# PRACTICAL SYNTHETIC
# ORGANIC CHEMISTRY

# PRACTICAL SYNTHETIC ORGANIC CHEMISTRY

## Reactions, Principles, and Techniques

Edited by

**STÉPHANE CARON**

**WILEY**

A JOHN WILEY & SONS, INC., PUBLICATION

Published by John Wiley & Sons, Inc., Hoboken, New Jersey
Published simultaneously in Canada

For general information on our other products and services or for technical support, please contact our Customer Care Department within the United States at (800) 762-2974, outside the United States at (317) 572-3993 or fax (317) 572-4002.

Wiley also publishes its books in a variety of electronic formats. Some content that appears in print may not be available in electronic formats. For more information about Wiley products, visit our web site at www.wiley.com.

*Library of Congress Cataloging-in-Publication Data:*

Practical synthetic organic chemistry : reactions, principles, and techniques / edited by Stéphane Caron.
    p. cm.
  Includes index.
    ISBN 978-0-470-03733-1 (pbk.)
  1. Organic compounds–Synthesis.   I. Caron, Stéphane.
  QD262.P688 2011
  547'.2–dc22

                    2011008293

Printed in the United States of America

oBooK ISBN: 978-1-118-09355-9
ePDF ISBN:  978-1-118-09357-3
ePub ISBN:   978-1-118-09356-6

10  9  8  7  6  5  4  3  2  1

# CONTENTS

# FOREWORD

While all practitioners in the art recognize that synthetic organic chemistry serves society, it is not always appreciated the other way around; namely that society does not associate the phenomenal rewards that are derived from our ability to assemble complex functional molecules. Indeed, the achievements of the pharmaceutical industry are underpinned by molecular design and synthesis of virtually all the healing drug substances that are currently on the market, which benefit mankind and cure human diseases.

There is, in my view, nothing nobler than this worthwhile task. However, synthesis of the complex functional architectures is not a trivial or routine process. This is always made more difficult on scaling up of the chemistries to comply with current standards of safety, robustness, costs, quality, and environmental impact. I, therefore, very much welcome this new textbook, Practical Synthetic Organic Chemistry: Reactions, Principles, and Techniques, which gives us an inspiring and pragmatic glimpse at the most commonly used and effective synthesis procedures on scale. This unique textbook has been written by experts in the field and provides an insight and critique of the key effective methods for molecular assembly. It will be a great starting point for anyone developing their career in this worthy industry.

The organization and chapter titles reflect a refreshing approach to explaining a complicated science covering all the major bond construction methods and functional control elements necessary for modern synthetic planning and execution.

The authors have incorporated chapters on important topics such as the effectiveness of rearrangement processes, radical reactions, access to chirality, and green chemistry, which are not captured as a group in more traditional texts. Furthermore, the educational component of this book highlighting nomenclature, pKa knowledge, solvent properties, and tips of the trade from the practicing chemist again adds value in providing a feast of information that otherwise is dispersed over many sources or is simply locked up in the personal database of knowledge of those experienced in the art.

The creative spirit of this area of science in bringing new functional molecules in our world for the first time for the betterment of us all is enduring and will continue to seed new discoveries for many years to come.

STEVEN V. LEY
*Cambridge, UK*
*2011*

# PREFACE

Despite the relative maturity of the field of organic chemistry, molecules continue to present synthetic chemists with significant challenges, especially as the expectations for synthetic, material, and energy efficiency continue to intensify. In response to these challenges, organic chemists have continued to develop innovative new synthetic methods. Many of these recently developed synthetic methods allow for achievement of ever more impressive chemo-, enantio-, diastereo-, and regioselectivity. Additionally, new synthetic methods are providing chemists with novel reactions and enabling synthetic strategies that would have previously been impractical or impossible.

While we are fascinated by the multitude of powerful synthetic methods that have been developed in recent years, these methods were not the direct inspiration for this book. Instead, the impetus for this book was actually an article request for a special issue of *Chemical Reviews* on process chemistry (Vol. 106, No. 7, July 2006). A group of Pfizer process chemists, at the Groton (CT) site, elected to write a review of oxidations performed on large scale. This review was divided among several authors and served as the basis for Chapter 10 in this book.[1] My personal contribution to the review article was the section covering C–O oxidations, and I was surprised by what I learned upon further investigation of this seemingly mundane topic. While a number of effective methods exist for the conversion of primary alcohols to aldehydes, only two general methods have been used consistently and successfully on manufacturing scale: Moffatt-type and TEMPO-mediated oxidations. Upon completion of this review, I asked myself if this would be true for other chemical transformations. Beyond simple curiosity, this question is relevant because search engines like SciFinder® and Reaxys™ can rapidly search through the chemical literature and identify large numbers of *potentially* relevant reactions/references. For example, a simple search for the nitration of an arene results in thousands of examples of this transformation. How does one sort through this incredible wealth of information?

---

[1]Caron, S.; Dugger, R. W.; Ruggeri, S. G.; Ragan, J. A.; Ripin, D. H. B. *Chem. Rev.* **2006**, *106*, 2943.

In this age of information overload, we thought that a simple collection of proven methodologies might be of value to practicing synthetic chemists. While many textbooks offer a compendium of the many synthetic methods that are available, we set out to create a guide that would help select the first few conditions to attempt for a given transformation. This might be of particular value when the substrate of interest is difficult to access in sufficient quantities for a large screen of reaction conditions. The authors agreed that the objective would not necessarily be to find the original report or the very best procedure for a given reaction, but rather to identify reaction conditions that work reliably well over the broadest possible range of substrates. It was a prerequisite for the references cited in this textbook that experimental data was provided, preferably on a multi-gram scale, and preferably using reaction conditions that would be amenable to scale-up. In the rare cases where such experimental data was not identified, the authors used their discretion to select reaction conditions that we would expect to scale up reasonably well. Not surprisingly, many of the examples cited in this textbook come from *Organic Syntheses* or from journals like *Organic Process Research and Development* and *The Journal of Organic Chemistry*. We also agreed that patent literature would be avoided, if possible, since it is not always easily accessed and/or interpreted. Finally, we made an effort to focus on references published after 1980.

In addition to the compilation of reliable methods for accomplishing various chemical transformations, we have tried to provide the reader with some information that may not be common knowledge outside of process chemistry groups. The second portion of this book focuses on this type of information. How can chiral building blocks be accessed most efficiently? What are the most reliable methods for preparing and naming common heterocycles? Why are some solvents favored over others? How does the synthesis of a drug evolve throughout development? What does it mean to have a green synthesis? We hope that readers will find some useful and practical information in this section of the book.

The completion of this endeavor would have never been possible without the commitment of several of my colleagues at Pfizer who agreed to the preparation and collective review of the chapters. We are also extremely grateful to our management for supporting the project. I would specifically like to thank Jodi Gaynor, Lisa Lentini, and Kim White for their administrative support. I would also like to thank Kris N. Jones for her assistance in editing the book. The commitment and support of our collaborators at Wiley-Interscience are very much appreciated. We especially appreciated their patience as our deadline came and passed.

Building each chapter of this manuscript served as an excellent chemistry refresher course for the authors, and we learned a lot of new chemistry in the process. We sincerely hope you enjoy the same experience as you read this book.

# CONTRIBUTORS

**Stéphane Caron,** Senior Director, Chemical R&D, Pfizer Worldwide R&D, Eastern Point Road, MS 8118D-4002, Groton CT, 06340, stephane.caron@pfizer.com.

**Juan Colberg,** API and Analytics Technology Leader, Technology & Strategic Sourcing, Pfizer Worldwide R&D, Eastern Point Road, MS 8118D-4025, Groton CT, 06340, Juan.colberg@pfizer.com.

**Pascal Dubé,** Principal Scientist, Chemical R&D, Pfizer Worldwide R&D, Eastern Point Road, MS 8118D-4060, Groton CT, 06340, pascal.dube@pfizer.com.

**Robert W. Dugger,** Research Fellow, Chemical R&D, Pfizer Worldwide R&D, Eastern Point Road, MS 8118D-4031, Groton CT, 06340, robert.w.dugger@pfizer.com.

**Heather N. Frost,** Chemistry Sourcing Lead, External Research Solutions Center of Emphasis, Pfizer Worldwide R&D, Eastern Point Rd, MS8200-4001, Groton, CT 06340, heather.n.frost@pfizer.com.

**Arun Ghosh,** Arch Chemicals, Process Technologies Group, 350 Knotter Drive, Cheshire CT, 06410, arun.ghosh@sbcglobal.net.

**Nathan D. Ide,** Principal Scientist, Chemical R&D, Pfizer Worldwide R&D, Eastern Point Road, MS 8118D-4058, Groton CT, 06340, nathan.d.ide@pfizer.com.

**Jade D. Nelson,** Associate Research Fellow, Chemical R&D, Pfizer Worldwide R&D, Eastern Point Road, MS 8118D-4047, Groton, CT 06340, jade.nelson@pfizer.com.

**Kristin E. Price,** Senior Scientist, Chemical R&D, Pfizer Worldwide R&D, Eastern Point Road, MS 8156-066, Groton, CT 06340, kristin.e.price@pfizer.com.

**John A. Ragan,** Associate Research Fellow, Chemical R&D, Pfizer Worldwide R&D, Eastern Point Road, MS 8118D-4010, Groton, CT 06340, john.a.ragan@pfizer.com.

**David H. Brown Ripin,** Scientific Director, Drug Access Team, Clinton Health Access Initiative, Boston, MA 02127, dripin@clintonHealthAccess.org.

**Sally Gut Ruggeri,** Associate Research Fellow, Research API, Pfizer Worldwide R&D, Eastern Point Road, MS 8156-078, Groton, CT 06340, sally.gut@pfizer.com.

**Rajappa Vaidyanathan,** Associate Research Fellow, Chemical R&D, Pfizer Worldwide R&D, Eastern Point Road, MS 8118D-4067, Groton, CT 06340, rajappa.vaidyanathan@pfizer.com.

**Carrie B. Wager,** Senior Principal Scientist, Chemical R&D, Pfizer Worldwide R&D, Eastern Point Road, MS 8118D-4017, Groton, CT 06340, carrie.b.wager@pfizer.com.

**Shu Yu,** Associate Research Fellow, Chemical R&D, Pfizer Worldwide R&D, Eastern Point Road, MS 8118D-4028, Groton, CT 06340, shu.yu@pfizer.com.

# 1

# ALIPHATIC NUCLEOPHILIC SUBSTITUTION

Jade D. Nelson

## 1.1 INTRODUCTION

Nucleophilic substitution reactions at an aliphatic center are among the most fundamental transformations in classical synthetic organic chemistry, and provide the practicing chemist with proven tools for simple functional group interconversion as well as complex target-oriented synthesis. Conventional $S_N2$ displacement reactions involving simple nucleophiles and electrophiles are well-studied transformations, are among the first concepts learned by chemistry students and provide a launching pad for more complex subject matter such as stereochemistry and physical organic chemistry.

A high level survey of the chemical literature provides an overwhelming mass of information regarding aliphatic nucleophilic substitution reactions. This chapter attempts to highlight those methods that stand out from the others in terms of scope, practicality, and scalability.

## 1.2 OXYGEN NUCLEOPHILES

### 1.2.1 Reactions with Water

***1.2.1.1 Hydrolysis of Alkyl Halides*** The reaction of water with an alkyl halide to form the corresponding alcohol is rarely utilized in target-oriented organic synthesis. Instead, conversion of alcohols to their corresponding halides is more common since methods for the synthesis of halides are less abundant. Nonetheless, alkyl halide hydrolysis can provide simple, efficient access to primary alcohols under certain circumstances. Namely, the hydrolysis of activated benzylic or allylic halides is a facile reaction, and following benzylic or allylic halogenation, provides a simple approach to the synthesis of this

*Practical Synthetic Organic Chemistry: Reactions, Principles, and Techniques*, First Edition.
Edited by Stéphane Caron.
© 2011 John Wiley & Sons, Inc. Published 2011 by John Wiley & Sons, Inc.

subset of alcohols. Typical conditions involve treatment of the alkyl halide with a mild base in an acetone-water or acetonitrile-water mixed solvent system. Moderate heating will accelerate the reaction, and is commonly employed.[1]

93%

### *1.2.1.2  Hydrolysis of gem-Dihalides*    Geminal dihalides can be converted to aldehydes or ketones via direct hydrolysis. The desired conversion can be markedly accelerated by heating in the presence of an acid or a base, or by including a nucleophilic amine promoter such as dimethylamine.[2]

86%

In the example below from Snapper and coworkers, a trichloro- intermediate was prepared from *p*-methoxystyrene via the Kharasch addition of 1,1,1-trichloroethane. Contact with silica gel effected elimination of the benzylic chloride as well as hydrolysis of the geminal dichloride moiety to yield the α,β-unsaturated methyl ketone in good overall yield.[3]

69%

### *1.2.1.3  Hydrolysis of 1,1,1-Trihalides*    1,1,1-Trihalides are at the appropriate oxidation state to serve as carboxylic acid precursors. These compounds react readily with water at acidic pH, providing the corresponding acids in high yield, and often at ambient temperature.[4] Trichloromethyl groups, rather than tribromo- or triiodo- analogs, are more often utilized due to superior access via nucleophilic displacement reactions by trichloromethyl anions.

[1]Lee, H. B.; Zaccaro, M. C.; Pattarawarapan, M.; Roy, S.; Saragovi, H. U.; Burgess, K. *J. Org. Chem.* **2004**, *69*, 701–713.
[2]Bankston, D. *Synthesis* **2004**, 283–289.
[3]Lee, B. T.; Schrader, T. O.; Martin-Matute, B.; Kauffman, C. R.; Zhang, P.; Snapper, M. L. *Tetrahedron* **2004**, *60*, 7391–7396.
[4]Martins, M. A. P.; Pereira, C. M. P.; Zimmermann, N. E. K.; Moura, S.; Sinhorin, A. P.; Cunico, W.; Zanatta, N.; Bonacorso, H. G.; Flores, A. C. F. *Synthesis* **2003**, 2353–2357.

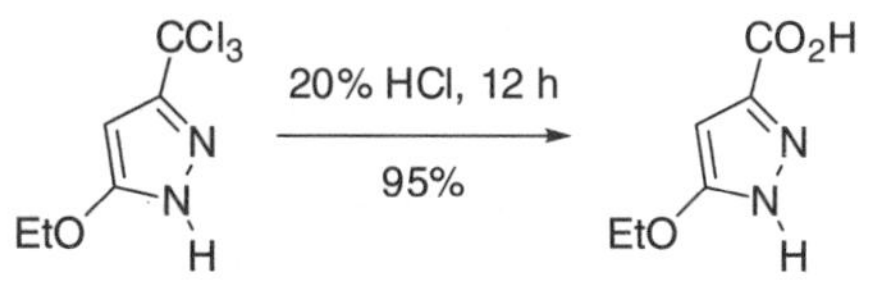

The trifluoromethyl group has seen increased application in the pharmaceutical industry in recent years due to its relative metabolic stability. Although trifluoromethyl groups are susceptible to vigorous hydrolytic conditions,[5] they are not frequently utilized as carboxylic acid precursors due to the relative expense of incorporating fluorine into building blocks. However, an increasingly common synthetic application of the $CF_3$ function is highlighted by the example below. Alkaline hydrolysis of a trifluoromethyl ketone provides the corresponding carboxylic acid in good yield. Here, the $CF_3$ group is not the point of nucleophilic attack by water. Instead, the strong inductive effect of the three highly electronegative fluorine atoms makes the trifluoromethyl anion an excellent leaving group, and attack occurs at the carbonyl carbon.[6]

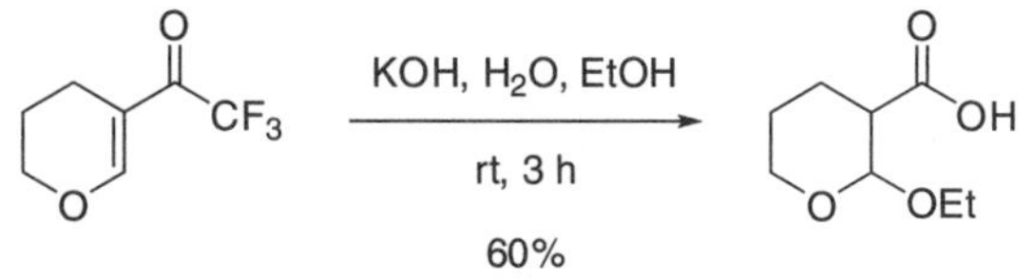

***1.2.1.4   Hydrolysis of Alkyl Esters of Inorganic Acids***    Alkaline hydrolysis of inorganic esters may proceed through competing mechanisms, as illustrated by the mesylate and boric acid monoester in the following scheme. Sulfonate hydrolysis favors the product of stereochemical inversion, via direct $S_N2$ attack at the carbon bearing the sulfonate. In contrast, the corresponding boron derivative is hydrolyzed under identical reaction conditions with retention of configuration, which is the result of formal attack by hydroxide at boron.[7]

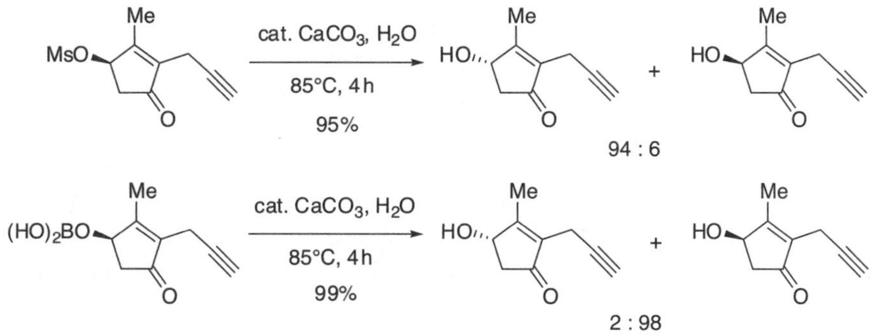

Enders and coworkers demonstrated that γ-sultone hydrolysis occurs exclusively via attack at carbon to provide γ-hydroxy sulfonates with a high degree of stereochemical

[5]Butler, D. E.; Poschel, B. P. H.; Marriott, J. G. *J. Med. Chem.* **1981**, *24*, 346–350.
[6]Hojo, M.; Masuda, R.; Sakaguchi, S.; Takagawa, M. *Synthesis* **1986**, 1016–1017.
[7]Danda, H.; Maehara, A.; Umemura, T. *Tetrahedron Lett.* **1991**, *32*, 5119–5122.

control.[8] In order to verify stereochemistry, the crude sulfonic acid was converted into the corresponding methyl sulfonate by treatment with diazomethane (see Section 1.2.3.5).

de ≥98%, ee ≥98%                                                    de ≥98%, ee ≥98%

***1.2.1.5  Hydrolysis of Diazo Ketones***    Treatment of simple α-diazo ketones with hydrochloric acid in aqueous acetone provides direct access to the corresponding alcohols.[9] In their 1968 paper, Tillett and Aziz describe an investigation into the kinetics of this transformation that utilized a number of diazoketones and mineral acids.[10] Although the reaction can be run under mild conditions, and is often high yielding, the preparation and handling of diazo compounds is a safety concern that may preclude their use on large scale.

***1.2.1.6  Hydrolysis of Acetals, Enol Ethers, and Related Compounds***    Acetals are highly susceptible to acid-catalyzed hydrolysis, typically providing the corresponding aldehydes under very mild conditions. Almost any acid catalyst can be employed, so the choice is usually dependent upon substrate compatibility. Solid-supported sulfonic acid catalysts such as Amberlyst-15[11] are an especially attractive option due to the relative ease of catalyst removal by simple filtration.[12]

Enol ethers of simple ketones may be similarly hydrolyzed by treatment with aqueous acid. A water-miscible organic cosolvent such as acetone or acetonitrile is often included to improve substrate solubility.[13] Moderate heating increases the rate of hydrolysis, but high temperatures are seldom required.

[8]Enders, D.; Harnying, W.; Raabe, G. *Synthesis* **2004**, 590–594.
[9]Pirrung, M. C.; Rowley, E. G.; Holmes, C. P. *J. Org. Chem.* **1993**, *58*, 5683–5689.
[10]Aziz, S.; Tillett, J. G. *J. Chem. Soc. B* **1968**, 1302–1307.
[11]Kunin, R.; Meitzner, E.; Bortnick, N. *J. Am. Chem. Soc.* **1962**, *84*, 305–306.
[12]Coppola, G. M. *Synthesis* **1984**, 1021–1023.
[13]Fuenfschilling, P. C.; Zaugg, W.; Beutler, U.; Kaufmann, D.; Lohse, O.; Mutz, J.-P.; Onken, U.; Reber, J.-L.; Shenton, D. *Org. Process Res. Dev.* **2005**, *9*, 272–277.

Dithioketene acetals may be hydrolyzed to thioesters under very mild conditions. Note that the strongly acidic reaction conditions employed in the example below resulted in concomitant β-dehydration and loss of the acid labile *N*-trityl protecting group.[9]

Orthoesters may also be hydrolyzed through treatment with aqueous acid, as exemplified in the scheme below.[14] Methanol is often included as a nucleophilic cosolvent that participates in the hydrolysis.

Under mildly acidic conditions, a terminal orthoester will provide the carboxylic acid ester.[15] However, prolonged exposure to aqueous acid will yield the carboxylic acid.

### 1.2.1.7 Hydrolysis of Silyl Enol Ethers
Silyl enol ethers are prone to hydrolysis at a rate generally consistent with their relative steric bulk. Trimethylsilyl (TMS) enol ethers are particularly labile, and may by hydrolyzed in the absence of an acid catalyst in some

[14]Kato, K.; Nouchi, H.; Ishikura, K.; Takaishi, S.; Motodate, S.; Tanaka, H.; Okudaira, K.; Mochida, T.; Nishigaki, R.; Shigenobu, K.; Akita, H. *Tetrahedron* **2006**, *62*, 2545–2554.
[15]Martynow, J. G.; Jozwik, J.; Szelejewski, W.; Achmatowicz, O.; Kutner, A.; Wisniewski, K.; Winiarski, J.; Zegrocka-Stendel, O.; Golebiewski, P. *Eur. J. Org. Chem.* **2007**, 689–703.

instances.[16] In contrast, bulky triisopropylsilyl (TIPS) enol ethers are generally stable enough to withstand an acidic aqueous workup and may even be purified via silica gel chromatography. In the scheme below, the TMS, *tert*-butyl dimethylsilyl (TBS) and TIPS enol ethers of cyclohexanone are desilylated by treatment with aqueous hydrochloric acid in tetrahydrofuran. The variation in reaction rate under these conditions is noteworthy.[17]

### 1.2.1.8  Hydrolysis of Silyl Ethers

**1.2.1.8  *Hydrolysis of Silyl Ethers***   Silyl ethers are generally less susceptible to acid-catalyzed hydrolysis than their enol ether counterparts. However, increasing the reaction time, reaction temperature or acid concentration can provide a simple, high-yielding method for the removal of silicon-based hydroxyl protecting groups, provided that the rest of the molecule is stable to such treatment.[18]

In the example below, the removal of two silyl ethers and one silyl enol ether was accomplished with aqueous acetic acid in THF at room temperature.[19] These milder conditions necessitated a longer reaction time, but conserved the acid sensitive methyl ester and tertiary alcohol functional groups.

[16]Keana, J. F. W.; Eckler, P. E. *J. Org. Chem.* **1976**, *41*, 2850–2854.
[17]Manis, P. A.; Rathke, M. W. *J. Org. Chem.* **1981**, *46*, 5348–5351.
[18]Sun, H.; Abboud, K. A.; Horenstein, N. A. *Tetrahedron* **2005**, *61*, 10462–10469.
[19]Collins, P. W.; Gasiecki, A. F.; Jones, P. H.; Bauer, R. F.; Gullikson, G. W.; Woods, E. M.; Bianchi, R. G. *J. Med. Chem.* **1986**, *29*, 1195–1201.

For acid-sensitive substrates, the use of fluoride ion (e.g., tetra-*n*-butylammonium fluoride, TBAF) provides an attractive alternative (see Section 1.5.1.7) and is increasingly employed as a first choice reagent.

Silyl ethers may be cleaved under aqueous alkaline conditions when the silicon center is less electron-rich. For example, *tert*-butyl diphenylsilyl (TBDPS or TPS) ethers are susceptible to base-promoted cleavage, despite their relative stability to acid. This reactivity is complementary to that of TBS ethers, allowing for selective silyl protection of similar functional groups.[20]

Silyl ethers may also be cleaved by nucleophilic alcohols in the presence of strong acid catalysts (see Section 1.2.2.6).

***1.2.1.9 Hydrolysis of Epoxides***   Epoxides can be efficiently ring-opened under acid catalysis in an aqueous environment to afford 1,2-diol products. In cases where the regiochemistry of attack by water is inconsequential, or is directed by steric and/or electronic bias within the substrate, simple Brønsted acid catalysts are utilized. In the example below, hydrolysis is promoted by the sulfonated tetrafluoroethylene copolymer Nafion-H to provide the racemic 1,2-*anti*-diol product via backside attack on the epoxide.[21]

### 1.2.2    Reactions with Alcohols

***1.2.2.1 Preparation of Ethers from Alkyl Halides***   The preparation of ethers from alcohols may be accomplished via treatment with an alkyl halide in the presence of a suitable base (the Williamson ether synthesis). For relatively acidic alcohols such as phenol derivatives, the use of potassium carbonate in acetone is a simple, low cost option.[22]

The rate of ether formation can be accelerated by increasing the reaction temperature, or more commonly by employing a more reactive alkyl halide. This is accomplished in the

[20]Hatakeyama, S.; Irie, H.; Shintani, T.; Noguchi, Y.; Yamada, H.; Nishizawa, M. *Tetrahedron* **1994**, *50*, 13369–13376.

[21]Olah, G. A.; Fung, A. P.; Meidar, D. *Synthesis* **1981**, 280–282.

[22]Garcia, A. L. L.; Carpes, M. J. S.; de Oca, A. C. B. M.; dos Santos, M. A. G.; Santana, C. C.; Correia, C. R. D. *J. Org. Chem.* **2005**, *70*, 1050–1053.

example below via conversion of benzyl chloride to benzyl iodide *in situ* by including potassium iodide in the reaction mixture (Finkelstein reaction; see Section 1.5.1.1).[23] Since the iodide ions are regenerated over the course of the reaction, a sub-stoichiometric quantity can often be used. However, a stoichiometric excess of iodide increases the concentration of the more reactive alkylating agent, and thus improves the kinetics of the alkylation reaction. Note that under the conditions utilized, the carboxylic acid is also converted to the corresponding benzyl ester. For more on the preparation of esters from carboxylic acids see Sections 1.2.3 and 2.29.

Henegar and coworkers reported the use of potassium isopropoxide in dimethyl carbonate for the large scale preparation of an ether intermediate in the synthesis of the commercial antidepressant (±)-reboxetine mesylate.[24]

***1.2.2.2    Preparation of Methyl Ethers from Dimethyl Carbonate or Dimethyl Sulfate***    Dimethyl carbonate is an ambident electrophile that typically reacts with soft nucleophiles at a methyl carbon and with hard nucleophiles at the central carbonyl. Via the former mechanism, dimethyl carbonate has been used for the methylation of phenols, leading to the formation of arylmethyl ethers. This is an especially attractive reagent for large scale applications, owing to the low cost, low toxicity and neglegible environmental impact of dimethyl carbonate,[25] however scope is limited due to the modest reactivity of the reagent. In the example below, Thiébaud and coworkers illustrate the selective methylation of a bis-phenol compound with dimethyl carbonate and potassium carbonate in the absence of solvent.[26]

[23]Bourke, D. G.; Collins, D. J. *Tetrahedron* **1997**, *53*, 3863–3878.
[24]Henegar, K. E.; Ball, C. T.; Horvath, C. M.; Maisto, K. D.; Mancini, S. E. *Org. Process Res. Dev.* **2007**, *11*, 346–353.
[25]Tundo, P.; Rossi, L.; Loris, A. *J. Org. Chem.* **2005**, *70*, 2219–2224.
[26]Ouk, S.; Thiebaud, S.; Borredon, E.; Le Gars, P. *Green Chem.* **2002**, *4*, 431–435.

Shen and coworkers discussed the "greenness" of their protocol for the preparation of arylmethyl ethers from their phenol precursors.[27] Thus, treatment of various phenols with dimethyl carbonate at 120°C in the recyclable ionic liquid 1-*n*-butyl-3-methylimidazolium-chloride [(BMIm)Cl] provides quantitative yields of aryl methyl ethers. The use of ionic liquids has been heavily debated in recent years, as these materials are relatively expensive and current methods for their manufacture pose a significant environmental impact. Nevertheless, opportunities for reuse and recycle make continued development in this area a worthwhile venture.

The preferred method for the preparation of methyl ethers via alkylation of alcohol precursors generally utilizes dimethyl sulfate $(MeO)_2SO_2$, rather than methyl iodide, due to the decreased worker-exposure risk that accompanies its lower volatility. Typically, the alcohol is treated with dimethyl sulfate and a suitable base in a polar aprotic solvent. Yields are generally high and product isolation relatively straightforward. The examples below are illustrative, and were both carried out on a large scale.[28,29] It should be noted that most alkylating agents—dimethylsulfate included—pose serious toxicological hazards and thus require cautious handling and environmental controls.

***1.2.2.3 Preparation of Ethers from Alkyl Sulfonates***    A widespread and practical strategy for the synthesis of ethers, especially on large scale, is through the intermediacy of alkyl sulfonates. Broad access to alcohol precursors, coupled with the low cost and high efficiency in converting these alcohols to electrophilic sulfonates has contributed to the popularity of this method. Mesylates or tosylates are used most often, as MsCl and TsCl are considerably less expensive than other sulfonyl halides. However, for demanding nucleophilic substitution reactions, the 4-bromobenzenesulfonate (brosylate) or 4-nitrobenzenesulfonate

[27]Shen, Z. L.; Jiang, X. Z.; Mo, W. M.; Hu, B. X.; Sun, N. *Green Chem.* **2005**, *7*, 97–99.

[28]Liu, Z.; Xiang, J. *Org. Process Res. Dev.* **2006**, *10*, 285–288.

[29]Prabhakar, C.; Reddy, G. B.; Reddy, C. M.; Nageshwar, D.; Devi, A. S.; Babu, J. M.; Vyas, K.; Sarma, M. R.; Reddy, G. O. *Org. Process Res. Dev.* **1999**, *3*, 121–125.

(nosylate, ONs) offer increased reactivity. In the following example, a substituted phenol was alkylated with a primary mesylate in good yield under representative conditions.[30]

75%

In a second example, alkylation was shown to proceed through an epoxide intermediate formed via intramolecular displacement of the tosylate by the adjacent hydroxyl. The zinc alkoxide was superior to the lithium derivative in terms of yield and reaction rate.[31]

67%

***1.2.2.4  Iodoetherification***    Treatment of olefins with iodine provides an electrophilic iodonium species that can be trapped with oxygen nucleophiles to provide ethers. A base is included to quench the HI that gets generated during the reaction. The intramolecular nucleophilic displacement occurs at a much higher rate than intermolecular trapping, so cyclic ethers are formed in high yield. In the example below, two additional stereocenters are created with high selectivity through treatment of an unsaturated *cis*-1,3-diol with iodine and sodium bicarbonate in aqueous ether at 0°C.[32]

79%

93 : 7

***1.2.2.5  Preparation of Silyl Ethers***    Despite the relatively high molar expense of substituted silicon compounds such as TBSCl, silyl ethers are often utilized as alcohol-protecting groups in synthesis because of their excellent compatibility with a broad range of common processing conditions, coupled with their relative ease of removal. Vanderplas and coworkers prepared the TBS ether below in high yield as part of a multi-kilogram synthesis of a β-3 adrenergic receptor agonist.[33] The preferred conditions for TBS protection of this secondary alcohol included imidazole base in dimethylformamide.

[30]Reuman, M.; Hu, Z.; Kuo, G.-H.; Li, X.; Russell, R. K.; Shen, L.; Youells, S.; Zhang, Y. *Org. Process Res. Dev.* **2007**, *11*, 1010–1014.
[31]Wu, G. G. *Org. Process Res. Dev.* **2000**, *4*, 298–300.
[32]Tamaru, Y.; Hojo, M.; Kawamura, S.; Sawada, S.; Yoshida, Z. *J. Org. Chem.* **1987**, *52*, 4062–4072.
[33]Vanderplas, B. C.; DeVries, K. M.; Fox, D. E.; Raggon, J. W.; Snyder, M. W.; Urban, F. *J. Org. Process Res. Dev.* **2004**, *8*, 583–586.

The example below highlights the strong steric bias of TBSCl toward reaction with the unhindered secondary alcohol over the more hindered secondary and tertiary alcohols.[34]

***1.2.2.6   Cleavage of Silyl Ethers with Alcohols***   Although the most common methods utilize fluoride ion (see Section 1.5.1.7), silyl ethers can be cleaved with alcohols promoted by strong acids. Watson and coworkers reported a high-yielding example of the use of ethanolic hydrogen chloride in their large scale synthesis of the tumor necrosis factor-α inhibitor candidate MDL 201449A.[35]

Selective cleavage of a TMS ether in the presence of the more sterically encumbered TBS derivative has been accomplished via treatment with a methanolic solution of oxalic acid dihydrate.[36]

[34]Van Arnum, S. D.; Carpenter, B. K.; Moffet, H.; Parrish, D. R.; MacIntrye, A.; Cleary, T. P.; Fritch, P. *Org. Process Res. Dev.* **2005**, *9*, 306–310.
[35]Watson, T. J. N.; Curran, T. T.; Hay, D. A.; Shah, R. S.; Wenstrup, D. L.; Webster, M. E. *Org. Process Res. Dev.* **1998**, *2*, 357–365.
[36]Maehr, H.; Uskokovic, M. R.; Adorini, L.; Reddy, G. S. *J. Med. Chem.* **2004**, *47*, 6476–6484.

***1.2.2.7 Transetherification*** The direct interconversion of ethers is not typically a practical transformation, and is thus rarely employed. The reaction can be accomplished, however, by treating the starting ether with an alcohol under the promotion of a strong mineral acid. The interconversion is an equilibrium process, so the alcohol is typically used as a cosolvent or in significant molar excess. A thorough study of the thermodynamics of interconversion of electronically diverse 1-phenylethyl ethers with aliphatic alcohols was reported by Jencks and coworkers.[37]

The following example from Pittman and coworkers highlights the efficiency of acetal exchange under Brønsted acid catalysis.[38] This reaction is formally a double transetherification, although it has the advantage of proceeding via an intermediate oxonium ion. In this instance, the reaction is rendered irreversible by running at a temperature where the liberated methanol is removed by distillation.

A wide variety of vinyl ethers can be prepared from ethyl vinyl ether via palladium-catalyzed transetherification, as exemplified below.[39] Additional substitution on the olefin, however, severely diminishes the efficiency of the transformation and limits scope. This reaction is not a nucleophilic aliphatic substitution, but is included here to illustrate this complementary, albeit limited methodology.

***1.2.2.8 Preparation of Epoxides*** Epoxides can be prepared in high yield through intramolecular nucleophilic displacement of halides by vicinal hydroxyls. In the example below, concentrated aqueous sodium hydroxide is used as the base in dichloromethane.[40] This solvent choice, when used in combination with highly concentrated NaOH solution, minimizes water content within the organic phase to prevent competitive epoxide hydrolysis to the corresponding 1,2-diol. The addition of a phase transfer catalyst, as well as aggressive agitation, facilitates movement of hydroxide ions into the dichloromethane.

[37]Rothenberg, M. E.; Richard, J. P.; Jencks, W. P. *J. Am. Chem. Soc.* **1985**, *107*, 1340–1346.
[38]Zhu, P. C.; Lin, J.; Pittman, C. U., Jr., *J. Org. Chem.* **1995**, *60*, 5729–5731.
[39]Weintraub, P. M.; King, C.-H. R. *J. Org. Chem.* **1997**, *62*, 1560–1562.
[40]Elango, S.; Yan, T.-H. *Tetrahedron* **2002**, *58*, 7335–7338.

Epoxides are often prepared from olefin precursors by sequential reaction with *N*-bromosuccinimide (NBS) and hydroxide in an aqueous environment. The NBS initially forms an electrophilic bromonium species, which is quickly converted to a bromohydrin via nucleophilic attack by water. As described in the following example, the bromohydrin can be converted to an epoxide in the same reaction vessel by addition of sodium hydroxide.[41] In this instance, the epoxide was not purified, but was further reacted to the azidohydrin by reaction with sodium azide at an elevated temperature. The opening of epoxides with nitrogen nucleophiles is also discussed in Section 1.4.1.6. More on the preparation of epoxides from carbon-carbon double bonds can be found in Section 3.10.4.

***1.2.2.9   Reaction of Alcohols with Epoxides***   Epoxides can be opened with nucleophilic alcohols in the presence of various activating agents. Despite a low degree of atom economy, *N*-bromosuccinimide (NBS) proved to be a very effective promoter in the example below.[42] A catalytic quantity of an acid catalyst is often sufficient, however.

In the example below from Loh and coworkers, a late stage intramolecular epoxide opening by an alcohol is catalyzed by camphorsulfonic acid in dichloromethane to provide the desired furan as a single diastereomer in good yield.[43] The γ-lactone resulting from intramolecular transesterification was observed as a minor byproduct.

95 : 5

[41]Loiseleur, O.; Schneider, H.; Huang, G.; Machaalani, R.; Selles, P.; Crowley, P.; Hanessian, S. *Org. Process Res. Dev.* **2006**, *10*, 518–524.
[42]Iranpoor, N.; Firouzabadi, H.; Chitsazi, M.; Ali Jafari, A. *Tetrahedron* **2002**, *58*, 7037–7042.
[43]Huang, J.-M.; Xu, K.-C.; Loh, T.-P. *Synthesis* **2003**, 755–764.

The specific case involving interconversion of two α,β-epoxy alcohols under basic conditions is known as the Payne rearrangement,[44] and is discussed in Section 7.2.2.3. The intermolecular ring-opening of epoxides by alcohol nucleophiles may be catalyzed by Lewis acids. The use of indium(III) chloride has proven to be particularly effective, as exemplified by Lee and coworkers in the following example.[45] Here, both sterics (geminal methyls) and electronics (positive charge stabilization) direct the isopropanol nucleophile to the benzylic position.

### 1.2.2.10    The Reaction of Alcohols with Diazo Compounds

In a series of publications, Moody and coworkers have demonstrated the utility of Rh(II)-catalyzed decomposition of α-diazo esters, sulfones, and phosphonates in the presence of various aliphatic and aromatic alcohols.[46,47] The intramolecular variant of this method has found utility in the synthesis of functionalized cyclic ethers. Widespread application of the intermolecular reaction, however, has been hindered by the need for a large stoichiometric excess of the alcohol component to limit competitive hydrogen abstraction that leads to reduced product.

This example highlights the superiority of the Rh(II) trifluoroacetamide catalyst, as the more commonly employed $Rh_2(OAc)_4$ afforded desired O-H insertion product in an inferior yield of 67% after 72 hours at reflux. Interestingly, photochemical decomposition of this diazo compound in isopropanol provided the product of hydrogen abstraction (i.e., reduction) exclusively in 90% yield. Complementary oxidative methods for the synthesis of related compounds are discussed in Section 10.3.4.

### 1.2.2.11    Preparation of Ethers via Dehydration of Alcohols

The direct formation of ethers from alcohol precursors is typically limited to the preparation of symmetrical ethers, as dehydration under the strongly acidic conditions employed generally occurs indiscriminately. In the example below, Ziyang and coworkers optimized a method for the production of solvent and one-time fuel additive methyl *tert*-butyl ether. In this

[44]Payne, G. B. *J. Org. Chem.* **1962**, *27*, 3819–3822.
[45]Lee, Y. R.; Lee, W. K.; Noh, S. K.; Lyoo, W. S. *Synthesis* **2006**, 853–859.
[46]Moody, C. J.; Taylor, R. J. *J. Chem. Soc., Perkin Trans.* **1989**, *1*, 721–731.
[47]Cox, G. G.; Miller, D. J.; Moody, C. J.; Sie, E. R. H. B.; Kulagowski, J. J. *Tetrahedron* **1994**, *50*, 3195–3212.

method, a heated stream of *tert*-butanol and methanol are passed through a column packed with the strongly acidic resin Amberlyst 15.[48]

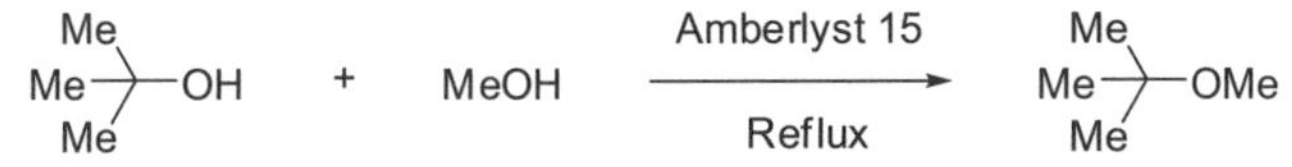

### 1.2.2.12    Addition of Alcohols to Boron, Phosphorous, and Titanium

Borate esters can be prepared by direct reaction of alcohols with boron halides in the presence of a suitable base. In the following example, a 1,3,5-triol is reacted with boron trichloride and pyridine to yield a novel "tripod borate ester" in high yield. This compound is of particular interest due to the considerable distortion from planar geometry generally observed in borate esters.[49]

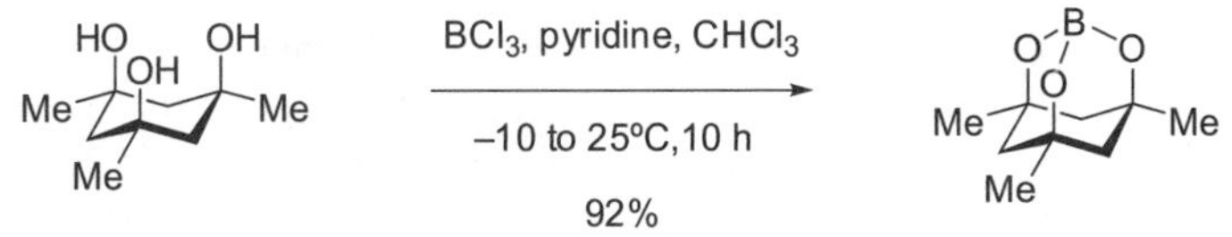

Phosphonate esters can be prepared in high yield by reaction of alcohols with diaryl halophosphates catalyzed by inorganic esters. As shown below, titanium(IV) *t*-butoxide was reported by Jones and coworkers to be the optimal catalyst.[50]

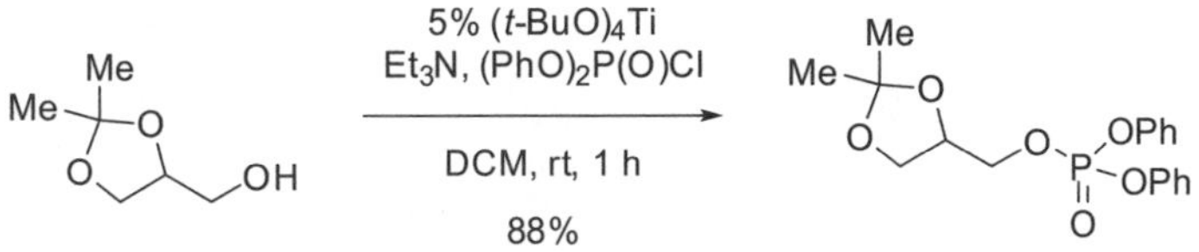

Phosphites and phosphonites are more commonly prepared by reaction of a metal alkoxide with a chlorophosphine derivative, as highlighted in the following example.[51] These reactions are exothermic, so cooling is generally required. Furthermore, most phosphites and phosphonites are air and moisture sensitive, so care must be taken to avoid degradation.

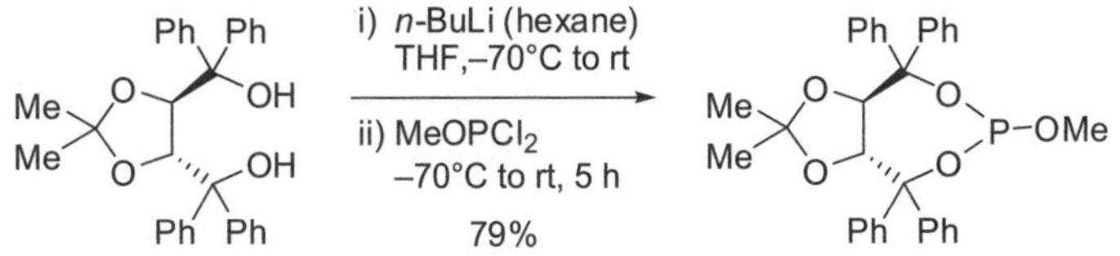

Due to the high oxophilicity of titanium(IV), titanates can be prepared through a process similar to the equilibrium driven ligand exchange observed with organometallic complexes. For example, heating titanium(IV) ethoxide in the presence of the TADDOL ligand shown

[48]Ziyang, Z.; Hidajat, K.; Ray, A. K. *J. Catal.* **2001**, *200*, 209–221.
[49]D'Accolti, L.; Fiorentino, M.; Fusco, C.; Capitelli, F.; Curci, R. *Tetrahedron Lett.* **2007**, *48*, 3575–3578.
[50]Jones, S.; Selitsianos, D.; Thompson, K. J.; Toms, S. M. *J. Org. Chem.* **2003**, *68*, 5211–5216.
[51]Seebach, D.; Hayakawa, M.; Sakaki, J.; Schweizer, W. B. *Tetrahedron* **1993**, *49*, 1711–1724.

below initially provides an equilibrium mixture of titanium alkoxides. Complete exchange can be achieved by removal of ethanol by distillation.[52]

### 1.2.3    Reactions with Carboxylates

***1.2.3.1   Alkylation of Carboxylic Acid Salts***    Carboxylic acids can serve as effective nucleophiles toward a variety of electrophilic reagents if first converted to their carboxylate salts via deprotonation with a suitable base. The relatively low pKa of carboxylic acids allows very mild bases to be employed. A reliable, scalable method for the preparation of methyl esters involves preparation of the lithium carboxylate with LiOH, followed by treatment with dimethyl sulfate, as exemplified below.[53] Note that methylation of the phenolic alcohol was avoided in this example through careful control of stoichiometry. Dimethyl sulfate may be replaced by iodomethane for most applications, however this is less preferred due to the increased volatility of the latter reagent (bp 42.4°C).

The most common method for the synthesis of esters from acids utilizes alkyl halides, owing to the relatively easy access to these electrophiles. Cabri and coworkers demonstrated a high-yielding *n*-propyl ester preparation via sodium carboxylate, which was obtained by treatment with $NaHCO_3$ in NMP.[54] Note that a tertiary amine was also quaternized under these conditions. For more on the alkylation of amines with alkyl halides, see Section 1.4.1.1.

[52]Von dem Bussche-Huennefeld, J. L.; Seebach, D. *Tetrahedron* **1992**, *48*, 5719–5730.

[53]Chakraborti, A. K.; Nandi, A. B.; Grover, V. *J. Org. Chem.* **1999**, *64*, 8014–8017.

[54]Cabri, W.; Roletto, J.; Olmo, S.; Fonte, P.; Ghetti, P.; Songia, S.; Mapelli, E.; Alpegiani, M.; Paissoni, P. *Org. Process Res. Dev.* **2006**, *10*, 198–202.

This approach to the conversion of carboxylic acids to esters is complementary to the Fischer esterification, whereby the carboxylic acid is heated in an alcohol solvent in the presence of a strong acid with concomitant generation of water (see Section 2.29.1).

***1.2.3.2  Iodolactonization***    Iodolactonization is a useful and practical method for the stereocontrolled preparation of lactones. In the example below from House and coworkers, the action of iodine upon the olefin provides an electrophilic intermediate iodonium ion. This highly reactive species is trapped in a stereospecific manner by the carboxylate nucleophile to provide a 1,2-anti relationship between the iodide and oxygen-bearing stereocenters.[55]

***1.2.3.3  Preparation of Silyl Esters***    The use of silyl esters as carboxylic acid protecting groups is an increasingly popular methodology due to the simplicity of preparation and relative ease of removal during workup. An attractive large scale method for trimethylsilyl ester preparation utilizes hexamethyldisilazane (HMDS) as both a base and TMS donor.[56] This method also proved effective in forming a silyl ether from the secondary alcohol precursor. For more on the formation of silyl ethers, see Section 1.2.2.5.

TPS = *tert*-butyl diphenylsilyl

The HMDS method described above is clearly advantageous if a TMS ester is desired. Unfortunately, the corresponding TES, TBS, or TIPS analogs are inaccessible with this methodology. As a result, silyl esters are more generally prepared by treatment of carboxylate salts with silyl chlorides or triflates. In the example below, *p*-methoxy cinnamic acid is converted to its TBS ester in high yield via treatment with *tert*-butyl dimethysilyl chloride and imidazole in DMF.[57]

The solvent employed for this reaction is often determined by the solubility of the intermediate carboxylate salt. Dichloromethane[58] and tetrahydrofuran[59] have proven successful in many cases. Triethylamine is commonly used as a base in place of imidazole,

[55]House, H. O.; Carlson, R. G.; Babad, H. *J. Org. Chem.* **1963**, *28*, 3359–3361.

[56]Smith, A. B.; Safonov, I. G.; Corbett, R. M. *J. Am. Chem. Soc.* **2002**, *124*, 11102–11113.

[57]Ponticello, G. S.; Freedman, M. B.; Habecker, C. N.; Holloway, M. K.; Amato, J. S.; Conn, R. S.; Baldwin, J. J. *J. Org. Chem.* **1988**, *53*, 9–13.

[58]Fache, F.; Suzan, N.; Piva, O. *Tetrahedron* **2005**, *61*, 5261–5266.

[59]McNamara, L. M. A.; Andrews, M. J. I.; Mitzel, F.; Siligardi, G.; Tabor, A. B. *J. Org. Chem.* **2001**, *66*, 4585–4594.

and 4-dimethylaminopyridine (DMAP) may be added as a nucleophilic catalyst to activate the silyl electrophile in case of a sluggish reaction. In the example below, a *tert*-butyldiphenylsilyl (TPS) ester was prepared in high yield.[60]

**1.2.3.4  Cleavage of Ethers with Acetic Anhydride**    The cleavage of ethers with acetic anhydride in the presence of ferric chloride was first reported by Knoevenagel in 1914 for the conversion of diethyl ether to ethyl acetate.[61] Ganem and coworkers followed years later with mechanistic studies and an examination of substrate scope for the method.[62] In the example below, benzyl-*n*-butyl ether is converted to *n*-butyl acetate in high yield after several hours at 80°C. The benzyl acetate byproduct was not quantified.

**1.2.3.5  Alkylation of Carboxylic Acids and Enols with Diazo Compounds**    One of the most efficient methods for the preparation of methyl esters from carboxylic acids involves reaction with a solution of diazomethane in ether.[63] Esterification is generally rapid and highly chemoselective. Therefore, products can typically be isolated by simple evaporation of the volatile ether solvent. This method is not typically applied on large scale due to the serious hazards associated with the preparation and handling of diazomethane. However, with specialized equipment and proper engineering controls in place, methyl ester formation via treatment with diazomethane is sometimes the method of choice, as exemplified by the example below.[64] These conditions are also effective for the methylation of sulfonic acids, as described in Section 1.2.1.4.

A safer, more practical alternative to the use of toxic and volatile diazomethane is provided by the commercially available reagent trimethylsilyl diazomethane. In the example

[60]Kim, W. H.; Hong, S. K.; Lim, S. M.; Ju, M.-A.; Jung, S. K.; Kim, Y. W.; Jung, J. H.; Kwon, M. S.; Lee, E. *Tetrahedron* **2007**, *63*, 9784–9801.
[61]Knoevenagel, E. *Justus Liebigs Ann. Chem.* **1914**, *402*, 111–148.
[62]Ganem, B.; Small, V. R., Jr., *J. Org. Chem.* **1974**, *39*, 3728–3730.
[63]Arndt, F. *Org. Synth.* **1935**, *15*, 3–5.
[64]Schmidt, R. R.; Frick, W. *Tetrahedron* **1988**, *44*, 7163–7169.

below, the enol form of a β-ketoaldehyde is converted to its methyl enol ether derivative in high yield and with excellent regioselectivity.[65] A discussion of the mechanism of this interesting transformation has been published.[66] For more on nucleophilic substitution reactions of enols and carbonyls, see Section 1.2.4.

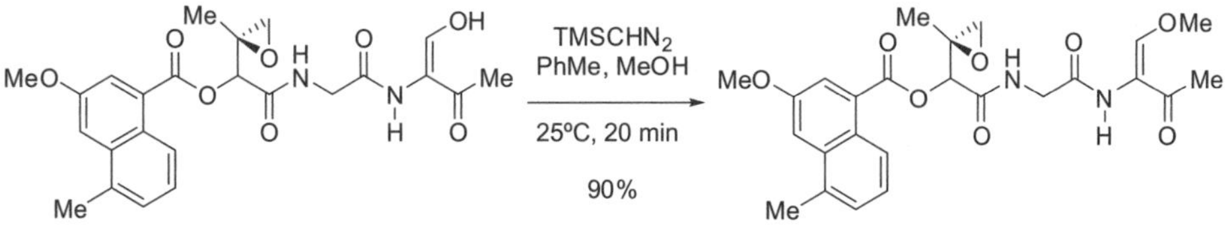

*1.2.3.6 Preparation of Mixed Organic-Inorganic Anhydrides*    The reaction of carboxylic acid salts with phosphonic acid halides can provide the corresponding carboxylic acid-phosphonic acid mixed anhydrides in good yield. In the example below, a diethylphosphonate ester was prepared from a triethylammonium carboxylate under very mild conditions.[67]

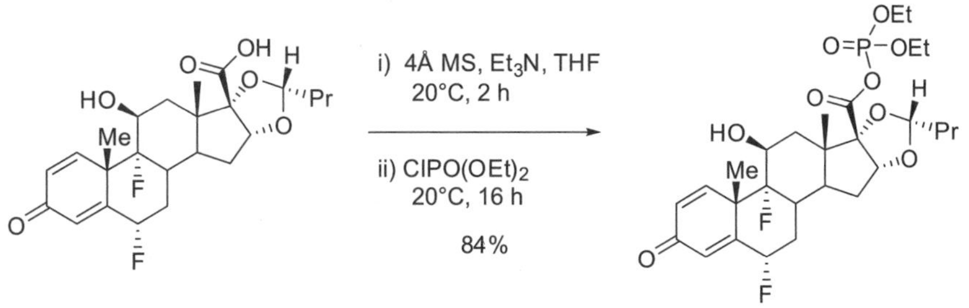

### 1.2.4  Reactions with Other Oxygen Nucleophiles

*1.2.4.1  Formation of Silyl Enol Ethers and Silyl Ketene Acetals*    The formation of silyl enol ethers has been a topic of considerable study, due to the broad utility of these compounds in carbon−carbon bond forming reactions. Silyl enol ethers are typically prepared by treatment of an enolizable carbonyl compound with a silyl halide or triflate in the presence of a suitable base. For unsymmetrical ketones, such as 2-methylcyclohexanone (shown below), regioisomeric enol ether products are possible, and methods for the selective preparation of both the kinetically favored and thermodynamically favored isomers have been developed.[68,69] Treatment of ketones with a silyl chloride in the presence of a weak nitrogen base such as triethylamine provides the thermodynamically favored silyl enol ether (more substituted olefin) at or near room temperature in excellent yield.[70]

[65]Coleman, R. S.; Tierney, M. T.; Cortright, S. B.; Carper, D. J. *J. Org. Chem.* **2007**, *72*, 7726–7735.

[66]Kuehnel, E.; Laffan, D. D. P.; Lloyd-Jones, G. C.; Martinez del Campo, T.; Shepperson, I. R.; Slaughter, J. L. *Angew. Chem., Int. Ed. Engl.* **2007**, *46*, 7075–7078.

[67]Procopiou, P. A.; Biggadike, K.; English, A. F.; Farrell, R. M.; Hagger, G. N.; Hancock, A. P.; Haase, M. V.; Irving, W. R.; Sareen, M.; Snowden, M. A.; Solanke, Y. E.; Tralau-Stewart, C. J.; Walton, S. E.; Wood, J. A. *J. Med. Chem.* **2001**, *44*, 602–612.

[68]D'Angelo, J. *Tetrahedron* **1976**, *32*, 2979–2990.

[69]Heathcock, C. H. *Modern Synthetic Methods* **1992**, *6*, 1–102.

[70]Takasu, K.; Ishii, T.; Inanaga, K.; Ihara, M. *Org. Synth.* **2006**, *83*, 193–199.

95%                    >97 : 3

The use of anhydrous zinc dichloride is key to formation of Danishefsky's diene.[71] The zinc additive plays a dual role of Lewis acid and dehydrative agent in this enol ether formation. The use of freshly fused $ZnCl_2$ is preferred and ensures a more robust outcome.

46%

In contrast to the conditions described above, enolization with a strong, bulky base at low temperature favors removal of the most sterically accessible α-proton to provide the kinetically favored enol ether upon silylation (less substituted olefin).[72] In the following example, the dianion of a β-hydroxy ketone was formed with 2.2 equivalents of LDA at −78°C. Silylation occurred with complete selectivity for the enol ether shown.

81%

In the following example from Aventis, a trimethylsilyl enol ether was prepared in good yield despite the presence of a rather labile epoxide. Slow addition of the starting ketone into a cold solution of LDA and a careful isolation protocol proved to be important.[73]

>83%

As exemplified by Albrecht and coworkers, treatment of a β-ketoester with two equivalents of LDA yields a dienolate intermediate. Quenching the dienolate with two molar equivalents of chlorotrimethylsilane provides the silyl enol ether as well as a silyl ketene acetal in excellent yield.[74]

[71]Danishefsky, S.; Kitahara, T.; Schuda, P. F. *Org. Synth.* **1983**, *61*, 147–151.
[72]Martin, V. A.; Albizati, K. F. *J. Org. Chem.* **1988**, *53*, 5986–5988.
[73]Larkin, J. P.; Wehrey, C.; Boffelli, P.; Lagraulet, H.; Lemaitre, G.; Nedelec, A.; Prat, D. *Org. Process Res. Dev.* **2002**, *6*, 20–27.
[74]Albrecht, U.; Nguyen, T. H. V.; Langer, P. *J. Org. Chem.* **2004**, *69*, 3417–3424.

i) LDA (2 eq), THF-hexane
   0°C, 2 h

ii) TMSCl, THF
    −78°C (1 h) to 0°C

98%

A second example of the preparation of a silyl ketene acetal from an enolizable ester is included below.[75] Note that the geometry of the newly formed double bond is not controlled, and a mixture of products is obtained.

i) LDA, THF-hexane
   −78°C, 1 h

ii) TMSCl, THF
    −78°C

iii) 0°C, 3 h

95%                    1.3 : 1

***1.2.4.2  Formation  of  Enol  Triflates***  The  increasing  utility  of  enol trifluoromethanesulfonates  (triflates)  in  metal-catalyzed  cross-coupling  reactions  has fueled the development of practical methods for their preparation. The synthesis of enol triflates is analogous to silyl enol ether formation. When the regioselectivity of enolization is not a concern, enol triflates may be prepared by trapping the enol tautomers of carbonyl compounds with triflic anhydride ($Tf_2O$) in the presence of a mild nitrogen base. These reactions are commonly carried out in dichloromethane to avoid interactions between the strongly electrophilic $Tf_2O$ and the solvent, although other non-Lewis basic media may be used. Due to the relative instability of enol triflates toward hydrolysis, manipulation during isolation and/or purification should be minimized and cold storage is recommended. In the following example, the enol triflate is prepared in good yield by treatment of the ketone with triflic anhydride and a bulky pyridine base.[76,77]

$Tf_2O$, 2,6-$t$-Bu-4-MePyr

70%

The use of phenyl- or $N$-(2-pyridyl)-triflimide in place of $Tf_2O$ can be advantageous, as described by Comins and coworkers.[78] In general, triflimides may be used to trap metallo enolates with excellent regiospecificity. Therefore, careful choice of enolization conditions can provide a vinyl triflate with a high degree of regioselection. The Comins pyridyl triflimide, which is more reactive than its phenyl variant, is a solid at ambient temperature and the byproducts are more easily removed from reaction mixtures.[79,80] In the example

[75]Mermerian, A. H.; Fu, G. C. *J. Am. Chem. Soc.* **2005**, *127*, 5604–5607.
[76]Scott, W. J.; Crisp, G. T.; Stille, J. K. *Org. Synth.* **1990**, *68*, 116–129.
[77]Cacchi, S.; Morera, E.; Ortar, G. *Org. Synth.* **1990**, *68*, 138–147.
[78]McMurry, J. E.; Scott, W. J. *Tetrahedron Lett.* **1983**, *24*, 979–982.
[79]Comins, D. L.; Dehghani, A. *Tetrahedron Lett.* **1992**, *33*, 6299–6302.
[80]Comins, D. L.; Dehghani, A.; Foti, C. J.; Joseph, S. P. *Org. Synth.* **1997**, *74*, 77–83.

below, an LDA-derived lithium enolate was trapped with Comins' chloropyridyl triflimide reagent to yield the vinyl triflate shown in high yield. It is notable that γ-deprotonation, leading to formation of the regioisomeric 1,3-diene, was not observed.

86%

### 1.2.4.3 Formation of Oxonium Salts

*1.2.4.3 Formation of Oxonium Salts* Trialkyloxonium salts are powerful alkylating agents with broad utility. These reagents are generally effective under very mild reaction conditions, and often alkylate modest nucleophiles when others fail; alkylation of over fifty functional groups has been reported with trialkyloxonium salts.[81] First reported by Meerwein (who was also responsible for the development of numerous applications), triethylammonium tetrafluoroborate provides access to ethyl enol ethers from carbonyl compounds (see Section 1.2.4.3). The synthesis of this reagent utilizes epichlorohydrin, boron trifluoride diethyletherate, and diethyl ether.[82] In an analogous manner, trimethylammonium tetrafluoroborate is prepared from dimethyl ether.[81]

*1.2.4.4 Reactions of Carbonyl Compounds with Oxonium Salts* Triethyloxonium tetrafluoroborate (Meerwein salt) is generally effective for the conversion of carbonyl compounds to their corresponding ethyl enol ether variants. For additional information regarding the synthesis of Meerwein salts, see Section 1.2.4.2. The reagent is an extremely powerful electrophile and care must be taken to ensure worker safety and minimize environmental impact during its use. That said, the example below illustrates the high yields that are achievable with $Et_3OBF_4$ under very mild conditions.[83]

[81]Curphey, T. J. *Org. Synth.* **1971**, *51*, 142–147.
[82]Meerwein, H. *Org. Synth.* **1966**, *46*, 113–115.
[83]Urban, F. J.; Anderson, B. G.; Orrill, S. L.; Daniels, P. J. *Org. Process Res. Dev.* **2001**, *5*, 575–580.

*1.2.4.5  Preparation of Hydroperoxides and Peroxyethers*  Care should be taken in handling all peroxides due to their propensity to decompose violently when exposed to shock or heat. As a result, peroxy compounds are rarely utilized as synthetic intermediates on large scale. However, well-studied hydroperoxide reagents are sometimes employed. A common method for hydroperoxide synthesis involves reaction of an electrophile with hydrogen peroxide under acidic or basic catalysis. Although most commonly utilized in oxidation reactions, urea hydrogen peroxide (UHP) is an alternative form of $H_2O_2$ that has proven useful in hydroperoxide synthesis. In the following example, moderate heating of an activated alkyl halide with UHP in DMF provides the hydroperoxide in excellent yield.[84] Some of the practical advantages of UHP include the fact that it is a relatively stable, inexpensive, and easily-handled solid.

In a manner analogous to the preparation of ethers from alcohols, peroxyethers can be prepared via treatment of hydroperoxides with a base in the presence of an alkyl halide or sulfonic acid ester (e.g., mesylate). These conditions are not always applicable however, as aliphatic peroxides containing α-hydrogens are prone to degradation under basic conditions. Nevertheless, when reactive electrophiles are used, appreciable yields and reaction rates can be seen at or near ambient temperature for many substrates. Reported methods involve the use of primary alkyl iodides in the presence of silver oxide[85,86] or, preferably, bromides or sulfonates with potassium hydroxide base. In the latter method, polyethylene glycol is a recommended cosolvent. The addition of crown ethers or PEG-400 is thought to promote nucleophilicity through chelation to potassium ions and sometimes leads to superior yields.[87]

For less reactive or sterically hindered peroxides, cesium hydroxide has been shown to provide acceptable reaction rates in peroxyether formation at ambient temperature, thus avoiding a thermal degradation threat.[88]

[84]Aoki, M.; Seebach, D. *Helv. Chim. Acta* **2001**, *84*, 187–207.

[85]Ito, T.; Tokuyasu, T.; Masuyama, A.; Nojima, M.; McCullough, K. J. *Tetrahedron* **2003**, *59*, 525–536.

[86]Hamada, Y.; Tokuhara, H.; Masuyama, A.; Nojima, M.; Kim, H.-S.; Ono, K.; Ogura, N.; Wataya, Y. *J. Med. Chem.* **2002**, *45*, 1374–1378.

[87]Bourgeois, M. J.; Montaudon, E.; Maillard, B. *Synthesis* **1989**, 700–701.

[88]Dussault, P. H.; Eary, C. T. *J. Am. Chem. Soc.* **1998**, *120*, 7133–7134.

3.6 : 1 *E/Z*          52%

***1.2.4.6   Alkylation of Oximes***   In most instances, oximes react with electrophiles such as aliphatic halides in a manner analogous to the corresponding reactions of alcohols. The example below outlines a typical procedure, in which an oxime is treated with an alkyl bromide and an inorganic base in a polar aprotic solvent.[89]

91%

Takemoto and coworkers have employed oxime nucleophiles in palladium-catalyzed allylic substitution reactions. Here, the electrophile is an intermediate palladium(II) $\pi$-allyl complex, and the base is potassium carbonate.[90] The reaction proceeds in good yield to give the product of O-alkylation near ambient temperature, and at a reasonable rate, despite the poor solubility of $K_2CO_3$ in dichloromethane. For more on metal-catalyzed allylic substitution with oxygen nucleophiles, see Section 6.8.3.

75%

The example below highlights the ambiguous nature of an oxime's ability to react at nitrogen, as opposed to oxygen, under certain conditions. In this instance, palladium(II) catalysis under solvent-free conditions provided the N-alkylated oxime as a mixture of geometrical isomers in moderate yield.

62%

In the example below, the oxygen atom of an oxime adds in a conjugate fashion to provide a stabilized anion for subsequent Horner-Wadsworth-Emmons reaction with nicotinaldehyde. Sodium hydride, which requires some additional handling precautions for large scale applications, is employed as the base in this instance.[91] This example is included to further

[89]Yamada, T.; Goto, K.; Mitsuda, Y.; Tsuji, J. *Tetrahedron Lett.* **1987**, *28*, 4557–4560.
[90]Miyabe, H.; Yoshida, K.; Reddy, V. K.; Matsumura, A.; Takemoto, Y. *J. Org. Chem.* **2005**, *70*, 5630–5635.
[91]Shen, Y.; Jiang, G.-F. *Synthesis* **2000**, 502–504.

illustrate the useful application of oxime nucleophiles. For more on nucleophilic addition of ROH to polarized carbon–carbon multiple bonds, see Section 3.2.2.

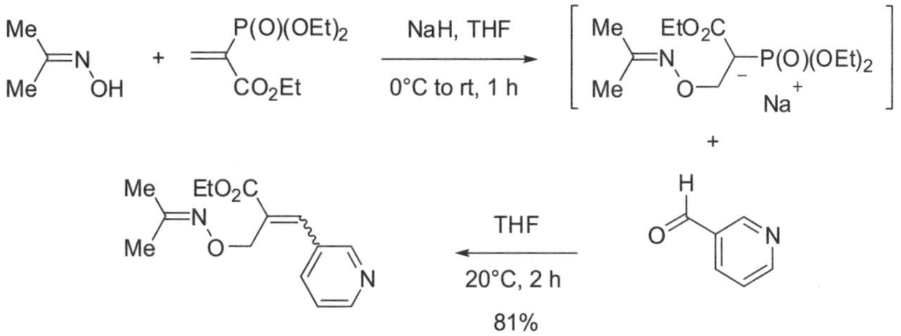

## 1.3  SULFUR NUCLEOPHILES

### 1.3.1  Reactions with Thiols

*1.3.1.1  Preparation of Thioethers*   Thioethers are generally prepared via reaction of thiols with electrophiles such as alkyl halides in the presence of an appropriate base. Thiols are not potent nucleophiles; however, $S_N2$ displacement reactions may frequently be accomplished near ambient temperature, as illustrated in the example below.[1] In this instance, sterically encumbered trityl thiol is alkylated in high yield and at a reasonable rate with a reactive benzylic bromide.

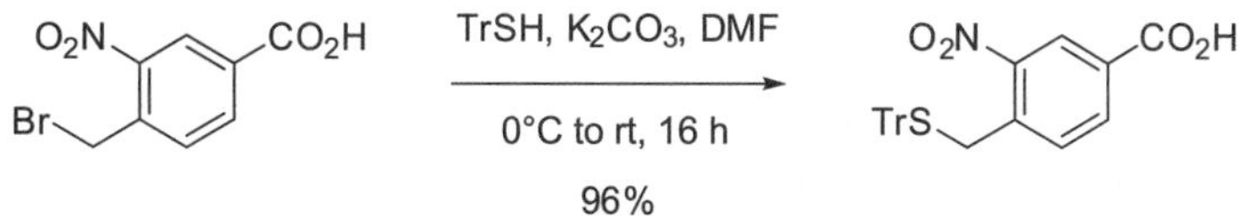

Cabri and coworkers described a highly efficient thioether formation in the synthesis of the ergot alkaloid derivative pergolide mesylate.[54] Direct $S_N2$ displacement of the primary mesylate shown below was accomplished with commercially available sodium methanethiolate, which can also be generated *in situ* from methanethiol and sodium hydroxide. The latter method is generally not preferred due to the challenges associated with handling toxic, malodorous methanethiol gas (bp 6°C).

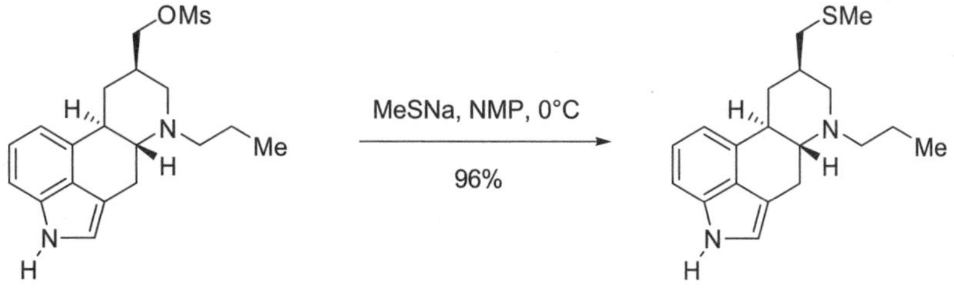

*1.3.1.2  Cleavage of Arylmethyl Ethers*   Another useful application of nucleophilic sulfur reagents involves the cleavage of arylmethyl ethers to afford the corresponding phenols.[92] In the example below, an intermediate in the synthesis of a VEGFR inhibitor candidate was prepared in high yield by treatment of a methoxybenzothiophene derivative with four equivalents of methionine in methanesulfonic acid.[93]

### 1.3.2   Reactions with Other Sulfur Nucleophiles

***1.3.2.1   Preparation of Thiols***   The preparation of thiols via direct displacement of alkyl halides with sodium sulfide is not straightforward due to low solubility of this sulfur reagent in most organic solvents. As a result, simple aliphatic thiols are most often prepared via the intermediacy of somewhat more elaborate precursors. In the example below an isothiuronium salt is prepared via reaction of an alkyl bromide with thiourea, followed by reaction with a nucleophilic amine.[94] This method relies on distillation of the thiol product for isolation and purification, and is therefore most applicable to the preparation of low molecular weight derivatives.

Perhaps more generally, aliphatic thiols are prepared by reaction of alkyl halides with thioacetic acid salts, followed by thioester cleavage under alkaline conditions.[95] The example below is representative, although use of commercially available thioacetic acid sodium salt would be preferred due to the obnoxious odor associated with free thioacids.

[92]Linderberg, M. T.; Moge, M.; Sivadasan, S. *Org. Process Res. Dev.* **2004**, *8*, 838–845.
[93]Scott, R. W.; Neville, S. N.; Urbina, A.; Camp, D.; Stankovic, N. *Org. Process Res. Dev.* **2006**, *10*, 296–303.
[94]Cossar, B. C.; Fournier, J. O.; Fields, D. L. *J. Org. Chem.* **1962**, *27*, 93–95.
[95]Canaria, C. A.; Smith, J. O.; Yu, C. J.; Fraser, S. E.; Lansford, R. *Tetrahedron Lett.* **2005**, *46*, 4813–4816.

Independent of the method employed for their construction, thiols are a challenge to store and manipulate due to their stench and propensity to form disulfides in a mildly oxidizing environment such as air. As a result, the more stable disulfide dimers are sometimes targeted as thiol precursors. An attractive method for nucleophilic disulfide cleavage involves treatment with Cleland's reagent (dithiothreitol, DTT), which liberates the thiol under mild aqueous conditions.[95] Although this method would undoubtedly apply to a broad range of synthetic organic challenges, the use of DTT is dominated by its applications in biochemistry as a means to reduce thiolated DNA.

A number of other chemical methods have been developed for disulfide cleavage, including the use of mild reducing agents such as phosphines, sodium borohydride, or zinc in acetic acid.[96] A single example is included here, but for more on the preparation of thiols via reductive cleavage of disulfides, see Section 9.7.5.

**1.3.2.2  Formation of Bunte Salts**  Salts of S-alkyl or S-aryl hydrogen thiosulfates are known as Bunte salts. They are frequently prepared as precursors to thiols or disulfides. Most commonly, Bunte salts are prepared from alkyl halides and sodium thiosulfate by heating in aqueous methanol.[97]

[96]Hartman, R. F.; Rose, S. D. *J. Org. Chem.* **2006**, *71*, 6342–6350.
[97]Tseng, C. C.; Handa, I.; Abdel-Sayed, A. N.; Bauer, L. *Tetrahedron* **1988**, *44*, 1893–1904.

In the example below, a pyridinium Bunte salt was prepared from the aryl thiol, then converted into the corresponding thiocyanate in good yield.[98] For a discussion on the preparation of aryl thiocyanates via electrophilic aromatic substitution, see Section 5.3.6.

### 1.3.2.3 Alkylation of Sulfinic Acid Salts

*1.3.2.3   Alkylation of Sulfinic Acid Salts*    Sulfinic acids react with common electrophiles upon treatment with a mild base. Various arene sulfinic acid derivatives are commercially available as their stable, crystalline salts, in contrast to the parent acids, which are somewhat prone to decomposition and dimerization. Sulfinic acid salts are generally nucleophilic at sulfur, as opposed to oxygen, although exceptions have been reported. In the example below, commercially available sodium benzenesulfinate is reacted with an allylic bromide in dimethylformamide to provide the sulfone in high yield.[99,100]

Another useful application of sulfinic acid nucleophiles is seen in palladium-catalyzed allylic substitution reactions. This mild methodology is useful for deprotection of allyl esters where standard acidic or alkaline hydrolysis would be incompatible with sensitive functionality, and is generally amenable to large scale. The following examples, from the research groups of Trost[101] and Hegedus[102] illustrate this reactivity toward π-allyl electrophiles. For more on metal-catalyzed allylic substitution, see Section 6.8.

[98]Alcaraz, M.-L.; Atkinson, S.; Cornwall, P.; Foster, A. C.; Gill, D. M.; Humphries, L. A.; Keegan, P. S.; Kemp, R.; Merifield, E.; Nixon, R. A.; Noble, A. J.; O'Beirne, D.; Patel, Z. M.; Perkins, J.; Rowan, P.; Sadler, P.; Singleton, J. T.; Tornos, J.; Watts, A. J.; Woodland, I. A. *Org. Process Res. Dev.* **2005**, *9*, 555–569.

[99]Nilsson, Y. I. M.; Andersson, P. G.; Baeckvall, J. E. *J. Am. Chem. Soc.* **1993**, *115*, 6609–6613.

[100]Sellen, M.; Baeckvall, J. E.; Helquist, P. *J. Org. Chem.* **1991**, *56*, 835–839.

[101]Trost, B. M.; Crawley, M. L.; Lee, C. B. *J. Am. Chem. Soc.* **2000**, *122*, 6120–6121.

[102]Sebahar, H. L.; Yoshida, K.; Hegedus, L. S. *J. Org. Chem.* **2002**, *67*, 3788–3795.

PhSO$_2$Na,
THAB, Cat.*

DCM, H$_2$O, rt

92%, 94% ee

TBSO — OAc — OAc  →  TBSO — OAc, H, SO$_2$Ph

*cat. = 2 mol% [n$^3$–C$_3$H$_5$PdCl]$_2$ + 6 mol% Ph$_2$P ··· NH  HN ··· PPh$_2$

PhSO$_2$Na,
[n$^3$–C$_3$H$_5$PdCl]$_2$, dppe

DMSO, 60°C, 0.5 h

64%

### 1.3.2.4 Attack by Sulfite Ion

Sulfite ion (SO$_3^{2-}$) is sufficiently nucleophilic to participate in S$_N$2 displacement reactions with aliphatic alkyl halides to provide the corresponding sulfonic acids after an acidic workup. The following example from Chen and coworkers is representative of typical reaction conditions and yields.[103]

Na$_2$SO$_3$, H$_2$O

Acetone, reflux

59%

Cl — CH$_2$ — C$_6$H$_4$ — CO$_2$Et, CO$_2$Et, NHAc  →  NaO$_3$S — CH$_2$ — C$_6$H$_4$ — CO$_2$Et, CO$_2$Et, NHAc

In the following example, the reaction of sodium sulfite with a symmetrical dichloride yielded the corresponding disulfonate intermediate. This was not isolated, but rather converted into a sultone by treatment with acidic Amberlite resin at an elevated temperature.[104]

i) Na$_2$SO$_3$, H$_2$O
Reflux

SO$_3$Na  SO$_3$Na

ii) Amberlite H$^+$
iii) 130–150°C

72–80%

2 × sultone

### 1.3.2.5 Preparation of Alkyl Thiocyanates

Thiocyanate salts (typically sodium or potassium) are ambident nucleophiles, capable of either sulfur- or nitrogen-centered S$_N$2 attack on electrophiles. Reaction at sulfur is more frequently observed, providing thiocyanate products (vs. isothiocyanates). In the following example, Sa and coworkers report on a high-yielding preparation of allylic thiocyanates from allylic bromide precursors.[105]

[103]Chen, Y. T.; Xie, J.; Seto, C. T. *J. Org. Chem.* **2003**, *68*, 4123–4125.
[104]Snoddy, A. O. *Org. Synth.* **1957**, *37*, 55–57.
[105]Sa, M. M.; Fernandes, L.; Ferreira, M.; Bortoluzzi, A. J. *Tetrahedron Lett.* **2008**, *49*, 1228–1232.

In addition to the primary reactivity described above, the authors also report measurable quantities of isothiocyanate products when the acrylate starting materials were alkyl substituted. Perhaps most interesting, the initially observed thiocyanate:isothiocyanate mixture (7:1) equilibrated to favor the latter compound (1:6) after prolonged periods at ambient temperature.

## 1.4 NITROGEN NUCLEOPHILES

### 1.4.1 Amine Alkylation

***1.4.1.1 Amine Alkylation with Alkyl Halides and Onium Salts***    The alkylation of nitrogen with alkyl halides is one of the most commonly utilized transformations in target-oriented synthetic chemistry. The inherently strong nucleophilicity of amines provides reliable, generally predictable reaction rates under mild conditions and in a variety of solvents. The nucleophilicity of an amine is strongly influenced by steric and inductive influences from substituents. A helpful discussion regarding the nucleophilicities of various primary and secondary amines has been published.[106] In the example below, the primary amine is first protected as a ketimine derivative through dehydrative condensation with methyl isobutylketone before introduction of the alkyl chloride. This process prevents the formation of a mixture of N-alkylation products. To enhance the alkylation rate, the chloride is converted to its corresponding iodide *in situ* via reaction with potassium iodide (Finkelstein reaction; see Section 1.5.1.1).[107] For further examples of ketimine formation, see Section 2.7.1).

[106]Brotzel, F.; Chu, Y. C.; Mayr, H. *J. Org. Chem.* **2007**, *72*, 3679–3688.
[107]Laduron, F.; Tamborowski, V.; Moens, L.; Horvath, A.; De Smaele, D.; Leurs, S. *Org. Process Res. Dev.* **2005**, *9*, 102–104.

For ease of handling, amines are often prepared and stored as crystalline, non-volatile salts of a mineral acid. In most instances, these salts may be utilized directly in alkylation processes by introducing of a stoichiometric quantity of a base. The salt is first broken to provide the corresponding "free base", which then undergoes alkylation.[108] These reactions are generally facile and efficient, providing the starting salt is sufficiently soluble in the reaction medium, and overalkylation is controlled through proper stoichiometry and reaction temperature.

The tendency of tertiary amines to react further with alkylating agents to provide quaternary ammonium salts is occasionally a complicating factor. There are examples, however, where tertiary amine quaternization has been utilized in a productive manner. In the Delépine reaction,[109,110] primary alkyl halides are converted to the corresponding primary amines via treatment with hexamethylenetetramine, followed by hydrolysis. This sequence provides an inexpensive and reliable entry into aliphatic amines that can be conveniently isolated and handled as hydrochloride salts. The example below is representative.[111]

(X = I, Cl)

When benzylic alkyl halides are treated in a similar fashion with excess hexamethylenetetramine, the product of hydrolysis is the aldehyde. This useful variation, known as the Sommelet reaction,[112] provides fairly general access to aromatic aldehydes from benzylic halide precursors.[113–118] For additional discussion on benzylic oxidation, see Section 10.3.8.1 and 10.4.1.5.

[108]Hashimoto, H.; Ikemoto, T.; Itoh, T.; Maruyama, H.; Hanaoka, T.; Wakimasu, M.; Mitsudera, H.; Tomimatsu, K. *Org. Process Res. Dev.* **2002**, *6*, 70–73.

[109]Delepine, M. *Bull. Soc. Chim. Fr.* **1895**, *13*, 352–355.

[110]Blazevic, N.; Kolbah, D.; Belin, B.; Sunjic, V.; Kajfez, F. *Synthesis* **1979**, 161–176.

[111]Nodiff, E. A.; Hulsizer, J. M.; Tanabe, K. *Chem. Ind.* **1974**, 962–963.

[112]Sommelet, M. *C. R. Acad. Sci. IIc* **1914**, *157*, 852–854.

[113]Angyal, S. J.; Morris, P. J.; Rassack, R. C.; Waterer, J. A. *J. Chem. Soc.* **1949**, 2704–2706.

[114]Angyal, S. J.; Rassack, R. C. *J. Chem. Soc.* **1949**, 2700–2704.

[115]Angyal, S. J.; Morris, P. J.; Tetaz, J. R.; Wilson, J. G. *J. Chem. Soc.* **1950**, 2141–2145.

[116]Angyal, S. J.; Barlin, G. B.; Wailes, P. C. *J. Chem. Soc.* **1953**, 1740–1741.

[117]Angyal, S. J.; Penman, D. R.; Warwick, G. P. *J. Chem. Soc.* **1953**, 1742–1747.

[118]Angyal, S. J.; Penman, D. R.; Warwick, G. P. *J. Chem. Soc.* **1953**, 1737–1739.

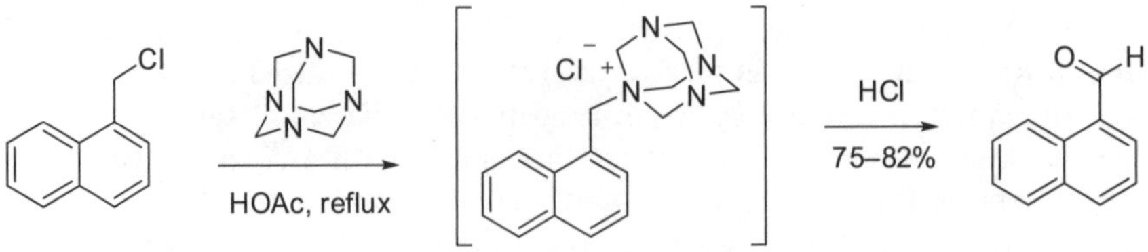

### 1.4.1.2 Amine Alkylation with Inorganic Esters

*1.4.1.2  Amine Alkylation with Inorganic Esters*    Amines typically react with inorganic esters in a manner analogous to their interaction with alkyl halides. As exemplified by the following scheme, reaction of an amine with an alkyl sulfonate (e.g., mesylate, tosylate, etc.) proceeds smoothly in a polar aprotic solvent to provide the substituted amine.[119] Mild to moderate heating is often employed to decrease reaction time. One equivalent of a sulfonic acid is generated over the course of the reaction, so the product will require neutralization unless another base is included in the reaction.

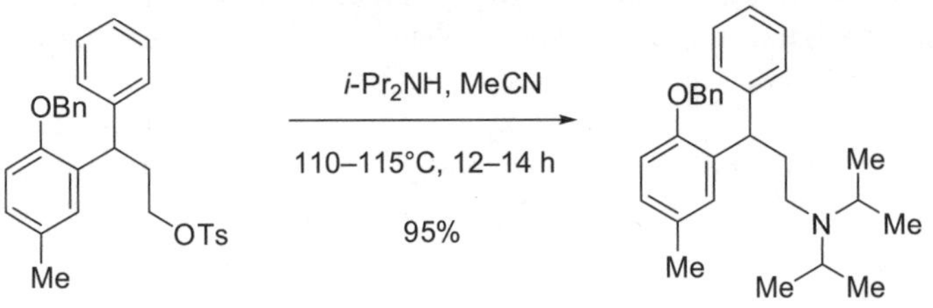

When cyclic sulfates are employed, initial attack typically occurs at the less sterically encumbered carbon center to provide the intermediate amino sulfate. In the example below, a second equivalent of amine is introduced to promote a second displacement at the more substituted carbon center. Note that considerably more forcing conditions are required for the second alkylation, and that this attack results in inversion of the stereocenter. The more sluggish reaction rate for the second addition often allows the use of two distinct amines in this stepwise reaction sequence.

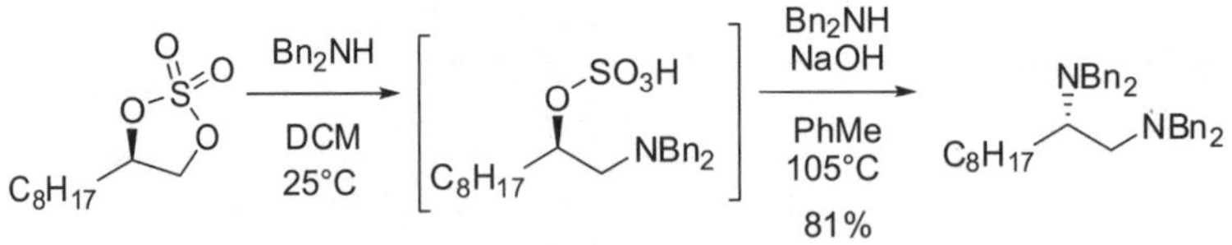

Trialkyloxonium tetrafluoroborate (Meerwein salts) have been used in the preparation of N-substituted pyrroles for pharmaceutical applications.[120] In the example below, the 2-oxazoline-5-one derived from phenylalanine is N-ethylated to produce the reactive dipole 1,3-oxazolium-5-oxide (munchnone). In the presence of dimethyl acetylenedicarboxylate, a [3 + 2] cycloaddition takes place to provide the N-ethyl pyrrole after carbon dioxide is lost via retro [4 + 2]. The authors note that extremely reactive alkylating agents

[119]Srinivas, K.; Srinivasan, N.; Reddy, K. S.; Ramakrishna, M.; Reddy, C. R.; Arunagiri, M.; Kumari, R. L.; Venkataraman, S.; Mathad, V. T. *Org. Process Res. Dev.* **2005**, *9*, 314–318.
[120]Hershenson, F. M.; Pavia, M. R. *Synthesis* **1988**, 999–1001.

are required for this transformation; methyl iodide, dimethyl sulfate, and benzyl bromide are ineffective.

### 1.4.1.3 Amine Alkylation with Alcohols

Alcohols may serve as useful precursors to amines, although an activation step is required. Unfortunately, the activation step often involves the use of reagents that add considerable expense and/or waste to the process. For instance, if the amine nucleophile is suitably acidic, as is the case for the Boc-protected sulfonamide below, Mitsunobu conditions may be employed.[121] Although this method can be effective for small scale applications, there is general agreement among process chemists that the phosphine, azodicarboxylate reagents and the byproducts they produce in the reaction decrease the utility of this method for large scale.

For more typical amine nucleophiles, such as simple unactivated primary amines, preactivation of the alcohol as its sulfonate ester (e.g., mesylate, tosylate, brosylate, etc.) is necessary.[122] This method is preferable for bulk production, due to the lower expense of sulfonyl ester halides, and the ease of removal of sulfonic acid byproducts via alkaline aqueous extraction. For more on the use of sulfonates in amine alkylation, see Section 1.4.1.2.

### 1.4.1.4 Amine Alkylation with Diazo Compounds

As discussed in Section 1.2.3.5, the use of diazo compounds of low molecular weight is not preferred for large scale applications. However, the reaction of activated amines with diazomethane to provide N-methylated

---

[121]Nishino, Y.; Komurasaki, T.; Yuasa, T.; Kakinuma, M.; Izumi, K.; Kobayashi, M.; Fujiie, S.; Gotoh, T.; Masui, Y.; Hajima, M.; Takahira, M.; Okuyama, A.; Kataoka, T. *Org. Process Res. Dev.* **2003**, *7*, 649–654.

[122]Conrow, R. E.; Dean, W. D.; Zinke, P. W.; Deason, M. E.; Sproull, S. J.; Dantanarayana, A. P.; DuPriest, M. T. *Org. Process Res. Dev.* **1999**, *3*, 114–120.

products is a clean, high-yielding transformation that can be carried out quite reliably on laboratory scale in fit-for-purpose glassware.[123]

### 1.4.1.5 Transamination

***1.4.1.5  Transamination***    The direct nucleophilic displacement of one aliphatic amine by another is not a widely employed transformation. The formal product of transamination can be obtained, however, under a range of reaction conditions. Although not technically a nucleophilic aliphatic substitution reaction, the example below highlights a formal transamination reaction. In this case, a series of elimination/conjugate addition reactions provide an efficient exchange of methylamine for *tert*-butylamine. Achieving a comparable transamination via direct $S_N2$ displacement with *tert*-butylamine would be very unlikely.[124]

***1.4.1.6  Amine Alkylation with Epoxides***    The ring-opening of epoxides with nitrogen nucleophiles is a facile, well-established transformation. The reaction will often proceed thermally, although the use of Lewis or Brønsted acid catalysis is common. In their kilogram-scale synthesis of the anti-influenza drug oseltamivir (active pharmaceutical ingredient in Tamiflu®), Rohloff and coworkers described a regioselective ring-opening of a key epoxide intermediate with sodium azide and ammonium chloride in aqueous ethanol. Although this reaction was run at an elevated temperature, the authors warned against exceeding 80°C due to the potential explosivity of azides.[125]

[123]Di Gioia, M. L.; Leggio, A.; Le Pera, A.; Liguori, A.; Napoli, A.; Siciliano, C.; Sindona, G. *J. Org. Chem.* **2003**, *68*, 7416–7421.
[124]Amato, J. S.; Chung, J. Y. L.; Cvetovich, R. J.; Reamer, R. A.; Zhao, D.; Zhou, G.; Gong, X. *Org. Process Res. Dev.* **2004**, *8*, 939–941.
[125]Rohloff, J. C.; Kent, K. M.; Postich, M. J.; Becker, M. W.; Chapman, H. H.; Kelly, D. E.; Lew, W.; Louie, M. S.; McGee, L. R.; Prisbe, E. J.; Schultze, L. M.; Yu, R. H.; Zhang, L. *J. Org. Chem.* **1998**, *63*, 4545–4550.

The thermal and shock sensitivity of some azide-containing compounds has limited their use in large scale synthesis. To address these issues for bulk production, Karpf and Trussardi have developed alternative conditions for the regioselective introduction of nitrogen to their substrate. The use of benzylamine, with magnesium bromide as a Lewis acid catalyst, provided the desired 1,2-amino alcohol in comparable yield and regioselectivity.[126]

85 : 15

For additional discussion on the use of azide nucleophiles, see Section 1.4.3.2. For another example of the reaction of an epoxide with sodium azide see, Section 1.2.2.8.

### *1.4.1.7  Preparation of 1° Amines via Hexamethyldisilazane*
The direct nucleophilic displacement of aliphatic halides by hexamethyldisilazane (HMDS) has been demonstrated.[127] The main advantages of this method for the synthesis of amines are the elimination of multi-alkylation issues and that TMS groups are easily removed by an acidic workup to liberate an ammonium salt. The method has not been widely adopted, however, due to the poor nucleophilicity of the sterically bulky HMDS. In most cases, reaction rates are slow and yields are modest. In the example below, the displacement of a secondary chloride is assisted via participation by the neighboring boron atom, thus accelerating the reaction rate significantly.[128] Note that participation by boron offsets the normal rate advantage offered by the primary bromide (vs. secondary chloride) and also provides a steric bias that influences the stereochemical course of the reaction.[129]

### *1.4.1.8  Preparation of Isocyanides ("Isonitriles")*
The first synthesis of an isocyanide was reported by Lieke in 1859; allyl iodide was reacted with silver cyanide to yield allyl isocyanide. A second method involves reaction of a primary amine with dichlorocarbene (generated by treatment of chloroform with a strong base), and is known as the Hofmann carbylamine reaction. Weber and coworkers reported an improvement to the latter method, wherein the use of phase transfer catalysis in dichloromethane-water solvent systems provides superior yields under milder conditions.[130,131]

[126]Karpf, M.; Trussardi, R. *J. Org. Chem.* **2001**, *66*, 2044–2051.
[127]Bestmann, H. J.; Woelfel, G. *Chem. Ber.* **1984**, *117*, 1250–1254.
[128]Matteson, D. S.; Schaumberg, G. D. *J. Org. Chem.* **1966**, *31*, 726–731.
[129]Wityak, J.; Earl, R. A.; Abelman, M. M.; Bethel, Y. B.; Fisher, B. N.; Kauffman, G. S.; Kettner, C. A.; Ma, P.; McMillan, J. L.;et al. *J. Org. Chem.* **1995**, *60*, 3717–3722.
[130]Weber, W. P.; Gokel, G. W. *Tetrahedron Lett.* **1972**, 1637–1640.
[131]Weber, W. P.; Gokel, G. W.; Ugi, I. K. *Angew. Chem., Int. Ed. Engl.* **1972**, *11*, 530–531.

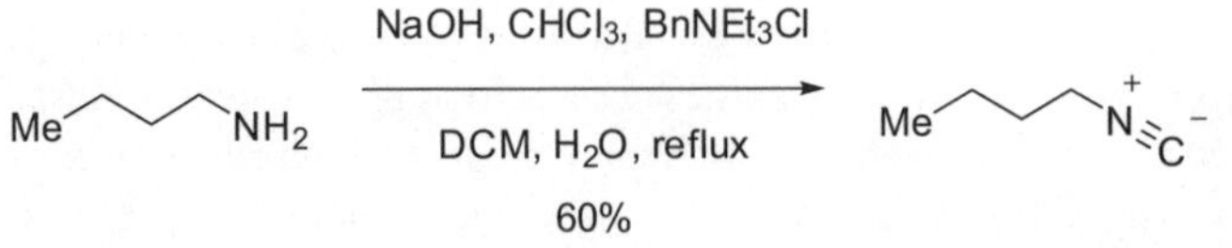

However, isocyanides can be more generally prepared via dehydration of *N*-formyl precursors.[132] A range of dehydrating agents is effective in the reaction, with phosphorous oxychloride (POCl$_3$) used in the representative example below.[133]

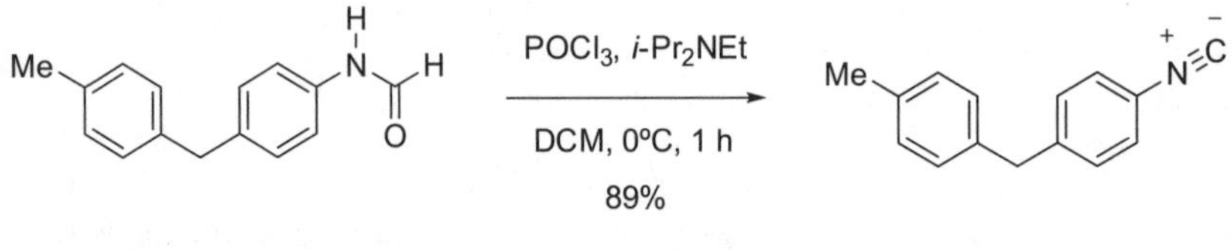

***1.4.1.9 Methylation of Amines, the Eschweiler-Clarke reaction*** A special case of imine reduction is demonstrated in the Eschweiler-Clarke reaction, in which an amine is condensed with formaldehyde in the presence of formic acid to yield the N-methylated product. The reaction is most simply carried out in water at reflux, as demonstrated in the synthesis of citalopram.[134]

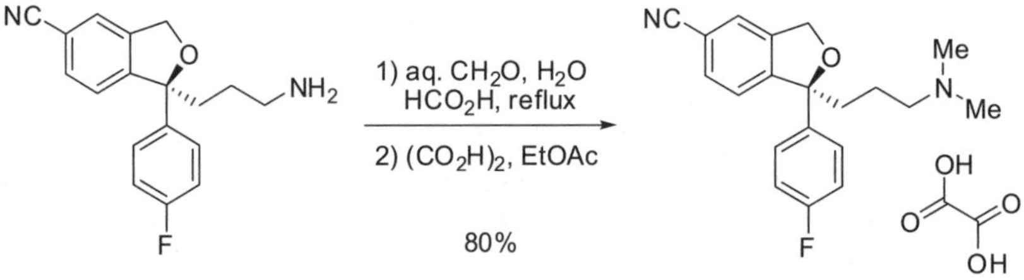

***1.4.1.10 Preparation Sulfenamides*** As illustrated by Walinsky and coworkers in their work toward the atypical antipsychotic drug ziprasidone, nucleophilic secondary amines react with disulfides via attack at sulfur to liberate sulfenamide products.[135] These hydrolytically labile compounds may serve as amine precursors[136] or intermediates in the preparation of heterocycles[137] such the benzisothiazole shown below.

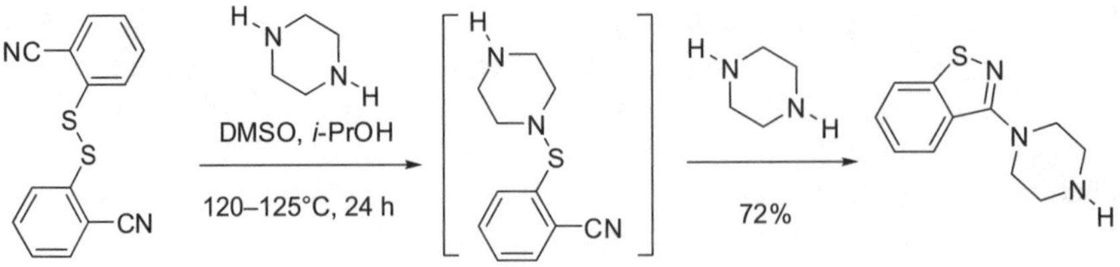

[132]Ugi, I.; Meyr, R. *Angew. Chem., Int. Ed. Engl.* **1958**, *70*, 702–703.

[133]Obrecht, R.; Herrmann, R.; Ugi, I. *Synthesis* **1985**, 400–402.

[134]Elati, C. R.; Kolla, N.; Vankawala, P. J.; Gangula, S.; Chalamala, S.; Sundaram, V.; Bhattacharya, A.; Vurimidi, H.; Mathad, V. T. *Org. Process Res. Dev.* **2007**, *11*, 289–292.

[135]Walinsky, S. W.; Fox, D. E.; Lambert, J. F.; Sinay, T. G. *Org. Process Res. Dev.* **1999**, *3*, 126–130.

[136]Wuts, P. G. M.; Jung, Y. W. *Tetrahedron Lett.* **1986**, *27*, 2079–2082.

[137]Davis, F. A. *Int. J. Sulfur Chem.* **1973**, *8*, 71–81.

### 1.4.2   N-Alkylation of Amides, Lactams, Imides, and Carbamates

*1.4.2.1   Alkylation with Alkyl Halides*   The alkylation of aryl amides with alkyl halides can be achieved under mild phase transfer conditions.[138] In the example below, benzamide is converted to its *N*-butyl derivative through reaction with *n*-butyl bromide in the presence of tetra-*n*-butylammonium hydrogensulfate. The high concentration of sodium hydroxide prevents hydrolytic side reactions from occurring by promoting clear separation between aqueous and organic phases. Benzene can typically be replaced by toluene or dichloromethane to provide secondary amides with equal efficiency. The primary concern with this protocol is competing formation of tertiary amides via double alkylation. However, this issue can be addressed by careful control of stoichiometry, as the second alkylation is generally much slower than the first.

79%

Unfortunately, yields drop off somewhat when aliphatic amides are reacted under these conditions, and attempted introduction of a second alkyl group is generally unsuccessful. Primary sulfonamides can also be alkylated with these conditions, although doubly alkylated products are primarily obtained due to the increased acidity of this substrate class.

In cases where the mild biphasic conditions described above are ineffective, amides can typically be alkylated by deprotonation with a strong base under anhydrous conditions, followed by combination with an alkyl halide. In the alkylation of δ-valerolactam below, sodium hexamethyldisilazide in THF is utilized.[139] The amide salts often exhibit poor solubility in solvents such as THF, however, so proper agitation, along with moderate heating may be required for successful reactions.

93%

Primary amides may be converted to iodo lactams via intramolecular alkylation of electrophilic iodonium species.[140,141] The nucleophilic component in the example below is not the primary amide itself, but rather the *N,O*-bis(trimethylsilyl)imidate derivative shown in brackets. Silylation of the starting amide promotes the N-alkylation pathway

[138]Gajda, T.; Zwierzak, A. *Synthesis* **1981**, 1005–1008.

[139]Roenn, M.; McCubbin, Q.; Winter, S.; Veige, M. K.; Grimster, N.; Alorati, T.; Plamondon, L. *Org. Process Res. Dev.* **2007**, *11*, 241–245.

[140]Knapp, S.; Levorse, A. T. *J. Org. Chem.* **1988**, *53*, 4006–4014.

[141]Knapp, S.; Rodriques, K. E.; Levorse, A. T.; Ornaf, R. M. *Tetrahedron Lett.* **1985**, *26*, 1803–1806.

and prevents competitive iodolactone formation, as originally reported by Ganem and coworkers.[142]

3 : 1

During their pilot scale synthesis of an oxazolidinone antibacterial candidate, Lu and coworkers carried out a multi-kilogram carbamate alkylation as the first operation in a highly efficient three step sequence.[143] In this example, an *N*-aryl carbamate was reacted with a primary alkyl chloride in the presence of lithium *t*-butoxide. The desired alkylation product was obtained in high yield under these conditions, which also lead to subsequent acetate cleavage and oxazolidinone formation. The use of lithium-derived bases was critical, as sodium and potassium counterions gave rise to isomeric byproducts, as reported by Perrault during the development of structurally similar linezolid.[144]

In the example below from Augustine and coworkers, potassium succinimide is reacted with α-chloroacetophenone in DMF to provide an *N*-alkyl succinimide that served as a precursor to enantiomerically enriched 2-amino-1-phenylethanol.[145] Consistent with their amide relatives, alkylation of imides most commonly occurs on nitrogen in the presence of a base, although O-alkylation can be a complicating side reaction. The use of a potassium counterion is generally effective for promoting nitrogen-centered attack.

67%

[142]Biloski, A. J.; Wood, R. D.; Ganem, B. *J. Am. Chem. Soc.* **1982**, *104*, 3233–3235.

[143]Lu, C. V.; Chen, J. J.; Perrault, W. R.; Conway, B. G.; Maloney, M. T.; Wang, Y. *Org. Process Res. Dev.* **2006**, *10*, 272–277.

[144]Perrault, W. R.; Pearlman, B. A.; Godrej, D. B.; Jeganathan, A.; Yamagata, K.; Chen, J. J.; Lu, C. V.; Herrinton, P. M.; Gadwood, R. C.; Chan, L.; Lyster, M. A.; Maloney, M. T.; Moeslein, J. A.; Greene, M. L.; Barbachyn, M. R. *Org. Process Res. Dev.* **2003**, *7*, 533–546.

[145]Tanielyan, S. K.; Marin, N.; Alvez, G.; Augustine, R. L. *Org. Process Res. Dev.* **2006**, *10*, 893–898.

### 1.4.2.2 Alkylation with Alkyl Sulfonates

*1.4.2.2  Alkylation with Alkyl Sulfonates*    Acetamide derivatives undergo intramolecular N-alkylation reactions with alkyl sulfonates (e.g., mesylates, tosylates, etc.) in good yield.[146] Sodium hydride in tetrahydrofuran was employed in this example, and is a common first choice for laboratory scale since the gaseous hydrogen byproduct is easily removed and relatively inert. Sodium hexamethyldisilazide (NaHMDS), which is safer and easier to handle on large scale, has also proven effective in some cases and should be considered.

Potassium *tert*-butoxide has also been successfully employed in amide alkylation reactions. In the example below, high yield of a bicyclic benzamide product was obtained via intramolecular mesylate displacement in THF.[147]

During development of synthetic methodology for the synthesis of oxacephem antibiotics, Chmielewski and coworkers reported an efficient intramolecular β-lactam alkylation with a primary tosylate.[148] In this instance, a very mild base (sodium carbonate) was sufficient to promote the alkylation, so long as tetra–*n*–butylammonium bromide was included.

### 1.4.3  Other Nitrogen Nucleophiles

*1.4.3.1  Nitrite Nucleophiles: Preparation of Nitro Compounds*    The preparation of aliphatic nitro compounds can be accomplished by treatment of alkyl halides with sodium or potassium nitrite. Moderate heating in a polar solvent such as acetone, acetonitrile, dimethylformamide or dimethylsulfoxide is generally required, and the formation of isomeric alkyl nitrites is competitive.[149]

[146]Hansen, M. M.; Bertsch, C. F.; Harkness, A. R.; Huff, B. E.; Hutchison, D. R.; Khau, V. V.; LeTourneau, M. E.; Martinelli, M. J.; Misner, J. W.; Peterson, B. C.; Rieck, J. A.; Sullivan, K. A.; Wright, I. G. *J. Org. Chem.* **1998**, *63*, 775–785.

[147]Avenoza, A.; Cativiela, C.; Busto, J. H.; Fernandez-Recio, M. A.; Peregrina, J. M.; Rodriguez, F. *Tetrahedron* **2001**, *57*, 545–548.

[148]Kaluza, Z.; Furman, B.; Krajewski, P.; Chmielewski, M. *Tetrahedron* **2000**, *56*, 5553–5562.

[149]Singh, P. N. D.; Mandel, S. M.; Sankaranarayanan, J.; Muthukrishnan, S.; Chang, M.; Robinson, R. M.; Lahti, P. M.; Ault, B. S.; Gudmundsdottir, A. D. *J. Am. Chem. Soc.* **2007**, *129*, 16263–16272.

71%

Silver nitrite is superior to its sodium relative in many cases, affording moderate to good yields of nitro compounds in an aqueous environment.[150] However, a significant excess of this expensive reagent is required to minimize competitive formation of alcohols.

***1.4.3.2 Azide Nucleophiles***    Due to a linear geometry that minimizes steric hindrance, azide ion is a highly effective nitrogen nucleophile. Azide has a nucleophilic constant $(n)$[151] that is approximately equivalent to that of $NH_3$, while possessing a basicity closer to that of acetate ion.[152] Furthermore, alkyl azides can be converted in a highly efficient manner to the corresponding amines via hydrogenolysis or via Staudinger reduction with a phosphine as the reducing agent (see Section 9.7.1.1). For these reasons, the use of azide ion as a nucleophile in organic synthesis is prevalent. In the example below, a primary alkyl tosylate is displaced by sodium azide in dimethylsulfoxide at 60°C to provide the primary alkyl azide in 95% yield.[153]

It should be noted that sodium azide and many related compounds are toxic substances and extreme care should be taken to minimize exposure while handling these reagents, reaction mixtures, and products. Furthermore, many azide compounds present an additional safety concern due to their tendency to aggressively decompose with concomitant nitrogen gas liberation. Many heavy metal azides are exceedingly shock sensitive, so handling of these compounds should be avoided. Finally, care must be taken during waste storage and disposal due to the propensity of azide ion to form explosive metal complexes with materials found in standard copper water pipes. For additional examples of C−N bond formation with azide nucleophiles, see Sections 1.2.2.8 and 1.4.1.6.

***1.4.3.3 Isocyanates and Isothiocyanates as Nucleophiles***    Isocyanate salts are commonly employed reagents for the introduction of urea or carbamate functionalities. Isocyanate ions are reasonably nucleophilic at nitrogen, but upon alkylation, the resulting N-substituted isocyanates are electrophilic at the central carbon atom. The example below illustrates how this dual reactivity can be utilized for the preparation of carbamates in a single reaction from potassium isocyanate, an alcohol, and an alkyl halide.[154]

[150]Ballini, R.; Barboni, L.; Giarlo, G. *J. Org. Chem.* **2004**, *69*, 6907–6908.

[151]Swain, C. G.; Scott, C. B. *J. Am. Chem. Soc.* **1953**, *75*, 141–147.

[152]Carey, F. A.; Sundberg, R. J. *Advanced Organic Chemistry, Part A: Structure and Mechanisms*. 4th ed.; Kluwer Academic/Plenum Publishers: Dordrecht, Netherlands, **2000**.

[153]Hoekstra, M. S.; Sobieray, D. M.; Schwindt, M. A.; Mulhern, T. A.; Grote, T. M.; Huckabee, B. K.; Hendrickson, V. S.; Franklin, L. C.; Granger, E. J.; Karrick, G. L. *Org. Process Res. Dev.* **1997**, *1*, 26–38.

[154]Argabright, P. A.; Rider, H. D.; Sieck, R. *J. Org. Chem.* **1965**, *30*, 3317–3321.

KNCO + EtOH + BnCl  →(100°C, 6 h / DMF)  70%

Magnus and coworkers reported a similar use of potassium isocyanate in their synthesis of NNRTI candidates for the treatment of HIV,[155] although in this case the carbonyl electrophile technically moves this example out of the scope of this chapter. For more on the addition of nitrogen nucleophiles to carbonyl carbon centers, see Chapter 2.

KOCN, AcOH, H$_2$O  →(60°C, 2.5 h / 96%)

Isothiocyanate ions afford the same general reactivity, although competitive S-alkylation may be a minor complication. In the following example, the authors we able to isolate methallyl isothiocyanate prepared by refluxing an acetone solution of ammonium isothiocyanate with methallyl chloride in 80% yield. This product was converted to a thiourea derivative via treatment with alcoholic ammonia, and to 2–butenamine via acidic hydrolysis.[156]

Acetone, reflux  →  80%

## 1.5  HALOGEN NUCLEOPHILES

### 1.5.1  Attack by Halides at Alkyl Carbon or Silicon

*1.5.1.1  Halide Exchange*  The exchange of one halide for another, known as the Finkelstein reaction,[157] is a thermodynamically driven process that can be influenced by the stoichiometric ratio of the two halides as well as relative solubilities of halide salts in the reaction medium. A halogen exchange reaction is commonly used to convert more stable, accessible, and inexpensive alkyl chlorides to their more reactive bromide or iodide analogs prior to an alkylation reaction. The Finkelstein reaction can frequently be carried out during an alkylation reaction, by adding a bromide or iodide salt to the reaction mixture. In the example below, an α-chloroamide is converted to the corresponding iodide through treatment with super-stoichiometric sodium iodide in acetone.[158]

[155]Magnus, N. A.; Confalone, P. N.; Storace, L.; Patel, M.; Wood, C. C.; Davis, W. P.; Parsons, R. L., Jr., *J. Org. Chem.* **2003**, *68*, 754–761.
[156]Young, W. G.; Webb, I. D.; Goering, H. L. *J. Am. Chem. Soc.* **1951**, *73*, 1076–1083.
[157]Finkelstein, H. *Ber. Dtsch. Chem. Ges.* **1910**, *43*, 1528–1532.
[158]Kropf, J. E.; Meigh, I. C.; Bebbington, M. W. P.; Weinreb, S. M. *J. Org. Chem.* **2006**, *71*, 2046–2055.

With the activated halide highlighted in the example above, exchange occurs at an acceptable rate at ambient temperature; however, moderate heating is commonly employed to accelerate the conversion. In the example below, the less-activated substrate required considerable heating and an extended reaction time to accomplish complete halide exchange.[159]

### 1.5.1.2 Preparation of Halides from Sulfonic Acid Esters

*1.5.1.2 Preparation of Halides from Sulfonic Acid Esters*   Alkyl halides can be prepared from aliphatic alcohol precursors via the intermediacy of sulfonic acid esters such as mesylates, tosylates, or triflates. Preferred conditions involve heating the sulfonate with a halide salt in a polar aprotic solvent such as acetone, acetonitrile, or dimethylformamide. The reaction generally proceeds via direct $S_N2$ displacement to provide the product of stereochemical inversion, although subsequent halide–halide exchange may occur to yield an epimeric mixture. In the example below from Robins and coworkers, clean inversion was achieved with lithium chloride in dimethylformamide.[160] The corresponding bromide could be prepared in 85% yield under analogous conditions with lithium bromide.

Interestingly, Lepore and coworkers have reported that certain sterically bulky arene-sulfonates can be converted to the corresponding chlorides with retention of configuration by treatment with titanium tetrachloride in dichloromethane at $-78°C$.[161] The method requires the use of an aryl sulfonic acid derivative that is capable of strong coordination to titanium. The authors propose an $S_Ni$-type mechanism to explain the observed stereochemical outcome.

[159]Jensen, A. E.; Kneisel, F.; Knochel, P. *Org. Synth.* **2003**, *79*, 35–42.

[160]Robins, M. J.; Nowak, I.; Wnuk, S. F.; Hansske, F.; Madej, D. *J. Org. Chem.* **2007**, *72*, 8216–8221.

[161]Lepore, S. D.; Bhunia, A. K.; Mondal, D.; Cohn, P. C.; Lefkowitz, C. *J. Org. Chem.* **2006**, *71*, 3285–3286.

With the steady increase in the number of fluorinated compounds being investigated in the pharmaceutical industry, numerous methods for the nucleophilic introduction of fluorine in a stereo- and regioselective manner have appeared.[162] Both alkali metal fluorides and tetra-*n*-butylammonium fluoride (TBAF) have been utilized successfully, but these reagents suffer from poor solubility and/or high basicity. A more general, albeit expensive, method for the direct nucleophilic displacement of halides (chlorides, bromides, and iodides) and sulfonates was introduced by DeShong and coworkers.[163] As highlighted by the example below, treatment of a diastereomerically pure secondary mesylate with tetra-*n*-butylammonium (triphenylsilyl) difluorosilicate (TBAT) in refluxing acetonitrile provided the secondary fluoride in good yield, albeit as a mixture of diastereomers. Interestingly, when the triflate-leaving group was utilized under identical conditions, the fluoride was obtained as a single syn-diastereoisomer in a slightly improved 76% yield. This result demonstrates that the substitution reaction occurs with clean inversion of stereochemistry and that erosion of diastereomeric purity does not occur via epimerization of the product under the reaction conditions.

***1.5.1.3   Preparation of Alkyl Halides from Alcohols***   Alcohols can serve as convenient precursors to halides, although direct displacement using hydrohalic acids is not always a practical operation due to competitive elimination and/or incompatibility with common functional groups. Instead, an activation step is carried out that renders the oxygen more reactive toward nucleophilic displacement. For large scale applications, the reagents of choice for conversion of an aliphatic alcohol to its corresponding chloride are typically thionyl chloride, oxalyl chloride, or methanesulfonyl chloride. Each reagent has advantages and disadvantages when cost, atom economy, and handling issues are considered. As a result,

[162]Bohm, H.-J.; Banner, D.; Bendels, S.; Kansy, M.; Kuhn, B.; Muller, K.; Obst-Sander, U.; Stahl, M. *Chembiochem* **2004**, *5*, 637–643.
[163]Pilcher, A. S.; Ammon, H. L.; DeShong, P. *J. Am. Chem. Soc.* **1995**, *117*, 5166–5167.

reagent choice is often determined on a case by case basis. The example below highlights the use of thionyl chloride in the presence of catalytic *N,N*-dimethylformamide (DMF) for the high-yielding conversion of a secondary benzylic alcohol to its chloride.[164] The reaction proceeds cleanly in heptane to afford the desired product with minimal manipulation. The workup for this reaction generally involves a slow, cautious addition of water to carefully quench residual thionyl chloride (HCl generation), followed by neutralization, phase separation, and concentration to remove volatiles.

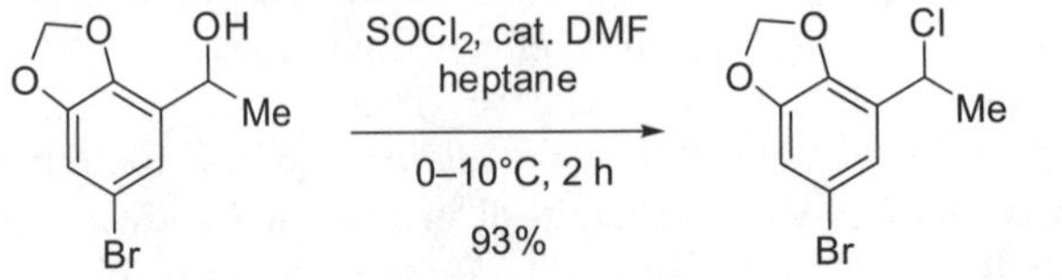

Alternatively, *N*-chlorosuccinimide (NCS) in combination with triphenylphosphine can be used to convert an alcohol to a chloride. The reaction on the chiral secondary alcohol shown below proceeded with inversion of stereochemistry and minimal erosion of enantiomeric excess.[165] In contrast, the use of the more common reagents described above would likely degrade the stereochemical integrity of the chiral center via reaction of the product with excess chloride.

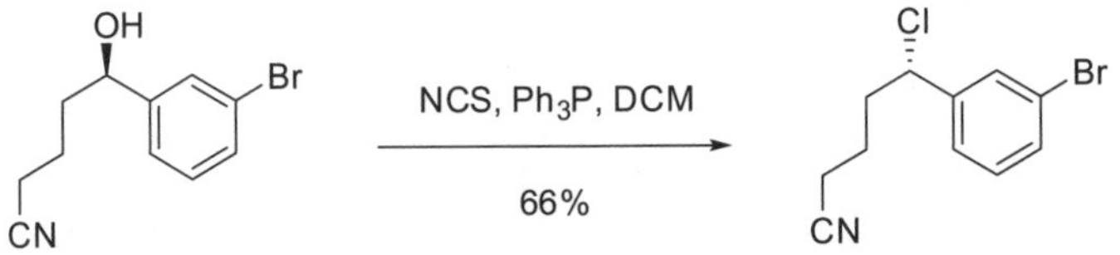

For the simple preparation of alkyl halides from alcohols, few methods are as straightforward and reliable as the Appel reaction.[166] Thus, treatment of an alcohol with carbon tetrabromide and triphenylphosphine provides the corresponding bromide, along with the stoichiometric triphenylphosphine oxide byproduct.[167] However, drawbacks associated with the use of this method include low atom economy, higher relative expense and additional challenges in purging phosphines and phosphine oxides from products. The latter can typically be accomplished via chromatography on silica gel or selective precipitation from a non-polar solvent.

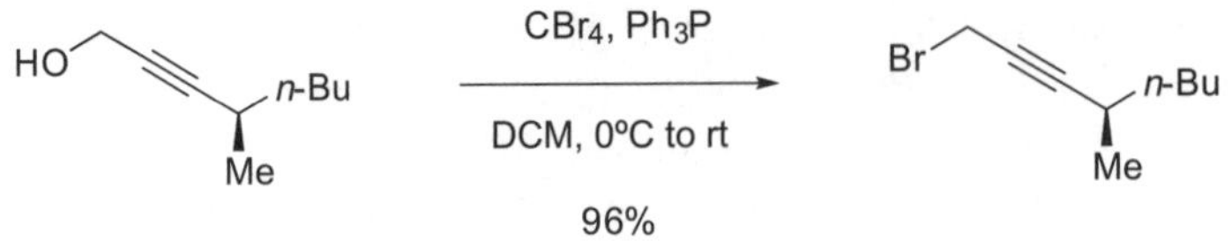

[164]Jacks, T. E.; Belmont, D. T.; Briggs, C. A.; Horne, N. M.; Kanter, G. D.; Karrick, G. L.; Krikke, J. J.; McCabe, R. J.; Mustakis, J. G.; Nanninga, T. N.; Risedorph, G. S.; Seamans, R. E.; Skeean, R.; Winkle, D. D.; Zennie, T. M. *Org. Process Res. Dev.* **2004**, *8*, 201–212.

[165]Merschaert, A.; Boquel, P.; Van Hoeck, J.-P.; Gorissen, H.; Borghese, A.; Bonnier, B.; Mockel, A.; Napora, F. *Org. Process Res. Dev.* **2006**, *10*, 776–783.

[166]Appel, R. *Angew. Chem., Int. Ed. Engl.* **1975**, *87*, 863–874.

[167]Woodin, K. S.; Jamison, T. F. *J. Org. Chem.* **2007**, *72*, 7451–7454.

**1.5.1.4  Preparation of Alkyl Halides from Ethers**    The action of the strong acids HI[92] and HBr[168] upon ethers provides one molar equivalent of an alkyl halide and another equivalent of an alcohol. If a stoichiometric excess of the acid is employed, the alcohol product is usually converted to its corresponding alkyl halide as well. This method is not commonly utilized to prepare alkyl halides from mixed ethers due to the lack of selectivity in the cleavage to halide and alcohol.

$$R^{\diagup O}{\diagdown}R' \xrightarrow{\ HX\ } R{-}X \ + \ R'{-}OH \ + \ R'{-}X \ + \ R{-}OH \xrightarrow{\ HX\ } R{-}X \ + \ R'{-}X$$

Conversely, there is broad utility for this method in the cleavage of alkyl aryl ethers (anisole derivatives), which liberate the alkyl halide and phenol preferentially (see Section 1.5.1.6). In this case, the goal is deprotection of the phenol rather than synthesis of the alkyl halide, although the latter is a consequence.

**1.5.1.5  Preparation of Alkyl Halides from Epoxides**    Epoxides are commonly prepared via intramolecular nucleophilic displacement of halides by vicinal alcohols (see Section 1.2.2.8). The inverse reaction can also be carried out; epoxides can be opened via nucleophilic attack by halide ions. Allevi reported that the treatment of a terminal epoxide with sodium iodide and acetic acid in tetrahydrofuran provides the primary alkyl iodide (via anti-Markovnikov attack) in high yield at room temperature.[169]

The ring-opening of epoxides with halide salts to produce halohydrins can also be promoted by either Lewis or Brønsted acids. The example below is representative of the stereochemical course of the reaction, which obeys the Fürst-Plattner rule,[170] providing the anti-halohydrin via backside attack on the epoxide.[171] The selectivity for the favored constitutional isomer is controlled in this case by steric factors, although electronic factors can also strongly influence the product distribution. The latter is especially true when strong Lewis or Brønsted acids are employed.

[168]Giles, M. E.; Thomson, C.; Eyley, S. C.; Cole, A. J.; Goodwin, C. J.; Hurved, P. A.; Morlin, A. J. G.; Tornos, J.; Atkinson, S.; Just, C.; Dean, J. C.; Singleton, J. T.; Longton, A. J.; Woodland, I.; Teasdale, A.; Gregertsen, B.; Else, H.; Athwal, M. S.; Tatterton, S.; Knott, J. M.; Thompson, N.; Smith, S. J. *Org. Process Res. Dev.* **2004**, *8*, 628–642.
[169]Allevi, P.; Anastasia, M. *Tetrahedron: Asymmetry* **2004**, *15*, 2091–2096.
[170]Furst, A.; Plattner, P. A. *Helv. Chim. Acta* **1949**, *32*, 275–283.
[171]Eipert, M.; Maichle-Moessmer, C.; Maier, M. E. *Tetrahedron* **2003**, *59*, 7949–7960.

In the interesting example below, from Seçen and coworkers, a cyclic epoxide was opened in a highly stereoselective manner to afford the 1,4-*syn*-haloacetate.[172] The authors proposed that this reaction may proceed through a semi-concerted transition state, where partial acylation of the epoxide occurs concurrently with $S_N2'$ attack of halide ion. Note that under these reaction conditions the acetonide was also converted to a diacetate.

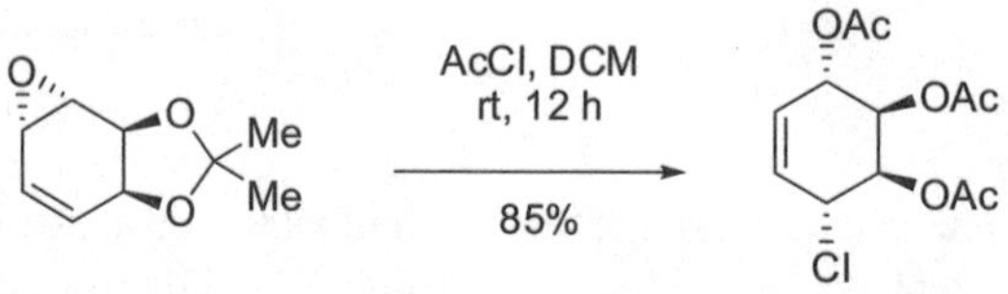

Another example of opening an allylic epoxide with halide ions is shown below.[173] In this case, halohydrin formation was promoted by magnesium bromide in cold acetonitrile. The ratio of isomers is presumably a reflection of increased carbocation stabilization at the allylic position, since sterics would appear to offer little influence.

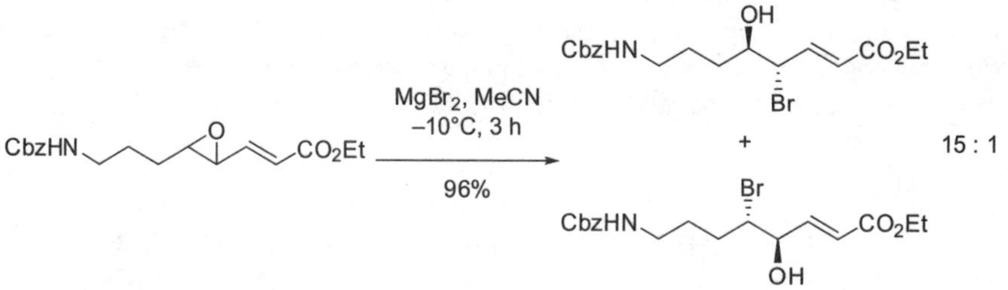

Epoxides can also be ring-opened in a highly stereoselective manner by hydrogen fluoride–pyridine complex.[174] It is notable that preferential attack by fluoride occurred away from the vinyl substituent in this example.

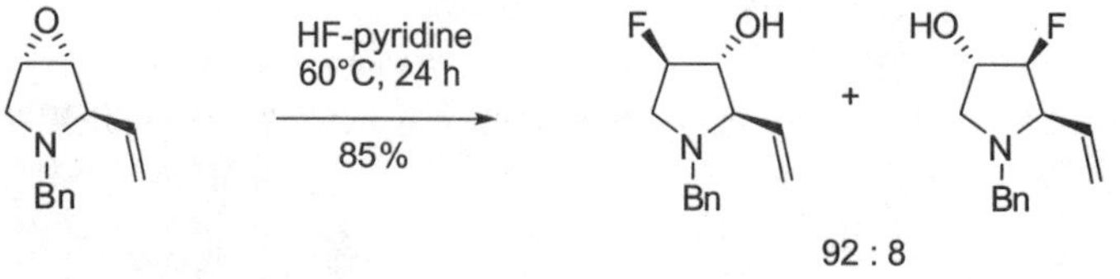

### 1.5.1.6  Cleavage of Alkyl Ethers with Halide Ion
The cleavage of arylmethyl ethers with halide ion to form the corresponding phenols is a common practice in the pharmaceutical industry. In the example below from Jacks and coworkers, hydrogen bromide in acetic acid gave good conversion and high yield of a catechol intermediate.[164]

[172]Baran, A.; Kazaz, C.; Secen, H. *Tetrahedron* **2004**, *60*, 861–866.

[173]Ha, J. D.; Kim, S. Y.; Lee, S. J.; Kang, S. K.; Ahn, J. H.; Kim, S. S.; Choi, J.-K. *Tetrahedron Lett.* **2004**, *45*, 5969–5972.

[174]Ayad, T.; Genisson, Y.; Broussy, S.; Baltas, M.; Gorrichon, L. *Eur. J. Org. Chem.* **2003**, 2903–2910.

Hydrogen chloride is generally less effective than hydrogen iodide or hydrogen bromide, although when used as a pyridine salt, high yields have been reported for demethylation of certain anisole derivatives. In the example below, the starting coumarin derivative was treated with pyridine hydrochloride in the absence of solvent at high temperature. The reagent cleaves two methyl ethers and an acetyl-protecting group in excellent yield, although the required reaction temperature was quite high.[175]

However, the use of strong mineral acids is often incompatible with organic substrates. Therefore, preferable methods for large scale applications often utilize reagents such as boron tribromide,[176] boron trichloride,[177] or aluminum trichloride[164,178] despite the challenges associated with safe handling of these materials. In these reactions, ionization of the departing alcohol is promoted by complexation of the boron or aluminum to the ether oxygen. The use of BBr$_3$ in dichloromethane has proven to be effective for the cleavage of a range of arylmethyl ethers.[179]

Arylmethyl ethers can also be converted to phenols by treatment with iodotrimethylsilane, which liberates iodomethane as a stoichiometric byproduct. This mild method can be further improved by utilizing the less expensive chlorotrimethylsilane in combination with sodium iodide.[180]

[175]Li, X.; Jain, N.; Russell, R. K.; Ma, R.; Branum, S.; Xu, J.; Sui, Z. *Org. Process Res. Dev.* **2006**, *10*, 354–360.
[176]Lim, C. W.; Tissot, O.; Mattison, A.; Hooper, M. W.; Brown, J. M.; Cowley, A. R.; Hulmes, D. I.; Blacker, A. J. *Org. Process Res. Dev.* **2003**, *7*, 379–384.
[177]Cai, S.; Dimitroff, M.; McKennon, T.; Reider, M.; Robarge, L.; Ryckman, D.; Shang, X.; Therrien, J. *Org. Process Res. Dev.* **2004**, *8*, 353–359.
[178]Haight, A. R.; Bailey, A. E.; Baker, W. S.; Cain, M. H.; Copp, R. R.; DeMattei, J. A.; Ford, K. L.; Henry, R. F.; Hsu, M. C.; Keyes, R. F.; King, S. A.; McLaughlin, M. A.; Melcher, L. M.; Nadler, W. R.; Oliver, P. A.; Parekh, S. I.; Patel, H. H.; Seif, L. S.; Staeger, M. A.; Wayne, G. S.; Wittenberger, S. J.; Zhang, W. *Org. Process Res. Dev.* **2004**, *8*, 897–902.
[179]Zhang, W.; Yamamoto, H. *J. Am. Chem. Soc.* **2007**, *129*, 286–287.
[180]Wack, H.; France, S.; Hafez, A. M.; Drury, W. J., III; Weatherwax, A.; Lectka, T. *J. Org. Chem.* **2004**, *69*, 4531–4533.

Perhaps the most attractive protocol for demethylation on large scale involves nucleophilic attack on the methyl group by a mild sulfur nucleophile such as methionine[93] or a simple aliphatic thiol.[92] This reaction is discussed in Section 1.3.1.2.

### *1.5.1.7   Cleavage of Silyl Ethers and Silyl Enol Ethers with Halide Ion*   Silicon-based protecting groups play a central role in complex target-oriented synthesis. As a result, numerous mild and effective methods have been developed for efficient cleavage of silicon oxygen bonds. Nevertheless, the use of tetra-*n*-butylammonium fluoride (TBAF) remains the most commonly employed reagent for this task. In their multi-kilogram scale preparation of vitamin $D_3$ analogs, Shimizu and coworkers were successful in removing two TBS-protecting groups in high yield with TBAF in refluxing THF.[181]

If mild reaction temperatures and appropriate stoichiometry are utilized, more labile silyl enol ethers may be selectively cleaved in the presence of a silyl ether with TBAF. The following example is particularly impressive, as the bulky triisopropylsilyl (TIPS) group resides at each position in the starting material.[182]

It should be noted that TBAF is strongly basic and is manufactured as a water-wet substance. As a result, the addition of a suitable acid (often acetic acid) is typical when

[181]Shimizu, H.; Shimizu, K.; Kubodera, N.; Mikami, T.; Tsuzaki, K.; Suwa, H.; Harada, K.; Hiraide, A.; Shimizu, M.; Koyama, K.; Ichikawa, Y.; Hirasawa, D.; Kito, Y.; Kobayashi, M.; Kigawa, M.; Kato, M.; Kozono, T.; Tanaka, H.; Tanabe, M.; Iguchi, M.; Yoshida, M. *Org. Process Res. Dev.* **2005**, *9*, 278–287.
[182]Outten, R. A.; Daves, G. D., Jr., *J. Org. Chem.* **1987**, *52*, 5064–5066.

base-sensitive substrates are involved. Furthermore, hydrogen fluoride is a highly toxic and corrosive substance, so extreme caution should be taken to minimize exposure during handling of TBAF or related materials.

For the desilylation of compounds where high pH must be avoided, the use of aqueous hydrofluoric acid may be considered. In the example below, two secondary TBS ethers are cleaved with high efficiency by utilizing 40% aqueous hydrofluoric acid in acetonitrile at ambient temperature.[183]

It should be noted that HF has been demonstrated to etch glass at low concentrations,[184–187] so care must be taken when choosing equipment for the laboratory or plant. Fluorinated polymers (e.g., Teflon-PTFE) have been shown to offer good corrosion resistance, and are generally the preferred materials of construction for HF-resistant laboratory equipment, while alloys of nickel and copper (e.g., Alloy 400, C71500, C70600, etc.) comprise the preferred materials of construction for commercial reactors that are routinely exposed to hydrofluoric acid.[188]

Other attractive options for large scale silyl ether cleavage are triethylamine trihydrofluoride,[189] ammonium bifluoride,[190,191] and hydrogen fluoride–pyridine complex.[65,192] As described in the example below, hydrogen fluoride–pyridine complex has been utilized for the selective cleavage of a primary silyl ether in the presence of a secondary silyl ether in high yield under very mild conditions. The cleavage of silyl ethers with acids or bases in an aqueous or alcoholic environment is discussed in Sections 1.2.1.8 and 1.2.2.6, respectively.

[183]Baumann, K.; Bacher, M.; Steck, A.; Wagner, T. *Tetrahedron* **2004**, *60*, 5965–5981.

[184]Spierings, G. A. C. M. *J. Mater. Sci.* **1993**, *28*, 6261–6273.

[185]Spierings, G. A. C. M. *J. Mater. Sci.* **1991**, *26*, 3329–3336.

[186]Spierings, G. A. C. M.; Van Dijk, J. *J. Mater. Sci.* **1987**, *22*, 1869–1874.

[187]Tso, S. T.; Pask, J. A. *J. Am. Ceram. Soc.* **1982**, *65*, 360–362.

[188]Schillmoller, C. M. *Chem. Eng. Prog.* **1998**, *94*, 49–54.

[189]Joubert, N.; Pohl, R.; Klepetarova, B.; Hocek, M. *J. Org. Chem.* **2007**, *72*, 6797–6805.

[190]Hu, X. E.; Kim, N. K.; Grinius, L.; Morris, C. M.; Wallace, C. D.; Mieling, G. E.; Demuth, T. P., Jr., *Synthesis* **2003**, 1732–1738.

[191]Seki, M.; Kondo, K.; Kuroda, T.; Yamanaka, T.; Iwasaki, T. *Synlett* **1995**, 609–611.

[192]Shin, Y.; Fournier, J.-H.; Brueckner, A.; Madiraju, C.; Balachandran, R.; Raccor, B. S.; Edler, M. C.; Hamel, E.; Sikorski, R. P.; Vogt, A.; Day, B. W.; Curran, D. P. *Tetrahedron* **2007**, *63*, 8537–8562.

82%

### 1.5.1.8 Cleavage of Carboxylic Acid and Sulfonic Acid Esters with Halide Ion

*1.5.1.8 Cleavage of Carboxylic Acid and Sulfonic Acid Esters with Halide Ion* In a series of seminal publications, Krapcho and coworkers reported that heating β-ketoesters, malonate esters, or α-cyanoesters in water-wet dimethylsulfoxide in the presence of chloride or cyanide ions provided the products of dealkoxycarbonylation.[193] A $B_{AL}2$ type mechanism (i.e., nucleophilic attack by chloride on the alkyl group of the ester to provide a sodium carboxylate) is generally accepted, although Krapcho demonstrated that certain dealkoxycarbonylations are more likely to proceed through a $B_{AC}2$ pathway. Dealkoxycarbonylation in the absence of salts was also reported, although both rate and yield improve with added chloride ion for most substrates.[194]

Nonetheless, the "Krapcho decarboxylation" has demonstrated utility in synthetic organic chemistry, as highlighted by the example below from Hoekstra and coworkers during development of a multi-kilogram synthesis of the anticonvulsant pregabalin.[153] The major disadvantage of the classic Krapcho conditions is that a temperature in excess of 130°C is generally required, which presents challenges for substrate compatibility and worker safety on large scale.

Through related transformations, nucleophilic cleavage of aliphatic esters has also been reported with cyanide[195–198] or thiolate anions.[199–203] These latter methods are typically employed for the ring-opening of cyclic esters (lactones) and therefore incorporate the nucleophiles into the product as nitriles or thioethers, respectively.

Sulfonic acid esters may also be converted to the parent sulfonic acids through nucleophilic attack by halide ion at carbon. In the example below, sodium iodide in acetone solvent provided the sodium sulfonate via a mild, high-yielding transformation.[8]

[193]Krapcho, A. P. *Synthesis* **1982**, 893–914.
[194]Krapcho, A. P.; Weimaster, J. F.; Eldridge, J. M.; Jahngen, E. G. E., Jr., ; Lovey, A. J.; Stephens, W. P. *J. Org. Chem.* **1978**, *43*, 138–147.
[195]Stanetty, P.; Froehlich, H.; Sauter, F. *Monatsh. Chem.* **1986**, *117*, 69–88.
[196]Anjanamurthy, C.; Rai, K. M. L. *Indian J. Chem. B Org* **1985**, *24B*, 502–504.
[197]Goya, S.; Takadate, A.; Tanaka, T.; Tsuruda, Y.; Ogata, H. *Yakugaku Zasshi.* **1980**, *100*, 819–825.
[198]Miller, A. W.; Stirling, C. J. M. *J. Chem. Soc., C* **1968**, 2612–2617.
[199]Kelly, T. R.; Dali, H. M.; Tsang, W. G. *Tetrahedron Lett.* **1977**, 3859–3860.
[200]Traynelis, V. J.; Love, R. F. *J. Org. Chem.* **1961**, *26*, 2728–2733.
[201]Kresze, G.; Schramm, W.; Cleve, G. *Chem. Ber.* **1961**, *94*, 2060–2072.
[202]Farrar, M. W. *J. Org. Chem.* **1958**, *23*, 1065–1066.
[203]Gresham, T. L.; Jansen, J. E.; Shaver, F. W.; Bankert, R. A.; Beears, W. L.; Prendergast, M. G. *J. Am. Chem. Soc.* **1949**, *71*, 661–663.

### 1.5.1.9 Preparation of Halides from α-Diazo Carbonyl Compounds

*1.5.1.9  Preparation of Halides from α-Diazo Carbonyl Compounds*    Treatment of α-diazo carbonyl compounds with a haloacid can efficiently provide the α-halo derivative. In the example below, a diazoketone is reacted with aqueous HBr in THF at 0°C to yield the corresponding bromoketone in high yield.[204]

α-Halo carbonyl compounds can also be prepared from α-amine precursors via diazotization *in situ* followed by reaction with a halide. Treatment of the amino carbonyl compound with a halide salt in the presence of sodium nitrite and in aqueous sulfuric acid provides the product at or near room temperature. When applied to α-amino carboxylic acids, this transformation has been shown to proceed in high yields, and with net retention of configuration, as illustrated in the example that follows.[205]

This stereochemical outcome is specific to α-amino acids, as the mechanism involves the intermediacy of an α-lactone. The analogous reaction with amino ketones, esters, or amides would proceed with less stereoselectivity. Complementary methods for the synthesis of α-halo carbonyl compounds are discussed in Section 10.3.3.

### 1.5.1.10 Preparation of Cyanamides

*1.5.1.10  Preparation of Cyanamides*    The reaction of an amine with cyanogen bromide yields cyanamide products of the general formula $R^1R^2N-CN$. When the reaction is carried out on a tertiary amine, one of the $C-N$ bonds of the initially formed quaternary amine is cleaved through nucleophilic attack at carbon by bromide ion (von Braun reaction).[206] The reaction may be carried out with a wide range of tertiary amines, although at least one substituent on the amine must be aliphatic, as aromatic groups are not cleaved after amine quaternization.[207,208] An illustrative example from Reinhoudt and coworkers is provided below.[209]

[204]Ramtohul, Y. K.; James, M. N. G.; Vederas, J. C. *J. Org. Chem.* **2002**, *67*, 3169–3178.

[205]Dutta, A. S.; Giles, M. B.; Gormley, J. J.; Williams, J. C.; Kusner, E. J. *J. Chem. Soc., Perkin Trans.* **1987**, *1*, 111–120.

[206]von Braun, J. *Ber. Dtsch. Chem. Ges.* **1900**, *33*, 1438–1452.

[207]Cooley, J. H.; Evain, E. J. *Synthesis* **1989**, 1–7.

[208]Hageman, H. *Organic Reactions (New York)* **1953**, *VII*, 198–262.

[209]Verboom, W.; Visser, G. W.; Reinhoudt, D. N. *Tetrahedron* **1982**, *38*, 1831–1835.

75%

## 1.6   CARBON NUCLEOPHILES

### 1.6.1   Attack by Carbon at Alkyl Carbon

*1.6.1.1   Direct Coupling of Alkyl Halides*   The direct coupling of alkyl halides via treatment with a metal or mixture of metals is known as the Wurtz reaction. Although it has been suggested that the initial stages of the Wurtz reaction are radical in nature,[210] the bond formation step is believed to consist of an $S_N2$ displacement of an alkyl halide with an aliphatic anion.[211] The reaction is not typically amenable to large scale applications, since specialized equipment is required. Furthermore, substrate scope is generally limited to homocoupling of activated halides and yields are typically low to moderate due to competitive alkene formation of the presumed radical intermediate. However, the homocoupled products can be useful, and might otherwise require multi-step synthesis.[212]

Unsymmetrical Wurtz couplings can be promoted by manganese and copper co-catalysts, as reported by Chan and Ma.[213] It is noteworthy that these reactions are carried out in an aqueous solvent system, thus avoiding the challenges associated with handling moisture sensitive reagents and intermediates. In addition, this methodology proved applicable to unactivated aliphatic alkyl bromides and iodides. In the following example, a significant quantity of 1,5-hexadiene byproduct was formed, which called for a stoichiometric excess of allyl bromide. However, the desired acid could be easily separated from this non-polar hydrocarbon. Note that for this particular substrate, a copper promoted $S_N2'$ displacement of the allylic bromide cannot be ruled out.

[210]Richards, R. B. *Trans. Faraday Soc.* **1940**, *36*, 956–960.
[211]LeGoff, E.; Ulrich, S. E.; Denney, D. B. *J. Am. Chem. Soc.* **1958**, *80*, 622–625.
[212]Fouquey, C.; Jacques, J. *Synthesis* **1971**, 306.
[213]Ma, J.; Chan, T.-H. *Tetrahedron Lett.* **1998**, *39*, 2499–2502.

For small scale applications, electrochemical Wurtz coupling methods may be applicable, although specialized equipment is required.[214] For more on the coupling of allylic halides, see Section 1.6.1.3.

$$e^-\,(\text{Zn anode})$$
$$\text{TBABF}_4,\ \text{TBAI}$$
$$\text{THF-TMU}$$
Reflux, 1 h
70%

### 1.6.1.2 Reactions of Organometallic Reagents with Alkyl Halides

The nucleophilic displacement of an alkyl halide with an organometallic reagent has been the subject of considerable research by the synthetic organic community. A wide variety of organometallic reagents have proven effective for this transformation, and their preparation is the subject of Chapter 12. A concern when choosing the appropriate methodology for a given system is the ability of many organometallic reagents to act as bases; deprotonation of the electrophile can lead to undesired elimination products. The relatively high basicity of alkyl lithium compounds makes their use in halide displacement reactions quite challenging. As a result, transmetallation from readily accessible but somewhat capricious alkyl lithiums to more predictable organometallic derivatives is common.

Perhaps the most highly utilized method was co-developed by the laboratories of Corey, House, Posner, and Whitesides.[215] This method consists of the reaction of a lithium dialkylcuprate (Gilman reagent), formed by treatment of an alkyl lithium with copper iodide in THF at $-78°$C, with a suitable alkyl halide to afford the desired product. The scope of this methodology is generally limited to primary or activated alkyl halides. A notable drawback to the use of Gilman reagents is that two equivalents of the alkyl lithium are required, but only one equivalent is incorporated into the product.

In 1983, Lipshutz and coworkers reported a valuable extension of this method: the facile coupling of higher order cyanocuprates with a range of alkyl halide electrophiles.[216] These reagents have an advantage over traditional Gilman reagents in that they can be prepared from more stable, easily accessible alkyl bromides or chlorides. However, they are somewhat less reactive, so higher than cryogenic temperatures are often utilized for the coupling step. For more on the use of cyanocuprates as nucleophiles, see Section 1.6.1.7. In the example below, a high yield of the substituted pyridine product is obtained after three hours at $-50°$C in tetrahydrofuran.

The displacement of alkyl halides with aliphatic Grignard reagents has also been reported, however these methods are often complicated by functional group incompatibility and/or elimination. The use of the non-Lewis basic solvent dichloromethane allowed Eguchi and coworkers to couple Grignard reagents with tertiary halides with

[214]Nedelec, J. Y.; Mouloud, H. A. H.; Folest, J. C.; Perichon, J. *J. Org. Chem.* **1988**, *53*, 4720–4724.

[215]Posner, G. H. *Organic Reactions (New York)* **1975**, *22*, 253–400.

[216]Lipshutz, B. H.; Parker, D.; Kozlowski, J. A.; Miller, R. D. *J. Org. Chem.* **1983**, *48*, 3334–3336.

moderate success.[217] The authors suspect competitive elimination was responsible for the lower yield observed with $t$–butyl chloride; elimination is prohibited by Bredt's rule for adamantyl chloride. A single electron transfer (radical) mechanism likely competes with an $S_N2$ pathway for these substrates.

### 1.6.1.3 Couplings of Allylic and Propargylic Halides

Allylic and propargylic halides are reactive electrophiles due to their ability to offer resonance stabilization to growing positive charge at the allylic/propargylic carbon during ionization of the carbon–halogen bond. This effect also allows for insertion of low valent metals into the carbon–halogen bond, which produces nucleophilic organometallic compounds. As a result, the addition of a metal in its lowest oxidation state, such as zinc, magnesium, or iron, to a solution of an allylic or propargylic halide will typically yield significant quantities of dimerization products (Wurtz coupling).[218]

The example below illustrates that this reductive dimerization is also effective with secondary allylic halides, although the rate and yield often decline for such systems.[219]

(meso / d,l)

For an additional example of the Wurtz coupling of an allylic halide, see Section 1.6.1.1.

### 1.6.1.4 Couplings of Organometallic Reagents with Sulfonate Esters

Most organometallic carbon nucleophiles (Grignards, cuprates, etc.) react with aliphatic sulfonate ester electrophiles in a manner analogous to that observed with alkyl halides

[217]Ohno, M.; Shimizu, K.; Ishizaki, K.; Sasaki, T.; Eguchi, S. *J. Org. Chem.* **1988**, *53*, 729–733.

[218]Rao, G. S. R. S.; Bhaskar, K. V. *J. Chem. Soc., Perkin Trans. 1* **1993**, 2813–2816.

[219]Voegtle, F.; Eisen, N.; Franken, S.; Buellesbach, P.; Puff, H. *J. Org. Chem.* **1987**, *52*, 5560–5564.

(see Section 1.6.1.2). In the example below, an aliphatic Grignard reagent is modified by the addition of copper(I) bromide to allow smooth displacement of a primary tosylate in tetrahydrofuran at ambient temperature.[220]

76%

Grignard reagents are frequently used without conversion to copper complexes in cases where allylic or propargylic carbon nucleophiles are involved. In the example below, a high yield of the alkylation product was obtained under mild conditions when an excess of 2-methallylmagnesium chloride was allowed to react with either the primary tosylate or iodide.[221] Benzenesulfonate derivatives typically outperform mesylates, as the latter contain acidic α-protons that may quench basic carbon nucleophiles.

THF, 0°C to rt, 16 h

R = OTs : 92%
R = I    : 83%

*1.6.1.5  Couplings Involving Alcohols*  Because hydroxide ion is a relatively poor leaving group, the direct $S_N2$ displacement at an alcohol-bearing carbon center is rare. Instead, alcohols are typically subjected to an activation step such as conversion to the corresponding halide[164] or inorganic ester[50–52,222](see Sections 1.5.1.3 and 1.7.1.2). A few methods for direct alcohol displacement have been reported, however. One particularly mild method, the treatment of benzylic alcohols with a catalytic quantity of InCl_3 (5 mol%) in the presence of allyl, propargyl, or alkynyl silanes, was reported by Baba and coworkers.[223]

81%

[220]Yadav, J. S.; Rao, K. V.; Prasad, A. R. *Synthesis* **2006**, 3888–3894.
[221]Marcos, I. S.; Pedrero, A. B.; Sexmero, M. J.; Diez, D.; Basabe, P.; Garcia, N.; Moro, R. F.; Broughton, H. B.; Mollinedo, F.; Urones, J. G. *J. Org. Chem.* **2003**, *68*, 7496–7504.
[222]Shan, Z.; Xiong, Y.; Zhao, D. *Tetrahedron* **1999**, *55*, 3893–3896.
[223]Yasuda, M.; Saito, T.; Ueba, M.; Baba, A. *Angew. Chem., Int. Ed. Engl.* **2004**, *43*, 1414–1416.

Aliphatic Grignard reagents have also been coupled with alcohols in the presence of titanium(IV) isopropoxide, although scope appears to be limited to activated (e.g. allylic) alcohols.[224]

### *1.6.1.6 Reactions of Organometallic Reagents with Allylic Esters and Carbonates*
Allylic esters and carbonates possess unique reactivity amongst aliphatic electrophiles in that they may be alkylated at either the 1- or 3-position of the allylic system via formal $S_N2$ or $S_N2'$ displacement of the leaving group. In the absence of a catalyst, the two modes of attack may be competitive, resulting in a product distribution that varies with substrate. In the example below, an allylic acetate was reacted with *n*-butyl magnesium bromide in the presence of CuCN to yield a mixture of regioisomers favoring the branched ($S_N2'$-type) alkylation product.[225]

More commonly, reactions of allylic electrophiles are catalyzed by an organometallic complex that provides a π-allyl intermediate via oxidative addition. The choice of metal and ligand set influences the sterics and electronics of the π-allyl intermediate, and therefore, the product distribution. The metal-catalyzed allylic substitution reaction falls outside the scope of this chapter, but is discussed thoroughly within Section 6.8.

### *1.6.1.7 Reactions of Organometallic Reagents with Epoxides*
The opening of epoxides with non-enolate carbon nucleophiles is most reliably carried out with higher order mixed organocuprate reagents "$R_2Cu(CN)Li_2$" generated *in situ* from two equivalents of the corresponding organolithium and one equivalent of copper cyanide, as reported by Lipshutz and coworkers. These reagents often outperform the related Gilman-type reagents "$R_2CuLi$" and "$RCu(CN)Li$", presumably due to decreased basicity and enhanced nucleophilicity. Furthermore, cyanocuprates exhibit increased thermal stability compared to classical Gilman reagents, which allows reactions to be carried out at or above ambient temperature. In the example below, the reagent generated from *n*-propyllithium and copper cyanide provided the desired ring-opened product in high yield under mild conditions in THF. The same transformation carried out with *n*-PrCu(CN)Li proceeded in only 15% to 30% yield, and required ethyl ether solvent to minimize competitive reagent complexation with the THF solvent.[226]

[224]Kulinkovich, O. G.; Epstein, O. L.; Isakov, V. E.; Khmel'nitskaya, E. A. *Synlett* **2001**, 49–52.
[225]Erdik, E.; Kocoglu, M. *Tetrahedron Lett.* **2007**, *48*, 4211–4214.
[226]Lipshutz, B. H.; Kozlowski, J.; Wilhelm, R. S. *J. Am. Chem. Soc.* **1982**, *104*, 2305–2307.

*1.6.1.8  Alkylation of Malonate and Acetoacetate Derivatives*    A carbon bearing at least one hydrogen atom and two to three electronegative substituents can be deprotonated to provide a stabilized carbanion that may be alkylated with a broad range of electrophiles under very mild conditions. The most common examples in this class are resonance stabilized salts of malonic acid derivatives (e.g., dimethylmalonate) and acetoacetate esters (e.g., ethyl acetoacetate). These compounds have found broad utility in transition metal-catalyzed allylic alkylation methodology pioneered by Tsuji[227,228] and Trost[229] as well as in the Knoevenagel condensation[230] to provide α,β-unsaturated carbonyl compounds. These reactions fall outside the scope of this chapter, but are discussed in Chapters 2 and 6, respectively.

Salts of β-dicarbonyl compounds can also be alkylated with alkyl halides,[231] epoxides,[232] and alkyl sulfonates.[233] In the representative example below, dibenzyl malonate was treated with sodium hydride in THF to provide a sodio-malonate that reacted with an alkyl bromide in high yield.[231]

In the second example, an enantiomerically enriched secondary tosylate is displaced by the sodium salt of di-*tert*-butyl malonate to provide the alkylation product in good yield. The product was formed with clean inversion of stereochemistry, in support of an $S_N2$ mechanism.

80%

*1.6.1.9  Alkylation of Aldehydes, Ketones, Nitriles, and Carboxylic Esters*    The α-alkylation of aldehyde enolates is plagued by self condensation byproducts, as aldehydes are reactive electrophiles. An elegant solution was introduced by Stork, as highlighted by the example below.[234] Condensation of an enolizable aldehyde with a primary amine provides the corresponding imine, which is far less prone to self condensation. Subjecting the imine to reaction with a base provides the metalated enamine, which can be reacted with mild electrophiles such as alkyl halides. Of

[227]Minami, I.; Shimizu, I.; Tsuji, J. *J. Organomet. Chem.* **1985**, *296*, 269–280.

[228]Tsuji, J. *Organic Synthesis with Palladium Compounds*; Springer-Verlag: Berlin, **1980**.

[229]Trost, B. M.; Verhoeven, T. R. *J. Am. Chem. Soc.* **1980**, *102*, 4730–4743.

[230]Davis, A. P.; Egan, T. J.; Orchard, M. G.; Cunningham, D.; McArdle, P. *Tetrahedron* **1992**, *48*, 8725–8738.

[231]Kaltenbronn, J. S.; Hudspeth, J. P.; Lunney, E. A.; Michniewicz, B. M.; Nicolaides, E. D.; Repine, J. T.; Roark, W. H.; Stier, M. A.; Tinney, F. J.;et al. *J. Med. Chem.* **1990**, *33*, 838–845.

[232]Takasu, H.; Tsuji, Y.; Sajiki, H.; Hirota, K. *Tetrahedron* **2005**, *61*, 8499–8504.

[233]Larcheveque, M.; Petit, Y. *Synthesis* **1991**, 162–164.

[234]Stork, G.; Dowd, S. R. *Org. Synth.* **1974**, *54*, 46–49.

particular note is that ethyl magnesium bromide is employed as a base in this particular example, and no addition of the Grignard reagent to the imine is observed. The resulting bromomagnesium counterion works in combination with the bulky *t*-butyl substituent on the imine nitrogen to provide excellent C- versus N-alkylation selectivity. A simple aqueous hydrochloric acid workup hydrolyzes the imine to release the aldehyde function for further elaboration.

In what has become the definitive example of stereocontrolled enolate alkylation under chiral phase transfer catalysis, Dolling and coworkers reported the following enantiose-lective methylation of an aryl ketone enolate.[235] The cinchona alkaloid-derived phase transfer catalyst blocks one face of the enolate to provide a substantial rate advantage over the racemic background alkylation. In their account of this work, Dolling and coworkers reported that several variables, including solvent polarity, reaction concentration, sodium hydroxide concentration, and catalyst loading were critical to optimizing yield and enantioselectivity.

Ketone enolates are readily generated by treatment with strong base.[68,69] When only one enolate is possible, as is the case with aryl ketones, a wide variety of bases can be used. In the example below, a chloroketone was treated with two equivalents of potassium *t*-butoxide in THF, to promote intramolecular alkylation and provide the cyclopropyl ketone in excellent yield.[236]

[235]Dolling, U. H.; Davis, P.; Grabowski, E. J. J. *J. Am. Chem. Soc.* **1984**, *106*, 446–447.
[236]Chang, S.-J.; Fernando, D. P.; King, S. *Org. Process Res. Dev.* **2001**, *5*, 141–143.

The generation and cyanation of ketone enolates obtained by treatment with lithium diisopropylamide in THF at −78°C has been reported by Kahne and Collum.[237] The use of *p*-toluenesulfonyl cyanide as the cyanide source proved to be critical, as did inverse addition of the enolate into a cold THF solution of *p*-TsCN. For more on the addition of carbon-based nucleophiles to nitriles see Section 2.39.

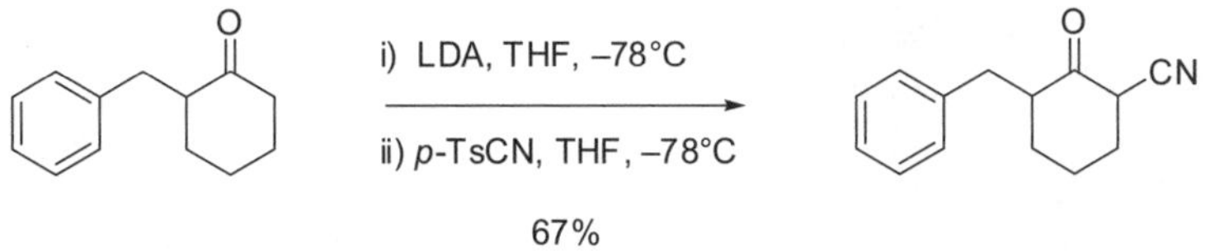

i) LDA, THF, −78°C

ii) *p*-TsCN, THF, −78°C

67%

It is also possible to deprotonate the position α to an aliphatic nitrile with LDA. In the example below, reaction of the deprotonated nitrile with an alkyl bromide provided the expected alkylation product in good yield.[238]

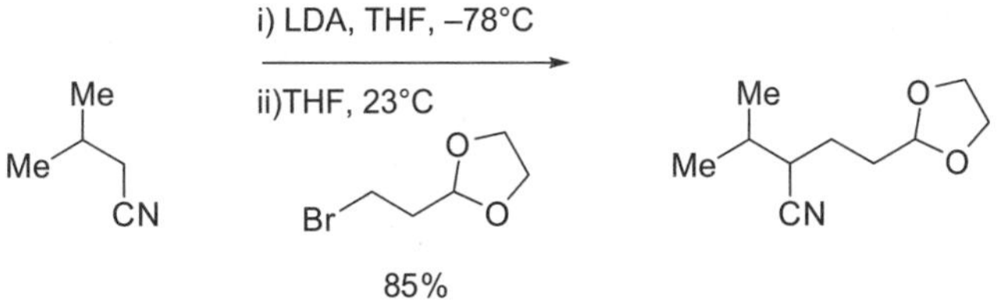

i) LDA, THF, −78°C

ii)THF, 23°C

85%

The alkylation of nitriles is often complicated by incomplete conversion and over-alkylation. In the following example from Taber and coworkers, however, a rather complex nitrile is further elaborated via deprotonation with LDA at −78°C, and subsequent treatment with an alkyl iodide.[239]

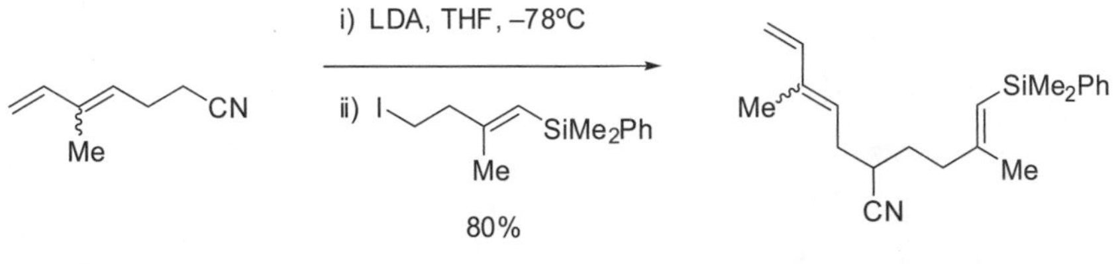

i) LDA, THF, −78°C

ii) 

80%

Taber and coworkers have also reported on the use of epoxides for the alkylation of nitrile enolates. In the example shown below, *n*-butyllithium is utilized for the deprotonation step, and an allylic epoxide serves as the electrophile. Of note is the apparent lack of regioisomeric products that could arise from attack at the other epoxide-bearing carbon or the less-substituted allylic terminus (S$_N$2′).[240]

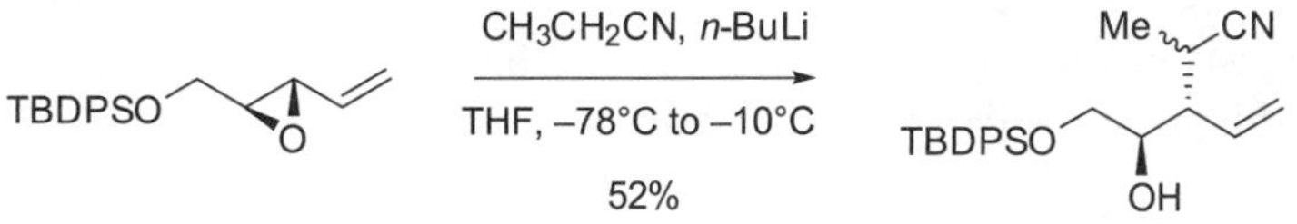

CH$_3$CH$_2$CN, *n*-BuLi

THF, −78°C to −10°C

52%

[237]Kahne, D.; Collum, D. B. *Tetrahedron Lett.* **1981**, *22*, 5011–5014.
[238]Wu, L.; Hartwig, J. F. *J. Am. Chem. Soc.* **2005**, *127*, 15824–15832.
[239]Taber, D. F.; Bhamidipati, R. S.; Yet, L. *J. Org. Chem.* **1995**, *60*, 5537–5539.

The use of strong bases derived from hexamethyldisilazane is becoming increasingly popular for both small and large scale applications. The lithium, sodium, and potassium hexamethyldisilazides are relatively stable, commercially available materials, and provide a convenient option for investigating counterion effects on rate and/or selectivity in a given reaction. In the example below, lithium hexamethyldisilazide (LiHMDS) in THF was employed for the selective formation of the kinetically favored (Z)-ester enolate at a temperature of $-23°C$. The subsequent addition of allyl iodide provided a 23:1 ratio of diastereoisomers in excellent yield.[241] Interestingly, potassium hexamethyldisilazide (KHMDS) also provided the (Z)-ester enolate, although the opposite diastereoisomer was favored during the alkylation. The authors propose an internal ester chelation mechanism for the potassium enolate that is not operable for the lithium enolate.

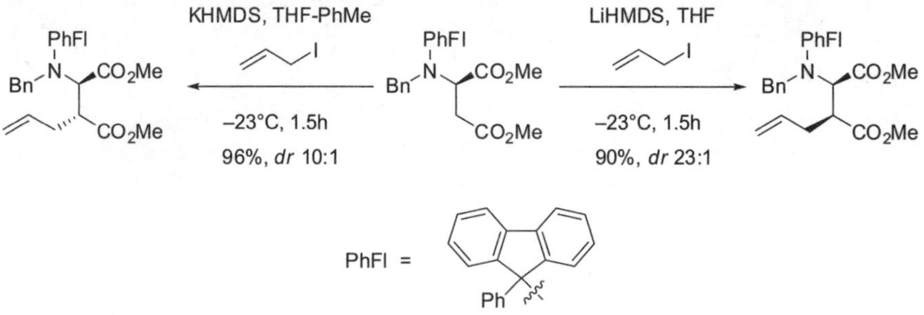

LiHMDS in THF/DMPU has been used for an ester enolate formation on multi-kilogram scale. The enolization and subsequent tosylate displacement were carried out below $-40°C$ in order to conserve the stereochemical integrity of the benzylic chiral center.[242]

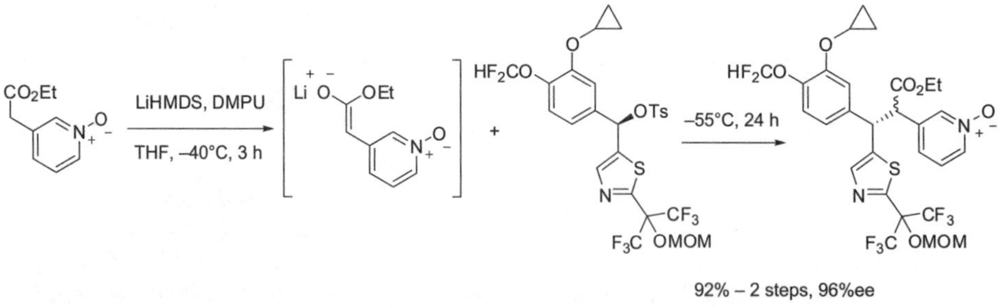

In a process to the HMG CoA reductase inhibitor simvastatin, lithium pyrrolidide has been used as a base for ester enolization, which allowed for the introduction of a challenging acyclic geminal methyl substituent.[243] Of significance is the lack of competitive amide methylation observed under these conditions. The authors were able to optimize reaction temperature and methyl iodide stoichiometry to minimize this side reaction, presumably aided by the high $pK_a$ of the amide $\alpha$-protons after initial lithium imidate formation.

[241]Humphrey, J. M.; Bridges, R. J.; Hart, J. A.; Chamberlin, A. R. *J. Org. Chem.* **1994**, *59*, 2467–2472.
[242]O'Shea, P. D.; Chen, C.-Y.; Chen, W.; Dagneau, P.; Frey, L. F.; Grabowski, E. J. J.; Marcantonio, K. M.; Reamer, R. A.; Tan, L.; Tillyer, R. D.; Roy, A.; Wang, X.; Zhao, D. *J. Org. Chem.* **2005**, *70*, 3021–3030.
[243]Askin, D.; Verhoeven, T. R.; Liu, T. M. H.; Shinkai, I. *J. Org. Chem.* **1991**, *56*, 4929–4932.

### 1.6.1.10  Alkylation of Carboxylic Acid Salts

The action of two equivalents of strong base on carboxylic acids produces a dianion that is capable of reaction with electrophiles to provide $\alpha$-substitution products (the Krapcho method).[244] Although the second deprotonation is considerably more challenging than the initial carboxylate salt formation, aggregate effects render the effective $pK_a$ of lithium carboxylates in THF solvent just a few units higher than the corresponding carboxylic esters.[245] In the example below, treatment of an enolizable carboxylic acid with two equivalents of lithium diisopropylamide (LDA) at $-75°C$ in THF provided the dianion. Subsequent addition of methyl iodide, followed by a hydrolytic workup, afforded the $\alpha$-methylation products in high yield as a mixture of diastereoisomers favoring axial substitution.[246]

### 1.6.1.11  Alkylation $\alpha$ to a Heteroatom

In a series of seminal publications, Corey and Seebach described the use of 1,3-dithianes as acyl anion synthons.[247–249] Soon after, Seebach introduced the term *umpolung* (German for "reversed polarity"), whereby a chemical transformation carried out on a functional group changes the inherent reactive polarity of that group. The use of dithianes as umpolung synthons has been reviewed.[250,251]

The example below clearly illustrates the umpolung concept, as the carbon in the 2-position of the starting 1,3-dithiane is derived from formaldehyde, which is typically electrophilic at carbon. Through conversion to the dithiane and deprotonation with *n*-butyllithium, this carbon becomes a nucleophilic center. Subsequent addition of 1-bromo-3-chloropropane provides the 3-chloropropyl alkylation product. In this example, another charge of *n*-BuLi addition led to intramolecular alkylation, which afforded dithiane-protected cyclobutanone in good yield. One major drawback of this methodology is the typical use of mercury(II) salts to promote conversion of the dithiane to the free

[244] Krapcho, A. P.; Dundulis, E. A. *Tetrahedron Lett.* **1976**, 2205–2208.
[245] Gronert, S.; Streitwieser, A. *J. Am. Chem. Soc.* **1988**, *110*, 4418–4419.
[246] Krapcho, A. P.; Dundulis, E. A. *J. Org. Chem.* **1980**, *45*, 3236–3245.
[247] Corey, E. J.; Seebach, D. *Angew. Chem., Int. Ed. Engl.* **1965**, *4*, 1075–1077.
[248] Corey, E. J.; Seebach, D. *Angew. Chem., Int. Ed. Engl.* **1965**, *4*, 1077–1078.
[249] Corey, E. J.; Seebach, D. *Org. Synth.* **1970**, *50*, 72–74.
[250] Groebel, B. T.; Seebach, D. *Synthesis* **1977**, 357–402.
[251] Seebach, D. *Synthesis* **1969**, *1*, 17–36.

carbonyl.[252,253] However, a handful of alternative methods for this conversion have been reported.[254,255] For more on the oxidation of sulfides, see Section 10.8.5.

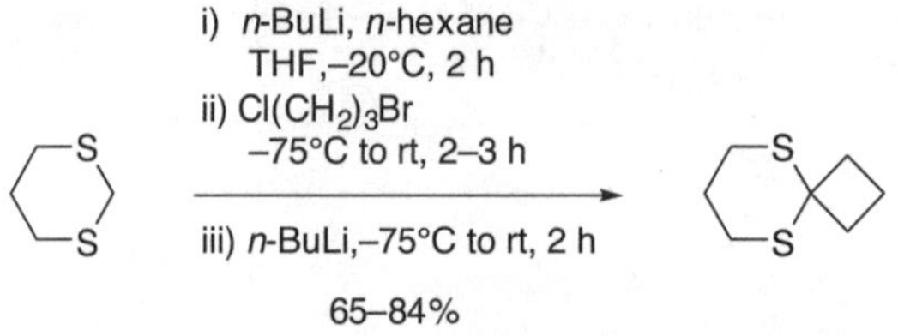

i) n-BuLi, n-hexane
THF,−20°C, 2 h
ii) Cl(CH₂)₃Br
−75°C to rt, 2–3 h
iii) n-BuLi,−75°C to rt, 2 h

65–84%

The dithiane alkylation reaction can also be used for the preparation of acyl silanes for further elaboration.[256] As described in the following scheme, the dithiane was deprotonated at the 2-position by n-BuLi in THF-hexanes and then reacted with chlorotrimethylsilane to afford the quaternary silane product in excellent yield.

n-BuLi, TMSCl
THF, hexanes

0°C to rt, 2 h
92%

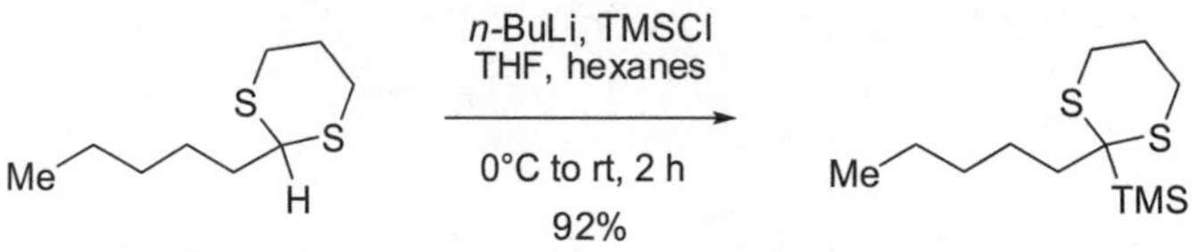

C-alkylation of trimethylsilyl cyanohydrins has emerged as an excellent method for homologation of aldehydes at the carbonyl carbon. As described in the example below, the TMS cyanohydrin prepared from furfuraldehyde and trimethylsilyl cyanide was treated with LDA at low temperature in THF to afford the carbon-centered anion. The addition of 4-bromo-1-butene followed by warming to room temperature provided the C-alkylated product in excellent yield. The cyanohydrin product was then subjected to triethylamine trihydrofluoride, which led to desilylation and loss of cyanide, and provide the ketone in high yield.[257]

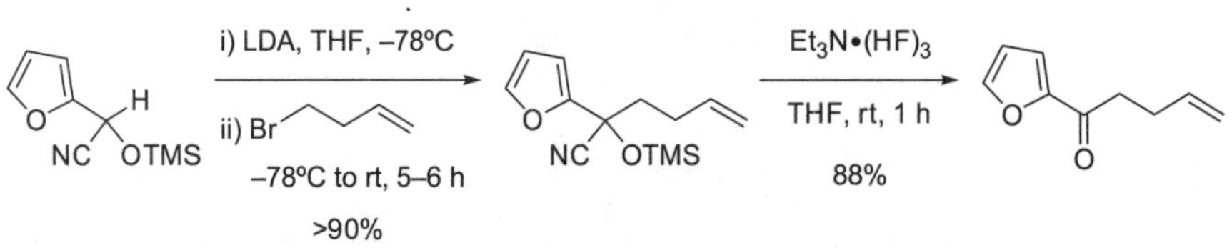

i) LDA, THF, −78°C

ii) Br
−78°C to rt, 5–6 h

>90%

Et₃N•(HF)₃

THF, rt, 1 h

88%

Hydrazones have also been shown to be effective acyl anion equivalents, as highlighted by the pioneering work of Baldwin[258–260] and coworkers. In the example below, the t-butyl

[252]Seebach, D.; Beck, A. K. *Org. Synth.* **1971**, *51*, 76–81.

[253]Manaviazar, S.; Frigerio, M.; Bhatia, G. S.; Hummersone, M. G.; Aliev, A. E.; Hale, K. J. *Org. Lett.* **2006**, *8*, 4477–4480.

[254]Firouzabadi, H.; Hazarkhani, H.; Hassani, H. *Phosphorus, Sulfur Silicon Relat. Elem.* **2004**, *179*, 403–409.

[255]Walsh, L. M.; Goodman, J. M. *Chem. Commun.* **2003**, 2616–2617.

[256]Huckins, J. R.; Rychnovsky, S. D. *J. Org. Chem.* **2003**, *68*, 10135–10145.

[257]Fischer, K.; Huenig, S. *J. Org. Chem.* **1987**, *52*, 564–569.

[258]Baldwin, J. E.; Adlington, R. M.; Bottaro, J. C.; Jain, A. U.; Kohle, J. N.; Perry, M. W. D.; Newington, I. M. *J. Chem. Soc., Chem. Commun.* **1984**, 1095–1096.

[259]Baldwin, J. E.; Adlington, R. M.; Bottaro, J. C.; Kolhe, J. N.; Perry, M. W. D.; Jain, A. U. *Tetrahedron* **1986**, *42*, 4223–4234.

[260]Baldwin, J. E.; Adlington, R. M.; Newington, I. M. *J. Chem. Soc., Chem. Commun.* **1986**, 176–178.

hydrazone of acetaldehyde was treated with *n*-butyllithium in THF to provide the expected lithiated intermediate at −20°C. Subsequent reaction with *o*-bromobenzylbromide at 15°C provided exclusively the product of C-alkylation in 77% yield.[261] Use of the sterically bulky *tert*-butyl substituent minimizes competitive N-alkylation.

Dieter and coworkers have extended this methodology through the use of organocuprates derived from lithiated hydrazones in reactions with Michael acceptors. This reaction is useful for the preparation of β-substituted carbonyl compounds.[262] The use of copper additives enhances the selectivity for C- versus N-alkylation of these ambident nucleophiles, and expands the substrate scope to include cyclic α,β-unsaturated ketones.

Alkoxy vinyl lithium reagents are another useful class of acyl anion equivalents. First reported by Baldwin and coworkers, the methyl enol ether of acetaldehyde can be deprotonated with *tert*-butyl lithium at the vinyl carbon-bearing oxygen.[263] The resulting vinyl anions can be alkylated with various electrophiles. Both *tert*-butyl lithium and alkoxy vinyl lithium reagents are highly reactive substances, which may limit practical use on a large scale; however a number of improvements to the original synthesis of these reagents have been made and several useful applications have been reported. In the example below from Denmark and coworkers, *n*-butyl vinyl ether is deprotonated with *tert*-butyllithium in tetrahydrofuran. Following transmetallation with CuCN the resulting vinyl cuprate was treated with allyl bromide to afford the alkylation product in good yield.[264]

Meyers and coworkers have communicated that the use of tetrahydropyran (THP) is superior to THF for the efficient preparation of these reagents via deprotonation by *t*-BuLi.[265] Meyers' group has also demonstrated the utility of α-ethoxyvinyl lithium in a diastereoselective conjugate addition to α,β-unsaturated oxazolines. Although not an aliphatic nucleophilic substitution reaction, an example is provided below for illustrative purposes.[266] The addition of carbon nucleophiles to polarized carbon−carbon multiple bonds is the subject of Section 3.5.

[261]Wang, S. F.; Warkentin, J. *Can. J. Chem.* **1988**, *66*, 2256–2258.

[262]Alexander, C. W.; Lin, S.-Y.; Dieter, R. K. *J. Organomet. Chem.* **1995**, *503*, 213–220.

[263]Baldwin, J. E.; Hoefle, G. A.; Lever, O. W., Jr., *J. Am. Chem. Soc.* **1974**, *96*, 7125–7127.

[264]Denmark, S. E.; Guagnano, V.; Dixon, J. A.; Stolle, A. *J. Org. Chem.* **1997**, *62*, 4610–4628.

[265]Shimano, M.; Meyers, A. I. *Tetrahedron Lett.* **1994**, *35*, 7727–7730.

[266]James, B.; Meyers, A. I. *Tetrahedron Lett.* **1998**, *39*, 5301–5304.

Alkoxyvinyl lithium derivatives have also found utility in nucleophilic additions to carbonyl centers such as carboxylic esters.[267]

### 1.6.1.12 Alkylation α to a Masked Carboxylic Acid

*1.6.1.12 Alkylation α to a Masked Carboxylic Acid* Elaboration at the α-position of carboxylic acid derivatives has been the subject of considerable research. Although much has been reported on the direct alkylation of carboxylate dianions, it is often preferable to employ carboxylic acid surrogates in bond forming operations and then deprotect later in a multi-step synthesis. In a series of seminal publications, Meyers and coworkers introduced 2-substituted–4,4-dimethyloxazolines as carboxylic acid synthons that can be easily elaborated via alkylation of their stabilized anion derivatives. The following example illustrates the deprotonation of 2-propyl oxazoline by *n*-butyllithium, followed by reaction with an alkyl iodide to provide the alkylated product.[268] The carboxylic acid moiety can be conveniently liberated under either acidic (3N HCl) or alkaline (MeI, then 1N NaOH) conditions. The latter method was utilized here to preserve the acid sensitive acetal functionality.

### 1.6.1.13 Alkylation at Alkynyl Carbon

*1.6.1.13 Alkylation at Alkynyl Carbon* Terminal alkynes can be converted to their acetylide anions by deprotonation with a strong base such as *n*-butyllithium. The acetylide formation and subsequent reaction with electrophiles are typically carried out in THF at temperatures near −78°C. In a few cases, additives such as HMPA or DMPU[269] are included to accelerate the overall reaction rate via coordination to lithium cations, resulting in deaggregation. An interesting study on the implications of HMPA additives on solution kinetics has been reported.[270] It should be noted that HMPA is a demonstrated health hazard and extreme care should be taken to avoid exposure during handling. If a strongly coordinating additive is desired, DMPU should be used preferentially, due to a superior safety profile. The following example is demonstrative of typical conditions for deprotonation and alkylation of terminal alkynes.[271]

[267]Li, Y.; Drew, M. G. B.; Welchman, E. V.; Shirvastava, R. K.; Jiang, S.; Valentine, R.; Singh, G. *Tetrahedron* **2004**, *60*, 6523–6531.

[268]Meyers, A. I.; Temple, D. L.; Nolen, R. L.; Mihelich, E. D. *J. Org. Chem.* **1974**, *39*, 2778–2783.

[269]Mukhopadhyay, T.; Seebach, D. *Helv. Chim. Acta* **1982**, *65*, 385–391.

[270]Collum, D. B.; McNeil, A. J.; Ramirez, A. *Angew. Chem., Int. Ed. Engl.* **2007**, *46*, 3002–3017.

[271]Patil, N. T.; Wu, H.; Yamamoto, Y. *J. Org. Chem.* **2007**, *72*, 6577–6579.

Epoxides are also suitable alkylating agents for lithium acetylides when the reaction is carried out in the presence of an oxophilic Lewis acid. In the example below, nucleophilic attack occurred at the less sterically encumbered position in the presence of boron trifluoride diethyl etherate.[272]

Trost and coworkers reported the use of catalytic diethylaluminum chloride as a Lewis acid on a closely related system to obtain the enantiomerically enriched alcohol in quantitative yield.[273]

Under analogous conditions, acetylides can be alkylated with oxetane via catalysis by $BF_3 \cdot OEt_2$, as described by Wessig and coworkers.[274]

Copper-promoted acetylinic alkylation reactions have also been reported. The mild conditions described in the example below are advantageous compared to methods that require cryogenic temperatures and pyrophoric reagents.[275]

[272]Reddy, M. S.; Narender, M.; Rao, K. R. *Tetrahedron* **2007**, *63*, 11011–11015.

[273]Trost, B. M.; Ball, Z. T.; Laemmerhold, K. M. *J. Am. Chem. Soc.* **2005**, *127*, 10028–10038.

[274]Wessig, P.; Mueller, G.; Kuehn, A.; Herre, R.; Blumenthal, H.; Troelenberg, S. *Synthesis* **2005**, 1445–1454.

[275]Alegret, C.; Santacana, F.; Riera, A. *J. Org. Chem.* **2007**, *72*, 7688–7692.

Acetylenes can be reacted with activated aziridines to afford homopropargylic amine derivatives in good yield.[276] It has been reported that potassium *tert*-butoxide base in dimethylsulfoxide is optimal for these systems.

*1.6.1.14 Alkylation of Cyanide Ion—Preparation of Nitriles* The use of cyanide anion for the preparation of nitriles (or amines via nitrile reduction) is a useful and commonly employed operation in target-directed synthesis. Despite the serious health hazards presented by hydrogen cyanide, sodium and potassium cyanide salts are utilized extensively as nucleophiles under basic conditions. Highly sensitive HCN detection monitors are available to provide an early warning to scientists working in the vicinity of cyanide-containing reactions. Diligence in the use of these detectors, in addition to personal protective equipment and proper engineering controls, is essential to maintaining a safe working environment.

In the example below, 2,4,6-triisopropylbenzyl chloride was converted to the corresponding benzylic nitrile in a biphasic solvent system via phase transfer catalysis.[277] This nucleophilic substitution reaction requires heating, presumably due to considerable steric obstruction from the isopropyl substituents. For some substrates, the more expensive, but also commercially available, tetrabutylammonium cyanide may provide a rate advantage over the catalytic method described here. However, this reagent is exceedingly hygroscopic, which complicates storage and handling.

Cyanide salts can also be reacted with alkyl sulfonates to provide the expected nitriles in high yield. In the example below, a primary alcohol was converted to its mesylate under standard conditions. The subsequent reaction of sodium cyanide in dimethylsulfoxide-toluene provided the nitrile in excellent overall yield on multi-gram scale.[278]

[276]Ding, C.-H.; Dai, L.-X.; Hou, X.-L. *Tetrahedron* **2005**, *61*, 9586–9593.

[277]Dozeman, G. J.; Fiore, P. J.; Puls, T. P.; Walker, J. C. *Org. Process Res. Dev.* **1997**, *1*, 137–148.

[278]Wu, G.; Wong, Y.; Steinman, M.; Tormos, W.; Schumacher, D. P.; Love, G. M.; Shutts, B. *Org. Process Res. Dev.* **1997**, *1*, 359–364.

Acetone cyanohydrin can also be employed as a convenient source of nucleophilic cyanide in the presence of a base. In the following example, an enantiomerically enriched secondary tosylate is converted to the corresponding nitrile in high yield and with clean inversion of stereochemistry. The authors reported that proper choice of base and solvent were critical to achieving high yield and reaction rate while preserving the stereochemical integrity of the product.[279] Optimal conditions included lithium hydroxide in a mixture of THF and 1,3-dimethyl-2-imidazolidinone (DMI).

In the example below, the action of tetrabutylammonium hydroxide on acetone cyano-hydrin produced tetrabutylammonium cyanide *in situ*. This underwent a smooth, high-yielding conjugate addition to the enone to afford the β-cyanoketone in high yield.[280] This methodology was successfully scaled to approximately 10 kilograms during the preparation of an endothelin antagonist candidate. Although this example does not satisfy the definition of a nucleophilic aliphatic substitution reaction, it is included here to illustrate the utility of *in situ* derived tetrabutylammonium cyanide, as the isolated compound is exceedingly hygroscopic and difficult to handle. For more discussion on the nucleophilic addition of cyanide to polarized carbon−carbon multiple bonds, see Section 3.5.6.

## 1.7 NUCLEOPHILIC SUBSTITUTION AT A SULFONYL SULFUR ATOM

### 1.7.1 Attack by Oxygen

*1.7.1.1 Hydrolysis of Sulfonic Acid Derivatives*  Sulfonyl halides are readily hydrolyzed to the parent sulfonic acids or sulfonate salts under acidic or basic conditions, respectively. In fact, most sulfonyl halides will hydrolyze in an aqueous environment in the absence of an additive, as the acidic products are themselves catalysts.[281]

[279]Hasegawa, T.; Kawanaka, Y.; Kasamatsu, E.; Ohta, C.; Nakabayashi, K.; Okamoto, M.; Hamano, M.; Takahashi, K.; Ohuchida, S.; Hamada, Y. *Org. Process Res. Dev.* **2005**, *9*, 774–781.
[280]Ellis, J. E.; Davis, E. M.; Brower, P. L. *Org. Process Res. Dev.* **1997**, *1*, 250–252.
[281]Hoffart, D. J.; Cote, A. P.; Shimizu, G. K. H. *Inorg. Chem.* **2003**, *42*, 8603–8605.

Sulfonic acid esters are also susceptible to hydrolysis. However, nucleophilic attack by water generally takes place at the carbon center, with concomitant displacement of the sulfonate moiety, as discussed in Section 1.2.1.4. Basic conditions are typically employed, and elimination is a concern. In a few rare cases, nucleophilic attack may occur at the sulfur center, as evidenced by net stereochemical retention at carbon in the following example.[282] The product is consistent with hydrolysis of the sulfonate, although steps were taken to ensure that water was not present in the reaction mixture.

Attack by water at the sulfur center is likely also operative during the hydrolysis of aryl sulfonate esters, although examples of this reaction are rare.[283,284]

Sulfonamides can be converted to the corresponding sulfonic acids via hydrolysis promoted by strong acids. Alkaline hydrolysis is typically ineffective, although a few special exceptions have been reported.[285] In the example below, *N*-methyl saccharin was treated with hydrochloric acid at reflux temperature to provide the methylammonium salt of the sulfonic acid resulting from complete hydrolysis of the mixed sulfonamide. The authors advise that the initial hydrolysis product is the *N*-methylsulfonamide, which reacts further with water to provide the sulfonic acid.[286]

***1.7.1.2  Formation of Sulfonate Esters***  Sulfonic acid esters are most commonly prepared by reaction of alcohols with sulfonyl halides under basic conditions. The use

[282]Chang, F. C. *Tetrahedron Lett.* **1964**, 305–309.

[283]Suh, J.; Suh, M. K.; Cho, S. H. *J. Chem. Soc., Perkin Trans.* **1990**, *2*, 685–688.

[284]Kano, K.; Nishiyabu, R.; Asada, T.; Kuroda, Y. *J. Am. Chem. Soc.* **2002**, *124*, 9937–9944.

[285]Cuvigny, T.; Larcheveque, M. *J. Organomet. Chem.* **1974**, *64*, 315–321.

[286]Meadow, J. R.; Cavagnol, J. C. *J. Org. Chem.* **1951**, *16*, 1582–1587.

of methanesulfonyl chloride (MsCl) and *p*-toluenesulfonyl chloride (TsCl) is commonplace, as these reagents are employed in the activation step during conversion of alcohols to other functionalities (for example, see Section 1.6.1.4). In the following example, a diol intermediate was differentially protected with TsCl and *p*-nitrobenzenesulfonyl chloride (NsCl) by taking advantage of the relative acidities of the two alcohols.[287] This selective protection sequence was used in the authors' synthesis of the muscarinic receptor antagonist tolterodine.

### 1.7.2  Attack by Nitrogen

*1.7.2.1  Formation of Sulfonamides*    The reaction of primary and secondary amines with sulfonyl halides is analogous to the corresponding reaction with oxygen nucleophiles, although the superior nucleophilicity of most amines makes the formation of sulfonamides a milder, more efficient transformation.[288] During the development of novel COX-2 inhibitors for the treatment of pain and related inflammatory disorders, Caron and coworkers reported the preparation of an aryl sulfonamide via treatment of an aniline derivative with tosyl chloride and pyridine in dichloromethane.

In the example below from Han and coworkers, the hydrochloride salt of (*R*)-phenyl-glycine methyl ester was treated with tosyl chloride in the presence of the mild base sodium bicarbonate to afford the sulfonamide in excellent yield at room temperature.[289] The authors were able to employ ethyl acetate as the solvent in this example, which is preferred for large scale applications due to its decreased environmental impact.

[287]De Castro, K. A.; Ko, J.; Park, D.; Park, S.; Rhee, H. *Org. Process Res. Dev.* **2007**, *11*, 918–921.

[288]Caron, S.; Vazquez, E.; Stevens, R. W.; Nakao, K.; Koike, H.; Murata, Y. *J. Org. Chem.* **2003**, *68*, 4104–4107.

[289]Han, Z.; Krishnamurthy, D.; Senanayake, C. H. *Org. Process Res. Dev.* **2006**, *10*, 327–333.

### 1.7.3 Attack by Halogen

***1.7.3.1  Formation of Sulfonyl Halides***    Sulfonic acids or their salts can be converted to the corresponding sulfonyl halides with methods analogous to the conversion of carboxylic acids or carboxylate salts to carboxylic acid halides (see Section 2.24). Sulfonyl chlorides are most commonly prepared by treatment with thionyl chloride[290] or phosphorus pentachloride,[291] with the former preferred for large scale applications due to the ease of reagent removal by distillation. In the example below, treatment of an acetonitrile suspension of a sulfonic acid sodium salt with thionyl chloride in the presence of DMF provided the sulfonyl chloride in excellent yield.[292]

Likewise, sulfonyl bromides can be prepared from sulfonic acid salts by utilizing phosphorous pentabromide,[293] or preferably, via treatment with triphenylphosphine and bromine or *N*-bromosuccinimide.[294] In the example below, the tetrabutylammonium salt of a sulfonic acid was converted to the corresponding sulfonyl bromide with triphenylphosphine and bromine. As sulfonyl bromides are generally unstable, the product was not isolated, but rather converted immediately to the isopropyl sulfonate ester by quenching the reaction with isopropanol and triethylamine.

A less frequently utilized method involves treatment of sulfonyl hydrazines with bromine.[295] As a relatively large number of sulfonyl chlorides are commercially available, a convenient method for the preparation of sulfonyl bromides involves treatment of sulfonyl chlorides with sodium sulfite, followed by bromine, as described in the example below.[293]

[290]Morikawa, A.; Sone, T.; Asano, T. *J. Med. Chem.* **1989**, *32*, 42–46.

[291]Ponticello, G. S.; Freedman, M. B.; Habecker, C. N.; Lyle, P. A.; Schwam, H.; Varga, S. L.; Christy, M. E.; Randall, W. C.; Baldwin, J. J. *J. Med. Chem.* **1987**, *30*, 591–597.

[292]Ikemoto, N.; Liu, J.; Brands, K. M. J.; McNamara, J. M.; Reider, P. J. *Tetrahedron* **2003**, *59*, 1317–1325.

[293]Block, E.; Aslam, M.; Eswarakrishnan, V.; Gebreyes, K.; Hutchinson, J.; Iyer, R.; Laffitte, J. A.; Wall, A. *J. Am. Chem. Soc.* **1986**, *108*, 4568–4580.

[294]Huang, J.; Widlanski, T. S. *Tetrahedron Lett.* **1992**, *33*, 2657–2660.

[295]Palmieri, G. *Tetrahedron* **1983**, *39*, 4097–4101.

i) $Na_2SO_3$, $H_2O$
0–20°C, 1 h

ii) $Br_2$, 0°C, 0.5 h

66%

Sulfonyl fluorides are also most easily prepared from their sulfonyl chloride analogs through halogen exchange with potassium fluoride[296] or TBAF.[297] In the representative example below, a sulfonyl fluoride was obtained in high yield via treatment of a sulfonyl chloride with KF in refluxing THF. The sulfonyl fluoride products typically require purification through distillation, which can be damaging to glass and is better carried out in specialized equipment.

$$n\text{-}C_{16}H_{32}SO_2Cl \quad \xrightarrow[\text{Reflux, 30 h}]{\text{KF, THF, } H_2O} \quad n\text{-}C_{16}H_{32}SO_2F$$

97%

### 1.7.4  Attack by Carbon

*1.7.4.1  Preparation of Sulfones*  The range of carbon-centered nucleophiles that can react with sulfonyl halide electrophiles to afford sulfone products with reasonable efficiency is rather limited. One complicating factor is that sulfone products typically have a lower pKa than the carbanion nucleophiles used to prepare them. Although stoichiometry may be adjusted to address this issue, the synthesis of alkyl sulfones is best accomplished via oxidation of sulfide or sulfoxide precursors (see Section 10.8.5). Nevertheless, the addition of certain carbon nucleophiles to sulfonyl halides has been accomplished under transition metal catalysis.[298] The example below from Labadie is representative and involves the use of a stoichiometric organostannane.

(Ph₃P)₄Pd, THF

65–70°C, 15 min

77%

[296]Hyatt, J. A.; White, A. W. *Synthesis* **1984**, 214–217.
[297]Sun, H.; DiMagno, S. G. *J. Am. Chem. Soc.* **2005**, *127*, 2050–2051.
[298]Labadie, S. S. *J. Org. Chem.* **1989**, *54*, 2496–2498.

# 2

# ADDITION TO CARBON–HETEROATOM MULTIPLE BONDS

Rajappa Vaidyanathan and Carrie Brockway Wager

## 2.1 INTRODUCTION

Additions to carbon–heteroatom multiple bonds will be presented in this chapter. The majority of the discussion will focus on additions to carbon–oxygen double bonds, carbon–nitrogen double bonds, and carbon–nitrogen triple bonds. Additions to carbon–sulfur double bonds are far less common, and will be briefly discussed in this chapter.

The polarity of the C=O, C=N and C≡N bonds makes prediction of their reactivity fairly simple. With the exception of isocyanides, the carbon atom is typically the more electropositive center, so nucleophiles generally attack the carbon atom, while the more electronegative heteroatom (O or N) generally attacks electrophiles. The relative stereochemistry of addition (*syn* or *anti*) across carbon–heteroatom double bonds is neither known nor of any consequence. However, the relative stereochemistry of addition across carbon–nitrogen triple bonds can be determined by examining the product (*E* or *Z* isomer).

The more interesting stereochemical aspect is the facial selectivity of nucleophilic attack onto C=O and C=N moieties. If $R^1$ and $R^2$ (the groups flanking the C=X center) are different, the resultant molecule after nucleophilic addition is chiral. In the absence of chirality in $R^1$, $R^2$, or the nucleophile, the products are racemic mixtures. The absolute configuration at the chiral center may be controlled by either of the R groups, the nucleophile, or more importantly, catalysts used to affect these reactions. The use of asymmetric catalysts to influence the stereochemical outcome of these reactions has revolutionized this area, and several examples will be presented.

In the specific case of nucleophilic attack on carbon–oxygen double bonds, substitiution or addition-derived products can arise. The mode of reactivity is determined by the nature of the groups flanking the carbonyl carbon. Aldehydes and ketones cannot undergo substitution

*Practical Synthetic Organic Chemistry: Reactions, Principles, and Techniques*, First Edition.
Edited by Stéphane Caron.
© 2011 John Wiley & Sons, Inc. Published 2011 by John Wiley & Sons, Inc.

reactions owing to the poor leaving group ability of H, alkyl, and aryl groups. However, carboxylic acids and their derivatives undergo substitution reactions since OH, OR, and $NR_2$ groups are better leaving groups.

The chapter is divided into additions to aldehydes or ketones (2.2–2.20), additions to carbon–nitrogen double bonds (2.21–2.23), additions to carboxylic acid derivatives (2.24–2.36), additions to nitriles (2.37–2.43), and miscellaneous C=X additions (2.44–2.48)

## 2.2 ADDITION OF WATER TO ALDEHYDES AND KETONES: FORMATION OF HYDRATES

The net addition of water across a carbonyl results in a hydrate. They are virtually useless from a synthetic standpoint, and hence are seldom intentionally synthesized. Hydrates *occur* readily in most aqueous solvent systems, but they are usually not isolable unless the carbonyl is sufficiently electron-deficient (e.g., di- and tri-carbonyl compounds, α-haloaldehydes, etc.).

A few instances where hydrates are formed are shown below. Both examples involve formation of electrophilic 1,2-dicarbonyl compounds, which result in the concomitant formation of a hydrate. In the first example, treatment of the fluoro-phenyl ketone with HBr in DMSO leads to the corresponding dibromide, which is then converted to the aldehyde during the workup.[1] In the second example, oxidative periodate-mediated cleavage of diethyl tartarate furnished 2 equivalents of ethyl glyoxaldehyde, which readily converts to a stable hydrated form in the presence of water.[2]

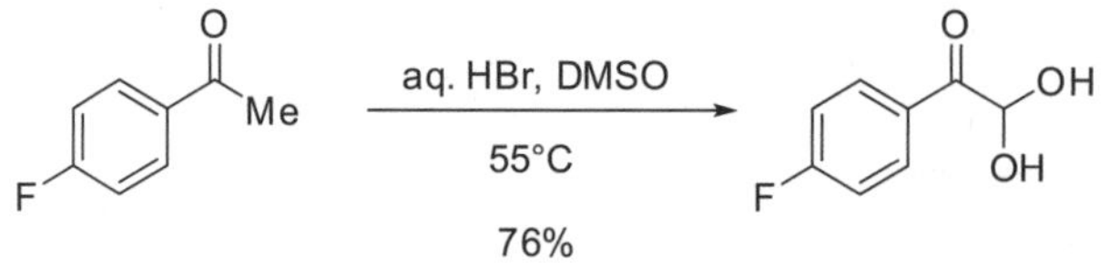

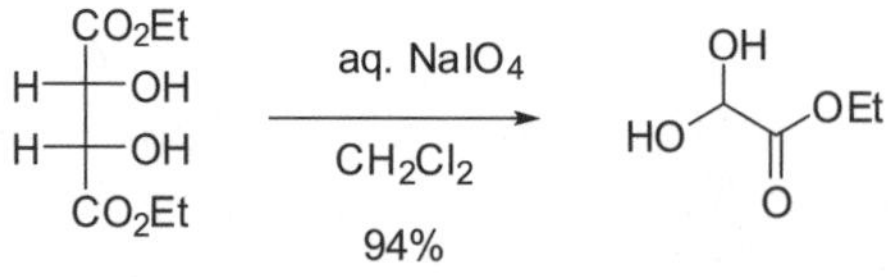

[1]Zhao, M. M.; McNamara, J. M.; Ho, G.-J.; Emerson, K. M.; Song, Z. J.; Tschaen, D. M.; Brands, K. M. J.; Dolling, U.-H.; Grabowski, E. J. J.; Reider, P. J.; Cottrell, I. F.; Ashwood, M. S.; Bishop, B. C. *J. Org. Chem.* **2002**, *67*, 6743–6747.

[2]Bailey, P. D.; Smith, P. D.; Pederson, F.; Clegg, W.; Rosair, G. M.; Teat, S. J. *Tetrahedron Lett.* **2002**, *43*, 1067–1070.

## 2.3  ADDITION OF BISULFITE TO ALDEHYDES AND KETONES

Bisulfite formation is a valuable, chemoselective derivatization procedure used for the protection and/or purification of carbonyl compounds (generally aldehydes), or for the removal of unwanted/residual aldehydes from reaction mixtures. In the most commonly utilized method, the aldehyde is treated with sodium bisulfite to afford the sodium salt of the bisulfite adduct. These charged adducts are often highly water soluble; however, careful solvent selection can allow for isolation or removal of the bisulfite adduct via precipitation. Regeneration of the adehyde can be accomplished by treatment with aqueous base or in some cases, by heating in water.[3]

Bisulfite adduct formation is heavily utilized in beer and wine production, where inhibition of oxidation of acetaldehyde (to form acetic acid) is critical. Brewing research scientists from the Sapporo Corporation formed the bisulfite adduct of acetaldehyde by mixing sodium bisulfite and acetaldehyde, thereby lengthening the shelf-life of their product.[4]

Hach and coworkers exploited this reaction to selectively isolate the β-isomer of thujone.[5] Treatment of thujone (as a mixture of isomers) in ethanol with a saturated aqueous solution of sodium bisulfite in the presence of sodium bicarbonate led to precipitation of the bisulfite adduct of the β-isomer. Decomposition of the bisulfite adduct by heating in water, followed by extraction with ether gave the desired thujone β-isomer.

Kjell and coworkers utilized an alternative approach for the regeneration of an aldehyde from the bisulfite adduct during their work to produce the oncolytic agent, LY231514•Na.[6] In this case, the alcohol impurities could be removed from the mixture by conversion of the aldehydes to the corresponding bisulfite adducts. Bisulfite adducts form slowly on hindered substrates, and as a result the less hindered aldehyde was enriched in the process. Anhydrous regeneration of the aldehyde was necessary in order to avoid inefficient extraction of the aldehyde from the aquous layer. This was achieved by treatment of the bisulfite adducts with an excess of TMSCl in acetonitrile.

[3]March, J. *Advanced Organic Chemistry: Reactions, Mechanisms, and Structure. 2nd ed.*; McGraw-Hill: New York, N.Y., **1977**.
[4]Kaneda, H.; Osawa, T.; Kawakishi, S.; Munekata, M.; Koshino, S. *J. Agric. Food Chem.* **1994**, *42*, 2428–2432.
[5]Hach, V.; Lockhart, R. W.; McDonald, E. C.; Cartlidge, D. M. *Can. J. Chem.* **1971**, *49*, 1762–1764.
[6]Kjell, D. P.; Slattery, B. J.; Semo, M. J. *J. Org. Chem.* **1999**, *64*, 5722–5724.

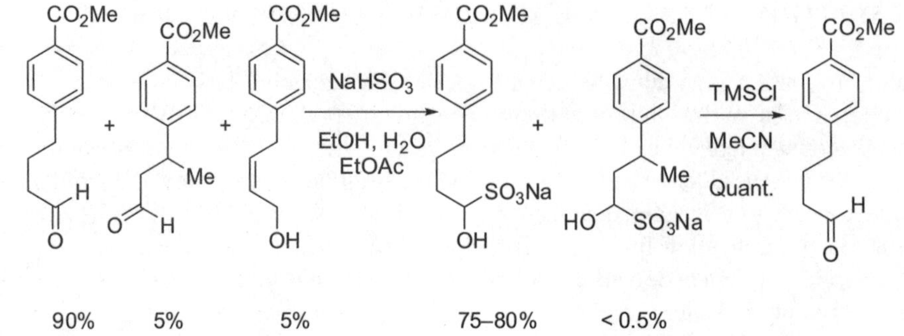
NaHSO3
EtOH, H2O
EtOAc
TMSCl
MeCN
Quant.
90%
5%
5%
75–80%
< 0.5%

Ragan and coworkers used sodium bisulfite for the *in situ* protection of a highly reactive aldehyde arising out of an ozonolysis sequence. The bisulfite adduct provided an easily characterized, isolable intermediate. The adduct was directly subjected to the next reductive amination step without the need to regenerate the aldehyde in a separate step.[7]

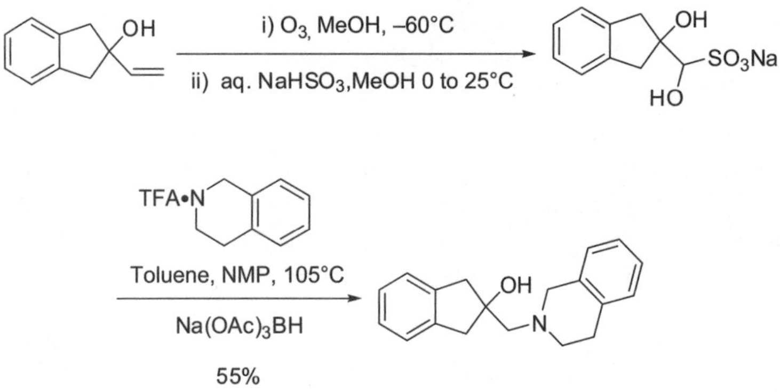
i) O3, MeOH, –60°C
ii) aq. NaHSO3,MeOH 0 to 25°C
TFA•N
Toluene, NMP, 105°C
Na(OAc)3BH
55%

Bisulfite adducts may also be formed by treatment of carbonyl compounds with aqueous sulfur dioxide (sulfurous acid). In the following example, treatment of a THF solution of the enamine with sulfurous acid gave the bisulfite adduct as a white solid in 64% yield on 10 kg scale.[8]

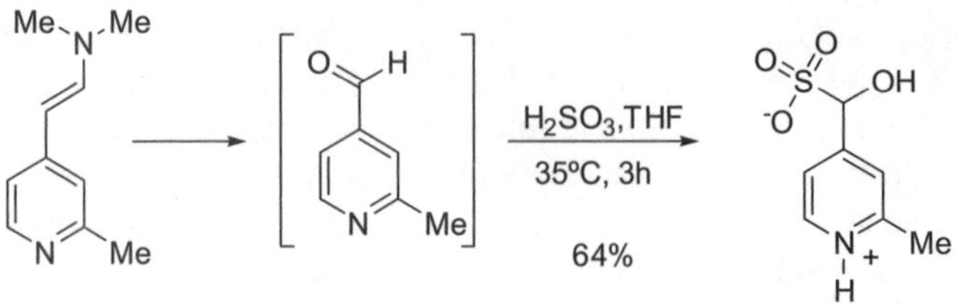
H2SO3,THF
35°C, 3h
64%

[7]Ragan, J. A.; am Ende, D. J.; Brenek, S. J.; Eisenbeis, S. A.; Singer, R. A.; Tickner, D. L.; Teixeira, J. J., Jr.; Vanderplas, B. C.; Weston, N. *Org. Process Res. Dev.* **2003**, *7*, 155–160.
[8]Ragan, J. A.; Jones, B. P.; Meltz, C. N.; Teixeira, J. J., Jr. *Synthesis* **2002**, 483–486.

## 2.4 THE ADDITION OF ALCOHOLS TO ALDEHYDES AND KETONES: ACETAL FORMATION

The reaction of carbonyl compounds with alcohols leads to acetal formation. Generally, these reactions are performed in the presence of excess alcohol, a Lewis or Brønsted acid, and often under dehydrating conditions. The water generated can either be azeotropically removed using a Dean-Stark apparatus, or by the use of a chemical trap (dehydrating reagent) such as an orthoformate or molecular sieves.

En route to quassinoid, the Heathcock group converted the enone to the acetal by treatment with *p*-toluenesulfonic acid and ethylene glycol in benzene at reflux.[9] In most cases, benzene can be substituted with toluene or trifluorotoluene.

In Corey's synthesis of ginkgolide, an acetal was obtained by treatment of the aldehyde precursor with a catalytic amount of *p*-toluenesulfonic acid in 5:1 methanol:trimethyl orthoformate at room temperature.[10] This is an excellent example of an acetal formation under relatively mild reaction conditions. Note that under these conditions, a hemiacetal was also converted to the mixed methyl acetal.

A particularly mild method for acetal formation was developed by Noyori and coworkers, wherein the carbonyl compound is treated with an alkoxytrimethylsilane in the presence of catalytic trimethylsilyl trifluoromethanesulfonate at temperatures as low as −78°C.[11] In this case, the high stability of the byproduct, hexamethydisiloxane, provides a major driving force for formation of the product acetal. A variant of this method was used by researchers at Merck to convert the ketone in the following example to the hydrobenzoin acetal.[12]

[9]Kerwin, S. M.; Paul, A. G.; Heathcock, C. H. *J. Org. Chem.* **1987**, *52*, 1686–1695.

[10]Corey, E. J.; Su, W. G. *J. Am. Chem. Soc.* **1987**, *109*, 7534–7536.

[11]Tsunoda, T.; Suzuki, M.; Noyori, R. *Tetrahedron Lett.* **1980**, *21*, 1357–1358.

[12]Tan, L.; Yasuda, N.; Yoshikawa, N.; Hartner, F. W.; Eng, K. K.; Leonard, W. R.; Tsay, F.-R.; Volante, R. P.; Tillyer, R. D. *J. Org. Chem.* **2005**, *70*, 8027–8034.

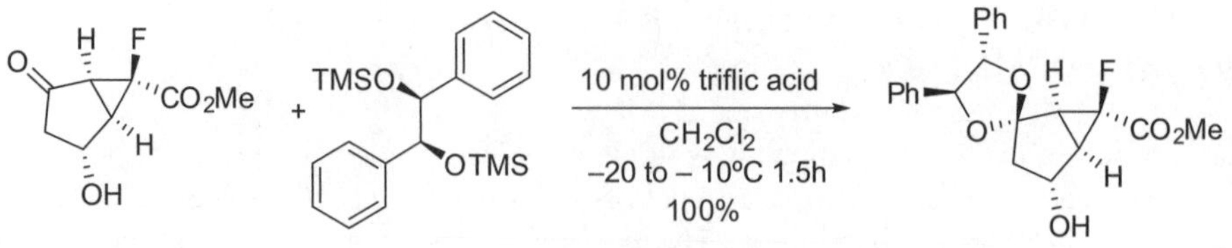

## 2.5   THE ADDITION OF THIOLS TO ALDEHYDES AND KETONES: S,S-ACETAL FORMATION

*S,S*-Acetal formation can be accomplished in a similar manner to the above *O,O*-acetal examples; in general, acidic dehydrating conditions are most effective. $BF_3 \cdot OEt_2$ is a commonly used catalyst for this reaction.[13]

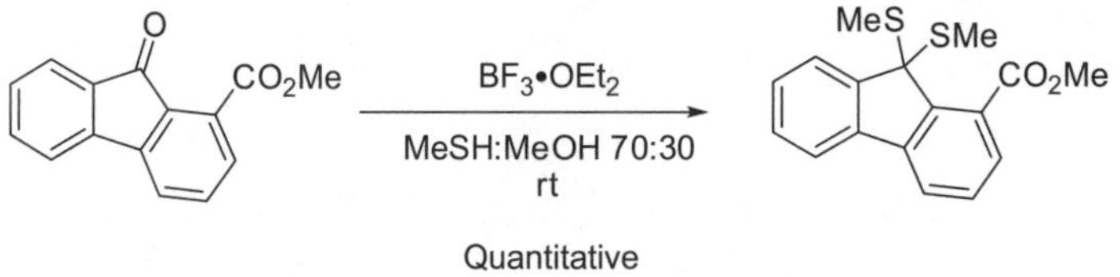

These reactions are often run using the thiol as the reaction solvent. Brønsted acids have been used to catalyze this reaction, as shown in the Valverde example below.[14] Glucose, dissolved in ethanethiol, is treated with HCl to afford the *S,S*-acetal after only one hour at room temperature. After precipitation with ethanol, ketal formation occurs after dissolution in acetone to give the desired product in a 74% yield over two steps.

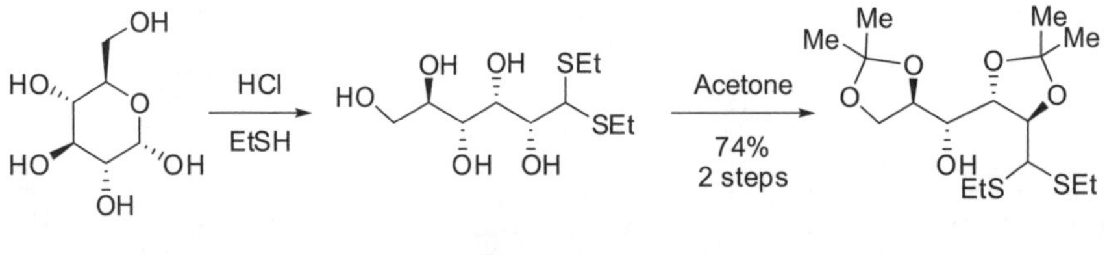

Djerassi and coworkers successfully synthesized the *S,S*-acetal of the a cortisone precursor in the following scheme under scrupulously anhydrous conditions.[15] The substrate in ethanethiol and dry dioxane was treated with an excess of anhydrous zinc chloride and sodium sulfate to give a good yield of the thioacetal product.

[13]Hojo, M.; Ichi, T.; Shibato, K. *J. Org. Chem.* **1985**, *50*, 1478–1482.
[14]Valverde, S.; Lopez, J. C.; Gomez, A. M.; Garcia-Ochoa, S. *J. Org. Chem.* **1992**, *57*, 1613–1615.
[15]Djerassi, C.; Batres, E.; Velasco, M.; Rosenkranz, G. *J. Am. Chem. Soc.* **1952**, *74*, 1712–1715.

A standard application of dithiane acetals was pioneered by Corey and Seebach.[16,17] Metallated *S,S*-acetals may be trapped by a variety of electrophiles, as described in the scheme. The dithiane was formed by treatment of the aldehydes in chloroform with propane-1,3-dithiol and either dry HCl gas or the Lewis acids anhydrous zinc dichloride or BF$_3$•OEt$_2$. The authors showed that after metallation with *n*-butyl lithiuim in cold THF, the acyl anion equivalents could then be trapped with an alkyl halides, aldehydes, or epoxides.

Aldehydes can be converted to a thioacetal, which can be metallated at C$_2$ and then react with electrophiles. This strategy converts the normally electrophilic aldehyde carbon into a nucleophile, and is referred to as "umpolung," or an inversion of electronic character. *S,S*-acetals can also be used to effect deoxygenation of the carbonyl by reduction of the C-S bonds. In their work to build compounds to bind the GABA neurotransmitter, Covey and coworkers found that their steroidal substrates were prone to decomposition, especially the aldehyde below.[18] In order to effect thioacetal formation in this substrate, the best strategy was to use ethanethiol in methylene chloride and boron trifluoride etherate. The *S,S*-acetal was then processed further with Raney Nickel to effect deoxygenation (reaction not shown).

[16]Corey, E. J.; Seebach, D. *Org. Synth.* **1970**, *50*, 72–74.
[17]Seebach, D.; Corey, E. J. *J. Org. Chem.* **1975**, *40*, 231–237.
[18]Hu, Y.; Zorumski, C. F.; Covey, D. F. *J. Med. Chem.* **1993**, *36*, 3956–3967.

Janda and coworkers also utilized boron trifluoride etherate in their conversion of the aldehyde to the thioacetal.[19] To a solution of the aldehyde and ethanethiol in dry methylene chloride was added boron trifluoride etherate. After overnight reaction, standard aqueous workup provided the cyclic *S,S*-acetal in excellent yield.

99%

Researchers at the Hebrew University found optimal conditions utilized hydrochloric acid in their conversion of a xylose-derived starting material to the *S,S*-acetal.[20] This reaction provided the desired protected xylose in good yield.

76%

After exploring standard methods for *S,S*-acetal formation, the laboratories of Dolors-Pujol found *p*-toluenesulfonic acid in hot toluene optimal for formation of the desired bis (phenylthio) acetal.[21]

80%

## 2.6   REDUCTIVE ETHERIFICATION

The reductive etherification of aldehydes and ketones occurs when they are treated with an acid (Lewis or Brønsted) followed by triethylsilane in the presence of an alcohol. The pathway involves hemi-acetal or ketal formation followed by oxocarbenium formation and reduction of the resultant intermediate by the silane.[22] Lewis acids that are often used for this reaction include $BF_3 \cdot OEt_2$ and bismuth tribromide. This method generally works well for the etherification of aldehydes, but can require some optimization for etherification of ketones, as ketone dimerization can be a complicating factor.

[19]Qi, L.; Yamamoto, N.; Meijler, M. M.; Altobell, L. J., III; Koob, G. F.; Wirsching, P.; Janda, K. D. *J. Med. Chem.* **2005**, *48*, 7389–7399.
[20]Gruzman, A.; Shamni, O.; Ben Yakir, M.; Sandovski, D.; Elgart, A.; Alpert, E.; Cohen, G.; Hoffman, A.; Katzhendler, Y.; Cerasi, E.; Sasson, S. *J. Med. Chem.* **2008**, *51*, 8096–8108.
[21]Capilla, A. S.; Sanchez, I.; Caignard, D. H.; Renard, P.; Pujol, M. D. *Eur. J. Med. Chem.* **2001**, *36*, 389–393.
[22]Tsunoda, T.; Suzuki, M.; Noyori, R. *Tetrahedron Lett.* **1979**, 4679–4680.

In the first example below, Bajwa and coworkers used bismuth tribromide and triethyl-silane in acetonitrile to deprotect the hydroxyl group and generate/reduce the oxocarbenium ion.[23]

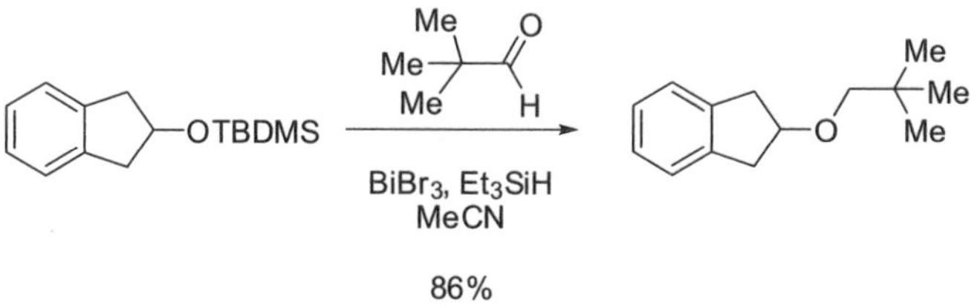

86%

In their synthesis of leucascandrolide A, the Floreancig group used very similar reaction conditions with trimethylallylsilane (instead of triethylsilane) to prepare the net allylated pyran.[24]

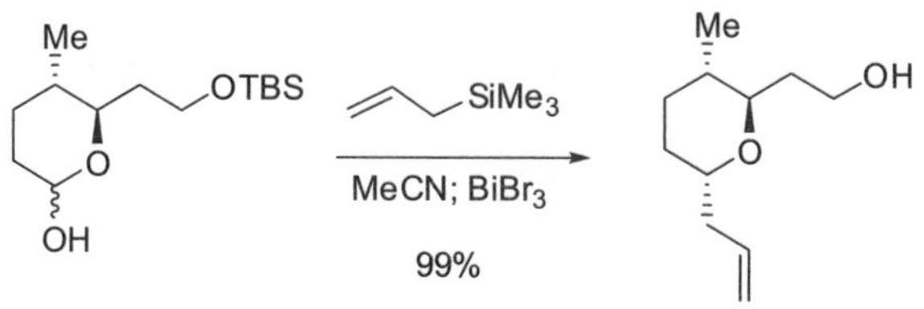

99%

Using $SnCl_4$, Mori and coworkers were able to form the pyran shown below.[25] Compared to other Lewis acids tested $SnCl_4$ minimized deleterious side reactions, such as elimination of the β-hydroxyl substituent. However, use of this stronger Lewis acid necessitates lower reaction temperatures to inhibit degradation of the starting materials.

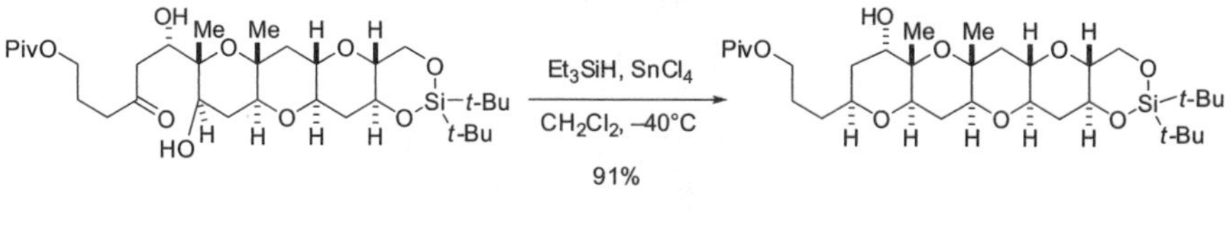

91%

As mentioned above, when standard reaction conditions are used with ketones, dimerization can be the main product. Olah found that use of an orthoformate can maximize the formation of the desired ether.[26] This reaction is initiated through the formation of the acetal or ketal with triethyl orthoformate in the presence of Nafion-H resin. This perfluorosulfonic acid resin catalyzes $O,O$-acetal formation and also the reduction with triethylsilane to give the ethyl ether below. Many other fluorinated sulfonic acids (super acids) can be used in this reaction, but Nafion-H performs well and is not as sensitive to moisture as alternatives. The main drawback to this method is that only simple orthoformates are commercially available.

[23]Bajwa, J. S.; Prasad, K.; Repic, O. *Org. Synth.* **2006**, *83*, 155–161.

[24]Jung, H. H.; Seiders, J. R.; Floreancig, P. E. *Angew. Chem., Int. Ed.* **2007**, *46*, 8464–8467.

[25]Furuta, H.; Hase, M.; Noyori, R.; Mori, Y. *Org. Lett.* **2005**, *7*, 4061–4064.

[26]Olah, G. A.; Yamato, T.; Iyer, P. S.; Prakash, G. K. S. *J. Org. Chem.* **1986**, *51*, 2826–2828.

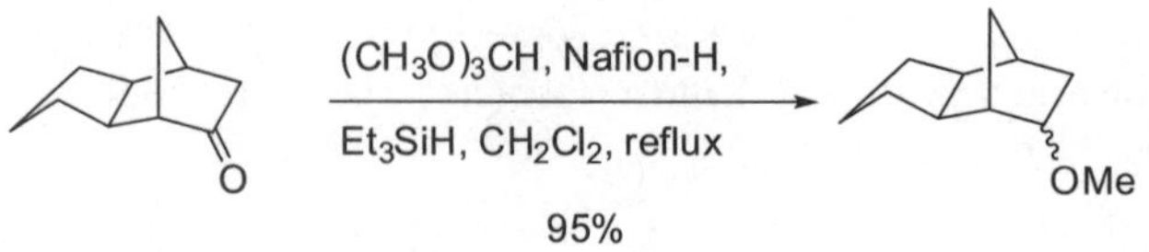

## 2.7  ATTACK BY NH$_2$, NHR, OR NR$_2$ (ADDITION OF NH$_3$, RNH$_2$, R$_2$NH)

### 2.7.1  The Addition of Amines to Aldehydes and Ketones: Imine and Oxime Formation

Imines and oximes are the replacement of the oxygen on an aldehyde or ketone with an -NR, -NH or -NOH. Imines and oximes have a wide range of reactivity. Some uses include the aza-Diels–Alder reaction (Section 3.14.2), Eschweiler–Clarke reaction (Section 1.4.1.9), and as an intermediate in reductive amination (Section 9.3.2). Common to imine and oxime formation is the use of dehydration conditions to drive the reaction to completion. This can be accomplished in a variety of ways, dictated by the sensitivity of the substrates. A few methods are shown below.

Banks and coworkers utilized a standard procedure for the primary imine formation shown below.[27] First, the HCl salt was broken with a solution of toluene and aqueous potassium carbonate. The toluene layer was dried by heating it to reflux with magnesium sulfate and with a Dean-Stark apparatus to remove water. As this solution was under reflux, a solution of the aldehyde in toluene was added while maintaining temperature. The reaction was maintained at toluene reflux and water collection measured to determine reaction completion. The imine was used in their next step without further purification.

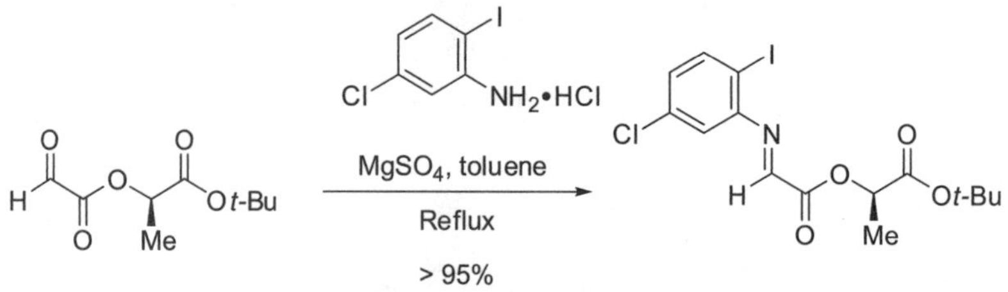

In their work towards avermectin, Cvetovich and coworkers found that standard conditions showed only a small amount of desired product and a significant amount of 1,4-addition byproduct.[28] As a result, they developed a Lewis acid approach to favor formation of the desired product. The procedure with ZnCl$_2$ and TMSONH$_2$ provided the desired oxime in good isolated yield.

[27]Banks, A.; Breen, G. F.; Caine, D.; Carey, J. S.; Drake, C.; Forth, M. A.; Gladwin, A.; Guelfi, S.; Hayes, J. F.; Maragni, P.; Morgan, D. O.; Oxley, P.; Perboni, A.; Popkin, M. E.; Rawlinson, F.; Roux, G. *Org. Process Res. Dev.* **2009**, *13*, 1130–1140.
[28]Cvetovich, R. J.; Leonard, W. R.; Amato, J. S.; DiMichele, L. M.; Reamer, R. A.; Shuman, R. F.; Grabowski, E. J. J. *J. Org. Chem.* **1994**, *59*, 5838–5840.

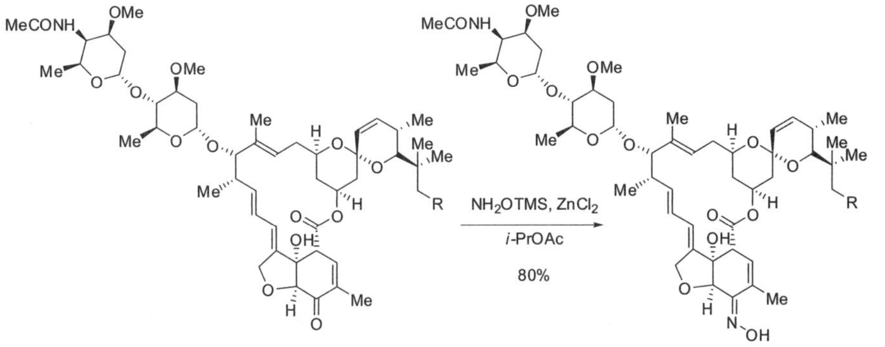

En route to a novel prostaglandin receptor agonist, Hida and coworkers utilized an oxime formation/reduction sequence for installation of the requisite amine.[29] Clean transformation to the *O*-methyl oxime was accomplished by dissolving the ketone in ethanol and adding equimolar amounts of the *O*-methyl hydroxyl amine salt and pyridine and heating to reflux. Acid/base aqueous workup and silica gel chromatography were necessary to isolate clean *O*-methyl hydroxyl amine. The oxime was transformed to the free amine by dissolution in diglyme with aluminum trichloride and then slow addition of an excess of sodium borohydride.

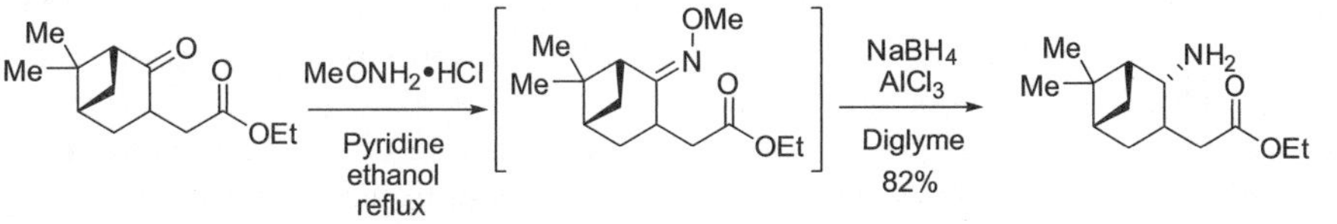

Another method for conversion of a ketone to a primary amine is shown below.[30]

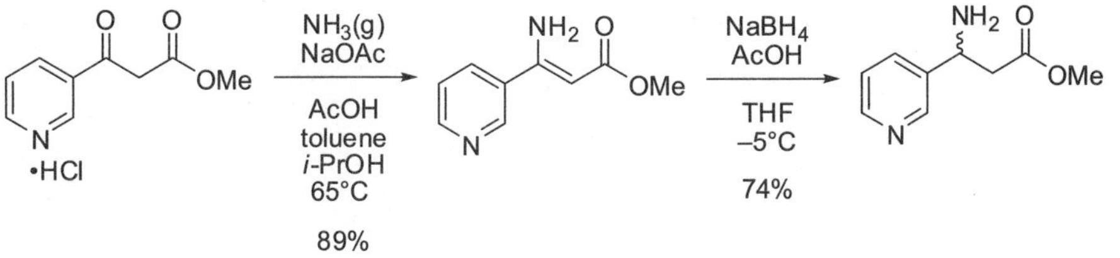

A chiral auxiliary controlled diastereoselective reductive amination used in the synthesis of leukocyte inhibitor DMP 777 is shown below.[31] Imine formation (2:1 *E:Z*) was

[29]Hida, T.; Mitsumori, S.; Honma, T.; Hiramatsu, Y.; Hashizume, H.; Okada, T.; Kakinuma, M.; Kawata, K.; Oda, K.; Hasegawa, A.; Masui, T.; Nogusa, H. *Org. Process Res. Dev.* **2009**, *13*, 1413–1418.

[30]Cohen, J. H.; Bos, M. E.; Cesco-Cancian, S.; Harris, B. D.; Hortenstine, J. T.; Justus, M.; Maryanoff, C. A.; Mills, J.; Muller, S.; Roessler, A.; Scott, L.; Sorgi, K. L.; Villani, F. J., Jr.,; Webster, R. R. H.; Weh, C. *Org. Process Res. Dev.* **2003**, *7*, 866–872.

[31]Storace, L.; Anzalone, L.; Confalone, P. N.; Davis, W. P.; Fortunak, J. M.; Giangiordano, M.; Haley, J. J., Jr.; Kamholz, K.; Li, H.-Y.; Ma, P.; Nugent, W. A.; Parsons, R. L., Jr.; Sheeran, P. J.; Silverman, C. E.; Waltermire, R. E.; Wood, C. C. *Org. Process Res. Dev.* **2002**, *6*, 54–63.

accomplished using titanium tetrachloride, triethylamine, and toluene. In the Raney Ni reduction, the *E* isomer was quickly reduced with good diastereoselectivity, while the less reactive *Z* isomer isomerized to the *E* isomer, which in turn was reduced to the desired product. The sequence was finished by standard hydrogenolysis to remove the chiral auxiliary.

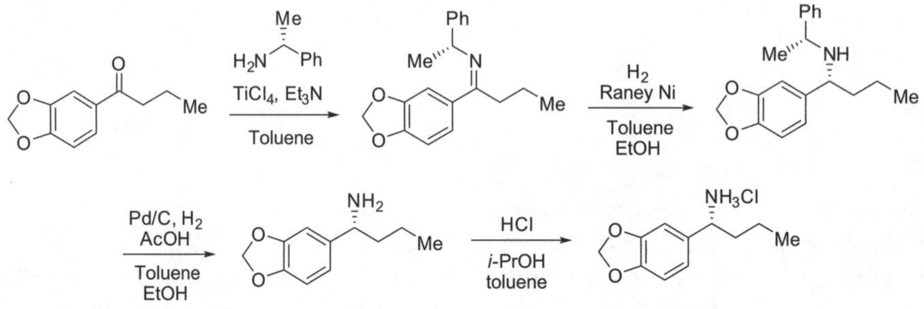

### 2.7.2  Redox Neutral Amination

Metal-catalyzed "redox neutral" methods for conversion of an alcohol into an amine have recently been developed. It is postulated that these complexes generate the carbonyl component *in situ*, followed by imine formation and reduction of the imine by the metal hydride species formed during the imine formation. Hence these reactions are sometimes referred to as "borrowing hydrogen" methods. These methods are particularly useful when aldehyde or ketone instability precludes standard reductive amination conditions. Two of the current methods are shown below, which use ruthenium and iridium complexes, respectively. In the first example, the Williams laboratories successfully applied a ruthenium *p*-cymene complex to the reaction shown.[32]

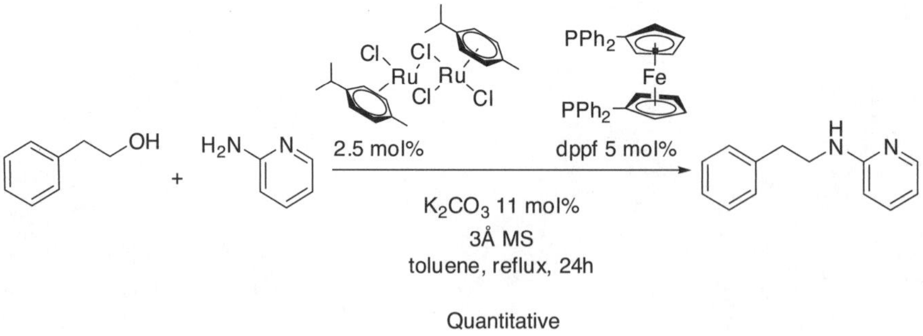

Quantitative

In the second example, Fujita and coworkers used the iridium pentamethyl-cyclopenta-dienyl dimer for net-double reductive amination to form *N*-benzylpiperidine.[33]

[32]Hamid, M. H. S. A.; Williams, J. M. J. *Chem. Commun. (Cambridge, U. K.)* **2007**, 725–727.
[33]Fujita, K.; Fujii, T.; Yamaguchi, R. *Org. Lett.* **2004**, *6*, 3525–3528.

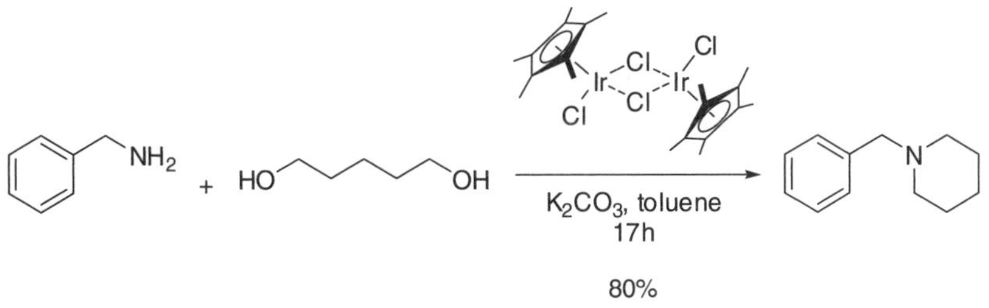

80%

## 2.8  FORMATION OF HYDRAZONES

Hydrazines condense with aldehydes and ketones to afford the corresponding hydrazones. Typically, hydrazine itself reacts only with aromatic ketones to produce hydrazones, whereas more advanced hydrazine derivatives exhibit broader substrate scope.[34]

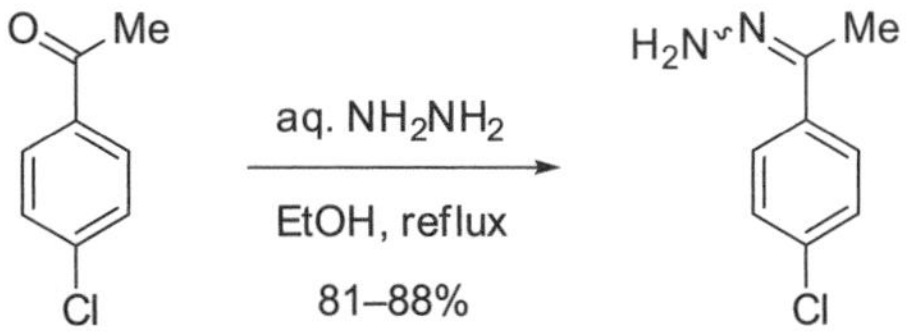

Reaction of hydrazine with aldehydes often leads to the bis-addition products (azines).[35] However, protected hydrazones can be synthesized by the treatment of aldehydes with protected hydrazines.[36]

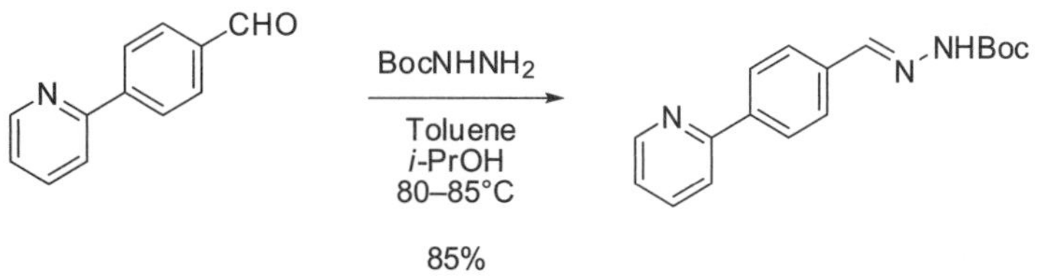

85%

The use of *p*-tosylhydrazine yields *p*-tosylhydrazones, which are substrates for the Shapiro reaction (see Section 9.3.4). This is a particularly useful method for the synthesis of olefins from ketones (see Section 8.2.8).[37]

[34]Nenajdenko, V. G.; Korotchenko, V. N.; Shastin, A. V.; Balenkova, E. S.; Brinner, K.; Ellman, J. A. *Org. Synth.* **2005**, *82*, 93–98.

[35]Newkome, G. R.; Fishel, D. L. *J. Org. Chem.* **1966**, *31*, 677–681.

[36]Xu, Z.; Singh, J.; Schwinden, M. D.; Zheng, B.; Kissick, T. P.; Patel, B.; Humora, M. J.; Quiroz, F.; Dong, L.; Hsieh, D.-M.; Heikes, J. E.; Pudipeddi, M.; Lindrud, M. D.; Srivastava, S. K.; Kronenthal, D. R.; Mueller, R. H. *Org. Process Res. Dev.* **2002**, *6*, 323–328.

[37]Faul, M. M.; Ratz, A. M.; Sullivan, K. A.; Trankle, W. G.; Winneroski, L. L. *J. Org. Chem.* **2001**, *66*, 5772–5782.

## 2.9 FORMATION OF OXIMES

The addition of hydroxylamine to aldehydes or ketones leads to oximes. In general, hydroxylamine is conveniently added as its hydrochloride salt; hence an equivalent amount of external base is required to neutralize the acid. Two representative examples illustrating the formation of an aldoxime[38] and a ketoxime[39] from the respective carbonyl compounds are given below.

## 2.10 THE FORMATION OF *GEM*-DIHALIDES FROM ALDEHYDES AND KETONES

Aldehydes and ketones can be converted to *gem*-dichlorides by treatment with $PCl_5$. This procedure was used to synthesize dichloromethylmesitylene from the corresponding aldehyde in excellent yield.[40]

[38]Boswell, G. E.; Licause, J. F. *J. Org. Chem.* **1995**, *60*, 6592–6594.

[39]Hobson, L. A.; Nugent, W. A.; Anderson, S. R.; Deshmukh, S. S.; Haley, J. J., III; Liu, P.; Magnus, N. A.; Sheeran, P.; Sherbine, J. P.; Stone, B. R. P.; Zhu, J. *Org. Process Res. Dev.* **2007**, *11*, 985–995.

[40]Yakubov, A. P.; Tsyganov, D. V.; Belen'kii, L. I.; Krayushkin, M. M. *Tetrahedron* **1993**, *49*, 3397–3404.

91%

In some instances, the transformation has been achieved using a mixture of $PCl_3$ and $PCl_5$, as exemplified by the synthesis of 2,2-dichloronorbornane from norcamphor.[41]

75%

There are very few examples of *gem*-dibromides synthesized using $PBr_5$. In the example below, norcamphor was converted to 2,2-dibromonorborane using $PBr_3$ and $Br_2$.[41]

53%

The most commonly synthesized dihalides are *gem*-difluorides. While these compounds were typically prepared using $SF_4$-HF and diethylaminosulfur trifluoride (DAST) in the past, safety concerns over these reagents led to the development and use of bis(2-methoxyethyl) aminosulfur trifluroide (Deoxo-Fluor) as the fluorinating agent of choice.[42]

95%

Mase and coworkers described the conversion of a ketone to a *gem*-difluoride using Deoxo-Fluor. While 2.5 equivalents of Deoxo-Fluor were originally required to achieve an 80% yield, the authors found that the addition of small amounts of Lewis acids such as $BF_3 \cdot OEt_2$ (5–10 mol%) led to much cleaner (albeit slower) reactions and higher yields, while allowing for the use of fewer equivalents (1.4) of Deoxo-Fluor.[43]

[41]Ashby, E. C.; Sun, X.; Duff, J. L. *J. Org. Chem.* **1994**, *59*, 1270–1278.

[42]Lal, G. S.; Pez, G. P.; Pesaresi, R. J.; Prozonic, F. M.; Cheng, H. *J. Org. Chem.* **1999**, *64*, 7048–7054.

[43]Mase, T.; Houpis, I. N.; Akao, A.; Dorziotis, I.; Emerson, K.; Hoang, T.; Iida, T.; Itoh, T.; Kamei, K.; Kato, S.; Kato, Y.; Kawasaki, M.; Lang, F.; Lee, J.; Lynch, J.; Maligres, P.; Molina, A.; Nemoto, T.; Okada, S.; Reamer, R.; Song, J. Z.; Tschaen, D.; Wada, T.; Zewge, D.; Volante, R. P.; Reider, P. J.; Tomimoto, K. *J. Org. Chem.* **2001**, *66*, 6775–6786.

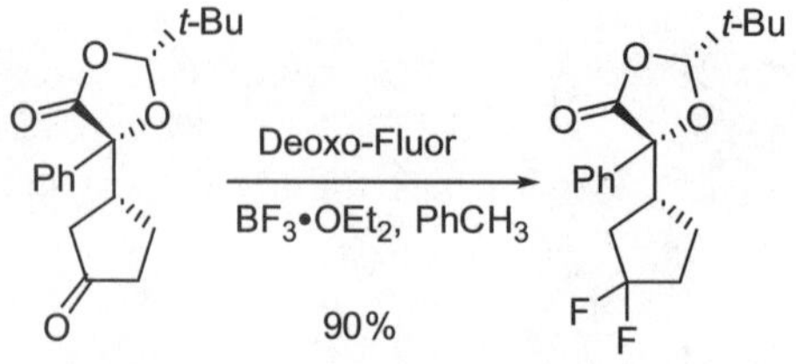

Another class of reagents that has come to the forefront in fluorination reactions are the XtalFluor reagents developed in the laboratories of Omegachem.[44] These reagents hold advantages over the others due to their safer preparation (no distillation of reagent), better safety profile, and the ability to use them in borosilicate (i.e., glass) vessels. Important for good reaction yield with this reagent is order of addition and use of the TEA•3HF complex. One example of fluorination with the XtalFluor reagent is shown below.

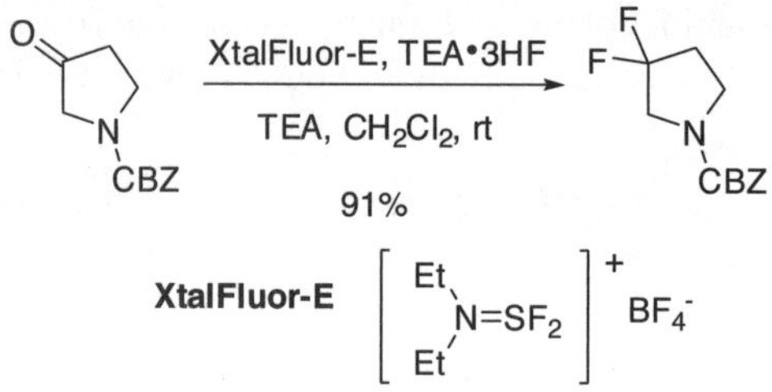

## 2.11   THE ALDOL REACTION

The aldol reaction is the addition of an enol or enolate nucleophile to an aldehyde, resulting in a β-hydroxycarbonyl product (see scheme below). The power of the aldol reaction lies in its formation of a carbon–carbon bond in a stereochemically predictable manner. Control of enolate geometry, catalyst selection, solvent choice and reaction conditions all lead to varied and generally predictable product diastereo- and (in some cases) enantioselectivities. Nearly every nuance of this powerful reaction has been studied and rationalized.[45,46] In the following sections are examples of the ways to unleash its potential and some lead references to aid in the application to other substrates of interest.

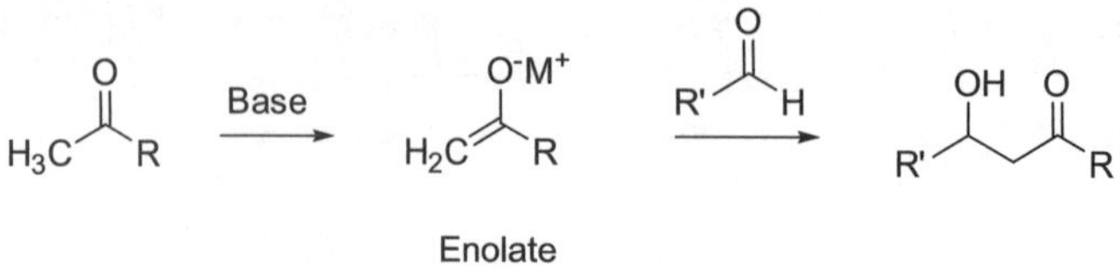

Enolate

[44]L'Heureux, A.; Beaulieu, F.; Bennett, C.; Bill, D. R.; Clayton, S.; La Flamme, F.; Mirmehrabi, M.; Tadayon, S.; Tovell, D.; Couturier, M. *J. Org. Chem.* 75, 3401–3411.
[45]Heathcock, C. H. *Science* **1981**, *214*, 395–400.
[46]Heathcock, C. H. *Science* **1995**, *267*, 116–118.

In the absence of any stereochemical control elements in enolate, aldehyde, or catalyst, the aldol reaction is non-selective in its stereochemical outcome and generates a racemic product. This construct is the key step in the synthesis of the Novartis cardiovascular drug fluvastatin.[47] In this case, use of excess base to form the keto-ester dianion ensures the desired regiochemical outcome.

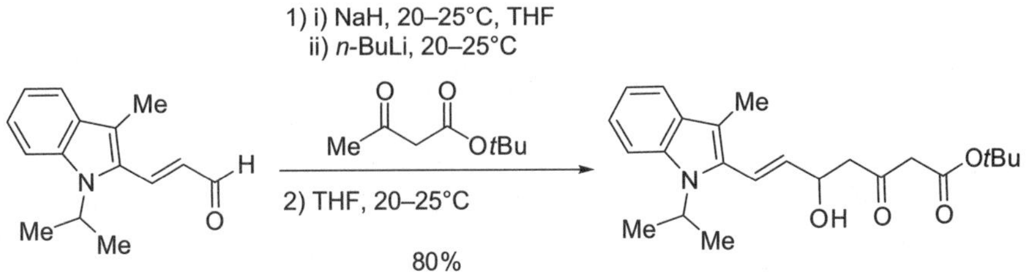

1) i) NaH, 20–25°C, THF
   ii) *n*-BuLi, 20–25°C

2) THF, 20–25°C

80%

Use of an ethyl ketone (or propionate if R′ = alkoxy) highlights the ability of the starting material to give rise to *syn* or *anti* aldol products. Addition of an *E* or *Z* enolate to the aldehyde gives complementary products with excellent selectivity, as shown below.[48]

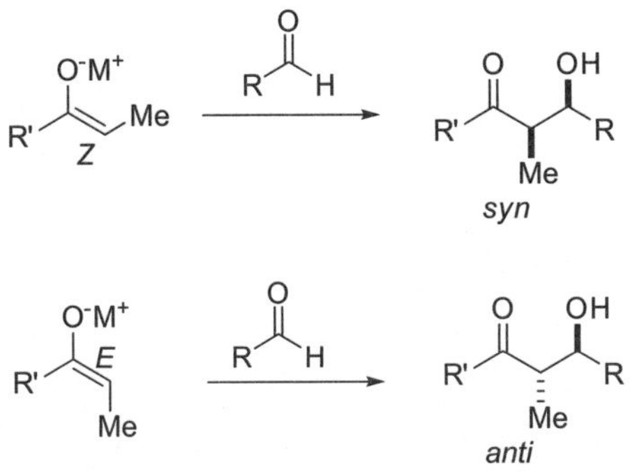

*syn*

*anti*

The Brown group[49–53] has demonstrated conditions for selective formation of either *E* or *Z*-enolborinates, which translate into *syn* or *anti* product geometries, respectively. These reactions are tuned through variation of the size of the alkyl borane and use of a complementary base to enable selection of the desired *E* or *Z* enolborinate (see below). Then, addition of the aldehyde and standard oxidative workup gives the β-hydroxy ketone product.

[47]Fuenfschilling, P. C.; Hoehn, P.; Mutz, J.-P. *Org. Process Res. Dev.* **2007**, *11*, 13–18.
[48]Heathcock, C. H. *Modern Synthetic Methods* **1992**, *6*, 1–102.
[49]Brown, H. C.; Dhar, R. K.; Bakshi, R. K.; Pandiarajan, P. K.; Singaram, B. *J. Am. Chem. Soc.* **1989**, *111*, 3441–3442.
[50]Brown, H. C.; Dhar, R. K.; Ganesan, K.; Singaram, B. *J. Org. Chem.* **1992**, *57*, 499–504.
[51]Brown, H. C.; Dhar, R. K.; Ganesan, K.; Singaram, B. *J. Org. Chem.* **1992**, *57*, 2716–2721.
[52]Brown, H. C.; Ganesan, K.; Dhar, R. K. *J. Org. Chem.* **1992**, *57*, 3767–3772.
[53]Brown, H. C.; Ganesan, K.; Dhar, R. K. *J. Org. Chem.* **1993**, *58*, 147–153.

Chirality in the starting materials can be exploited to relay asymmetric induction. Chiral centers at the 2- and 3-positions ($\alpha$ and $\beta$) in the electrophile, have the greatest effect on facial selectivity of addition. If both the $\alpha$- and $\beta$-positions are substituted in the aldehyde, the size of the nucleophile determines whether the stereocenter at the 2- or 3-positions is dominant in predicting stereochemical outcome.[54]

Evans has studied the stereochemical outcome of additions to aldehydes containing both $\alpha$ and $\beta$ substituents. In substrates where Felkin and 1,3-anti selectivity act in concert, high Felkin selectivity is observed (example 1); however, when the Felkin ($\alpha$ substituent) and 1,3-anti ($\beta$ substituent) predictors are opposed, the stereochemical outcome became dependent on secondary effects. Smaller enolsilanes (Me vs. $t$-Bu) and less polar solvents (toluene vs. $CH_2Cl_2$) were found to allow the 1,3-induction to dominate, leading to useful levels of 1,3-anti (anti-Felkin) selectivity (example 2). However, with the $t$-butyl enolsilane, Felkin selectivity dominated (96:4 in $CH_2Cl_2$, 88:12 in toluene).

[54]Evans, D. A.; Dart, M. J.; Duffy, J. L.; Yang, M. G.; Livingston, A. B. *J. Am. Chem. Soc.* **1995**, *117*, 6619–6620.

Heathcock documented the use of the α stereocenter in aldol selectivity nearly 30 years ago.[55] Generation of the (Z)-lithium enolate of the ketone led to good selectivity for the *anti: syn* bond construction. Through this early method for stereoselective bond construction Heathcock synthesized several macrolides.

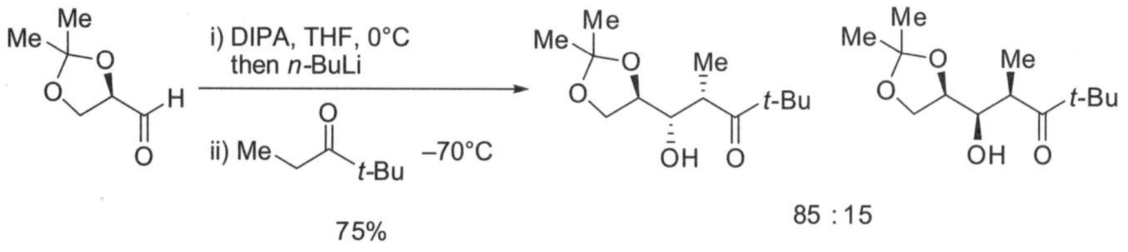

As a result of the power of this synthetic transformation, many authors have documented both empirical results and theoretical models to explain how electrophile substitution, enolate geometry, and variance in Lewis acid catalysts lead to different stereochemical outcomes.[56–61] More complex examples that exhibit the use of substrate control are shown below. In the first example, both aldehyde substituents and a non-chelating Lewis acid act in concert to provide high diastereoselectivity.[54]

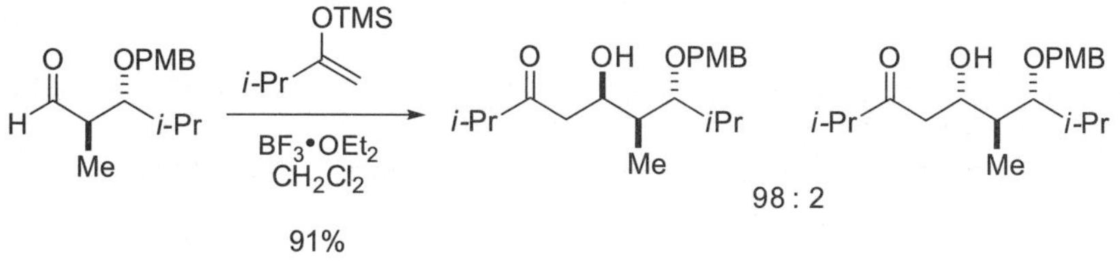

Examples where the α and β stereocenters are not working in tandem are considered non-reinforcing or mismatched. This is shown in the Novartis synthesis by Mickel and coworkers of a late stage discodermolide intermediate, which overcomes this issue via control of the Lewis acid. This example is shown later in this section.

When substituents are present only in the enolate nucleophile, modest selectivity is generally observed. In the example below, the Z enolate generates the aldol product with good relative (*syn*) facial selectivity (86:14 syn:anti) but poor enolate facial selectivity (65:21 ratio within the *syn* product distribution).[60]

[55]Heathcock, C. H.; Young, S. D.; Hagen, J. P.; Pirrung, M. C.; White, C. T.; VanDerveer, D. *J. Org. Chem.* **1980**, *45*, 3846–3856.

[56]Evans, D. A.; Cee, V. J.; Siska, S. J. *J. Am. Chem. Soc.* **2006**, *128*, 9433–9441.

[57]Evans, D. A.; Dart, M. J.; Duffy, J. L.; Rieger, D. L. *J. Am. Chem. Soc.* **1995**, *117*, 9073–9074.

[58]Evans, D. A.; Dart, M. J.; Duffy, J. L.; Yang, M. G. *J. Am. Chem. Soc.* **1996**, *118*, 4322–4343.

[59]Evans, D. A.; Duffy, J. L.; Dart, M. J. *Tetrahedron Lett.* **1994**, *35*, 8537–8540.

[60]Evans, D. A.; Yang, M. G.; Dart, M. J.; Duffy, J. L. *Tetrahedron Lett.* **1996**, *37*, 1957–1960.

[61]Evans, D. A.; Yang, M. G.; Dart, M. J.; Duffy, J. L.; Kim, A. S. *J. Am. Chem. Soc.* **1995**, *117*, 9598–9599.

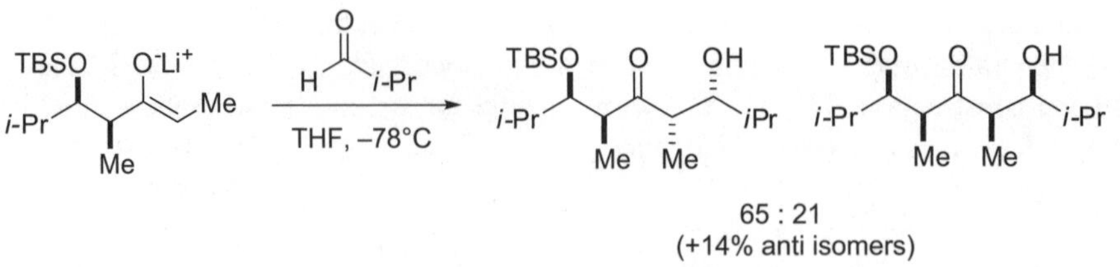

65 : 21
(+14% anti isomers)

Finally, even though the example below is considered only partially matched (due to the stereochemistry of the PMB ether in the aldehyde), good selectivity is achieved.[60]

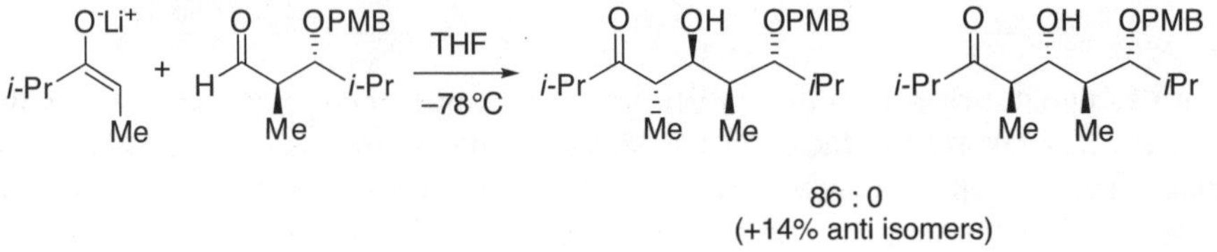

86 : 0
(+14% anti isomers)

The examples demonstrate how substrate geometry can be harnessed to obtain a desired polyol product. An expansion on this idea is the use of chiral auxiliaries to influence the stereochemical outcome of the reaction; the auxiliary is removed post-reaction when its bias is no longer needed. In the case below, Mapp and Heathcock[62] used the oxazolidinone chiral auxiliary developed in the Evans lab[63,64] in their synthesis of myxalamide A. An exchange resin (IRA-743) was used to remove residual dialkylboronates after the initial aldol reaction; it was also found that the aldol product was somewhat sensitive to silica gel chromatography, and higher yields were realized by directly reducing to the diols, which were isolated in excellent yield.

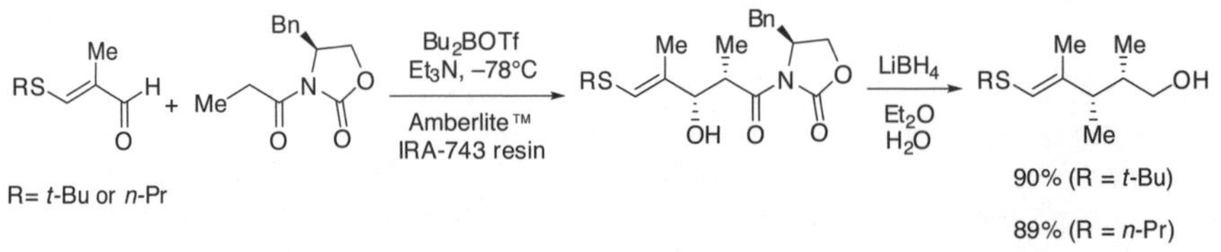

90% (R = t-Bu)

89% (R = n-Pr)

Finally, with use of a chiral Lewis acid, asymmetric aldol reactions are achievable with achiral substrates. One example of this approach is the use of a chiral enolborinate.[65–68] The concept is shown below in an example from Paterson's synthesis of tirandamycin.

[62]Mapp, A. K.; Heathcock, C. H. *J. Org. Chem.* **1999**, *64*, 23–27.
[63]Gage, J. R.; Evans, D. A. *Org. Synth.* **1990**, *68*, No pp given.
[64]Gage, J. R.; Evans, D. A. *Org. Synth.* **1990**, *68*, No pp given.
[65]Cowden, C. J.; Paterson, I. *Org. React. (New York)* **1997**; Vol. 51, p 1–200.
[66]Paterson, I.; Lister, M. A.; McClure, C. K. *Tetrahedron Lett.* **1986**, *27*, 4787–4790.
[67]Paterson, I.; Lister, M. A.; Ryan, G. R. *Tetrahedron Lett.* **1991**, *32*, 1749–1752.
[68]Paterson, I.; Mackay, A. C. *Tetrahedron Lett.* **2001**, *42*, 9269–9272.

In the example below, the chiral enolborinate is used to dictate the desired stereochemical outcome, overcoming the intrinsic bias of the reaction components. Mickel's team at Novartis used the Paterson asymmetric aldol conditions for the construction of this late stage discodermolide intermediate. As can be seen in the table below, by fine tuning the amounts of each component they were able to optimize formation of the desired isomer, **3a**, while minimizing byproduct formation.[69]

| Entry | **1** (equiv) | DIP-Cl (equiv) | Et$_3$N (equiv) | Yield of **3a** (%) | **3a/3b** | **3a/4** | **3a/2** |
|---|---|---|---|---|---|---|---|
| 1 | 6.6 | 5.4 | 6.6 | 55 | 3.9/1 | 25/1 | 28/1 |
| 2 | 4.0 | 3.0 | 4.0 | 56 | 3.9/1 | 30/1 | 30/1 |
| 3 | 3.3 | 2.7 | 3.3 | 48 | 4/1 | 28/1 | 35/1 |
| **4** | **3.3** | **2.5** | **3.3** | **68** | **3/1** | **22/1** | **22/1** |
| 5 | 2.4 | 2.0 | 2.4 | 49 | 3.6/1 | 28/1 | 5/1 |

In the example below, Keck and Krishnamurthy developed the catalytic asymmetric aldol with a ketene silyl acetal as the nucleophile. With benzyloxyacetaldehyde, the C$_2$ symmetry of the BINOL and titanium catalyst system enables the aldol reaction to proceed with high enantioselectivity.[70]

[69]Mickel, S. J.; Daeffler, R.; Prikoszovich, W. *Org. Process Res. Dev.* **2005**, *9*, 113–120.
[70]Keck, G. E.; Krishnamurthy, D. *J. Am. Chem. Soc.* **1995**, *117*, 2363–2364.

The catalyst system described above represents a class that is considered 'privileged,' in that it can be used for numerous asymmetric transformations, for example, aldol, allylation, and Diels–Alder reactions.[71,72]

### 2.11.1   Ketene and Silyl Enol Ether Addition to Aldehydes

One standard method for ensuring clean aldol reactions is through pre-formation of the nucleophile via ketene and silyl enol ethers. When traditional enolate aldol conditions are employed, other acidic functionality in the reactants can interfere with the desired reaction; it is necessary to rely on the inherent pKa differences of the substrates to ensure desired reaction. For example, if the aldehyde has an α hydrogen that is more acidic than the reacting ketone enolate, proton transfer can occur, leading to undesired reactions such as aldehyde self-condensation. Another reason for use of silyl enol ether or ketene silyl acetal nucleophiles is their utility with Lewis acid catalysts, which complements the basic conditions of lithium enolate aldols. This can be a desirable alternative with certain base-sensitive substrates. Although the initial product is a β-silyloxy aldol, the silyl ether is frequently cleaved during aqueous workup. These nucleophiles have been used extensively in both methodological and total synthesis applications of the aldol reaction.

### 2.11.2   Silyl Enol Ether Addition to Aldehydes: The Mukaiyama Aldol

In the example below, Kelly and Vanderplas developed a titanium-mediated silyl enol ether aldol reaction that provided the desired product on large scale in good yield. This method was found to be more successful than either lithium enolate or enamine alternatives.[73]

In another example, Kuroda and coworkers utilized boron trifluoride etherate as the Lewis acid of choice for addition of an enol ether to an α,β-unsaturated aldehyde.[74]

[71] Krauss, R.; Koert, U. *Org. Synth. Highlights IV* **2000**, 144–154.
[72] Wabnitz, T.; Reiser, O. *Org. Synth. Highlights IV* **2000**, 155–165.
[73] Kelly, S. E.; Vandeplas, B. C. *J. Org. Chem.* **1991**, *56*, 1325–1327.
[74] Kuroda, H.; Hanaki, E.; Izawa, H.; Kano, M.; Itahashi, H. *Tetrahedron* **2004**, *60*, 1913–1920.

Through careful choice of enolsilane geometry, Lewis acid, and reaction conditions, the Evans group was able to achieve excellent diastereoselectivity for key bond formations in the sequence below, which lead to the synthesis of both 6-deoxyerythronolide B and oleandolide.[75]

The next two examples show catalytic asymmetric variants of the Mukaiyama aldol reaction. Hisashi Yamamoto and coworkers developed a catalyst derived from tryptophan that works exceptionally well for a variety of silyl enol ethers in condensations with benzaldehyde.[76]

The Evans group was able to expand the scope of this asymmetric transformation with their cationic scandium pybox catalyst system.[77] As seen in the example below, with the *gem*-dimethyl silyl enol ether and ethyl glyoxalate, good yield and excellent enantio-selectivity are achieved. The key for asymmetric induction in this catalyst system is the two point coordination of the catalyst to the aldehyde. This manifests itself in the necessity for an α-Lewis basic functional group in the aldehyde starting material. Another key difference in this catalyst system is its ability to effectively catalyze reaction with electron deficient aldehydes.

[75]Evans, D. A.; Kim, A. S.; Metternich, R.; Novack, V. J. *J. Am. Chem. Soc.* **1998**, *120*, 5921–5942.
[76]Ishihara, K.; Kondo, S.; Yamamoto, H. *J. Org. Chem.* **2000**, *65*, 9125–9128.
[77]Evans, D. A.; Masse, C. E.; Wu, J. *Org. Lett.* **2002**, *4*, 3375–3378.

85%
95% ee

Catalyst =

### 2.11.3   Ketene Silyl Acetal and Thio-Acetal Addition to Aldehydes

The examples below illustrate ketene silyl acetal additions with a variety of substrates and Lewis acids. In the first, Keck and coworkers used an attenuated titanium species in order to realize a good yield for formation of the desired hydroxyl-ester.[78] Use of this Lewis acid was based on the lability of the p-methoxybenzyl ether. Low temperatures are critical with pre-formed nucleophiles (e.g., silyl ketene acetals); in the example below a pre-cooled solution of Lewis acid is added to a cold solution of aldehyde, followed by addition of the nucleophile. The reactions need to be quenched at a low temperature and allowed to warm up to room temperature for optimal stereoselectivity and yield.

95%
41 : 1 dr

In the second example, stereochemistry is again directed through chelation-controlled addition of the ketene silyl acetal. In this case the more robust benzyl ether-protecting group allows for use of the stronger catalyst titanium tetrachloride.[79]

94%
97 : 3 dr

Finally, Shiina and Fukui extended this methodology by incorporating a third stereo-center and setting the stereocenters via an open-chained (Felkin–Ahn) model that capitalizes on the steric bulk of the substrates to dictate facial selectivity.[80]

[78]Keck, G. E.; Welch, D. S.; Vivian, P. K. *Org. Lett.* **2006**, *8*, 3667–3670.
[79]Zanato, C.; Pignataro, L.; Hao, Z.; Gennari, C. *Synthesis* **2008**, 2158–2162.
[80]Fukui, H.; Shiina, I. *Org. Lett.* **2008**, *10*, 3153–3156.

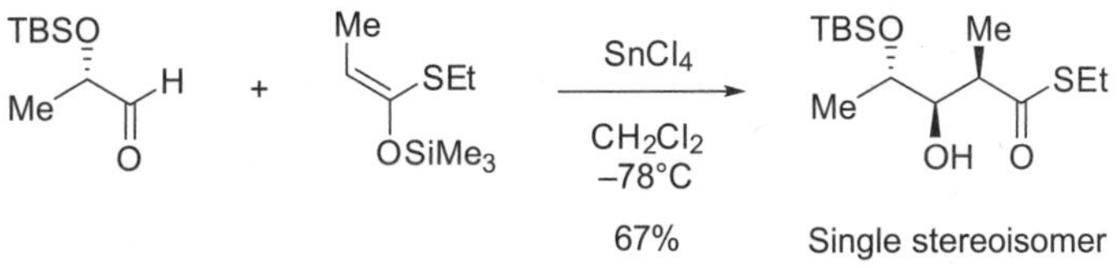

67%    Single stereoisomer

The remainder of this section presents examples of catalytic asymmetric aldol reactions using ketene silyl acetals. In their work towards a cholesterol absorption inhibitor, coworkers Wu and Tormos from Schering-Plough found that a valine-derived sulfonamide catalyst performed well to effect the reaction with the protected cyclohexyl silyl ketene acetal shown below.[81]

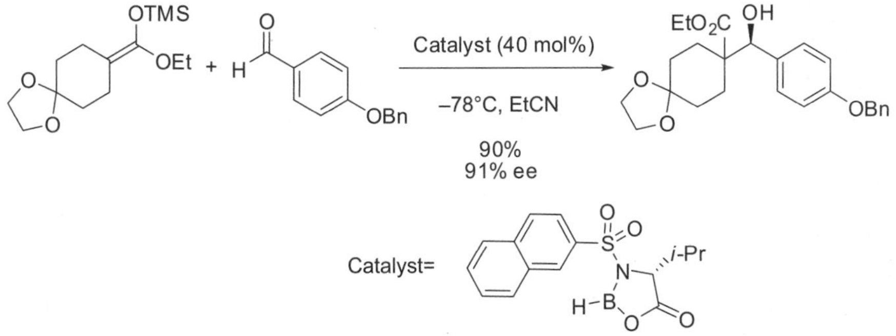

90%
91% ee

Catalyst=

Huang and coworkers used a catalyst developed in the Masamune labs in their synthesis of the marine anti-cancer natural product psymberin.[82] In this process, the catalyst is preformed at low temperature and then the ketene acetal is added. Finally, the aldehyde is added over a few hours.

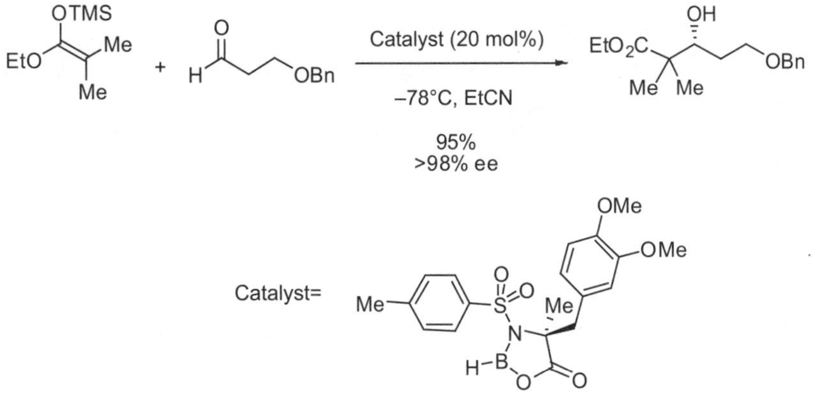

95%
>98% ee

Catalyst=

[81]Wu, G.; Tormos, W. *J. Org. Chem.* **1997**, *62*, 6412–6414.
[82]Huang, X.; Shao, N.; Palani, A.; Aslanian, R.; Buevich, A. *Org. Lett.* **2007**, *9*, 2597–2600.

Thioketene acetals (formed from thioesters) are also frequently utilized. Bulky alkyl groups (e.g., *t*-butyl, phenyl) generally lead to better selectivity. In the next schemes are examples of an asymmetric system developed by the Evans group. The first highlights application of the pybox tin and copper catalysts in the synthesis of phorboxazole B and an extended methodology evaluation.[83,84] In the final scheme is a related example from Larionov and Meijere in their work toward the belactosins.[85]

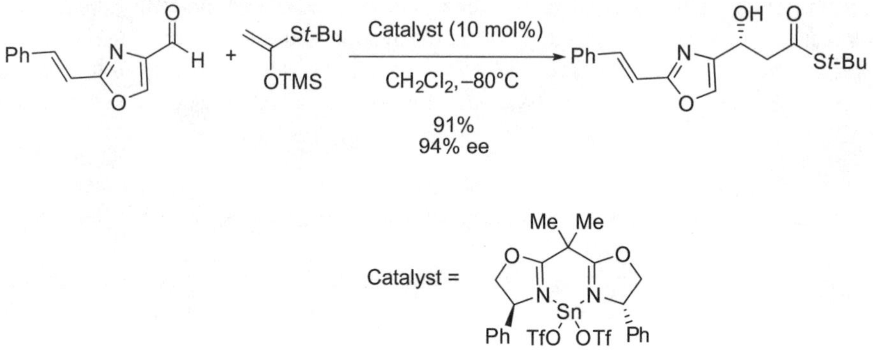

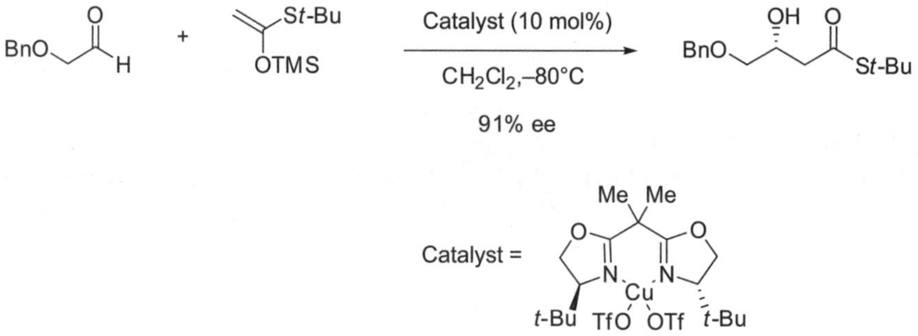

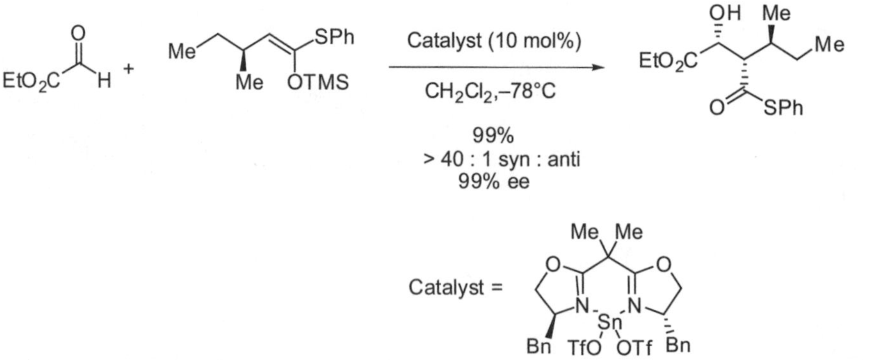

[83]Evans, D. A.; Fitch, D. M.; Smith, T. E.; Cee, V. J. *J. Am. Chem. Soc.* **2000**, *122*, 10033–10046.

[84]Evans, D. A.; Kozlowski, M. C.; Murry, J. A.; Burgey, C. S.; Campos, K. R.; Connell, B. T.; Staples, R. J. *J. Am. Chem. Soc.* **1999**, *121*, 669–685.

[85]Larionov, O. V.; De Meijere, A. *Org. Lett.* **2004**, *6*, 2153–2156.

The simple system below, initially put forward by the Keck labs, does not necessitate a Lewis basic center functional group at the α center and uses inexpensive and easily handled reagents, but reaction times can be longer. These reactions generally work better in the presence of molecular sieves.[86]

## 2.12 ALLYLORGANOMETALLICS: STANNANE, BORANE, AND SILANE

In this section we will discuss the more common types of allylation reactions, focusing on nucleophiles containing silicon, boron, and tin. Allylation provides a few advantages over the classic aldol reaction:

1. An alkene is embedded in the resultant product providing a handle for further elaboration.
2. The nucleophiles (e.g., allylsilanes) are generally stable species amenable to storage.
3. The Lewis acidic metal embedded in the nucleophile leads to enhanced reactivity via coordination with the Lewis basic carbonyl.
4. The ability to place chiral ligands on the metal center, which translates to a net enantioselective reaction after removal of the ligand in the reaction workup.

Transmetallation can change the reactivity and stereochemical outcome. A complication with use of these nucleophiles, however, is removal of the resultant stannous, borate or silyl byproducts. With careful operations these byproducts can be effectively purged, but this can be a challenge (e.g., coordination of borane byproducts to the desired product, removal of toxic organostannane reagents, etc.).

### 2.12.1  Allylsilane Additions

The first class of allyl organometallic reagents described are allylsilanes. Although attenuated in reactivity compared to other allylorganometallics, these reagents offer significant advantages in terms of reduced toxicity and ease of handling.

[86]Dalgard, J. E.; Rychnovsky, S. D. *Org. Lett.* **2004**, *6*, 2713–2716.

Jervis and Cox used this classic allylsilane reaction to build the tetrahydrofuran core of aureonitol.[87] By tethering the allylsilane, this intramolecular reaction proceeds with good stereoselectivity and yield to give the desired tetrahydrofuran.

Chemler and Roush developed a trifluorocrotylsilane reagent that gives good selectivity for the elusive *anti:anti* bond construction.[88] Selectivity in this system is dictated by the reactive trifluorosilane. It is proposed that the reaction proceeds via a Zimmerman–Traxler-like transition state where the silicon of the allylsilane coordinates with both the aldehyde and β-hydroxyl group in the substrate to lock the reaction conformation.

In the next two examples, work from the Panek laboratories shows the versatility of chiral crotylsilane reactions. In the first, Kesavan, Panek, and Porco performed this enantioselective crotylation to give, after treatment with methyl sulfate, the desired *syn* methyl ether in good yield and with excellent diastereoselectivity.[89]

In the second example, Arefolov and Panek used a titanium-catalyzed crotylsilane addition in their approach to the total synthesis of discodermolide.[90]

[87]Jervis, P. J.; Cox, L. R. *J. Org. Chem.* **2008**, *73*, 7616–7624.
[88]Chemler, S. R.; Roush, W. R. *J. Org. Chem.* **1998**, *63*, 3800–3801.
[89]Kesavan, S.; Panek, J. S.; Porco, J. A., Jr., *Org. Lett.* **2007**, *9*, 5203–5206.
[90]Arefolov, A.; Panek, J. S. *J. Am. Chem. Soc.* **2005**, *127*, 5596–5603.

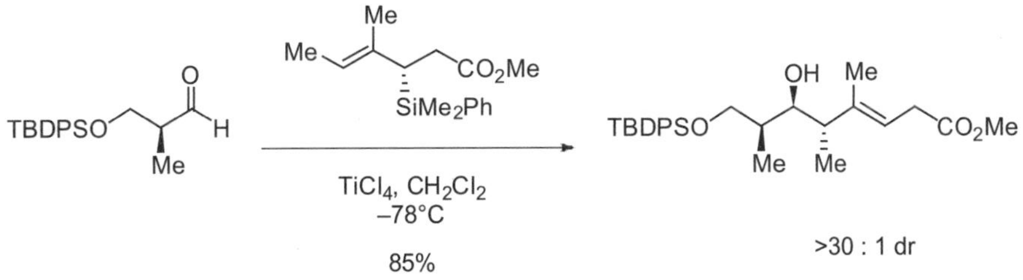

Denmark and coworkers developed a Lewis base–catalyzed allylsilylation using the chiral bisphosphoramide shown below.[91,92] This reaction was used in a sequence for the synthesis of the natural product papulacandin.

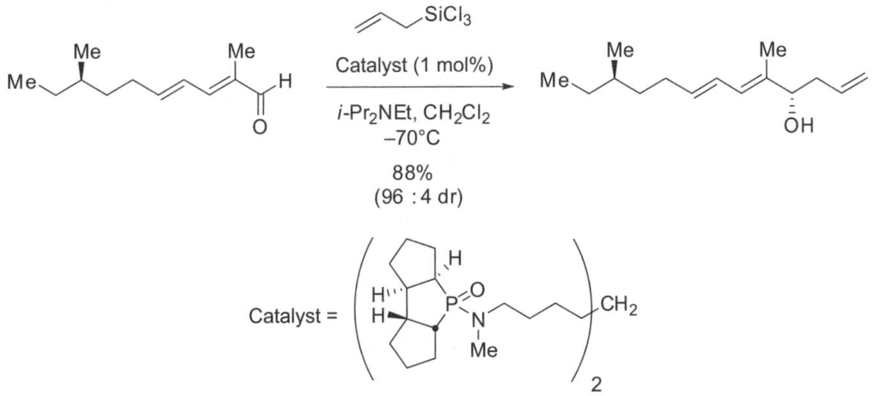

## 2.12.2  Allylborane Additions

One strength of allylborane additions to carbonyls is excellent reactivity and selectivity due to the oxophilicity of boron. Another relies on the fact that many chiral ligands can be tethered to the borane nucleophile. A downside to this class of reagents is difficulty sometimes encountered in releasing the desired products from borane-derived byproducts.

Ramachandran and Chatterjee used a difluoro allylborane reagent for the synthesis of the *gem*-difluoro alkene shown below.[93] The reagent was prepared from trifluoroethanol. They found good yields in difluoroallylation of a variety of substrates. One conclusion of their work was the finding that the benzyl-protected allyl borane had much better reactivity than the tosyl variant initially investigated. They also were able to extend this methodology to a variety of aliphatic, aromatic, and substituted aldehyde substrates.

[91]Denmark, S. E.; Regens, C. S.; Kobayashi, T. *J. Am. Chem. Soc.* **2007**, *129*, 2774–2776.
[92]Denmark, S. E.; Pham, S. M.; Stavenger, R. A.; Su, X.; Wong, K.-T.; Nishigaichi, Y. *J. Org. Chem.* **2006**, *71*, 3904–3922.
[93]Ramachandran, P. V.; Chatterjee, A. *Org. Lett.* **2008**, *10*, 1195–1198.

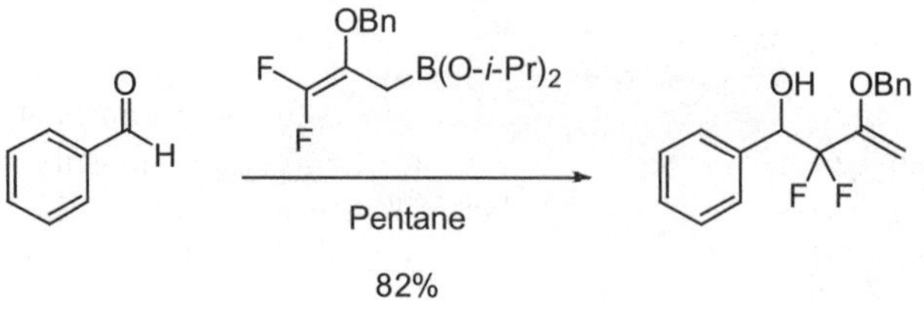

82%

Thadani and Batey developed the (Z)-crotyl trifluoroborate salt shown below, which has the distinct advantage of tolerating a biphasic aqueous solvent system.[94] Phase transfer reagents are necessary for this transformation but the allyl trifluoroborate reagent can be stored for long periods of time and is stable to air and water.

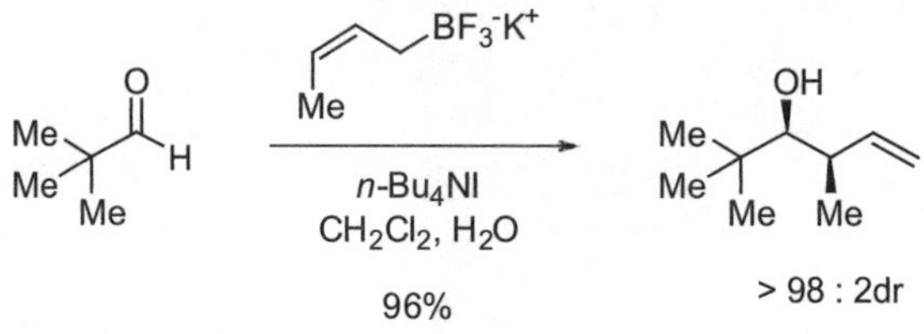

96%

> 98 : 2dr

Smith and Zheng used classic Roush crotylboration conditions in a key bond-forming sequence for their salicylihalamide A synthesis.[95] These reactions were performed by adding a solution of the allylborane to a low temperature slurry of aldehyde and molecular sieves. One advantage of this series of reagents is that in general the crotylborane nucleophile overrides the inherent facial selectivity of the chiral aldehyde.

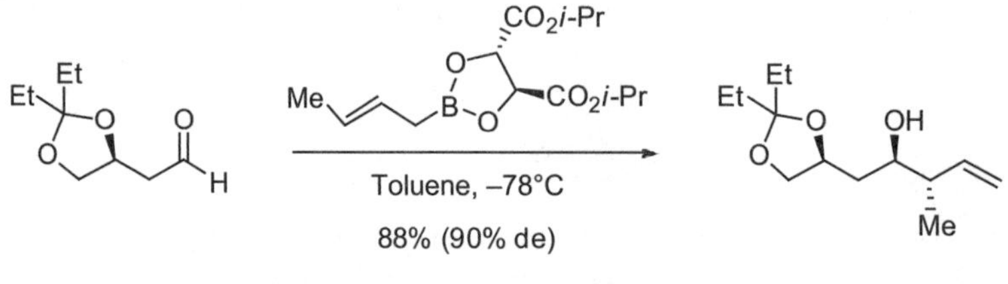

Toluene, −78°C

88% (90% de)

Va and Roush modified the crotylborane nucleophile with the γ-silylallylborinate shown below.[96]

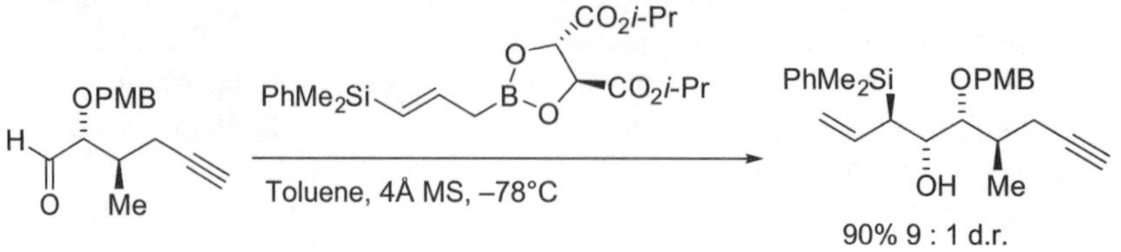

Toluene, 4Å MS, −78°C

90% 9 : 1 d.r.

[94]Thadani, A. N.; Batey, R. A. *Org. Lett.* **2002**, *4*, 3827–3830.
[95]Smith, A. B.; Zheng, J. *Tetrahedron* **2002**, *58*, 6455–6471.
[96]Va, P.; Roush, W. R. *Tetrahedron* **2007**, *63*, 5768–5796.

### 2.12.3 Allylstannane Additions

Even with the detriment of tin toxicity, the literature is full of examples of various allyl tin additions to aldehydes. This is due to the fact that the reagents' toxicity is balanced with higher reactivity at lower temperatures and their ability to coordinate with Lewis bases and transmetallate with the Lewis acid reactants. The latter qualities can lead to excellent stereoselectivity.

In their synthesis of mycalamide A, Rawal and coworkers utilized prenyl stannane with zinc dibromide to catalyze the transformation.[97] After pre-complexation of the aldehyde in methylene chloride, prenylation proceeded with excellent yield and greater than 50:1 diastereoselectivity. Allylation with this and other metal variants is a common method for the installation of the *gem*-dimethyl functionality.

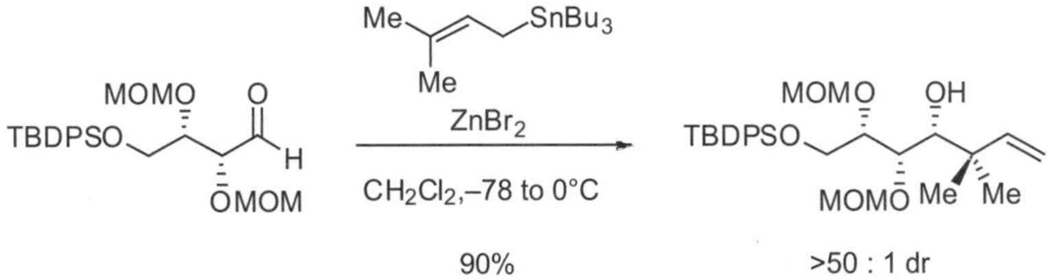

The Cossy laboratories demonstrated a non-chelation controlled crotylstannane reaction.[98] In their synthesis of spongidepsin, through use of an *E*-crotyl stannane, the silyl-protected aldehyde and $BF_3 \cdot OEt_2$, they set the *syn, syn* bond construction required for the natural product.

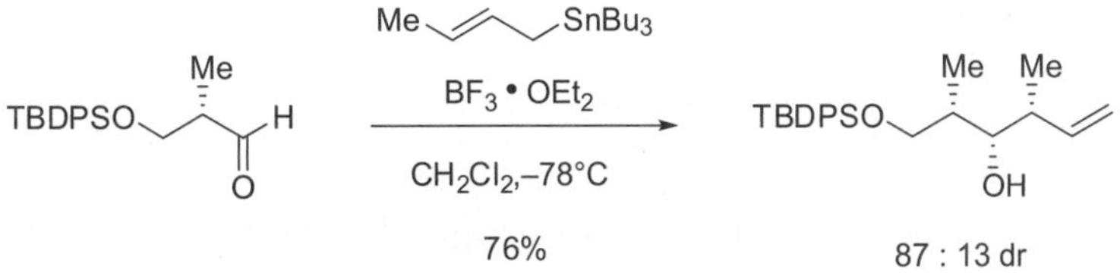

Using chiral stannanes, Marshall and Welmaker obtained good diastereoselectivity in the reaction with the enal ester shown below.[99] Key for good selectivity in this transformation was use of the unsaturated aldehyde and silyl protection in the chiral stannane. Also of note is that use of the enantio- and regio-divergent reagent (scheme below) gave the opposite product enantiomer.

[97]Sohn, J.-H.; Waizumi, N.; Zhong, H. M.; Rawal, V. H. *J. Am. Chem. Soc.* **2005**, *127*, 7290–7291.
[98]Ferrie, L.; Reymond, S.; Capdevielle, P.; Cossy, J. *Org. Lett.* **2006**, *8*, 3441–3443.
[99]Marshall, J. A.; Welmaker, G. S. *J. Org. Chem.* **1994**, *59*, 4122–4125.

79%

>95 : 5 dr
>95% ee

77%

The final example shows a standard application of catalytic asymmetric allylstanylation. In the synthesis of bryostatin A, Keck and coworkers using utilized asymmetric allylation for several key stereoselective bond formations.[78] In the example below, pre-formation of the catalyst using BINOL, titanium isopropoxide, molecular sieves and trifluoroacetic acid in methylene chloride was followed by the addition of the aldehyde and allylstannane. With this reaction system the desired product was obtained in excellent enantioselectivity and yield.

97%

## 2.13  ADDITION OF ORGANOMETALLIC REAGENTS TO CARBONYLS

Grignard and organolithium additions are one of the classic methods for carbon–carbon bond formation. They perform well in addition to carbonyl compounds due to their ability to coordinate with the carbonyl or transmetallate to form other organometallic compounds. The highly basic nature of these reagents should be taken into consideration when designing a synthesis, but there are examples where relatively sensitive functionality (e.g., a stereocenter adjacent to a carbonyl) can be tolerated.

There are a few challenges with carrying out effective organolithium or organomagnesium (Grignard) reactions. Both nucleophiles are very basic: for example, butyllithium has a pKa of 48 and phenylmagnesium bromide a pKa of approximately 35.[100,101] Therefore,

---

[100]Bordwell, F. G. *Acc. Chem. Res.* **1988**, *21*, 456–463.

[101]Silverman, G. S.; Rakita, P. E.;Editors *Handbook of Grignard Reagents. [In: Chem. Ind. (Dekker), 1996; 64]* 1996.

care has to be taken to ensure reaction of the desired components; proton transfer can easily compete with the desired reaction. In organolithium additions, unintended cross-coupling such as the Wurtz reaction (see Section 1.6.1.1) should be anticipated; this particular side reaction can be circumvented by avoiding alkyl iodides. Appropriate care must be taken for the exclusion of water and oxygen. Other obstacles exist in the formation of Grignard reagents.[102] Grignard reagents are formed from the reaction of magnesium metal and an organohalide, typically in an ethereal solvent (to aid solubility). Given the biphasic nature of the metal and organohalide, care must be taken upon reaction due to the ability to have an induction period followed by rapid exothermic reaction. One method to manage this is the use of activators (e.g., iodine, dibromomethane) to help minimize the onset period. Other techniques include the use of activated magnesium turnings and starting with a partial charge, instead of a full charge of the organohalide. The reaction of organolithium and organomagnesium reagents with aldehydes and ketones is presented in this section. Reactions of these compounds with carboxylic acid derivatives and nitriles are presented in Sections 2.35 and 2.39, respectively.

### 2.13.1  Organolithium Additions

Zhang and coworkers prepared 2-lithiofuran by metallation with *n*-butyllithium, and this reagent was in turn added to a non-enolizable aryl ketone.[103] Upon reaction completion the mixture was quenched with an aqueous ammonium chloride solution and standard workup ensued to give an excellent yield of the desired alcohol.

Caron and Do[104] explored the effect of temperature, solvent, and character of the electrophile in their work toward the optimization of reactions with dibromobenzene. Choice of reaction solvent and low temperatures were key to optimizing the yield for this lithiation-addition sequence.

[102]Knochel, P.; Dohle, W.; Gommermann, N.; Kneisel, F. F.; Kopp, F.; Korn, T.; Sapountzis, I.; Vu, V. A. *Angew. Chem., Int. Ed.* **2003**, *42*, 4302–4320.

[103]Zhang, P.; Kern, J. C.; Terefenko, E. A.; Fensome, A.; Unwalla, R.; Zhang, Z.; Cohen, J.; Berrodin, T. J.; Yudt, M. R.; Winneker, R. C.; Wrobel, J. *Bioorg. Med. Chem.* **2008**, *16*, 6589–6600.

[104]Caron, S.; Do, N. M. *Synlett* **2004**, 1440–1442.

Moseley and coworkers at AstraZeneca used similar conditions for their synthesis of several neurokinin antagonists.[105] Both Grignard and aryllithium reagents were investigated, with the latter providing cleaner addition products.

Modeling work in the Corey lab, Pierce and coworkers developed an asymmetric variant of an organolithium addition in their work on the HIV candidate efavirenz.[106] In this reaction sequence, the ephedrine-derived ligand, butyl lithium and the cyclopropyl acetylene were premixed and cooled to −50°C prior to addition of the trifluoromethyl ketone. Enantioselectivity of this transformation was optimized through temperature control, variation of substituents at the aniline, and careful consideration of the ligand chirality.

### 2.13.2  Organomagnesium Additions

Katzenellenbogen and Robertson constructed tamoxifen and its analogs through the phenyl Grignard addition to the ketone below.[107] In their synthesis, the phenyl Grignard

[105]Bowden, S. A.; Burke, J. N.; Gray, F.; McKown, S.; Moseley, J. D.; Moss, W. O.; Murray, P. M.; Welham, M. J.; Young, M. J. *Org. Process Res. Dev.* **2004**, *8*, 33–44.

[106]Pierce, M. E.; Parsons, R. L., Jr.; Radesca, L. A.; Lo, Y. S.; Silverman, S.; Moore, J. R.; Islam, Q.; Choudhury, A.; Fortunak, J. M. D.; Nguyen, D.; Luo, C.; Morgan, S. J.; Davis, W. P.; Confalone, P. N.; Chen, C.-y.; Tillyer, R. D.; Frey, L.; Tan, L.; Xu, F.; Zhao, D.; Thompson, A. S.; Corley, E. G.; Grabowski, E. J. J.; Reamer, R.; Reider, P. J. *J. Org. Chem.* **1998**, *63*, 8536–8543.

[107]Robertson, D. W.; Katzenellenbogen, J. A. *J. Org. Chem.* **1982**, *47*, 2387–2393.

was generated by subjection of a portion of the bromobenzene to magnesium turnings in diethyl ether. After initiation commenced, the remainder of the bromide was added slowly so as to maintain a gentle reflux of the solvent. The phenyl Grignard solution was then added slowly to a cooled solution of the diaryl ketone. After aqueous workup, the reaction mixture was concentrated to give an oil that upon subjection to alcoholic HCl generated the desired triaryl alkene.

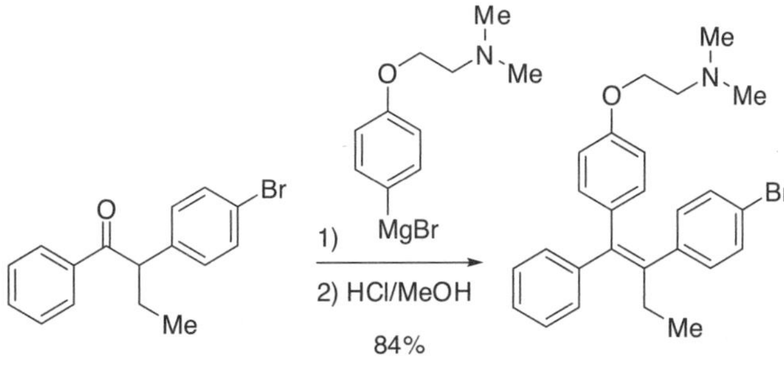

## 2.14   ADDITION OF CONJUGATED ALKENES TO ALDEHYDES: THE BAYLIS–HILLMAN REACTION

The addition of the α-carbon of an α,β-unsaturated carbonyl compound to an aldehyde is referred to as the Baylis–Hillman reaction. The product of this reaction is an allylic alcohol. Nucleophiles such as tertiary amines and phosphines catalyze this reaction.[108] The use of asymmetric catalysts can lead to optically enriched products.

A classic example, by Basavaiah and Suguna Hyma is shown below.[109] Hexanal and methyl acrylate are mixed neat with a catalytic amount of DABCO to give a good yield of the desired allylic alcohol. As illustrated in this example, neat or very concentrated reaction conditions are frequently beneficial for this reaction.

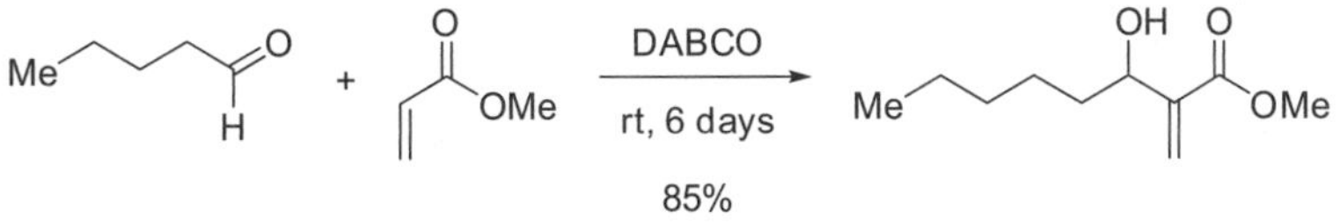

Dunn and coworkers at Pfizer used the Baylis–Hillman reaction as the first step for the large scale production of sampatrilat.[110] Use of aqueous acetontrile with 2 equivalents of aldehyde and an equivalent of 3-quinuclidinol was found to give complete and clean conversion to the allyic alcohol. During the process run, the 3-quinuclidinol was dropped to a catalytic amount due to cost considerations. This optimization work (excess aldehyde in a polar solvent) supports the second order kinetics observed in the

[108]Basavaiah, D.; Rao, P. D.; Hyma, R. S. *Tetrahedron* **1996**, *52*, 8001–8062.
[109]Basavaiah, D.; Suguna Hyma, R. *Tetrahedron* **1996**, *52*, 1253–1258.
[110]Dunn, P. J.; Hughes, M. L.; Searle, P. M.; Wood, A. S. *Org. Process Res. Dev.* **2003**, *7*, 244–253.

laboratories of McQuade.[111] Using this process, 80 kilograms of the allylic alcohol were produced.

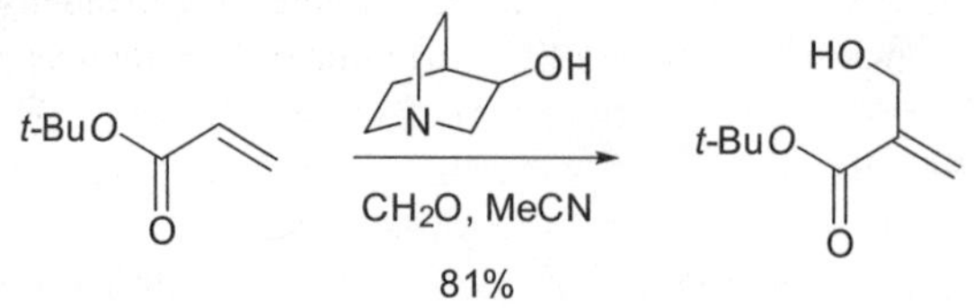

81%

Key examples for the asymmetric Baylis–Hillman reaction include use of an optically active Lewis acid complex as well as chiral auxiliaries and optically active catalysts. Chen and coworkers used chiral lanthanide complexes to successfully perform the addition; unfortunately, high enantioselectivities were observed only with electron-poor aromatic aldehydes.[112] Leahy and Hatekeyama obtained success with a wider scope of substrates in their chiral auxiliary and chiral catalyst approaches, as shown below. Use of the Oppolzer sultam with a catalytic amount of DABCO gave excellent yield and high enantioselectivity.[113]

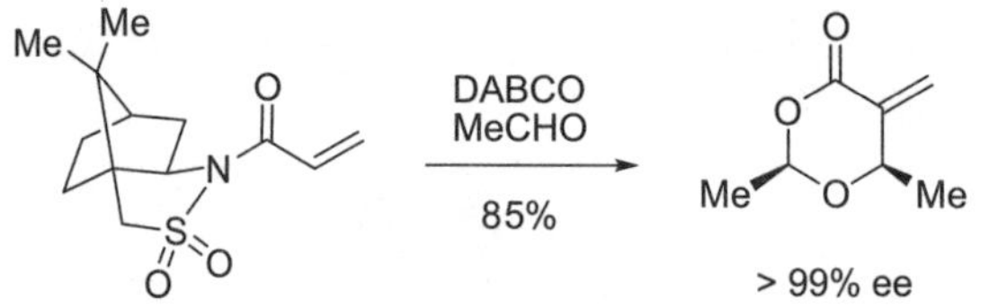

85%

> 99% ee

Hatakeyama and coworkers expanded chiral Baylis–Hillman methodology through use of the activated alkene and the chiral quinidine.[114] Although yields are moderate and in most cases a small amount of the lactone dimer is observed, the aldehyde scope is broad and the enantioselectivity is excellent.

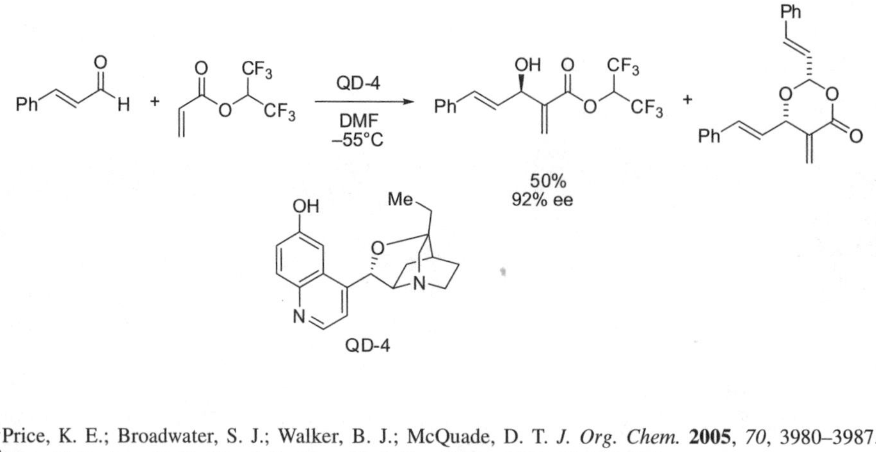

50%
92% ee

QD-4

[111]Price, K. E.; Broadwater, S. J.; Walker, B. J.; McQuade, D. T. *J. Org. Chem.* **2005**, *70*, 3980–3987.
[112]Yang, K.-S.; Lee, W.-D.; Pan, J.-F.; Chen, K. *J. Org. Chem.* **2003**, *68*, 915–919.
[113]Brzezinski, L. J.; Rafel, S.; Leahy, J. W. *Tetrahedron* **1997**, *53*, 16423–16434.
[114]Iwabuchi, Y.; Nakatani, M.; Yokoyama, N.; Hatakeyama, S. *J. Am. Chem. Soc.* **1999**, *121*, 10219–10220.

## 2.15   THE REFORMATSKY REACTION

The Reformatsky reaction is the reaction of a zinc enolate with an aldehyde or ketone. Zinc enolates are less basic than magnesium or lithium enolates. This decreased basicity allows a wider range of functional group toleration in the reaction substrates. One challenge for this reaction is the requirement of activated zinc. Multiple methods are published that accomplish this task, and a few variants are shown below.

Hallinan and coworkers used classic conditions for the formation of the requisite di-flouro compound shown below.[115] Starting with freshly etched (rinsed with strong acid) zinc in THF, the zinc enolate was formed from the α-bromo ester, then treated with the aldehyde.

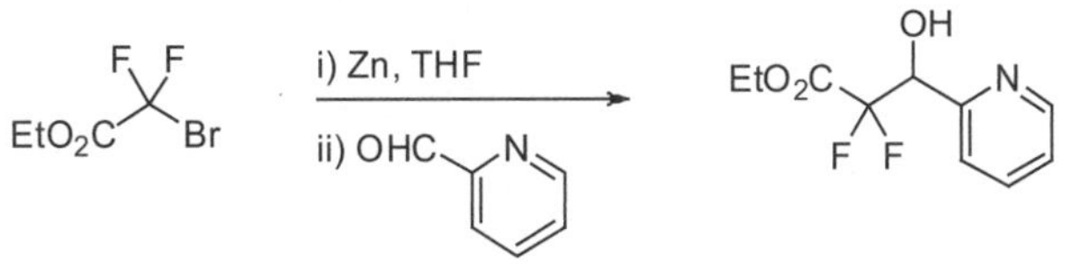

93%

Siegel and Scherkenbeck found a mixture of copper chloride, zinc powder, and sulfuric acid to be effective for the Reformatsky reaction shown below.[116] For this reaction, the zinc–copper couple was formed by making a suspension of copper chloride and zinc powder in THF. To this suspension was added a catalytic amount of sulfuric acid and the solution was heated. The ketone was added followed by the α-bromo-ester. Following this procedure the group obtained the desired hydroxy-ester (as a mixture of diastereomers) in excellent yield.

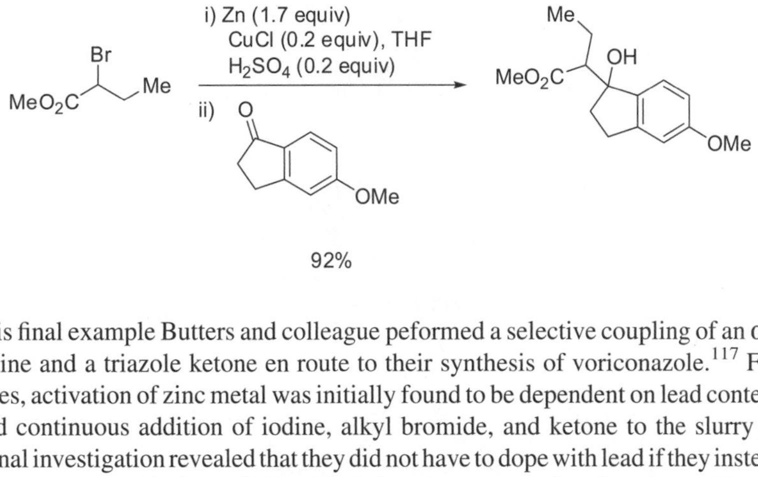

92%

In this final example Butters and colleague peformed a selective coupling of an α-bromo pyrimidine and a triazole ketone en route to their synthesis of voriconazole.[117] For these substrates, activation of zinc metal was initially found to be dependent on lead content in the zinc and continuous addition of iodine, alkyl bromide, and ketone to the slurry of zinc. Additional investigation revealed that they did not have to dope with lead if they instead had a continuous stream of iodine added with their substrates to the zinc slurry. Temperature

[115]Hallinan, E. A.; Hagen, T. J.; Husa, R. K.; Tsymbalov, S.; Rao, S. N.; vanHoeck, J. P.; Rafferty, M. F.; Stapelfeld, A.; Savage, M. A.; Reichman, M. *J. Med. Chem.* **1993**, *36*, 3293–3299.

[116]Scherkenbeck, T.; Siegel, K. *Org. Process Res. Dev.* **2005**, *9*, 216–218.

[117]Butters, M.; Ebbs, J.; Green, S. P.; MacRae, J.; Morland, M. C.; Murtiashaw, C. W.; Pettman, A. J. *Org. Process Res. Dev.* **2001**, *5*, 28–36.

control was also important to maintain selectivity; optimal selectivity was observed at reaction temperatures at or below 5°C.

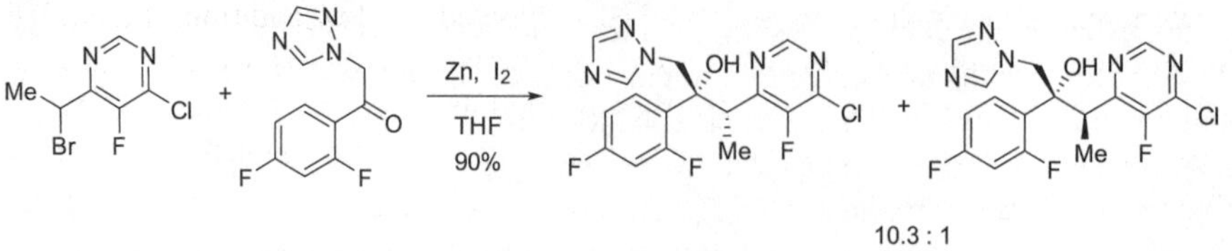

10.3 : 1

## 2.16   WITTIG REACTION

The reaction of a Wittig reagent (phosphorous ylide) with an aldehyde or ketone gives an alkene product and triphenylphospine oxide. *E* vs. *Z* olefin selectivity is a complex function of ylide stability, counterion effects, solvent, etc., and the subject has been well reviewed.[118]

Taber and Nelson used potassium hydride in paraffin as the base in the otherwise standard Wittig reaction shown below.[119]

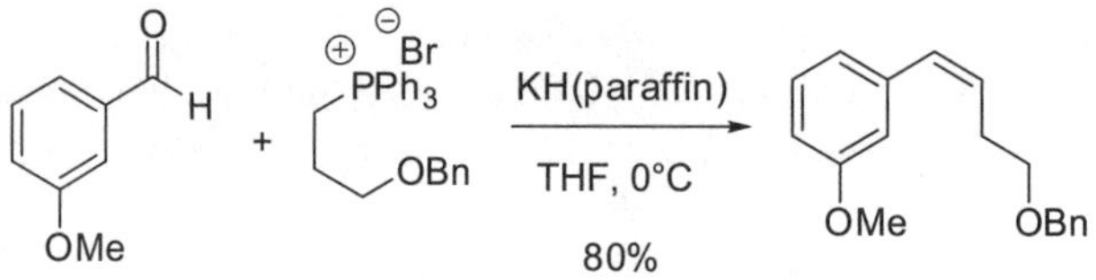

Stuk and colleagues used a pre-formed Wittig ylide in the synthesis of pagoclone.[120] Heating the ylide with a hydroxy-isoindolinone in refluxing xylenes for 24 hours enabled smooth conversion, via the ring-opened aldehyde, to the initial enone product, which then underwent a Michael addition of the amide nitrogen to reform the isoindolinone. A solvent exchange from xylenes into 2-propanol allowed for the isolation of a good yield of the desired product.

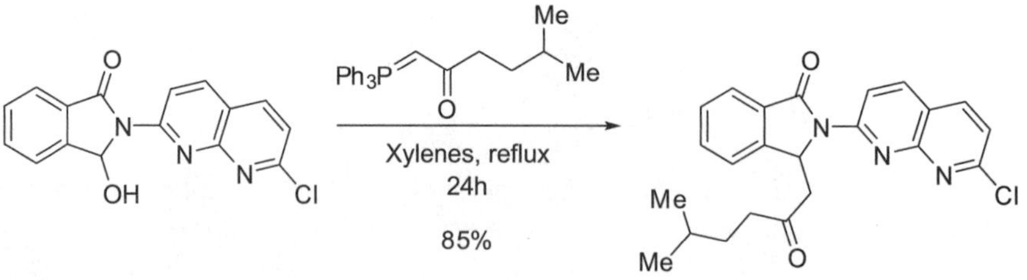

After exploring multiple methods for the olefination of the ketone below, Ainge and colleagues found the Wittig reaction to be most effective.[121] They investigated several

[118]Maryanoff, B. E.; Reitz, A. B. *Chem. Rev.* **1989**, *89*, 863–927.

[119]Taber, D. F.; Nelson, C. G. *J. Org. Chem.* **2006**, *71*, 8973–8974.

[120]Stuk, T. L.; Assink, B. K.; Bates, R. C., Jr.; Erdman, D. T.; Fedij, V.; Jennings, S. M.; Lassig, J. A.; Smith, R. J.; Smith, T. L. *Org. Process Res. Dev.* **2003**, *7*, 851–855.

[121]Ainge, D.; Ennis, D.; Gidlund, M.; Stefinovic, M.; Vaz, L.-M. *Org. Process Res. Dev.* **2003**, *7*, 198–201.

variants: use of the pre-formed ylide, different orders of addition, and different bases and different solvents. For this substrate, the reaction was optimized for overall yield and the ratio of exocyclic versus endocyclic olefin. The optimized procedure involved pre-mixing the ketone and phosphonium bromide in THF, followed by slow addition of a cold THF solution of potassium *t*-butoxide. After reaction completion, it was necessary to remove the reaction solvent in order to precipitate the triphenylphosphine oxide byproduct. Washing the product with water and solvent exchange into *t*-butyl alcohol allowed for near quantitative isolation of product.

In the next two reactions the use of a stabilized Wittig reagent is demonstrated. In the first, Prasad Raju and coworkers utilized the stabilized-Wittig reaction in their synthesis of lacidipine.[122] Generation of the Wittig ylide was accomplished with sodium hydroxide at 0°C. After reaction completion, heptanes were added to the solution to allow for precipitation of triphenylphosphine oxide. Following filtration, the product was used without further purification in the subsequent reaction to form the isolated dihydropyridine product.

In the second example, Chen and coworkers generated the reactive aldehyde by treating the primary alcohol with pyridine–sulfur trioxide complex and an excess of pyridine (to remove excess sulfuric acid in the complex).[123] The aldehyde was then treated with the stabilized Wittig ylide to give the α,β-unsaturated ester as a single olefin isomer.

[122]Prasada Raju, V. V. N. K. V.; Ravindra, V.; Mathad, V. T.; Dubey, P. K.; Pratap Reddy, P. *Org. Process Res. Dev.* **2009**, *13*, 710–715.
[123]Chen, L.; Lee, S.; Renner, M.; Tian, Q.; Nayyar, N. *Org. Process Res. Dev.* **2006**, *10*, 163–164.

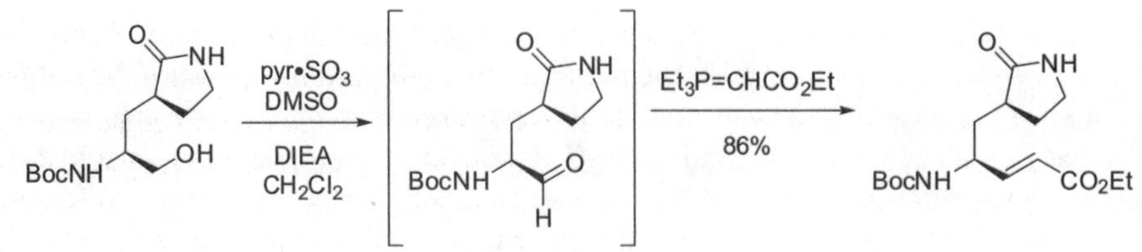

## 2.17 HORNER–WADSWORTH–EMMONS REACTION

The Horner–Wadsworth–Emmons reaction involves reaction of a phosphonate ester with an aldehyde or ketone. Due to the stabilized nature of the ylide, the *E*-olefin geometry is formed predominantly. In the synthesis of cefovecin, Norris and coworkers used the Horner–Wadsworth–Emmons reaction as a key bond-forming step.[124] In their process, the group built the phosphonate ester via the α-chloro ester, shown below. Iodide-catalyzed phosphite displacement and treatment with Hunig's base in methylene chloride resulted in an intramolecular Horner–Wadsworth–Emmons reaction.

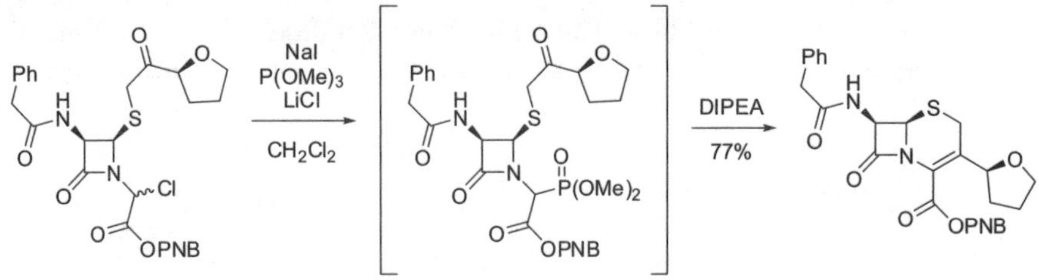

For Alimaridinov and coworkers at Wyeth, the Horner–Wadsworth–Emmons reaction proved critical for synthesis of the desired *E* alkene in their synthesis of the trifluoromethyl alkene shown below.[125] Key to reaction success were the use of $K_3PO_4$ (NaH was less effective) and aging the reaction in ethanol. Potassium phosphate provided moderate *E:Z* selectivity (83:17), whereas NaH was essentially non-selective (53:47). Further aging of the reaction in ethanol (20 h) provided enhancement of the *E:Z* selectivity to 95:5.

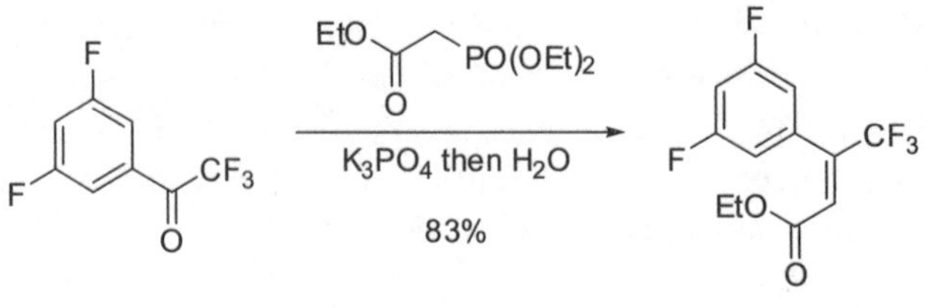

[124]Norris, T.; Nagakura, I.; Morita, H.; McLachlan, G.; Desneves, J. *Org. Process Res. Dev.* **2007**, *11*, 742–746.
[125]Alimardanov, A.; Nikitenko, A.; Connolly, T. J.; Feigelson, G.; Chan, A. W.; Ding, Z.; Ghosh, M.; Shi, X.; Ren, J.; Hansen, E.; Farr, R.; MacEwan, M.; Tadayon, S.; Springer, D. M.; Kreft, A. F.; Ho, D. M.; Potoski, J. R. *Org. Process Res. Dev.* **2009**, *13*, 1161–1168.

Despite obtaining four different isomeric alkenes in their first attempt at the reaction shown below, Zanka and coworkers were able to develop conditions for the preparation of the desired α,β-unsaturated ester.[126] When the initial NaH/DMSO reaction provided a complex mixture, the group examined different solvents and bases to fine tune the reaction. This optimization work led them to a combination of sodium hydroxide and potassium carbonate in DME, which provided an excellent yield of the desired product and minimal amounts of isomeric byproducts.

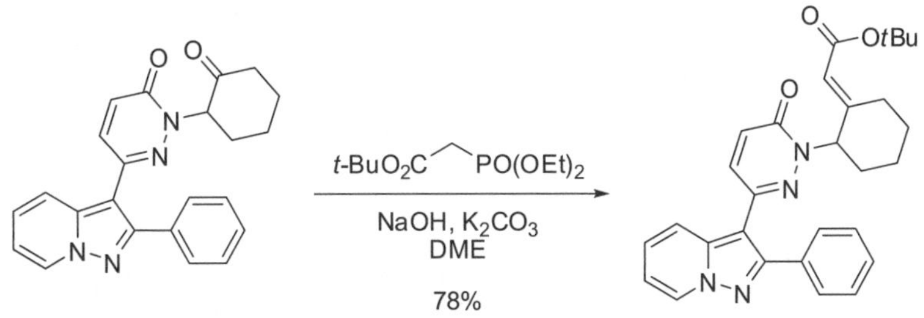

A modification of the Horner–Wadsworth–Emmons reaction initially developed by Still and Gennari, allows for the isolation of the Z-alkene instead. In their synthesis of the enantiomer of kallolide B, Marshall and coworkers used the Still modification of the Horner–Wadsworth–Emmons to append their furan intermediate to the (Z)-conjugated ester in excellent yield.[127]

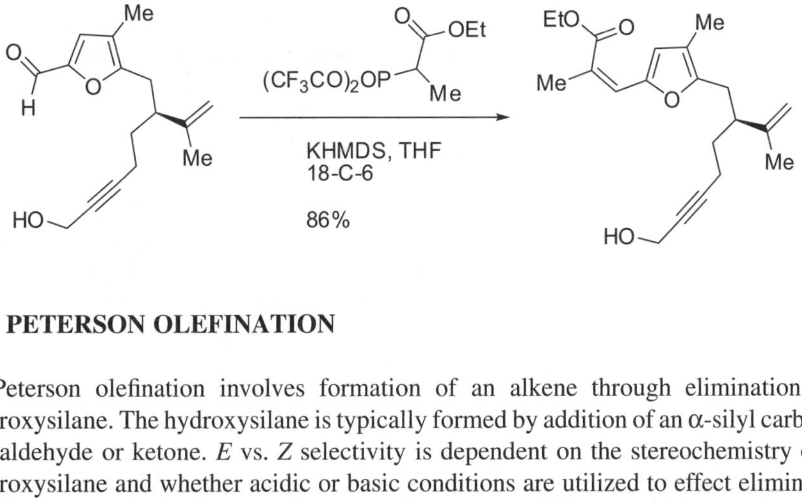

## 2.18 PETERSON OLEFINATION

The Peterson olefination involves formation of an alkene through elimination of a β-hydroxysilane. The hydroxysilane is typically formed by addition of an α-silyl carbanion to an aldehyde or ketone. E vs. Z selectivity is dependent on the stereochemistry of the β-hydroxysilane and whether acidic or basic conditions are utilized to effect elimination.

Key to the Denmark and Yang synthesis of brasilenyne was a Peterson olefination for the installation of the sensitive Z-enyne.[128] The lithiated silyl precursor was formed by treating a cold THF solution of the disilyl alkyne with n-butyllithium. To this solution was added the aldehyde and upon slow warming the desired enyne was formed in good yield and 6:1 Z:E selectivity.

[126]Zanka, A.; Itoh, N.; Kuroda, S. *Org. Process Res. Dev.* **1999**, *3*, 394–399.
[127]Marshall, J. A.; Bartley, G. S.; Wallace, E. M. *J. Org. Chem.* **1996**, *61*, 5729–5735.
[128]Denmark, S. E.; Yang, S.-M. *J. Am. Chem. Soc.* **2002**, *124*, 15196–15197.

Tius and Harrington utilized a Peterson olefination early in their formal synthesis of roseophilin.[129] The *t*-butylimine was first α-silylated with LDA and TMSCl. A second lithiation and condensation with isobutyraldehyde formed the α,β-unsaturated imine, and treatment with aqueous oxalic acid in THF delivered the desired aldehyde in 71% isolated yield.

In their synthesis of maritimol, Deslongchamps and coworkers utilized a Peterson olefination to enable the formation of the required Z-olefin.[130] The reactive boronate was formed by treatment of a lithiated solution of trimethylsilylacetonitrile with triisopropyl-borate. To this cooled reagent was added a solution of aldehyde, providing a 79% yield of the desired alkene.

## 2.19 JULIA–LYTHGOE OLEFINATION

Julia–Lythgoe olefinations involve α-metallation of an alkyl sulfone and addition to an aldehyde or ketone. The resulting alcohol is activated (e.g., by acylation) and then reductively eliminated to form the olefin (see Section 9.5.2).

[129]Harrington, P. E.; Tius, M. A. *Org. Lett.* **1999**, *1*, 649–651.
[130]Toro, A.; Nowak, P.; Deslongchamps, P. *J. Am. Chem. Soc.* **2000**, *122*, 4526–4527.

Keck and coworkers used modified Julia–Lythgoe conditions in their synthesis of rhizoxin.[131] They found that their target triene product was incompatible with standard Na(Hg) reduction conditions due to reduction to form a 1,5-diene. Pursuing observations of Kende and other workers, they studied the use of samarium diiodide as an alternative reductant with good success. They found that the β-acetoxy sulfone adduct could be eliminated to a vinyl sulfone, which was efficiently reduced by $SmI_2$ in a THF-DMPU-MeOH solvent mixture to selectively give the desired $E$-olefin. In deuterium-labeling studies they found that whereas $SmI_2$ appears to reduce β-acetoxy sulfones directly to olefins, the standard Na(Hg) reduction appears to proceed via a vinyl sulfone (i.e., a stepwise elimination/reduction sequence).

Danishefsky and coworkers utilized a standard Julia–Lythgoe olefination in their total synthesis of indolizomycin.[132] The lithiated allylic sulfone was added to the aldehyde and acetylated to form a mixture of diastereomers in high yield (86%), and reduction with Na(Hg) generated the desired $E,E,E$-triene with excellent efficiency (89%).

The Kocienski modification of the Julia–Lythgoe olefination allows for the removal of the one-electron reductant through use of a benzothiazole sulfone. The first step is the same as with the standard Julia–Lythgoe protocol. Deprotonation α to the sulfone and addition to the aldehyde occur to generate a metal alkoxide, which adds to the thiazole to generate an intermediate that collapses to give the desired olefin.

[131]Keck, G. E.; Savin, K. A.; Weglarz, M. A. *J. Org. Chem.* **1995**, *60*, 3194–3204.
[132]Kim, G.; Chu-Moyer, M. Y.; Danishefsky, S. J.; Schulte, G. K. *J. Am. Chem. Soc.* **1993**, *115*, 30–39.

Waykole and coworkers at Novartis found two procedural changes that are key for achieving the Kocienski modification of the Julia–Lythgoe olefination for their substrates: heating the reaction post-sulfone addition to the aldehyde and addition of an additive such as TMSCl or BF$_3$•Et$_2$O.[133] Initial experiments showed less than 10% conversion to the desired olefin, although the aldehyde was consumed. While heating a failed reaction to remove solvent, the group noticed desired product formed, suggesting that elimination (rather than addition) was rate-limiting. This led them to identify the successful modification, which provided the desired alkene in 45% yield on 170 g scale.

Hobson and coworkers utilized a variant of the Julia–Lythgoe olefination in which a phenyltetrazolyl sulfone serves as the nucleophile.[134] Examination of several reaction parameters indicated that aging temperature was critical to *E:Z* selectivity. The optimized process involved metallation with LiHMDS at $-70°$C, then warming to $-10°$C and aging for 1 hour prior to quench. With this protocol an average yield of 80% with 99.9 area % purity was realized on 27 kg scale ($>$100:1 *E:Z* selectivity).

## 2.20   TEBBE METHYLENATION

The Tebbe reagent, which can be made from trimethylaluminum and dicyclopentyl titanocene, is used for methylenation of an aldehyde, ketone, ester, or amide.[135] There

[133]Xu, D. D.; Waykole, L.; Calienni, J. V.; Ciszewski, L.; Lee, G. T.; Liu, W.; Szewczyk, J.; Vargas, K.; Prasad, K.; Repic, O.; Blacklock, T. J. *Org. Process Res. Dev.* **2003**, *7*, 856–865.

[134]Hobson, L. A.; Akiti, O.; Deshmukh, S. S.; Harper, S.; Katipally, K.; Lai, C. J.; Livingston, R. C.; Lo, E.; Miller, M. M.; Ramakrishnan, S.; Shen, L.; Spink, J.; Tummala, S.; Wei, C.; Yamamoto, K.; Young, J.; Parsons, R. L. *Org. Process Res. Dev.* **2010**, *14*, 441–458.

[135]Cannizzo, L. F.; Grubbs, R. H. *J. Org. Chem.* **1985**, *50*, 2386–2387.

are a few advantages to the Tebbe Lewis acid reagent. The reagent is highly reactive and can be used with sterically demanding substrates. It can also be used to olefinate carbonyls with an α stereocenter without epimization. Another added benefit is the trade of phosphine oxide side products common to the Wittig and Horner–Wadsworth–Emmons olefinations for titanium oxide. A modification was made by Petasis in the production of the titanocene reagent by replacing trimethyl aluminum (original Tebbe) with methyl lithium. This reagent procedure produces a reagent with similar reactivity but better handling and storage properties.[136]

The two examples below showcase the ability of the Tebbe reagent to accomplish methylenation of hindered and sensitive substrates. In the first example, the Sinay group appended their hindered glycoside using the Tebbe reagent.[137] The acetate was dissolved in 5:1 pyridine:THF, cooled to −60°C, treated with the Tebbe reagent and warmed to 0°C. Following quench and workup the desired product was isolated in excellent yield.

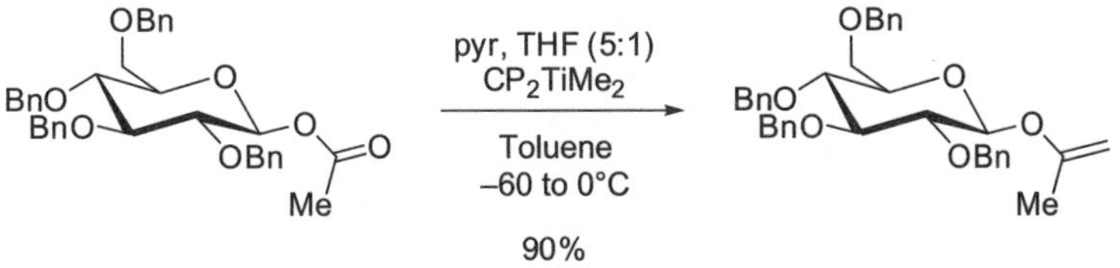

In the second example, Howell and Blauvelt were able to methylenate a challenging β-lactone substrate, shown below.[138] The lactone substrate in toluene was treated with the Tebbe reagent and after precipitation of titanium byproducts, the product was purified by silica gel chromatography. This is an impressive reaction because it has an α-stereocenter, is hindered by the trityl group, and is sensitive to ring opening.

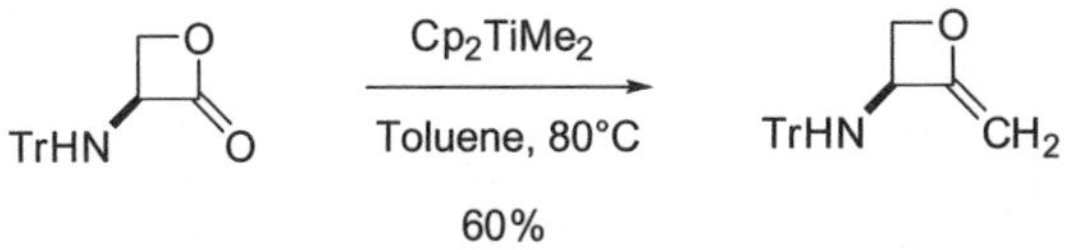

Using the Petasis modification of Tebbe reagent formation, the Merck team utilized methyl magnesium chloride and dichlorocyclopentadienyl titanocene for effective production of the necessary dimethyltitanocene reagent.[139] Payack and coworkers used the Petasis-modified Tebbe reaction for the preparation of aprepitant. The olefination was clean in this case but excess Tebbe reagent caused product deterioration if it was not quenched immediately. This issue was overcome by addition of a small amount of sacrificial ester to intercept excess Tebbe reagent. The reaction was quenched, and filtered through diatomaceous earth (to remove titanium byproducts) to provide a 91% yield of the desired enol ether.

[136]Petasis, N. A.; Bzowej, E. I. *J. Am. Chem. Soc.* **1990**, *112*, 6392–6394.

[137]Marra, A.; Esnault, J.; Veyrieres, A.; Sinay, P. *J. Am. Chem. Soc.* **1992**, *114*, 6354–6360.

[138]Blauvelt, M. L.; Howell, A. R. *J. Org. Chem.* **2008**, *73*, 517–521.

[139]Payack, J. F.; Huffman, M. A.; Cai, D.; Hughes, D. L.; Collins, P. C.; Johnson, B. K.; Cottrell, I. F.; Tuma, L. D. *Org. Process Res. Dev.* **2004**, *8*, 256–259.

## 2.21 THE MANNICH REACTION

The Mannich reaction is the condensation of a carbonyl compound, an amine and a carbon nucleophile. The nitrogen nucleophile may be a primary or secondary amine, or ammonia; however, anilines do not participate in the Mannich reaction. The carbon nucleophile is typically a second carbonyl compound with an enolizable α-proton. The reaction proceeds via activation of the aldehyde via formation of a Schiff base with the nitrogen nucleophile, followed by attack of the carbon nucleophile to afford the product, as described in the scheme below.

The simplest manifestation of the Mannich reaction entails stirring a nucleophilic amine, paraformaldehyde and an enolizable carbon nucleophile in the presence of an acid at an elevated temperature. A few representative examples are given below.[140–142]

[140]Fujima, Y.; Ikunaka, M.; Inoue, T.; Matsumoto, J. *Org. Process Res. Dev.* **2006**, *10*, 905–913.

[141]Sole, D.; Vallverdu, L.; Solans, X.; Font-Bardia, M.; Bonjoch, J. *J. Am. Chem. Soc.* **2003**, *125*, 1587–1594.

[142]Kellogg, R. M.; Nieuwenhuijzen, J. W.; Pouwer, K.; Vries, T. R.; Broxterman, Q. B.; Grimbergen, R. F. P.; Kaptein, B.; La Crois, R. M.; de Wever, E.; Zwaagstra, K.; van der Laan, A. C. *Synthesis* **2003**, 1626–1638.

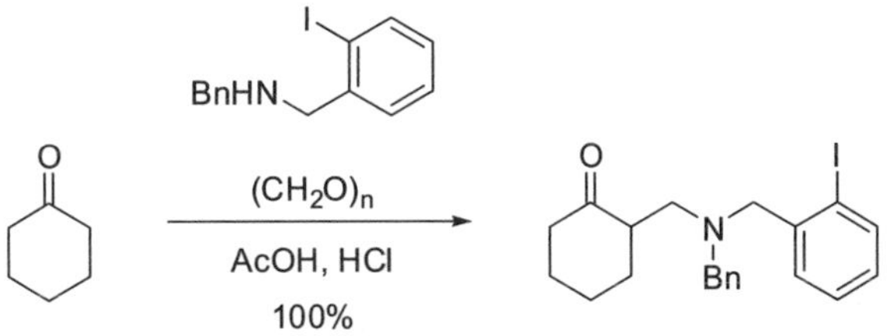

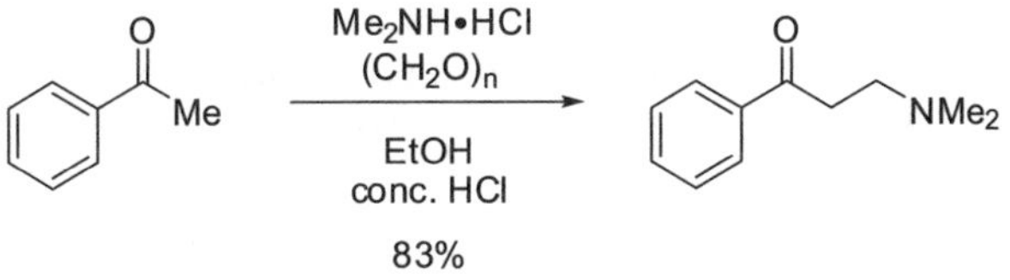

Several excellent examples of intra-molecular Mannich condensations can be found in Heathcock's synthesis of lycopodium alkaloids. In the following scheme, treatment of the amino bis-acetal starting material with HCl led to acetal hydrolysis followed by an intramolecular Mannich reaction to produce the tricyclic amino-ketone in 63% yield.[143]

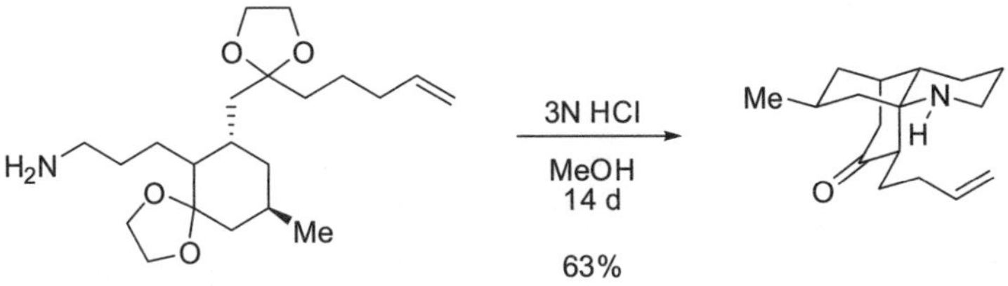

Recent advances in the development of asymmetric versions of the Mannich reaction have been reviewed by List and coworkers.[144]

## 2.22   THE STRECKER REACTION

The Strecker synthesis can be viewed as a variant of the Mannich reaction (2.21), where cyanide plays the role of the carbon nucleophile. The original Strecker reaction was a "one-step" method using ammonia as the nitrogen source, where all three components were mixed together at the start of the reaction. More recent versions of this reaction involve pre-formation of the imine (using amines), followed by addition of the cyanide nucleophile.

The Strecker reaction is perhaps the oldest known non-biosynthetic method to synthesize α-amino acids. These compounds are formed by hydrolysis of the intermediate α-aminonitrile.

[143]Heathcock, C. H.; Kleinman, E. F.; Binkley, E. S. *J. Am. Chem. Soc.* **1982**, *104*, 1054–1068.

[144]Mukherjee, S.; Yang, J. W.; Hoffmann, S.; List, B. *Chem. Rev. (Washington, DC, U. S.)* **2007**, *107*, 5471–5569.

One of the simplest methods to carry out the Strecker reaction is to combine the carbonyl compound and the amine with a cyanide source (preferably KCN or NaCN), and stir the mixture until the product is formed. A nice example of the use of the Strecker reaction was provided by Mehrotra and coworkers.[145]

A slight variation of this procedure was used for the reaction of (*R*)-phenylglycinol with KCN and glutaraldehyde.[146]

The use of alkali metal cyanides invariably necessitates the use of water, and this can be deleterious in some cases. For example, in the synthesis of carfentanil, it was found that the α-aminonitrile (formed via the reaction of *N*-benzyl-4-piperidone with aniline and KCN in *aqueous* acetic acid) underwent a retro-Strecker reaction when heated in the presence of aqueous acid. Under these conditions, only 71% of the desired product was isolated; however, the use of trimethylsilyl cyanide (TMSCN) in *glacial* acetic acid (anhydrous conditions) furnished the product in 81% yield. The difference in yields was more pronounced when 2-fluoroaniline was used as the amine component—the aqueous Strecker reaction afforded the product in 12% yield, while the ahydrous modification yielded 72% of the desired product.[147]

[145]Mehrotra, M. M.; Heath, J. A.; Smyth, M. S.; Pandey, A.; Rose, J. W.; Seroogy, J. M.; Volkots, D. L.; Nannizzi-Alaimo, L.; Park, G. L.; Lambing, J. L.; Hollenbach, S. J.; Scarborough, R. M. *J. Med. Chem.* **2004**, *47*, 2037–2061.
[146]Bonin, M.; Grierson, D. S.; Royer, J.; Husson, H. P. *Org. Synth.* **1992**, *70*, 54–59.
[147]Feldman, P. L.; Brackeen, M. F. *J. Org. Chem.* **1990**, *55*, 4207–4209.

Lewis acids are also known to promote the formation of α-aminonitriles from ketones. A diastereoselective Strecker reaction was performed on the ketone below using $NH_3$, $Ti(Oi\text{-}Pr)_4$, and TMSCN.[12]

Several asymmetric modifications of the Strecker reaction have been developed and utilized to synthesize a host of optically enriched, structurally diverse, and unnatural α-amino acids. These non-proteinogenic amino acids have found extensive use as key building blocks in the pharmaceutical industry. Recent developments in the area of catalytic enantioselective Strecker reactions have been chronicled in a review by Gröger.[148]

## 2.23  HYDROLYSIS OF CARBON–NITROGEN DOUBLE BONDS

The carbon–nitrogen double bond can be hydrolyzed to a carbon–oxygen double bond by the reaction of the substrate under aqueous acidic, aqueous basic, or oxidative conditions. The oxidative methods will not be covered here; some examples are the use of PDC (pyridinium dichromate), IBX (o-iodoxybenzoic acid), Dess–Martin periodinane and ozone (see Chapter 10 for C−N oxidation strategies). The ease of hydrolysis varies with the substitution on the nitrogen and also on the ability of the substrate to tolerate changes in pH. In general, it is more difficult to convert hydrazones and oximes back to the respective aldehyde and ketone than imines. In the first example, after the Friedel–Crafts reaction with the sesamol nitrile, the resultant imine is converted to the ketone by heating an acidic aqueous solution for a few hours.[149]

[148]Groeger, H. *Chem. Rev.* **2003**, *103*, 2795–2827.
[149]Hastings, J. M.; Hadden, M. K.; Blagg, B. S. J. *J. Org. Chem.* **2008**, *73*, 369–373.

It can be difficult to maintain the imine functionality unless rigorous care is taken to maintain an anhydrous environment. As with other generally sensitive functionality such as acid chlorides, some compounds in this class are more hydrolytically stable and can even be chromatographed.

A common method for the hydrolysis of hydrazones or oximes involves the use of a transfer agent. In the examples below, acetone and formaldehyde are used to trap the hydrazine or hydroxylamine when they are released from the substrate of interest. This works well when two factors are present: (1) the substrate can tolerate exposure to an abundance of the sacrificial aldehyde or ketone (formaldehyde or acetone in this case) without deleterious side reactions, and (2) there is an easy method for separation of the acetyl or formyl byproduct after the reaction.

In the first example, the Fuchs group used boron trifluoride etherate and acetone to effect hydrolysis of the tosyl hydrazone.[150] The tosyl hydrazone byproduct was easily removed by precipation with hexanes or through degradation and separation by stirring with a basic aqueous wash.

94%

Another option is the use of a formaldehyde trap in a binary aqueous hydrochloric acid/tetrahydrofuran solution.[151] This reaction was stirred at room temperature with a 35% HCl solution until completion. The product may then be isolated by extraction.

79%

When functionality is present that is sensitive to strongly acidic or basic reagents, there are other options available for oxime or hydrazone hydrolysis. In the examples below, mildly acidic reagents (i.e., pH 4–7) are used for hydrolysis. Ziegler was able to successfully degrade the Enders RAMP hydrazone using copper acetate in THF and water.[152] The subsequent ketone was isolated in moderate yield after distillation. In the second example below, the hydrazone was converted to the aldehyde in excellent yield by treatment with $CuCl_2$.[153]

[150]Sacks, C. E.; Fuchs, P. L. *Synthesis* **1976**, 456–457.
[151]Severin, T.; Lerche, H. *Synthesis* **1982**, 305–307.
[152]Ziegler, F. E.; Becker, M. R. *J. Org. Chem.* **1990**, *55*, 2800–2805.
[153]Caron, S.; Do, N. M.; Sieser, J. E.; Arpin, P.; Vazquez, E. *Org. Process Res. Dev.* **2007**, *11*, 1015–1024.

Likewise, Mitra was able to reveal the latent ketone in the alkenyl hydrazone below. This substrate and others with THP and acetal functionality were stirred in THF/water (10/1 v/v) with silica gel at room temperature.[154] The reaction was then concentrated to dryness and the product (adsorbed onto the silica gel) was chromatographed to give the desired product.

One final example shows the use of dichloroamine-T (*N,N*-dichloro-4-toluenesulfona-mide, DCT). This slightly acidic solution allowed the removal of the oxime without scrambling of the (*E*)-α,β unsaturated system.[155]

## 2.24 CONVERSION OF CARBOXYLIC ACIDS TO ACYL CHLORIDES

This section will deal with the synthesis of acid chlorides from carboxylic acids; the synthesis of acyl bromides and iodides will not be discussed in this chapter, while the synthesis of acyl fluorides is discussed in Section 2.25. One of the most common methods to activate carboxylic acids towards nucleophilic attack at the carboxyl center is to convert them to acyl halides, generally acyl chlorides. The reaction is typically performed by

[154]Mitra, R. B.; Reddy, G. B. *Synthesis* **1989**, 694–698.
[155]Gupta, P. K.; Manral, L.; Ganesan, K. *Synthesis* **2007**, 1930–1932.

treatment of the carboxylic acid with a halogenating agent such as thionyl chloride or oxalyl chloride, optionally with a catalytic amount of dimethylformamide. Vilsmeier reagent (derived from reaction of thionyl chloride or oxalyl chloride with dimethylformamide) has also been used. Representative examples and relative merits of these protocols are discussed below.

### 2.24.1 Procedures Using Oxalyl Chloride in the Absence of DMF

It is generally possible to treat a carboxylic acid with oxalyl chloride in an organic solvent to form the acid chloride. The advantage of using oxalyl chloride is that byproducts of the reaction (CO, $CO_2$, and HCl) are volatile and can be easily removed from the reaction mixture. In any event, they rarely interfere with subsequent reactions of the acid halides, rendering oxalyl chloride one of the top reagents of choice to effect this transformation. A representative example is given below.

Interestingly, DBU has been found to be an effective catalyst for this transformation. Treatment of the carboxylic acid with 2 equivalents of oxalyl chloride and 0.06 equivalents of DBU in THF at room temperature led to clean conversion to the acid chloride (yield not mentioned). Subsequent reaction of the acid chloride with hydroxylamine furnished the hydroxamic acid in 61% overall yield after recrystallization.[156]

### 2.24.2 Procedures Using Thionyl Chloride in the Absence of DMF

In general, treatment of carboxylic acids with thionyl chloride affords the corresponding acyl chlorides. The byproducts of the reaction are $SO_2$ and HCl. The reaction can be carried out in the absence of an organic co-solvent (i.e., in neat thionyl chloride).[157] The excess thionyl chloride can be conveniently removed from the reaction mixture by distillation due to its relatively low boiling point (79°C), thereby avoiding complications in subsequent steps.

[156]Frampton, G. A.; Hannah, D. R.; Henderson, N.; Katz, R. B.; Smith, I. H.; Tremayne, N.; Watson, R. J.; Woollam, I. *Org. Process Res. Dev.* **2004**, *8*, 415–417.

[157]Sano, T.; Sugaya, T.; Inoue, K.; Mizutaki, S.-i.; Ono, Y.; Kasai, M. *Org. Process Res. Dev.* **2000**, *4*, 147–152.

The reaction has also been carried out in organic solvents such as THF, as exemplified below.[158]

This transfomation may also be effected in the presence of a base. In the following example, the acid chloride was synthesized by treating the carboxylic acid with thionyl chloride and triethylamine in toluene/MTBE. The acid chloride was taken on to the Boc-protected amine via a Curtius rearrangement (see Section 7.2.3.2) in 73% overall yield.[159]

It is to be noted that the use of $SOCl_2$ in MTBE in the absence of a base is particularly hazardous. The HCl liberated in reactions using $SOCl_2$ could lead to decomposition of MTBE and generate *iso*-butylene gas.[160]

### 2.24.3  Procedures Using a Halogenating Agent and DMF

A small amount of DMF typically catalyzes the reaction of a carboxylic acid with a chlorinating agent such as oxalyl chloride or thionyl chloride. This is presumably due to the *in situ* generation of the Vilsmeier reagent (see Section 2.24.4). One major drawback of utilizing DMF in chlorodehydroxylation reactions is the unwanted co-production of dimethylcarbamoyl chloride (DMCC). This compound is a known animal carcinogen, and a potential human carcinogen.[161] Due to these toxicological concerns, adequate containment measures must be taken while performing reactions involving DMF and chlorinating agents, especially in the pharmaceutical industry. With that note of caution, here are a few examples of reactions using this reagent combination.

In the following example, oxalyl chloride was used in the presence of catalytic DMF to convert the acid to the corresponding acyl chloride.[162]

[158]Stoner, E. J.; Stengel, P. J.; Cooper, A. J. *Org. Process Res. Dev.* **1999**, *3*, 145–148.

[159]Varie, D. L.; Beck, C.; Borders, S. K.; Brady, M. D.; Cronin, J. S.; Ditsworth, T. K.; Hay, D. A.; Hoard, D. W.; Hoying, R. C.; Linder, R. J.; Miller, R. D.; Moher, E. D.; Remacle, J. R.; Rieck, J. A., III; Anderson, D. D.; Dodson, P. N.; Forst, M. B.; Pierson, D. A.; Turpin, J. A. *Org. Process Res. Dev.* **2007**, *11*, 546–559.

[160]Grimm, J. S.; Maryanoff, C. A.; Patel, M.; Palmer, D. C.; Sorgi, K. L.; Stefanick, S.; Webster, R. R. H.; Zhang, X. *Org. Process Res. Dev.* **2002**, *6*, 938–942.

[161]Levin, D. *Org. Process Res. Dev.* **1997**, *1*, 182.

[162]Slade, J. S.; Vivelo, J. A.; Parker, D. J.; Bajwa, J.; Liu, H.; Girgis, M.; Parker, D. T.; Repic, O.; Blacklock, T. *Org. Process Res. Dev.* **2005**, *9*, 608–620.

In the following example, clean conversion to the acid chloride was achieved using $SOCl_2$ in the presence of catalytic DMF in toluene. Since the acid chloride was being converted to the amide under Schotten–Baumann conditions (using aqueous NaOH) in the next step, it was not necessary to remove the excess thionyl chloride after the acid chloride formation.[163]

### 2.24.4 Procedures Using Vilsmeier Reagent

The reaction of DMF with a halogenating agent such as $SOCl_2$, $POCl_3$ or oxalyl chloride leads to formation of Vilsmeier reagent (*N,N*-dimethylchloromethyleneammonium chloride). This reagent is commercially available and has been used in the conversion of carboxylic acids to acid chlorides. A representative example is provided below.[164]

## 2.25 SYNTHESIS OF ACYL FLUORIDES FROM CARBOXYLIC ACIDS

One of the most commonly used reagents for the conversion of carboxylic acids to acyl fluorides is diethylaminosulfur trifluoride (DAST).[165] The reaction is carried out at near-ambient conditions.

[163]Burks, J. E., Jr.; Espinosa, L.; LaBell, E. S.; McGill, J. M.; Ritter, A. R.; Speakman, J. L.; Williams, M.; Bradley, D. A.; Haehl, M. G.; Schmid, C. R. *Org. Process Res. Dev.* **1997**, *1*, 198–210.
[164]Koch, G.; Kottirsch, G.; Wietfeld, B.; Kuesters, E. *Org. Process Res. Dev.* **2002**, *6*, 652–659.
[165]Wipf, P.; Wang, Z. *Org. Lett.* **2007**, *9*, 1605–1607.

80%

One major drawback with DAST is its well-documented thermal instability. From this perspective, bis(2-methoxyethyl)aminosulfur trifluoride (Deoxo-Fluor) is a better choice due to its enhanced thermal stability.[42]

96%

Efficient conversion of acids to acyl fluorides can also be achieved using 2,4,6-trifluoro-1,3,5-triazine (cyanuric fluoride).[166]

97%

## 2.26    FORMATION OF AMIDES FROM CARBOXYLIC ACIDS

Amides may be formed either by direct coupling of carboxylic acids (by activation *in situ*) with amines, or in a two step process via activation of the carboxylic acid followed by reaction of an amine with the activated intermediate.

### 2.26.1    Direct Coupling of Carboxylic Acids and Amines

Carbodiimides such as DCC (dicyclohexylcarbodiimide) are the most commonly utilized class of reagents for the direct coupling of acids and amines. The major drawback with these reagents is the formation of urea byproducts that are typically separable only by chromatography. This problem can be alleviated by the use of 1-[3-(dimethylamino)propyl]-3-ethylcarbodiimide hydrochloride (EDC),[167] wherein the urea byproduct is water soluble and can be conveniently removed through extractive workup or a water wash. In most cases, these reactions require catalysis by 1-hydroxybenzotriazole (HOBt). Thus, the need to use a stoichiometric amount of EDC, coupled with the known hazards of HOBt,[168] makes this protocol less desirable for large-scale applications.

---

[166]Suaifan, G. A. R. Y.; Mahon, M. F.; Arafat, T.; Threadgill, M. D. *Tetrahedron* **2006**, *62*, 11245–11266.
[167]Ormerod, D.; Willemsens, B.; Mermans, R.; Langens, J.; Winderickx, G.; Kalindjian, S. B.; Buck, I. M.; McDonald, I. M. *Org. Process Res. Dev.* **2005**, *9*, 499–507.
[168]Dunn, P. J.; Hoffmann, W.; Kang, Y.; Mitchell, J. C.; Snowden, M. J. *Org. Process Res. Dev.* **2005**, *9*, 956–961.

A particularly convenient and "green" protocol for the synthesis of amides involves the direct reaction of a carboxylic acid and an amine in the presence of catalytic boric acid with azeotropic removal of water. This reaction was shown to be successful with a variety of primary and secondary aliphatic amines as well as anilines.[169]

### 2.26.2   Via acid Chlorides

A common approach to the synthesis of amides is via the reaction of amines with acid chlorides in the presence of a base. When sodium hydroxide is used as the base, the reaction is called the Schotten–Baumann reaction. Organic bases such as triethylamine have also been employed in this transformation. In some cases, the reactions are accelerated by the addition of a small quantity of 4-(dimethylamino)pyridine (DMAP). This is a convenient and fairly general method for amide formation when the acid chlorides are readily available (see Section 2.24 for the preparation of acid chlorides from carboxylic acids).

Despite the convenience and practicality it affords, this method has a few limitations. The co-production of HCl in the coupling reaction can lead to undesired side reactions such as unwanted removal of certain acid-labile protecting groups. Moreover, the base used to scavenge the HCl could lead to racemization of enolizable acid chlorides, especially when applied to peptide synthesis where epimerizable centers are abundant. In the following example, the use of a mild base (sodium bicarbonate) led to smooth amidation while preserving stereochemistry and avoiding hydrolysis of the ester functionalities.[170]

[169]Tang, P. *Org. Synth.* **2005**, *81*, 262–272.
[170]Ager, D. J.; Babler, S.; Erickson, R. A.; Froen, D. E.; Kittleson, J.; Pantaleone, D. P.; Prakash, I.; Zhi, B. *Org. Process Res. Dev.* **2004**, *8*, 72–85.

In the following example, the acid chloride was formed in quantitative yield by treatment of the carboxylic acid with thionyl chloride. The amidation reaction was then performed using imidazole as the base.[158] The use of imidazole circumvented self-condensation and consequent polymerization of the intermediate acid chloride—a problem observed when several other bases were used.

Imidazole, EtOAc, 0°C to rt

87–95%

### 2.26.3  Via Acyl Imidazoles (Imidazolides)

A convenient and widely used reagent for the formation of amides is $N,N'$-carbonyldiimidazole (CDI). The carboxylic acid is activated with CDI to produce the intermediate acyl imidazole, which is then coupled with the desired amine in the subsequent step. This protocol was successfully used to synthesize a key intermediate in the manufacture of sildenafil.[171]

i) CDI, EtOAc
reflux

ii)

90%

Since imidazole is produced in this sequence of reactions, there is no need to utilize additional base for the coupling reaction. Moreover, salts of amines may be directly used without the need to generate the free-base prior to the reaction. The byproducts of amidations using CDI, namely, imidazole and carbon dioxide, are fairly innocuous. It has been shown that the $CO_2$ evolved in the activation step catalyzes the subsequent amidation step.[172,173]

[171]Dale, D. J.; Dunn, P. J.; Golightly, C.; Hughes, M. L.; Levett, P. C.; Pearce, A. K.; Searle, P. M.; Ward, G.; Wood, A. S. *Org. Process Res. Dev.* **2000**, *4*, 17–22.
[172]Williams, I.; Kariuki, B. M.; Reeves, K.; Cox, L. R. *Org. Lett.* **2006**, *8*, 4389–4392.
[173]Vaidyanathan, R.; Kalthod, V. G.; Ngo, D. P.; Manley, J. M.; Lapekas, S. P. *J. Org. Chem.* **2004**, *69*, 2565–2568.

One minor drawback of the CDI protocol is that acyl imidazoles are slightly less reactive than the corresponding acid chlorides, and hence couplings involving either sterically hindered carboxylic acids or weakly nucleophilic amines tend to be sluggish. This problem may be overcome by the use of a catalyst such as 2-hydroxy-5-nitropyridine. The safety hazards and efficiencies of such catalysts have been thoroughly investigated.[174]

### 2.26.4   Using Anhydrides

Symmetrical anhydrides are rarely used as the acyl component in amide formation reactions because only one half of the acid is used in the coupling reaction while the other half is "wasted." This method is attractive only when the anhydride is cheap and commercially available (e.g., acetic anhydride). The reaction of primary amines with cyclic anhydrides leads to imides. The phthalimido group is commonly used as a protecting group for amines, and is installed via reaction with phthalic anhydride.[175]

Mixed anhydrides have been widely used to form amides. Mixed anhydrides can be categorized into two types, namely, mixed carboxylic anhydrides and mixed carbonic anhydrides.

The most widely used example of a mixed carboxylic anhydride is pivalic anhydride. The carboxylic acid is treated with pivaloyl chloride (trimethylacetyl chloride) in the presence of a base (tertiary amine), and the resultant anhydride is coupled with an amine in the next step. The steric hindrance provided by the *t*-butyl group forces the amine to react preferentially at the distal carboxyl center, and thus regioselectivity is obtained.[170] Other mixed carboxylic anhydrides are plagued with regioselectivity issues, and hence are less commonly used.

[174]Bright, R.; Dale, D. J.; Dunn, P. J.; Hussain, F.; Kang, Y.; Mason, C.; Mitchell, J. C.; Snowden, M. J. *Org. Process Res. Dev.* **2004**, *8*, 1054–1058.
[175]Meffre, P.; Durand, P.; Le Goffic, F. *Org. Synth.* **1999**, *76*, 123–132.

NMM = *N*-methylmorpholine

Mixed carbonic anhydrides are extensively used for amide formation. These compounds can be readily accessed by treatment of a carboxylic acid with an alkyl chloroformate in the presence of a base (typically a tertiary amine). Regiochemistry of amine attack is primarily dictated by electronic factors in these cases. The reaction preferentially occurs at the more electrophilic carboxylic site rather than the carbonate site. The byproducts from these reactions, namely the hydrochloride salt of the base, $CO_2$ and the alcohol from the chloroformate, are fairly innocuous and render this protocol attractive for large scale applications.

In a typical reaction, the chloroformate is added to a mixture of the carboxylic acid and the base to form the mixed anhydride, which is treated with an amine in the next step to afford the amide.[176]

Interestingly, when the same protocol was applied for the following transformation, significant amounts of the symmetrical anhydride were formed in the activation step due to reaction of the carboxylic acid with the activated species. This problem was circumvented by modifying the order of addition: a mixture of the carboxylic acid and base was added to a solution of *iso*-butylchloroformate in toluene. The choice of base is noteworthy. *N,N*-Dimethylbenzylamine was extracted from the aqueous layers after the workup, and recycled.[177]

[176]Chen, J.; Corbin, S. P.; Holman, N. J. *Org. Process Res. Dev.* **2005**, *9*, 185–187.
[177]Prashad, M.; Har, D.; Hu, B.; Kim, H.-Y.; Girgis, M. J.; Chaudhary, A.; Repic, O.; Blacklock, T. J.; Marterer, W. *Org. Process Res. Dev.* **2004**, *8*, 330–340.

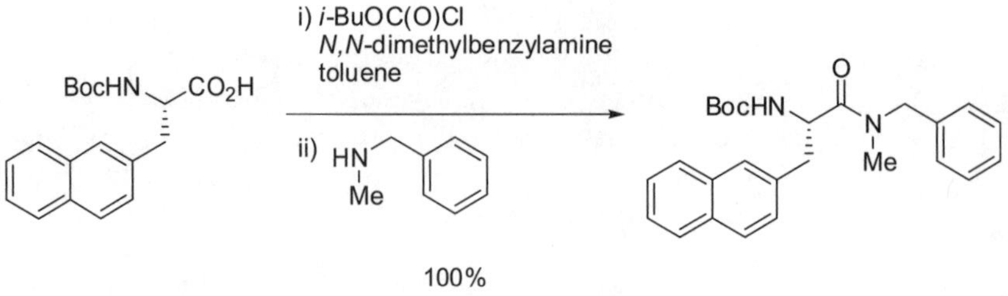

100%

Another method to form mixed carbonic anhydrides is to treat the carboxylic acid with 2-ethoxy-1-ethoxycarbonyl-1,2-dihydroquinoline (EEDQ). The main disadvantage of this method compared to the chloroformate protocol is the production of stoichiometric quinoline as the byproduct.

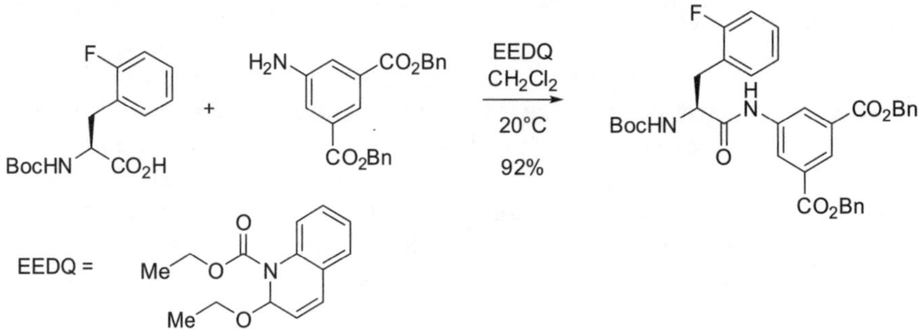

Several other reagents have been developed for the formation of amide bonds, and the literature in this area has been extensively reviewed.[178] Most of these reagents such as phosphonium salts and uronium salts are very effective; however, they are unattractive for large scale applications because of extensive byproduct formation. These methods may be applicable in extreme cases should the aforementioned protocols fail to produce the desired results.

## 2.27   FORMATION OF AMIDES FROM ESTERS

The direct coupling of alkyl esters with an amine is an attractive method to form amides. The reaction may be catalyzed by a mild Brønsted acid such as 2-hydroxypyridine.[179]

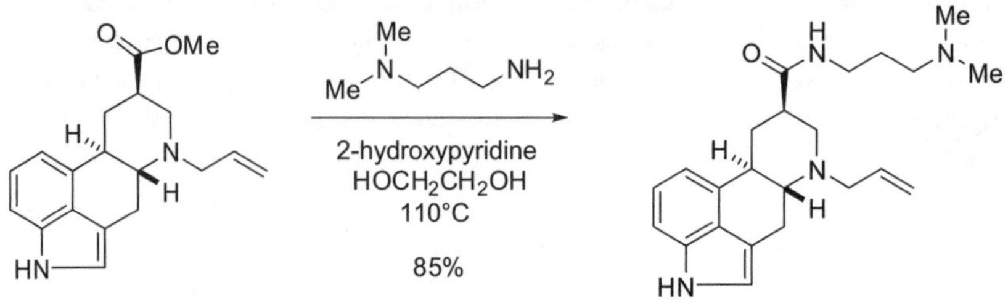

85%

[178]Montalbetti, C. A. G. N.; Falque, V. *Tetrahedron* **2005**, *61*, 10827–10852.
[179]Ashford, S. W.; Henegar, K. E.; Anderson, A. M.; Wuts, P. G. M. *J. Org. Chem.* **2002**, *67*, 7147–7150.

The coupling of esters with amines to form amides is also catalyzed by Lewis acids such as magnesium halides. This procedure has been used to amidate the ester at the 2-position of indoles and pyridines, presumably due to complexation of the metal with the nitrogen atom. It has also been shown to work with other unactivated esters.[180]

## 2.28 HYDROLYSIS OF ACYL HALIDES

The hydrolysis of acyl halides leads to carboxylic acids. This is a reaction that typically works extremely and almost equally well in the absence of a skilled chemist or in the presence of a careless one. Most acyl halides are formed from the corresponding carboxylic acids, and the reverse reaction is rarely attempted deliberately. Several acid halides are water-sensitive, and therefore need to be stored under anhydrous conditions to prevent hydrolysis. Nevertheless, if one desperately needs to effect this transformation, it can be easily accomplished by treating the acyl halide with water (or hydroxide, in some rare instances). Note that the reaction could be exothermic and will lead to the co-production of 1 equivalent of HX.

## 2.29 CONVERSION OF CARBOXYLIC ACIDS TO ESTERS

Carboxylic acids can be converted to esters in two different ways. Alkylation of the carboxylate oxygen with a suitable electrophile will lead to esters. In this case, both of the oxygen atoms in the carboxylic acid are retained in the ester formed, and this type of transformation is covered in Section 1.2.2.1. A complementary method is via treatment of the acid with an oxygen nucleophile in the presence of an appropriate catalyst or activating agent. The oxygen atom from the nucleophile is incorporated into the ester formed, and this type of transformation is covered in this section.

[180]Guo, Z.; Dowdy, E. D.; Li, W. S.; Polniaszek, R.; Delaney, E. *Tetrahedron Lett.* **2001**, *42*, 1843–1845.

### 2.29.1 Fisher Esterification

The conversion of carboxylic acids to esters via treatment with alcohols in the presence of a protic acid catalyst is referred to as the Fisher esterification reaction. This is essentially an equilibrium reaction, and is generally driven to the right by the use of one of the reagents in excess, or removal of water (either by distillation or the use of a suitable desiccant/dehydrating agent).

In the case of esterification reactions to form simple esters such as methyl and ethyl esters, the alcohol is used in large excess, usually as the solvent.[153,181] Note that in the second example below, the strongly acidic conditions led to the removal of the Boc protecting group as well.

When the alcohol cannot be used as the solvent, the reaction is generally driven to completion by azeotropic removal of water as depicted in the example below.[182]

The use of catalytic $SOCl_2$ in the presence of an alcohol is another convenient method for the preparation of esters.[183] The thionyl chloride reacts with methanol to form HCl, and thus serves as a convenient source of anhydrous HCl. Acetyl chloride is also frequently used for this purpose.

[181]Boesch, H.; Cesco-Cancian, S.; Hecker, L. R.; Hoekstra, W. J.; Justus, M.; Maryanoff, C. A.; Scott, L.; Shah, R. D.; Solms, G.; Sorgi, K. L.; Stefanick, S. M.; Turnheer, U.; Villani, F. J.; Walker, D. G. *Org. Process Res. Dev.* **2001**, *5*, 23–27.

[182]Furuta, K.; Gao, Q.-z.; Yamamoto, H. *Org. Synth.* **1995**, *72*, 86–94.

[183]Gurjar, M. K.; Murugaiah, A. M. S.; Reddy, D. S.; Chorghade, M. S. *Org. Process Res. Dev.* **2003**, *7*, 309–312.

The Fisher esterification protocols are effective when primary or secondary alcohols are used. Tertiary alcohols typically undergo elimination under these conditions, and hence esters of tertiary alcohols are seldom prepared using this method (see 2.29.2). While this procedure is operationally simple, the need for strong acids is a limitation when acid-labile functionality is present in the molecule(s).

### 2.29.2    Widmer's Method for the Synthesis of *t*-Butyl Esters

A particularly convenient method for the formation of *t*-butyl esters is Widmer's method, wherein a carboxylic acid is treated with *N,N*-dimethylformamide di-*tert*-butyl acetal. While the original procedure utilized benzene as the reaction solvent,[184] toluene has been found to be an acceptable alternative.[185]

### 2.29.3    Via Acid Chlorides

Carboxylic acids can easily be transformed to the corresponding esters via acid chlorides (see Section 2.24 for the conversion of carboxylic acids to acid chlorides). As in the case of amidation reactions, esterifications via acid chlorides generate an equivalent of HCl, and hence require an equivalent of base in order to proceed to completion. These reactions are generally catalyzed by the addition of small quantities of DMAP.[182]

[184]Widmer, U. *Synthesis* **1983**, 135–136.

[185]Tagat, J. R.; McCombie, S. W.; Nazareno, D. V.; Boyle, C. D.; Kozlowski, J. A.; Chackalamannil, S.; Josien, H.; Wang, Y.; Zhou, G. *J. Org. Chem.* **2002**, *67*, 1171–1177.

### 2.29.4   Via Acyl Imidazoles (Imidazolides)

Carboxylic acids can be converted to the corresponding esters through a two-step sequence that proceeds via an acyl imidazole. In a typical reaction, the acid is activated with *N,N*-carbonyldiimidazole (CDI) to afford the acyl imidazole, which is then treated with an alcohol to furnish the ester.[186] Since the byproduct in this transformation is imidazole, no additional base is required.

### 2.29.5   Using Carbodiimides

The use of carbodiimides is a convenient method for the synthesis of esters from acids. The most commonly used carbodiimides are DCC[187] and EDC.[188] Under these conditions, sterically demanding substrates as well as weakly nucleophilic alcohols such as phenols react to produce the corresponding esters. However, since most of the urea byproducts arising from the carbodiimides are separable only by chromatography, this method is less favored than the ones listed above.

DMTr = Dimethoxytrityl

[186]Couturier, M.; Le, T. *Org. Process Res. Dev.* **2006**, *10*, 534–538.

[187]Bringmann, G.; Breuning, M.; Henschel, P.; Hinrichs, J. *Org. Synth.* **2003**, *79*, 72–83.

[188]de Koning, M. C.; Ghisaidoobe, A. B. T.; Duynstee, H. I.; Ten Kortenaar, P. B. W.; Filippov, D. V.; van der Marel, G. A. *Org. Process Res. Dev.* **2006**, *10*, 1238–1245.

### 2.29.6  Via Anhydrides

As discussed in the amidation Section (2.26.4), symmetrical anhydrides are rarely used as the acyl component in ester formation reactions because only one half of the anhydride gets incorporated into the product. Alcoholysis of symmetrical anhydrides is typically performed only when the anhydride is inexpensive and readily available (e.g., acetic anhydride). This kind of transformation is generally useful for the protection of alcohols as esters.

Most commonly, the alcohol is treated with an anhydride in the presence of a base to afford the ester. A catalytic amount of DMAP is often added to enhance the reaction rate. Two representative examples using priopionic anhydride[189] and *iso*-butyric anhydride[190] are given below.

Several Lewis acid catalysts have been shown to promote the alcoholysis of anhydrides. For example, it has been demonstrated that the addition of very small quantities (< 0.01 equiv) of $Bi(OTf)_3$ is an effective catalyst.[191] Under these conditions, sterically demanding alcohols could be acylated efficiently. Similarly, relatively less reactive anhydrides such as pivalic anhydride also underwent efficient alcoholysis.

[189]Sofiyev, V.; Navarro, G.; Trauner, D. *Org. Lett.* **2008**, *10*, 149–152.

[190]Barnes, D. M.; Christesen, A. C.; Engstrom, K. M.; Haight, A. R.; Hsu, M. C.; Lee, E. C.; Peterson, M. J.; Plata, D. J.; Raje, P. S.; Stoner, E. J.; Tedrow, J. S.; Wagaw, S. *Org. Process Res. Dev.* **2006**, *10*, 803–807.

[191]Orita, A.; Tanahashi, C.; Kakuda, A.; Otera, J. *J. Org. Chem.* **2001**, *66*, 8926–8934.

Symmetrical diols can be selectively mono-acylated with anhydrides using catalytic Yb (OTf)$_3$.[192] For instance, *meso*-hydrobenzoin was selectively mono-acylated by treatment with acetic anhydride in the presence of 0.1 mol% Yb(OTf)$_3$.

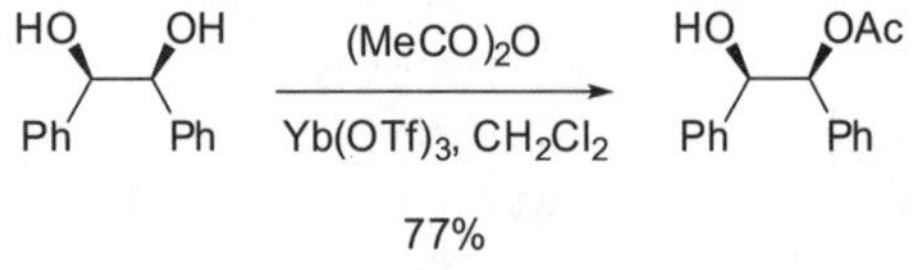

77%

## 2.30  HYDROLYSIS OF AMIDES

Hydrolysis of amides leads to the corresponding carboxylic acid and the amine. However, this reaction is seldom used in multi-step syntheses due to the harsh conditions typically required. Hydrolysis of amides is generally accomplished under acidic, basic, or oxidative conditions. The choice of conditions for a particular molecule is dictated largely by the nature of functional groups contained therein. When the hydrolysis is carried out under acidic or basic conditions, the desired product (either the amine or the carboxylic acid component) may be conveniently isolated by a pH-controlled extractive workup.

### 2.30.1  Under Acidic Conditions

Heating an amide under strongly acid conditions promotes hydrolysis to the carboxylic acid. In the following example, the chiral auxiliary phenylalaninol was removed by heating the amide in sulfuric acid.[193] Note that in this example, stereochemistry at the α center was unaffected.

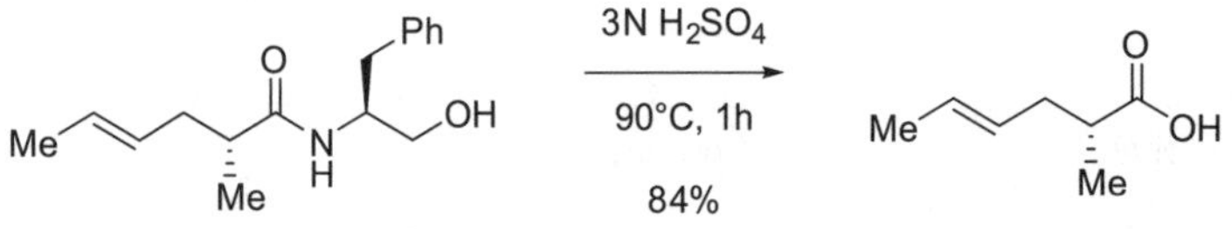

### 2.30.2  Under Basic Conditions

In the example shown below, both of the amides are hydrolyzed under basic conditions to give the corresponding amino acid.[194] The reaction mixture was heated at 100°C to effect hydrolysis.

[192]Clarke, P. A.; Kayaleh, N. E.; Smith, M. A.; Baker, J. R.; Bird, S. J.; Chan, C. *J. Org. Chem.* **2002**, *67*, 5226–5231.
[193]Rao, A. V. R.; Dhar, T. G. M.; Bose, D. S.; Chakraborty, T. K.; Gurjar, M. K. *Tetrahedron* **1989**, *45*, 7361–7370.
[194]Vanderplas, B. C.; DeVries, K. M.; Fox, D. E.; Raggon, J. W.; Snyder, M. W.; Urban, F. J. *Org. Process Res. Dev.* **2004**, *8*, 583–586.

In the following example, removal of the silyl-protecting groups and chiral auxiliary was accomplished in a single pot via sequential treatment with TBAF and $n$-Bu$_4$NOH.[195] The hydrolysis reaction was complete within 24 hours in refluxing $t$-butyl alcohol-water.

While hydroxide-based methods are commonly employed for amide hydrolysis, the high temperatures required could lead to complications when thermally labile functionality is present in the molecule. A particularly benign alternative developed by Gassman and coworkers involves treatment of the amide with 6 equivalents of potassium $t$-butoxide and 2 equivalents of water at room temperature.[196] The reaction of potassium $t$-butoxide with water leads to finely divided anhydrous hydroxide, which is poorly solvated and strongly nucleophilic. In the example below, the hindered amide was quantitatively hydrolyzed under these conditions to the corresponding amine (desired product).[197]

### 2.30.3   Under Oxidative Conditions

Sodium peroxide has also been used to hydrolyze amides.[198] In the following examples, treatment of the di-amide with 1.1 equivalent of Na$_2$O$_2$ at 50°C led to selective deamidation at the less hindered position to afford the mono-acid. Complete hydrolytic conversion to the diacid was achieved using an excess of Na$_2$O$_2$ at higher temperatures.[199] While these conditions seem mild, one needs to exercise caution while working with Na$_2$O$_2$, especially at elevated temperatures.

[195]Su, Q.; Dakin, L. A.; Panek, J. S. *J. Org. Chem.* **2007**, *72*, 2–24.

[196]Gassman, P. G.; Hodgson, P. K. G.; Balchunis, R. J. *J. Am. Chem. Soc.* **1976**, *98*, 1275–1276.

[197]Li, H.-Y.; DeLucca, I.; Boswell, G. A.; Billheimer, J. T.; Drummond, S.; Gillies, P. J.; Robinson, C. *Bioorg. Med. Chem.* **1997**, *5*, 1345–1361.

[198]Vaughn, H. L.; Robbins, M. D. *J. Org. Chem.* **1975**, *40*, 1187–1189.

[199]Dandekar, S. A.; Greenwood, S. N.; Greenwood, T. D.; Mabic, S.; Merola, J. S.; Tanko, J. M.; Wolfe, J. F. *J. Org. Chem.* **1999**, *64*, 1543–1553.

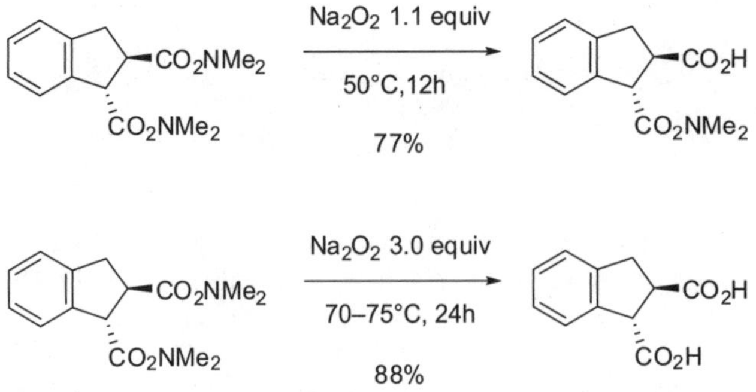

### 2.30.4 Miscellaneous Methods

Amides can also be converted to the corresponding carboxylic acids via the White rearrangement. In the following scheme, the amide was selectively hydrolyzed to the corresponding acid in the presence of the methyl ester functionality. This transformation was accomplished by a conversion of the amide to the $N$-nitrosoamide, by treatment with $N_2O_4$. This was then heated in dioxane to afford a mixture of the desired acid and its 1-phenethyl ester. Hydrogenation of the crude reaction mixture converted the ester to the acid, which was isolated as the 1-adamantamine salt in 96% overall yield.[200]

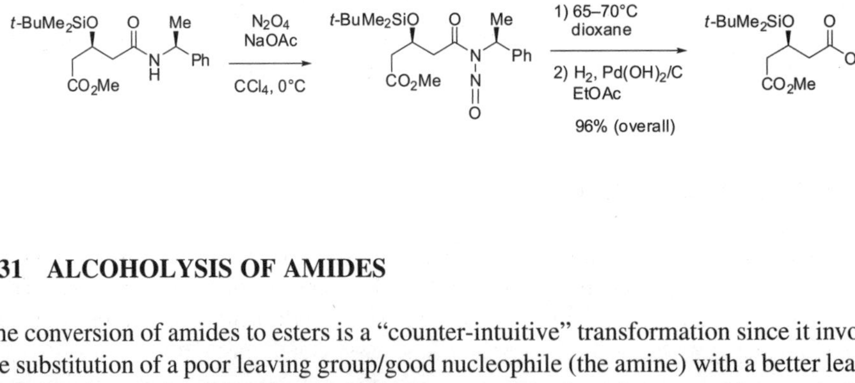

### 2.31 ALCOHOLYSIS OF AMIDES

The conversion of amides to esters is a "counter-intuitive" transformation since it involves the substitution of a poor leaving group/good nucleophile (the amine) with a better leaving group/weaker nucleophile (the alcohol). As a general rule, primary amides are easier to cleave than secondary or tertiary amides.

Alcoholysis of amides is most commonly carried out under acidic conditions, wherein protonation of the amide nitrogen enhances its leaving group ability. If alcohol addition is intramolecular (to yield a lactone), the reaction can be performed under relatively mild conditions as exemplified below.[201]

[200]Karanewsky, D. S.; Malley, M. F.; Gougoutas, J. Z. *J. Org. Chem.* **1991**, *56*, 3744–3747.
[201]Kolla, N.; Elati, C. R.; Arunagiri, M.; Gangula, S.; Vankawala, P. J.; Anjaneyulu, Y.; Bhattacharya, A.; Venkatraman, S.; Mathad, V. T. *Org. Process Res. Dev.* **2007**, *11*, 455–457.

More forcing conditions (i.e., strong acids and excess alcohol) are needed for intermolecular reactions. In most cases, the acid employed is generally strong enough to afford the ester via a Fisher esterification pathway (Section 2.29.1) even if the corresponding carboxylic acid is formed as an intermediate.[202]

Secondary and tertiary amides can be converted to esters by treatment with a stoichiometric amount of triflic anhydride and pyridine. The iminum triflate initially formed is converted to the ester upon addition of the alcohol.[203]

## 2.32  HYDROLYSIS OF ESTERS

The ester functionality is commonly found in natural products and key building blocks. In multi-step syntheses, the ester group serves as a "masked" carboxylic acid moiety. One of the most frequently used methods to convert an ester to a carboxylic acid is via hydrolysis. Ester hydrolysis is generally performed under acidic or basic conditions.

### 2.32.1  Under Basic Conditions

The product of base-promoted hydrolysis of esters is the carboxylate salt, which is then converted to the corresponding acid via an acidic workup. Generally, alkali metal hydroxides are the bases of choice to effect this reaction.[204]

[202]Imamura, S.; Ichikawa, T.; Nishikawa, Y.; Kanzaki, N.; Takashima, K.; Niwa, S.; Iizawa, Y.; Baba, M.; Sugihara, Y. *J. Med. Chem.* **2006**, *49*, 2784–2793.

[203]Charette, A. B.; Chua, P. *Synlett* **1998**, 163–165.

[204]Deussen, H.-J.; Jeppesen, L.; Schaerer, N.; Junager, F.; Bentzen, B.; Weber, B.; Weil, V.; Mozer, S. J.; Sauerberg, P. *Org. Process Res. Dev.* **2004**, *8*, 363–371.

In an interesting article, Chan and coworkers reported the dependence of relative rates of hydrolysis of alkyl carboxylate and sulfonate esters on the pH of the reaction mixture.[205] Carboxylate esters are hydrolyzed slowly at lower pH (pH 5–6), while the rate of hydrolysis increases substantially with increasing hydroxide concentration, that is, pH. Sulfonate esters, on the other hand, exhibit a higher but constant rate of hydrolysis up to ~pH 12. The pH dependence on rate of hydrolysis can be exploited to enable selective hydrolysis of sulfonate esters in the presence of carboxylate esters.

### 2.32.2   Under Acidic Conditions

In general, esters that are capable of forming stabilized carbocations (benzhydryl esters) or that readily eliminate to form olefins (*t*-butyl esters) are cleaved under acidic conditions. In the following example, treatment of the *t*-butyl ester with aqueous HCl leads to hydrolysis of both the ester and imine functionalities to furnish the amino acid as the HCl salt in excellent yield.[206]

This transformation may also be accomplished under anhydrous conditions using methanesulfonic acid or trifluoroacetic acid. For instance, in the synthesis of aztreonam, the benhydryl ester was selectively hydrolyzed in the presence of other sensitive functionality by treatment with methanesulfonic acid and anisole (used to trap the benzhydryl cation).[207]

[205]Chan, L. C.; Cox, B. G.; Sinclair, R. S. *Org. Process Res. Dev.* **2008**, *12*, 213–217.

[206]Patterson, D. E.; Xie, S.; Jones, L. A.; Osterhout, M. H.; Henry, C. G.; Roper, T. D. *Org. Process Res. Dev.* **2007**, *11*, 624–627.

[207]Singh, J.; Denzel, T. W.; Fox, R.; Kissick, T. P.; Herter, R.; Wurdinger, J.; Schierling, P.; Papaioannou, C. G.; Moniot, J. L.; Mueller, R. H.; Cimarusti, C. M. *Org. Process Res. Dev.* **2002**, *6*, 863–868.

84%

## 2.33  TRANSESTERIFICATION

Transesterification is conceptually similar to ester hydrolysis. The main difference is that the nucleophile in a transesterification is an alcohol instead of water. The transformation is an equilibrium-driven reaction, and can thus be impacted by stoichiometry of the two competing alcohols. Acid or base catalysts are commonly employed to increase the rate of transesterification.

This reaction is particularly useful for the conversion of low molecular weight esters to their higher homologues, wherein the lower boiling alcohol is removed from the reaction mixture by distillation. In the following example, the methyl ester was converted to the *n*-butyl ester by treatment with 1-butanol in the presence of catalytic Ti(*i*-PrO)$_4$.[208]

91%

The conversion of methyl *N*-(2,6-dimethylphenyl)alaninate to the corresponding 2-methoxyethyl ester was carried out by heating it in the presence of DBU and 2-methoxyethanol at 135°C. It is noteworthy that the reaction was incomplete when DMAP, a commonly employed acylation catalyst, was used.[209]

96%

The next two examples illustrate contrasting conditions to accomplish transesterification reactions. In the first case, the hydroxy ester was treated with *p*-TsOH to furnish the desired lactone.[210] In the second case, the reverse reaction, that is, conversion of the lactone to the

[208] Atkins, R. J.; Banks, A.; Bellingham, R. K.; Breen, G. F.; Carey, J. S.; Etridge, S. K.; Hayes, J. F.; Hussain, N.; Morgan, D. O.; Oxley, P.; Passey, S. C.; Walsgrove, T. C.; Wells, A. S. *Org. Process Res. Dev.* **2003**, *7*, 663–675.
[209] Park, O.-J.; Lee, S.-H.; Park, T.-Y.; Chung, W.-G.; Lee, S.-W. *Org. Process Res. Dev.* **2006**, *10*, 588–591.
[210] Mackey, S. S.; Wu, H.; Matison, M. E.; Goble, M. *Org. Process Res. Dev.* **2005**, *9*, 174–178.

hydroxy ester was desired. This was realized by reaction with ethanol in the presence of sodium ethoxide.[211]

76%

99%

## 2.34   ALKYL THIOL ADDITION TO ESTERS

Thioesters are more reactive than esters. The larger sulfur atom allows for less delocalization of charge across the carbonyl in a thioester as compared to an ester. As a result, thioesters are often generated to enable milder reaction conditions when the ester variant proves impervious to reaction. Key to these reactions are effective formation of thiolate ion (generally alkyl lithium reagents are used as bases), good purging or "positive pressure" systems to avoid thiol oxidation side products, and post-reaction scrubbers or low-reaction temperature to avoid the strong smell associated with some of these reagents. Many thiols can be smelled at the parts per million (ppm) level.

Pesti and coworkers have developed a method using aluminium trichloride and a silylated thiol for transestification.[212] They successfully performed this potentially problematic reaction on large scale with a few key procedural adaptations. In the first sequence, the silylated thiol is formed by cold, slow addition of chlorotrimethylsilane to a THF solution of *n*-butyllithium and propanethiol. After addition, the reaction was sparged with nitrogen to remove the butane side product and then diluted with heptane. The resulting slurry was filtered into a solution of the ester. To this solution was added aluminium trichloride and the reaction was heated. After reaction completion and standard workup, the thioester was isolated in excellent yield.

91%

[211]Raw, A. S.; Jang, E. B. *Tetrahedron* **2000**, *56*, 3285–3290.

[212]Pesti, J. A.; Yin, J.; Zhang, L.-h.; Anzalone, L.; Waltermire, R. E.; Ma, P.; Gorko, E.; Confalone, P. N.; Fortunak, J.; Silverman, C.; Blackwell, O. J.; Chung, J. C.; Hrytsak, M. D.; Cooke, M.; Powell, L.; Ray, C. *Org. Process Res. Dev.* **2004**, *8*, 22–27.

Warm and coworkers utilized a lactone to thiolactone conversion, which proceeded well for the synthesis of biotin.[213] Key to this reaction was the avoidance of oxidation side products common with reactions of thiols, through the use of rigorous nitrogen purging and dihydroquinone. With use of dihydroquinone as an antioxidant, careful attention to the quality of nitrogen used in multiple purges of the reaction, and use of a small positive pressure on the reactor, the reaction proceeded smoothly.

When Damon and Coppola had difficulty with removal of a hindered oxazolidone auxiliary, they found that thiolate addition could cleanly remove the chiral auxiliary.[214] In their hands, other methods for chiral auxiliary removal gave various byproducts that involved the unraveling of the oxazolidinone. Using a mixture of 2 equivalents of benzyl mercaptan and 1.5 equivalents of *n*-butyllithium they saw clean conversion to the desired thioester with excellent recovery of the chiral auxiliary.

## 2.35  ADDITION OF ORGANOMETALLIC REAGENTS TO CARBOXYLIC ACID DERIVATIVES

Generally, when an organolithium or organomagnesium reagent is added to an ester, controlled mono-addition to result in the ketone is difficult; frequently over addition to form a tertiary alcohol takes place. This was the case in Alibe's grandisol synthesis.[215] Initial attempts for single addition of methyllithium to give the methyl ketone from their cyclobutane lactone met with low yield (13%). The authors instead incorporated the over addition product in their route and were able to prepare grandisol in this manner. Using 5.8 equivalents of methyllithium in THF at −78°C, the diol was formed in excellent yield.

[213]Warm, A.; Naughton, A. B.; Saikali, E. A. *Org. Process Res. Dev.* **2003**, *7*, 272–284.
[214]Damon, R. E.; Coppola, G. M. *Tetrahedron Lett.* **1990**, *31*, 2849–2852.
[215]Alibes, R.; Bourdelande, J. L.; Font, J.; Parella, T. *Tetrahedron* **1996**, *52*, 1279–1292.

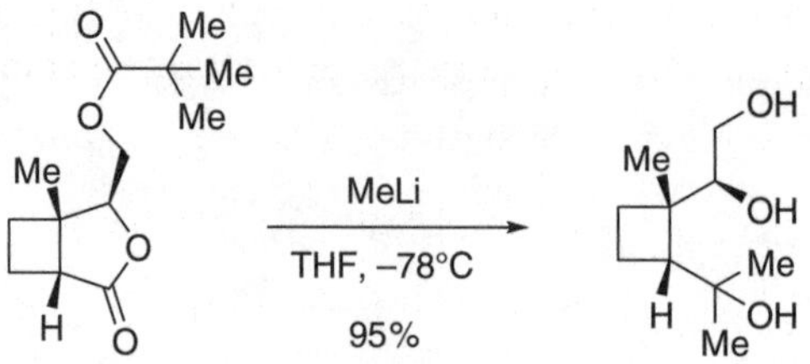

Similarly, in the example below, treatment of the diester with excess of the organo-magnesium reagent led to the product diol in excellent yield.[216]

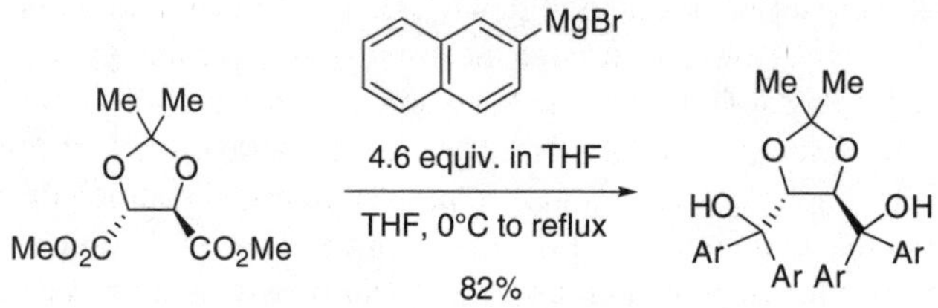

Mono-addition of organolithium or organomagnesium reagents to esters can be achieved by strict attention to reaction temperature, slow dosing rate of the organometallic reagent and cold quench of the resultant mixture, as seen in a few examples below. Mulzer's second generation synthesis of epothilone[217] utilized a mono-addition of methyllithium to the lactone shown below, and the hemi-ketal/hydroxyl ketone equilibrium mixture was carried forward through a Wittig olefination reaction to give the thiazole product in excellent yield.

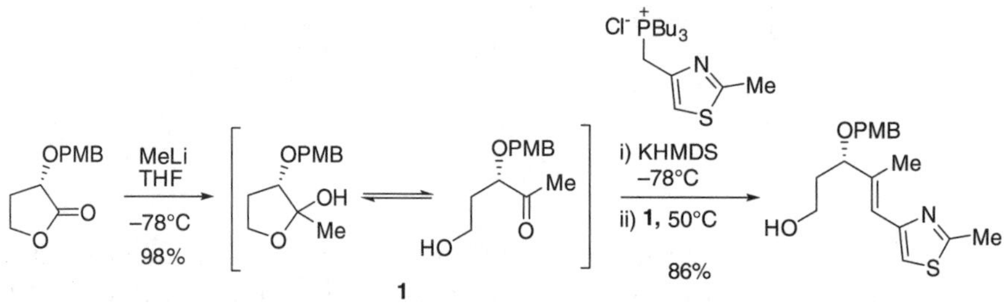

Another example from the Bach labs is shown below.[218] The addition of the lithiated thiazole proceeds well with standard conditions at −78°C in diethyl ether.

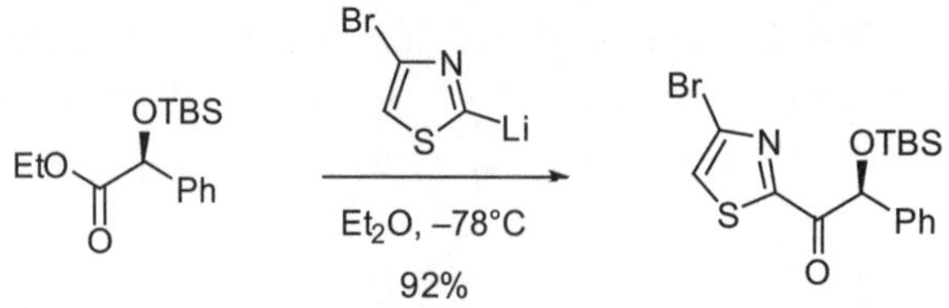

[216]Beck, A. K.; Gysi, P.; La Vecchia, L.; Seebach, D. *Org. Synth.* **1999**, *76*, 12–22.
[217]Mulzer, J.; Mantoulidis, A.; Oehler, E. *J. Org. Chem.* **2000**, *65*, 7456–7467.
[218]Delgado, O.; Heckmann, G.; Mueller, H. M.; Bach, T. *J. Org. Chem.* **2006**, *71*, 4599–4608.

In the following example, selective monoaddition of cyclohexylmagnesium bromide to the pyruvate moiety was achieved by carrying out the reaction at low temperature.[219]

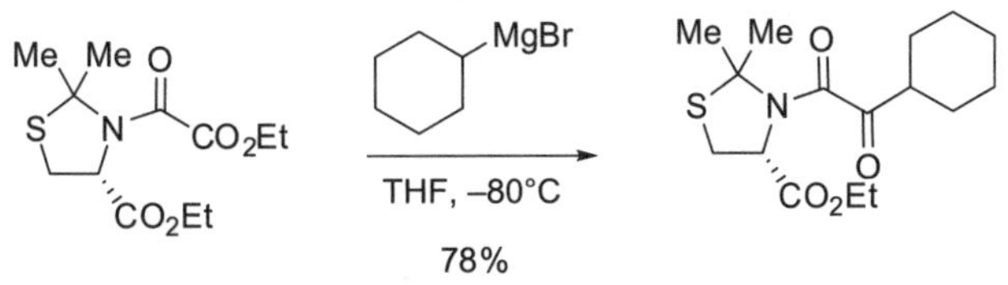

THF, −80°C

78%

A useful method to ensure mono-addition of either an organolithium or organomagnesium reagent to form the ketone is through the use of a Weinreb amide. In these substrates, after addition of 1 equivalent of the organometallic reagent, the substrate is locked into a tetrahedral intermediate via coordination of the metal and the pendant N-O ether of the Weinreb amide. This constrains the system to prevent a second addition of the nucleophile and leads to good reaction control.

An excellent example of the chemoselective addition that can be achieved is shown below. After lithium halogen exchange on the bromopyridine the reaction proceeds with addition to the Weinreb amide leaving the α-substituted ketone intact for further elaboration.[220]

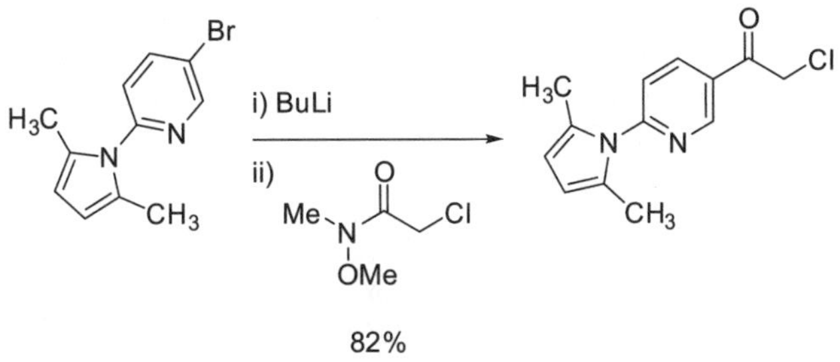

82%

It is important to note that with enolizable substrates, organomagnesium reagents are generally preferred over their alkyllithium counterparts. Urban and Jasys utilized this strategy in their work towards an aspartyl protease inhibitor.[221] In the presence of the Boc-protected amine and magnesium metal, the substrate was first deprotonated with either methyl or benzylmagnesium bromide until trace methyl or benzyl ketone was detected, then the bromo dioxolane was added to form the the corresponding organomagnesium reagent *in situ*. On scale, methylmagnesium bromide was substituted with the benzyl variant to give toluene rather than methane as a byproduct. It was also noted that in contrast to MeMgBr, the BnMgBr did not add to the Weinreb amide.

[219]Zhao, L.; Huang, W.; Liu, H.; Wang, L.; Zhong, W.; Xiao, J.; Hu, Y.; Li, S. *J. Med. Chem.* **2006**, *49*, 4059–4071.
[220]Scott, R. W.; Fox, D. E.; Wong, J. W.; Burns, M. P. *Org. Process Res. Dev.* **2004**, *8*, 587–592.
[221]Urban, F. J.; Jasys, V. J. *Org. Process Res. Dev.* **2004**, *8*, 169–175.

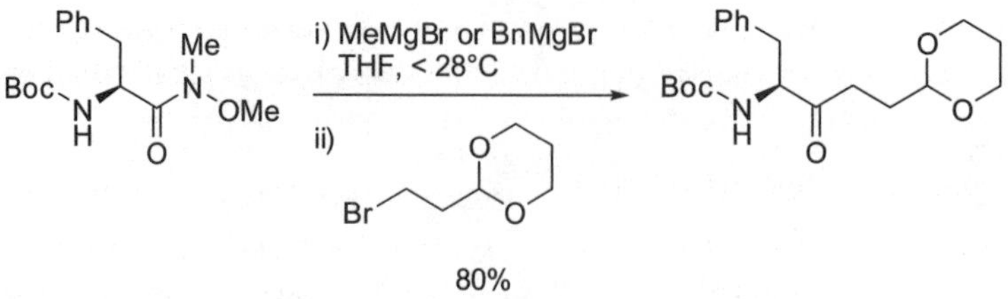

80%

Key in the synthesis of the selective estrogen receptor modulator raloxifen by Schmid and coworkers from Lilly is the addition shown below.[222] Organomagnesium addition (as described by Grese)[223] to the Weinreb amide furnished the desired ketone in good yield with no observed epimerization of the stereocenter.

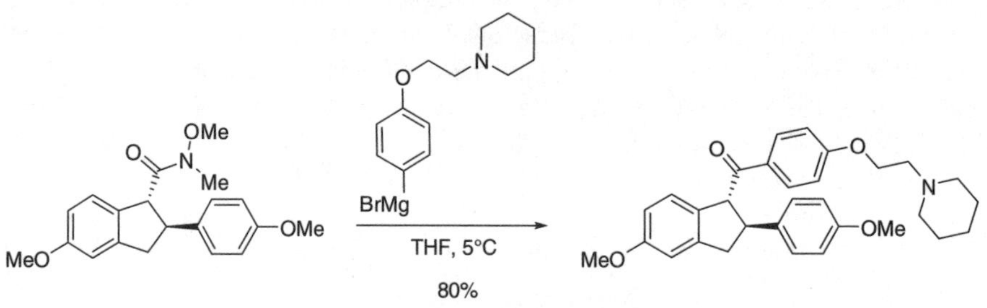

80%

Gallou and coworkers demonstrated the addition of organometallics to traditionally less reactive substrates via the use of lithium triarylmagnesiate complexes.[224] A key feature of this process is the non-cryogenic conditions employed. The complex is formed by initial reaction of isopropylmagnesium chloride followed by addition of $n$-butyllithium. Addition to DMF forms the homologated aldehyde.

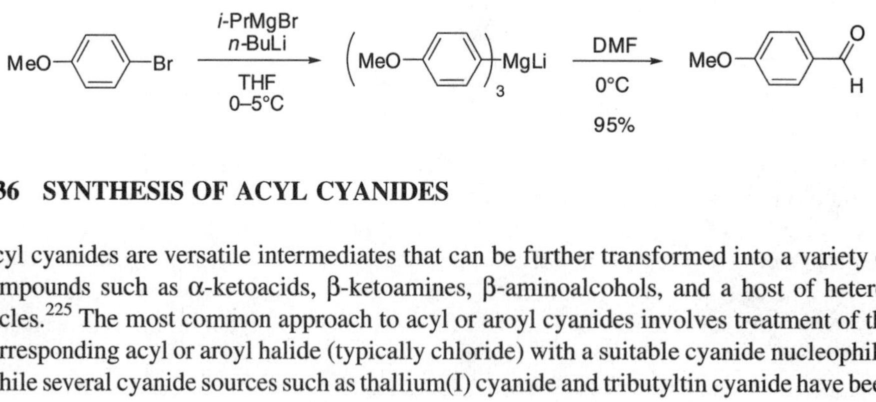

95%

## 2.36   SYNTHESIS OF ACYL CYANIDES

Acyl cyanides are versatile intermediates that can be further transformed into a variety of compounds such as α-ketoacids, β-ketoamines, β-aminoalcohols, and a host of hetero-cycles.[225] The most common approach to acyl or aroyl cyanides involves treatment of the corresponding acyl or aroyl halide (typically chloride) with a suitable cyanide nucleophile. While several cyanide sources such as thallium(I) cyanide and tributyltin cyanide have been

[222]Schmid, C. R.; Glasebrook, A. L.; Misner, J. W.; Stephenson, G. A. *Bioorg. Med. Chem. Lett.* **1999**, *9*, 1137–1140.

[223]Grese, T. A.; Pennington, L. D.; Sluka, J. P.; Adrian, M. D.; Cole, H. W.; Fuson, T. R.; Magee, D. E.; Phillips, D. L.; Rowley, E. R.; Shetler, P. K.; Short, L. L.; Venugopalan, M.; Yang, N. N.; Sato, M.; Glasebrook, A. L.; Bryant, H. U. *J. Med. Chem.* **1998**, *41*, 1272–1283.

[224]Gallou, F.; Haenggi, R.; Hirt, H.; Marterer, W.; Schaefer, F.; Seeger-Weibel, M. *Tetrahedron Lett.* **2008**, *49*, 5024–5027.

[225]Huenig, S.; Schaller, R. *Angew. Chem.* **1982**, *94*, 1–15.

used for this transformation, the most practical reagents from a preparative standpoint are trimethylsilyl cyanide, copper(I) cyanide and, to a lesser extent, alkali metal cyanides.

### 2.36.1    Using Trimethylsilyl Cyanide

Trimethylsilyl cyanide is one of the most convenient sources of cyanide since it is a relatively stable and non-volatile liquid (bp = 119°C). In most cases, the reaction to synthesize an acyl cyanide involves treatment of the acyl halide with neat TMSCN at reflux.[226,227] This method is particularly attractive for the synthesis of acyl cyanides (i.e., cyanide derivatives of aliphatic carboxylic acids).

In some cases, the addition of trace amounts of aluminum chloride catalyzes the reaction of acid chlorides with TMSCN.[228]

### 2.36.2    Using Copper(I) Cyanide

Copper(I) cyanide is one of the most extensively used reagents to effect this transformation. The example below illustrates the conversion of 3-methyl-2-butenoyl chloride to the corresponding cyanide by treatment with copper(I) cyanide in acetonitrile.[229]

[226]Bischofberger, N.; Waldmann, H.; Saito, T.; Simon, E. S.; Lees, W.; Bednarski, M. D.; Whitesides, G. M. *J. Org. Chem.* **1988**, *53*, 3457–3465.

[227]Kaila, N.; Janz, K.; Huang, A.; Moretto, A.; DeBernardo, S.; Bedard, P. W.; Tam, S.; Clerin, V.; Keith, J. C., Jr.; Tsao, D. H. H.; Sushkova, N.; Shaw, G. D.; Camphausen, R. T.; Schaub, R. G.; Wang, Q. *J. Med. Chem.* **2007**, *50*, 40–64.

[228]Bartmann, E.; Krause, J. *J. Fluorine Chem.* **1993**, *61*, 117–122.

[229]Jung, M. E.; Min, S.-J. *J. Am. Chem. Soc.* **2005**, *127*, 10834–10835.

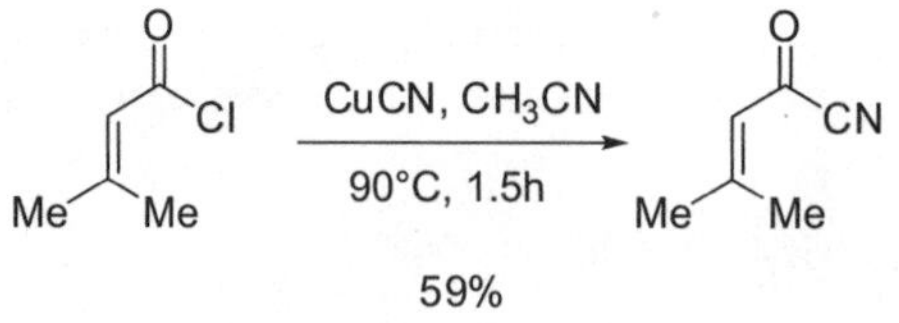

59%

In some instances, the transformation is sluggish with acyl chlorides. This problem has been circumvented by the addition of an iodide source (to convert the acyl chloride to the acyl iodide *in situ*) prior to treatment with CuCN.[230]

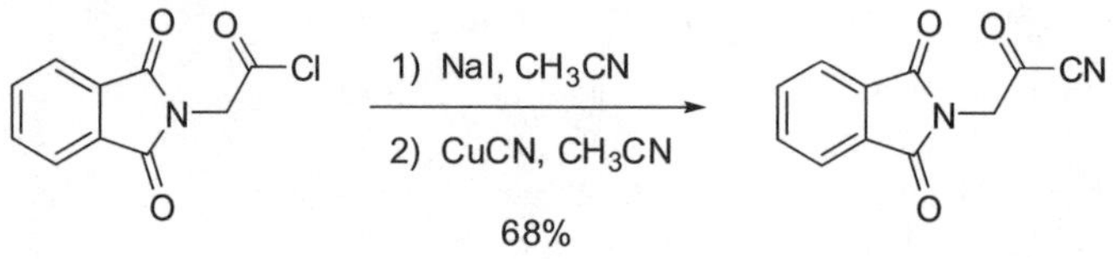

68%

CuCN with $P_2O_5$ has also been used successfully to synthesize acyl cyanides from acyl chlorides.[228]

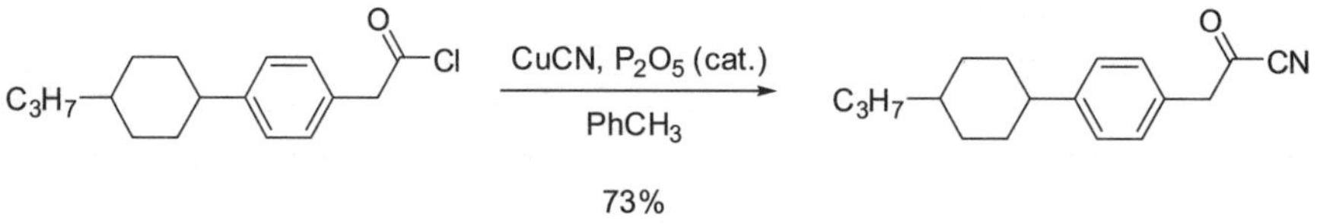

73%

### 2.36.3 Miscellaneous Methods

In an interesting article,[231] potassium hexacyanoferrate(II) has been used as the cyanide source in conjunction with catalytic quantities of AgI, PEG400 and KI for the conversion of aroyl chlorides to aroyl cyanides (there is no mention of the utility of this reaction for the synthesis of acyl cyanides). AgI is believed to enhance the reaction rate by promoting the release of cyanide ion from $K_4[Fe(CN)_6]$. KI presumably increases the solubility of AgI and increases the concentration of iodide in solution. It is also conceivable that iodide ion leads to the formation of the acyl iodide *in situ*. PEG400 is believed to function as a phase transfer catalyst.

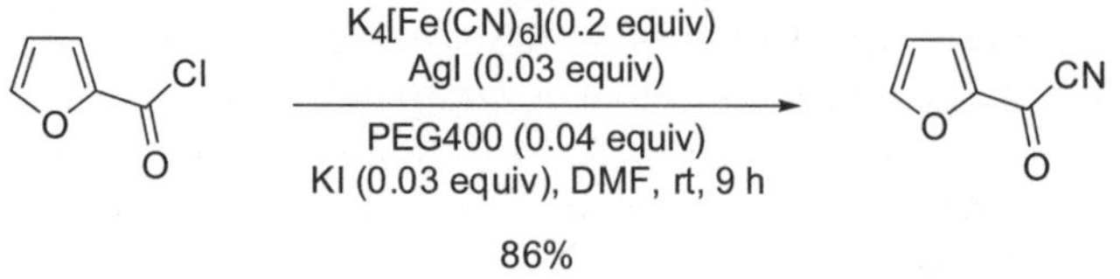

86%

[230]Nozaki, K.; Sato, N.; Takaya, H. *Tetrahedron: Asymmetry* **1993**, *4*, 2179–2182.
[231]Li, Z.; Shi, S.; Yang, J. *Synlett* **2006**, 2495–2497.

Reactions using alkali metal cyanides are usually catalyzed by the addition of a phase-transfer catalyst.[232]

60%

## 2.37   THE RITTER REACTION

The Ritter reaction involves the nucleophilic attack of the nitrogen atom of a nitrile onto a carbocation, followed by the addition of water and subsequent tautomerization to afford an *N*-alkyl amide. This reaction is particularly useful for the preparation of amides bearing secondary or tertiary alkyl groups on the nitrogen.

In a classic Ritter reaction, the carbocation is generated from the corresponding alcohol via treatment with a strong acid. Over the years, the scope of the reaction has been widened to include electrophiles such as epoxides, and a wide range of mild Lewis acids as promoters.

The most common procedure for this reaction involves the addition of acid (typically $H_2SO_4$) to a mixture of the nitrile and the alcohol in an appropriate solvent. In the following example, the *t*-butyl amide was obtained in excellent yield from the nitrile (with minimal hydrolysis of the enol ether functionality).[233]

78%

While this addition protocol can work in most cases, a violent reaction was observed when acrylonitrile and cyclohexanol were used. This was presumably due to the long induction period arising out of the low sulfuric acid to nitrile:alcohol ratio until all the sulfuric acid had been added. The problem was circumvented by maintaining a high and constant sulfuric acid to cyclohexanol ratio throughout the addition. The acrylamide has also been prepared using cyclohexene and acrylonitrile using an analogous procedure.[234]

[232]Koenig, K. E.; Weber, W. P. *Tetrahedron Lett.* **1974**, 2275–2278.
[233]Effenberger, F.; Jaeger, J. *J. Org. Chem.* **1997**, *62*, 3867–3873.
[234]Chang, S.-J. *Org. Process Res. Dev.* **1999**, *3*, 232–234.

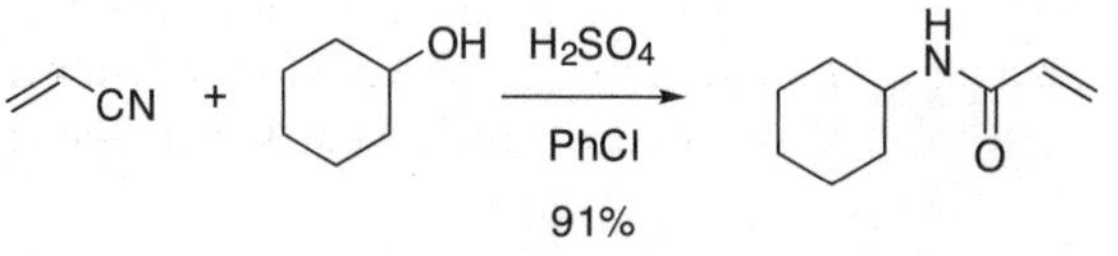

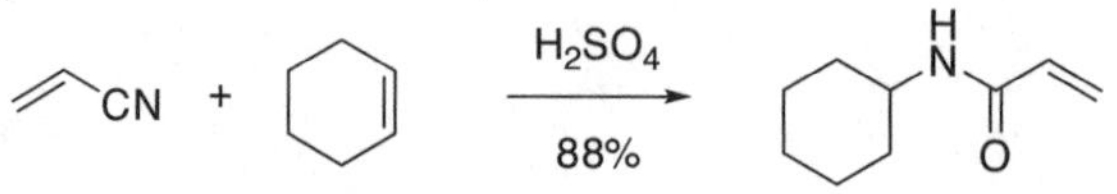

91%

88%

An interesting modification of the classical Ritter reaction conditions utilizes *t*-butyl acetate as the source of the *t*-butyl cation, and a *catalytic* amount of sulfuric acid.[235]

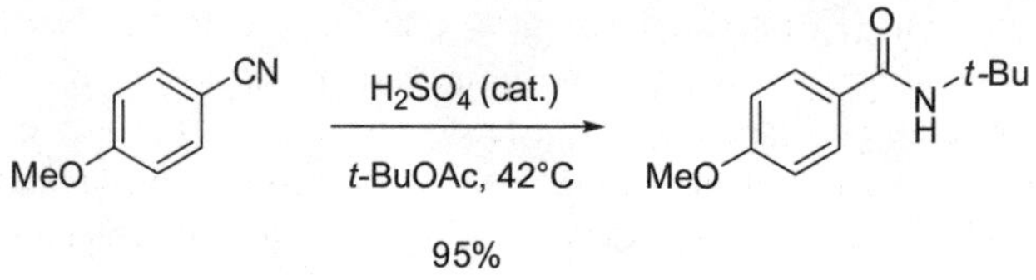

95%

One of the best applications of the Ritter reaction is in the synthesis of optically pure *cis*-1-amino-2-alcohols. The optically pure epoxides or diols were synthesized from the corresponding olefins via an asymmetric epoxidation or dihydroxylation reaction, and were subjected to a Ritter reaction to furnish the desired amino alcohol in excellent ee's.[236] The *cis* diasteromer arises from preferential formation of the intermediate *cis*-oxazoline (95:5 *cis/trans*) prior to hydrolysis.

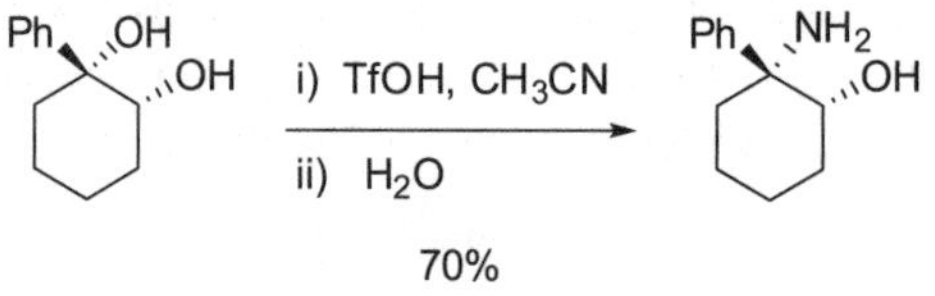

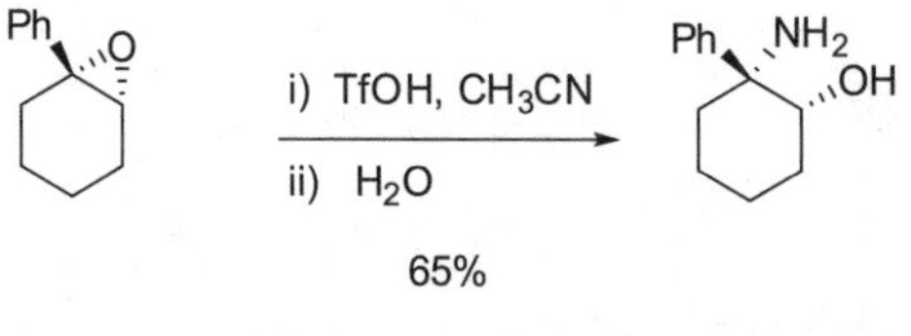

70%

65%

The Ritter reaction can be used to directly synthesize formamides from alcohols by using TMSCN as the nitrile source.[237]

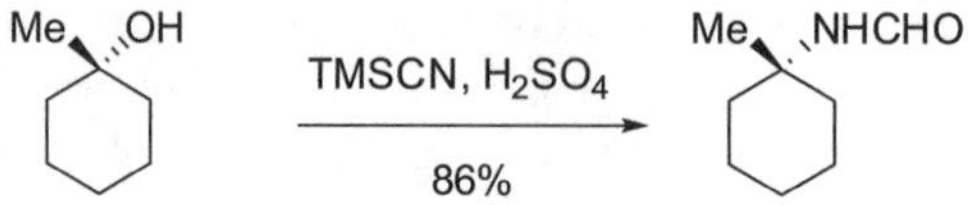

86%

[235]Reddy, K. L. *Tetrahedron Lett.* **2003**, *44*, 1453–1455.

[236]Senanayake, C. H.; Larsen, R. D.; DiMichele, L. M.; Liu, J.; Toma, P. H.; Ball, R. G.; Verhoeven, T. R.; Reider, P. J. *Tetrahedron: Asymmetry* **1996**, *7*, 1501–1506.

[237]Chen, H. G.; Goel, O. P.; Knobelsdorf, J. *Tetrahedron Lett.* **1996**, *37*, 8129–8132.

Functionalized nitriles have also been used in the Ritter reaction. For example, Ritter reaction of tertiary alcohols using chloroacetonitrile leads to N-*tert*-alkylchloroacetamides, which can then be further functionalized, or cleaved with thiourea to produce *tert*-alkylamines.[238]

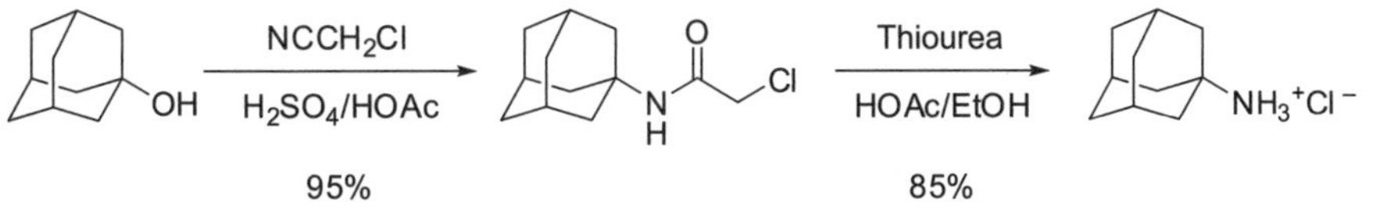

95%                                    85%

## 2.38   THORPE REACTION

The addition of an anion generated at the α-carbon of a nitrile onto the CN carbon of another nitrile is known as the Thorpe reaction. The reaction is performed by treating the nitrile(s) with an appropriate base. Conceptually, this reaction can be visualized as the nitrile version of the aldol reaction. The product of the reaction is typically a β-iminonitrile (or its tautomer, the enaminonitrile), which undergoes hydrolysis during workup to generate a β-ketonitrile.[239]

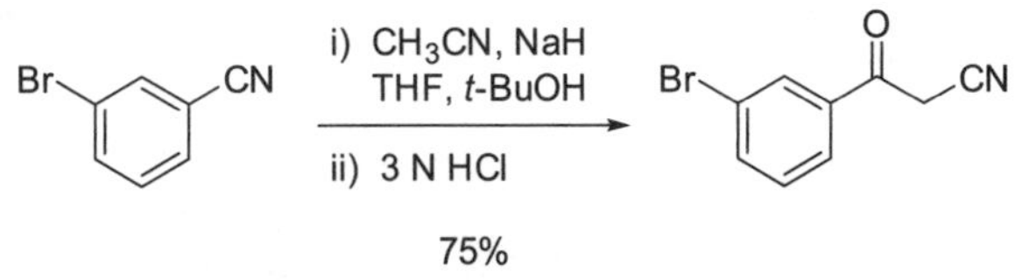

75%

The intramolecular version of this reaction is called the Thorpe–Zeigler reaction. When the reaction mixture is heated during the acidic workup, the β-ketonitrile could undergo hydrolysis, and subsequent decarboxylation to afford the corresponding ketone.[240] Thus, this reaction constitutes a useful method to synthesize cyclic ketones from acyclic nitriles.

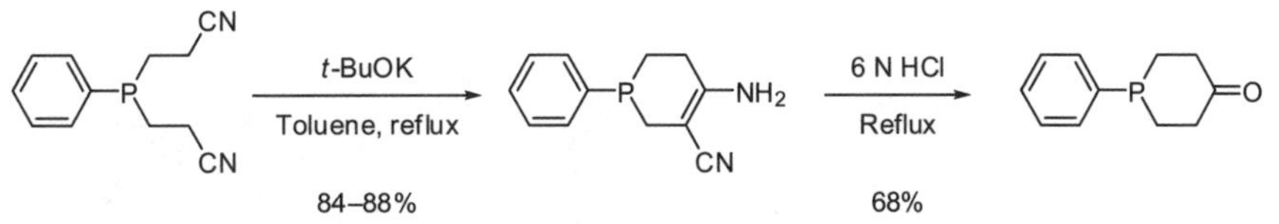

84–88%                              68%

The following transformation is an excellent example of a one-pot alkylation of a nitrile with 5-bromovaleronitrile, followed by a Thorpe–Zeigler reaction. Hydrolysis of the enaminonitrile, followed by decarboxylation of the resulting carboxylic acid, leads to the cyclohexanone derivative shown below.[241]

[238]Jirgensons, A.; Kauss, V.; Kalvinsh, I.; Gold, M. R. *Synthesis* **2000**, 1709–1712.

[239]Takayama, K.; Iwata, M.; Hisamichi, H.; Okamoto, Y.; Aoki, M.; Niwa, A. *Chem. Pharm. Bull.* **2002**, *50*, 1050–1059.

[240]Snider, T. E.; Morris, D. L.; Srivastava, K. C.; Berlin, K. D. *Org. Syn.* **1973**. *53*, 98–103.

[241]Hashimoto, A.; Przybyl, A. K.; Linders, J. T. M.; Kodato, S.; Tian, X.; Deschamps, J. R.; George, C.; Flippen-Anderson, J. L.; Jacobson, A. E.; Rice, K. C. *J. Org. Chem.* **2004**, *69*, 5322–5327.

91%    99%

This reaction is the subject of an exhaustive review,[242] which contains several other examples of cyclizations of nitriles onto other carbonyls as well.

## 2.39  ADDITION OF ORGANOMETALLIC REAGENTS TO NITRILES

Addition of an organomagnesium reagent to a nitrile most often results in isolation of a ketone.[243] While an imine is initially formed via reaction of the organomagnesium reagent with the nitrile, a standard aqueous workup leads to isolation of the ketone.

42%

Isolation of the ketimine intermediate takes a bit more finesse. Pickard and Tolbert showed how it could be done by quench of the initial addition product (the ketimine) into methanol.[244] After filtration of this solution and addition of HCl, the HCl salt of the imine was isolated.

51%

## 2.40  CONVERSION OF NITRILES TO AMIDES, ESTERS, AND CARBOXYLIC ACIDS

Nitriles can be hydrolyzed to provide the corresponding amides or carboxylic acids, under appropriate conditions. In most instances, either acids or bases are used to effect the transformation. As the conversion of nitriles to carboxylic acids under these conditions proceeds via the corresponding amides, it is often possible to isolate the amide "intermediate" by proper choice of reaction conditions. Since nitriles can be converted to esters under similar conditions, this transformation will also be discussed in this section. Enzymes (nitrilases) have also been utilized to hydrolyze nitriles.

[242]Fleming, F. F.; Shook, B. C. *Tetrahedron* **2002**, *58*, 1–23.
[243]Jones, R. L.; Pearson, D. E.; Gordon, M. *J. Org. Chem.* **1972**, *37*, 3369–3370.
[244]Pickard, P. L.; Tolbert, T. L. *J. Org. Chem.* **1961**, *26*, 4886–4887.

### 2.40.1   Hydrolysis of Nitriles Under Acidic Conditions

In the following example, the nitrile underwent facile hydrolysis to the corresponding carboxylic acid when heated in the presence of concentrated HCl. It is noteworthy that prolonged heating at elevated temperatures led to the decarboxylated product.[245]

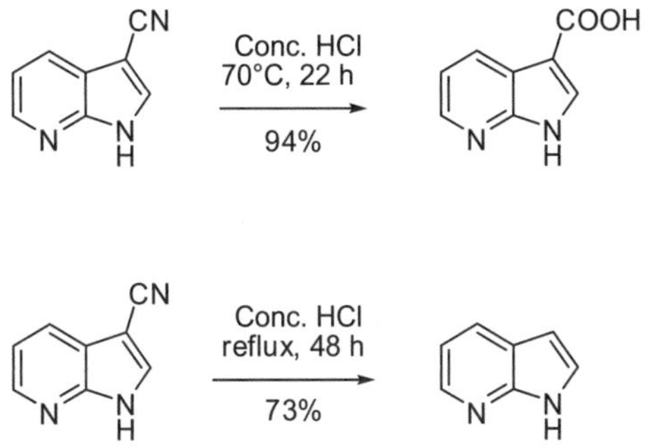

Nitriles can be converted to esters in the presence of an acid and the desired alcohol. In the following scheme, the same nitrile was converted to the acid and the ester by slightly varying the reaction conditions. Treatment of the nitrile with sulfuric acid afforded the acid, while running the reaction in excess methanol furnished the methyl ester.[246]

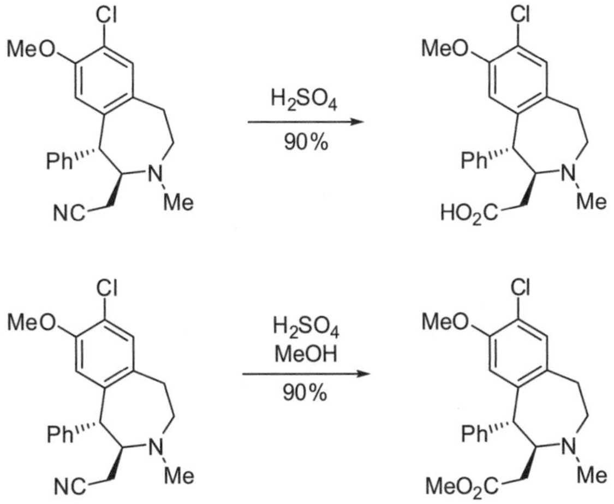

It is also possible to selectively "hydrolyze" a secondary nitrile in the presence of a tertiary nitrile. In the following example,[247] treatment of the bis-nitrile with ethanol saturated with HCl led to the formation of the imidate hydrochloride of the secondary nitrile. This was then

[245] Allegretti, M.; Anacardio, R.; Cesta, M. C.; Curti, R.; Mantovanini, M.; Nano, G.; Topai, A.; Zampella, G. *Org. Process Res. Dev.* **2003**, *7*, 209–213.

[246] Wu, G.; Wong, Y.; Steinman, M.; Tormos, W.; Schumacher, D. P.; Love, G. M.; Shutts, B. *Org. Process Res. Dev.* **1997**, *1*, 359–364.

[247] Caron, S.; Vazquez, E. *Org. Process Res. Dev.* **2001**, *5*, 587–592.

hydrolyzed with toluene/water to furnish the ethyl ester. Subsequent hydrolysis ultimately led to the carboxylic acid (not shown in scheme). Interestingly, the use of methanol instead of ethanol led to a mixture of the secondary ester and the diester.

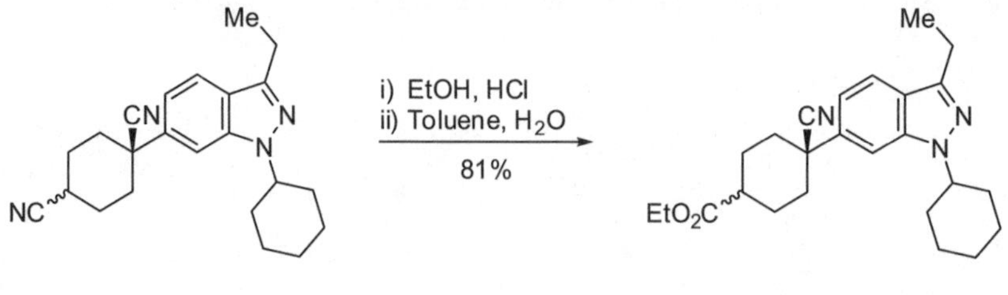

Sulfuric acid is the reagent of choice if the amide is the desired hydrolysis product. In the following scheme, the hindered tertiary nitrile underwent facile transformation to the amide upon treatment with sulfuric acid.[248]

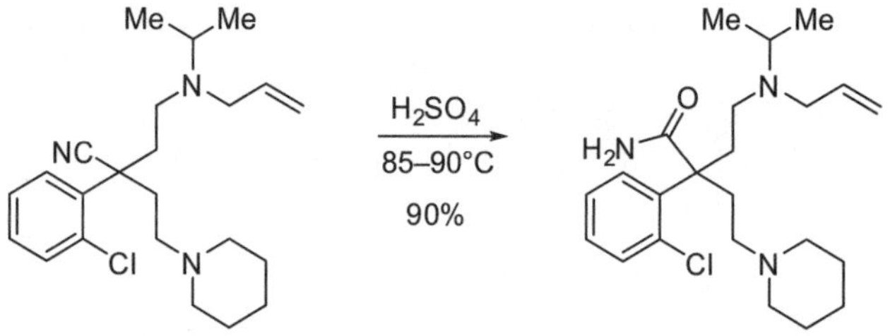

There is one potential complication when sulfuric acid is used to hydrolyze nitriles that contain an electron-rich aromatic ring, as sulfonation of the aromatic ring might occur under these conditions. The ring sulfonation was avoided in the example depicted below by carefully controlling the sulfuric acid stoichiometry and by using a sacrificial solvent. Treatment of the nitrile with 50 equivalents of 98% sulfuric acid at 50°C followed by an aqueous quench led to ring sulfonation and nitrile hydrolysis. In contrast, virtually quantitative hydrolysis (with minimal ring sulfonation) was observed using 5 to 6 equivalents of sulfuric acid in toluene at 70°C after 2 hours. Some sulfonation of toluene was observed, though.[249]

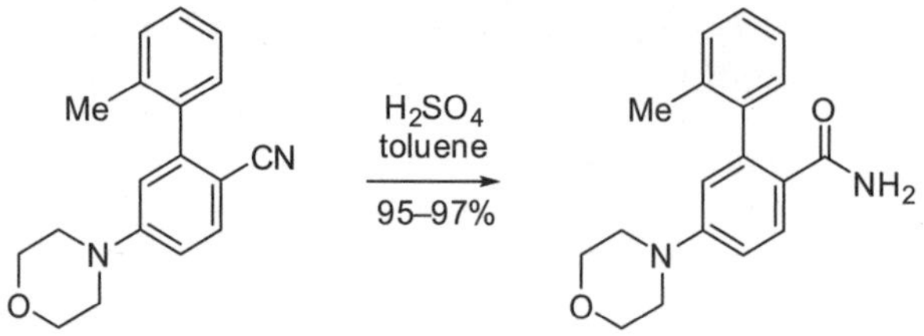

[248] Awasthi, A. K.; Paul, K. *Org. Process Res. Dev.* **2001**, *5*, 528–530.
[249] Harrington, P. J.; Johnston, D.; Moorlag, H.; Wong, J.-W.; Hodges, L. M.; Harris, L.; McEwen, G. K.; Smallwood, B. *Org. Process Res. Dev.* **2006**, *10*, 1157–1166.

### 2.40.2  Hydrolysis of Nitriles Under Basic Conditions

Nitriles can also be easily transformed to the corresponding carboxylic acids by treatment with base. An example is provided below.[250]

### 2.40.3  Hydrolysis of Nitriles under Oxidative Conditions

A particularly mild and convenient method for the conversion of nitriles to amides is the Katrizky protocol[251] using $K_2CO_3$ and hydrogen peroxide. These conditions led to the conversion of the secondary nitrile to the amide within 1 hour at 50–70°C.[252]

### 2.40.4  Enzymatic Hydrolysis of Nitriles

The use of an enzyme is perhaps the mildest method to hydrolyze nitriles. In the following example, 3-hydroxyglutaronitrile was desymmetrized using an enzyme to produce (*R*)-4-cyano-3-hydroxybutyric acid.[253] While enzymes are used primarily to impart or enhance optical purity in substrates via kinetic resolution, they can also be used to effect non-selective hydrolysis of sensitive molecules under mild conditions.

[250]Chandrasekhar, B.; Prasad, A. S. R.; Eswaraiah, S.; Venkateswaralu, A. *Org. Process Res. Dev.* **2002**, *6*, 242–245.

[251]Katritzky, A. R.; Pilarski, B.; Urogdi, L. *Synthesis* **1989**, 949–950.

[252]Hasegawa, T.; Kawanaka, Y.; Kasamatsu, E.; Ohta, C.; Nakabayashi, K.; Okamoto, M.; Hamano, M.; Takahashi, K.; Ohuchida, S.; Hamada, Y. *Org. Process Res. Dev.* **2005**, *9*, 774–781.

[253]Bergeron, S.; Chaplin, D. A.; Edwards, J. H.; Ellis, B. S. W.; Hill, C. L.; Holt-Tiffin, K.; Knight, J. R.; Mahoney, T.; Osborne, A. P.; Ruecroft, G. *Org. Process Res. Dev.* **2006**, *10*, 661–665.

## 2.41   CONVERSION OF NITRILES TO THIOAMIDES

Treatment of nitriles with hydrogen sulfide leads to the thioamide, akin to the formation of amides by reaction of water with nitriles. While simple nitriles (such as 4-bromobenzonitrile) can be converted to thioamides under rather mild conditions,[254] deactivated nitriles necessitate slightly harsher conditions, as evidenced in the last two examples below.[255]

## 2.42   THE ADDITION OF AMMONIA OR AMINES TO NITRILES

Ammonia and amines add to nitriles to produce amidines. For most primary and secondary amines the reaction is sluggish unless either the nitrile is sufficiently electrophilic or the nucleophilicity of the amine is appropriately enhanced.

For example, trichloroacetonitrile reacts with aminoacetaldehyde dimethylacetal to furnish the amidine in virtually quantitative yield under mild conditions.[256]

[254]Ikemoto, N.; Liu, J.; Brands, K. M. J.; McNamara, J. M.; Reider, P. J. *Tetrahedron* **2003**, *59*, 1317–1325.

[255]Rudolph, J.; Chen, L.; Majumdar, D.; Bullock, W. H.; Burns, M.; Claus, T.; Dela Cruz, F. E.; Daly, M.; Ehrgott, F. J.; Johnson, J. S.; Livingston, J. N.; Schoenleber, R. W.; Shapiro, J.; Yang, L.; Tsutsumi, M.; Ma, X. *J. Med. Chem.* **2007**, *50*, 984–1000.

[256]Galeazzi, E.; Guzman, A.; Nava, J. L.; Liu, Y.; Maddox, M. L.; Muchowski, J. M. *J. Org. Chem.* **1995**, *60*, 1090–1092.

In the case of 4-chloroaniline (a relatively poor nucleophile), the addition of a strong base (NaHMDS) was necessary to promote addition to the nitrile.[257]

In an interesting example, Reid and coworkers used lithium hexamethyldisilazide as an ammonia surrogate to efficiently synthesize the amidine from a substituted benzonitrile.[258]

The addition of hydroxylamine to nitriles leads to amidoximes. A nice, simple procedure to effect this transformation has been described by Jendralla and coworkers.[259]

[257]Lange, J. H. M.; van Stuivenberg, H. H.; Coolen, H. K. A. C.; Adolfs, T. J. P.; McCreary, A. C.; Keizer, H. G.; Wals, H. C.; Veerman, W.; Borst, A. J. M.; de Looff, W.; Verveer, P. C.; Kruse, C. G. *J. Med. Chem.* **2005**, *48*, 1823–1838.

[258]Reid, C. M.; Ebikeme, C.; Barrett, M. P.; Patzewitz, E.-M.; Mueller, S.; Robins, D. J.; Sutherland, A. *Bioorg. Med. Chem. Lett.* **2008**, *18*, 5399–5401.

[259]Jendralla, H.; Seuring, B.; Herchen, J.; Kulitzscher, B.; Wunner, J.; Stueber, W.; Koschinsky, R. *Tetrahedron* **1995**, *51*, 12047–12068.

## 2.43  ALKYL THIOL ADDITION TO NITRILES

Addition of an alkyl thiol or thiol to a nitrile results in a thioate or thioamide, respectively.  Pertinent discussion regarding thioates (and thioesters) can be found in section 2.34.

Webber and coworkers achieved clean thioamide formation by addition of hydrogen sulfide to the nitrile in the presence of a ketone and dimethyl acetal.[260] In this procedure, the substrate, triethylamine and pyridine were added to a sealed tube and cooled to 0°C. The reaction was purged with argon, hydrogen sulfide was added, the reaction was sealed, and the reaction was warmed to room temperature for two days. Product isolation by chromatography provided a good yield of the desired thioamide.

96%

In the example below[261] the nitrile and mercaptan were combined in dry dioxane and saturated with HCl gas. After 5 days of reaction, the product was precipitated by addition of ether.

87%

## 2.44  THE ADDITION OF ALCOHOLS TO ISOCYANATES

The reaction of alcohols with isocyantes is a practical way to generate carbamates and is often a key step in the Curtius rearrangement (see Section 7.2.3.2). This type of reaction is typically straightforward due to the reactivity of isocyantes. In general, the main constraint is the use of a reaction solvent that is not reactive; typical solvents are toluene or THF. In the first two examples below, the isocyanates are generated via Curtius rearrangement and then used *in situ*. The first reaction utilizes potassium *t*-butoxide in THF.[159] A toluene solution of the isocyanate was added to a cold stirring solution of the *t*-butoxide in THF. The reaction was warmed to room temperature for 40 minutes. After extraction with MTBE/water, the

[260]Webber, S. E.; Tikhe, J.; Worland, S. T.; Fuhrman, S. A.; Hendrickson, T. F.; Matthews, D. A.; Love, R. A.; Patick, A. K.; Meador, J. W.; et al. *J. Med. Chem.* **1996**, *39*, 5072–5082.
[261]Lunt, E.; Newton, C. G.; Smith, C.; Stevens, G. P.; Stevens, M. F. G.; Straw, C. G.; Walsh, R. J. A.; Warren, P. J.; Fizames, C.; et al. *J. Med. Chem.* **1987**, *30*, 357–366.

organic phase was displaced with heptanes and concentrated to give nearly 6 kilograms of the desired Boc-protected amine.

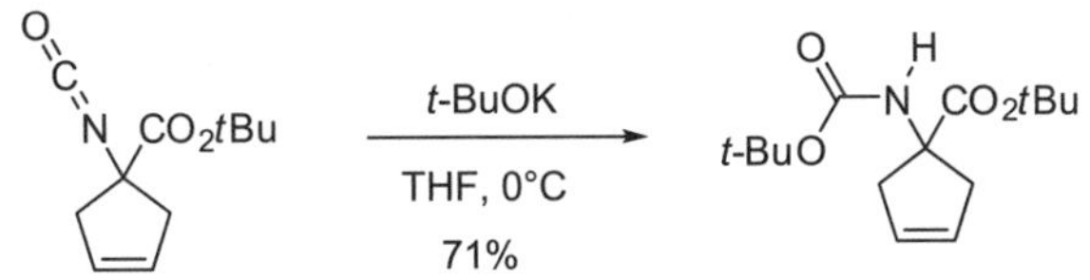

In the second example[262] the isocyanate is treated with ethanol in toluene to provide the desired carbamate as a solid.

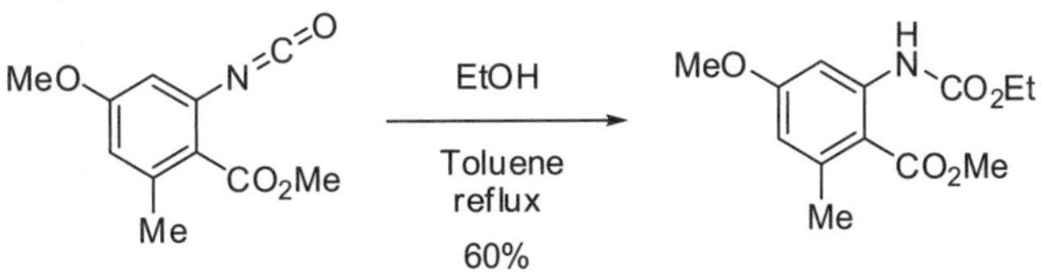

After process optimization by Masui and coworkers, carbamate formation proceeded smoothly on the sensitive chlorosulfonyl isocyanate shown below.[263] This reaction was performed by adding 1 equivalent of *t*-butyl alcohol to the isocyanate in toluene at 0°C. Sulfonamide conversion was accomplished by pyridine-catalyzed displacement with ammonia. The product was isolated as a colorless precipitate in excellent yield.

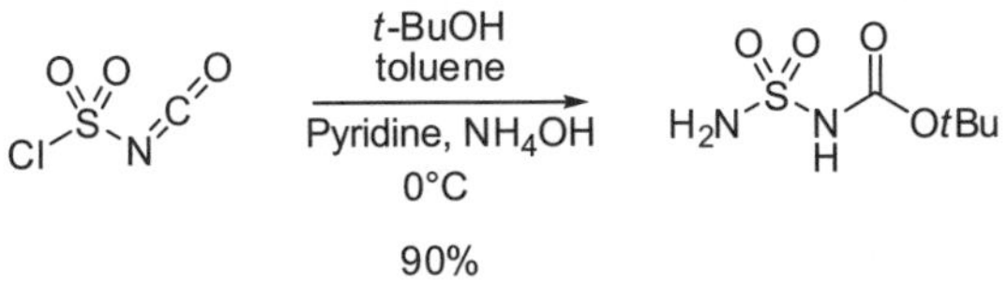

## 2.45   THE ADDITION OF AMINES AND AMIDES TO ISOCYANATES

The addition of amines to isocyanates leads to ureas. This is a good method for the synthesis of substituted ureas by appropriate choice of the amine and isocyanate. The reaction occurs under mild conditions, and the product ureas frequently precipitate out of the reaction mixture.[264]

[262]Tam, T. F.; Coles, P. *Synthesis* **1988**, 383–386.

[263]Masui, T.; Kabaki, M.; Watanabe, H.; Kobayashi, T.; Masui, Y. *Org. Process Res. Dev.* **2004**, *8*, 408–410.

[264]Bankston, D.; Dumas, J.; Natero, R.; Riedl, B.; Monahan, M.-K.; Sibley, R. *Org. Process Res. Dev.* **2002**, *6*, 777–781.

92%

Amides can also add to isocyanates, but the reaction typically requires catalytic quantities of base. Storace and coworkers reported the addition of a β-lactam to an isocyanate in the presence of catalytic lithium *t*-butoxide. The choice of reaction conditions was noteworthy—the catalyst had to be a poor nucleophile that was sufficiently basic to deprotonate the lactam. Additionally, the resulting anion needed to react faster with the isocyanate than undergo decomposition via elimination of phenolate. The base of choice turned out to be *t*-butoxide, while the strongly coordinating lithium ion imparted the desired anion stability.[31]

90%

## 2.46   THE FORMATION OF XANTHATES

Common uses of xanthates are as an intermediate in the Chugaev elimination reaction (see Section 8.2.4) and in the Barton–McCombie deoxygenation (see Section 9.4.3.1). Xanthates are typically formed by deprotonation of alcohols with a suitable base, treatment with carbon disulfide and then addition of an alkylating agent.[265] The first two examples show one of the standard procedures for xanthate formation.

i) NaH, imidazole
ii) CS$_2$

iii) MeI

84%

Boyd and coworkers formed the xanthate below, which was a starting material for a key deoxygenation.[266] To a solution of the substrate in THF at 0°C was added both carbon disulfide and methyl iodode, followed by 60% NaH dispersion in oil. After reaction

[265]Chen, W.; Xia, C.; Wang, J.; Thapa, P.; Li, Y.; Nadas, J.; Zhang, W.; Zhou, D.; Wang, P. G. *J. Org. Chem.* **2007**, *72*, 9914–9923.
[266]Boyd, S. A.; Mantei, R. A.; Hsiao, C. N.; Baker, W. R. *J. Org. Chem.* **1991**, *56*, 438–442.

completion, the mixture was quenched into an ice water solution. Standard workup provided the desired product in good yield.

84%

Another method for xanthate formation is shown by Dehmel and coworkers in their work toward histone deacetylases, a potentially useful class of chemotherapeutic agents.[267] Sodium methoxide and carbon disulfide were premixed in THF and then treated with the $\alpha$-bromo ketone. An aqueous dichloromethane workup provided a good yield of the desired xanthate.

93%

## 2.47   THE ADDITION OF AMINES TO CARBON DIOXIDE

The reaction of amines with carbon dioxide leads to the formation of carbamate salts or carbamic acids. Most primary and secondary aliphatic amines readily absorb $CO_2$ from air to form these adducts. The reaction is generally reversible, and the amine can be regenerated by simply heating the reaction mixture.[268] The reversible nature of this reaction has been exploited for the development of a transient protection strategy for amines.

Katritzky and coworkers have utilized carbon dioxide to simultaneously protect an indole nitrogen and activate the 2-position.[269] Hudkins and coworkers have used this reaction as the basis for synthesizing 2-vinyl-1$H$-indoles via 1-carboxy-2-(tributylstannyl)indole.[270]

[267]Dehmel, F.; Weinbrenner, S.; Julius, H.; Ciossek, T.; Maier, T.; Stengel, T.; Fettis, K.; Burkhardt, C.; Wieland, H.; Beckers, T. *J. Med. Chem.* **2008**, *51*, 3985–4001.
[268]Hampe, E. M.; Rudkevich, D. M. *Chem. Commun. (Cambridge, U. K.)* **2002**, 1450–1451.
[269]Katritzky, A. R.; Faid-Allah, H.; Marson, C. M. *Heterocycles* **1987**, *26*, 1333–1344.
[270]Hudkins, R. L.; Diebold, J. L.; Marsh, F. D. *J. Org. Chem.* **1995**, *60*, 6218–6220.

Another example of this concept can be found in the work of Williams and coworkers, where $CO_2$ was utilized to protect the pyrrolidine nitrogen, thereby inhibiting attack on the carbapenem.[271]

If the reaction of an amine with $CO_2$ is carried out in the presence of an alkylating agent and a base, it is possible to trap the intermediate adduct and produce carbamates. For example, the three-component coupling between valine ester, benzylchloride, and $CO_2$ led to the corresponding carbamate in excellent yield.[272]

## 2.48   THE ADDITION OF AMINES TO CARBON DISULFIDE

Amines add to carbon disulfide (analogous to the reaction with $CO_2$, Section 2.48) to form dithiocarbamic acids or their salts. The reaction is facile, and occurs under fairly mild conditions.[273]

[271]Williams, J. M.; Brands, K. M. J.; Skerlj, R. T.; Jobson, R. B.; Marchesini, G.; Conrad, K. M.; Pipik, B.; Savary, K. A.; Tsay, F.-R.; Houghton, P. G.; Sidler, D. R.; Dolling, U.-H.; DiMichele, L. M.; Novak, T. J. *J. Org. Chem.* **2005**, *70*, 7479–7487.

[272]Salvatore, R. N.; Shin, S. I.; Nagle, A. S.; Jung, K. W. *J. Org. Chem.* **2001**, *66*, 1035–1037.

[273]Kruse, L. I.; Kaiser, C.; DeWolf, W. E.; Finkelstein, J. A.; Frazee, J. S.; Hilbert, E. L.; Ross, S. T.; Flaim, K. E.; Sawyer, J. L. *J. Med. Chem.* **1990**, *33*, 781–789.

Again, similar to the reaction of amines with $CO_2$ (Section 2.48) and the reaction of alcohols with $CS_2$ (Section 2.47), the intermediate adduct can be trapped with an alkylating agent to furnish $S$-alkyldithiocarbamates.[274]

[274]Vaillancourt, V. A.; Larsen, S. D.; Tanis, S. P.; Burr, J. E.; Connell, M. A.; Cudahy, M. M.; Evans, B. R.; Fisher, P. V.; May, P. D.; Meglasson, M. D.; Robinson, D. D.; Stevens, F. C.; Tucker, J. A.; Vidmar, T. J.; Yu, J. H. *J. Med. Chem.* **2001**, *44*, 1231–1248.

# 3

# ADDITION TO CARBON–CARBON MULTIPLE BONDS

JOHN A. RAGAN

## 3.1 INTRODUCTION

Functionalization of olefins, acetylenes, and 1,3-dienes constitutes some of the most important transformations in organic synthesis. A rough approximation is that 10 of the 98 Nobel Prizes awarded in chemistry between 1901 and 2010 have included in significant portion the development or utilization of such reactions in the prizewinner's research. The list of chemists in this analysis includes several leading and historical figures in organic synthesis (e.g., Alder, Brown, Corey, Diels, Noyori, Sharpless, and Woodward). The chapter is organized according to the nature of the elements being added to an olefin (or acetylene). Sections 3.2 to 3.5 cover addition of H-X, and 3.6 to 3.15 cover addition of X-Y (where X and Y are non-hydrogen elements).

## 3.2 HYDROGEN–HALOGEN ADDITION (HYDROHALOGENATION)

### 3.2.1 Hydrohalogenation of Olefins

Direct addition of HX to an olefin is known, and in some cases is a preparatively useful method for the synthesis of alkyl halides. Polarized olefins (e.g., $\alpha,\beta$-unsaturated carbonyl compounds, terminal olefins, styrene derivatives, or 1,1-disubstituted olefins) are particularly good substrates for this reaction. However, non-polarized olefins can yield problematic mixtures of regioisomers. Note that in general, alkyl halides are more readily prepared from alcohols (see Section 1.5.1.3). The examples that follow are arranged in the sequence of addition of HF, HCl, HBr, and HI.

*Practical Synthetic Organic Chemistry: Reactions, Principles, and Techniques*, First Edition.
Edited by Stéphane Caron.
© 2011 John Wiley & Sons, Inc. Published 2011 by John Wiley & Sons, Inc.

For hydrofluorinating alkenes and other substrates, Olah developed a mixture of anhydrous HF and pyridine (*Olah reagent*).[1] The corrosive nature of this reagent is an issue, however, and other methods for installation of fluorine are often preferred (e.g., fluoride ion displacement of alkyl sulfonates, Section 1.5.1.2).[2,3]

Hydrochlorination can be an efficient process. In the first example, Markovnikov addition to a trisubstituted olefin is achieved with TMSCl and water.[4] In the second example, HCl addition occurs in a conjugate fashion with concomitant acetal formation of the aldehyde.[5]

Carreira has developed a cobalt-catalyzed hydrochlorination of unactivated olefins with *p*-toluenesulfonyl chloride and phenylsilane serving as the source of HCl.[6] This method provides remarkable levels of regioselectivity with either terminal or trisubstituted olefins, as shown in the two examples below. The catalytic cycle proceeds via a Co(II)/Co(III) radical process.

[1]Olah, G. A.; Welch, J. T.; Vankar, Y. D.; Nojima, M.; Kerekes, I.; Olah, J. A. *J. Org. Chem.* **1979**, *44*, 3872–3881.
[2]Percy, J. M. *Top. Curr. Chem.* **1997**, *193*, 131–195.
[3]Resnati, G. *Tetrahedron* **1993**, *49*, 9385–9445.
[4]Boudjouk, P.; Kim, B.-K.; Han, B.-H. *Synth. Commun.* **1996**, *26*, 3479–3484.
[5]Kawashima, M.; Fujisawa, T. *Bull. Chem. Soc. Jpn.* **1988**, *61*, 3377–3379.
[6]Gaspar, B.; Carreira, E. M. *Angew. Chem., Int. Ed. Engl.* **2008**, *47*, 5758–5760.

Hydrobromination of methyl acrylate is similarly effected; use of bromine-free HBr is an important factor for this reaction. A radical inhibitor (hydroquinone) is utilized to quench any radicals formed, which can lead to *anti*-Markovnikov addition products.[7]

$$
\text{CH}_2=\text{CHCO}_2\text{Me} \xrightarrow[\substack{0-25°C \\ 80-84\%}]{\substack{\text{HBr} \\ \text{Et}_2\text{O}}} \text{Br-CH}_2\text{CH}_2\text{CO}_2\text{Me}
$$

Landini has utilized a tetralkylphosphonium salt as a phase-transfer catalyst for the hydrohalogenation (HCl, HBr, and HI) of terminal olefins; this allows for the use of aqueous (48%) HBr in place of HBr gas. The regioselectivity reflects Markovnikov's rule. The reaction is run neat in olefin at 115°C.[8]

$$
\xrightarrow[\substack{86\%}]{\substack{\text{aq.HBr} \\ C_{16}H_{33}PBu_3Br}}
$$

Hydroiodination can be effected with HI, or more practically, by *in situ* formation of HI from an iodide salt (e.g., KI and $H_3PO_4$).[9]

$$
\xrightarrow[\substack{88-90\%}]{\substack{\text{KI} \\ H_3PO_4}}
$$

Terminal olefins can be selectively converted to the primary halide by hydrozirconation followed by trapping of the C–Zr bond with an electrophilic halogen source (further examples of hydrozirconation can be found in Section 12.2.12). The examples below utilize iodine to generate the alkyl iodide.[10,11]

$$
\xrightarrow[\substack{2)\ I_2 \\ 90\%}]{\substack{1)\ Cp_2ZrHCl \\ PhH,25°C,\ 9\ h}}
$$

$$
\xrightarrow[\substack{2)\ I_2,\ toluene}]{\substack{1)\ Cp_2ZrHCl \\ PhCH_3}}
$$

[7]Mozingo, R.; Patterson, L. A. *Org. Synth.* **1940**, *20*, 64–65.
[8]Landini, D.; Rolla, F. *J. Org. Chem.* **1980**, *45*, 3527–3529.
[9]Stone, H.; Schechter, H. *Org. Synth.* **1951**, *31*, 66–67.
[10]Evans, D. A.; Bender, S. L.; Morris, J. *J. Am. Chem. Soc.* **1988**, *110*, 2506–2526.
[11]Meijboom, R.; Moss, J. R.; Hutton, A. T.; Makaluza, T.-A.; Mapolie, S. F.; Waggie, F.; Domingo, M. R. *J. Organomet. Chem.* **2004**, *689*, 1876–1881.

### 3.2.2   Hydrohalogenation of Alkynes

Preparation of vinyl halides by addition of HX to acetylenes is not generally the most practical route to these compounds. An exception is polarized alkynes, such as propynoic acid derivatives; these substrates will react with halide salts to generate β-haloacrylic acid derivatives.[12] The reaction shows useful levels of Z-selectivity with esters, amides, and nitriles, but not with ketones.

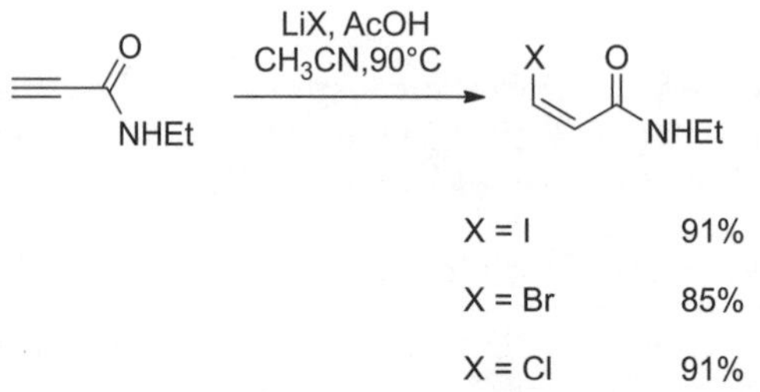

| | |
|---|---|
| X = I | 91% |
| X = Br | 85% |
| X = Cl | 91% |

For unactivated acetylenes, addition of HCl can be effected with triethylammonium hydrogen dichloride ($Et_3NH^+\ HCl_2^-$), although isomeric mixtures are typically observed.[13] More useful is the addition of HBr to terminal acetylenes with tetraethylammonium hydrogen dibromide, which generates the 1-alkyl-1-bromo olefins in useful yields on 5–10 g scale.[14]

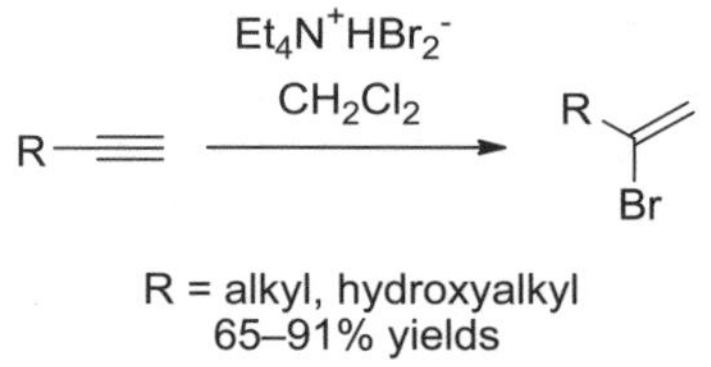

R = alkyl, hydroxyalkyl
65–91% yields

Hydrozirconation of alkynes followed by trapping of the C−Zr bond with an electrophilic halogen source can be useful for the regio- and stereoselective preparation of vinyl halides. The reaction is generally selective for the *syn*-addition of H-X, and can be remarkably sensitive to steric differences in the acetylene substrate.[15] The high regioselectivity has been shown to arise via Zr-catalyzed equilibration of the isomers; the initial kinetic selectivity is modest. The examples shown below are taken from total syntheses of discodermolide[16] and FK506,[17] respectively, and demonstrate the utility of this reaction for the stereo- and regioselective preparation of trisubstituted olefins.

[12]Ma, S.; Lu, X.; Li, Z. *J. Org. Chem.* **1992**, *57*, 709–713.

[13]Cousseau, J.; Gouin, L. *J. Chem. Soc., Perkin Trans.* **1977**, *1*, 1797–1801.

[14]Cousseau, J. *Synthesis* **1980**, 805–806.

[15]Hart, D. W.; Blackburn, T. F.; Schwartz, J. *J. Am. Chem. Soc.* **1975**, *97*, 679–680.

[16]Arefolov, A.; Panek, J. S. *J. Am. Chem. Soc.* **2005**, *127*, 5596–5603.

[17]Nakatsuka, M.; Ragan, J. A.; Sammakia, T.; Smith, D. B.; Uehling, D. E.; Schreiber, S. L. *J. Am. Chem. Soc.* **1990**, *112*, 5583–5601.

## 3.3 HYDROGEN–OXYGEN ADDITION

### 3.3.1 Addition of H−OH (Hydration)

***3.3.1.1 Hydration of Olefins***   Direct hydration of olefins to generate alcohols is usually not a practical process, but indirect, metal-mediated methods are common. Three of the most commonly used are covered in this section: hydroboration, hydrosilylation, and oxymercuration.

*Hydroboration*   Hydroboration is a useful indirect method for hydration of olefins via oxidation of the C−B bond resulting from the initial hydroboration. This is a widely utilized method for *anti*-Markovnikov addition of water to an olefin (steric and electronic factors favor boron addition to the less-substituted carbon, and thus lead to formation of the less-substituted alcohol after oxidation).

Borane • THF is commercially available, and can serve as a convenient source of borane on laboratory scale. There are potential handling and stability issues with this reagent, however.[18,19] These can be avoided through *in situ* formation of borane. One of the most useful methods for this is the reaction of sodium borohydride with $BF_3$ • $Et_2O$, as shown in the example below.[20]

Dialkylboranes such as disiamylborane (($Me_2CHCHMe)_2BH$, prepared *in situ* by reaction of $BH_3$ • THF with 2-methyl-2-butene) can exhibit useful levels of regio- and stereocontrol. In the following example, a triene is selectively hydroborated at the terminal olefin to give the primary alcohol following oxidation.[21]

[18]am Ende, D. J.; Vogt, P. F. *Org. Process Res. Dev.* **2003**, *7*, 1029–1033.

[19]Potyen, M.; Josyula, K. V. B.; Schuck, M.; Lu, S.; Gao, P.; Hewitt, C. *Org. Process Res. Dev.* **2007**, *11*, 210–214.

[20]Rathke, M. W.; Millard, A. A. *Org. Synth.* **1978**, *58*, 32–36.

[21]Leopold, E. J. *Org. Synth.* **1986**, *64*, 164–174.

Diasteroselectivity is another useful feature of hydroborations, with *syn* addition of the borane reagent from the less-hindered face of the olefin predominating. The following example illustrates this for a bicyclic olefin, with the borane reagent approaching from the less-hindered face. This procedure also illustrates the use of sodium perborate, a safe and convenient alternative to hydrogen peroxide for oxidation of the intermediate organoborane.[22]

Enantioselective hydroboration is a challenging transformation. In selected cases, it can be achieved with the use of chiral, non-racemic dialkylborane reagents. In the example below, the two olefins are enantiotopic, and the chiral reagent selectively reacts with one olefin to generate the product alcohol with useful levels of enantioselectivity after an oxidative workup (albeit in modest yield, 27–31% overall for the alkylation/hydroboration sequence from cyclopentadienyl sodium).[23]

Acyclic olefins can also be good substrates for diastereoselective hydroboration, as first noted by Still.[24] The following example from Evans is representative.[25]

[22] Kabalka, G. W.; Maddox, J. T.; Shoup, T.; Bowers, K. R. *Org. Synth.* **1996**, *73*, 116–122.
[23] Partridge, J. J.; Chadha, N. K.; Uskokovic, M. R. *Org. Synth.* **1985**, *63*, 44–56.
[24] Still, W. C.; Barrish, J. C. *J. Am. Chem. Soc.* **1983**, *105*, 2487–2489.
[25] Evans, D. A.; Kim, A. S.; Metternich, R.; Novack, V. J. *J. Am. Chem. Soc.* **1998**, *120*, 5921–5942.

Hydroboration of olefins with catecholborane catalyzed by rhodium or iridium has been developed by Evans and coworkers, and provides products complementary to the uncatalyzed reaction in terms of regio- and stereocontrol. For example, the catalyzed hydroboration shown below provides the *syn* product (93:7 selectivity), whereas the uncatalyzed hydroboration with 9-BBN provides predominantly the *anti* product (87:13 selectivity, not shown).[26]

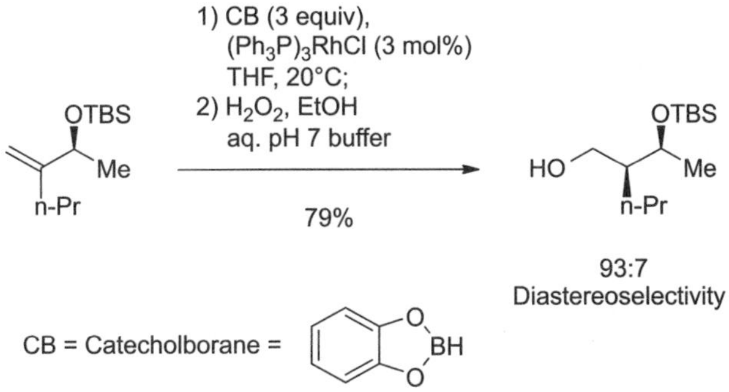

Cyclic olefins also exhibit this complementarity, with the 1,3-*anti* product favored with catalyzed hydroboration, versus the 1,2-*anti* product in the uncatalyzed variant.

Coordinating substituents such as a phosphinite or an amide have a dramatic impact on the stereo- and regiochemical outcome. For example, allylic cyclohexenols provide the 1,3-*anti* product with a non-coordinating group (TBSO), whereas the 1,2-*syn* product predominates with a coordinating group (Ph$_2$PO). Amides also serve as directing groups in this reaction.[26]

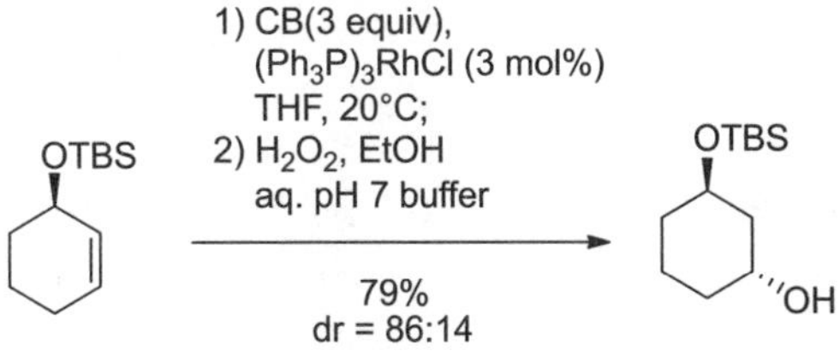

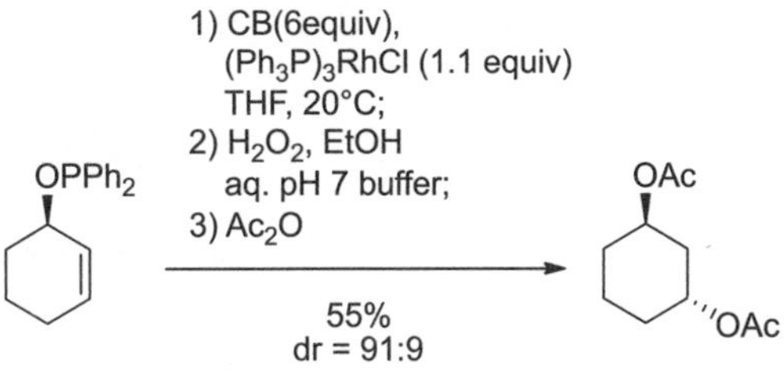

[26]Evans, D. A.; Fu, G. C.; Hoveyda, A. H. *J. Am. Chem. Soc.* **1992**, *114*, 6671–6679.

*Hydrosilylation*    Hydrosilylation of an olefin followed by oxidation of the C–Si bond achieves the overall hydration of the olefin, similar to hydroboration/oxidation. In the following intramolecular example this reaction is highly selective for the 2,3-*syn* triol.[27]

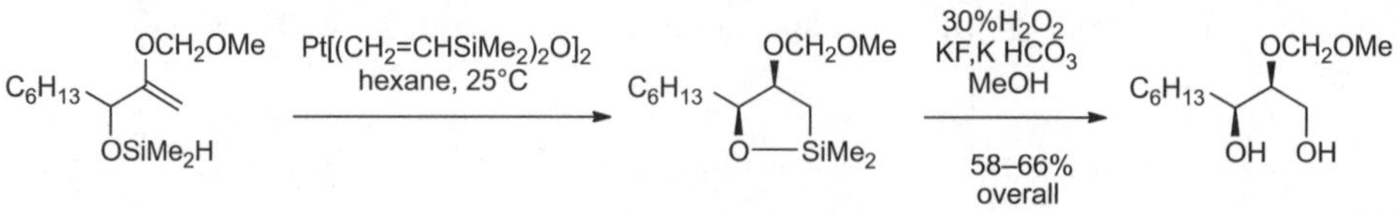

*Oxymercuration*    Oxymercuration of olefins followed by reduction of the C–HgX bond effects net hydration of the olefin. The regioselectivity of this method is generally excellent, providing the Markovnikov addition product, and it is thus complementary to hydroboration (see Section 3.3.1.1.1). The generation of a stoichiometric organomercurial intermediate and resulting mercury waste streams makes this method impractical on large scale, but for laboratory applications it can be useful.

In the example shown, 1-methylcyclohexene is oxymercurated and reduced with sodium borohydride to generate 1-methylcyclohexanol.[28]

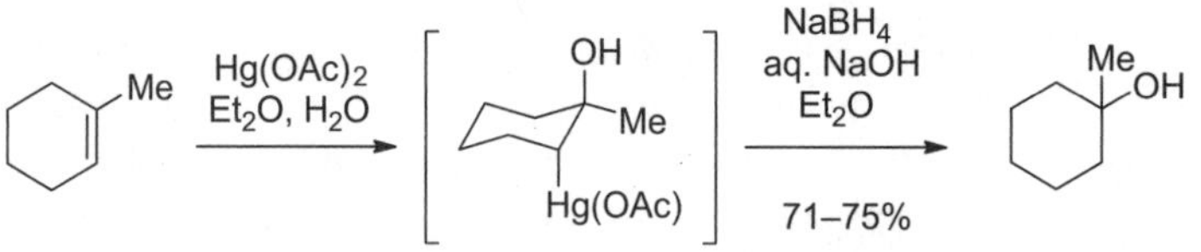

### 3.3.1.2  *Hydration of Acetylenes*

Addition of water to an acetylene formally generates an enol, which tautomerizes to the carbonyl form. Although this reaction can be catalyzed by acid, its most common variant utilizes Hg(II) salts as catalysts. Any water-soluble mercuric salt can typically be used; HgO and $HgSO_4$ are frequently utilized. Addition occurs according to Markovnikov's rule, such that terminal acetylenes generate the methyl ketone, as shown below for the preparation of 1-acetylcyclohexanol.[29] Au(III) complexes have also been reported to effect a similar hydration of acetylenes, thus avoiding the use of toxic mercury salts.[30]

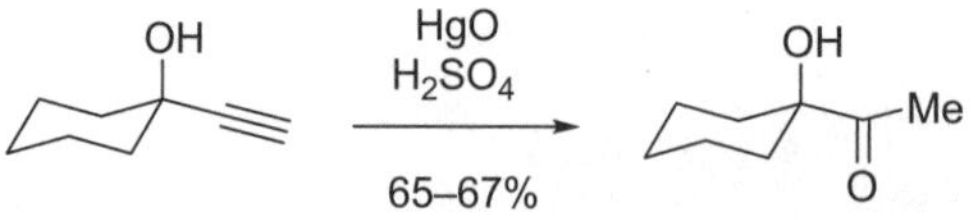

Olah has reported a Nafion-H resin impregnated with HgO that effects this transformation in 80–93% yields for terminal olefins.[31] The catalyst is prepared by combining Nafion-H resin and HgO in a 4:1 w/w ratio in aqueous sulfuric acid, stirring at ambient temperature, filtering, and drying.

[27] Tamao, K.; Nakagawa, Y.; Ito, Y. *Org. Synth.* **1996**, *73*, 94–109.

[28] Jerkunica, J. M.; Traylor, T. G. *Org. Synth.* **1973**, *53*, 94–97.

[29] Stacy, G. W.; Mikulec, R. A. *Org. Synth.* **1955**, *35*, 1–3.

[30] Casado, R.; Contel, M.; Laguna, M.; Romero, P.; Sanz, S. *J. Am. Chem. Soc.* **2003**, *125*, 11925–11935.

[31] Olah, G. A.; Meidar, D. *Synthesis* **1978**, 671–672.

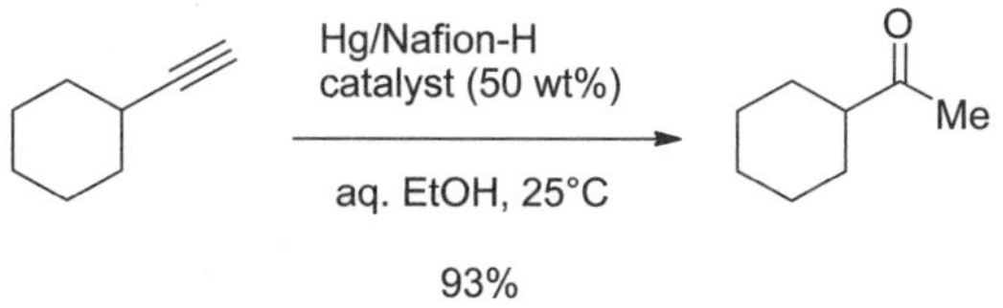

93%

Terminal alkynes can also be hydrated to give the alternative regioisomer (i.e., the aldehyde). Wakatsuki and coworkers have reported the Ru-catalyzed hydration of terminal acetylenes to provide the *anti*-Markovnikov product, an attractive alternative to stoichiometric hydroboration/oxidation (see below).[32] The catalyst (RuCpCl(dppm))[33] is commercially available.

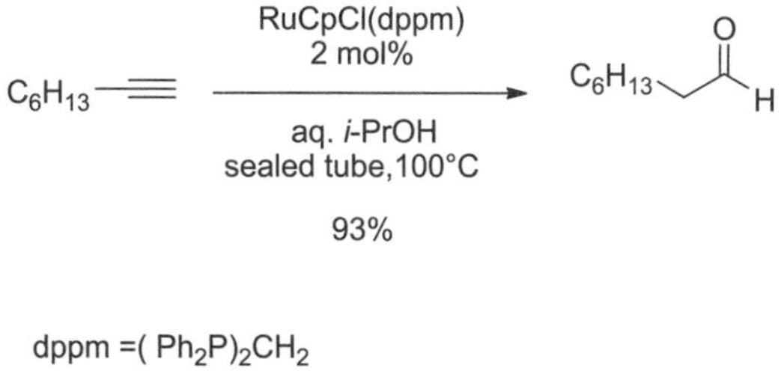

93%

dppm =( $Ph_2P)_2CH_2$

Hydroboration of alkynes also provides the *anti*-Markovnikov hydration product, via oxidation of the initially formed terminal vinyl organoborane. The example shown below converts 1-hexyne to hexanal.[34]

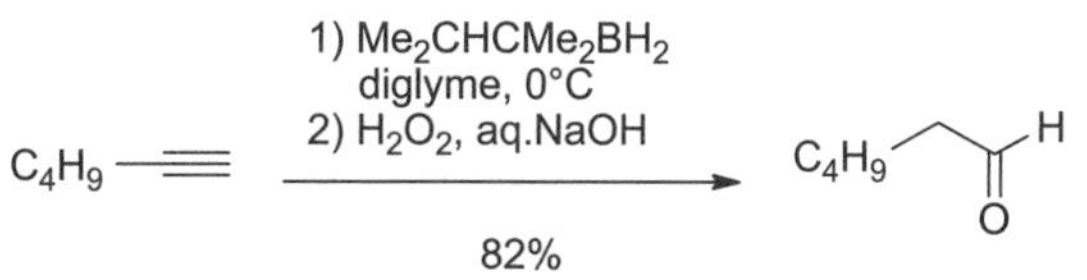

82%

Internal (unsymmetrical) acetylenes generally give mixtures of regioisomers with mercury or gold catalysts.[35] There is a report of regioselective addition of methanol to methylisopropylacetylene catalyzed by a cationic gold(I) complex,[36] although no yield and limited experimental details are provided.

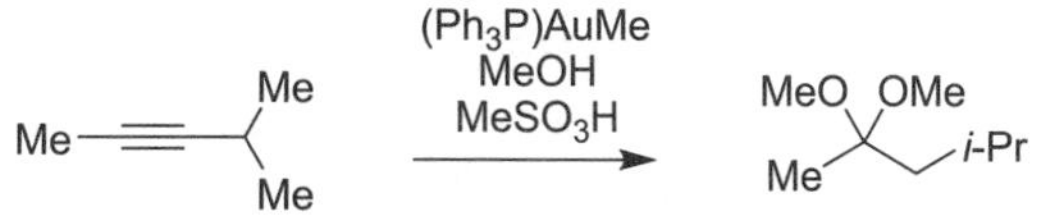

It is reported that the regioselective hydration of an unsymmetrical alkyne can be achieved indirectly via hydroboration, although no experimental details are provided.[37]

[32]Suzuki, T.; Tokunaga, M.; Wakatsuki, Y. *Org. Lett.* **2001**, *3*, 735–737.

[33]Ashby, G. S.; Bruce, M. I.; Tomkins, I. B.; Wallis, R. C. *Aust. J. Chem.* **1979**, *32*, 1003–1016.

[34]Zweifel, G.; Brown, H. C. *J. Am. Chem. Soc.* **1963**, *85*, 2066–2072.

[35]Norman, R. O. C.; Parr, W. J. E.; Thomas, C. B. *J. Chem. Soc., Perkin Trans.* **1976**, *1*, 1983–1987.

[36]Teles, J. H.; Brode, S.; Chabanas, M. *Angew. Chem., Int. Ed. Engl.* **1998**, *37*, 1415–1418.

[37]Trost, B. M.; Martin, S. J. *J. Am. Chem. Soc.* **1984**, *106*, 4263–4265.

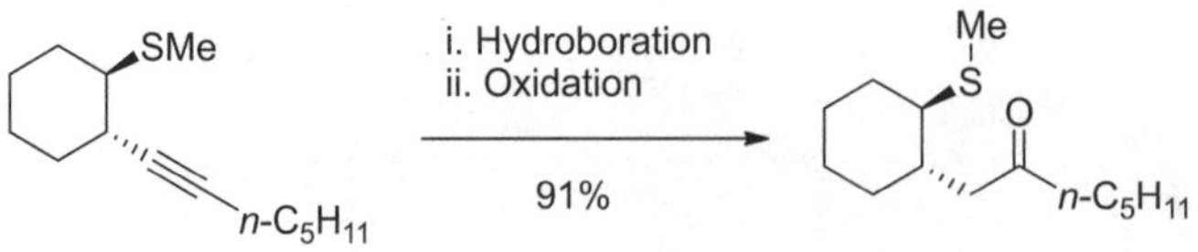

### 3.3.2 Addition of H–OR (Hydroalkoxylation)

*3.3.2.1 Addition of H–OR to Olefins* Polarized olefins can be reacted with alcohols under more practical conditions. One such example is shown below, in which catalytic sodium is used to generate the catalytic alkoxide.[38] An earlier reference for the addition of simpler, less expensive alcohols to acrylate esters (e.g., methanol, ethanol) is also provided.[39]

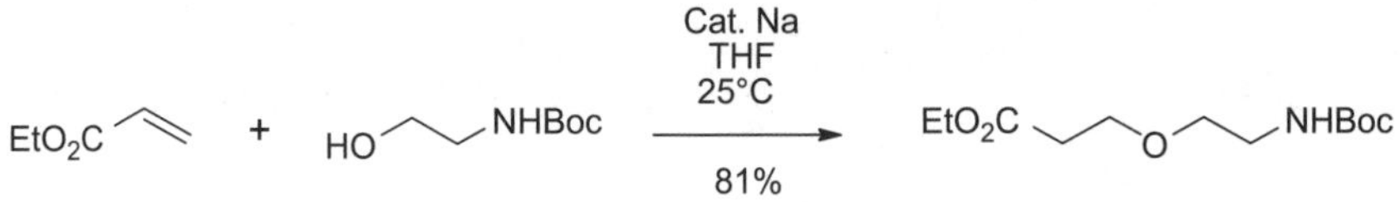

Unsaturated lactones can also serve as electrophiles for this conjugate addition, as in the following example.[40]

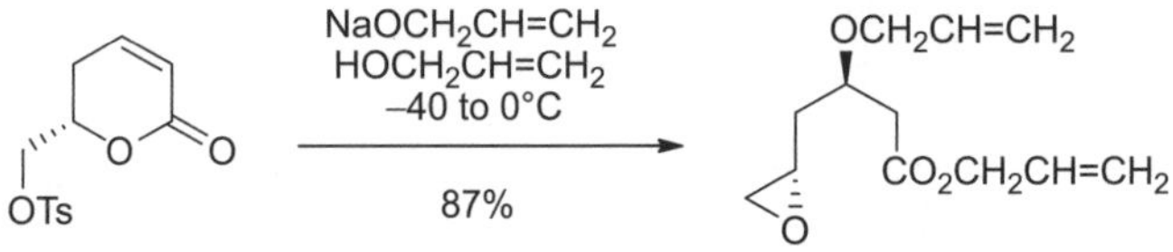

As in the oxymercuration of olefins to form alcohols (see Section 3.3.1.1.3), mercury (II) salts can effect addition of alcohols to olefins (alkoxymercuration) to generate the corresponding ethers following reduction of the C–HgX bond with NaBH$_4$.[41,42] Due to the generation of stoichiometric mercury metal as a byproduct, this method is not practical on any significant scale, however; appropriate caution should be utilized even on modest laboratory scale.

Highly polarized olefins can be directly reacted with an alcohol and acid catalyst. Two such examples can be found in alcohol-protecting-group chemistry. Addition of an alcohol to dihydropyran generates the corresponding tetrahydropyranyl (THP) ether,[43] an excellent blocking group for strong base chemistry (note that a chiral center is generated in the product, such that a chiral alcohol will generate a pair of diastereomeric THP ethers).

[38]Reddy, D. S.; Vander Velde, D.; Aube, J. *J. Org. Chem.* **2004**, *69*, 1716–1719.
[39]Rehberg, C. E.; Dixon, M. B.; Fisher, C. H. *J. Am. Chem. Soc.* **1946**, *68*, 544–546.
[40]Roth, B.; Baccanari, D. P.; Sigel, C. W.; Hubbell, J. P.; Eaddy, J.; Kao, J. C.; Grace, M. E.; Rauckman, B. S. *J. Med. Chem.* **1988**, *31*, 122–129.
[41]Brown, H. C.; Kurek, J. T.; Rei, M. H.; Thompson, K. L. *J. Org. Chem.* **1984**, *49*, 2551–2557.
[42]Brown, H. C.; Kurek, J. T.; Rei, M. H.; Thompson, K. L. *J. Org. Chem.* **1985**, *50*, 1171–1174.
[43]Earl, R. A.; Townsend, L. B. *Org. Synth.* **1981**, *60*, 81–87.

p-TsOH

78–92%

Isobutylene can also be condensed with an alcohol under strongly acidic conditions to generate the corresponding *tert*-butyl ether.[44]

Addition of carboxylic acids to olefins is also known. Formation of *tert*-butyl esters from isobutylene falls in this category, and represents another example of protecting-group chemistry.[45] Intramolecular addition of carboxylic acids to olefins can also be a useful route to lactones. This transformation is frequently effected by iodolactonization (see Section 1.2.3.2) followed by reduction of the carbon–iodine bond; but in certain cases, the transformation can be realized directly by treatment with acid, such as toluenesulfonic acid in the example below (note that in this example the preference for Markovnikov addition and the kinetic preference for 5- vs. 6-membered ring formation are aligned to provide a single regioisomer).[46] Triflic acid in nitromethane has also been utilized to generate several 5- and 6-membered lactones.[47]

p-TsOH

Toluene
reflux, 1h

75%

### 3.3.2.2 Addition of H–OR to Acetylenes

Alcohols will also add to polarized acetylenes to generate β-alkoxy acrylate derivatives, as shown in the example below.[48] Phenols will also add in a similar fashion.[49]

MeOH, Et$_3$N
Et$_2$O, 25°C

≡—CO$_2$Me

90%

MeO $\diagup$ CO$_2$Me

PhOH
N-Me-morpholine
THF, 0°C

≡—CO$_2$Me

82%

PhO $\diagup$ CO$_2$Me

[44]Ireland, R. E.; O'Neil, T. H.; Tolman, G. L. *Org. Synth.* **1983**, *61*, 116–121.

[45]Strube, R. E. *Org. Synth.* **1957**, *37*, 34–36.

[46]Barrero, A. F.; Altarejos, J.; Alvarez-Manzaneda, E. J.; Ramos, J. M.; Salido, S. *Tetrahedron* **1993**, *49*, 6251–6262.

[47]Coulombel, L.; Dunach, E. *Synth. Commun.* **2005**, *35*, 153–160.

[48]Ireland, R. E.; Wipf, P.; Xiang, J. N. *J. Org. Chem.* **1991**, *56*, 3572–3582.

[49]Glorius, F.; Neuhurger, M.; Pfaltz, A. *Helv. Chim. Acta* **2001**, *84*, 3178–3196.

## 3.4 HYDROGEN–NITROGEN ADDITION (HYDROAMINATION)

### 3.4.1 Hydroamination of Olefins

Direct hydroamination of unactivated olefins is challenging, and is an active area of research in organometallic catalysis.[50] Mercury salts will catalyze the addition of amines to olefins (similar to the oxymercuration discussed in Section 3.3.1.1.3). Markovnikov's Rule predicts the regioselectivity of this reaction. As with oxymercuration, a significant drawback to this approach is the generation of mercury metal in the subsequent reduction of the carbon–mercury bond.[51] Hydroboration followed by treatment of the organoborane with a chloroamine is an indirect approach to this transformation.[52] Another variant on this approach is shown below, in which the intermediate organoborane is oxidized with hydroxylamine-$O$-sulfonic acid to provide the corresponding primary amine[20] (note that this method for *in situ* generation of borane is also useful for hydroborations to generate alcohols, as shown earlier in Section 3.3.1.1.1).

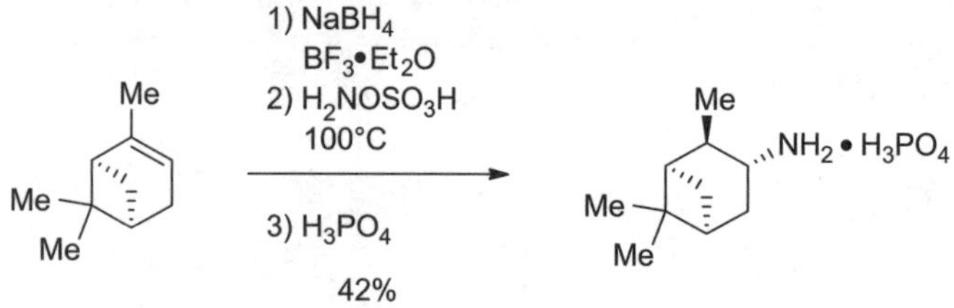

Alkyl azides can also be coupled with alkylboranes to affect the net hydroamination of the olefin substrate. The two examples below are from H. C. Brown[53] and David Evans.[54] In the first example, the chiral dichloroalkylborane is prepared by asymmetric hydroboration of 1-methylcyclopentene with di- or mono-isopinocampheylborane, and formation of the dichloroborane by treatment with HCl-Me$_2$S. The second example, from Evans' synthesis of echinocandin D, is particularly interesting as it involves a diastereoselective hydroboration followed by internal coupling of an alkylazide to form the pyrrolidine.

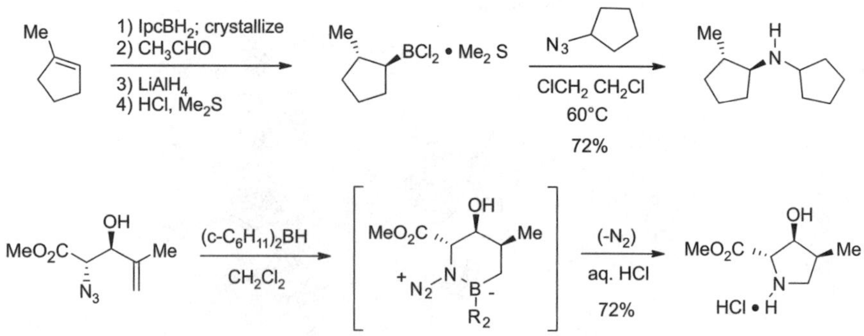

[50]Ryu, J.-S.; Li, G. Y.; Marks, T. J. *J. Am. Chem. Soc.* **2003**, *125*, 12584–12605.
[51]Griffith, R. C.; Gentile, R. J.; Davidson, T. A.; Scott, F. L. *J. Org. Chem.* **1979**, *44*, 3580–3583.
[52]Kabalka, G. W.; McCollum, G. W.; Kunda, S. A. *J. Org. Chem.* **1984**, *49*, 1656–1658.
[53]Brown, H. C.; Salunkhe, A. M.; Singaram, B. *J. Org. Chem.* **1991**, *56*, 1170–1175.
[54]Evans, D. A.; Weber, A. E. *J. Am. Chem. Soc.* **1987**, *109*, 7151–7157.

In contrast to unactivated olefins, amines readily add to highly polarized olefins such as α,β-unsaturated esters. Care must be taken to avoid polymerization, but with proper choice of conditions this can be a useful synthesis of β-aminoesters, as shown in the example below, where two molecules of ethyl acrylate condense with 3-hydroxy-1-aminopropane (this is the first step in a preparation of azetidine).[55] Cupric acetate has been used as a catalyst for mono-addition of 2-chloroaniline to acrylonitrile.[56]

Additions to α,β-unsaturated amides are also useful, and the following experimental is offered on more typical laboratory scale and in ethanol solvent (3.6 g product, isolated via chromatography).[57]

MacMillan has applied imidazolidinone organic catalysis to the conjugate addition of O-silyl hydroxylamines to enals.[58]

### 3.4.2 Hydroamination of Acetylenes and Allenes

Hydroamination of alkynes (and allenes) can be a useful approach to the synthesis of alkyl amines.[59] A variety of catalysts for effecting this transformation have been developed.

[55]Wadsworth, D. H. *Org. Synth.* **1973**, *53*, 13–16.
[56]Heininger, S. A. *Org. Synth.* **1958**, *38*, 14–16.
[57]Gutierrez-Garcia, V. M.; Lopez-Ruiz, H.; Reyes-Rangel, G.; Juaristi, E. *Tetrahedron* **2001**, *57*, 6487–6496.
[58]Chen, Y. K.; Yoshida, M.; MacMillan, D. W. C. *J. Am. Chem. Soc.* **2006**, *128*, 9328–9329.
[59]Pohlki, F.; Doye, S. *Chem. Soc. Rev.* **2003**, *32*, 104–114.

Doye and coworkers have reported dimethyltitanocene as a catalyst for the intermolecular hydroamination of acetylenes.[60] In the example below, nitrogen adds to the carbon β to the aromatic ring; the initial imine product is then hydrogenated to generate the primary amine through hydrogenation of the CN bond and hydrogenolysis of the benzyhydryl group.

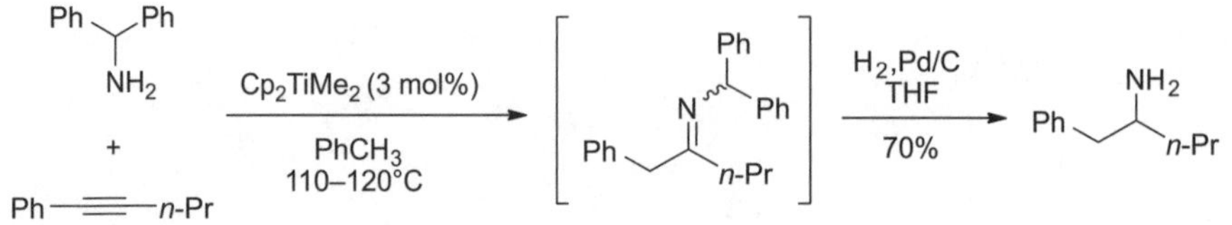

Bergman has reported that tetrakis(amido) titanium complexes effect intramolecular hydroaminations of alkynes and allenes; the reaction temperature is somewhat reduced with this catalyst.[61]

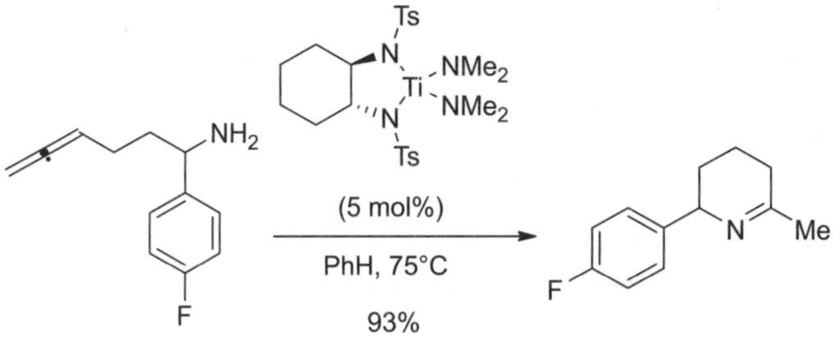

In the final example, Livinghouse utilized an amido titanocene complex to effect the intramolecular hydroamination of an acetylene in his synthesis of monomorine.[62]

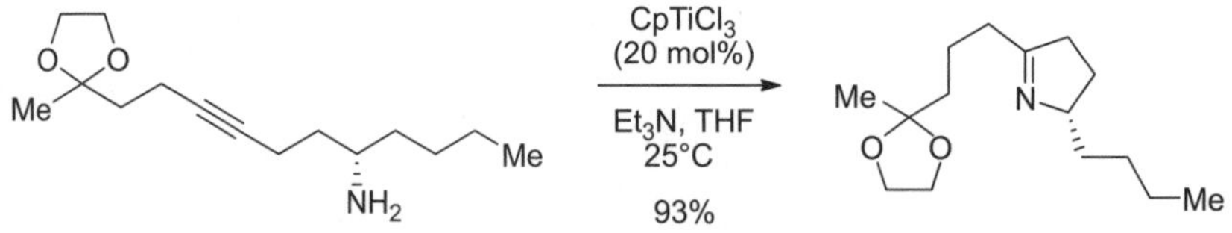

Gold complexes have also been used as catalysts for hydroamination of alkynes. Both intra-[63] and intermolecular[64] examples are shown below.

[60]Haak, E.; Siebeneicher, H.; Doye, S. *Org. Lett.* **2000**, *2*, 1935–1937.
[61]Ackermann, L.; Bergman, R. G. *Org. Lett.* **2002**, *4*, 1475–1478.
[62]McGrane, P. L.; Livinghouse, T. *J. Org. Chem.* **1992**, *57*, 1323–1324.
[63]Fukuda, Y.; Utimoto, K. *Synthesis* **1991**, 975–978.
[64]Mizushima, E.; Hayashi, T.; Tanaka, M. *Org. Lett.* **2003**, *5*, 3349–3352.

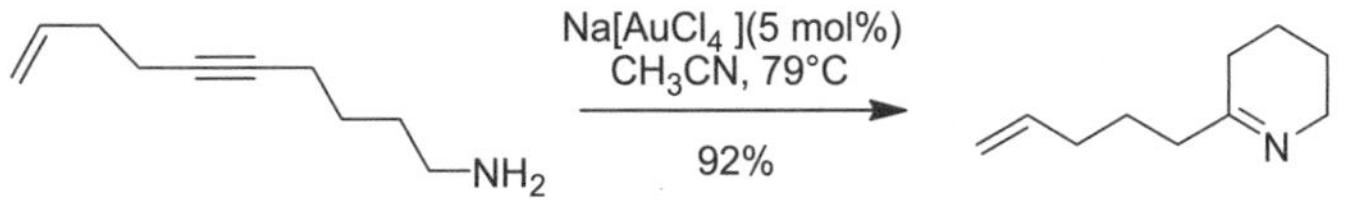

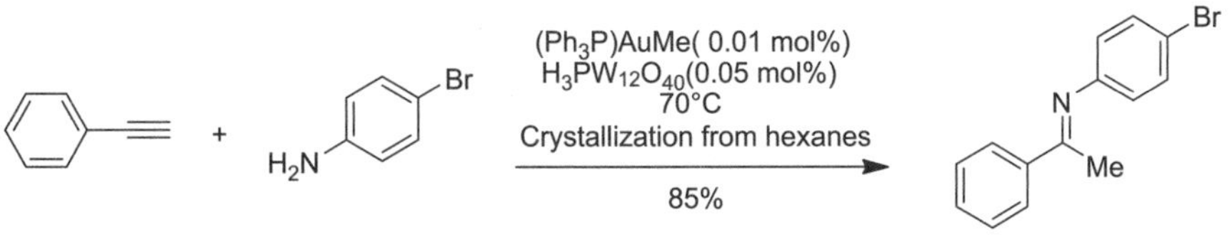

Direct addition of amines to polarized acetylenes is generally an efficient process, as shown in the example below where proline ethyl ester adds to ethyl propynoate.[65]

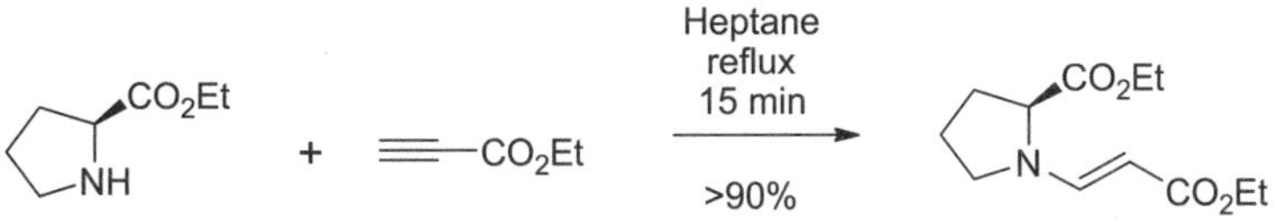

## 3.5    HYDROGEN–CARBON ADDITION (HYDROALKYLATION)

### 3.5.1    Direct Hydrogen–Alkyl Addition

There are few preparatively useful reactions for the direct hydroalkylation of unactivated olefins. (Note that indirect methods that proceed via hydrometallation, e.g., carboalumination, are covered in Chapter 12). One notable exception is the cascade cyclization of polyenes to generate steroid and terpene-like structures. The majority of these reactions fall in the category of carbon–carbon addition to an olefin (see Section 3.15.8). The following cyclization, a Friedel-Crafts addition to a conjugated dienone, provides an example of a formal hydroalkylation of an olefin.[66]

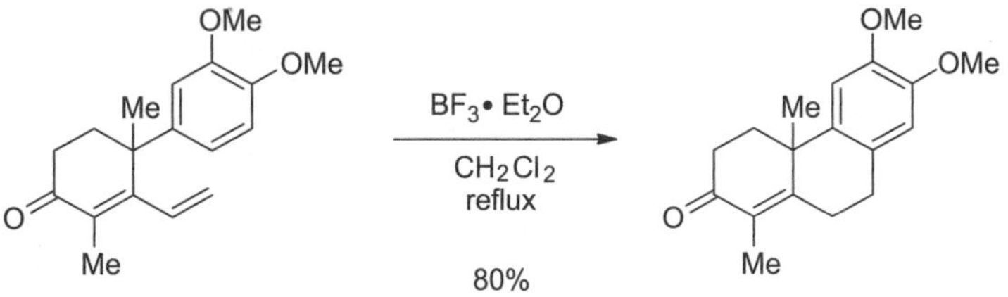

A similar example was utilized in the cyclization shown below.[67] This reaction can also be viewed as a Lewis acid-mediated ene reaction (see Section 3.5.2).

[65]Walter, P.; Harris, T. M. *J. Org. Chem.* **1978**, *43*, 4250–4252.
[66]Majetich, G.; Liu, S.; Fang, J.; Siesel, D.; Zhang, Y. *J. Org. Chem.* **1997**, *62*, 6928–6951.
[67]Schmidt, C.; Chishti, N. H.; Breining, T. *Synthesis* **1982**, 391–393.

Another example of hydroalkylation is the dimerization of styrene to generate 1-methyl-3-phenylindane.[68]

### 3.5.2   Hydrogen–Allyl Addition (Alder Ene Reaction)

The Alder ene reaction, a [2,3]-sigmatropic process, effects the net addition of hydrogen and an allyl group across an olefin. The reaction can be effected thermally, as exemplified below.[69]

Lewis acid catalysis has also been utilized to effect ene reactions at lower temperatures. The following cyclization is representative, and proceeds with high diastereoselectivity (>30:1).[70]

An enantioselective ene cyclization was reported in the synthesis of (−)-α-kainic acid.[71] Although two equivalents of the ligand and metal reagents are required, the reaction proceeds in good yield and enantioselectivity. Note that appropriate precautions should be taken when handling perchlorate salts.

[68]Rosen, M. J. *Org. Synth.* **1955**, *35*, 83–84.
[69]Rondestvedt, C. S., Jr., *Org. Synth.* **1951**, *31*, 85–87.
[70]Tietze, L. F.; Beifuss, U. *Org. Synth.* **1993**, *71*, 167–174.
[71]Xia, Q.; Ganem, B. *Org. Lett.* **2001**, *3*, 485–487.

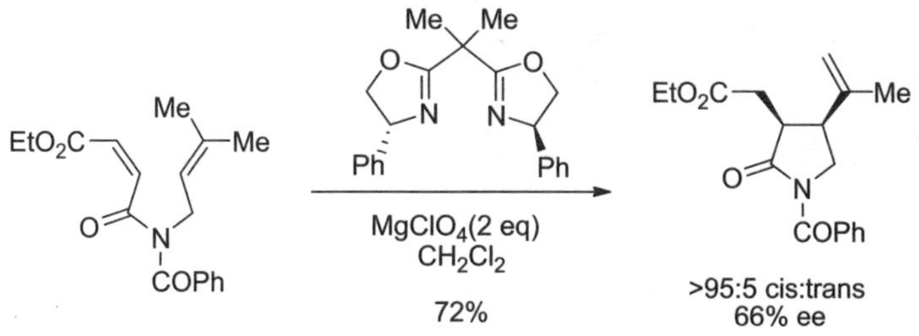

>95:5 cis:trans
66% ee

72%

A similar transformation can occur wherein the ene component is the enol of a carbonyl compound. This transformation is known as a Conia reaction.[72] The following example is from Paquette's synthesis of modhephene. While efficient (85% yield on 500 mg scale), the reaction does require extreme temperatures.[73]

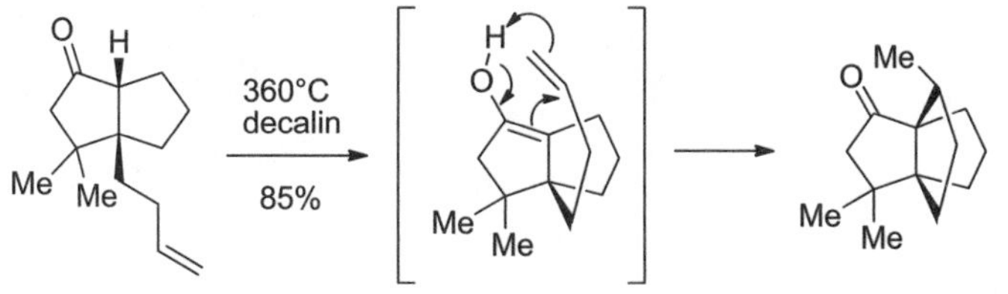

### 3.5.3   Hydrogen–Malonate/Enolate Addition (Michael Reaction)

Active methylene compounds such as malonates and β-keto esters readily add to activated olefins (enones, acrylates). This can be coupled with a subsequent saponification/decarboxylation to give the acid corresponding to conjugate addition of an acetic acid dianion equivalent.

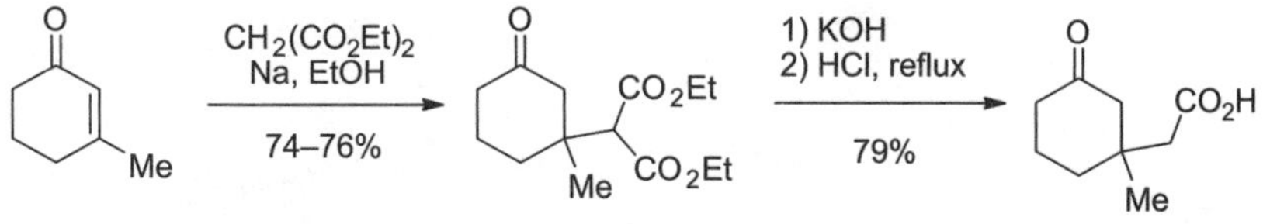

Malonate anions also add to acetylenes, albeit with modest E/Z selectivity. In the following example, triethylamine catalyzes the addition to alkynoate with 3:1 E/Z selectivity; the paper includes several examples with enones, which exhibit more modest selectivity (1-2:1). Phoramidites (e.g., HMPT = $(Me_2N)_3P$) also catalyze this reaction.[74]

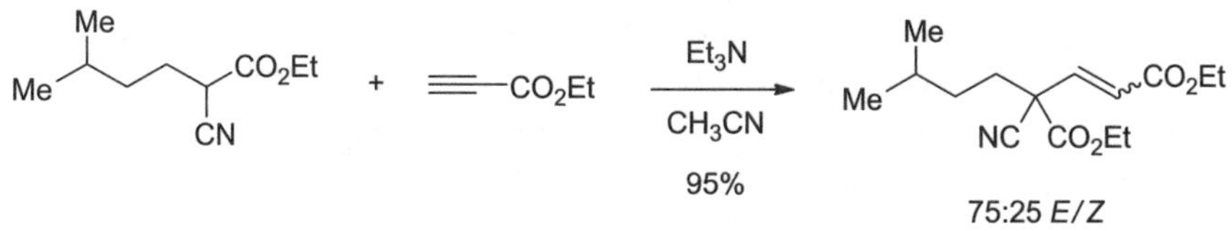

75:25 *E/Z*

95%

[72]Clarke, M. L.; France, M. B. *Tetrahedron* **2008**, *64*, 9003–9031.

[73]Schostarez, H.; Paquette, L. A. *Tetrahedron* **1981**, *37*, 4431–4435.

[74]Grossman, R. B.; Comesse, S.; Rasne, R. M.; Hattori, K.; Delong, M. N. *J. Org. Chem.* **2003**, *68*, 871–874.

Heathcock and coworkers have extensively studied the stereoselectivity of enolate additions to Michael acceptors. The two examples shown below exhibit high regio- and stereoselectivity; the latter is not the case for many other examples. Both examples involve addition of a pre-formed lithium enolate of a propionamide to an $\alpha,\beta$-unsaturated ketone.[75]

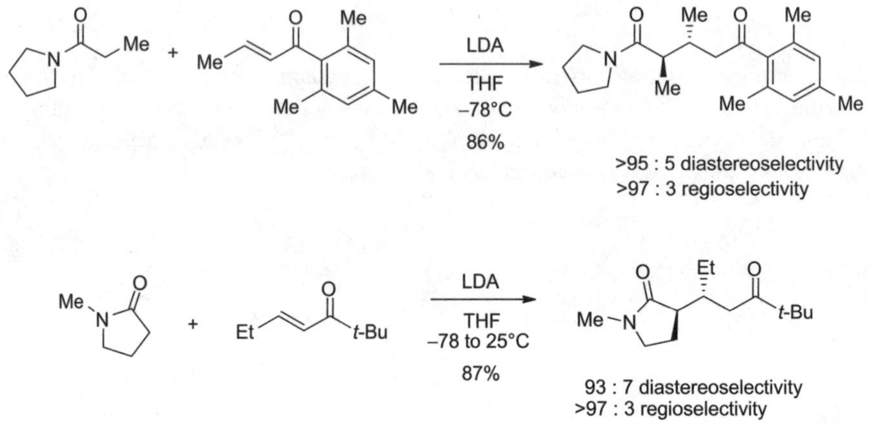

>95 : 5 diastereoselectivity
>97 : 3 regioselectivity

>93 : 7 diastereoselectivity
>97 : 3 regioselectivity

Enol silanes can also add to activated olefins (Mukaiyama–Michael addition). Several useful variants of this reaction have been developed; a preparatively useful enantioselective example from the Evans lab is shown below.[76]

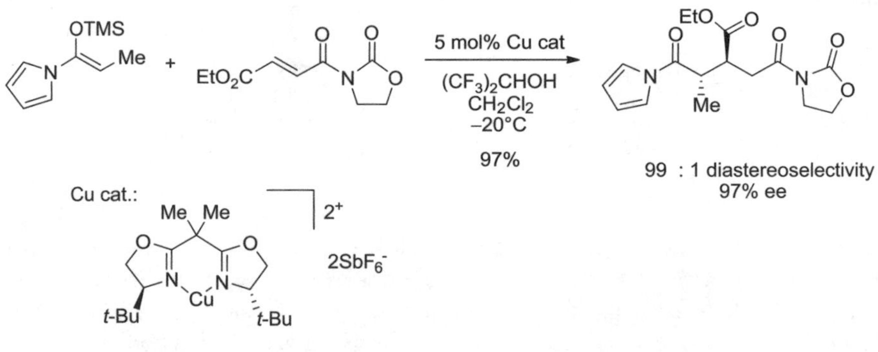

99 : 1 diastereoselectivity
97% ee

Electron-rich aromatic rings can also function as nucleophiles in these reactions. For example, MacMillan has applied organocatalysis to effect addition of furan[77] and pyrrole[78] nucleophiles to enals. An example from the second reference is shown below.

[75]Oare, D. A.; Henderson, M. A.; Sanner, M. A.; Heathcock, C. H. *J. Org. Chem.* **1990**, *55*, 132–157.
[76]Evans, D. A.; Scheidt, K. A.; Johnston, J. N.; Willis, M. C. *J. Am. Chem. Soc.* **2001**, *123*, 4480–4491.
[77]Brown, S. P.; Goodwin, N. C.; MacMillan, D. W. C. *J. Am. Chem. Soc.* **2003**, *125*, 1192–1194.
[78]Paras, N. A.; MacMillan, D. W. C. *J. Am. Chem. Soc.* **2001**, *123*, 4370–4371.

Nitroalkanes can also serve as effective nucleophiles in conjugate additions, as shown in the following example for an $\alpha,\beta$-unsaturated ester.[79]

Trialkyl phosphines can catalyze a vinylogous Baylis–Hillman reaction, which is essentially the addition of the $\alpha$-carbon of an enone to a Michael acceptor. The two examples below come from Roush[80] and Krische.[81] Each cyclization proceeds with high diastereoselectivity (10:1 and >19:1, respectively).

[79]Hanessian, S.; Yun, H.; Hou, Y.; Tintelnot-Blomley, M. *J. Org. Chem.* **2005**, *70*, 6746–6756.
[80]Frank, S. A.; Mergott, D. J.; Roush, W. R. *J. Am. Chem. Soc.* **2002**, *124*, 2404–2405.
[81]Wang, L.-C.; Luis, A. L.; Agapiou, K.; Jang, H.-Y.; Krische, M. J. *J. Am. Chem. Soc.* **2002**, *124*, 2402–2403.

### 3.5.4 Hydrogen–Alkyl Addition, Stork Enamine Reaction

When condensed with secondary amines (e.g., pyrrolidine), enolizable aldehydes and ketones generate enamines, which are excellent nucleophiles for conjugate addition to Michael acceptors such as methyl vinyl ketone. The following two examples are representative.[82,83] In both cases, Mannich cylization of the initial product generates a cyclohexenone product.

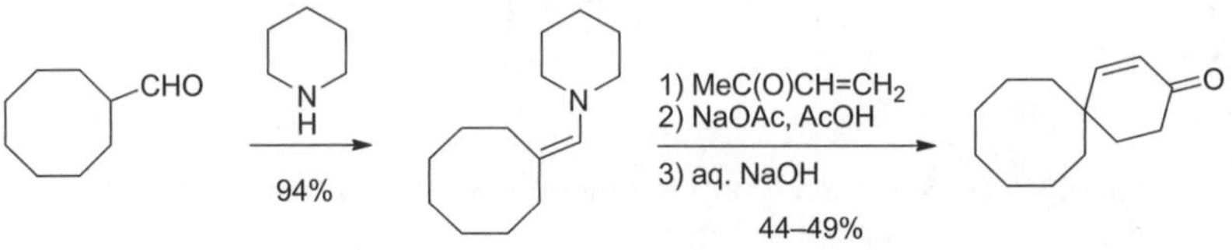

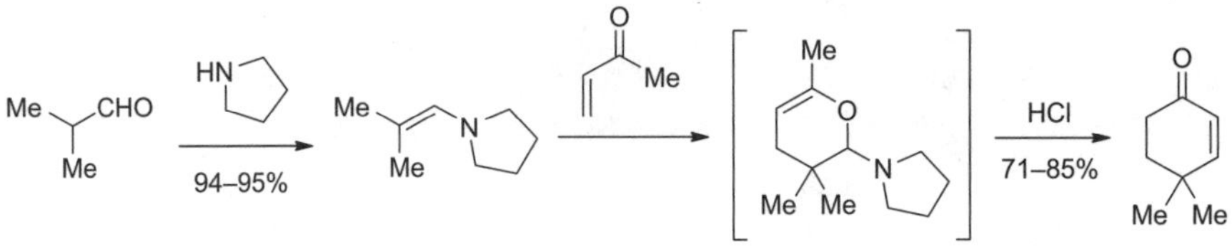

Use of a chiral secondary amine such as (S)-proline can generate chiral, non-racemic products, as in the classic preparation of the Wieland-Miescher ketone shown below.[84]

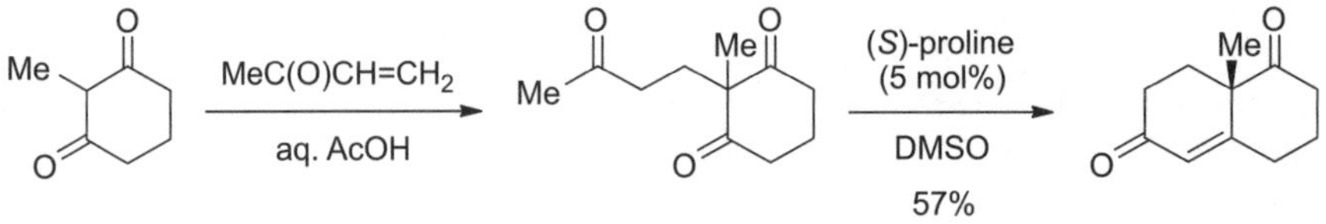

### 3.5.5 Hydrogen–Alkyl Addition, Metal-Catalyzed

The first example in this category is narrow in scope, being limited to malonate nucleophiles, but is remarkably atom economical and efficient (e.g., no solvent, and just 2 mol% of an inexpensive catalyst).[85]

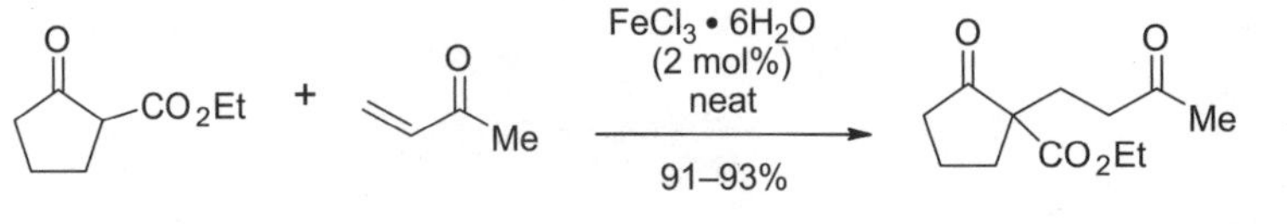

Addition of cuprate reagents ($R_2CuLi$) to electron-deficient olefins is well known; copper-catalyzed addition of Grignard reagents effects a similar transformation. The example below comes from a synthesis of (−)-8-phenylmenthol.[86]

[82]Kane, V. V.; Jones, M., Jr., *Org. Synth.* **1983**, *61*, 129–133.
[83]Chan, Y.; Epstein, W. W. *Org. Synth.* **1973**, *53*, 48–52.
[84]Buchschacher, P.; Fuerst, A. *Org. Synth.* **1985**, *63*, 37–43.
[85]Christoffers, J. *Org. Synth.* **2002**, *78*, 249–253.
[86]Ort, O. *Org. Synth.* **1987**, *65*, 203–214.

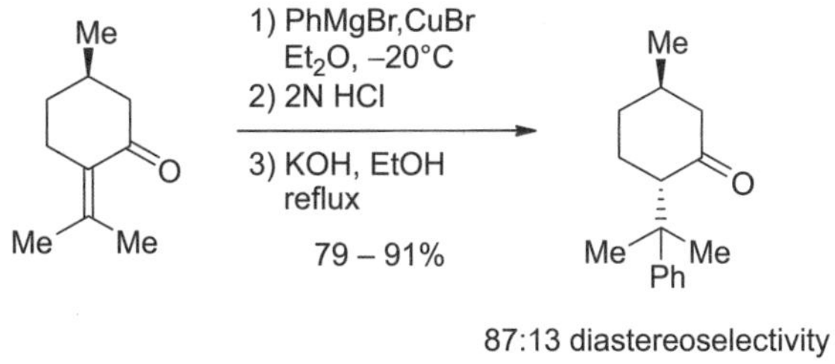

87:13 diastereoselectivity

Another useful variant is shown below, and utilizes catalytic $MnCl_2$ in addition to CuCl.[87] Note that the decreased diastereoselectivity is due to the absence of the KOH-EtOH equilibration step.

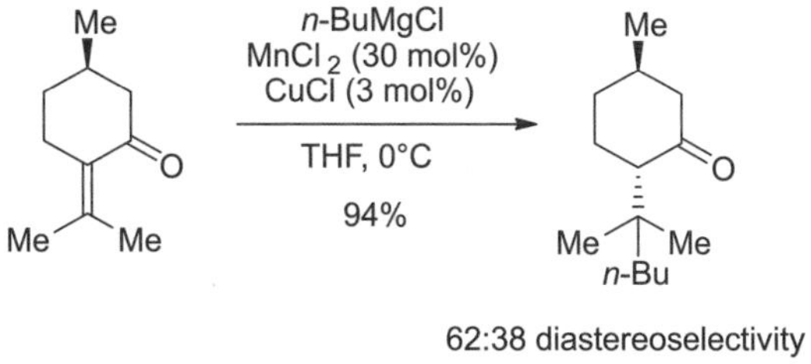

62:38 diastereoselectivity

Several metal-mediated conjugate additions have been developed more recently, similar to conjugate addition of cuprate or Grignard reagents. Nickel catalyzes the conjugate addition of vinylzirconium reagents to enones, as shown below.[88]

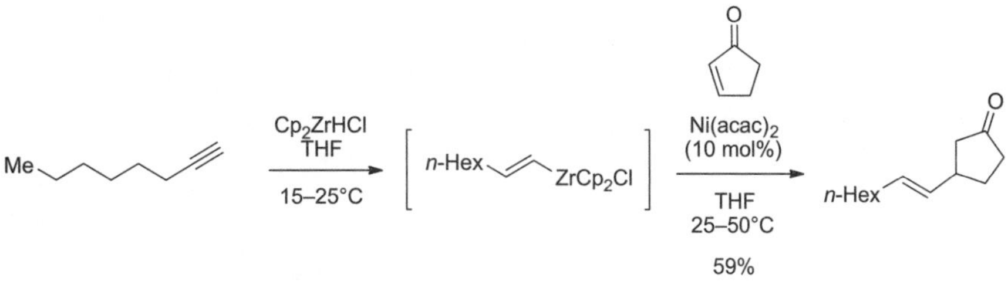

Copper can catalyze the conjugate addition of alkylzirconium reagents to enones. The following example proceeds in 73% yield on 1 g scale (the initial product, 3-butyl-4,4-dimethylcyclohexanone, was isolated as the crystalline dinitrophenylsulfonyl-hydrazone).[89]

[87] Alami, M.; Marquais, S.; Cahiez, G. *Org. Synth.* **1995**, *72*, 135–146.
[88] Sun, R. C.; Okabe, M.; Coffen, D. L.; Schwartz, J. *Org. Synth.* **1993**, *71*, 838–.
[89] Wipf, P.; Xu, W.; Smitrovich, J. H.; Lehmann, R.; Venanzi, L. M. *Tetrahedron* **1994**, *50*, 1935–1954.

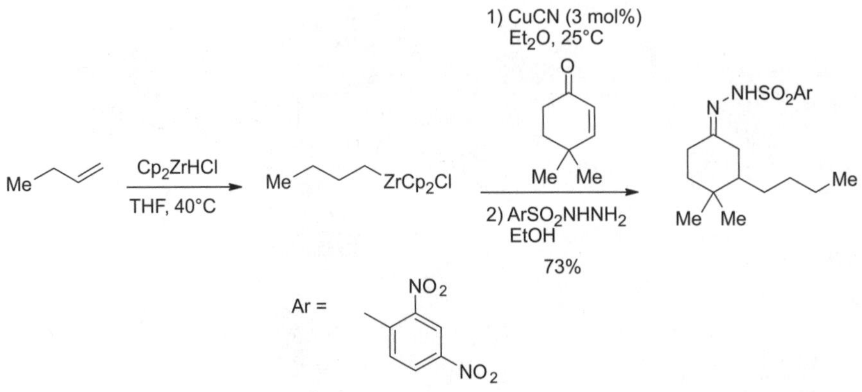

The following example is similar to the preceding one, but does not involve isolation of the intermediate vinylzirconium reagent.[90]

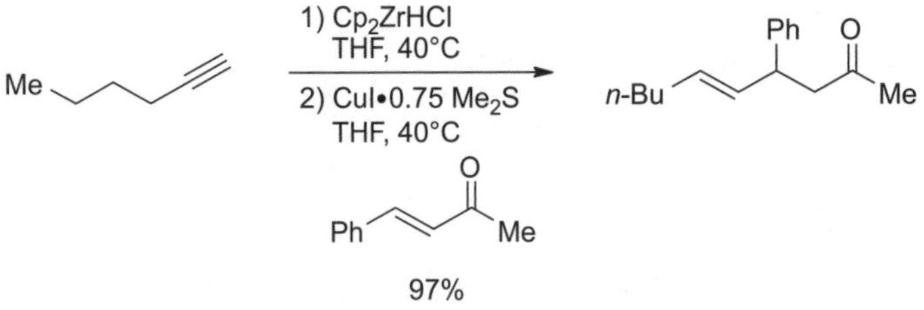

### 3.5.6   Hydroformylation/Hydroacylation

Addition of H-C(O)X can be a useful process, and is frequently executed with a metal catalyst using CO and $H_2$ (for X = H), or CO and XH (where X = OR, NR$_2$, etc.). A Rh-catalyzed hydroformylation (X = H) is shown below.[91]

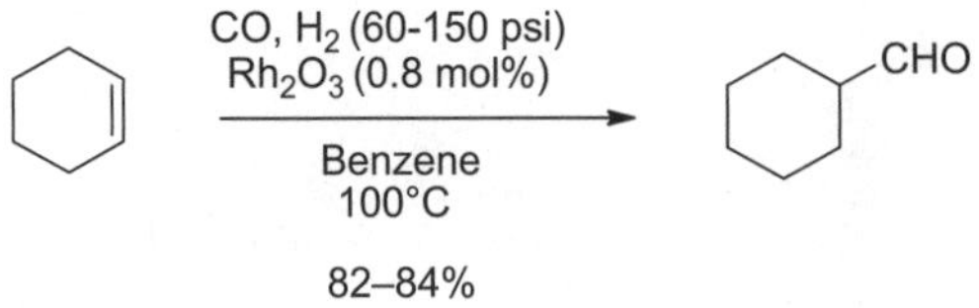

An application of this reaction to the synthesis of tolterodine is shown below.[92]

[90]El-Batta, A.; Hage, T. R.; Plotkin, S.; Bergdahl, M. *Org. Lett.* **2004**, *6*, 107–110.
[91]Pino, P.; Botteghi, C. *Org. Synth.* **1977**, *57*, 11–16.
[92]Botteghi, C.; Corrias, T.; Marchetti, M.; Paganelli, S.; Piccolo, O. *Org. Process Res. Dev.* **2002**, *6*, 379–383.

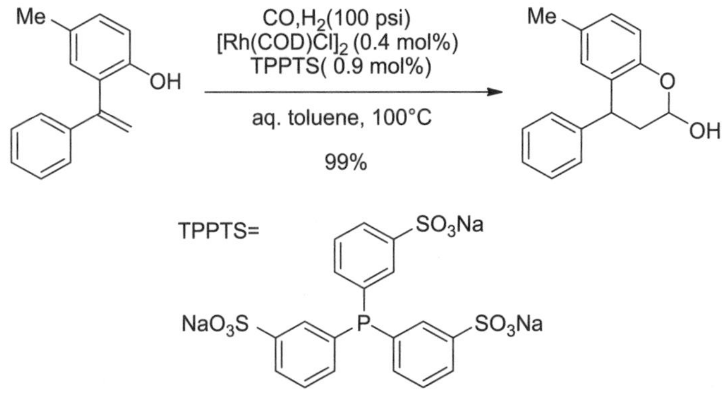

A variant of this reaction couples a reductive amination with the initial hydroformylation to effect the net addition of H-CH$_2$NR$_2$ across a terminal olefin with high regioselectivity.[93]

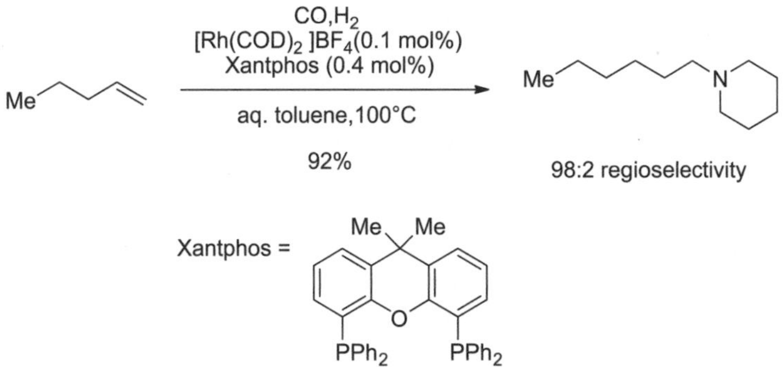

Hydrocyanation of enones can be achieved via the conjugate addition of Et$_2$AlCN, as developed by Nagata.[94] The example shown below traps the initial enolate with TMSCl and then NBS to provide the α-bromoketone. Replacing NBS with aqueous perchloric acid generates the corresponding ketone (appropriate caution should be exercised when handling perchloric acid).[95]

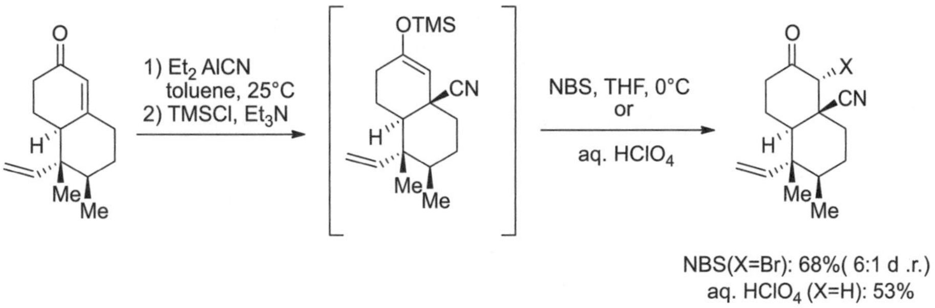

NBS(X=Br): 68%( 6:1 d .r.)
aq. HClO$_4$ (X=H): 53%

[93] Ahmed, M.; Seayad, A. M.; Jackstell, R.; Beller, M. *J. Am. Chem. Soc.* **2003**, *125*, 10311–10318.

[94] Nagata, W.; Yoshioka, M. *Organic Reactions (New York)* **1977**, *25*, 255–476.

[95] Ihara, M.; Katsumata, A.; Egashira, M.; Suzuki, S.; Tokunaga, Y.; Fukumoto, K. *J. Org. Chem.* **1995**, *60*, 5560–5566.

An indirect way to effect hydroacylation of an enone is a conjugate reduction and subsequent trapping of the enolate with a suitable acylating reagent. The example below is from Mander.[96]

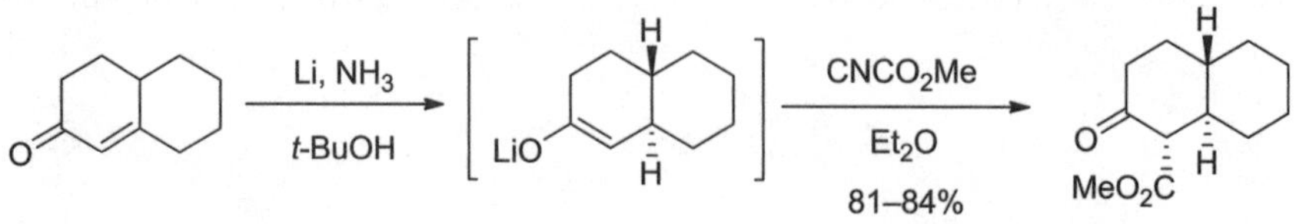

The Stetter reaction is formally a hydroacylation of an olefin; it proceeds via a deprotonated cyanohydrin, which functions as an acyl anion equivalent. Conjugate addition of this anion to a Michael acceptor constitutes hydroalkylation of the electron-deficient olefin, as shown in the example below with acrylonitrile.[97]

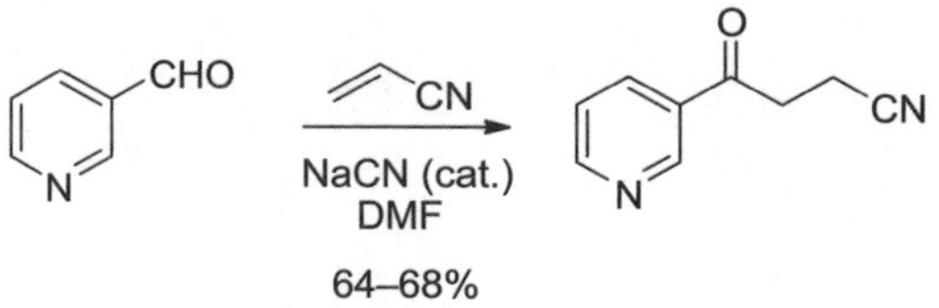

A particularly useful variant of the Stetter reaction uses a thiazolium salt in place of the cyanide catalyst, and thus is much safer to operate. Two examples are shown below, the first demonstrating preparation of 2,5,8-nonanetrione,[98] and the second providing access to a densely substituted cyclopentenone utilized by Tius in a formal total synthesis of roseophilin.[99]

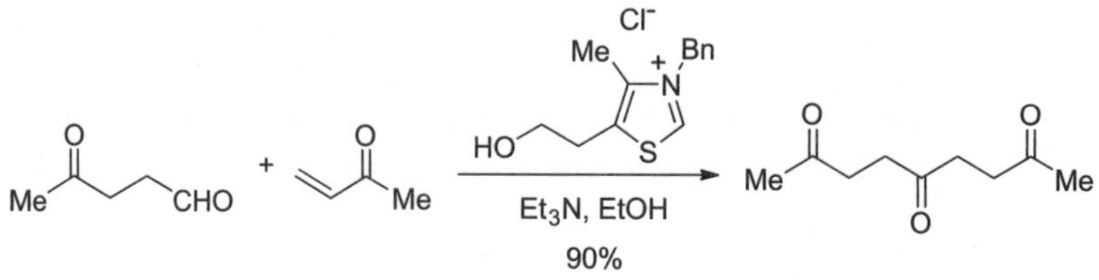

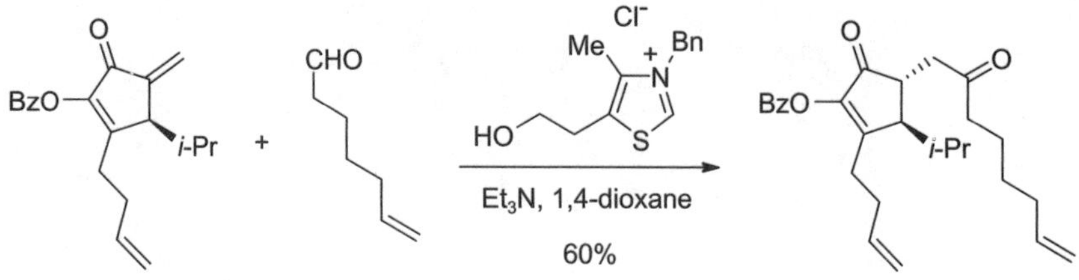

[96]Crabtree, S. R.; Mander, L. N.; Sethi, S. P. *Org. Synth.* **1992**, *70*, 256–264.
[97]Stetter, H.; Kuhlmann, H.; Lorenz, G. *Org. Synth.* **1980**, *59*, 53–58.
[98]Sant, M. P.; Smith, W. B. *J. Org. Chem.* **1993**, *58*, 5479–5481.
[99]Harrington, P. E.; Tius, M. A. *Org. Lett.* **1999**, *1*, 649–651.

### 3.5.7   Nazarov Cyclization

The Nazarov cyclization is an electrocyclic, cationic process, which provides a powerful approach to cyclopentenones.[100,101]

Denmark has developed a silicon-directed Nazarov cyclization that offers improvements in reaction rate and regioselectivity relative to the non-directed reaction. The following example is taken from Stille's synthesis of $\Delta^{9(12)}$-capnellene, in which he utilized this reaction in both five-membered ring cyclizations.[102]

The following example was utilized in a synthesis of the hydroazulene core of guanacastepene A.[103]

Conia has developed a fragmentation-recombination Nazarov cyclization of $\alpha,\beta$-unsaturated esters; the following example was reported by J. D. White in *Organic Syntheses*.[104]

[100]Habermas, K. L.; Denmark, S. E.; Jones, T. K. *Organic Reactions (New York)* **1994**, *45*, 1–158.

[101]Pellissier, H. *Tetrahedron* **2005**, *61*, 6479–6517.

[102]Crisp, G. T.; Scott, W. J.; Stille, J. K. *J. Am. Chem. Soc.* **1984**, *106*, 7500–7506.

[103]Chiu, P.; Li, S. *Org. Lett.* **2004**, *6*, 613–616.

[104]Schwartz, K. D.; White, J. D.; Tosaki, S.-y.; Shibasaki, M. *Org. Synth.* **2006**, *83*, 49–54.

Scandium and copper catalysts have been reported to effectively catalyze the Nazarov cyclization. Moderate to good asymmetric induction has been observed with chiral bis-oxazoline ligands (61–88% ee with 20–100 mol% catalyst loads).[105–107]

The palladium-catalyzed Nazarov cyclization shown below occurs under particularly mild conditions (25°C, aqueous acetone).[108] The reaction retains its conrotatory stereo-specificity despite the reduced temperature.

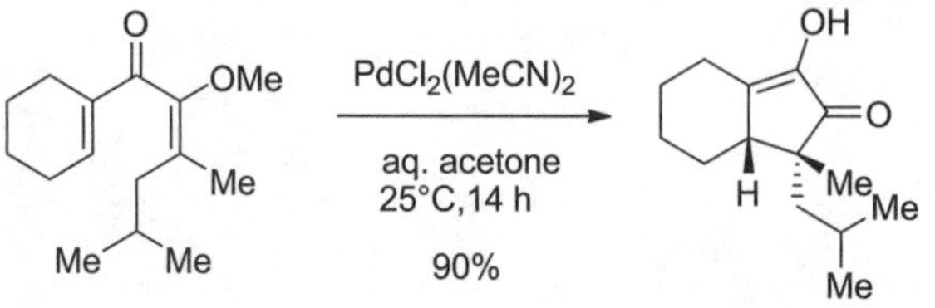

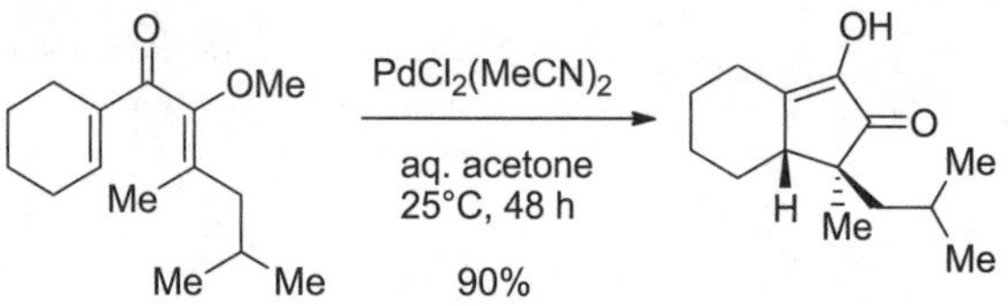

Cascade polyene cyclizations induced by an initial Nazarov cyclization have also been reported. These involve a combination of C–H and C–C additions across olefins, and are discussed in Section 3.15.8.

### 3.5.8    Radical-Mediated C–H Addition

Intramolecular radical cyclizations have been utilized effectively in numerous total syntheses. These reactions can be efficient, but their overall attractiveness frequently suffers from the use of toxic organostannane reagents to serve as radical propagators (e.g., $Bu_3SnH$). The example below shows cyclization of an α-silyl radical, and was utilized in a total synthesis of talaromycin A.[109]

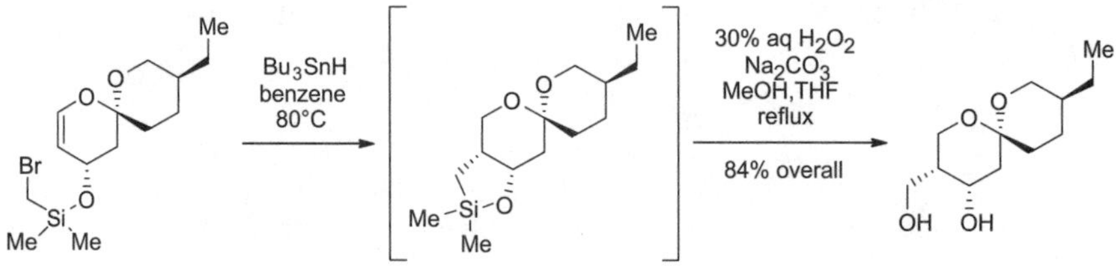

[105]Liang, G.; Gradl, S. N.; Trauner, D. *Org. Lett.* **2003**, *5*, 4931–4934.

[106]Aggarwal, V. K.; Belfield, A. J. *Org. Lett.* **2003**, *5*, 5075–5078.

[107]He, W.; Sun, X.; Frontier, A. J. *J. Am. Chem. Soc.* **2003**, *125*, 14278–14279.

[108]Bee, C.; Leclerc, E.; Tius, M. A. *Org. Lett.* **2003**, *5*, 4927–4930.

[109]Crimmins, M. T.; O'Mahony, R. *J. Org. Chem.* **1989**, *54*, 1157–1161.

In some cases, the quantity of alkyltin reagent can be reduced to catalytic quantities through use of a stoichiometric reducing agent such as a trialkylsilane.[110] The following example from the synthesis of (−)-malyngolide uses 10 mol% tributyltin chloride and sodium borohydride as the stoichiometric reductant.[111] Not surprisingly, no diastereoselectivity is observed for the newly formed stereocenter.

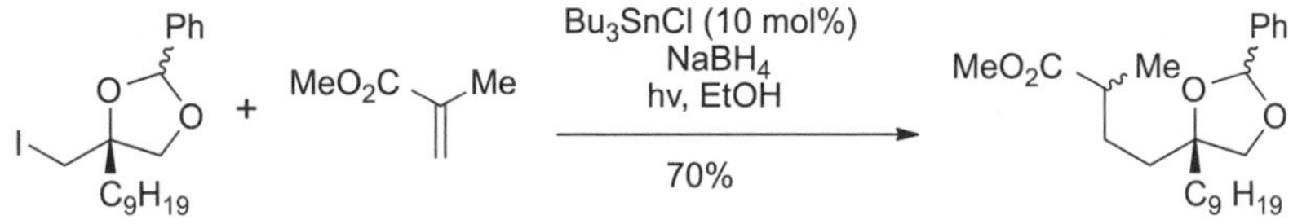

## 3.6   HALOGEN–HALOGEN ADDITION

Addition of chlorine or bromine to olefins is generally an efficient reaction; an example from the steroid literature is shown below.[112] In the second example, formation of the 5α-bromo-6β-chloro isomer provides access to mixed-dihalides, and also served as proof of the mechanism (i.e., α-bromononium ion formation followed by chloride attack from the β-face at the less-substituted 6-position).[113]

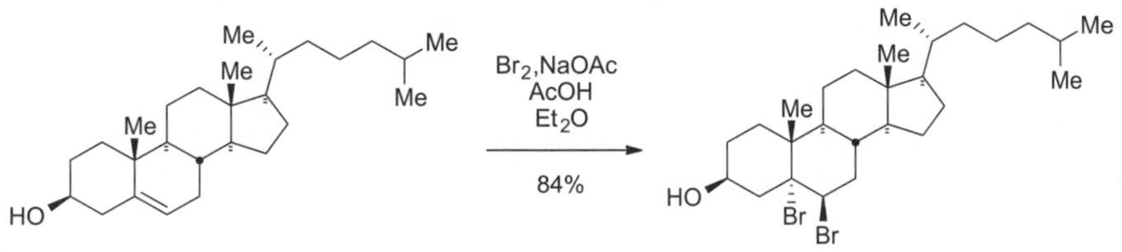

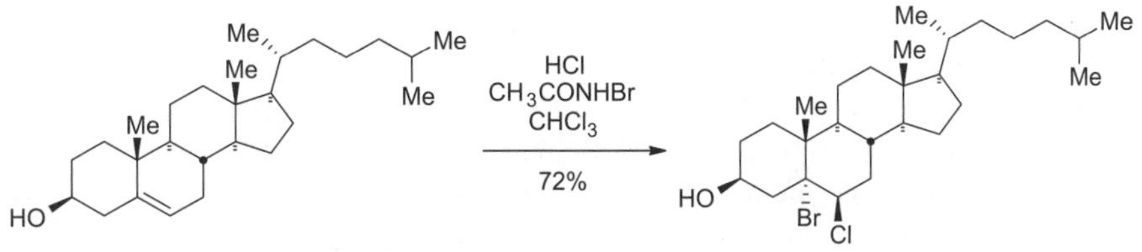

## 3.7   HYDROXY–HALOGEN ADDITION

Bromohydrins can be prepared from olefins by bromination in aqueous media. In the example below, NBS in aqueous acetone provided the bromohydrin in 95% yield on 40 g scale.[114]

[110]Hays, D. S.; Scholl, M.; Fu, G. C. *J. Org. Chem.* **1996**, *61*, 6751–6752.

[111]Giese, B.; Rupaner, R. *Liebigs Ann. Chem* **1987**, 231–233.

[112]Fieser, L. F. *Org. Synth.* **1955**, *35*, 43–49.

[113]Ziegler, J. B.; Shabica, A. C. *J. Am. Chem. Soc.* **1952**, *74*, 4891–4894.

[114]Larsen, R. D.; Davis, P.; Corley, E. G.; Reider, P. J.; Lamanec, T. R.; Grabowski, E. J. J. *J. Org. Chem.* **1990**, *55*, 299–304.

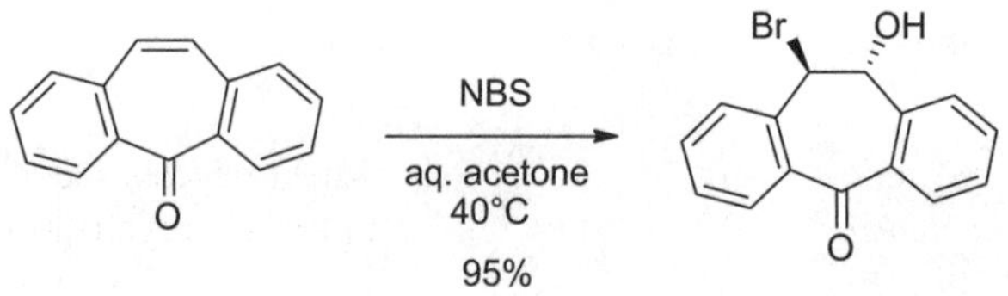

When performed in an alcohol solvent, a similar reaction delivers the analogous alkyl ether.[115] The regioselectivity in this reaction is noteworthy, and presumably reflects attack of methanol on the bromonium ion carbon distal from the electron-withdrawing alcohol functionality.

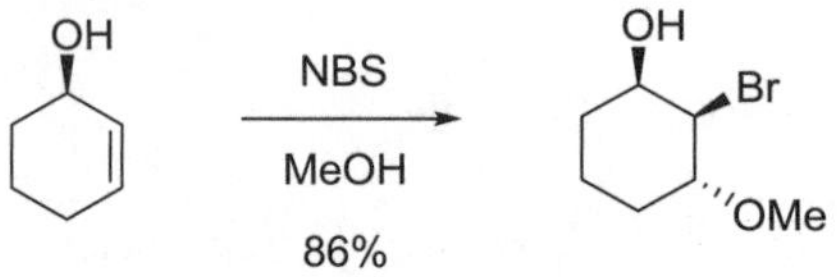

Intramolecular cyclization of olefins containing oxygen nucleophiles can be similarly effected to form cyclic ethers. Iodine is frequently used as the electrophile in these cyclizations, although other electrophiles have been used (e.g., PhSeBr, Hg(OAc)$_2$, NBS). With carboxylic acids, the product is a lactone.[116]

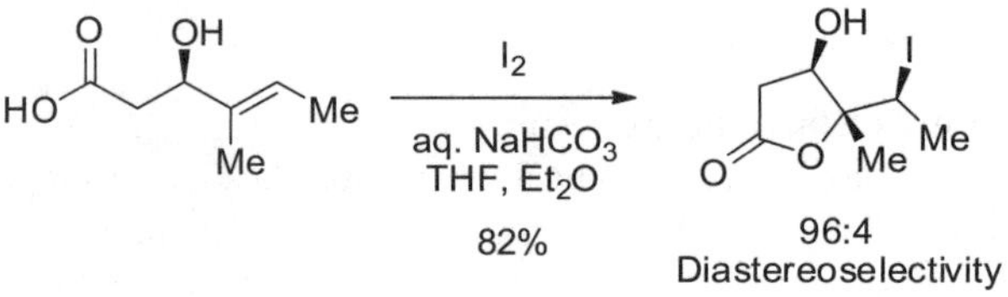

With alcohol substrates, the product is a cyclic ether, such as a tetrahydrofuran, as shown below for a spiro-substituted diol (see also Section 1.2.2.4).[117] Levels of diastereoselection in these cyclizations can be quite high with proper choice of substrate.[118] Note that with this and the preceding example, diethyl ether is used as a solvent; less volatile and easier handled solvents such as MTBE or 2-methyl-THF are preferred, and may very well work in place of Et$_2$O in these types of reactions.

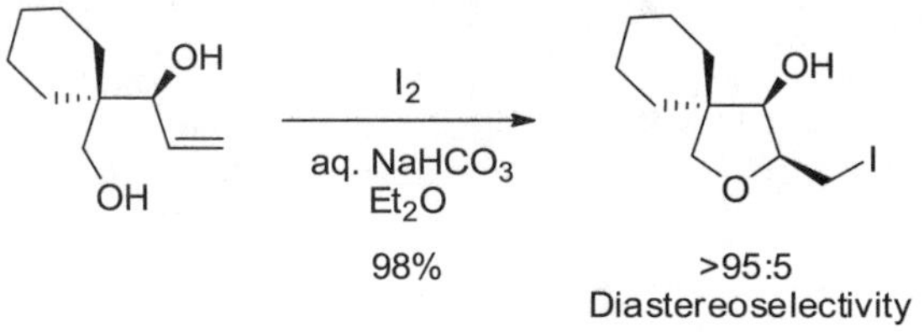

[115]Yang, C. H.; Wu, J. S.; Ho, W. B. *Tetrahedron* **1990**, *46*, 4205–4216.

[116]Chamberlin, A. R.; Dezube, M.; Dussault, P.; McMills, M. C. *J. Am. Chem. Soc.* **1983**, *105*, 5819–5825.

[117]Tamaru, Y.; Hojo, M.; Kawamura, S.; Sawada, S.; Yoshida, Z. *J. Org. Chem.* **1987**, *52*, 4062–4072.

[118]Rychnovsky, S. D.; Bartlett, P. A. *J. Am. Chem. Soc.* **1981**, *103*, 3963–3964.

## 3.8  AMINO–HALOGEN ADDITION

Properly configured amino olefins will cyclize to form heterocycles under conditions leading to bromonium ion intermediates. The following example shows cyclization of an aziridine to form a bicyclic pyrrolidine. The analogous cyclizations of the *trans*-aziridine and the piperidine homologue are higher yielding, but give a mixture of diastereomers.[119]

Free radical addition of dialkyl-*N*-chloramines is also known, as shown in the following example with diethyl-*N*-chloramine.[120]

## 3.9  CARBON–HALOGEN ADDITION

### 3.9.1  Alkyl–Halogen Addition

Carbohalogenation of olefins is not a transformation of wide utility. One notable exception to this generalization is atom-transfer radical cyclization, largely developed by Curran and coworkers. One advantage of this method relative to the tin hydride approach (see Section 3.5.8) is that catalytic quantities of hexaalkyldistannane ($R_3SnSnR_3$) reagents are frequently adequate to initiate the radical chain. The examples below show additions of alkyl iodides to both olefins[121] and acetylenes.[122]

[119]Chen, G.; Sasaki, M.; Li, X.; Yudin, A. K. *J. Org. Chem.* **2006**, *71*, 6067–6073.
[120]Neale, R. S.; Marcus, N. L. *J. Org. Chem.* **1967**, *32*, 3273–3284.
[121]Curran, D. P.; Chang, C. T. *J. Org. Chem.* **1989**, *54*, 3140–3157.
[122]Curran, D. P.; Chen, M. H.; Kim, D. *J. Am. Chem. Soc.* **1989**, *111*, 6265–6276.

Two other preparatively useful examples are shown below. The first is a CuBr-catalyzed addition of a diazonium salt to acrylic acid.[123] The second is a free radical addition of chloroform to a terminal olefin.[124]

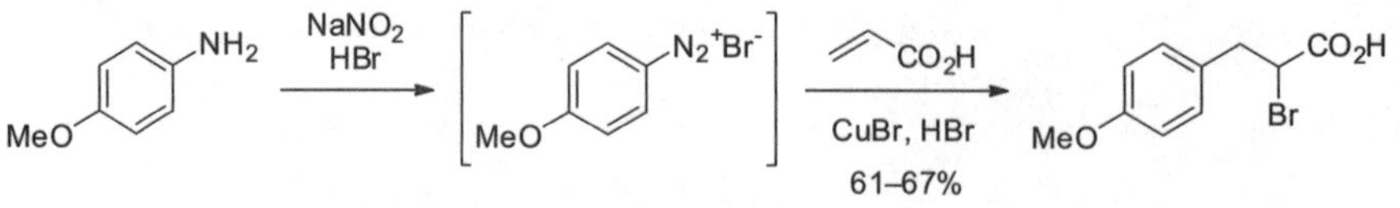

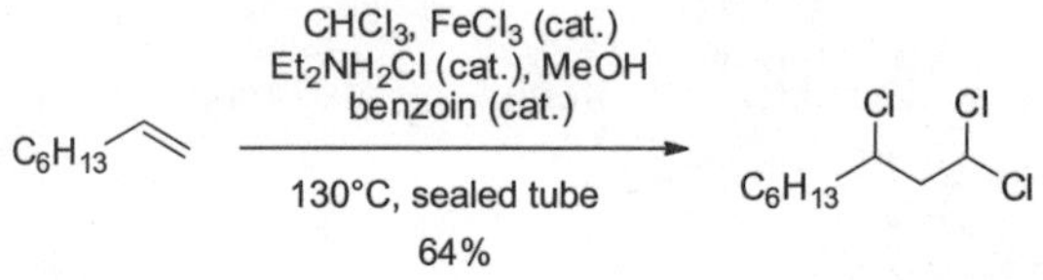

## 3.9.2 Acyl–Halogen Addition

Addition of an acyl (or aldehyde) group and a halogen atom across an olefin is a synthetically useful type of Friedel–Crafts acylation. The product β-halo carbonyl compound is frequently eliminated to the olefin in a subsequent step, as in the following preparation of ethyl 3,3-diethoxypropanoate.[125]

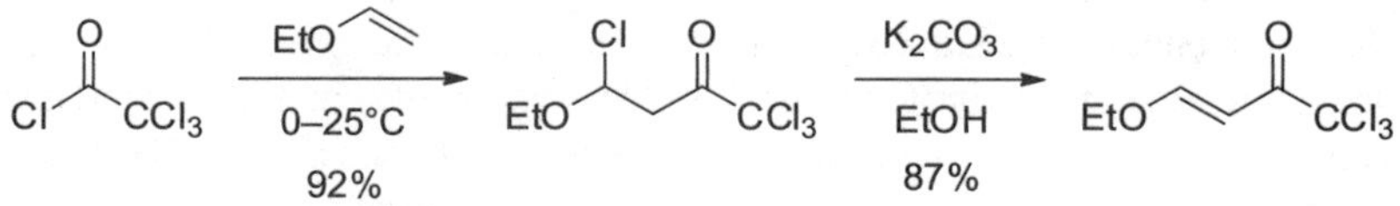

Intramolecular variants are also known, as shown in the following example from Paquette's lab. In this particular case, the anticipated Prins cyclization did not occur, presumably due to stereoelectronic constraints.[126]

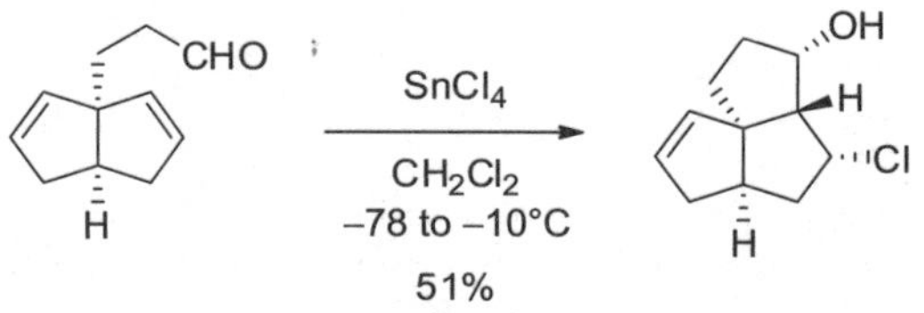

The following examples show the intramolecular cyclization of an enone onto an aldehyde or another enone with placement of an iodide on the β-carbon.[127]

[123]Cleland, G. H. *Org. Synth.* **1971**, *51*, 1–4.
[124]Vofsi, D.; Asscher, M. *Org. Synth.* **1965**, *45*, 104–107.
[125]Tietze, L. F.; Voss, E.; Hartfiel, U. *Org. Synth.* **1990**, *69*, 238–244.
[126]Cheney, D. L.; Paquette, L. A. *J. Org. Chem.* **1989**, *54*, 3334–3347.
[127]Yagi, K.; Turitani, T.; Shinokubo, H.; Oshima, K. *Org. Lett.* **2002**, *4*, 3111–3114.

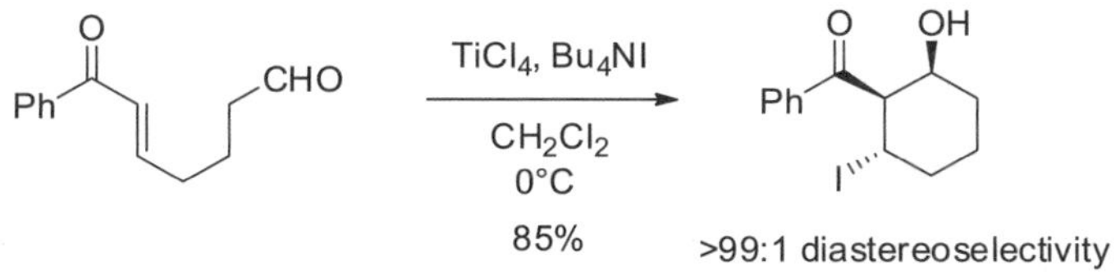

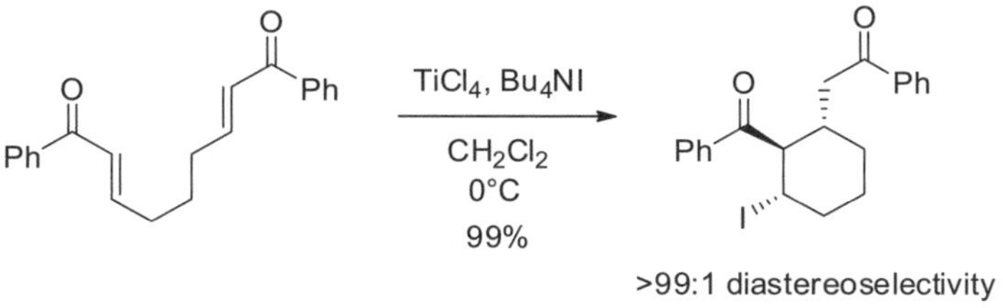

85%    >99:1 diastereoselectivity

99%

>99:1 diastereoselectivity

## 3.10  OXYGEN–OXYGEN ADDITION

### 3.10.1  Dihydroxylation of Olefins

Dihydroxylation of olefins is an important transformation in organic synthesis, and several methods have been developed to effect this oxidation. Although osmium tetroxide is most frequently used, it suffers from the handling and waste stream issues associated with osmium. Nonetheless, this reagent is particularly useful on laboratory scale for its predictability and tremendously wide scope of substrates. An example using the "Upjohn procedure" (aqueous N-methylmorpholine-N-oxide as the stoichiometric oxidant) is shown below.[128] For water-soluble product diols, *in situ* protection with $PhB(OH)_2$ is a useful variation.[129–131]

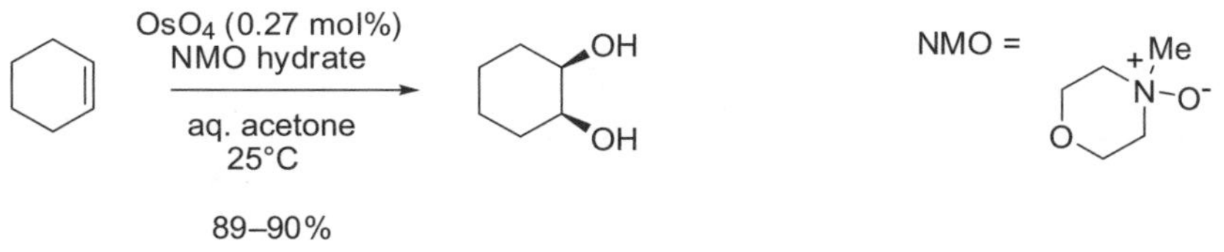

89–90%

Although less general than osmium-mediated dihydroxylation, potassium permanganate offers a more practical solution when the substrate allows use of this reagent.[132] This transformation was recently utilized in a large scale preparation of a nicotine partial agonist; the diol product was then converted to a piperidine via an oxidative cleavage/reductive amination sequence.[133]

[128]VanRheenen, V.; Cha, D. Y.; Hartley, W. M. *Org. Synth.* **1978**, *58*, 43–52.
[129]Iwasawa, N.; Kato, T.; Narasaka, K. *Chemistry Letters* **1988**, 1721–1724.
[130]Gypser, A.; Michel, D.; Nirschl, D. S.; Sharpless, K. B. *J. Org. Chem.* **1998**, *63*, 7322–7327.
[131]Sakurai, H.; Iwasawa, N.; Narasaka, K. *Bull. Chem. Soc. Jpn.* **1996**, *69*, 2585–2594.
[132]Ogino, T.; Mochizuki, K. *Chemistry Letters* **1979**, 443–446.
[133]Bashore, C. G.; Vetelino, M. G.; Wirtz, M. C.; Brooks, P. R.; Frost, H. N.; McDermott, R. E.; Whritenour, D. C.; Ragan, J. A.; Rutherford, J. L.; Makowski, T. W.; Brenek, S. J.; Coe, J. W. *Org. Lett.* **2006**, *8*, 5947–5950.

Catalytic ruthenium represents another alternative to osmium for effecting olefin dihydroxylation. The following example was optimized for a 50 kg pilot plant run.[134]

One of the most important catalytic asymmetric transformations in organic chemistry is the Sharpless asymmetric dihydroxylation (AD) of olefins. The OsO$_4$-ligand-salt complex is commercially available (AD-mix-α and β), making this a particularly convenient transformation, although the osmium waste stream and handling issues must still be dealt with. The following example provides a quantitative yield of product with 84% ee; after recrystallization, a 52–57% yield of 97% ee material is obtained.[135,136]

[134]Couturier, M.; Andresen, B. M.; Jorgensen, J. B.; Tucker, J. L.; Busch, F. R.; Brenek, S. J.; Dube, P.; Ende, D. J.; Negri, J. T. *Org. Process Res. Dev.* **2002**, *6*, 42–48.

[135]Oi, R.; Sharpless, K. B. *Org. Synth.* **1996**, *73*, 1–12.

[136]Sharpless, K. B.; Amberg, W.; Beller, M.; Chen, H.; Hartung, J.; Kawanami, Y.; Lubben, D.; Manoury, E.; Ogino, Y.; et al. *J. Org. Chem.* **1991**, *56*, 4585–4588.

### 3.10.2  Keto-Hydroxylation of Olefins

With proper choice of conditions, olefins can also be converted to α-hydroxy ketones with $KMnO_4$.[137] Note that the outcome of these oxidations is dependent on both solvent and pH. In the example shown, the use of aqueous acetone and a weak acid was credited with controlling the path of oxidation (i.e., diol vs. hydroxy-ketone vs. diketone).

RuO$_4$ (in this case generated catalytically from $RuCl_3$ and Oxone, $2KHSO_5 \cdot KHSO_4 \cdot K_2SO_4$) effects a similar transformation, with good levels of regioselectivity with certain unsymmetrical disubstituted olefins.[138] Styrene derivatives and enones are particularly good substrates for this oxidation.

### 3.10.3  Dihydroxylation of Acetylenes

In a reaction similar to the dihydroxylation of olefins, acetylenes can be oxidized to α-diketones. One of the simplest procedures uses $KMnO_4$ in aqueous acetone (example below).[139] $H_5IO_6$ in acetic acid effects the analogous transformation of diphenylacetylene with similar efficiency.[140]

[137]Srinivasan, N. S.; Lee, D. G. *Synthesis* **1979**, 520–521.
[138]Plietker, B. *J. Org. Chem.* **2004**, *69*, 8287–8296.
[139]Srinivasan, N. S.; Lee Donald, G. *J. Org. Chem.* **1979**, *44*, 1574.
[140]Gebeyehu, G.; McNelis, E. *J. Org. Chem.* **1980**, *45*, 4280–4283.

A variety of other oxidants have been reported to accomplish this transformation, SO$_3$-dioxane complex provides good results,[141] while elevated temperatures (>100°C) are required for oxidation in DMSO with PdCl$_2$[142] and I$_2$.[143,144]

In the example below, an alkoxyacetylene is oxidized to an α-ketoester; since the substrate is generated from addition of LiCCOEt to a ketone, the acetylene anion is serving as an acyl anion equivalent.[145]

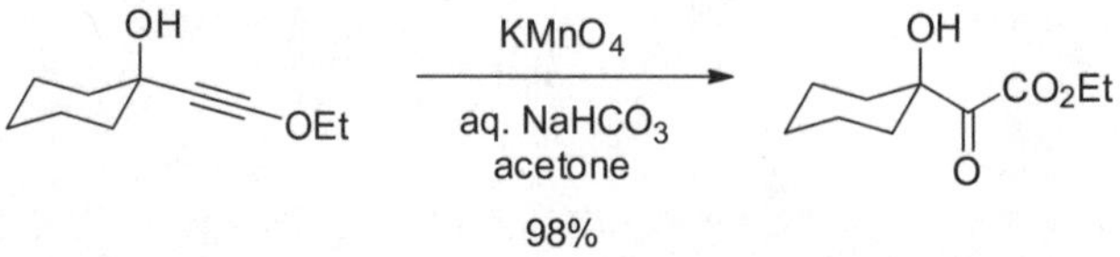

### 3.10.4  Epoxidation

Peracids (RCO$_3$H) are a useful reagent class for effecting epoxidations, with *meta*-chloroperbenzoic acid (*m*-CPBA) and magnesium monoperoxyphthalate (MMPP) being particularly convenient for laboratory scale oxidations.

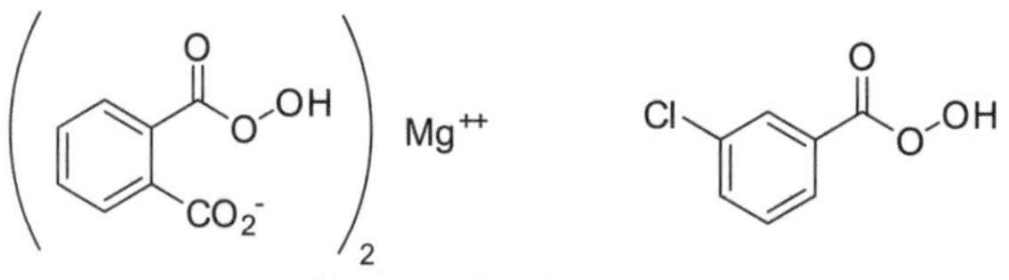

MMPP                    *m*-CPBA

It should be noted that high concentrations of non-aqueous *m*-CPBA have been reported to be explosive,[146] and are generally not commercially available. Several commercial suppliers sell the reagent at 75–80%, of which the remainder is the benzoic acid derivative and water. This material can be used as received, or purified to the crystalline peracid before use.[147] Two examples of its use are shown below. The first example provides high α-diastereoselectivity for the epoxidation of cholesterol, and in his detailed procedure, Fieser reports that other peracids were less stereoselective.[148] The second example also exhibited high diastereoselectivity with *m*-CPBA (>99:1), but little selectivity with MMPP (53:47).

[141]Rogatchov, V. O.; Filimonov, V. D.; Yusubov, M. S. *Synthesis* **2001**, 1001–1003.

[142]Yusubov, M. S.; Zholobova, G. A.; Vasilevsky, S. F.; Tretyakov, E. V.; Knight, D. W. *Tetrahedron* **2002**, *58*, 1607–1610.

[143]Yusybov, M. S.; Filimonov, V. D. *Synthesis* **1991**, 131–132.

[144]Yusubov, M. S.; Filimonov, V. D.; Vasilyeva, V. P.; Chi, K.–W. *Synthesis* **1995**, 1234–1236.

[145]Tatlock, J. H. *J. Org. Chem.* **1995**, *60*, 6221–6223.

[146]Urben, P.; *Bretherick's Handbook of Reactive Chemical Hazards, 6th ed*; Butterworth-Heinemann Ltd: Oxford, **1999**.

[147]Schwartz, N. N.; Blumbergs, J. H. *J. Org. Chem.* **1964**, *29*, 1976–1979.

[148]Fieser, L. F.; Fieser, M. *Reagents for Organic Synthesis*; Wiley: New York, **1967**.

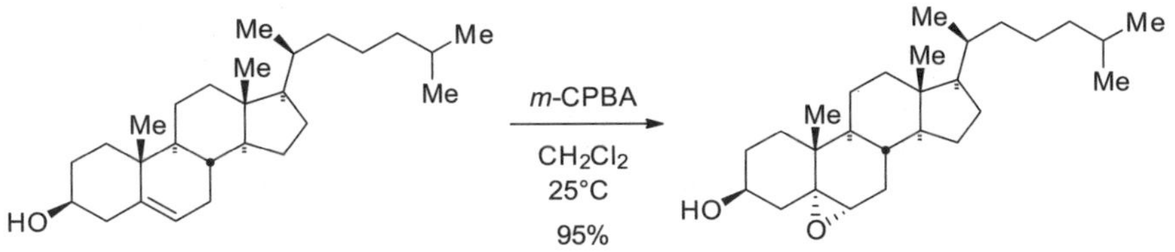

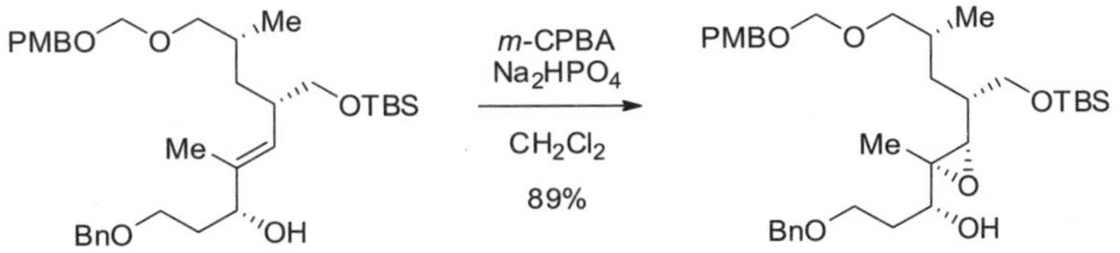

An example of an allylic alcohol epoxidation with MMPP is shown below in which there is no diastereoselectivity issue; the example was run on 12 g scale.[149]

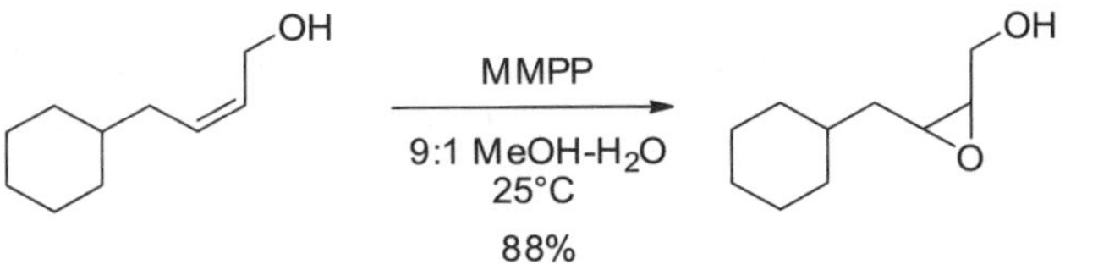

Also convenient for laboratory scale epoxidations are dioxirane reagents, generally derived by the *in situ* oxidation of ketones with Oxone (larger scale examples are known using the *in situ* formation method as well). The example shown below uses tetrahydrothiopyran-4-one (the active catalyst is the derived sulfone).[150] Numerous examples from the acetone-derived dioxirane (3,3-dimethyldioxirane, DMDO) are also known.[151–153]

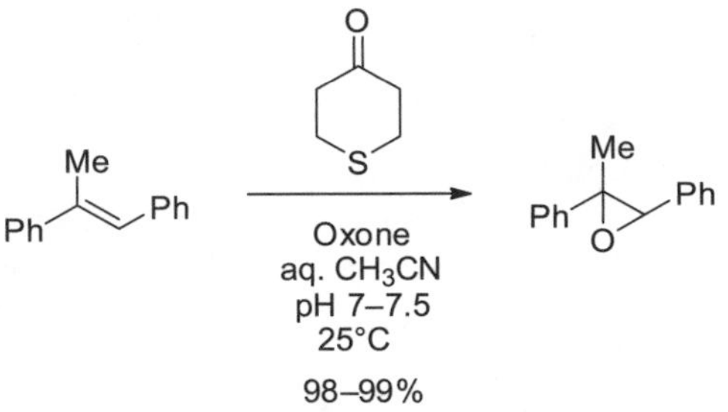

[149]Krysan, D. J.; Haight, A. R.; Menzia, J. A.; Welch, N. *Tetrahedron* **1994**, *50*, 6163–6172.

[150]Yang, D.; Yip, Y.-C.; Jiao, G.-S.; Wong, M.-K. *Org. Synth.* **2002**, *78*, 225–233.

[151]Roberge, J. Y.; Beebe, X.; Danishefsky, S. J. *J. Am. Chem. Soc.* **1998**, *120*, 3915–3927.

[152]Murray, R. W. *Chem. Rev.* **1989**, *89*, 1187–1201.

[153]Adam, W.; Rao, P. B.; Degen, H.-G.; Levai, A.; Patonay, T.; Saha-Moeller, C. R. *J. Org. Chem.* **2002**, *67*, 259–264.

Several groups have reported asymmetric variants of dioxirane-mediated epoxidations.[154–156] The example below is an improved preparation of one of Shi's chiral catalysts; the ketone catalyst is prepared in four steps from *D*-fructose.[157]

Metal-mediated epoxidations represent one of the first practical examples of catalytic enantioselective transformations, and are ubiquitous in epoxidation chemistry. The following example represents a procedure that has undergone significant optimization in the Sharpless lab.[158]

This reaction is also useful for the kinetic resolution of racemic allylic alcohol substrates. The example below is from Roush's synthesis of ( + )-olivose.[159] Note that the yield of 27–33% represents 54–66% of theory. A 30% recovery (60% of theory) of optically enriched starting alcohol is also obtained from the reaction (72% ee, (*S*)-enantiomer). Vanadium-catalyzed epoxidations (and kinetic resolution) of homoallylic alcohols have also been reported.[160]

[154]Wu, X.-Y.; She, X.; Shi, Y. *J. Am. Chem. Soc.* **2002**, *124*, 8792–8793.

[155]Yang, D.; Wong, M.-K.; Yip, Y.-C.; Wang, X.-C.; Tang, M.-W.; Zheng, J.-H.; Cheung, K.-K. *J. Am. Chem. Soc.* **1998**, *120*, 5943–5952.

[156]Denmark, S. E.; Wu, Z. *Synlett* **1999**, 847–859.

[157]Nieto, N.; Molas, P.; Benet-Buchholz, J.; Vidal-Ferran, A. *J. Org. Chem.* **2005**, *70*, 10143–10146.

[158]Hill, J. G.; Sharpless, K. B.; Exon, C. M.; Regenye, R. *Org. Synth.* **1985**, *63*, 66–78.

[159]Roush, W. R.; Brown, R. J. *J. Org. Chem.* **1983**, *48*, 5093–5101.

[160]Zhang, W.; Yamamoto, H. *J. Am. Chem. Soc.* **2007**, *129*, 286–287.

Jacobsen has developed a class of Mn-salen catalysts for the enantioselective epoxidation of olefins. This catalyst class offers a wider range of substrate scope than the Sharpless epoxidation, as the requirement for an allylic alcohol in the substrate olefin is obviated. The following example was utilized in the synthesis of the HIV protease inhibitor indinavir.[161] The catalyst preparation is also described in an earlier *Organic Syntheses* procedure.[162]

Related metal-salen catalysts developed by Jacobsen are useful for the hydrolytic kinetic resolution of epoxides.[163] As noted earlier for the Sharpless kinetic resolution, the 36–37% yield represents 72–74% of theory.

[161]Larrow, J. F.; Roberts, E.; Verhoeven, T. R.; Ryan, K. M.; Senanayake, C. H.; Reider, P. J.; Jacobsen, E. N. *Org. Synth.* **1999**, *76*, 46–56.
[162]Larrow, J. F.; Jacobsen, E. N. *Org. Synth.* **1998**, *75*, 1–11.
[163]Stevenson, C. P.; Nielsen, L. P. C.; Jacobsen, E. N.; McKinley, J. D.; White, T. D.; Couturier, M. A.; Ragan, J. *Org. Synth.* **2006**, *83*, 162–169.

A final class of epoxidations that differs in mechanism from those described previously is the nucleophilic epoxidation of enones. This can be effected with basic hydrogen peroxide, and proceeds via a conjugate addition/elimination mechanism.[164]

### 3.10.5   Singlet Oxygen Addition to Dienes

Photochemical-mediated addition of singlet oxygen to dienes can be an efficient route to endoperoxides, which can then be reduced to diols or converted to other functional groups. *Meso*-tetraphenylporphine is frequently used as the photosensitizing agent. The following example from Carl Johnson's group proceeds in 76% yield on 20 g scale.[165] In a useful extension of this chemistry, Toste has demonstrated the desymmetrization of endoperoxides with chiral quinuclidine bases (Kornblum DeLeMare rearrangement).[166]

### 3.11   OXYGEN–NITROGEN ADDITION

Sharpless has developed the osmium-mediated aminohydroxylation of olefins with chloramines to provide either sulfonamide or carbamate-protected *cis*-aminoalcohols. For sulfonamides, chloramine-T is used in the presence of 5 mol% phase-transfer catalyst.[167]

[164]Felix, D.; Wintner, C.; Eschenmoser, A. *Org. Synth.* **1976**, *55*, 52–56.
[165]Johnson, C. R.; Golebiowski, A.; Steensma, D. H. *J. Am. Chem. Soc.* **1992**, *114*, 9414–9418.
[166]Staben, S. T.; Xin, L.; Toste, F. D. *J. Am. Chem. Soc.* **2006**, *128*, 12658–12659.
[167]Herranz, E.; Sharpless, K. B. *Org. Synth.* **1983**, *61*, 85–93.

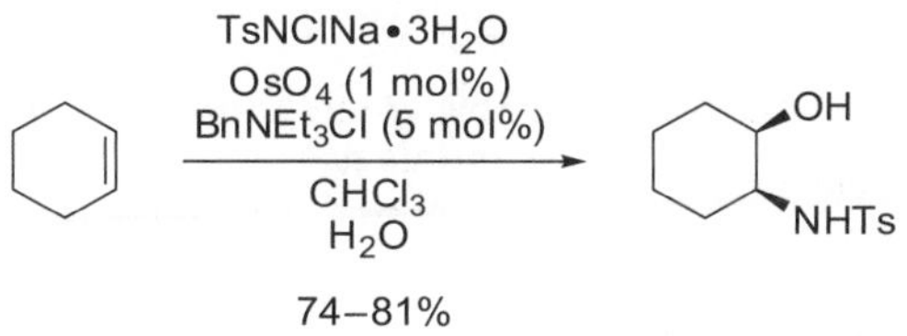
TsNClNa•3H2O
OsO4 (1 mol%)
BnNEt3Cl (5 mol%)
CHCl3
H2O
74–81%
OH
NHTs

For preparation of carbamates, an *N*-chloro sodium salt is utilized in the presence of silver nitrate. The *N*-chloro carbamate salt is prepared by a known method from *t*-butyl hypochlorite, ethyl carbamate, and base.[168]

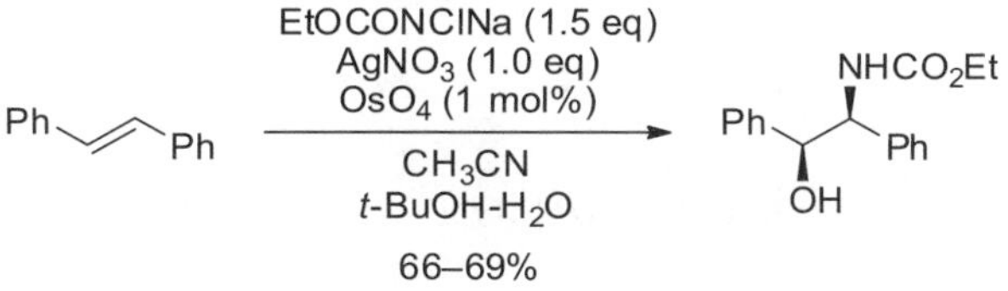
EtOCONClNa (1.5 eq)
AgNO3 (1.0 eq)
OsO4 (1 mol%)
CH3CN
t-BuOH-H2O
66–69%
Ph
Ph
NHCO2Et
Ph
Ph
OH

This method has been utilized as shown below for the preparation of oxazolidinones from styrenes. Regioselectivity for the benzylamine isomer ranges from good to modest (91:9 to 50:50), but the desired isomer cyclizes more rapidly to the oxazolidinone, which facilitates separation (uncyclized material can be separated by acid extraction). The example shown proceeded in 73% yield and 90–93% ee. An experiment on 2 g scale is described, and the authors indicate that scale-up to kilograms has been performed.[169]

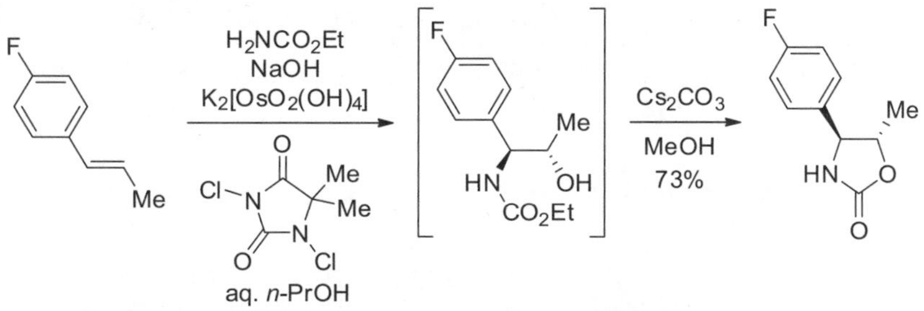
F
Me
H2NCO2Et
NaOH
K2[OsO2(OH)4]
aq. n-PrOH
Cs2CO3
MeOH
73%
Me
HN
OH
CO2Et
HN
O
O

Merck scientists have utilized an asymmetric epoxidation followed by a Ritter reaction to provide the net *syn*-aminohydroxylation of olefins. The sequence shown below was developed for the synthesis of the protease inhibitor indinavir. The preparation of the starting epoxide is described in the same *Organic Syntheses* procedure.[161]

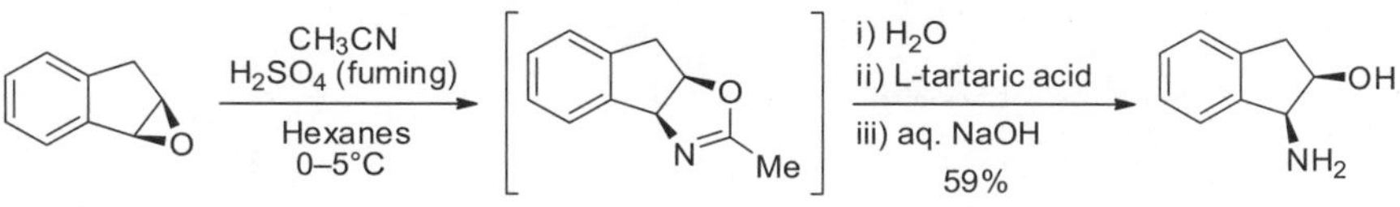
CH3CN
H2SO4 (fuming)
Hexanes
0–5°C
i) H2O
ii) L-tartaric acid
iii) aq. NaOH
59%
OH
NH2

[168]Herranz, E.; Sharpless, K. B. *Org. Synth.* **1983**, *61*, 93–97.

[169]Barta, N. S.; Sidler, D. R.; Somerville, K. B.; Weissman, S. A.; Larsen, R. D.; Reider, P. J. *Org. Lett.* **2000**, *2*, 2821–2824.

Copper reagents will effect the coupling of oxaziridines and olefins to generate oxazolines, which constitutes a net aminohydroxylation of the olefin.[170] Experimental evidence supports a stepwise, cationic mechanism for this reaction.

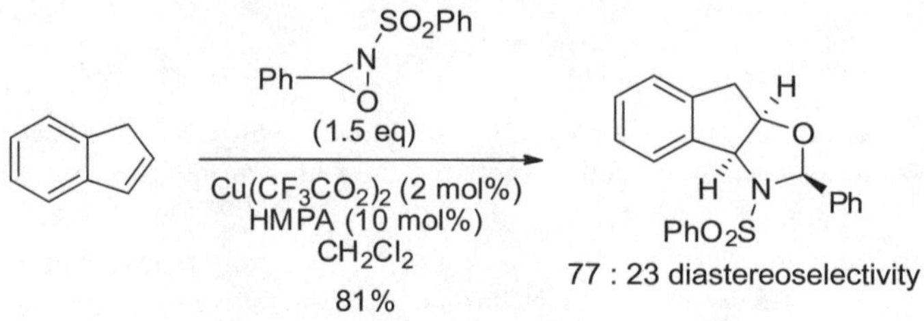

77 : 23 diastereoselectivity

## 3.12    NITROGEN–NITROGEN ADDITION

### 3.12.1    Aziridination

Rhodium-catalyzed oxidation of olefins with sulfonamides in the presence of a stoichiometric oxidant provides an efficient synthesis of aziridines. The following examples are from Du Bois;[171] Doyle has reported a similar catalyst system for the addition of p-toluenesulfonamide ($TsNH_2$) to olefins catalyzed by $Rh_2(cap)_4$ (cap = caprolactam).[172]

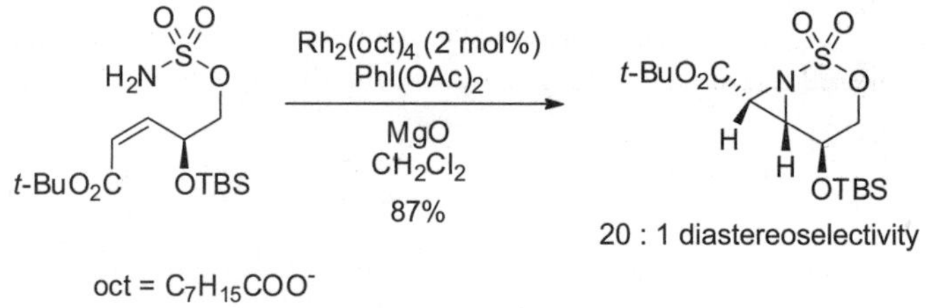

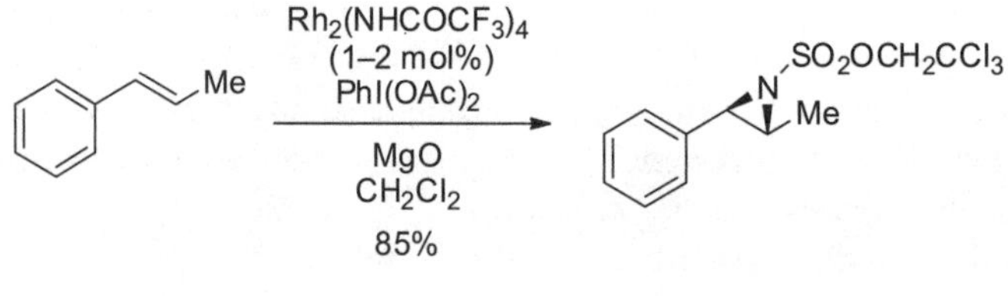

$oct = C_7H_{15}COO^-$

20 : 1 diastereoselectivity

### 3.12.2    N–N Addition to Olefins

A practical, general method for the diamination of olefins remains elusive. Osmium-mediated diaminations are known, as shown below.[173] Reactions with fumarate esters

[170]Michaelis, D. J.; Shaffer, C. J.; Yoon, T. P. *J. Am. Chem. Soc.* **2007**, *129*, 1866–1867.
[171]Guthikonda, K.; Wehn, P. M.; Caliando, B. J.; Du Bois, J. *Tetrahedron* **2006**, *62*, 11331–11342.
[172]Catino, A. J.; Nichols, J. M.; Forslund, R. E.; Doyle, M. P. *Org. Lett.* **2005**, *7*, 2787–2790.
[173]Chong, A. O.; Oshima, K.; Sharpless, K. B. *J. Am. Chem. Soc.* **1977**, *99*, 3420–3426.

showed the diamination to be stereospecific. The practicality of this method is limited by the need for a stoichiometric osmium reagent, the use of LiAlH$_4$ for converting the initial metal complex to the free diamine, and the use of less desirable solvents.

Palladium-catalyzed diamination of conjugated dienes and trienes is known, and utilizes a diaziridinone reagent to generate cyclic ureas.[174] While of more limited substrate scope, this method offers considerable advantages in terms of handling and waste disposal. A similar Pd(II)-catalyzed diamination of dienes has been reported,[175] and utilizes ureas as the diamine source. A stoichiometric oxidant is required (O$_2$ or benzoquinone).

Palladium-catalyzed intramolecular diaminations to generate *N*-sulfonyl ureas are known (example below).[176] An asymmetric variant has also been reported.[177] Intramolecular copper-catalyzed diaminations to generate sulfonylureas are also known.[178]

Vicinal diamines can also be prepared indirectly from diols, via bis-opening of the cyclic sulfate (available in two steps from the corresponding diol). While indirect, this method translates the practicality of Sharpless' asymmetric dihydroxylation methodology (see Section 3.10.1) into diamines.[179] The example shown is for a symmetrical diamine; use of two different amines (the first for formation of the aziridinium ion, and the second to open this intermediate) was also reported.

[174]Du, H.; Zhao, B.; Shi, Y. *J. Am. Chem. Soc.* **2007**, *129*, 762–763.
[175]Bar, G. L. J.; Lloyd-Jones, G. C.; Booker-Milburn, K. I. *J. Am. Chem. Soc.* **2005**, *127*, 7308–7309.
[176]Streuff, J.; Hoevelmann, C. H.; Nieger, M.; Muniz, K. *J. Am. Chem. Soc.* **2005**, *127*, 14586–14587.
[177]Muniz, K.; Nieger, M. *Synlett* **2003**, 211–214.
[178]Zabawa, T. P.; Kasi, D.; Chemler, S. R. *J. Am. Chem. Soc.* **2005**, *127*, 11250–11251.
[179]Richardson, P. F.; Nelson, L. T. J.; Sharpless, K. B. *Tetrahedron Lett.* **1995**, *36*, 9241–9244.

### 3.12.3 Triazines from Azide–Olefin Cycloaddition

Cycloaddition of alkylazides with olefins generates triazines, and is formally an example of an olefin diamination. The following example provides access to a variety of useful 2-aminoglycosides (via the 1,2-aziridine intermediate).[180] Appropriate care should be taken when handling any alkyl azide reagent.

### 3.12.4 N–N Addition to Dienes (1,4)

While limited in scope, the following diene–diazodicarboxylate cycloaddition is an example of a preparatively useful 1,4-diamination of a diene; the product is a precursor to bicyclo [2.1.0]pentane.

### 3.13 CARBON–OXYGEN ADDITION

### 3.13.1 Carbon–Oxygen Addition to Olefins (1,2)

#### 3.13.1.1 [2 + 2] Cycloadditions of Olefins and Carbonyl Compounds

*Oxetane Formation (Olefin + Carbonyl)* The photochemical reaction of aldehydes and electron-rich olefins (Paterno–Büchi reaction)[181] has been utilized for the synthesis of substituted oxetanes. With proper disposition of substituents, moderate to good levels of diastereo- and regiocontrol can be realized with this reaction, as shown in the example below.[182]

>90 : 10 diastereoselectivity
>95 : 5 regioselectivity

[180]Dahl, R. S.; Finney, N. S. *J. Am. Chem. Soc.* **2004**, *126*, 8356–8357.
[181]Buchi, G.; Inman, C. G.; Lipinsky, E. S. *J. Am. Chem. Soc.* **1954**, *76*, 4327–4331.
[182]Bach, T.; Joedicke, K.; Kather, K.; Froehlich, R. *J. Am. Chem. Soc.* **1997**, *119*, 2437–2445.

Cycloaddition of aldehydes and furans (also known as the photoaldol reaction) has received particular attention in the context of natural product synthesis, as exemplified below from a total synthesis of avenaciolide.[183] The photocycloaddition is run neat in 10 to 15 equivalents of furan, and the product is carried on without further purification to the crystalline lactol.

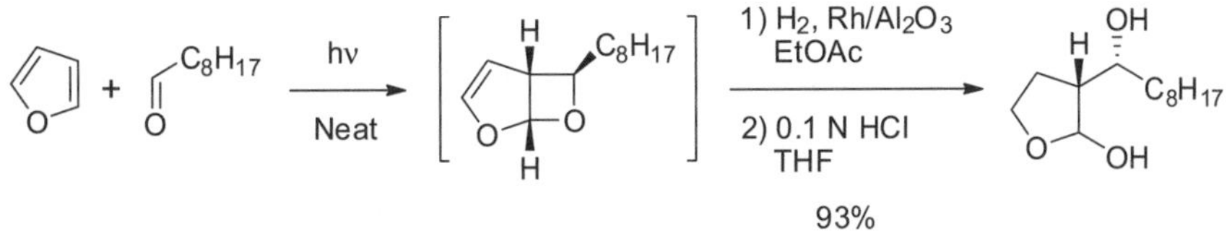

β-*Lactone Formation (Ketene + Carbonyl)*    Nelson has reported the cinchona alkaloid-catalyzed [2 + 2] cycloaddition of ketenes and aldehydes, as shown below for the synthesis of a *cis*-α,β-disubstituted β-lactone.[184] Note that this is an extension of Wynberg's 1982 report of the quinidine-catalyzed reaction of ketene and chloral to form the β-lactone in 95% chemical and optical yield.[185]

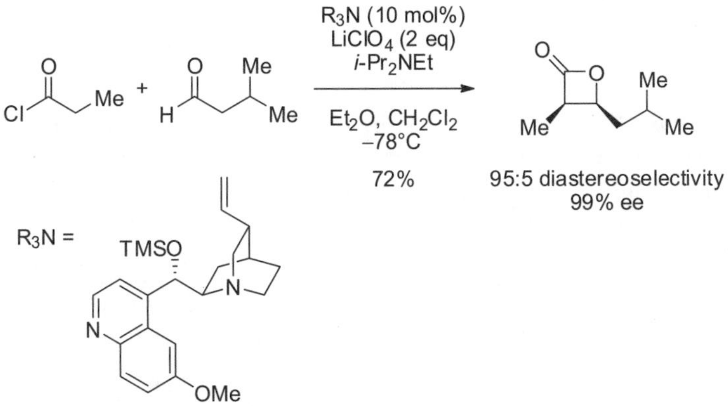

The following example shows a chiral aluminum amide catalyst that effects the [2 + 2] cycloaddition of ketenes and aldehydes to form β-lactones with good enantioselectivity. The catalyst is formed *in situ* by the reaction of $Me_3Al$ and the bis-sulfonamide; the latter reagent is available in two steps from valinol.[186] Similarly, $C_2$-symmetric Cu(II) complexes catalyze the [2 + 2] cycloaddition of TMS-ketene with aldehydes.[187]

[183]Schreiber, S. L.; Hoveyda, A. H. *J. Am. Chem. Soc.* **1984**, *106*, 7200–7202.
[184]Zhu, C.; Shen, X.; Nelson, S. G. *J. Am. Chem. Soc.* **2004**, *126*, 5352–5353.
[185]Wynberg, H.; Staring, E. G. J. *J. Am. Chem. Soc.* **1982**, *104*, 166–168.
[186]Nelson, S. G.; Mills, P. M.; Ohshima, T.; Shibasaki, M. *Org. Synth.* **2005**, *82*, 170–178.
[187]Evans, D. A.; Janey, J. M. *Org. Lett.* **2001**, *3*, 2125–2128.

Romo has developed a SnCl$_4$-mediated [2 + 2] cycloaddition route to *cis*-α,β-disubstituted β-lactones from aldehydes and thiopyridyl ketene acetals.[188] The *cis*-selectivity is in contrast to his ZnCl$_2$-mediated coupling, which provides predominantly the *trans* diastereomer (and presumably occurs via a cationic, stepwise mechanism).

**3.13.1.2  Nitrone-Olefin [3 + 2] Cycloadditions**   The [3 + 2] cycloaddition of nitrones and olefins effects the net addition of carbon and oxygen across an olefin, and has seen wide application in synthesis.[189] The first example utilizes achiral nitrone and olefin, and demonstrates the intrinsic relative facial selectivity (20:1 *trans:cis*) arising from an exo transition state.[190]

Highly diastereoselective chiral nitrone cycloadditions with enol ethers have been applied to the preparation of carbohydrates (high diastereoselectivity refers to the absolute facial selectivity of the nitrone; the modest 4:1 relative facial selectivity is of no consequence to the downstream chemistry). The example below is taken from the synthesis of daunosamine.[191]

[188]Wang, Y.; Zhao, C.; Romo, D. *Org. Lett.* **1999**, *1*, 1197–1199.
[189]Confalone, P. N.; Huie, E. M. *Organic Reactions (New York)* **1988**, *36*, 1–173.
[190]Iida, H.; Watanabe, Y.; Tanaka, M.; Kibayashi, C. *J. Org. Chem.* **1984**, *49*, 2412–2418.
[191]DeShong, P.; Dicken, C. M.; Leginus, J. M.; Whittle, R. R. *J. Am. Chem. Soc.* **1984**, *106*, 5598–5602.

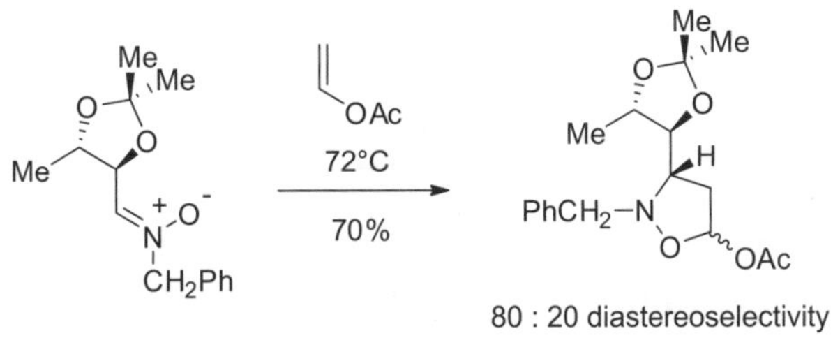

80 : 20 diastereoselectivity

The following example shows an enantioselective nitrone-olefin cycloaddition using a $C_2$-symmetric nickel catalyst.[192] The catalyst ligand preparation has been reported in *Organic Syntheses*.[193]

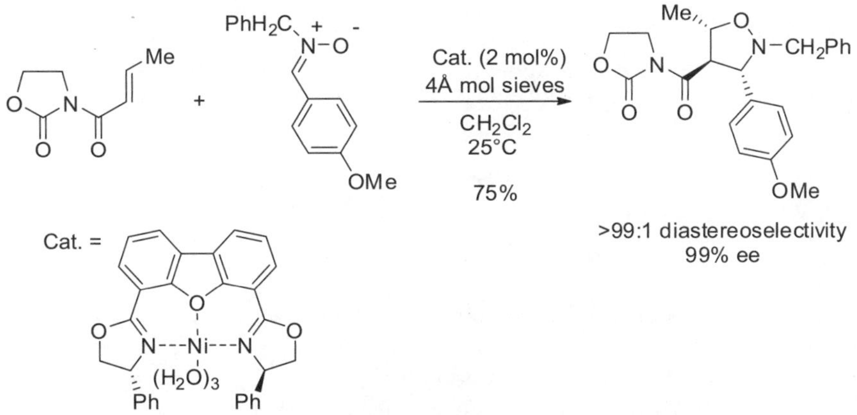

>99:1 diastereoselectivity
99% ee

The Kinugasa reaction involves coupling of a terminal alkyne with a nitrone to form a β-lactam.[194] The mechanism involves an initial [3 + 2] cycloaddition to form a dihydroisoxazole, which then rearranges to the β-lactam product. Asymmetric variants have been developed. The first example from Fu utilizes a Cu(I)-bis(azaferrocene) catalyst system.[195]

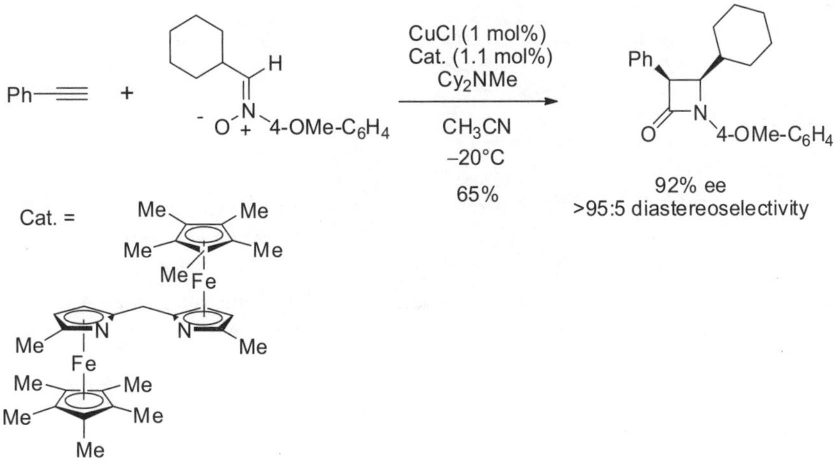

65%

92% ee
>95:5 diastereoselectivity

[192]Kanemasa, S.; Oderaotoshi, Y.; Tanaka, J.; Wada, E. *J. Am. Chem. Soc.* **1998**, *120*, 12355–12356.
[193]Iserloh, U.; Oderaotoshi, Y.; Kanemasa, S.; Curran, D. P. *Org. Synth.* **2003**, *80*, 46–56.
[194]Kinugasa, M.; Hashimoto, S. *J. Chem. Soc., Chem. Commun.* **1972**, 466–467.
[195]Lo, M. M. C.; Fu, G. C. *J. Am. Chem. Soc.* **2002**, *124*, 4572–4573.

Tang has developed a Cu(II)-tris(oxazoline) catalyst system as shown below.[196]

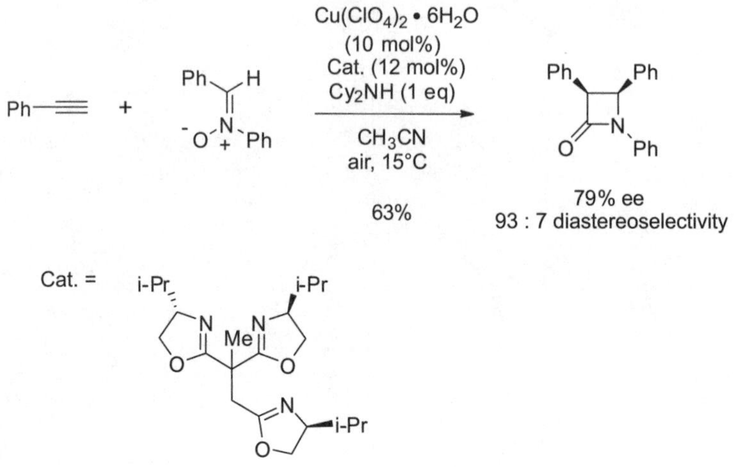

63%

79% ee
93 : 7 diastereoselectivity

### 3.13.1.3 Non-Cycloaddition Carbon–Oxygen Additions

The free radical oxy-acylation of terminal olefins to provide butyrolactones can be effected by an *in situ* generated $Mn(OAc)_3 \cdot H_2O$ reagent (prepared from $Mn(OAc)_3 \cdot 4H_2O$, $KMnO_4$ and $Ac_2O$ in AcOH). The yield is >95% based on consumed olefin, and 66% based on potassium permanganate.[197]

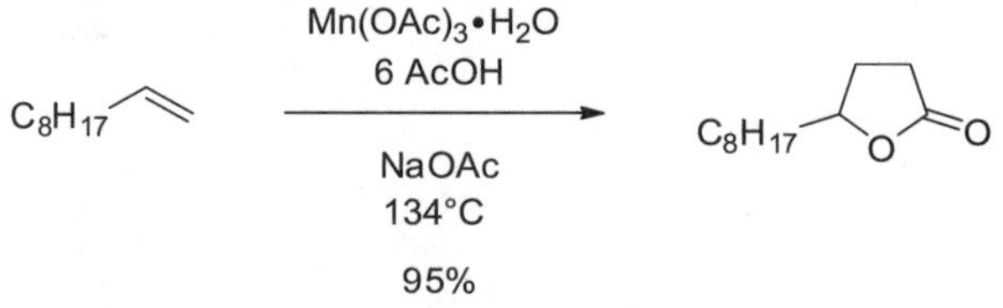

95%

A multi-step protocol for addition of carbon to the β-position of an enone with subsequent α-hydroxylation is shown below.[198] It involves copper-catalyzed addition of MeMgCl to the enone (see Section 3.5.5) followed by enolate trapping with $Ac_2O$ to generate an intermediate enol acetate. Epoxidation with buffered peracetic acid generates the corresponding epoxide, which hydrolyzes to generate the α-hydroxyketone in 95% yield.

[196]Ye, M.-C.; Zhou, J.; Tang, Y. *J. Org. Chem.* **2006**, *71*, 3576–3582.
[197]Heiba, E. I.; Dessau, R. M.; Williams, A. L.; Rodewald, P. G. *Org. Synth.* **1983**, *61*, 22–24.
[198]Ohta, T.; Zhang, H.; Torihara, Y.; Furukawa, I. *Org. Process Res. Dev.* **1997**, *1*, 420–424.

## 3.13.2    Carbon–Oxygen Addition to Dienes (1,4)

### 3.13.2.1    *Hetero Diels–Alder Cycloaddition*    The [4 + 2] cycloaddition of an aldehyde and diene (hetero ç) constitutes the 1,4 addition of carbon and oxygen across a diene. Danishefsky was an early pioneer of this methodology,[199] as shown in the two examples below.[200,201]

Jacobsen has developed a chromium catalyst for effecting hetero Diels–Alder reactions with high levels of enantioselectivity.[202]

[199]Danishefsky, S. J. *Aldrichimica Acta* **1986**, *19*, 59–69.

[200]Danishefsky, S. J.; Myles, D. C.; Harvey, D. F. *J. Am. Chem. Soc.* **1987**, *109*, 862–867.

[201]Danishefsky, S. J.; Pearson, W. H.; Harvey, D. F.; Maring, C. J.; Springer, J. P., *J. Am. Chem. Soc.* **1985**, *107*, 1256–1268.

[202]Chavez, D. E.; Jacobsen, E. N.; Grabowski, E. J. J.; Kubryk, M. *Org. Synth.* **2005**, *82*, 34–42.

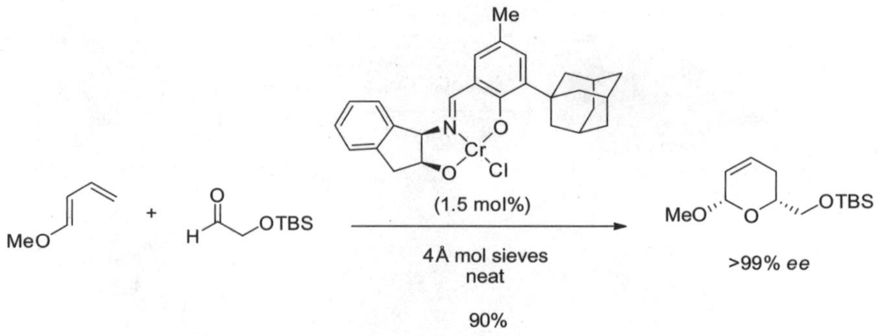

90%

## 3.14   CARBON–NITROGEN ADDITION

### 3.14.1   Carbon–Nitrogen Addition to Olefins

Ketenes and imines undergo a formal [2 + 2] cycloaddition to form β-lactams (Staudinger reaction). The relative diastereofacial selectivity can be quite high for these reactions, as shown in the example below[203] (a single diastereomer was detected, suggesting >95:5 diastereoselectivity).

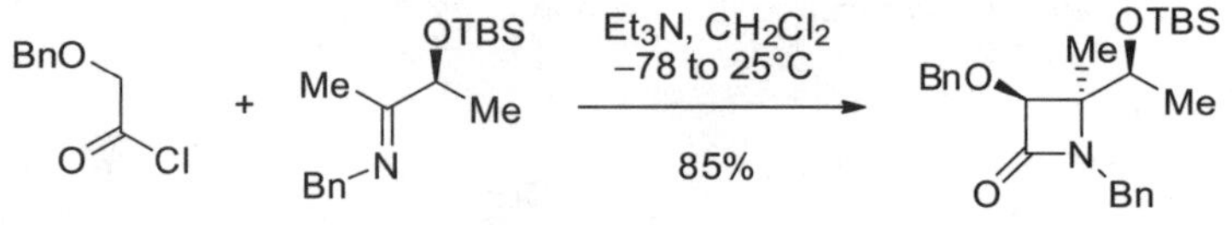

Chiral auxiliary-mediated Staudinger reactions have also been utilized in the synthesis of carbacephem intermediates, as shown below. In this example, the product crystallizes from 2-propanol in 75% yield on 140 g scale.[204]

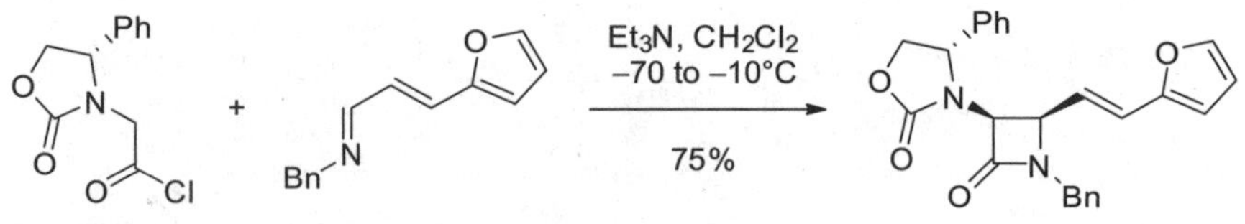

Variants using chiral catalysts have also been developed, as shown below for the chiral bifunctional catalysis system developed by Lectka.[205]

[203]Palomo, C.; Aizpurua, J. M.; Garcia, J. M.; Galarza, R.; Legido, M.; Urchegui, R.; Roman, P.; Luque, A.; Server-Carrio, J.; Linden, A. *J. Org. Chem.* **1997**, *62*, 2070–2079.

[204]Kumar, Y.; Tewari, N.; Nizar, H.; Rai, B. P.; Singh, S. K. *Org. Process Res. Dev.* **2003**, *7*, 933–935.

[205]France, S.; Wack, H.; Hafez, A. M.; Taggi, A. E.; Witsil, D. R.; Lectka, T. *Org. Lett.* **2002**, *4*, 1603–1605.

BQ (10 mol%)
In(OTf)$_3$ (10 mol%)
proton sponge

Toluene,
−78 to 25°C

95%                    98% *ee*

BQ =

### 3.14.2   Carbon–Nitrogen Addition to Dienes

The [4 + 2] cycloaddition of dienes and imines falls into this reaction class, and is a form of a hetero-Diels–Alder reaction. The following example utilizes the *N*-benzylimine derived from benzylamine and formaldehyde.[206]

PhCH$_2$NH$_2$•HCl

aq. H$_2$C=O
H$_2$O
25°C

91–92%

## 3.15   CARBON–CARBON ADDITION

The addition of a carbon atom to each end of an olefin encompasses a wide variety of cycloaddition reactions and represents one of the most important classes of bond-forming methods in organic chemistry. The wide variety of cycloaddition reactions makes it difficult to be comprehensive in this section, but several of the most important classes will be exemplified.

### 3.15.1   [4 + 2] Cycloaddition: Diels–Alder Reaction

The Diels–Alder reaction, a cycloaddition of a conjugated diene with an olefin (dienophile) to provide a cyclohexene product, is one of the highest profile C−C bond-forming reactions in organic synthesis.[207] Its utility arises in large part from the specificity exhibited with regards to regioselectivity, stereospecifity, and relative (facial) stereoselectivity. In this section examples are subdivided into several categories based on the structural elements of each reaction component.

[206]Grieco, P. A.; Larsen, S. D. *Org. Synth.* **1990**, *68*, 206–209.
[207]Trost, B. M.; Fleming, I. *Comprehensive Organic Synthesis*; Pergamon Press: New York, **1991**; Vol. *5*, Chapters 4.1–4.5.

**3.15.1.1  Intermolecular, Non-Substituted**  The following example comes from the preparation of 2-cyclohexene-1,4-dione. The initial cycloadduct's enone is selectively reduced with Zn/AcOH, then heated to effect a retro-Diels–Alder reaction to generate the title compound. Cyclopentadiene is a particularly reactive diene (due to its constraint to the s-*cis* conformation), and the cycloaddition thus occurs under particularly mild conditions.[208]

An intramolecular example in which an anthracene moiety serves as the diene is provided below[209] (see Section 3.15.1.9 for intramolecular examples in which relative stereochemistry is explored).

**3.15.1.2  Intermolecular, Heteroatom-Substituted Dienophile**  Diels–Alder cycloadditions are accelerated through the combination of electron-deficient dienophiles and electron-rich dienes (the opposite combination applies to inverse electron-demand Diels–Alder reactions). As such, dienophiles conjugated to an electron-withdrawing group (e.g., $CO_2R$, · COR, CN, $NO_2$) have seen wide application. In addition to providing rate acceleration, they offer the possibility of post-cycloaddition manipulation. In the following example from Paquette, the sulfone-substituent on the dienophile provides a handle for post-cycloaddition manipulations such as alkylation and/or reductive cleavage.[210] Note that the direct cycloaddition of this diene to an unactivated olefin (e.g., cyclopentene) is not an efficient process.

[208]Oda, M.; Kawase, T.; Okada, T.; Enomoto, T. *Org. Synth.* **1996**, *73*, 253–261.
[209]Ciganek, E. *J. Org. Chem.* **1980**, *45*, 1497–1505.
[210]Lin, H. S.; Paquette, L. A. *Org. Synth.* **1989**, *67*, 163–169.

***3.15.1.3  Intermolecular, Heteroatom-Substituted Diene***   Oxygen substitution on the diene component accelerates the rate of cycloaddition, is a useful regiochemistry control element, and provides a handle for subsequent synthetic manipulations. The following example from Danishefsky is representative.[211] An improved preparation of a more highly substituted 1,3-dioxydiene has also been reported.[212]

Amino-substituted dienes can be even more reactive, as shown in the following example from Rawal.[213] The preparation of the diene is also reported in the same volume of *Organic Syntheses*.[214]

***3.15.1.4  Intermolecular, Aqueous Media***   Dramatic rate-acceleration of Diels–Alder cycloadditions in water has been noted by several labs,[215–217] and was utilized in the synthesis of vernolepin in the example below.[218]

[211]Danishefsky, S.; Kitahara, T.; Schuda, P. F. *Org. Synth.* **1983**, *61*, 147–151.

[212]Myles, D. C.; Bigham, M. H. *Org. Synth.* **1992**, *70*, 231–239.

[213]Kozmin, S. A.; He, S.; Rawal, V. H. *Org. Synth.* **2002**, *78*, 160–168.

[214]Kozmin, S. A.; He, S.; Rawal, V. H. *Org. Synth.* **2002**, *78*, 152–159.

[215]Hopff, H.; Rautenstrauch, C. W. (I. G. Farbenindustrie AG). *Addition products of butadiene with vinyl methyl ketone, maleic acid, etc.*, (1941) US 2262002 19411111 Patent language unavailable. Application: US 1939-274040 19390516. CAN 36:6302 AN 1942:6302 CAPLUS.

[216]Rideout, D. C.; Breslow, R. *J. Am. Chem. Soc.* **1980**, *102*, 7816–7817.

[217]Braun, R.; Schuster, F.; Sauer, J. *Tetrahedron Lett.* **1986**, *27*, 1285–1288.

[218]Yoshida, K.; Grieco, P. A. *J. Org. Chem.* **1984**, *49*, 5257–5260.

### 3.15.1.5 Intermolecular, Aromatic Product

With heteroatom-substituted dienes and/or dienophiles, the Diels–Alder product may eliminate the elements of HX (X = heteroatom), and the resulting dihydroaromatic species may undergo further oxidation to provide the fully aromatic product. The following example utilizes an enamine dienophile to provide a mixture of hexahydroisoquinolines, which upon treatment with Pd/C undergoes further 2-electron oxidation to the tetrahydroisoquinoline as a single regioisomer.[219] Note that the use of an electron-rich dienophile with an electron-deficient diene constitutes a reversal of the normal electronics of the reaction, and is referred to as an inverse electron demand Diels–Alder reaction (see Section 3.15.1.7).

Danheiser has developed a cycloaddition of vinylketenes with electron-rich acetylenes in which the former species is generated by electrocyclic opening of cyclobutenones. The initial cyclobutanone [2 + 2] adduct undergoes a cascade of two electrocyclic openings and closures to generate a highly substituted phenol product. Because the initial cycloaddition is [2 + 2], it could arguably be included in Section 3.15.2. But as the final product is formally the product of a [4 + 2] cycloaddition, it is included in this section. The example below is taken from the synthesis of mycophenolic acid, and proceeds in 73% yield on 1 g scale.[220]

[219]Danishefsky, S.; Cavanaugh, R. *J. Org. Chem.* **1968**, *33*, 2959–2962.
[220]Danheiser, R. L.; Gee, S. K.; Perez, J. J. *J. Am. Chem. Soc.* **1986**, *108*, 806–810.

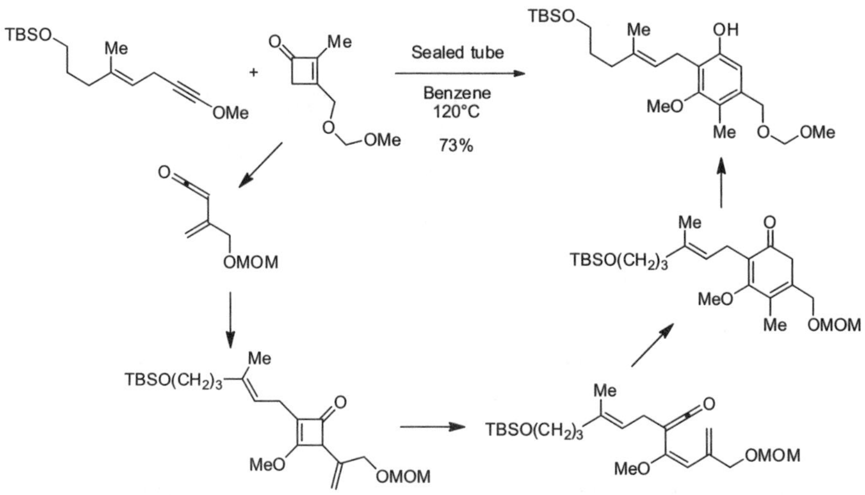

### 3.15.1.6 Intermolecular, Lewis Acid Catalyzed

*3.15.1.6  Intermolecular, Lewis Acid Catalyzed*  Lewis acids can serve as effective catalysts for Diels–Alder cycloadditions through coordination with the carbonyl oxygen and lowering of the dienophile's lowest unoccupied molecular orbital (LUMO). Two examples from the pharmaceutical industry are shown below. The first was utilized in the asymmetric synthesis of an NMDA receptor antagonist.[221]

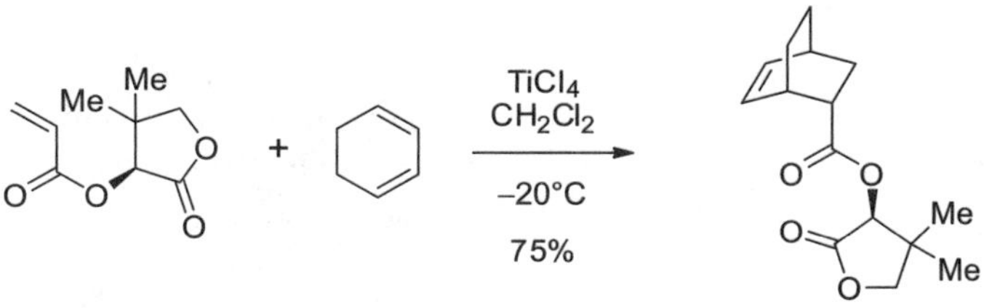

97 : 3 diastereoselectivity

The second example involves a cyclopropylidene dienophile.[222]

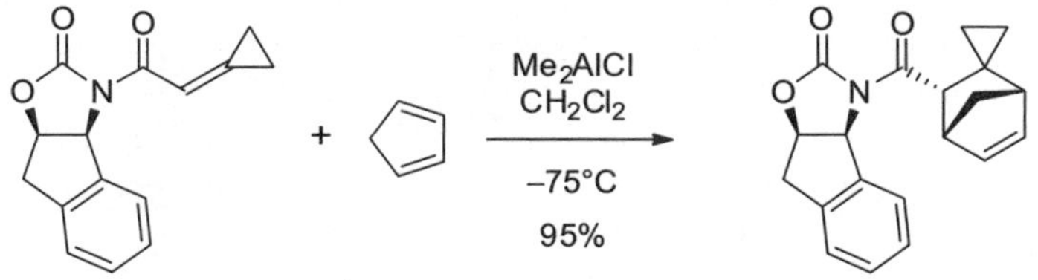

36:1 endo-exo selectivity

[221]Hansen, M. M.; Bertsch, C. F.; Harkness, A. R.; Huff, B. E.; Hutchison, D. R.; Khau, V. V.; LeTourneau, M. E.; Martinelli, M. J.; Misner, J. W.; Peterson, B. C.; Rieck, J. A.; Sullivan, K. A.; Wright, I. G. *J. Org. Chem.* **1998**, *63*, 775–785.
[222]Kuethe, J. T.; Zhao, D.; Humphrey, G. R.; Journet, M.; McKeown, A. E. *J. Org. Chem.* **2006**, *71*, 2192–2195.

*3.15.1.7 Intermolecular, Inverse-Electron Demand* The normal electronic configuration in a Diels–Alder reaction is an electron-rich diene and an electron-deficient dienophile. This distribution can be inverted to an electron-deficient diene and electron-rich dienophile as shown in the following two examples from Boger's lab. In each case, a subsequent extrusion occurs to generate an aromatic product ($CO_2$[223] or $N_2$ followed by elimination of pyrrolidinone).[224]

*3.15.1.8 Intermolecular, Benzyne as Dienophile* Benzyne can serve as the dienophile in a Diels–Alder reaction when generated *in situ*, as in the following example.[225] As in the Boger example cited previously, a $CO_2$ extrusion occurs after the initial cycloaddition.

[223]Boger, D. L.; Mullican, M. D. *Org. Synth.* **1987**, *65*, 98–107.

[224]Boger, D. L.; Panek, J. S.; Yasuda, M. *Org. Synth.* **1988**, *66*, 142–150.

[225]Ashworth, I. W.; Bowden, M. C.; Dembofsky, B.; Levin, D.; Moss, W.; Robinson, E.; Szczur, N.; Virica, J. *Org. Process Res. Dev.* **2003**, *7*, 74–81.

### 3.15.1.9 Intramolecular Examples

Intramolecular Diels–Alder (IMDA) cycloadditions have been studied extensively, particularly in the context of complex natural product total synthesis.[226,227] The following example is from an early series of studies in the Roush lab directed at the synthesis of the nargenicins.[228] Note that the *cis*-ring fusion is undesired relative to the target structures, and extensive studies were pursued by both Roush and Boeckman to alter the stereochemical outcome with bromo[229] or trimethylsilyl substituents[230] on the diene moiety.

The following example is from Boeckman's synthesis of the cyclohexene subunit of (+)-tetronolide.[231,232]

Overman utilized a silyl-tethered IMDA cyclization in his synthesis of (+)-aloperine. The Diels–Alder cyclization is quite efficient, as the 76% overall yield encompasses five individual transformations (BOC removal, *N*-silylation, cycloaddition, desilylation/rearrangement, and oxidation of the C–Si bond).[233]

[226]Takao, K.; Munakata, R.; Tadano, K. *Chem. Rev.* **2005**, *105*, 4779–4807.

[227]Ciganek, E. *Organic Reactions (New York)* **1984**, *32*, 1–374.

[228]Coe, J. W.; Roush, W. R. *J. Org. Chem.* **1989**, *54*, 915–930.

[229]Roush, W. R.; Kageyama, M.; Riva, R.; Brown, B. B.; Warmus, J. S.; Moriarty, K. J. *J. Org. Chem.* **1991**, *56*, 1192–1210.

[230]Boeckman, R. K., Jr., ; Barta, T. E. *J. Org. Chem.* **1985**, *50*, 3421–3423.

[231]Boeckman, R. K., Jr., ; Wrobleski, S. T. *J. Org. Chem.* **1996**, *61*, 7238–7239.

[232]Boeckman, R. K., Jr., ; Estep, K. G.; Nelson, S. G.; Walters, M. A. *Tetrahedron Lett.* **1991**, *32*, 4095–4098.

[233]Brosius, A. D.; Overman, L. E.; Schwink, L. *J. Am. Chem. Soc.* **1999**, *121*, 700–709.

Deslongchamps has published extensive stereochemical studies and synthetic applications of the transannular Diels–Alder cyclization. The example shown below is taken from his studies on the synthesis of kempane diterepenes.[234]

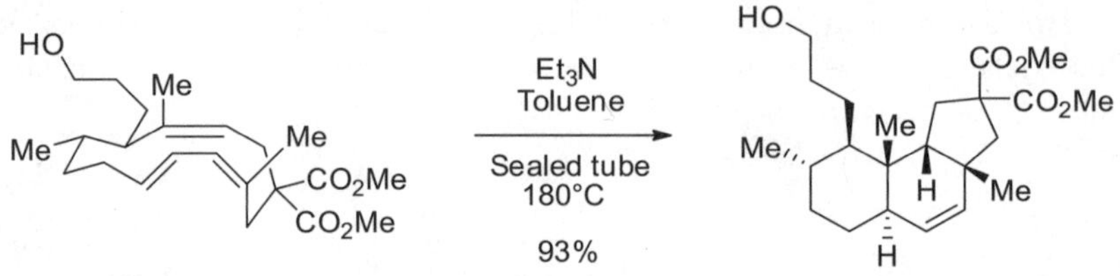

3.15.1.10   *Asymmetric Examples*   Asymmetric Diels–Alder cycloadditions have been effected with both chiral-auxiliary and chiral-catalyst methods. The following two examples show chiral-auxiliary methods. The first example is from the Evans lab.[235]

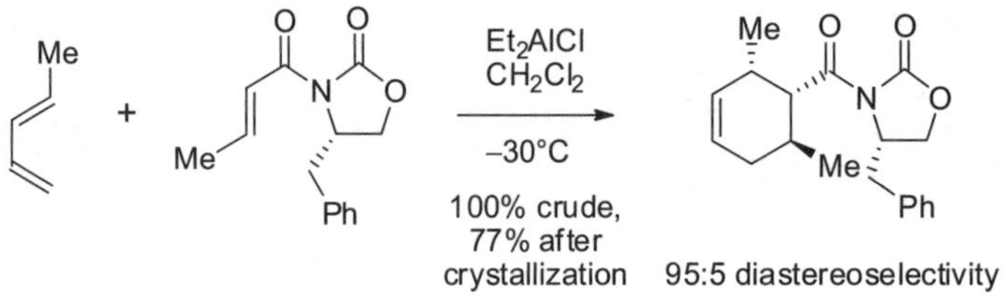

The second example is from Oppolzer, and was utilized in his enantiospecific total synthesis of loganin.[236] The high crystallinity of the sultam auxiliary is an attractive feature of this methodology.

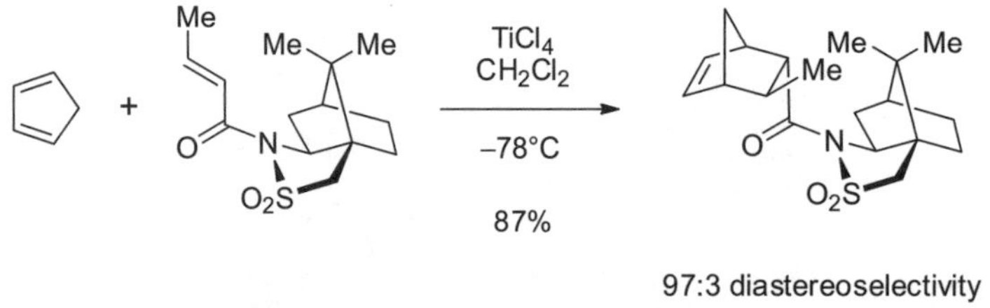

Asymmetric catalysts have also been developed for effecting enantioselective Diels–Alder cycloadditions. The first example is from Corey, one of the early pioneers in asymmetric Diels–Alder reactions.[237,238] The *in situ* catalyst synthesis from the bissulfonamide (prepared in one step from the diamine) and a 10 g Diels–Alder cycloaddition have been reported.[239] The same authors also provide a synthesis of the chiral diamine in two

[234]Caussanel, F.; Wang, K.; Ramachandran, S. A.; Deslongchamps, P. *J. Org. Chem.* **2006**, *71*, 7370–7377.

[235]Evans, D. A.; Chapman, K. T.; Bisaha, J. *J. Am. Chem. Soc.* **1988**, *110*, 1238–1256.

[236]Vandewalle, M.; Van der Eycken, J.; Oppolzer, W.; Vullioud, C. *Tetrahedron* **1986**, *42*, 4035–4043.

[237]Corey, E. J.; Imwinkelried, R.; Pikul, S.; Xiang, Y. B. *J. Am. Chem. Soc.* **1989**, *111*, 5493–5495.

[238]Corey, E. J.; Ensley, H. E. *J. Am. Chem. Soc.* **1975**, *97*, 6908–6909.

[239]Pikul, S.; Corey, E. J. *Org. Synth.* **1993**, *71*, 30–38.

steps from (PhCO)₂ (benzil) and ammonium acetate via resolution with tartaric acid.[240] The diamine is commercially available.

Yamamoto has reported a boron-based catalyst for effecting enantioselective Diels–Alder cycloadditions. The reference for the example below includes the preparation of the diacid pre-catalyst, available in two steps from tartaric acid.[241]

### 3.15.2  [2 + 2] Cycloaddition

Olefins can undergo photochemically mediated [2 + 2] cycloadditions. One of the olefin partners is frequently electron deficient, such as an enone. The following intramolecular example was utilized in the synthesis of isocomene,[242,243] and is preparatively useful (76% yield on 6 g scale, albeit using a large volume of solvent: 500 mL/g).

[240]Pikul, S.; Corey, E. J. *Org. Synth.* **1993**, *71*, 22–29.

[241]Furuta, K.; Gao, Q.-z.; Yamamoto, H. *Org. Synth.* **1995**, *72*, 86–94.

[242]Pirrung, M. C. *J. Am. Chem. Soc.* **1981**, *103*, 82–87.

[243]Pirrung, M. C. *J. Am. Chem. Soc.* **1979**, *101*, 7130–7131.

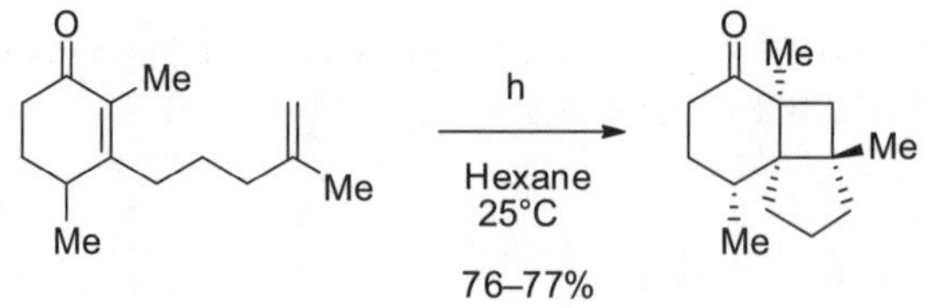

76–77%

Allenes will also add to enones in a photocycloaddition to give exo-methylene-substituted cyclobutanes.[244]

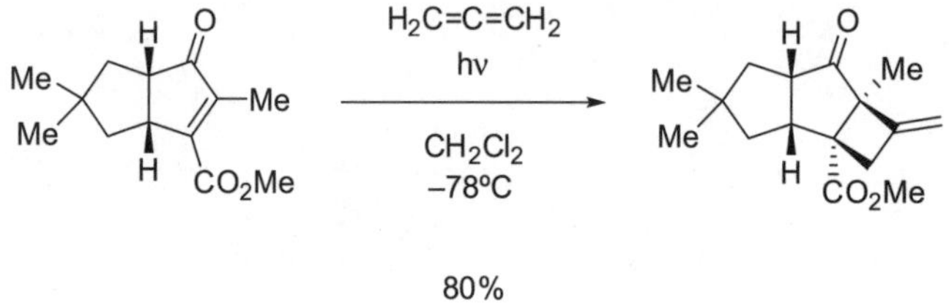

80%

Cyclobutenes are also available by photochemical [2 + 2] cycloadditions between an acetylene and an electron-deficient olefin. The following is from a synthesis of xanthocidin, and proceeds in 57% yield on 5 g scale.[245]

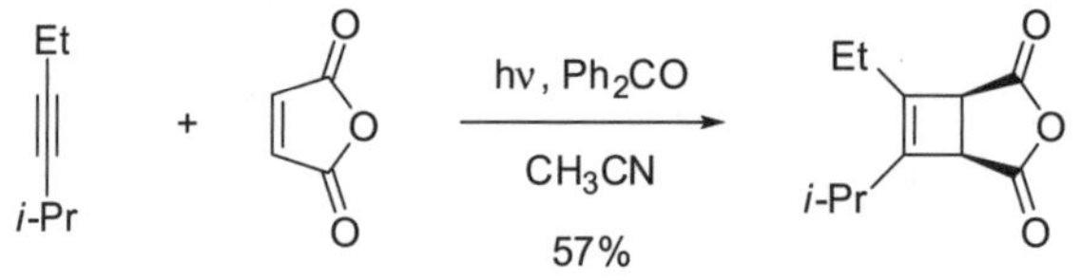

57%

Cyclobutanones can also be formed by [2 + 2] cycloaddition between olefins and ketenes (also known as the Staudinger ketene cycloaddition). Intramolecular variants are particularly useful for formation of bicyclic ring systems, as exemplified below.[246] The substrate was designed to test the kinetic preference for 5- vs. 6-membered tether length; complete regioselectivity for the 5-membered ring was observed.

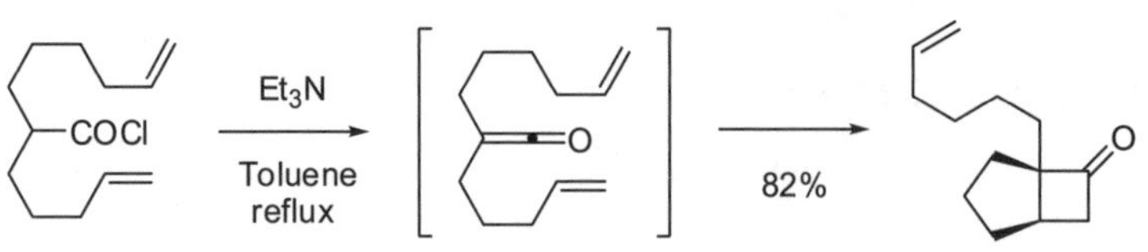

82%

[244]Tobe, Y.; Yamashita, D.; Takahashi, T.; Inata, M.; Sato, J.; Kakiuchi, K.; Kobiro, K.; Odaira, Y. *J. Am. Chem. Soc.* **1990**, *112*, 775–779.
[245]Smith, A. B., III; Boschelli, D. *J. Org. Chem.* **1983**, *48*, 1217–1226.
[246]Belanger, G.; Levesque, F.; Paquet, J.; Barbe, G. *J. Org. Chem.* **2005**, *70*, 291–296.

Intermolecular examples are also known, frequently with dichloroketene, as exemplified below from the synthesis of homogynolide B.[247]

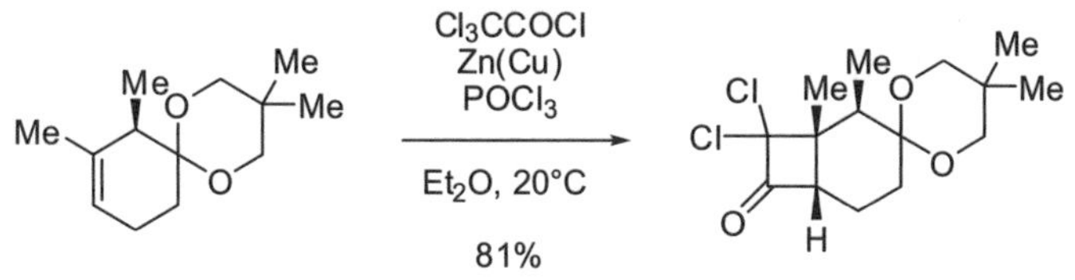

81%

### 3.15.3    [3 + 2] Cycloaddition

Several methods for effecting [3 + 2] annulations have been developed. The order of presentation below is arbitrary; the preferred method for a given application will largely depend on the specific functionality present in the target.

Danheiser has developed a [3 + 2] annulation reaction between electron-deficient olefins and trimethylsilylallenes; the example shown below proceeds in 71% yield on 5 g scale.[248]

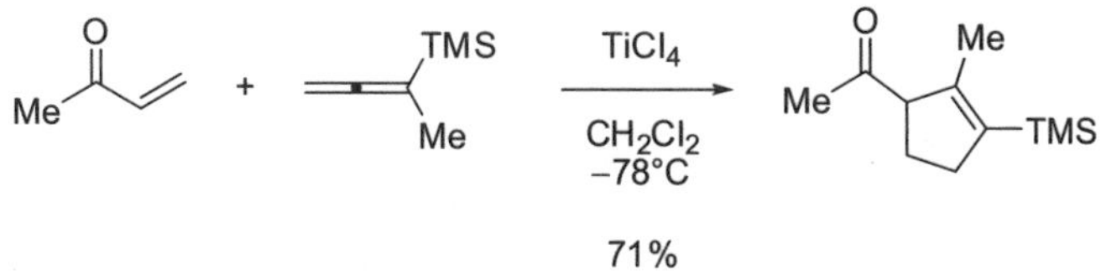

71%

Trost has developed a [3 + 2] cycloaddition utilizing a Pd-mediated coupling of electron-deficient olefins with allylsilane/allylic acetate nucleophiles.[249] The example shown below is from Paquette's lab.[250]

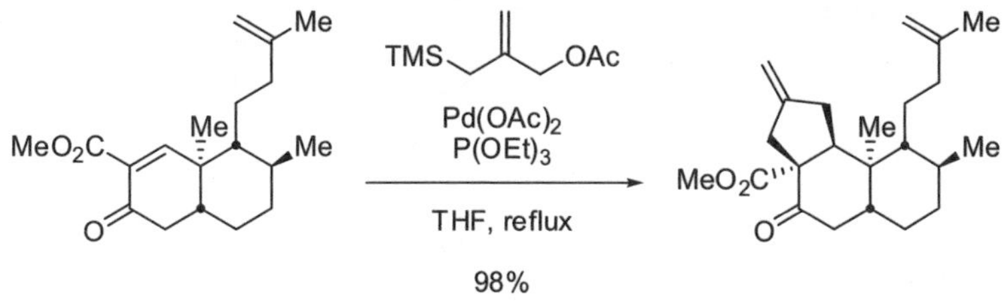

98%

[247]Brocksom, T. J.; Coelho, F.; Depres, J.-P.; Greene, A. E.; Freire de Lima, M. E.; Hamelin, O.; Hartmann, B.; Kanazawa, A. M.; Wang, Y. *J. Am. Chem. Soc.* **2002**, *124*, 15313–15325.

[248]Danheiser, R. L.; Carini, D. J.; Fink, D. M.; Basak, A. *Tetrahedron* **1983**, *39*, 935–947.

[249]Trost, B. M. *Angewandte Chemie International Edition in English* **1986**, *98*, 1–20.

[250]Paquette, L. A.; Sauer, D. R.; Cleary, D. G.; Kinsella, M. A.; Blackwell, C. M.; Anderson, L. G. *J. Am. Chem. Soc.* **1992**, *114*, 7375–7387.

A similar example is shown below.[251]

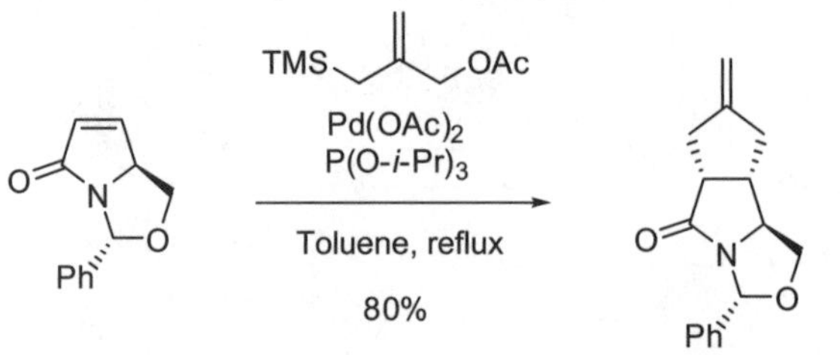

The Pauson–Khand cyclization involves a [2 + 2 + 1] cycloaddition of an olefin, an acetylene, and carbon monoxide; the latter two components arise from a cobalt complex of the acetylene, generated either *in situ* or as an isolated intermediate.[252,253] The following example comes from the synthesis of methyl deoxynorpentalenolactone H.[254]

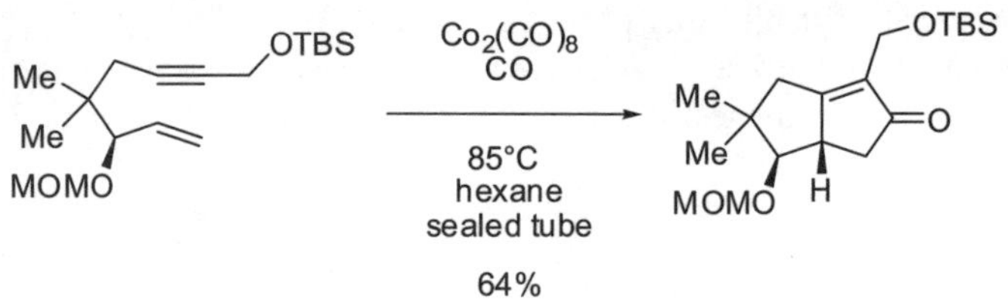

A second example shown below is from Schreiber's synthesis of epoxydictymene.[255,256] It utilizes a cobalt-stabilized propargylic cation cyclization to generate the 8-membered carbocycle, followed by a Pauson–Khand cyclization of the isolated cobalt complex. The olefin component of the Pauson–Khand cyclization was present in the propargylic cation precursor (identification of conditions to provide regiocontrol in the acetal ionization required considerable optimization).

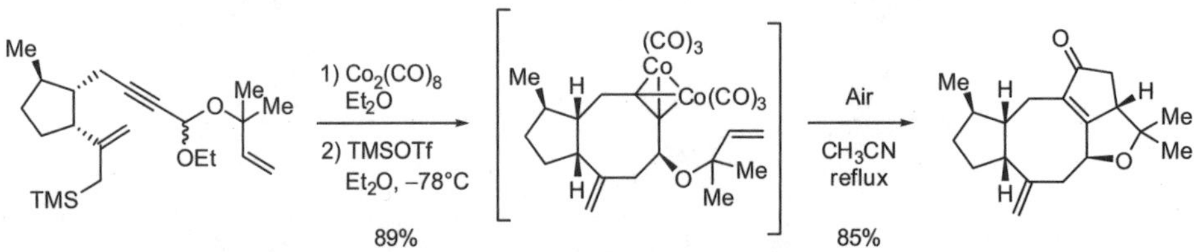

Brummond has developed a Pauson–Khand cyclization of allenes, as shown in the example below. The cyclization product constitutes the carbon skeleton of guanacastepene A.[257]

[251] Jao, E.; Bogen, S.; Saksena, A. K.; Girijavallabhan, V. *Tetrahedron Lett.* **2003**, *44*, 5033–5035.

[252] Brummond, K. M.; Kent, J. L. *Tetrahedron* **2000**, *56*, 3263–3283.

[253] Schore, N. E. *Organic Reactions (New York)* **1991**, *40*, 1–90.

[254] Magnus, P.; Slater, M. J.; Principe, L. M. *J. Org. Chem.* **1989**, *54*, 5148–5153.

[255] Jamison, T. F.; Shambayati, S.; Crowe, W. E.; Schreiber, S. L. *J. Am. Chem. Soc.* **1997**, *119*, 4353–4363.

[256] Jamison, T. F.; Shambayati, S.; Crowe, W. E.; Schreiber, S. L. *J. Am. Chem. Soc.* **1994**, *116*, 5505–5506.

[257] Brummond, K. M.; Gao, D. *Org. Lett.* **2003**, *5*, 3491–3494.

Trialkylphosphines can mediate a [3 + 2] cycloaddition of allenic esters with electron-deficient olefins, as shown in the examples below.[258] The reaction proceeds with high diastereospecificity with respect to the olefin geometry.

Krische has extended this approach to the intramolecular [3 + 2] annulation of ynones with enones mediated by trialkylphosphines. The reaction requires reasonably electrophilic olefins (enones and thioesters cyclize efficiently, whereas esters do not).[259]

Azomethine ylides can be utilized to generate pyrrolidines from addition to electron-deficient olefins. The following example utilizes N-methylglycine and paraformaldehyde to generate an azomethine ylide, which undergoes a [3 + 2] cycloaddition to generate an N-methylpyrrolidine with 4:1 diastereoselectivity.[79]

[258]Zhang, C.; Lu, X. *J. Org. Chem.* **1995**, *60*, 2906–2908.
[259]Wang, J.-C.; Ng, S.-S.; Krische, M. J. *J. Am. Chem. Soc.* **2003**, *125*, 3682–3683.

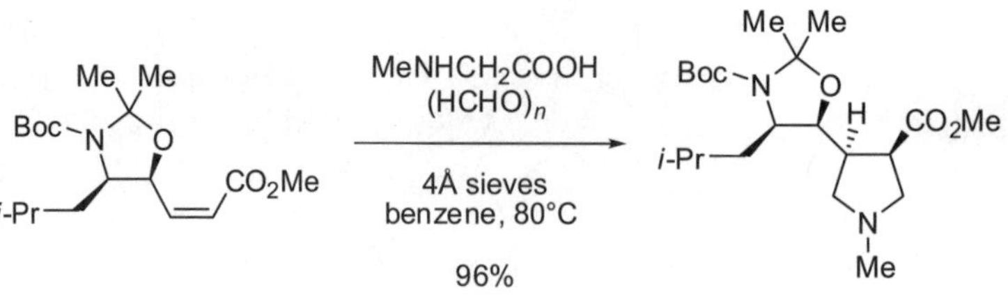

96%

Azomethine ylides can also be generated by high temperature thermolysis of aziridines. The following example was executed on 500 g scale and utilized the unusual solvent DW-therm, a mixture of triethoxyalkylsilanes (DW-therm is a heat transfer fluid available from Huber Corporation of Germany, www.huber-online.com).[260]

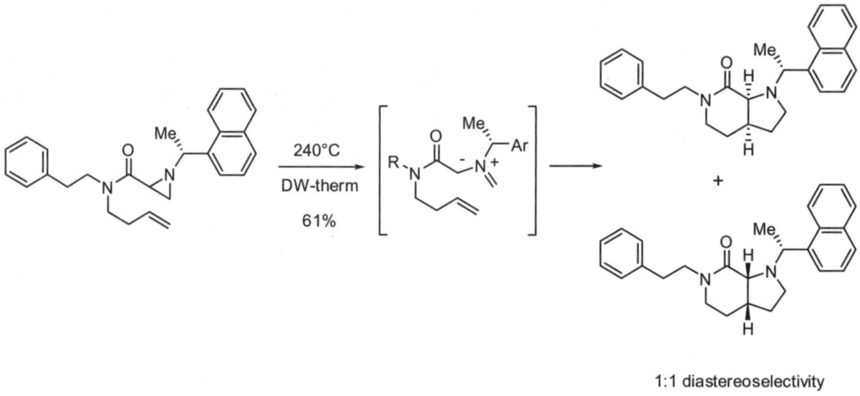

1:1 diastereoselectivity

### 3.15.4   Carbene Addition (Cyclopropanation)

The Simmons–Smith reaction is an efficient route to cyclopropanes. It involves the Zn-mediated addition of a dihalomethane to an olefin. The following example was executed on 60 g scale and proceeded in 84% yield.[261] Some nitrile reduction to the aldehyde was observed (approximately 8%), which was minimized by optimization of reaction conditions.

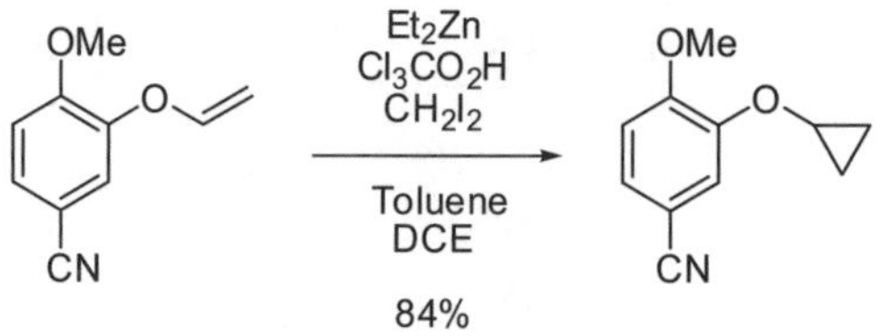

84%

Several enantioselective variants of the Simmons–Smith reaction are known. The transition metal-mediated coupling of diazoesters with olefins can be highly selective.

[260]Shieh, W.-C.; Chen, G.-P.; Xue, S.; McKenna, J.; Jiang, X.; Prasad, K.; Repic, O.; Straub, C.; Sharma, S. K. *Org. Process Res. Dev.* **2007**, *11*, 711–715.
[261]Frey, L. F.; Marcantonio, K. M.; Chen, C.-y.; Wallace, D. J.; Murry, J. A.; Tan, L.; Chen, W.; Dolling, U. H.; Grabowski, E. J. J. *Tetrahedron* **2003**, *59*, 6363–6373.

Substrate scope is limited to α-diazo carbonyl reagents, however, which means that only carbonyl-substituted cyclopropanes are available via this route. The example shown below proceeds in 91% yield on 35 g scale with a catalyst load of just 0.12 mol%.[262]

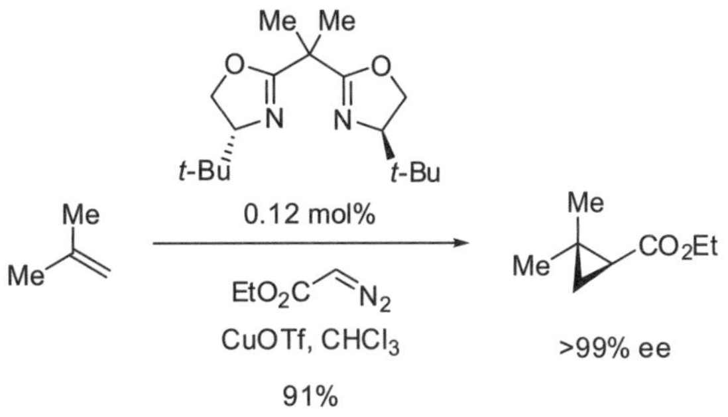

Intramolecular variants of the transition metal-catalyzed cyclopropanation can be useful for preparation of bicyclic cyclopropanes. The preparation of Doyle's chiral dirhodium catalyst used in the example below is described in an *Organic Syntheses* procedure.[263] It is also commercially available in smaller quantities (CAS 131796-58-2).

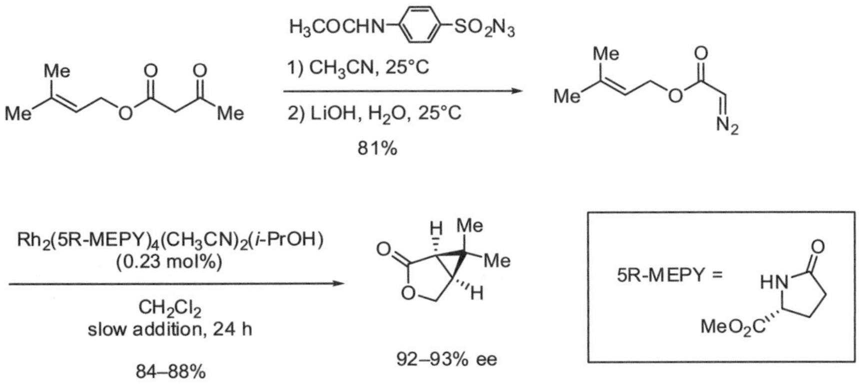

Efficient catalysts for the cyclopropanation of allylic alcohols have been developed that avoid the use of diazo esters (and thus allow unsubstituted cyclopropanes to be prepared). The following two examples are from Charette[264] and Denmark,[265] respectively.

[262]Evans, D. A.; Woerpel, K. A.; Hinman, M. M.; Faul, M. M. *J. Am. Chem. Soc.* **1991**, *113*, 726–728.

[263]Doyle, M. P.; Winchester, W. R.; Protopopova, M. N.; Kazala, A. P.; Westrum, L. J. *Org. Synth.* **1996**, *73*, 13–24.

[264]Charette, A. B.; Molinaro, C.; Brochu, C. *J. Am. Chem. Soc.* **2001**, *123*, 12168–12175.

[265]Denmark, S. E.; O'Connor, S. P. *J. Org. Chem.* **1997**, *62*, 584–594.

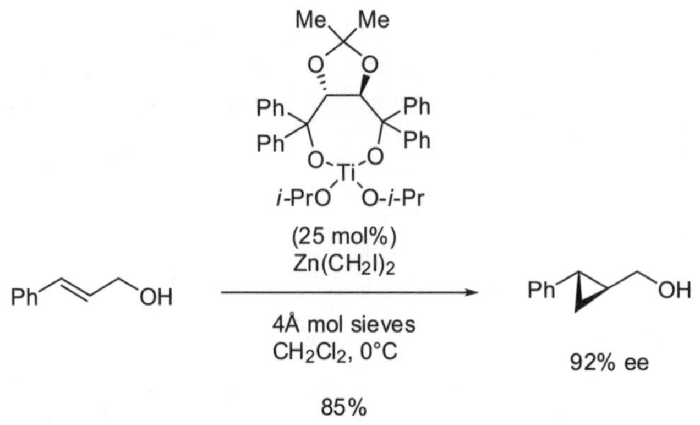

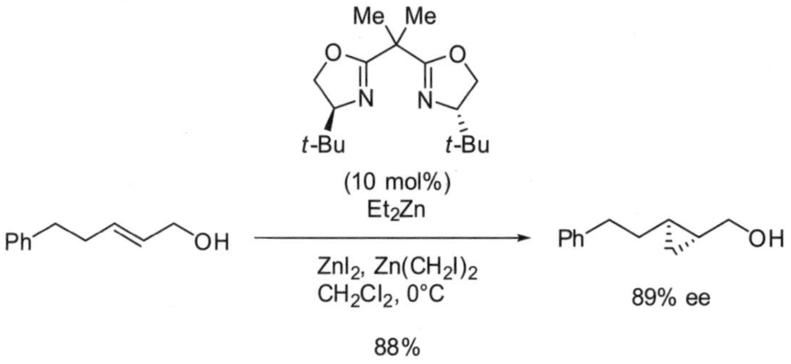

85%

88%

Dihalocarbenes will also add to olefins, and generally do not require metal catalysis. The following example prepares a crystalline bis-cyclopropane tetrabromide from dibromo-carbene addition to 1,4-cyclohexadiene, and was utilized as an intermediate in Danheiser's synthesis of anatoxin *a*.[266]

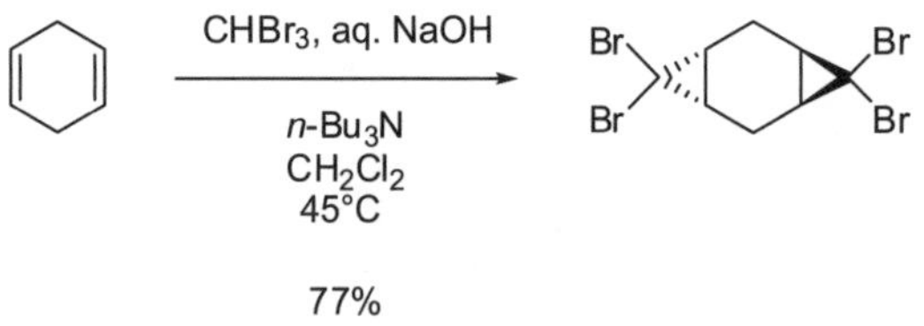

77%

Corey has developed epoxidation of carbonyl compounds (and cyclopropanation of electron-deficient olefins) with the anion of DMSO. The reaction proceeds by an addition/elimination mechanism, although the net effect is that of carbene addition to the olefin. Heathcock utilized this cyclopropanation in a synthesis of isovelleral; the reaction shown was run on a 3 g scale.[267]

[266]Danheiser, R. L.; Morin, J. M., Jr., ; Salaski, E. J. *J. Am. Chem. Soc.* **1985**, *107*, 8066–8073.
[267]Thompson, S. K.; Heathcock, C. H. *J. Org. Chem.* **1992**, *57*, 5979–5989.

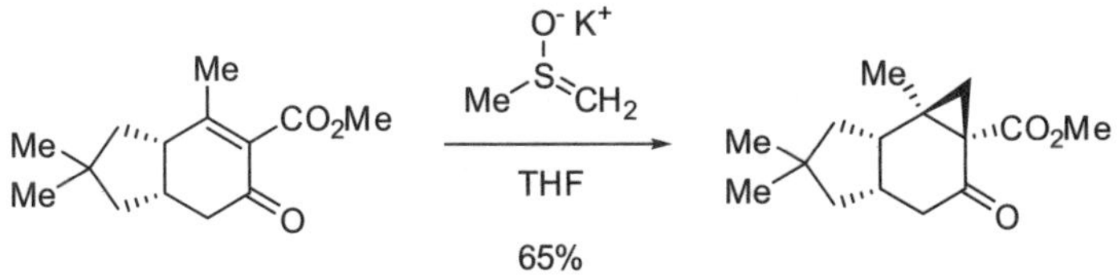

### 3.15.5   [4 + 3] Cycloadditions

Although not formally a C–C bis-addition to an olefin, [4 + 3] cycloadditions will be briefly discussed, given their analogy to other cycloadditions described in this section. Cycloadditions in which the 4-carbon component is a heterocycle (e.g., furan or pyrrole) are well known, and include the oxyallyl cation/furan cycloaddition to form oxabicyclo[3.2.1] octenes.[268,269] The interested reader is directed to these reviews. The examples in this section are limited to cycloadditions that generate products in which only carbon atoms comprise the bicyclic core.

The example below generates an α-bromoketone tricyclic adduct through the cycloaddition of an oxyallylcation and cyclopentadiene.[270]

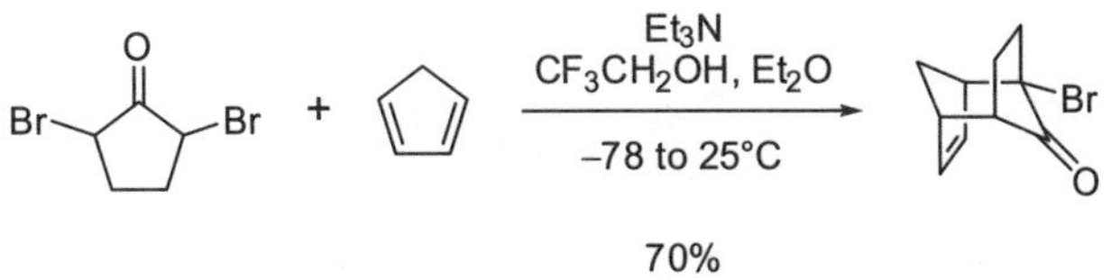

West has reported an intramolecular [4 + 3] cycloaddition in which a diene traps a Nazarov cationic intermediate to form tricyclic products such as that shown below.[271]

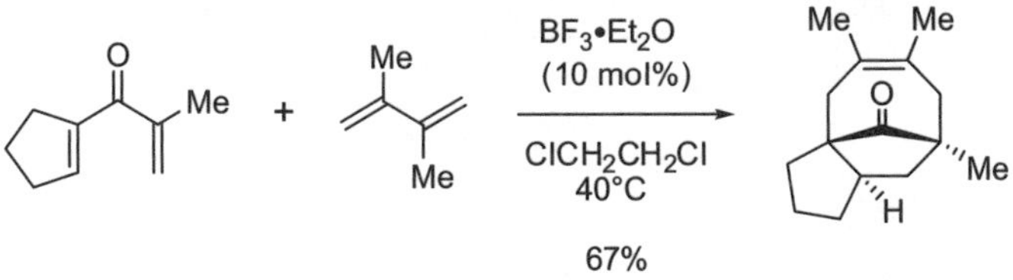

The following example is an apparent [4 + 3] cycloaddition, which actually entails a sequence of anionic additions/rearrangements.[272] The stereospecificity is attributed to the

[268]Hosomi, A. T.,Y. in *Comprehensive Organic Synthesis*; Pergamon Press: Oxford, **1991**; Vol. *5*, Ch. 5.1, p. 593.
[269]Noyori, R.; Hayakawa, Y. *Organic Reactions (New York)* **1983**, *29*, 163–344.
[270]Harmata, M.; Wacharasindhu, S. *Org. Lett.* **2005**, *7*, 2563–2565.
[271]Wang, Y.; Schill, B. D.; Arif, A. M.; West, F. G. *Org. Lett.* **2003**, *5*, 2747–2750.
[272]Takeda, K.; Nakajima, A.; Takeda, M.; Yoshii, E. *Org. Synth.* **1999**, *76*, 199–213.

intermediacy of a divinylcyclopropane, which undergoes anion-accelerated Cope rearrangement to generate the 7-membered carbocycle. The cyclopropane arises from initial 1,2-addition to the acylsilane followed by Brook rearrangement/cyclopropanation.

### 3.15.6    Conjugate Addition-Alkylation

As discussed in Section 3.5.5, cuprate reagents are effective for the 1,4-addition of alkyl groups to activated olefins. If the resulting enolate is trapped with an alkylating agent, then the net addition of two carbon atoms across the olefins is achieved. The following example couples a cuprate addition to an enone with a conjugate addition to a second enone, followed by an aldol reaction (net Robinson annulation).[273] The α-silyl MVK (methyl vinyl ketone) reagent was developed specifically for this type of reaction sequence.

The following example shows a similar sequence, but with enolate trapping with an alkyl triflate.[274] For details on the generation of the vinyl cuprate reagent and the enolate transmetallation, see Chapter 12.

[273]Boeckman, R. K., Jr., ; Blum, D. M.; Ganem, B. *Org. Synth.* **1978**, *58*, 158–163.
[274]Lipshutz, B. H.; Wood, M. R. *J. Am. Chem. Soc.* **1994**, *116*, 11689–11702.

## 3.15.7  Bis-Alkoxycarbonylation

Bis-alkoxycarbonylation of olefins can be effected with $CuCl_2$ and Pd/C in methanol, as shown in the following example.[275]

## 3.15.8  Cascade Cyclizations

Although limited in scope to olefin substrates possessing the requisite arrangement of initiator and terminator, cationic cascade cyclizations of polyenes as pioneered by W. S. Johnson represent a powerful strategy for construction of steroids and steroid-like carbon skeletons. In the following example, a fluorine atom is installed on one of the olefins to serve as a cation-stabilizing group, a strategy that improves the efficiency of the cascade cyclization.[276] This remarkable transformation forms four C–C sigma bonds in a single synthetic operation; the 70% yield translates to an average yield of 91.5% per bond forming event!

[275]Jolliffe, K.; Paddon-Row, M. N. *Tetrahedron* **1995**, *51*, 2689–2698.
[276]Johnson, W. S.; Plummer, M. S.; Reddy, S. P.; Bartlett, W. R. *J. Am. Chem. Soc.* **1993**, *115*, 515–521.

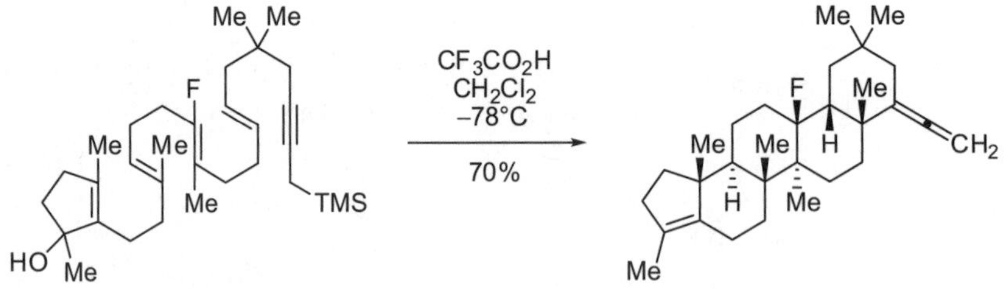

The following example was executed on a 5 g scale. The initial cation is formed by addition of the terminal olefin to a Pummerer-type intermediate generated from phenyl methyl sulfoxide and trifluoroacetic anhydride, which leads to sulfur incorporation into the cyclized product.[277]

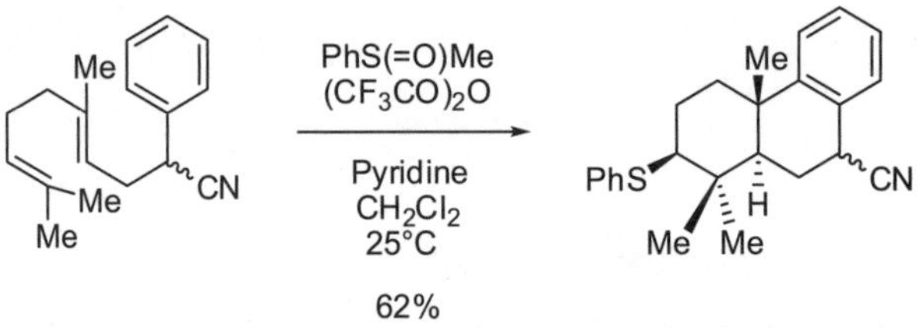

Cascade radical cyclizations have also been used to dramatic effect in natural product syntheses in which multiple C–C bonds are formed in a tandem series of radical cyclizations. The following example from Snider utilizes a Mn(III) oxidation of a β-ketoester to generate the initial radical, which then undergoes two sequential radical cyclizations to generate the bicyclic product.[278]

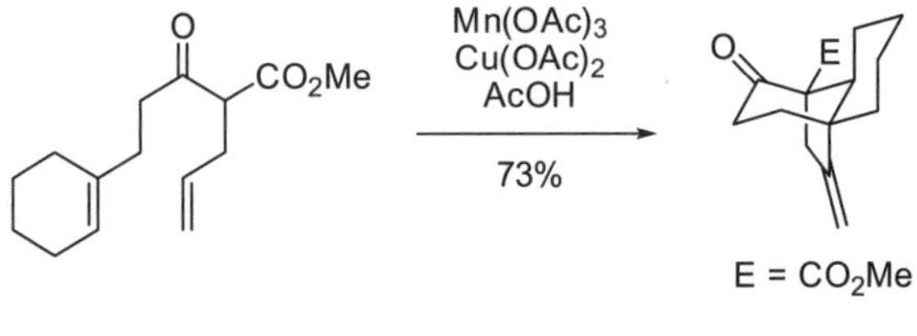

Lanthanide catalysts can serve to enhance the selectivity and efficiency of a similar cascade cyclization, as shown below.[279]

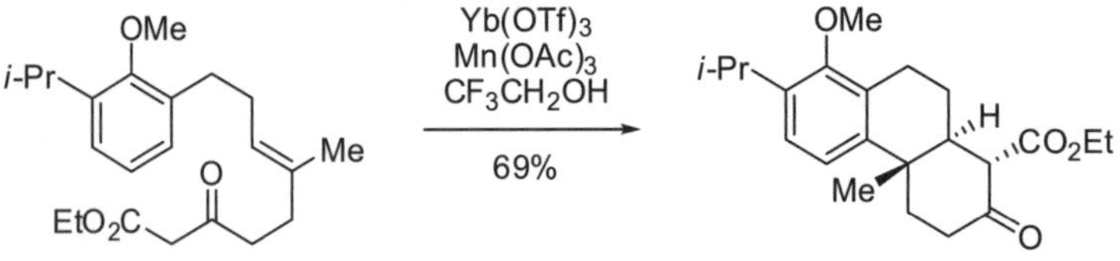

[277]Burnell, R. H.; Caron, S. *Can. J. Chem.* **1992**, *70*, 1446–1454.
[278]Snider, B. B.; Dombroski, M. A. *J. Org. Chem.* **1987**, *52*, 5487–5489.
[279]Yang, D.; Ye, X.-Y.; Xu, M.; Pang, K.-W.; Cheung, K.-K. *J. Am. Chem. Soc.* **2000**, *122*, 1658–1663.

The following example shows a rather extreme extension of this strategy. Although of modest efficiency (35% yield), the formation of four C–C bonds in a single reaction is impressive.[280]

The following cascade radical cyclization was utilized as the final step in Curran's synthesis of hirsutene. The isolated yield (65%) suffers somewhat from the volatility of the product hydrocarbon; the crude yield was estimated to be approximately 80%.[281]

Cascade polyene cyclizations initiated by a Nazarov cyclization have been reported, as exemplified below.[282]

[280]Snider, B. B.; Kiselgof, J. Y.; Foxman, B. M. *J. Org. Chem.* **1998**, *63*, 7945–7952.
[281]Curran, D. P.; Rakiewicz, D. M. *Tetrahedron* **1985**, *41*, 3943–3958.
[282]Bender, J. A.; Arif, A. M.; West, F. G. *J. Am. Chem. Soc.* **1999**, *121*, 7443–7444.

# 4

# NUCLEOPHILIC AROMATIC SUBSTITUTION

STÉPHANE CARON AND ARUN GHOSH

## 4.1 INTRODUCTION

Nucleophilic aromatic substitution, which can operate through several different reaction mechanisms, is considered one of the preferred methods to derivatize simple arenes. The scope of this reaction is guided by three basic principles: electron deficiency at the reactive carbon on the aromatic system, nature of the leaving group to be displaced, and reactivity of the nucleophile.[1] In general, more electron-deficient arenes will undergo more facile aromatic nucleophilic substitution in an addition/elimination sequence. Aryl halides, specifically fluorides, and diazonium compounds have proven to be the most successful substrates for this reaction. While the typical order of reactivity for an aliphatic nucleophilic substitution follows $I^- > Br^- > Cl^- \gg F^-$, this trend is generally reversed for the nucleophilic aromatic substitution. The electron withdrawing ability of the fluoride atom makes aryl fluorides better substrates for the addition of the nucleophile. Primary and secondary amines, as well as alkoxides, are usually excellent nucleophiles for the reaction. A few types of carbon nucleophiles, including cyanide and malonate derivatives, are also commonly used.

The preparation of *ortho*-arynes will also be briefly discussed in this chapter. In this case the elimination of a leaving group obviously precedes the addition of a nucleophile in what is formally a nucleophilic aromatic substitution. This methodology is not as common as the $S_N$–Ar reaction because of the very high reactivity and instability of the aryne generated. However, it has been used in cycloaddition reactions to access relatively complex polycycles.

---

[1]Buncel, E.; Dust, J. M.; Terrier, F. *Chem. Rev.* **1995**, *95*, 2261–2280.

---

*Practical Synthetic Organic Chemistry: Reactions, Principles, and Techniques*, First Edition.
Edited by Stéphane Caron.
© 2011 John Wiley & Sons, Inc. Published 2011 by John Wiley & Sons, Inc.

The chapter is constructed based on the type of nucleophile and nature of the product generated. This account should not be considered a comprehensive review in this area. Furthermore, transition metal promoted couplings are discussed separately in Chapter 6.

## 4.2   OXYGEN NUCLEOPHILES

The aryl ether linkage is abundant in natural products as well as in several pharmaceutical targets. The nucleophilic aromatic substitution of an arene with an oxygen nucleophile is one of the most powerful methods to access this class of molecules. Synthetic studies on the glycopeptide antibiotic vancomycin clearly demonstrated the utility of this reaction.[2–4]

**Vancomycin aglycone**

While several approaches have been considered for the preparation of the aryl ethers of vancomycin, the examples shown demonstrate the utility of the method for such a complex natural product.[5–7]

### 4.2.1   Preparation of Phenols

Phenols can be synthesized via nucleophilic aromatic substitutions from several starting materials with either hydroxide or water as the nucleophile. For example, aryl fluorides can be displaced by hydroxide under fairly mild conditions.[8]

[2]Nicolaou, K. C.; Boddy, C. N. C.; Brase, S.; Winssinger, N. *Angew. Chem., Int. Ed. Engl.* **1999**, *38*, 2097–2152.
[3]Burgess, K.; Lim, D.; Martinez, C. I. *Angew. Chem., Int. Ed. Engl.* **1996**, *35*, 1077–1078.
[4]Zhu, J. *Synlett* **1997**, 133–144.
[5]Bois-Choussy, M.; Beugelmans, R.; Bouillon, J.-P.; Zhu, J. *Tetrahedron Lett.* **1995**, *36*, 4781–4784.
[6]Boger, D. L.; Zhou, J. *J. Org. Chem.* **1996**, *61*, 3938–3939.
[7]Evans, D. A.; Barrow, J. C.; Watson, P. S.; Ratz, A. M.; Dinsmore, C. J.; Evrard, D. A.; DeVries, K. M.; Ellman, J. A.; Rychnovsky, S. D.; Lacour, J. *J. Am. Chem. Soc.* **1997**, *119*, 3419–3420.
[8]Hankovszky, H. O.; Hideg, K.; Lovas, M. J.; Jerkovich, G.; Rockenbauer, A.; Gyor, M.; Sohar, P. *Can. J. Chem.* **1989**, *67*, 1392–1400.

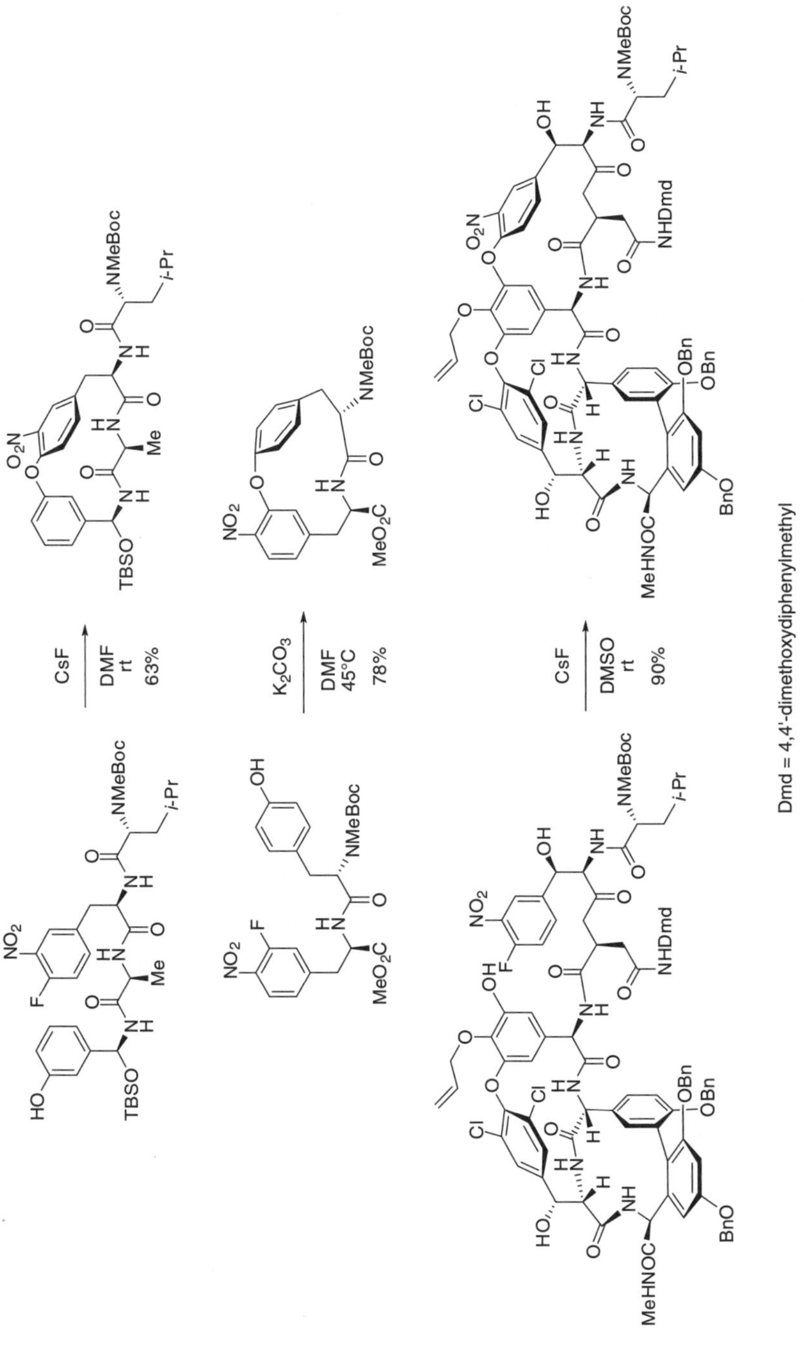

CsF

DMF
rt
63%

K₂CO₃

DMF
45°C
78%

CsF

DMSO
rt
90%

Dmd = 4,4'-dimethoxydiphenylmethyl

In cases where arylsulfonic acids prove to be an inexpensive and readily available starting materials, their conversion to phenols can be accomplished by reaction with hydroxide at high temperature.[9] Finally, the substitution of a diazonium salt provides another alternative in the case where the starting aniline is readily accessible. In the example provided, water from the aqueous sulfuric acid acts as the nucleophile.[10] It is always preferable to generate the diazonium *in situ* rather than attempt its isolation due to their known instability and potential shock sensitivity.

### 4.2.2   Preparation of Aryl Ethers

Aryl ethers are an important class of compounds and have most often been prepared using copper-catalyzed Ullman couplings (see Section 6.11). However, these compounds can also be accessed via nucleophilic aromatic substitutions, which usually require elevated reaction temperatures unless the arene starting material is very electron-deficient. Displacement of an aryl fluoride with an alkoxide is well precedented. The presence of an electron-withdrawing group *para* or *ortho* to the fluoride renders this reaction more facile.[11] An impressive example came from GlaxoSmithKline where difluorobenzene starting materials were functionalized by sequential addition of two alcohols. The deactivating nature of the first ether introduced allowed for mono-substitution with the first alcohol and more forcing conditions permitted formation of the second ether.[12]

[9]Campayo, L.; Jimenez, B.; Manzano, T.; Navarro, P. *Synthesis* **1985**, 197–200.

[10]Singh, B.; Bacon, E. R.; Lesher, G. Y.; Robinson, S.; Pennock, P. O.; Bode, D. C.; Pagani, E. D.; Bentley, R. G.; Connell, M. J.;et al. *J. Med. Chem.* **1995**, *38*, 2546–2550.

[11]Umemoto, T.; Adachi, K.; Ishihara, S. *J. Org. Chem.* **2007**, *72*, 6905–6917.

[12]Kim, A.; Powers, J. D.; Toczko, J. F. *J. Org. Chem.* **2006**, *71*, 2170–2172.

Aryl bromides can also participate in this reaction although the substrate scope is not as broad.[13] Chloropyridines are generally excellent substrates for this reaction, especially when the chloride is positioned at the 2-, 4-, or 6-position. Conversely, 3- or 5-substituted pyridines are poorly reactive. The example presented demonstrates once again that mono-addition is possible since the ether generated renders the product more electron-rich, and therefore less reactive.[14]

### 4.2.3  Preparation of Diaryl Ethers

Diaryl ethers can be accessed by nucleophilic aromatic substitution. As a rule of thumb, it is preferable to have the most electron-deficient arene contain the leaving group, while the "phenol" provides the electron-rich partner. The reaction is generally conducted using a mild base such as a carbonate, in a polar aprotic solvent with heat. This synthetic method is now complemented by the metal catalyzed cross-coupling of aryl halides (see Section 6.11).

[13]Wilson, J. M.; Cram, D. J. *J. Org. Chem.* **1984**, *49*, 4930–4943.
[14]Henegar, K. E.; Ashford, S. W.; Baughman, T. A.; Sih, J. C.; Gu, R.-L. *J. Org. Chem.* **1997**, *62*, 6588–6597.

Activated bromides[15] and fluorides[16] generally provide high-yielding reactions. In the last example shown below, chemoselectivity of the phenol over the aniline was achieved when reacting with a 2-chloropyridine, an excellent class of substrate for this reaction. The authors also noticed an appreciable rate difference when using powdered, as opposed to granular $K_2CO_3$.[17]

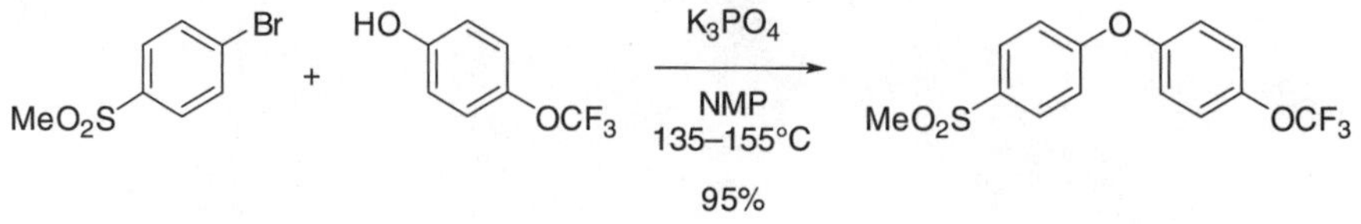

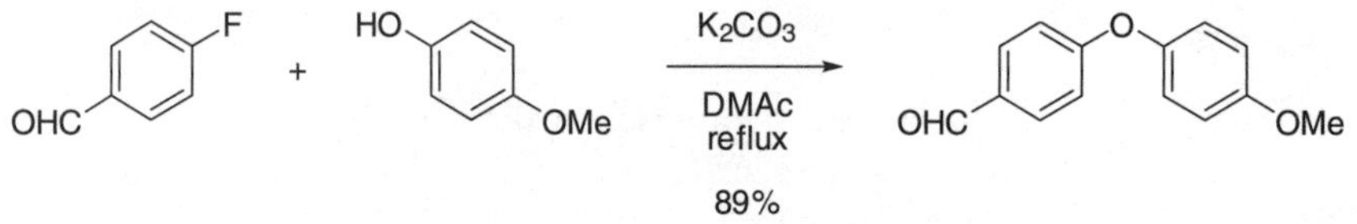

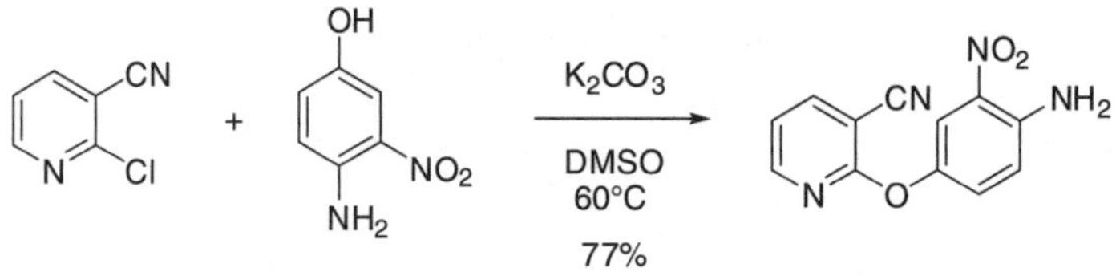

95%

89%

77%

## 4.3 SULFUR NUCLEOPHILES

### 4.3.1 Preparation of Aryl Thioethers

Thiols will participate efficiently in nucleophilic aromatic substitutions and are generally more reactive than the corresponding alcohols, allowing for a lower reaction temperature. It is convenient to utilize the thiolate directly if it is commercially available to avoid potential disulfide formation and minimize the odor associated with free thiols.[18]

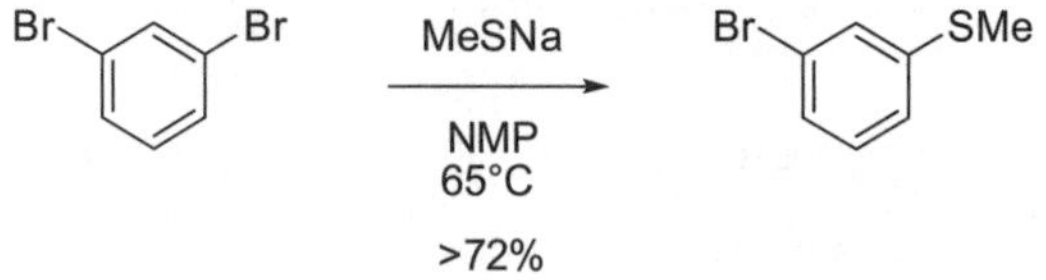

>72%

[15]Chang, S.-J.; Fernando, D.; Fickes, M.; Gupta, A. K.; Hill, D. R.; McDermott, T.; Parekh, S.; Tian, Z.; Wittenberger, S. J. *Org. Process Res. Dev.* **2002**, *6*, 329–335.

[16]Yeager, G. W.; Schissel, D. N. *Synthesis* **1991**, 63–68.

[17]Ruggeri, S. G.; Bill, D. R.; Bourassa, D. E.; Castaldi, M. J.; Houck, T. L.; Ripin, D. H. B.; Wei, L.; Weston, N. *Org. Process Res. Dev.* **2003**, *7*, 1043–1047.

[18]Lipton, M. F.; Mauragis, M. A.; Maloney, M. T.; Veley, M. F.; VanderBor, D. W.; Newby, J. J.; Appell, R. B.; Daugs, E. D. *Org. Process Res. Dev.* **2003**, *7*, 385–392.

When necessary, the thiol can be used in the presence of a mild base. Because of its lower p$K_a$ and higher nucleophilicity, the reaction can be conducted in an alcohol solvent.[19] Furthermore, introduction of a hindered $t$-BuS moiety is feasible[20] and it has been shown that a nitro moiety is a sufficient leaving group to participate is this reaction.[21]

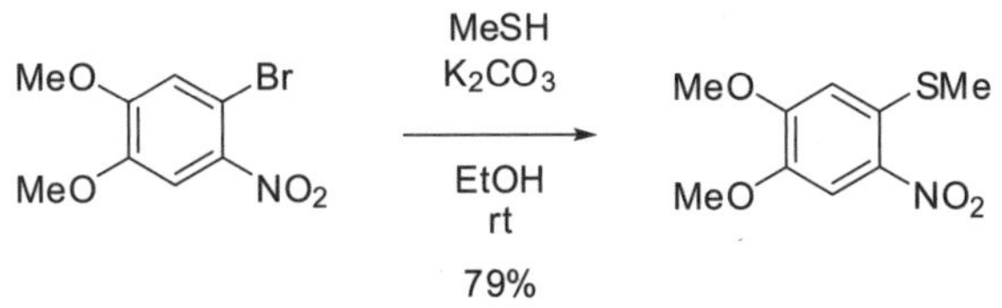

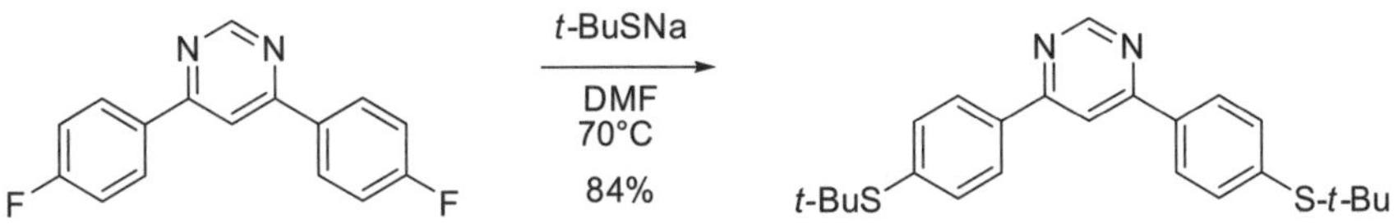

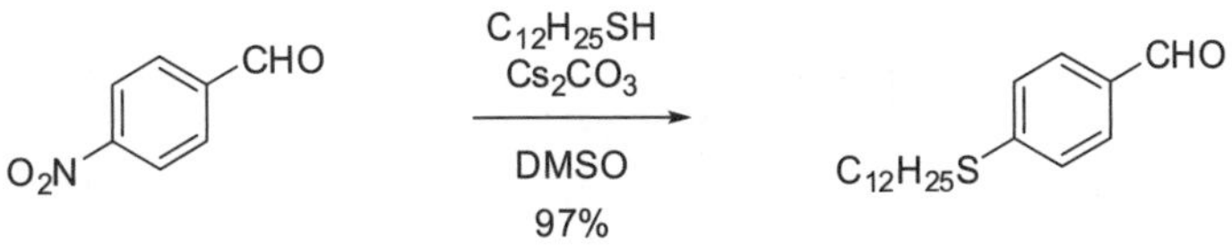

## 4.3.2  Preparation of Diaryl Thioethers

Arylthiols are effective partners in the formation of diaryl thioethers by nucleophilic aromatic substitutions. As with alkylthiols, arylthiols are excellent nucleophiles, leading to faster reactions compared to the corresponding phenols. As shown below, displacement of an aryl fluoride[22a] or an activated aryl bromide[22b] with thiophenol can occur under fairly mild conditions.

[19]Dillard, R. D.; Yen, T. T.; Stark, P.; Pavey, D. E. *J. Med. Chem.* **1980**, *23*, 717–722.
[20]Wheelhouse, R. T.; Jennings, S. A.; Phillips, V. A.; Pletsas, D.; Murphy, P. M.; Garbett, N. C.; Chaires, J. B.; Jenkins, T. C. *J. Med. Chem.* **2006**, *49*, 5187–5198.
[21]Kondoh, A.; Yorimitsu, H.; Oshima, K. *Tetrahedron* **2006**, *62*, 2357–2360.
[22](a) Becker, D. P.; Villamil, C. I.; Barta, T. E.; Bedell, L. J.; Boehm, T. L.; DeCrescenzo, G. A.; Freskos, J. N.; Getman, D. P.; Hockerman, S.; Heintz, R.; Howard, S. C.; Li, M. H.; McDonald, J. J.; Carron, C. P.; Funckes-Shippy, C. L.; Mehta, P. P.; Munie, G. E.; Swearingen, C. A. *J. Med. Chem.* **2005**, *48*, 6713–6730. (b) Li, Z.; Yang, Q.; Qian, X. *Tetrahedron* **2005**, *61*, 8711–8717.

### 4.3.3 Other Sulfur Nucleophiles

Other sulfur-substituted arenes have been prepared by nucleophilic aromatic substitutions. For example, a sulfinate anion reacts at the sulfur center, generating directly the corresponding sulfone.[23] Sodium sulfide can react with electron-deficient arene to generate sodium thiolates directly, albeit in moderate yields. The regioselectivity observed reflects the activating nature of a nitro moiety at the *ortho* position.[24]

## 4.4 NITROGEN NUCLEOPHILES

Substituted anilines are commonly seen in pharmaceutical agents, agrochemicals, dyes, and many other useful materials. Their preparation by nucleophilic aromatic substitution of an amine, which is generally an excellent nucleophile, to an electron-deficient arene containing

[23]Ulman, A.; Urankar, E. *J. Org. Chem.* **1989**, *54*, 4691–4692.
[24]Gupta, R. R.; Kumar, R.; Gautam, R. K. *J. Heterocycl. Chem.* **1984**, *21*, 1713–1715.

a leaving group is well precedented. This class of compounds has attracted much attention and led to the development of new synthetic methods such as palladium and copper-mediated aryl aminations (see Section 6.10). Additionally, alkylation or reductive amination of anilines, often obtained from a previous nitration, remains an attractive synthetic approach. The availability of the prerequisite starting materials often dictates which strategy might be preferred.

### 4.4.1  Preparation of Aryl Amines

Primary and secondary amines react readily with appropriately functionalized electron-deficient arenes to provide the desired anilines. An excess of the starting amine is often utilized if it is an inexpensive commodity, since they are stronger bases than the products, and hence can effectively scavenge the acid generated. In cases where the amine is more valuable, an additional non-nucleophilic base, such as a carbonate or a tertiary amine can be employed to neutralize the acid generated in the process. For primary amines, disubstitution is avoided as the resulting product is a less nucleophilic aniline.[25] When a substrate contains multiple leaving groups, introduction of the first amine renders the product more electron-rich, thereby slowing down the second nucleophilic substitution.[26]

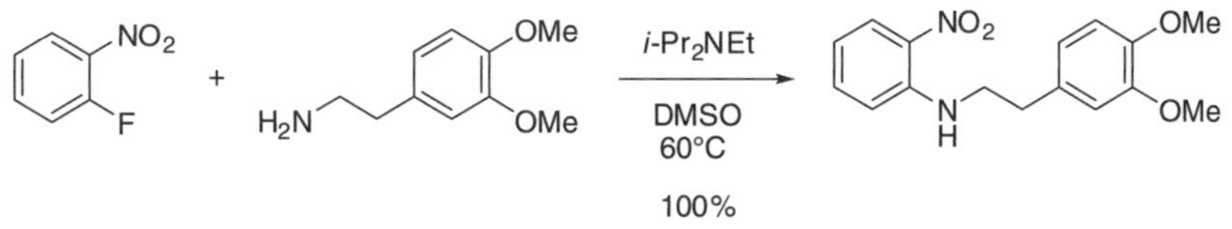

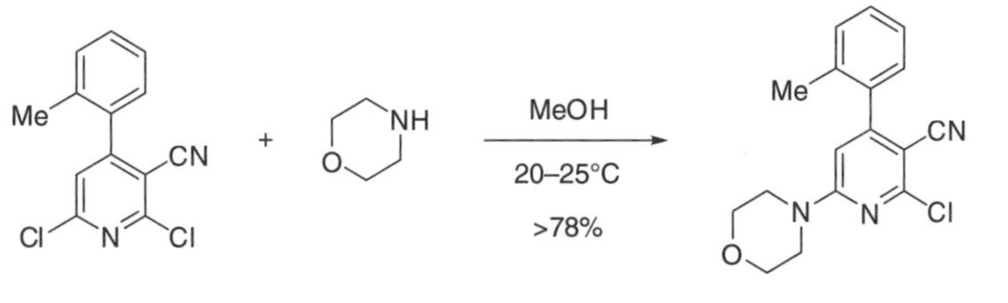

In the case of the trichloropyrimidine shown below, the first two substitutions were accomplished readily using $S_N$–Ar, but the introduction of a third amine required the generation of a lithium amide in order to avoid harsh reaction conditions.[27] The second example shows the addition of a pyrrolidine that proceeded in very high yield where tetramethylguanidine was used as the base in a fairly complex system.[28]

[25]Beaulieu, P. L.; Hache, B.; Von Moos, E. *Synthesis* **2003**, 1683–1692.

[26]Harrington, P. J.; Johnston, D.; Moorlag, H.; Wong, J.-W.; Hodges, L. M.; Harris, L.; McEwen, G. K.; Smallwood, B. *Org. Process Res. Dev.* **2006**, *10*, 1157–1166.

[27]Mauragis, M. A.; Veley, M. F.; Lipton, M. F. *Org. Process Res. Dev.* **1997**, *1*, 39–44.

[28]Beylin, V.; Boyles, D. C.; Curran, T. T.; Macikenas, D.; Parlett, R. V., IV; Vrieze, D. The Preparation of Two, Preclinical Amino-quinazolinediones as Antibacterial Agents. *Org. Process Res. Dev.* **2007**, *11*, 441–449.

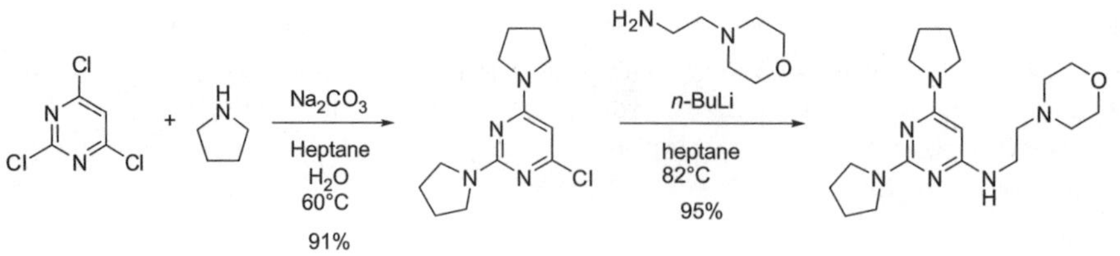

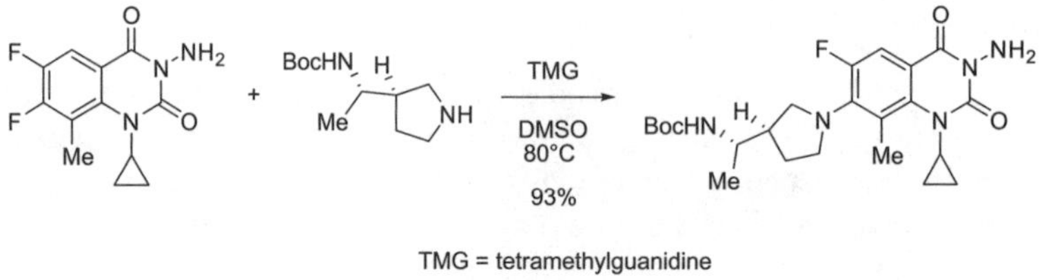

91%

95%

93%

TMG = tetramethylguanidine

### 4.4.2   Preparation of Diaryl Amines

Because they are not as nucleophilic as alkylamines, anilines require more forcing conditions to participate in a nucleophilic aromatic substitution.[29]

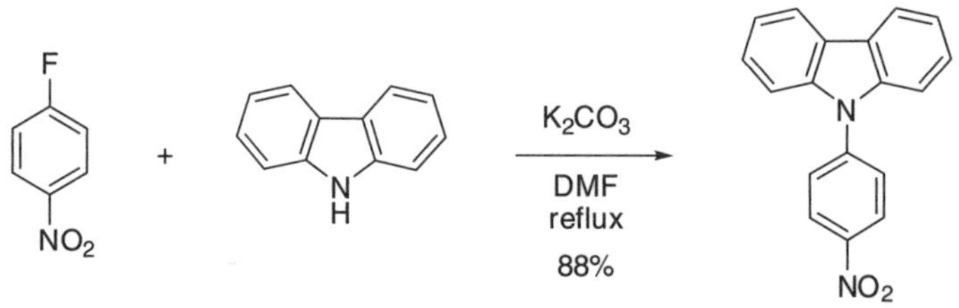

88%

One method to circumvent this problem is to deprotonate the aniline as shown in the example below.[30] It is worth mentioning that in this reaction, a mixture of both substrates is added to a suspension of the lithium amide and deprotonation occurs more rapidly than addition of the lithium amide. In some instances where the product of the nucleophilic aromatic substitution is a better base than the starting aniline, it is possible to conduct the reaction under acidic conditions.[31]

[29]Jian, H.; Tour, J. M. *J. Org. Chem.* **2003**, *68*, 5091–5103.

[30]Davis, E. M.; Nanninga, T. N.; Tjiong, H. I.; Winkle, D. D. *Org. Process Res. Dev.* **2005**, *9*, 843–846.

[31]Denni-Dischert, D.; Marterer, W.; Baenziger, M.; Yusuff, N.; Batt, D.; Ramsey, T.; Geng, P.; Michael, W.; Wang, R.-M. B.; Taplin, F., Jr., ; Versace, R.; Cesarz, D.; Perez, L. B. *Org. Process Res. Dev.* **2006**, *10*, 70–77.

### 4.4.3 Other Nitrogen Nucleophiles

Several other nitrogen nucleophiles can participate in nucleophilic aromatic substitution. For example, a pyrimidone will react with an activated aryl fluoride.[32] Hydrazine is also an excellent nucleophile and can be introduced under relatively mild conditions.[33]

Hydrazones will undergo intramolecular cyclization to 1*H*-indazoles when a leaving group is present at the *ortho* position. It has been demonstrated that a fluoride[34] requires less forcing conditions than a mesylate[35] to promote cyclization. Finally, an elegant sequence involving two nucleophilic aromatic substitutions was developed at Abbott for the preparation of a quinolone antibiotic. The first substitution generated the quinolone by cyclization of

[32]De Napoli, L.; Messere, A.; Montesarchio, D.; Piccialli, G. *J. Org. Chem.* **1995**, *60*, 2251–2253.
[33]Fleck, T. J.; Chen, J. J.; Lu, C. V.; Hanson, K. J. *Org. Process Res. Dev.* **2006**, *10*, 334–338.
[34]Caron, S.; Vazquez, E. *Org. Process Res. Dev.* **2001**, *5*, 587–592.
[35]Caron, S.; Vazquez, E. *Synthesis* **1999**, 588–592.

a vinylogous amide on an aryl fluoride. This was followed by introduction of an azetidine side chain in a second substitution.[36]

## 4.5 HALOGEN NUCLEOPHILES

### 4.5.1 Reaction of Diazonium Salts

Introduction of a halogen atom onto an aromatic ring by nucleophilic aromatic substitution is complementary to halogenation of arenes through electrophilic aromatic substitution discussed in Section 5.4. One of the most common methods is the Sandmeyer reaction, where an aniline is diazotized and decomposed in the presence of a halide. The diazonium salt is obtained by treatment of the sodium nitrite under acidic conditions or with an alkyl nitrite. While the salt can be isolated, it is preferable from a safety standpoint to keep this reactive intermediate in solution and to carry it through the next step. When setting up this reaction, it is prudent to have sufficient venting and a large headspace, as a large amount of gas evolves over the course of the reaction.

In general, formation of fluoroarenes are lower yielding than other aryl halides and often requires isolation of the diazonium salt.[37]

For the formation of chlorides, copper(II) chloride is often used as the halide source.[38] As part of a methodology evaluation, the Sandmeyer reaction has been conducted on

[36]Barnes, D. M.; Christesen, A. C.; Engstrom, K. M.; Haight, A. R.; Hsu, M. C.; Lee, E. C.; Peterson, M. J.; Plata, D. J.; Raje, P. S.; Stoner, E. J.; Tedrow, J. S.; Wagaw, S. *Org. Process Res. Dev.* **2006**, *10*, 803–807.

[37]Munson, P. M.; Thompson, W. J. *Synth. Commun.* **2004**, *34*, 759–766.

[38]Knapp, S.; Ziv, J.; Rosen, J. D. *Tetrahedron* **1989**, *45*, 1293–1298.

7-amino-1-indanone and 8-amino-tetralone with several halide sources. The preparation of 8-bromo-1-tetralone is exemplified.[39]

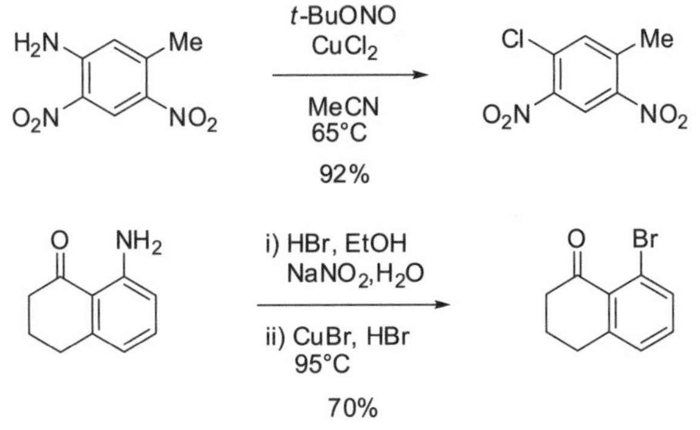

92%

70%

For the preparation of aryl iodides, potassium iodide is generally the reagent selected.[40,41]

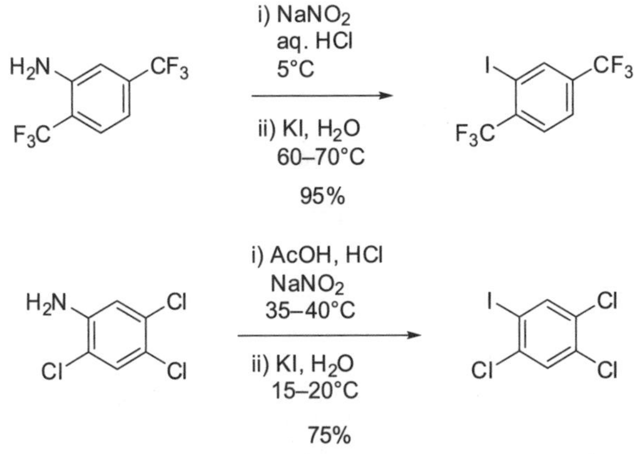

95%

75%

## 4.5.2  Preparation of 2-Halopyridines and Derivatives

Another type of halogenation through nucleophilic substitution is the conversion of a pyrone-type structure to a halopyridine. The oxygen atom generally reacts with either a phosphorous or sulfur reagent to provide an activated ester that is displaced by a halide.

[39]Nguyen, P.; Corpuz, E.; Heidelbaugh, T. M.; Chow, K.; Garst, M. E. *J. Org. Chem.* **2003**, *68*, 10195–10198.
[40]Satyanarayana, K.; Srinivas, K.; Himabindu, V.; Reddy, G. M. *Org. Process Res. Dev.* **2007**, *11*, 842–845.
[41]Singh, J.; Kim, O. K.; Kissick, T. P.; Natalie, K. J.; Zhang, B.; Crispino, G. A.; Springer, D. M.; Wichtowski, J. A.; Zhang, Y.; Goodrich, J.; Ueda, Y.; Luh, B. Y.; Burke, B. D.; Brown, M.; Dutka, A. P.; Zheng, B.; Hsieh, D.-M.; Humora, M. J.; North, J. T.; Pullockaran, A. J.; Livshits, J.; Swaminathan, S.; Gao, Z.; Schierling, P.; Ermann, P.; Perrone, R. K.; Lai, M. C.; Gougoutas, J. Z.; DiMarco, J. D.; Bronson, J. J.; Heikes, J. E.; Grosso, J. A.; Kronenthal, D. R.; Denzel, T. W.; Mueller, R. H. *Org. Process Res. Dev.* **2000**, *4*, 488–497.

For the preparation of chlorides, phosphorous oxychloride[42] or thionyl chloride are most commonly used.[43] When phosphorous oxychloride is used, proper precautions should be taken in the workup (see Section 20.3.4.6).

Phosphorous oxybromide can be utilized for the preparation of bromides but is generally not as efficient as phosphorous oxychloride for the synthesis of chlorides.[44] Another common method is the *in situ* generation of phosphorous pentabromide, although this method might be less practical on small scale due to the challenges with handling bromine.[45]

An attractive method for this transformation utilizes phosphorous pentoxide in the presence of a bromide source. This procedure offers some advantage in the workup as the product resides in the toluene layer while the phosphoric acid generated can easily be removed via an aqueous wash.[46]

[42]Connolly, T. J.; Matchett, M.; McGarry, P.; Sukhtankar, S.; Zhu, J. *Org. Process Res. Dev.* **2006**, *10*, 391–397.
[43]Ple, P. A.; Green, T. P.; Hennequin, L. F.; Curwen, J.; Fennell, M.; Allen, J. Lambert-van der Brempt, C.; Costello, G. *J. Med. Chem..* **2004**, *47*, 871.
[44]Ricci, G.; Ruzziconi, R.; Giorgio, E. *J. Org. Chem.* **2005**, *70*, 1011–1018.
[45]Ager, D. J.; Erickson, R. A.; Froen, D. E.; Prakash, I.; Zhi, B. *Org. Process Res. Dev.* **2004**, *8*, 62–71.
[46]Kato, Y.; Okada, S.; Tomimoto, K.; Mase, T. *Tetrahedron Lett.* **2001**, *42*, 4849–4851.

Preparation of aryl fluorides or iodides by these methods is rather uncommon.

## 4.6 CARBON NUCLEOPHILES

### 4.6.1 Cyanide as a Nucleophile

Cyanide is an excellent nucleophile and can participate in nucleophilic aromatic substitutions. This method is complementary, but often inferior to, the metal-mediated couplings that typically utilize zinc cyanide (see Section 6.14). For substrates that are highly activated, sodium cyanide can be employed.[47] In cases where the starting material is a chloride, copper(I) cyanide is most often utilized.[48] This reaction most likely proceeds through an electron-transfer mechanism and usually requires a polar aprotic solvent and elevated reaction temperature.

### 4.6.2 Malonates as Nucleophiles

Malonate derivatives have been utilized in nucleophilic aromatic substitutions of electron-deficient arenes. In some cases, the salt of the malonate is available and can be used directly.[49] In most cases, sodium hydride is utilized to generate the enolate at lower temperatures in tetrahydrofuran and the temperature is increased for the substitution to proceed.[50]

[47]Cheung, E.; Rademacher, K.; Scheffer, J. R.; Trotter, J. *Tetrahedron* **2000**, *56*, 6739–6751.
[48]Skeean, R. W.; Goel, O. P. *Synthesis* **1990**, 628–630.
[49]Gurjar, M. K.; Murugaiah, A. M. S.; Reddy, D. S.; Chorghade, M. S. *Org. Process Res. Dev.* **2003**, *7*, 309–312.
[50]Butters, M.; Ebbs, J.; Green, S. P.; MacRae, J.; Morland, M. C.; Murtiashaw, C. W.; Pettman, A. J. *Org. Process Res. Dev.* **2001**, *5*, 28–36.

Methyl cyanoacetate is also an effective nucleophile and can allow for the preparation of benzylic nitriles after hydrolysis and decarboxylation.[51]

### 4.6.3   Other Carbon Nucleophiles

While less common, other carbon nucleophiles can participate in nucleophilic aromatic substitution. One such example is the preparation of indoles through intramolecular cyclization, which is believed to proceed through a benzyne mechanism.[52]

Sulfone anions can also act as nucleophiles, as shown in the following example.[53]

[51]Selvakumar, N.; Rajulu, G. G. *J. Org. Chem.* **2004**, *69*, 4429–4432.
[52]Kudzma, L. V. *Synthesis* **2003**, 1661–1666.
[53]Sommer, M. B.; Begtrup, M.; Boegesoe, K. P. *J. Org. Chem.* **1990**, *55*, 4817–4821.

Furthermore, the reaction of secondary nitrile anions with aryl fluorides has been demonstrated, even with electron-rich arenes.[34,54]

## 4.7 *ORTHO*-ARYNES

Arynes are extremely active intermediates that react rapidly with a number of reagents, most notably alkenes and alkynes in dipolar cycloadditions. The chemistry for the preparation of arynes and their reactivity has been extensively reviewed.[55,56] Two major methods exist for the preparation of arynes. The first method consists of generation of an anion *ortho* to a leaving group. Typically, this is accomplished by *ortho*-metallation[57] or halogen-metal exchange.[58]

While diazotization and decomposition of an anthranilic acid derivatives has historically been an important method for the generation of arynes, it is now less commonly used due to its impracticality and potential safety hazard associated with the large volume of gas generated.[59]

[54]Caron, S.; Wojcik, J. M.; Vazquez, E. *Org. Synth.* **2003**, *79*, 209–215.
[55]Dyke, A. M.; Hester, A. J.; Lloyd-Jones, G. C. *Synthesis* **2006**, 4093–4112.
[56]Pellissier, H.; Santelli, M. *Tetrahedron* **2003**, *59*, 701–730.
[57]Sutherland, H. S.; Higgs, K. C.; Taylor, N. J.; Rodrigo, R. *Tetrahedron* **2001**, *57*, 309–317.
[58]Coe, J. W.; Wirtz, M. C.; Bashore, C. G.; Candler, J. *Org. Lett.* **2004**, *6*, 1589–1592.
[59]Pu, L.; Grubbs, R. H. *J. Org. Chem.* **1994**, *59*, 1351–1353.

# 5

# ELECTROPHILIC AROMATIC SUBSTITUTION

Stéphane Caron

## 5.1 INTRODUCTION

Electrophilic aromatic substitution reactions are the most commonly used methods to derivatize simple aromatic substrates.[1] The parameters that control the outcome of these reactions are generally well understood. Electron-rich arenes will often undergo facile reactions while electron-poor arenes require more forcing conditions or will not react at all. The factors that control the regiochemical outcome of the reaction on substituted arenes follow the general rule that electron-rich substituents direct to the position *para* and to a lesser extent *ortho*, while electron-withdrawing groups will favor *meta* substitution. Halogens are also *ortho/para* directors, but reduce the rate of substitution relative to more electron-rich substituents. The most electron-rich functional group usually has the dominant effect. Most of the reactions presented in this chapter proceed at room temperature or above and many require the presence of a protic or Lewis acid in stoichiometric amount. The most commonly utilized Lewis acid is $AlCl_3$, but many others are acceptable as well.

This chapter is organized based on the nature of the electrophile and then subdivided by the type of products obtained from the substitution. Many reactions with little practical use have been omitted.

---

[1]Olah, G. A. *Interscience Monographs on Organic Chemistry: Friedel–Crafts Chemistry*; Wiley-Interscience: New York, N. Y., 1973.

---

*Practical Synthetic Organic Chemistry: Reactions, Principles, and Techniques*, First Edition.
Edited by Stéphane Caron.
© 2011 John Wiley & Sons, Inc. Published 2011 by John Wiley & Sons, Inc.

## 5.2   NITROGEN ELECTROPHILES

### 5.2.1   Nitration

Nitration is the most utilized reaction for incorporation of a nitrogen substituent on an aromatic ring. A desirable feature of this reaction is that it directly functionalizes an arene C$-$H bond and the reagent used, nitric acid, is extremely cheap. Another advantage of the nitration reaction is that the active nitrating species, the nitronium ion ($NO_2^+$), is very reactive such that most substrates will be mono-nitrated. The reaction will usually stop after a single nitration because the nitro group is highly deactivating.

The most common and practical way to perform a nitration is to dissolve the substrate in sulfuric acid (often an exothermic process in and of itself) followed by slow addition of concentrated $HNO_3$ or a cold mixture of $H_2SO_4$ and $HNO_3$. When the reaction is performed in this manner, the nitronium ion is generated in the presence of the substrate and is rapidly consumed, thus minimizing safety concerns. Nitrations are highly exothermic and the exotherm can be controlled by the rate of addition of the $HNO_3$. Typical nitrations are often dose-controlled processes.[2,3]

A modification of the procedure above is to incorporate acetic anhydride as a water scavenger. The reaction can be performed using $Ac_2O$ as either the solvent without $H_2SO_4$[4] or as an additive. In the second case below (115 kg scale), the yield dropped significantly when the reaction was conducted only in acetic acid without acetic anhydride.[5]

[2]Hutt, M. P.; MacKellar, F. A. *J. Heterocycl. Chem.* **1984**, *21*, 349–352.

[3]Marterer, W.; Prikoszovich, W.; Wiss, J.; Prashad, M. *Org. Process Res. Dev.* **2003**, *7*, 318–323.

[4]Maehr, H.; Smallheer, J. *J. Am. Chem. Soc.* **1985**, *107*, 2943–2945.

[5]Giles, M. E.; Thomson, C.; Eyley, S. C.; Cole, A. J.; Goodwin, C. J.; Hurved, P. A.; Morlin, A. J. G.; Tornos, J.; Atkinson, S.; Just, C.; Dean, J. C.; Singleton, J. T.; Longton, A. J.; Woodland, I.; Teasdale, A.; Gregertsen, B.; Else, H.; Athwal, M. S.; Tatterton, S.; Knott, J. M.; Thompson, N.; Smith, S. J. *Org. Process Res. Dev.* **2004**, *8*, 628–642.

Nitronium triflate is a superior nitrating agent for less reactive substrates, and can be generated with tetraalkylammonium nitrate and triflic anhydride[6] or triflic acid in the presence of fuming nitric acid.[7] Because of the nature of these nitrations, a very unreactive solvent is required and dichloromethane is often selected. The product is then isolated through an extractive workup.

### 5.2.2   Nitrosation

The most common method for the preparation of a nitroso derivative is by slow addition of aqueous sodium nitrite to an arene dissolved in a carboxylic acid. One advantage of carrying out the reaction in this manner is that if the product is a solid, it will often crystallize from the reaction mixture and can be easily isolated by filtration. The three examples shown are representative of this reaction. In the first example, the reaction proceeded at 0°C,[8] in the second, at room temperature,[9] and the third example required 90°C, due to the lack of reactivity of the substrate.[10]

[6]Shackelford, S. A.; Anderson, M. B.; Christie, L. C.; Goetzen, T.; Guzman, M. C.; Hananel, M. A.; Kornreich, W. D.; Li, H.; Pathak, V. P.; Rabinovich, A. K.; Rajapakse, R. J.; Truesdale, L. K.; Tsank, S. M.; Vazir, H. N. *J. Org. Chem.* **2003**, *68*, 267–275.

[7]Coe, J. W.; Brooks, P. R.; Vetelino, M. G.; Wirtz, M. C.; Arnold, E. P.; Huang, J.; Sands, S. B.; Davis, T. I.; Lebel, L. A.; Fox, C. B.; Shrikhande, A.; Heym, J. H.; Schaeffer, E.; Rollema, H.; Lu, Y.; Mansbach, R. S.; Chambers, L. K.; Rovetti, C. C.; Schulz, D. W.; Tingley, F. D., III; O'Neill, B. T. *J. Med. Chem.* **2005**, *48*, 3474–3477.

[8]Maleski, R. J. *Synth. Commun.* **1993**, *23*, 343–348.

[9]Liu, Y.; McWhorter, W. W., Jr., *J. Am. Chem. Soc.* **2003**, *125*, 4240–4252.

[10]Erickson, R. H.; Hiner, R. N.; Feeney, S. W.; Blake, P. R.; Rzeszotarski, W. J.; Hicks, R. P.; Costello, D. G.; Abreu, M. E. *J. Med. Chem.* **1991**, *34*, 1431–1435.

Alkyl nitrites are an alternative class of reagent that is particularly useful if the starting material is not soluble in a carboxylic acid. The most common alkyl nitrites are *n*-butyl and *i*-amyl nitrite.[11] An extractive workup is usually required for isolation of the desired product when these conditions are utilized.

### 5.2.3  Diazonium Coupling

The reaction between a diazonium salt and an arene, known as a diazonium coupling, is not a very common reaction although the products from this reaction often lead to materials with interesting optical properties. When performing this reaction, it is recommended that proper safety precautions be used in handling the diazonium salts, since they are known to be very energetic and can decompose easily.[12]

[11]Marchal, A.; Melguizo, M.; Nogueras, M.; Sanchez, A.; Low, J. N. *Synlett* **2002**, 255–258.
[12]Ulman, A.; Willand, C. S.; Kohler, W.; Robello, D. R.; Williams, D. J.; Handley, L. *J. Am. Chem. Soc.* **1990**, *112*, 7083–7090.

## 5.3  SULFUR ELECTROPHILES

### 5.3.1  Sulfonation

While the sulfonation reaction is not utilized as often as the chlorosulfonation, this transformation allows for the direct formation of a sulfonic acid. Sulfonation is conducted in fuming sulfuric acid (oleum), a highly corrosive reagent. On a reactive substrate, the reaction is generally carried out around 0°C. Ideally, the product is precipitated by pouring the reaction mixture into cold water and isolated by filtration.[13] Alternatively, the reaction can be quenched and the crude sulfonic acid can be isolated and purified by generation of the sodium salt.[14]

Another method to access the sulfonic acid is similar to a halosulfonation (Section 5.3.2) reaction, but the sulfonic acid is obtained by not employing a large excess of the halosulfonic acid.[15]

Not isolated

### 5.3.2  Halosulfonation

The halosulfonation reaction is probably the most commonly employed method for introduction of a sulfur substituent onto an arene. The substrate is normally added directly to the halosulfonic acid, the mixture is heated until reaction completion, then quenched into cold water. The initial product of the reaction is the sulfonic acid, which is converted to the sulfonyl chloride by reaction with the excess halosulfonic acid.[16,17]

[13]Dorogov, M. V.; Filimonov, S. I.; Kobylinsky, D. B.; Ivanovsky, S. A.; Korikov, P. V.; Soloviev, M. Y.; Khahina, M. Y.; Shalygina, E. E.; Kravchenko, D. V.; Ivachtchenko, A. V. *Synthesis* **2004**, 2999–3004.

[14]Ikemoto, N.; Liu, J.; Brands, K. M. J.; McNamara, J. M.; Reider, P. J. *Tetrahedron* **2003**, *59*, 1317–1325.

[15]Urban, F. J.; Jasys, V. J.; Raggon, J. W.; Buzon, R. A.; Hill, P. D.; Eggler, J. F.; Weaver, J. D. *Synth. Commun.* **2003**, *33*, 2029–2043.

[16]Borror, A. L.; Chinoporos, E.; Filosa, M. P.; Herchen, S. R.; Petersen, C. P.; Stern, C. A.; Onan, K. D. *J. Org. Chem.* **1988**, *53*, 2047–2052.

[17]Moore, R. M., Jr., *Org. Process Res. Dev.* **2003**, *7*, 921–924.

Proper precautions must be taken when using chlorosulfonic acid, which is highly toxic.

### 5.3.3 Sulfurization

The sulfurization reaction is not commonly employed. It can be performed with a variety of sulfur sources such as $SCl_2$, $S_2Cl_2$, or even elemental sulfur. In some cases, a Lewis acid is employed but it is not always necessary with activated substrates. This method has been used in an efficient synthesis of a benzothiophene.[18] Disulfides can also be generated in a single step by treatment with sulfur monochloride in the presence of $TiCl_4$.[19]

A variation on the sulfurization is the Herz reaction, in which an aniline reacts with $SCl_2$ to generate a thiazothionium halide (Herz compound).[20,21] One positive aspect of the Herz compounds is that they can be further derivatized. For instance, they will react with sodium hydroxide to generate the ortho thioaniline, which makes the two-step sequence very attractive for the preparation of such products.

[18]Engman, L. *J. Heterocycl. Chem.* **1984**, *21*, 413–416.
[19]Pastor, S. D. *J. Org. Chem.* **1984**, *49*, 5260–5262.
[20]Belica, P. S.; Manchand, P. S. *Synthesis* **1990**, 539–540.
[21]Girard, G. R.; Bondinell, W. E.; Hillegass, L. M.; Holden, K. G.; Pendleton, R. G.; Uzinskas, I. *J. Med. Chem.* **1989**, *32*, 1566–1571.

### 5.3.4  Sulfinylation

There are only a few reports on the sulfinylation of arenes and they generally lead to the symmetrical diaryl sulfoxide. The reaction is executed by treating an aromatic compound with thionyl chloride in the presence of a Lewis acid. The example shown proved to be fairly general, especially for electron-rich systems.[22]

### 5.3.5  Sulfonylation

The sulfonylation reaction is similar to the sulfinylation reaction. A key advantage is that it is straightforward to synthesize asymmetrical sulfones, since the reaction can be performed using an alkyl or aryl sulfonyl chloride and an arene.[23] It has also been shown that Lewis acids with triflate counterions are superior to the corresponding chlorides for catalysis of the reaction.[24] The reaction usually requires high temperature and is not suitable for sensitive substrates.

[22]Yadav, J. S.; Reddy, B. V. S.; Rao, R. S.; Kumar, S. P.; Nagaiah, K. *Synlett* **2002**, 784–786.
[23]Singh, R. P.; Kamble, R. M.; Chandra, K. L.; Saravanan, P.; Singh, V. K. *Tetrahedron* **2001**, *57*, 241–247.
[24]Repichet, S.; Le Roux, C.; Hernandez, P.; Dubac, J.; Desmurs, J.-R. *J. Org. Chem.* **1999**, *64*, 6479–6482.

### 5.3.6 Thiocyanation

The thiocyanation reaction is another method for the introduction of sulfur to an arene. The reaction is typically performed with a thiocyanate source in the presence of an oxidizing agent. For example, ammonium thiocyanate is used in conjunction with bromine.[25] An alternative method, which might be more practical on a small scale, generates the reactive intermediate from sodium thiocyanate and N-bromosuccinimide.[26] The resulting product can easily be hydrolyzed to the thiol for further functionalization.

## 5.4 HALOGENATION

### 5.4.1 Fluorination

Electrophilic fluorination of arenes is generally not a very common reaction. Several of the reagents used for electrophilic fluorination are not practical or have a very small safety margin.[27] Furthermore, the regioselectivity observed in electrophilic fluorination is usually very poor. For instance, the fluorination of anisole using N-fluoro pentachloropyridinium triflate gives a 1 : 1 ratio of the *ortho* and *para* regioisomers.[28] Aryl fluorides are usually accessed through a Balz–Schiemann reaction using a nucleophilic fluoride source (Section 4.5.1).

[25]Hirokawa, Y.; Harada, H.; Yoshikawa, T.; Yoshida, N.; Kato, S. *Chem. Pharm. Bull. (Tokyo).* **2002**, *50*, 941–959.
[26]Toste, F. D.; De Stefano, V.; Still, I. W. J. *Synth. Commun.* **1995**, *25*, 1277–1286.
[27]Lal, G. S.; Pez, G. P.; Pesaresi, R. J.; Prozonic, F. M.; Cheng, H. *J. Org. Chem.* **1999**, *64*, 7048–7054.
[28]Umemoto, T.; Fukami, S.; Tomizawa, G.; Harasawa, K.; Kawada, K.; Tomita, K. *J. Am. Chem. Soc.* **1990**, *112*, 8563–8575.

## 5.4.2    Chlorination

The electrophilic chlorination of arenes is not as straightforward as the corresponding bromination or iodination, because the regioselectivity observed is not always as high and the simplest chlorinating reagent, $Cl_2$, is a gas. Nonetheless, chlorination of anilines has been reported using $N$-chlorosuccinimide.[29] Sulfuryl chloride has been utilized as a chlorinating agent but usually provides poor regioselectivity. However, it is a good reagent choice when regiochemistry is not an issue.[30] It has been reported that catalytic amounts of diphenylsulfide in the presence of $AlCl_3$ improve the regioselectivity of the reaction.[31]

## 5.4.3    Bromination

The bromination of arenes is the most common aromatic electrophilic halogenation. Aryl bromides are synthetically useful since they are readily converted to organomagnesium or organolithium species (see Chapter 12) as well as being excellent substrates for a number of metal-catalyzed processes (see Chapter 6). There are many methods available for the preparation of aryl bromides. The two reagents that are most practical and easily scalable are 1,3-dibromo-5,5-dimethylhydantoin (DBDMH also named dibromantin) and $N$-bromosuccinimide (NBS). DBDMH is an effective reagent with both electron-deficient[32] and electron-rich[33] arenes. In some cases, NBS proved to be a superior reagent to bromine in the bromination of *ortho*-nitroanilines.[34]

[29]Nickson, T. E.; Roche-Dolson, C. A. *Synthesis* **1985**, 669–670.

[30]Masilamani, D.; Rogic, M. M. *J. Org. Chem.* **1981**, *46*, 4486–4489.

[31]Watson, W. D. *J. Org. Chem.* **1985**, *50*, 2145–2148.

[32]Leazer, J. L., Jr., ; Cvetovich, R.; Tsay, F.-R.; Dolling, U.; Vickery, T.; Bachert, D. *J. Org. Chem.* **2003**, *68*, 3695–3698.

[33]Connolly, T. J.; Matchett, M.; McGarry, P.; Sukhtankar, S.; Zhu, J. *Org. Process Res. Dev.* **2004**, *8*, 624–627.

[34]Manley, P. W.; Acemoglu, M.; Marterer, W.; Pachinger, W. *Org. Process Res. Dev.* **2003**, *7*, 436–445.

Bromine, while impractical on small scale because of its low boiling point, is a very useful brominating agent on large scale because of its low cost and the generation of HBr as the sole byproduct. The bromination is usually carried out in acetic acid at room temperature.[35] Another reagent proven to be superior to bromine on a very electron-poor arene is sodium bromate.[36] Unfortunately, the strongly acidic conditions required for this reaction do not make it practical for substrates with sensitive functionality.

### 5.4.4 Iodination

The iodination reaction is not as common as bromination. Since iodide is not as electronegative as other halogens, diiodination can be observed on very reactive substrates.[37] Aryl iodides can also be accessed using iodo pyridinium chloride, which is easier from a handling

[35]Tilley, J. W.; Coffen, D. L.; Schaer, B. H.; Lind, J. *J. Org. Chem.* **1987**, *52*, 2469–2474.
[36]Groweiss, A. *Org. Process Res. Dev.* **2000**, *4*, 30–33.
[37]Wariishi, K.; Morishima, S.; Inagaki, Y. *Org. Process Res. Dev.* **2003**, *7*, 98–100.

perspective than iodine.[38] Finally, another proven alternative is the use of iodine in the presence of iodic acid.[39]

An interesting procedure for the iodination of an arene is the use of *N*-chlorosuccinimide (NCS) in the presence of hydriodic acid and potassium iodide.[40] This procedure has the advantage that the cost of NCS is significantly lower then NIS.

## 5.5   CARBON ELECTROPHILES

### 5.5.1   Friedel–Crafts Alkylation

The reaction of an alkyl halide or alkene with an arene in the presence of an acid is usually referred to as a Friedel–Crafts alkylation. Overall, this reaction is not as useful as its acylation counterpart, mainly because of the lower reactivity of the alkylating agent and the harsher conditions often necessary for the electrophilic substitution to proceed. While much of the older literature will cite the use of carbon disulfide or deactivated arenes such as nitrobenzene as solvents, these solvents are now generally avoided because of the safety hazards

[38]Atkins, R. J.; Banks, A.; Bellingham, R. K.; Breen, G. F.; Carey, J. S.; Etridge, S. K.; Hayes, J. F.; Hussain, N.; Morgan, D. O.; Oxley, P.; Passey, S. C.; Walsgrove, T. C.; Wells, A. S. *Org. Process Res. Dev.* **2003**, *7*, 663–675.
[39]Huth, A.; Beetz, I.; Schumann, I.; Thielert, K. *Tetrahedron* **1987**, *43*, 1071–1074.
[40]Herrinton, P. M.; Owen, C. E.; Gage, J. R. *Org. Process Res. Dev.* **2001**, *5*, 80–83.

associated with them and have often been replaced by dichloromethane or by the arene itself if it is a commodity.

The Friedel–Crafts alkylation proceeds best when performed in an intramolecular fashion as exemplified in the preparation of a fluoroindanone in sulfuric acid at elevated temperature.[41] In another example an excess of toluene is used in the presence of a dichloride freshly prepared from the diol to effect intermolecular alkylation followed by intramolecular cyclization in 91% yield.[42]

Another method by which to perform the Friedel–Crafts alkylation is to utilize an alkene as the electrophilic reagent. This method is often utilized to prepare *tert*-butyl arenes using isobutylene.[43] The only inconvenience with this procedure is the necessity for a reaction vessel which can be pressurized, since isobutylene is a gas.

Once again, this reaction is more general in the intramolecular sense. Benzomorphane[44] and benzazepine[45] derivatives could be prepared by initial protonation of an olefin with a strong acid followed by nucleophilic attack of the arene at the more stable tertiary carbocation.

[41]Sircar, I.; Duell, B. L.; Cain, M. H.; Burke, S. E.; Bristol, J. A. *J. Med. Chem.* **1986**, *29*, 2142–2148.

[42]Faul, M. M.; Ratz, A. M.; Sullivan, K. A.; Trankle, W. G.; Winneroski, L. L. *J. Org. Chem.* **2001**, *66*, 5772–5782.

[43]Aeilts, S. L.; Cefalo, D. R.; Bonitatebus, P. J., Jr.,; Houser, J. H.; Hoveyda, A. H.; Schrock, R. R. *Angew. Chem., Int. Ed. Engl.* **2001**, *40*, 1452–1456.

[44]Grauert, M.; Bechtel, W. D.; Weiser, T.; Stransky, W.; Nar, H.; Carter, A. J. *J. Med. Chem.* **2002**, *45*, 3755–3764.

[45]Varlamov, A.; Kouznetsov, V.; Zubkov, F.; Chernyshev, A.; Alexandrov, G.; Palma, A.; Vargas, L.; Salas, S. *Synthesis* **2001**, 849–854.

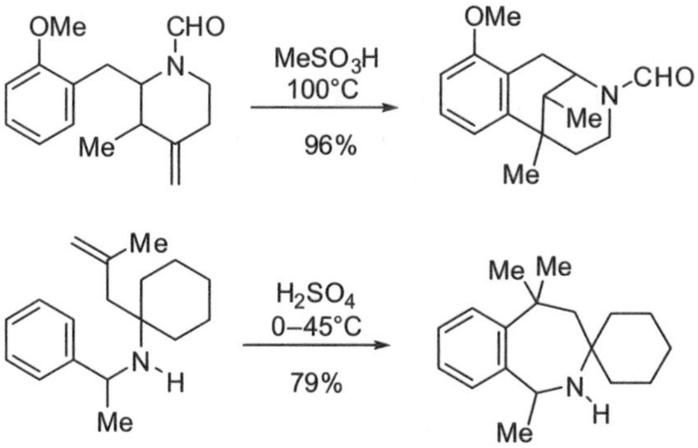

## 5.5.2 Friedel–Crafts Arylation

The Friedel–Crafts arylation, also known as the Scholl reaction, involves the generation of a doubly benzylic carbocation followed by intramolecular formation of the diphenyl bond. The carbocation is usually generated from an alcohol precursor upon treatment with a protic or Lewis acid. Often, the starting material is dissolved in acetic acid followed by addition of sulfuric acid, which induces the cation formation and cyclization.[46]

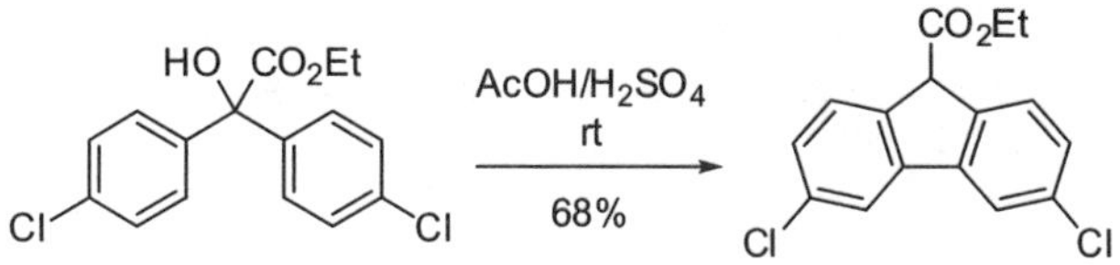

A spectacular example of a cascade Friedel–Crafts alkylation and arylation is shown below. Starting from hexaphenylbenzene, six alkylations and arylations are performed in a single operation in excellent yield.[47]

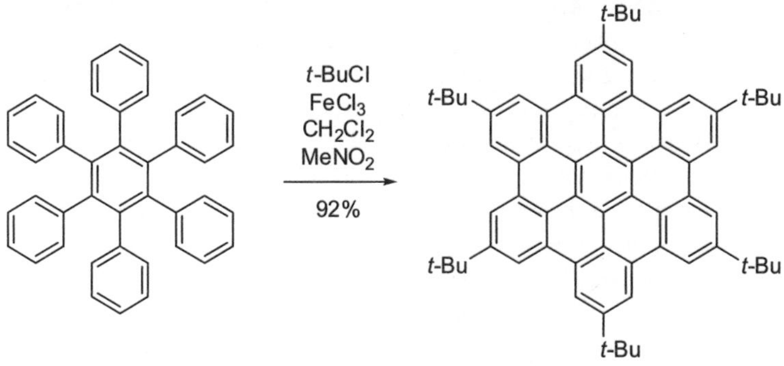

[46]Levy, A.; Rakowitz, A.; Mills, N. S. *J. Org. Chem.* **2003**, *68*, 3990–3998.
[47]Rathore, R.; Burns, C. L. *J. Org. Chem.* **2003**, *68*, 4071–4074.

### 5.5.3    Claisen Rearrangement

The aryl Claisen rearrangement generally requires much harsher conditions than the alkyl variant (Section 7.3.2.4) since aromaticity is lost in the transition state. Below are three representative examples highlighting the high temperature usually required for this transformation and the occasional low regioselectivity that can result.[48,49] It is interesting that in the third example, modification from a methyl ester to a diethyl amide changed the product distribution from a ratio of 87 : 13 to 96 : 4.[50] The predominant regioisomer arises from two sequential [3.3] sigmatropic rearrangements, first a Claisen, followed by a Cope rearrangement

### 5.5.4    Formylation

The formylation of arenes is a particularly useful electrophilic aromatic substitution because of the synthetic utility of the resulting aldehyde. A large number of procedures are available to effect this transformation including the reaction of a disubstituted formamide in the presence of an activating agent such as phosphorous oxychloride or oxalyl chloride (Vilsmeier–Haack reaction), reaction with hexamethylenetetramine (HMT) in the presence of an acid (Duff reaction), reaction of a 1,1-dichloroether in the presence of a Lewis acid, reaction of zinc cyanide in the presence of HCl (Gatterman reaction), reaction of carbon monoxide and HCl in the presence of $AlCl_3$ and CuCl (Gatterman–Koch reaction), and the reaction with a dichlorocarbene generated by treatment of chloroform with sodium hydroxide (Reimer–Tiemann reaction).

One of the most frequently utilized procedures for the formylation reaction is the Vilsmeier–Haack reaction. DMF is usually used as the formamide and is activated with

[48]Burks, J. E., Jr., ; Espinosa, L.; LaBell, E. S.; McGill, J. M.; Ritter, A. R.; Speakman, J. L.; Williams, M.; Bradley, D. A.; Haehl, M. G.; Schmid, C. R. *Org. Process Res. Dev.* **1997**, *1*, 198–210.

[49]Manchand, P. S.; Micheli, R. A.; Saposnik, S. J. *Tetrahedron* **1992**, *48*, 9391–9398.

[50]Patterson, J. W. *J. Org. Chem.* **1995**, *60*, 4542–4548.

POCl$_3$.[51] In the formylation of resorcinol, use of POCl$_3$ was shown to be slightly superior to use of oxalyl chloride.[52] Another convenient way to carry out the formylation is to utilize the Vilsmeier reagent, which is a commercially available solid. The last example below shows the preparation of an iminium chloride, which was further reacted with a nucleophile.[53] While there are many reports of the use of 1,1-dichloroethers in the presence of a Lewis acid to generate benzaldehyde derivatives,[54] the toxicity of the starting dichloroether makes this procedure less preferable.

One of the most practical formylation procedures is the Duff reaction where HMT is used as the formaldehyde surrogate to generate an iminium species in the presence of acid. A nice feature of the Duff reaction is that it is not as exothermic as the Vilsmeier–Haack and often necessitates higher temperatures in order to proceed. This can be an advantage from a process safety standpoint.[55,56]

[51]Black, D. S. C.; Kumar, N.; Wong, L. C. H. *Synthesis* **1986**, 474–476.

[52]Mendelson, W. L.; Hayden, S. *Synth. Commun.* **1996**, *26*, 603–610.

[53]Manley, J. M.; Kalman, M. J.; Conway, B. G.; Ball, C. C.; Havens, J. L.; Vaidyanathan, R. *J. Org. Chem.* **2003**, *68*, 6447–6450.

[54]Boswell, G. E.; Licause, J. F. *J. Org. Chem.* **1995**, *60*, 6592–6594.

[55]Roth, B.; Baccanari, D. P.; Sigel, C. W.; Hubbell, J. P.; Eaddy, J.; Kao, J. C.; Grace, M. E.; Rauckman, B. S. *J. Med. Chem.* **1988**, *31*, 122–129.

[56]Lindoy, L. F.; Meehan, G. V.; Svenstrup, N. *Synthesis* **1998**, 1029–1032.

The Reimer–Tiemann reaction allows for introduction of an aldehyde *ortho* to a phenol. The reaction proceeds through the generation of dichlorocarbene by deprotonation of chloroform followed by α elimination. The initially formed dichloride hydrolyzes to the aldehyde. Generally, the reaction is low yielding and precautions must be taken when adding the chloroform to the strongly basic reaction mixture.[57]

### 5.5.5 Hydroxyalkylation

Aldehydes and derivatives thereof can also be cyclized under Friedel–Crafts conditions. In the first example below, an aldehyde is cyclized to a benzylic alcohol by treatment with a Lewis acid.[58] A dimethyl acetal can be cyclized using methane sulfonic acid and the resulting ether reduced using $BH_3 \cdot t\text{-}BuNH_2$.[59] In the last example, formation of an oxocarbenium ion from a dimethylacetal in the presence of an alcohol resulted in the desired isochromane after cyclization.[60]

[57]Baker, R.; Castro, J. L. *J. Chem. Soc., Perkin Trans. 1* **1990**, 47–65.

[58]Achmatowicz, O.; Szechner, B. *J. Org. Chem.* **2003**, *68*, 2398–2404.

[59]Draper, R. W.; Hou, D.; Iyer, R.; Lee, G. M.; Liang, J. T.; Mas, J. L.; Vater, E. J. *Org. Process Res. Dev.* **1998**, *2*, 186–193.

[60]DeNinno, M. P.; Perner, R. J.; Morton, H. E.; DiDomenico, S., Jr., *J. Org. Chem.* **1992**, *57*, 7115–7118.

## 5.5.6   Haloalkylation

Haloalkylation is the outcome of a hydroxyalkylation of an arene followed by *in situ* conversion of the resulting benzylic alcohol to a halide. Benzylic chlorides have been obtained in very high yield using this method.[61]

## 5.5.7   Aminoalkylation

The aminoalkylation of an arene through electrophilic aromatic substitution is possible by generation of an iminium ion. The reaction proceeds with both acyl[62] and alkyl[63] iminium ions.

The most powerful version of the aminoalkylation is the intramolecular variant known as the Pictet–Spengler reaction. A β-amino arene is treated with an aldehyde to generate an imine, which is protonated and undergoes the electrophilic aromatic substitution as shown in the example below.[64] A variation of the reaction that leads directly to a *N*-folmyl tetrahydroisoquinoline has also been reported.[65]

[61]Ladd, D. L.; Weinstock, J. *J. Org. Chem.* **1981**, *46*, 203–206.
[62]Maulding, D. R.; Lotts, K. D.; Robinson, S. A. *J. Org. Chem.* **1983**, *48*, 2938–2939.
[63]O'Neill, P. M.; Mukhtar, A.; Stocks, P. A.; Randle, L. E.; Hindley, S.; Ward, S. A.; Storr, R. C.; Bickley, J. F.; O'Neil, I. A.; Maggs, J. L.; Hughes, R. H.; Winstanley, P. A.; Bray, P. G.; Park, B. K. *J. Med. Chem.* **2003**, *46*, 4933–4945.
[64]Ornstein, P. L.; Arnold, M. B.; Augenstein, N. K.; Paschal, J. W. *J. Org. Chem.* **1991**, *56*, 4388–4392.
[65]Maryanoff, B. E.; Rebarchak, M. C. *Synthesis* **1992**, 1245–1248.

Another variation of the aminoalkylation is the Bischler–Napieralski reaction wherein an amide is converted to an imidoyl chloride, which undergoes electrophilic aromatic substitution followed by elimination to provide an acyclic imine (dihydroisoquinoline).[66]

### 5.5.8   Thioalkylation

The thioalkylation of arenes through electrophilic aromatic substitution is not a very common reaction. The reaction usually utilizes a sulfoxide as a starting material, which is activated to generate the oxosulfenium ion and ultimately a sulfenium ion, which will react with the arene.[67] Sometimes, the reaction is also observed as a side product from Moffatt-type oxidations (see Section 10.4.1.2).[68]

### 5.5.9   Friedel–Crafts Acylation

The most utilized Friedel–Crafts reaction is the Friedel–Crafts acylation. Some of the aspects that render this reaction practical are the fact that a number of acid chlorides are commercially available or easily prepared (Section 2.24). The reaction will usually stop after mono-acylation since the newly introduced carbonyl is highly deactivating, and the activated intermediate, either an acyl cation or a complex of a Lewis acid with the carbonyl, is sufficiently reactive such that the reaction can be carried out at a reasonable temperature and with acceptable reaction rates. Furthermore, the product of the reaction is an aryl ketone, which can easily be further derivatized in a number of ways. Several Lewis acids can be employed. Two of the most frequently utilized reagents are $AlCl_3$ and $TiCl_4$ and the selection is often a personal preference between the addition of a solid ($AlCl_3$) or a liquid ($TiCl_4$). On smaller scale, $TiCl_4$ is often preferred since it is available as a solution in dichloromethane. Generally, $AlCl_3$ is mixed with the acid chloride in dichloromethane followed by addition of the arene. In the case of an electron-rich arene,[69] the reaction might require cooling in order to control the exotherm. For less reactive substrates, the reaction can be performed at higher temperatures.[70] For unreactive substrates, where dichloromethane does not provide the

[66]von Nussbaum, F.; Schumann, S.; Steglich, W. *Tetrahedron* **2001**, *57*, 2331–2335.

[67]Veeraraghavan, S.; Jostmeyer, S.; Myers, J'n.; Wiley, J. C., Jr. *J. Org. Chem.* **1987**, *52*, 1355–1357.

[68]Bailey, P. D.; Cochrane, P. J.; Irvine, F.; Morgan, K. M.; Pearson, D. P. J.; Veal, K. T. *Tetrahedron Lett.* **1999**, *40*, 4593–4596.

[69]Alabaster, R. J.; Cottrell, I. F.; Marley, H.; Wright, S. H. B. *Synthesis* **1988**, 950–952.

[70]Tang, P. W.; Maggiulli, C. A. *J. Org. Chem.* **1981**, *46*, 3429–3432.

appropriate temperature range, dichloroethane can serve as a substitute since the Friedel–Crafts alkylation is much slower than the acylation. However, dichloromethane is a preferable solvent from a safety standpoint.

The choice of Lewis acid and acylating agent can have an effect on the regioselectivity of the acylation. For instance, much higher regioselectivity was observed using $AlBr_3$ than with $AlCl_3$.[71]

The Friedel–Crafts acylation can also proceed from other activated carboxylic acid derivatives. For instance, $P_2O_5$ can allow for the direct condensation of arene and carboxylic acid, particularly for intramolecular cases.[72]

Anhydrides can also be used for the acylation. In the first example below, dichlorobenzene was used in excess (3 equivalents) such that no additional solvent is needed.[73] In cases where it might not be desirable to use the arene as a solvent, a mixture of dichloromethane and nitromethane has proven to be an efficient combination although the use of nitromethane is usually not recommended.[74]

[71]Caron, S.; Do, N. M.; Arpin, P.; Larivee, A. *Synthesis* **2003**, 1693–1698.

[72]Chandrasekhar, B.; Prasad, A. S. R.; Eswaraiah, S.; Venkateswaralu, A. *Org. Process Res. Dev.* **2002**, *6*, 242–245.

[73]Quallich, G. J.; Williams, M. T.; Friedmann, R. C. *J. Org. Chem.* **1990**, *55*, 4971–4973.

[74]Melillo, D. G.; Larsen, R. D.; Mathre, D. J.; Shukis, W. F.; Wood, A. W.; Colleluori, J. R. *J. Org. Chem.* **1987**, *52*, 5143–5150.

Another formal acylation is the Hoesch reaction, wherein a nitrile is used as the electrophile. This reaction is seldom used since nitriles are not very reactive. The reaction can be achieved when performed in an intramolecular fashion.[75] For intermolecular cases, pre-complexation of a phenol allowed for the reaction to proceed.[76]

### 5.5.10   Fries Rearrangement

The Fries rearrangement is one of the most useful electrophilic aromatic substitution reactions since it regioselectively introduces an acyl group *ortho* to a phenol. An acylated phenol is treated with a Lewis acid under conditions similar to the Friedel–Crafts reaction to induce the rearrangement which usually proceeds in high yield as shown in the example below.[77]

[75]He, H.-Z.; Kwon, C.-H. *Synth. Comm.* **2003**, *33*, 2437–2440.
[76]Toyoda, T.; Sasakura, K.; Sugasawa, T. *J. Org. Chem.* **1981**, *46*, 189–191.
[77]Nguyen Van, T.; Kesteleyn, B.; De Kimpe, N. *Tetrahedron* **2001**, *57*, 4213–4219.

The acylation and rearrangement can easily be accomplished in a single operation. The acylation proceeds at low temperature and once disappearance of the phenol is confirmed, the reaction mixture can be heated to induce the rearrangement.[78]

## 5.5.11   Carboxylation

The carboxylation of an arene through electrophilic aromatic substitution is seldom employed and the desired product is usually obtained through a halogenation followed by generation of an organometallic species, which is reacted with an electrophile such as $CO_2$. The Kolbe–Schmitt reaction is an example of an electrophilic carboxylation, which is specific for the introduction of a carboxylic acid at the *ortho* position of a phenol. The reaction suffers from the fact that it requires elevated temperature and $CO_2$ under high pressure, making it inconvenient for general application. This reaction was recently demonstrated in a microreactor.[79]

## 5.5.12   Amidation

Formation of an amide through electrophilic aromatic substitution is not straightforward. The reaction of phenoxides with isocyanates has been reported to proceed in low yield with high regioselectivity in the presence of $AlCl_3$.[80] It has also been reported that $BCl_3$ can perform as a better Lewis acid for this transformation.[81]

[78]Caron, S.; Vazquez, E. *Synthesis* **1999**, 588–592.
[79]Hessel, V.; Hofmann, C.; Loeb, P.; Loehndorf, J.; Loewe, H.; Ziogas, A. *Org. Process Res. Dev.* **2005**, *9*, 479–489.
[80]Balduzzi, G.; Bigi, F.; Casiraghi, G.; Casnati, G.; Sartori, G. *Synthesis* **1982**, 879–881.
[81]Piccolo, O.; Filippini, L.; Tinucci, L.; Valoti, E.; Citterio, A. *Tetrahedron* **1986**, *42*, 885–891.

### 5.5.13    Thioamidation and Thioesterification

Electron-rich arenes can react under Friedel–Crafts conditions with isothiocyanates to provide a thioamide. It has been shown that for phenol derivatives, the reaction does not proceed through an acylation-rearrangment mechanism (like a Fries rearrangement) and that a polar solvent is required for the electrophilic substitution to proceed. In $CH_2Cl_2$, acylation of the phenol was observed.[82] An alternative to $AlCl_3$ in $MeNO_2$ is the commercially available AcOH complex of $BF_3$. This method has proven to be advantageous in some cases and works with alkyl isothiocyanates.[83]

Methyl thioesters can be prepared by reaction of an arene with $CS_2$ and MeI in the presence of $AlCl_3$. Interestingly, the reaction is far superior when using an alkyl ether rather than a phenol as the starting material.[84] The reaction proceeds with high selectivity for the *para* regioisomer.

### 5.5.14    Cyanation

Preparation of cyanoarenes through electrophilic aromatic substitution is not commonly utilized. This transformation has been reported for *ortho*-substitution of anilines with trichloroacetonitrile in the presence of one equivalent of $BCl_3$. Interestingly, the reaction can also be accomplished with an alkylthiocyanate. Depending on the method utilized for the quench of the reaction, the nitrile or thioester can be obtained. The methodology also works with phenols as starting materials.[85] Less direct methods such as the elimination of oxime

[82]Jagodzinski, T. *Org. Prep. Proced. Int.* **1990**, *22*, 755–760.
[83]Sosnicki, J.; Jagodzinski, T.; Krolikowska, M. *J. Heterocycl. Chem.* **1999**, *36*, 1033–1041.
[84]Dieter, R. K.; Lugade, A. G. *Synthesis* **1988**, 303–306.
[85]Adachi, M.; Sugasawa, T. *Synth. Commun.* **1990**, *20*, 71–84.

derivatives (Section 8.6.2) or cross coupling reactions of aryl halides with cyanide are usually preferred (see Section 6.14).

# 6

# SELECTED METAL-MEDIATED CROSS-COUPLING REACTIONS

STÉPHANE CARON, ARUN GHOSH, SALLY GUT RUGGERI, NATHAN D. IDE, JADE D. NELSON, AND JOHN A. RAGAN

## 6.1  INTRODUCTION

In the past few decades, significant improvements have been achieved in metal-mediated processes, often utilizing catalytic quantities of a transition metal. Extensive mechanistic insight has been obtained in some cases. A plethora of novel synthetic methodologies have emerged from research in both academic and industrial laboratories. In pharmaceutical companies, these processes are now widely utilized for the synthesis of new drug candidates in medicinal chemistry as well as for the preparation of active pharmaceutical ingredients (API) on small and up to commercial scale.

Predicting the efficiency of these reactions can be difficult since yields can vary based on the electronic and steric nature of the substrates, the choice of ligands, catalysts, and other additives. While electron-withdrawing groups facilitate nucleophilic aromatic substitution and to some extent the ability of an arene to undergo oxidative addition with a metal, the nature of the halogen or pseudo-halide is more important in the cross-coupling reactions. The reactivity of the halide starting material follows a reverse order for the oxidative insertion of the metal where I > Br > Cl >> F. This trend for metal insertion is generally based on the increasing C$-$X bond strength and a number of reviews have been published in recent years that cover different aspects of these processes.[1] This chapter will primarily consider cross-coupling reactions catalyzed by palladium, copper, and a few other metals with particular attention on identifying generally reliable reaction conditions with an emphasis on heterocyclic substrates. The last section of the chapter also discusses metathesis reactions, a very different but important type of metal-catalyzed cross-coupling reaction.

---

[1]Corbet, J.-P.; Mignani, G. *Chem. Rev.* **2006**, *106*, 2651–2710.

---

*Practical Synthetic Organic Chemistry: Reactions, Principles, and Techniques*, First Edition.
Edited by Stéphane Caron.
© 2011 John Wiley & Sons, Inc. Published 2011 by John Wiley & Sons, Inc.

## 6.2  ORGANOBORON REAGENTS: THE SUZUKI–MIYAURA COUPLING

The palladium-catalyzed cross-coupling reaction between organoborane derivatives and an electrophile, known as the Suzuki–Miyaura reaction,[2–4] is one of the most widely explored cross-coupling reactions. The Suzuki–Miyaura reaction has received special attention since its mild reaction conditions make it amenable to a broad scope of substrates. In addition, the byproduct boric acid is much more environmentally friendly than the tin byproduct from Stille couplings. A base is required in this reaction and probably serves two purposes, generating the corresponding borate as a superior nucleophile that facilitates the transmetalation and neutralizing the acid generated in the reaction.[5] The byproducts are often easily removed by an aqueous wash in the workup. The reaction tolerates a wide range of solvents, and works in homogeneous or biphasic conditions.

The organoboranes used in this reaction are usually boronic acids that are crystalline, air, and moisture stable and can be stored indefinitely. Borate complexes generated *in situ* have also been used. In some cases, boronic acids have been derivatized as a diethanolamine complex[6] or a N-methyliminodiacetic acid (MIDA) boronate[7] for added stability. All states of carbon hybridization have been exemplified for this cross-coupling. While acetylenic boronic acid derivatives are prepared *in situ* due to poor stability, the alkyl derivatives are much less reactive and often require electron-rich organoboranes (e.g., 9-BBN derivatives) and/or electron-rich ligands (e.g., $PCy_3$) to participate in the reaction.[8] Trifluoroborate reagents have also emerged as a means of adding stability and improving reactivity in some cases.[9] The most common coupling partners are aryl and alkenyl boronic acids, which are easily prepared by addition of an organolithium or magnesium reagent to a trialkylborate or by hydroboration of an unsaturated organic molecule followed by hydrolysis. The often used pinacol arylborates are prepared by a palladium-catalyzed coupling of bis(pinacolato) diboron[10] or pinacolborane[11] with the corresponding arylhalides, or by an iridium-catalyzed direct borylation of simple arenes (see Section 12.2.2.3).[12]

### 6.2.1  Preparation of Biaryls

Generation of biaryls is the predominant use for the Suzuki–Miyaura reaction. Aryl iodides and bromides are known to be the most reactive organic halides but for large scale synthesis, aryl chlorides[13] are often preferred because they are cheaper. New catalyst systems allow for mild reaction conditions and low catalyst loadings.[14,15] The early literature on the reaction

[2]Miyaura, N.; Yamada, K.; Suzuki, A. *Tetrahedron Lett.* **1979**, 3437–3440.

[3]Martin, R.; Buchwald, S. L. *Acc. Chem. Res.* **2008**, *41*, 1461–1473.

[4]Bellina, F.; Carpita, A.; Rossi, R. *Synthesis* **2004**, 2419–2440.

[5]Matos, K.; Soderquist, J. A. *J. Org. Chem.* **1998**, *63*, 461–470.

[6]Caron, S.; Hawkins, J. M. *J. Org. Chem.* **1998**, *63*, 2054–2055.

[7]Gillis, E. P.; Burke, M. D. *Aldrichimica Acta* **2009**, *42*, 17–27.

[8]Littke, A. F.; Dai, C.; Fu, G. C. *J. Am. Chem. Soc.* **2000**, *122*, 4020–4028.

[9]Molander, G. A.; Biolatto, B. *J. Org. Chem.* **2003**, *68*, 4302–4314.

[10]Ishiyama, T.; Murata, M.; Miyaura, N. *J. Org. Chem.* **1995**, *60*, 7508–7510.

[11]Murata, M.; Watanabe, S.; Masuda, Y. *J. Org. Chem.* **1997**, *62*, 6458–6459.

[12]Ishiyama, T.; Takagi, J.; Ishida, K.; Miyaura, N.; Anastasi, N. R.; Hartwig, J. F. *J. Am. Chem. Soc.* **2002**, *124*, 390–391.

[13]Littke, A. F.; Fu, G. C. *Angew. Chem., Int. Ed. Engl.* **2002**, *41*, 4176–4211.

[14]Barder, T. E.; Walker, S. D.; Martinelli, J. R.; Buchwald, S. L. *J. Am. Chem. Soc.* **2005**, *127*, 4685–4696.

[15]Marion, N.; Navarro, O.; Mei, J.; Stevens, E. D.; Scott, N. M.; Nolan, S. P. *J. Am. Chem. Soc.* **2006**, *128*, 4101–4111.

usually utilized Pd(PPh$_3$)$_4$ as the catalyst of choice, but a more stable palladium (II) source, such as Pd(OAc)$_2$ in the presence of a phosphine ligand, is generally preferred.

While aryl iodides are often unstable and fairly expensive, they are very good partners in the Suzuki–Miyaura coupling. A convergent route to LY 451395, an AMPA potentiator for the treatment of cognitive deficits associated with Alzheimer's disease, was developed by Eli Lilly.[16] A large-scale Suzuki–Miyaura coupling was used in the final step to produce the desired biaryl using palladium black as the catalyst. Rigorous exclusion of dissolved oxygen and introduction of potassium formate as a mild reducing agent to minimize the concentration of Pd(II) efficiently suppressed formation of the undesired homocoupled product.

Aryl bromides are the most commonly utilized electrophilic starting materials in this cross-coupling. An intermediate in the synthesis of Losartan, an angiotensin II receptor antagonist developed by DuPont and Merck,[17] was prepared by a cross-coupling that was tolerant of both aryl halide and primary alcohol functionalities. The second example showed below also demonstrates the mild nature of the reaction, where a sulfonyl urea is not hydrolyzed under the cross-coupling conditions.[18]

[16]Miller, W. D.; Fray, A. H.; Quatroche, J. T.; Sturgill, C. D. *Org. Process Res. Dev.* **2007**, *11*, 359–364.

[17]Larsen, R. D.; King, A. O.; Chen, C. Y.; Corley, E. G.; Foster, B. S.; Roberts, F. E.; Yang, C.; Lieberman, D. R.; Reamer, R. A.;et al. *J. Org. Chem.* **1994**, *59*, 6391–6394.

[18]Heitsch, H.; Wagner, A.; Yadav-Bhatnagar, N.; Griffoul-Marteau, C. *Synthesis* **1996**, 1325–1330.

For the multi-kilogram synthesis of a pharmaceutical intermediate at Abbott, a biphenyl subunit was assembled from 4-trifluoromethoxy phenylboronic acid and bromofluorobenzonitrile using a Suzuki–Miyaura reaction.[12] Complete conversion could be achieved in 15 hours using 0.025 mol% of $(Ph_3P)_2PdCl_2$. With 0.06 mol% of catalyst, the reaction was complete (98% yield) within 6 hours using sodium bicarbonate as the base in a biphasic toluene and water. This process was well designed to achieve high turnover numbers because of the nature of the coupling partners. The electron-withdrawing groups of the aryl halide greatly facilitate oxidative insertion and the electron-rich nature of the boronic acid facilitates the transmetalation as well.

Electron-rich phosphine ligands, often $t$-$Bu_3P$ or $Cy_3P$, generally provide enhanced catalytic activity. The Merck process group has utilized such a catalytic system in a reaction where the boronic acid is activated with a fluoride source.[19]

Heterocyclic aryl bromides also participate efficiently in the Suzuki–Miyaura reaction. A variety of substituted 2-bromoimidazoles were further functionalized using a Suzuki–Miyaura coupling reaction to obtain the corresponding 2-arylimidazoles.[20]

8 other examples
41–99%

Control of the palladium levels in the isolated product can be problematic, especially when the resulting product can serve as a ligand for the metal. The Abbott group developed an efficient four-step process for the synthesis of multi-kilogram quantities of a COX-2

[19]Cameron, M.; Foster, B. S.; Lynch, J. E.; Shi, Y.-J.; Dolling, U.-H. *Org. Process Res. Dev.* **2006**, *10*, 398–402.
[20]Lee, S.-H.; Yoshida, K.; Matsushita, H.; Clapham, B.; Koch, G.; Zimmermann, J.; Janda Kim, D. *J. Org. Chem.* **2004**, *69*, 8829–8835.

inhibitor that involved a high-yielding Suzuki–Miyaura coupling in the penultimate step.[21] To alleviate the high level of Pd contamination, the coupled product was treated with a Deloxan resin that typically reduced Pd levels from 200 ppm to 20–40 ppm.

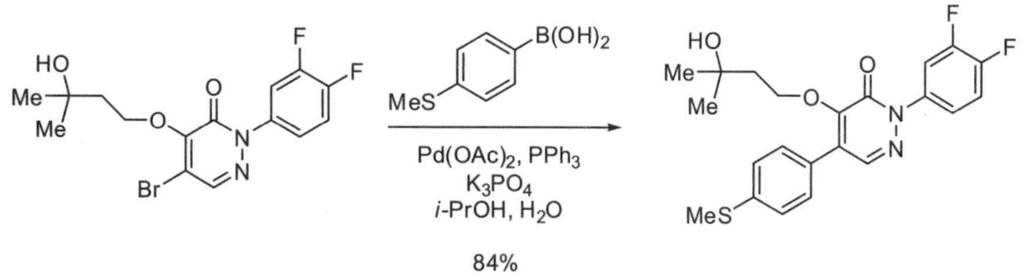

84%

Dialkyl aryl boranes participate efficiently in the Suzuki–Miyaura reaction. In the example shown below, the initial conditions for the preparation of the biaryl involved a cross-coupling reaction between 3-bromophenylsulfone and 3-pyridyldiethyl borane, which reproducibly delivered the free base or hydrochloride salt in 80% yield. However, the catalyst loading was significantly reduced by changing the solvent from tetrahydrofuran to toluene and the base from sodium hydroxide to potassium carbonate. The coupled product was reproducibly isolated as the methanesulfonic acid salt in 92–94% yield on 200 kg scale.[22]

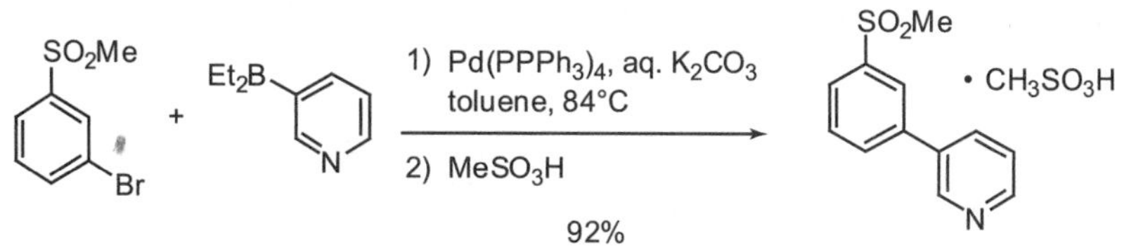

92%

Aryl chlorides are far less reactive toward the oxidative addition of the metal and usually require electron-deficient substrates and electron-rich ligands in order to proceed. In the example shown below, the reaction originally required the use of 2 to 2.5 equivalents of boronic acid to completely consume the chloroarene. The efficiency of the process was improved by using the boronic ester produced *in situ* from isopropyl borate, which resulted in faster reactions requiring less of the expensive boronic acid reagent. The rate acceleration was attributed to the presence of a trace amount of isopropyl alcohol in the reaction mixture. Subsequently, the reaction conditions were further modified to incorporate isopropyl alcohol as solvent and used only 1.1 equivalent of the boronic acid to yield the coupled product.[23] An *N*-heterocyclic carbene ligand also proved to be efficient for this reaction.[24]

[21]Kerdesky, F. A. J.; Leanna, M. R.; Zhang, J.; Li, W.; Lallaman, J. E.; Ji, J.; Morton, H. E. *Org. Process Res. Dev.* **2006**, *10*, 512–517.

[22]Lipton, M. F.; Mauragis, M. A.; Maloney, M. T.; Veley, M. F.; VanderBor, D. W.; Newby, J. J.; Appell, R. B.; Daugs, E. D. *Org. Process Res. Dev.* **2003**, *7*, 385–392.

[23]Chung, J. Y. L.; Cvetovich, R. J.; McLaughlin, M.; Amato, J.; Tsay, F.-R.; Jensen, M.; Weissman, S.; Zewge, D. *J. Org. Chem.* **2006**, *71*, 8602–8609.

[24]Chung, J. Y. L.; Cai, C.; McWilliams, J. C.; Reamer, R. A.; Dormer, P. G.; Cvetovich, R. J. *J. Org. Chem.* **2005**, *70*, 10342–10347.

IMes =

A highly chemo/regioselective Suzuki–Miyaura coupling of a tetrachloropyrimidine was described in a concise synthesis of fused benzo[4,5]furano heterocycles. Under the optimized conditions [Pd(OAc)$_2$, PPh$_3$, and K$_3$PO$_4$ in CH$_3$CN/H$_2$O] the desired product was obtained in 87% yield. Notably, the use of more electron-rich ligands, such as dppf or tricyclohexylphosphine, resulted in very poor regioselectivity.[25]

Dialkylbiarylphosphine ligands have shown enhanced reactivity in the Suzuki–Miyaura couplings of highly functionalized aryl chlorides or heterocyclic chlorides/bromides. The Buchwald group has relied on the increased reactivity and stability of metal catalysts using these ligands. Thus, cross-coupling reactions of aryl- and heteroaryl chlorides with potassium aryl- and heteroaryltrifluoroborates have been accomplished with the supporting ligand 2-(2′,6′-dimethoxybiphenyl)-dicyclohexylphosphine (S-Phos) in good to excellent yield.[3] Substrates that are sterically demanding and contain a variety of functional groups can be prepared following this protocol. Couplings of several trifluoroborate salts with aryl- and heteroaryl chlorides proceed in good yields.[26]

[25]Liu, J.; Fitzgerald, A. E.; Mani, N. S. *J. Org. Chem.* **2008**, *73*, 2951–2954.
[26]Barder, T. E.; Buchwald, S. L. *Org. Lett.* **2004**, *6*, 2649–2652.

93%

Electrophiles containing free amine groups can be problematic substrates. One strategy to deal with this issue is to temporarily protect the amine *in situ*, as demonstrated by researchers at Pfizer. A protocol where the benzaldehyde imine was generated *in situ* allowed the Suzuki–Miyaura coupling to proceed efficiently. The imine of the product was hydrolyzed as part of the reaction workup.[27]

99%

While the majority of Suzuki–Miyaura couplings utilize an aryl halide as a starting material, aryl triflates also participate in the reaction. Scientists at Johnson & Johnson have successfully accomplished the coupling of a challenging substrate using $PdCl_2$(dppf) as the catalyst. The addition of an excess of the dppf ligand improved the yield as well as the substrate scope. A dppf/Pd ratio of 1.5 in conjunction with $PdCl_2$(dppf) as the palladium source was crucial for broader applicability to a diverse range of substrates.[28]

90%

While the triflate moiety is an acceptable coupling group, it suffers from the disadvantage of being fairly unstable along with the difficulties associated with working with triflic anhydride as a reagent for its preparation. In the preparation of a pharmaceutical intermediate at Pfizer, several alternatives to a triflate were investigated, namely the mesylate, tosylate, phosphate, benzenesulfonate, 4-fluorobenzenesulfonate, and 4-chlorobenzenesulfonate esters. Among them, the 4-fluorobenzenesulfonate ester was found to be the most suitable substitute for the triflate group in the Suzuki–Miyaura reaction. Of the catalyst systems

[27]Caron, S.; Massett, S. S.; Bogle, D. E.; Castaldi, M. J.; Braish, T. F. *Org. Process Res. Dev.* **2001**, *5*, 254–256.
[28]Dvorak, C. A.; Rudolph, D. A.; Ma, S.; Carruthers, N. I. *J. Org. Chem.* **2005**, *70*, 4188–4190.

studied, $(PPh_3)_2 PdCl_2$ in the presence of triphenylphosphine turned out to be the most reliable and robust.[29] Imidazoylsulfonates have also proven to be suitable coupling partners.[30]

### 6.2.2    Preparation of Alkynyl-Substituted Arenes

Coupling of alkynes using the Suzuki–Miyaura coupling is very rare, as this overall reaction is usually achieved by a Sonogashira coupling (see Section 6.7). The Molander group has demonstrated that tetrafluoroborate salts can lead to productive cross-couplings as shown in the example below.[31]

### 6.2.3    Preparation of Vinyl-Substituted Arenes

Researchers at Boehringer Ingelheim developed a practical approach for the synthesis of 1,1-diarylsubstituted alkenes with a Suzuki–Miyaura reaction. A variety of boronic acids and substituted acrylonitriles were used to define the scope of the reaction.[32] In their work, they

[29]Jacks, T. E.; Belmont, D. T.; Briggs, C. A.; Horne, N. M.; Kanter, G. D.; Karrick, G. L.; Krikke, J. J.; McCabe, R. J.; Mustakis, J. G.; Nanninga, T. N.; Risedorph, G. S.; Seamans, R. E.; Skeean, R.; Winkle, D. D.; Zennie, T. M. *Org. Process Res. Dev.* **2004**, *8*, 201–212.
[30]Albaneze-Walker, J.; Raju, R.; Vance, J. A.; Goodman, A. J.; Reeder, M. R.; Liao, J.; Maust, M. T.; Irish, P. A.; Espino, P.; Andrews, D. R. *Org. Lett.* **2009**, *11*, 1463–1466.
[31]Molander, G. A.; Katona, B. W.; Machrouhi, F. *J. Org. Chem.* **2002**, *67*, 8416–8423.
[32]Taylor, S. J.; Netherton, M. R. *J. Org. Chem.* **2006**, *71*, 397–400.

elected to use the salt of tri-*tert*-butyl phosphine[33] a more stable and easily handled form of the phosphine ligand.

The strategy of using a vinylboronic acid, usually prepared by hydroboration of an alkyne, is also an effective approach to styrene derivatives. The Suzuki–Miyaura coupling reactions of aryl and heteroaryl halides with vinylboronic acids proceeds in good to excellent yield using S-Phos as a ligand that allows reactions to be performed at low catalyst levels, even in the preparation of hindered aryl halides.[14]

One of the most impressive examples of the Suzuki–Miyaura cross-coupling has been in the synthesis of carbapenem derivatives, which are known to be highly sensitive to reaction conditions. In the example below, a vinyltriflate was coupled with a complex boronic acid to afford the fully elaborated carbapenem in excellent yield.[34]

[33]Netherton, M. R.; Fu, G. C. *Org. Lett.* **2001**, *3*, 4295–4298.

[34]Yasuda, N.; Huffman, M. A.; Ho, G.-J.; Xavier, L. C.; Yang, C.; Emerson, K. M.; Tsay, F.-R.; Li, Y.; Kress, M. H.; Rieger, D. L.; Karady, S.; Sohar, P.; Abramson, N. L.; DeCamp, A. E.; Mathre, D. J.; Douglas, A. W.; Dolling, U.-H.; Grabowski, E. J. J.; Reider, P. J. *J. Org. Chem.* **1998**, *63*, 5438–5446.

### 6.2.4 Preparation of Dienes

1,3-Dienes can be prepared via the Suzuki–Miyaura reaction. An impressive example was demonstrated in the synthesis of a pharmaceutical intermediate where two vinylic coupling partners were joined to provide a sensitive product.[35]

> 59%

### 6.2.5 Preparation of Alkyl-Substituted Arenes

Following its work in palladium-catalyzed cross-coupling,[36] the Fu group has demonstrated the Suzuki–Miyaura cross-coupling reactions of a series of unactivated primary and secondary alkyl halides through the use of nickel catalysts aided by an amino alcohol ligand. This catalyst/ligand combination has circumvented the notorious problem with β-hydride elimination that is usually seen with this class of substrates.[37]

With proper tuning of the amino alcohol ligands, the reaction can be successfully carried out using unactivated secondary alkyl chlorides.

### 6.2.6 Preparation of Alkyl-Substituted Alkenes

Elaboration of an alkene is possible via the Suzuki–Miyaura coupling. Often, the alkyl borane utilized is the 9-BBN derivative that arises from hydroboration of the corresponding

[35]Yu, M. S.; Lopez De Leon, L.; McGuire, M. A.; Botha, G. *Tetrahedron Lett.* **1998**, *39*, 9347–9350.
[36]Netherton, M. R.; Fu, G. C. *Top. Organomet. Chem.* **2005**, *14*, 85–108.
[37]Gonzalez-Bobes, F.; Fu, G. C. *J. Am. Chem. Soc.* **2006**, *128*, 5360–5361.

alkene (see Section 12.2.2.1). An impressive application of this methodology came from the Shionogi laboratory in the preparation of carbapanem derivatives.[38]

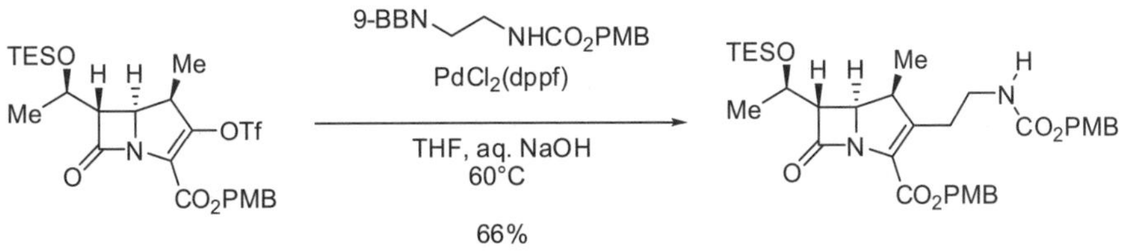

66%

## 6.2.7 Preparation of Alkanes

The coupling of sp$^3$ to sp$^3$ carbons by the Suzuki–Miyaura coupling is very rare. One exception is the coupling of cyclopropanes, which have more sp$^2$ character. The example below shows the formation of contiguous cyclopropanes using this strategy.[39]

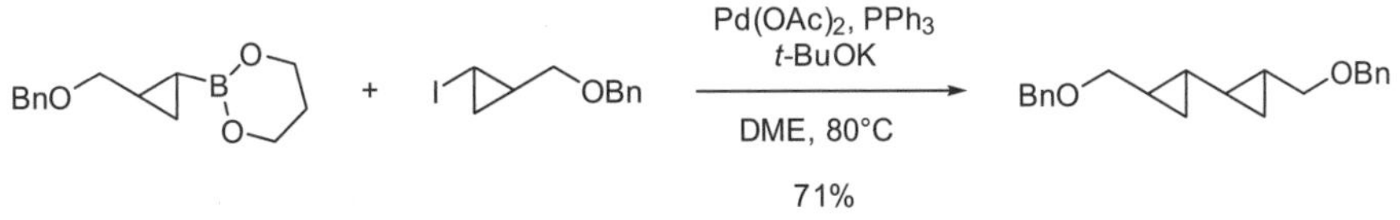

71%

The Fu group has developed a nickel-mediated Suzuki–Miyaura reaction catalyzed by chiral 1,2-diamines that enables the coupling of racemic secondary bromides with 9-BBN adducts to afford branched alkanes with good yield enantioselectivity controlled by the ligand making it such that all of the racemic starting material is consummed.[40]

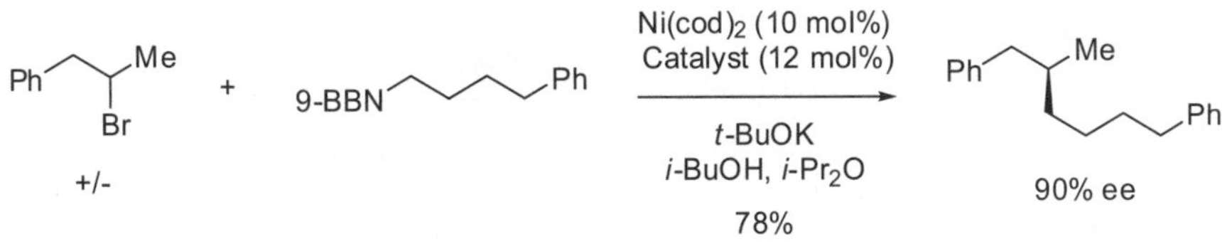

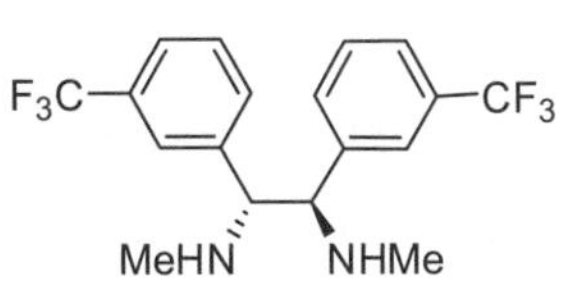

[38]Narukawa, Y.; Nishi, K.; Onoue, H. *Tetrahedron* **1997**, *53*, 539–556.
[39]Charette, A. B.; De Freitas-Gil, R. P. *Tetrahedron Lett.* **1997**, *38*, 2809–2812.
[40]Saito, B.; Fu, G. C. *J. Am. Chem. Soc.* **2008**, *130*, 6694–6695.

## 6.3 ORGANOTIN REAGENTS: THE STILLE COUPLING (MIGITA–STILLE REACTION)

The Stille coupling, also known as the Migita–Stille reaction, was independently reported in the late 1970's[41,42] and has been the subject of many reviews.[43,44] It has mainly been utilized for the coupling of sp and sp$^2$ carbons where an alkyl, vinyl or aryl halide is reacted with an organotin reagent in the presence of a palladium catalyst. Due to the use of toxic organostannanes and the poor atom economy of the tin reagents, this reaction is not ideal from a "green" or process chemistry perspective. However, the highly covalent nature of the Sn$-$C bond compared to other organometallic reagents (e.g., Li-, Mg-, Al-, Zn-) makes this reagent less nucleophilic, more stable, and very tolerant of a variety of different functional groups. Furthermore, the Stille coupling is one of the most reliable and predictable cross-coupling reactions and remains one of the preferred synthetic methods at small laboratory scale, especially in the field of natural product synthesis. Residual tin byproducts are often a problem with this method. Addition of a fluoride source[45] as part of the reaction quench and extractive workup has been used, although the efficiency of the tin removal is usually specific to the nature of the product. More often than not, the Stille coupling requires difficult chromatographic purification in order to eliminate unwanted tin derivatives in the product.

### 6.3.1 Preparation of Biaryls

The Stille coupling is an efficient method for the preparation of biaryls, although it is not as practical as the Suzuki–Miyaura reaction. In the example shown below, an activated aryl chloride is coupled with an arylstannane to afford the desired biphenyl in high yield. As part of the workup, an aqueous solution of KF was utilized to reduce the levels of alkyltin byproducts.[46]

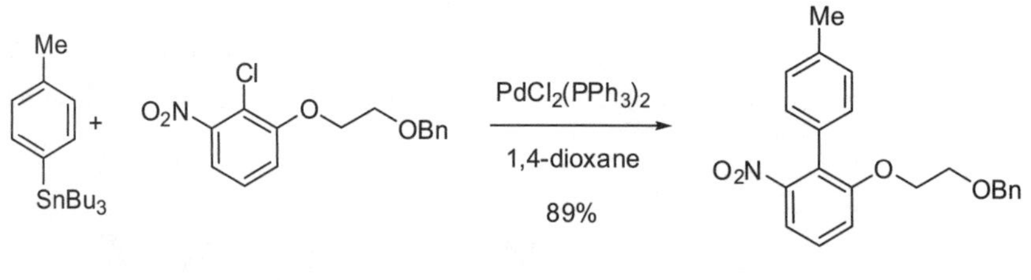

Another example of a Stille coupling of an iodide was reported by the Pfizer process group. While several other cross-coupling methods were investigated, the organostannane

[41]Kosugi, M.; Sasazawa, K.; Shimizu, Y.; Migita, T. *Chemistry Letters* **1977**, 301–302.

[42]Stille, J. K.; Lau, K. S. Y. *Acc. Chem. Res.* **1977**, *10*, 434–442.

[43]Stille, J. K. *Angew. Chem., Int. Ed. Engl.* **1986**, *98*, 504–519.

[44]Farina, V.; Krishnamurthy, V.; Scott, W. J. *Organic Reactions(New York)*; Wiley: 1997; Vol. 50, p 1–652.

[45]Hoshino, M.; Degenkolb, P.; Curran, D. P. *J. Org. Chem.* **1997**, *62*, 8341–8349.

[46]Morimoto, H.; Shimadzu, H.; Kushiyama, E.; Kawanishi, H.; Hosaka, T.; Kawase, Y.; Yasuda, K.; Kikkawa, K.; Yamauchi-Kohno, R.; Yamada, K. *J. Med. Chem.* **2001**, *44*, 3355–3368.

was the only reagent that reliably provided the desired product. A number of different workups were investigated to control the residual tin levels.[47]

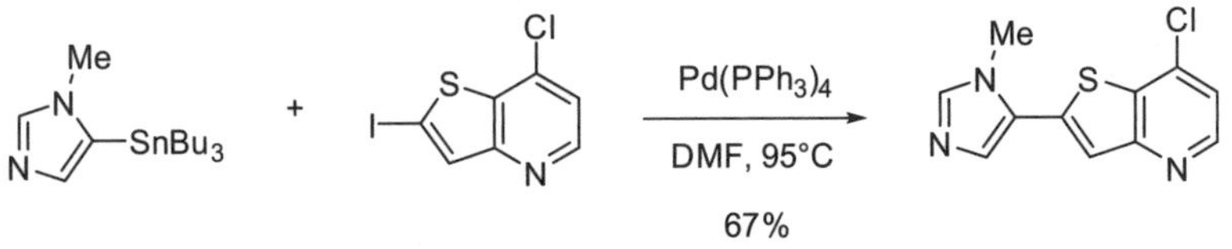

67%

The Stille coupling can also proceed starting from the aryl triflate[48] or 4-fluorophenyl-sulfonate,[49] which are easily accessed from the corresponding phenol.

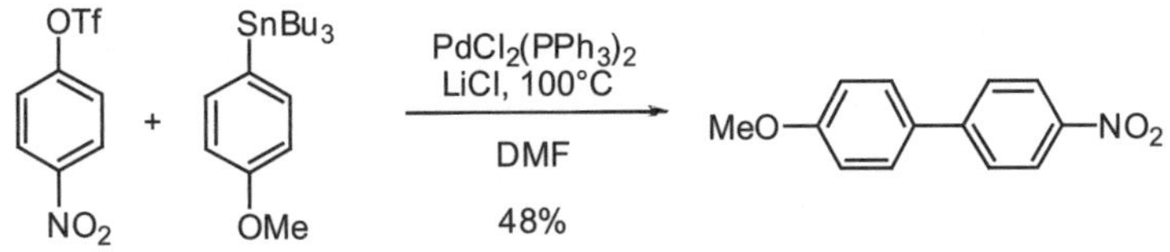

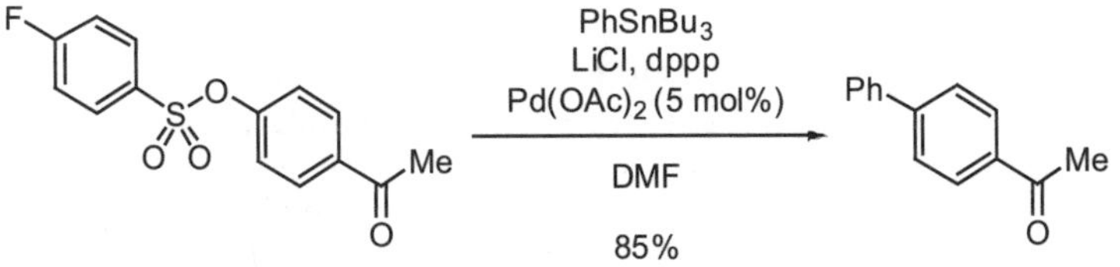

48%

85%

### 6.3.2  Preparation of Vinyl-Substituted Arenes

Vinyl stannanes will undergo efficient cross-couplings with aryl and heteroaryl halides. In the example shown below, the Stille coupling proceeded at room temperature and in high yield on a sensitive substrate.[50]

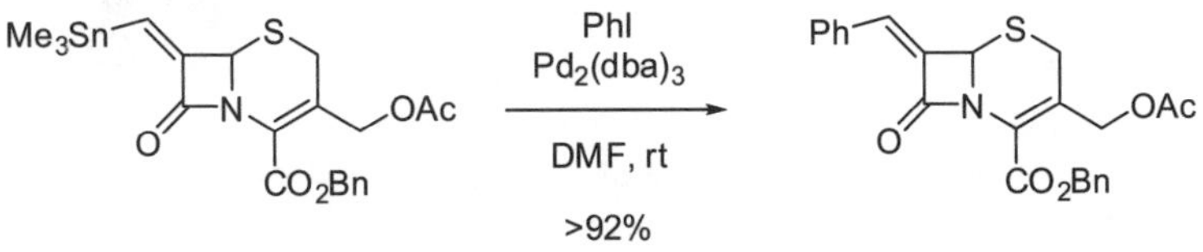

>92%

[47]Ragan, J. A.; Raggon, J. W.; Hill, P. D.; Jones, B. P.; McDermott, R. E.; Munchhof, M. J.; Marx, M. A.; Casavant, J. M.; Cooper, B. A.; Doty, J. L.; Lu, Y. *Org. Process Res. Dev.* **2003**, *7*, 676–683.

[48]Stille, J. K.; Echavarren, A. M.; Williams, R. M.; Hendrix, J. A. *Org. Synth.* **1993**, *71*, 97–106.

[49]Badone, D.; Cecchi, R.; Guzzi, U. *J. Org. Chem.* **1992**, *57*, 6321–6323.

[50]Buynak, J. D.; Doppalapudi, V. R.; Frotan, M.; Kumar, R.; Chambers, A. *Tetrahedron* **2000**, *56*, 5709–5718.

Tributylvinyl tin is frequently used for vinylation of aryl halides and triflates. Lithium chloride is often added as an additive to the reaction to facilitate the transmetalation step by exchanging a Pd–O bond for a more active Pd–Cl bond. The two cases shown below exemplify the level of tolerance for other functional groups in this cross-coupling.[51,52]

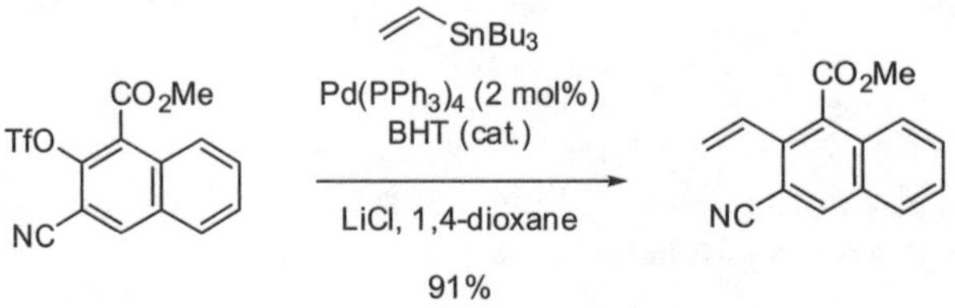

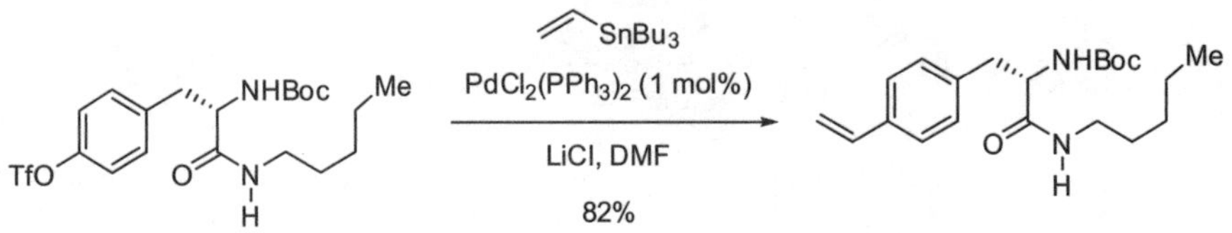

When a simple phenyl group needs to be introduced, phenyltributyltin can be utilized. Enol tosylates have been utilized in a few cases and have the advantage of being more stable than the corresponding vinyl triflates.[53]

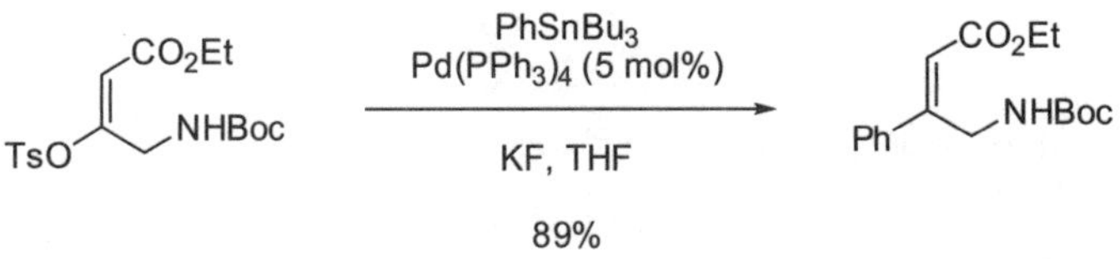

### 6.3.3  Preparation of 1,3-Dienes

1,3-Dienes have been prepared via Stille coupling between the vinyl triflate derived from the corresponding ketone and commercially available vinyltributyl tin. In the first example below, triphenylarsine was utilized as a ligand,[54] while the second example shows the use of Pd(PPh3)4 as the palladium source without an additional ligand.[55]

[51]Albert, J. S.; Ohnmacht, C.; Bernstein, P. R.; Rumsey, W. L.; Aharony, D.; Alelyunas, Y.; Russell, D. J.; Potts, W.; Sherwood, S. A.; Shen, L.; Dedinas, R. F.; Palmer, W. E.; Russell, K. *J. Med. Chem.* **2004**, *47*, 519–529.

[52]Larsen, S. D.; Barf, T.; Liljebris, C.; May, P. D.; Ogg, D.; O'Sullivan, T. J.; Palazuk, B. J.; Schostarez, H. J.; Stevens, F. C.; Bleasdale, J. E. *J. Med. Chem.* **2002**, *45*, 598–622.

[53]Steinhuebel, D.; Baxter, J. M.; Palucki, M.; Davies, I. W. *J. Org. Chem.* **2005**, *70*, 10124–10127.

[54]Pal, K. *Synthesis* **1995**, 1485–1487.

[55]Scott, W. J.; Crisp, G. T.; Stille, J. K. *Org. Synth.* **1990**, *68*, 116–129.

While the use of stannanes is usually not preferred at large scale, the Stille coupling has been used extensively in the preparation of complex and sensitive β-lactams. One of the first examples was reported by Bristol-Myers Squibb chemists for their orally active antibiotic, *cis*-cefprozil.[56] A Stille coupling was used to avoid the limitations of the earlier synthesis, which utilized a Wittig reaction. The requisite Z-alkene was obtained by coupling a stable *cis*-vinylstannane reagent with 3-trifloxycephem.[57] Later, the same group reported the use of fluorosulfonate as a triflate replacement and NMP as the solvent to perform the Stille coupling under ligandless conditions, producing a variety of semi-synthetic β-lactam products.[58,59]

### 6.3.4   Preparation of Alkyl-Substituted Alkenes

The Stille coupling has also been utilized in the preparation of alkyl-substituted carbapenems. Merck utilized this approach for the conversion of a vinyl triflate to the corresponding hydroxymethyl analog.[60]

[56]Naito, T.; Hoshi, H.; Aburaki, S.; Abe, Y.; Okumura, J.; Tomatsu, K.; Kawaguchi, H. *J. Antibiot. (Tokyo).* **1987,** *40,* 991–1005.

[57]Farina, V.; Baker, S. R.; Sapino, C., Jr., *Tetrahedron Lett.* **1988,** *29,* 6043–6046.

[58]Baker, S. R.; Roth, G. P.; Sapino, C. *Synth. Commun.* **1990,** *20,* 2185–2189.

[59]Roth, G. P.; Sapino, C. *Tetrahedron Lett.* **1991,** *32,* 4073–4076.

[60]Yasuda, N.; Yang, C.; Wells, K. M.; Jensen, M. S.; Hughes, D. L. *Tetrahedron Lett.* **1999,** *40,* 427–430.

A shorter and superior synthesis of the same carbapenem was accomplished using the cross-coupling of the enol triflate with the fully elaborated side chain, utilizing a stannatrane as the heteroalkyl transfer reagent.[61] It is worth mentioning that the toxic HMPA was replaced by Hunig's base and NMP as the solvent.

An allyl chloride has been used as the coupling partner in the preparation of a cephalo-sporin. Trifurylphosphine was selected as the preferred ligand for this transformation.[62]

## 6.4 CROSS-COUPLING REACTIONS WITH ORGANOSILICON COMPOUNDS

While much of cross-coupling chemistry, both in academia and in industry, is dominated by the Stille and Suzuki–Miyaura reactions, couplings that utilize organosilicon compounds have advanced to the point that they are a viable alternative to these well established methods. In addition to being efficient reactions, cross-coupling reactions with organo-silicon compounds can offer advantages in terms of waste disposal and atom economy.[63,64]

[61]Jensen, M. S.; Yang, C.; Hsiao, Y.; Rivera, N.; Wells, K. M.; Chung, J. Y. L.; Yasuda, N.; Hughes, D. L.; Reider, P. J. *Org. Lett.* **2000**, *2*, 1081–1084.
[62]Farina, V.; Baker, S. R.; Benigni, D. A.; Hauck, S. I.; Sapino, C., Jr., *J. Org. Chem.* **1990**, *55*, 5833–5847.
[63]Denmark, S. E.; Sweis, R. F. *Organosilicon Compounds in Cross-Coupling Reactions. 2nd ed;* Wiley-VCH: Weinheim, Germany, 2004; Vol. 1.
[64]Spivey, A. C.; Gripton, C. J. G.; Hannah, J. P. *Curr. Org. Synth.* **2004**, *1*, 211–226.

The utility of this approach was shown by Denmark and Yang during their synthesis of (+)-brasilenyne.[65] In this case, the cyclic silyl ether was activated by fluoride, transmetalated with palladium, and coupled to the pendant iodide to form the nine-membered ring in 61% yield.

During their total synthesis of herboxidiene/GEX 1A, Panek and Zhang utilized a similar method for the synthesis of the diene functionality.[66] In this case, the yields were variable (50–71%), but the conditions provided the diene exclusively as the (E,E) isomer.

## 6.5 ORGANOZINC REAGENTS: NEGISHI COUPLING

During the late 1970's, the Negishi group first reported the systematic screening of various alkynylmetals as nucleophiles in cross-couplings, and identified Zn, B, and Sn as three superior metals. These are the three most widely used metals today, associated with the Negishi, Suzuki–Miyaura and Stille couplings, respectively. Carbon nucleophiles derived from organozinc reagents have long been known. They are much less reactive than organolithium and organomagnesium reagents and the modest nucleophilicity of organozinc reagents has been exploited to achieve excellent chemoselectivity in coupling reactions.[67] One of the challenges with the Negishi coupling is that the required organozinc reagents are typically more difficult to prepare than the corresponding boronic acids and organostannanes utilized in the Suzuki–Miyaura and Stille reactions (see Section 12.2.11).

[65]Denmark, S. E.; Yang, S.-M. *J. Am. Chem. Soc.* **2002**, *124*, 15196–15197.
[66]Zhang, Y.; Panek, J. S. *Org. Lett.* **2007**, *9*, 3141–3143.
[67]Negishi, E.-i.; Zeng, X.; Tan, Z.; Qian, M.; Hu, Q.; Huang, Z. *Palladium- or Nickel-Catalyzed Cross-Coupling with Organometals Containing Zinc, Aluminum, and Zirconium: The Negishi Coupling. 2nd ed;* Wiley-VCH: Weinheim, Germany, 2004; Vol. 2.

### 6.5.1   Preparation of Biaryls

Synthesis of biaryls using the Negishi coupling has been the most prominent use of this reaction, and proceeds with aryl iodides, bromides, triflates, and even some chlorides.

Cross-coupling between the *in situ* prepared 2-pyridylzinc chloride and 5-iodo-2-chloropyrimidine catalyzed by Pd(PPh$_3$)$_4$ was used during the synthesis of 2-chloro-5-(pyridin-2-yl)pyrimidine, an intermediate in the synthesis of a selective PDE-V inhibitor developed by Johnson & Johnson. In this case, the yield increased from 55% with the bromide to >80% with the iodide.[68]

Eli Lilly scientists reported the preparation of a potent 5-HT$_{1a}$ agonist using an efficient Negishi cross-coupling strategy.[69] Although the yield is modest, it was unnecessary to protect the indole nitrogen of the starting material. In this case, the organozinc reagent was prepared by deprotonation/transmetalation with two equivalents of *n*-butyllithium, and the catalyst was generated from a stable palladium(II) source reduced *in situ* with catalytic *n*-butyllithium.

Novartis scientists investigated different synthetic routes to a selective inhibitor of the phosphodiesterase PDE$_4$D isoenzyme. An important refinement of the key Negishi aryl–aryl coupling involved premixing of the arylbromide and the catalyst, which was then added to the arylzinc intermediate. The workup of this reaction was also thoroughly optimized to minimize the levels of residual metals in the final product.[70]

[68]Perez-Balado, C.; Willemsens, A.; Ormerod, D.; Aelterman, W.; Mertens, N. *Org. Process Res. Dev.* **2007**, *11*, 237–240.
[69]Anderson, B. A.; Becke, L. M.; Booher, R. N.; Flaugh, M. E.; Harn, N. K.; Kress, T. J.; Varie, D. L.; Wepsiec, J. P. *J. Org. Chem.* **1997**, *62*, 8634–8639.
[70]Manley, P. W.; Acemoglu, M.; Marterer, W.; Pachinger, W. *Org. Process Res. Dev.* **2003**, *7*, 436–445.

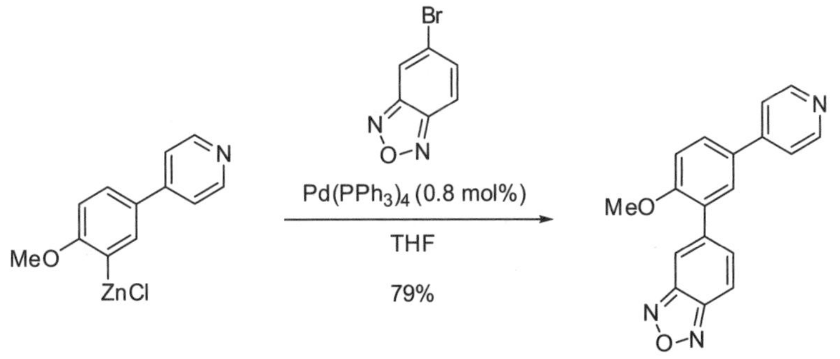

When catalytic amounts of Zn or Cd salts are present, the Ni- or Pd-catalyzed coupling of aryl halides (X = Cl, Br, I) with aryl Grignard reagents can provide unsymmetrical biaryls.[71] This double metal-catalyzed synthesis of biaryls avoids the stoichiometric preparation of arylzinc reagents. It is noteworthy that addition to the nitrile can be avoided by slow addition of the organomagnesium reagent.

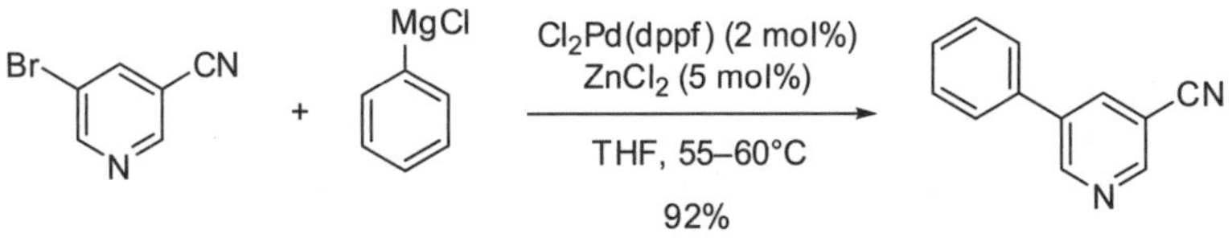

An efficient process for the preparation of adapalene is shown below. The key step employs a palladium-zinc double metal-catalyzed coupling, which avoids the use of stoichiometric zinc and greatly improves the purification process.[72]

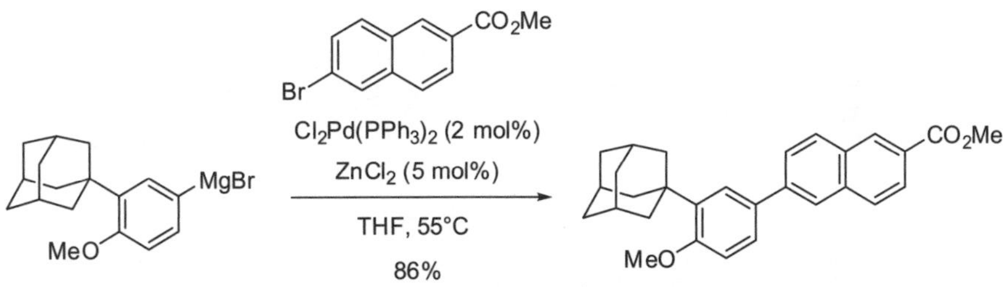

Readily available aryl chlorides have been used as substrates in Negishi-type cross-coupling reactions with arylzinc reagents via either Ni- or Pd-catalysis to produce unsymmetrical biaryls in an efficient manner. Since a wide range of functional groups (e.g., nitrile, ketone, ester) tolerate arylzinc compounds, this methodology results in the direct synthesis of biaryls from aryl chlorides containing those functionalities. In the case presented below, the Ni(0) catalyst is prepared *in situ* and mixed with the organozinc

[71]Miller, J.; Farrell, R. P. *Tetrahedron Lett.* **1998**, *39*, 7275–7278.
[72]Liu, Z.; Xiang, J. *Org. Process Res. Dev.* **2006**, *10*, 285–288.

reagent prior to introduction of the aryl chloride, which affords the biphenyl product in high yield.[73]

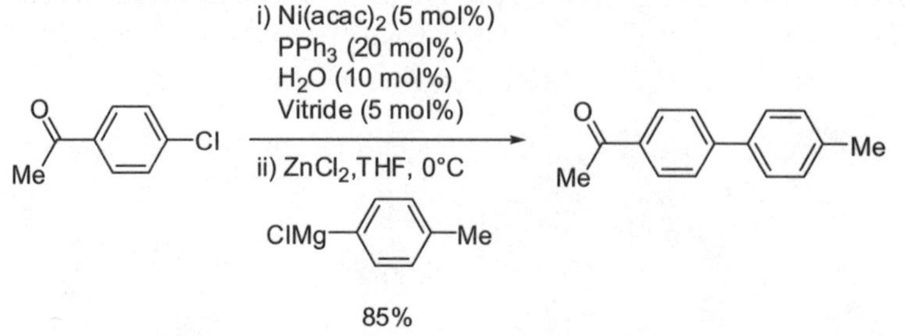

85%

Vitride = Red-Al = bis(2-methoxyethoxy) aluminum hydride

Aryl triflates can also be utilized in the Negishi coupling. In the example below, a chemoselective halogen-metal exchange followed by transmetalation provides the organozinc reagent, which is then coupled with the aryl triflate.[74]

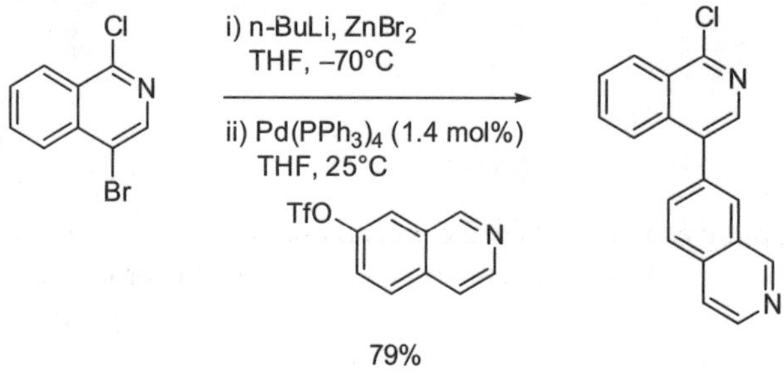

79%

The cross-coupling of a pyridyl zinc reagent and a pyridyl triflate under Negishi conditions can be used for the efficient and high-yielding synthesis of 4-, 5-, and 6-methyl-2,2'-bipyridines.[75]

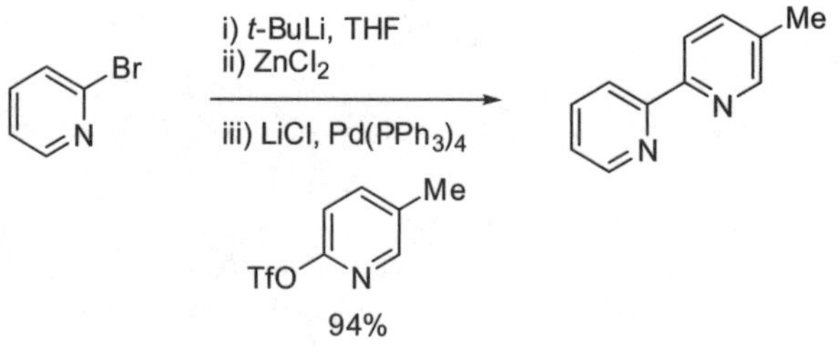

94%

[73]Miller, J. A.; Farrell, R. P. *Tetrahedron Lett.* **1998**, *39*, 6441–6444.

[74]Denni-Dischert, D.; Marterer, W.; Baenziger, M.; Yusuff, N.; Batt, D.; Ramsey, T.; Geng, P.; Michael, W.; Wang, R.-M. B.; Taplin, F., Jr., ; Versace, R.; Cesarz, D.; Perez, L. B. *Org. Process Res. Dev.* **2006**, *10*, 70–77.

[75]Smith, A. P.; Savage, S. A.; Love, J. C.; Fraser, C. L. *Org. Synth.* **2002**, *78*, 51–62.

### 6.5.2 Preparation of Aryl–Alkyl Bonds

Because of the ready availability of benzylic halides and their straightforward conversion to benzylic organozinc reagents, the Negishi reaction has been utilized for their coupling to aryl halides. The example below shows the introduction of a benzyl group to the 2-position of a pyridine that proceeded in excellent yield.[76]

This method has also been employed starting from the methylthioether, although an electron-deficient arene is required for the reaction to proceed.[77]

### 6.5.3 Preparation of 1,3-Dienes

The Negishi coupling can be used for the preparation of 1,3-dienes. In the example below, dibromostyrene undergoes a selective coupling to provide the bromodiene in high yield.[78]

The vinylzinc reagent can also be accessed through *in situ* transmetalation of an organoaluminum reagent as shown in the reaction of the vinyl iodide below.[79]

[76]Khatib, S.; Tal, S.; Godsi, O.; Peskin, U.; Eichen, Y. *Tetrahedron* **2000**, *56*, 6753–6761.
[77]Angiolelli, M. E.; Casalnuovo, A. L.; Selby, T. P. *Synlett* **2000**, 905–907.
[78]Ogasawara, M.; Ikeda, H.; Hayashi, T. *Angew. Chem., Int. Ed. Engl.* **2000**, *39*, 1042–1044.
[79]Negishi, E.; Takahashi, T.; Baba, S. *Org. Synth.* **1988**, *66*, 60–66.

### 6.5.4 Preparation of Ketones

The Negishi coupling provides a mild alternative for the preparation of benzylic ketones. In the example below, the benzylic bromide is coupled with the acid chloride in the presence of zinc and a palladium(0) catalyst.[80]

78%

## 6.6 ORGANOMAGNESIUM REAGENTS: KUMADA COUPLING

The transition-metal-catalyzed reaction of an organomagnesium reagent with a vinyl or aryl halide, published by Kumada and Corriu in 1972, was one of the first reported examples of a cross-coupling reaction.[81,82] The reaction, generally utilizing nickel or palladium catalysis has found many industrial applications.[83] One of the major limitations of the Kumada reaction is the use of highly reactive Grignard reagents. Due to their high nucleophilicity and basicity, the scope of this reaction is often limited to less-functionalized or relatively less-reactive substrates. However, the reaction benefits from the fact that many organomagnesium reagents are commercially available.

### 6.6.1 Preparation of Biaryls

In cases where an aryl Grignard is easily accessible, the Kumada coupling presents an attractive option for the preparation of biaryls. In the example below, the cross-coupling proceeded in high efficiency using the organomagnesium reagent derived from bromoanisole.[84]

97%

[80]Brandt, T. A.; Caron, S.; Damon, D. B.; Di Brino, J.; Ghosh, A.; Griffith, D. A.; Kedia, S.; Ragan, J. A.; Rose, P. R.; Vanderplas, B. C.; Wei, L. *Tetrahedron* **2009**, *65*, 3292–3304.

[81]Kumada, M. *Pure Appl. Chem.* **1980**, *52*, 669–679.

[82]Corriu, R. J. P.; Masse, J. P. *J. Chem. Soc., Chem. Commun.* **1972**, 144.

[83]Banno, T.; Hayakawa, Y.; Umeno, M. *J. Organomet. Chem.* **2002**, *653*, 288–291.

[84]Mewshaw, R. E.; Edsall, R. J., Jr., ; Yang, C.; Manas, E. S.; Xu, Z. B.; Henderson, R. A.; Keith, J. C., Jr., ; Harris, H. A. *J. Med. Chem.* **2005**, *48*, 3953–3979.

## 6.6.2    Preparation of Styrene Derivatives

Styrene derivatives have been prepared using the Kumada coupling. The aryl halide is generally added to the vinyl Grignard reagent as opposed to the aryl Grignard with the vinyl halide. In the example below, a dibromobenzofuran was regioselectively functionalized using a nickel catalyst.[85]

## 6.6.3    Preparation of Aryl–Alkyl Bonds

Aryl halides can be converted to the corresponding alkyl derivatives by treatment with a Grignard reagent and a catalyst, usually nickel. In the example below, a 3,6-dimethylcarbazole was prepared from the dibromo precursor.[86]

A very efficient Kumada cross-coupling was demonstrated during the development of an environmentally benign scalable process for the synthesis of the asymmetric phase-transfer catalyst (R)-3,5-dihydro-4H-dinaphth[2,1-c:1'2'-e]azepine.[87]

In the preparation of a key precursor used in the synthesis of a thymidylate synthase inhibitor, an iodoarene was methylated with methylmagnesium bromide under palladium catalysis. While the crude yield was excellent, recrystallization resulted in a 58% yield.[88]

[85]Bach, T.; Bartels, M. *Synthesis* **2003**, 925–939.
[86]Buck, J. R.; Park, M.; Wang, Z.; Prudhomme, D. R.; Rizzo, C. J. *Org. Synth.* **2000**, *77*, 153–161.
[87]Ikunaka, M.; Maruoka, K.; Okuda, Y.; Ooi, T. *Org. Process Res. Dev.* **2003**, *7*, 644–648.
[88]Marzoni, G.; Varney, M. D. *Org. Process Res. Dev.* **1997**, *1*, 81–84.

## 6.7  METAL-CATALYZED COUPLING OF ALKYNES (SONOGASHIRA COUPLING)

Nearly a decade after the first cross-coupling between an sp and an $sp^2$ carbon was reported by Stephens and Castro using stoichiometric copper (I) acetylide, Cassar and Heck independently demonstrated the cross-coupling of aryl and vinyl halides with terminal acetylenes via palladium catalysis. Mechanistically, this procedure could be considered as a Heck reaction on an alkyne substrate. A major improvement on this protocol was reported by Sonogashira and Hagihara, who found that the presence of CuI as a co-catalyst and an amine base allowed the reaction to proceed under very mild conditions. The operational simplicity and functional group tolerance of this procedure has resulted in it becoming the most common method for cross-coupling between a vinyl or aryl halide/triflate and a terminal acetylene.[89,90]

### 6.7.1  Reaction with Aryl Halides

The example below shows the robust preparation of a disubstituted alkyne using 2-bromopyridine and a terminal alkyne, which was prepared by alkylation with propargyl bromide and used without purification.[91]

When a terminal alkyne is the desired product, the reaction is not conducted with acetylene but rather with a protected derivative of acetylene, typically the trimethylsilyl derivative[92] or 2-methyl-3-butyn-2-ol, an inexpensive alternative that can be deprotected under basic conditions.[93] In general, electron-rich alynes tend to work well in the transmetalation step while electron-poor substrates often fail to react.

[89]Chinchilla, R.; Najera, C. *Chem. Rev.* **2007**, *107*, 874–922.

[90]Plenio, H. *Angew. Chem., Int. Ed. Engl.* **2008**, *47*, 6954–6956.

[91]Li, H.; Xia, Z.; Chen, S.; Koya, K.; Ono, M.; Sun, L. *Org. Process Res. Dev.* **2007**, *11*, 246–250.

[92]Andresen, B. M.; Couturier, M.; Cronin, B.; D'Occhio, M.; Ewing, M. D.; Guinn, M.; Hawkins, J. M.; Jasys, V. J.; LaGreca, S. D.; Lyssikatos, J. P.; Moraski, G.; Ng, K.; Raggon, J. W.; Stewart, A. M.; Tickner, D. L.; Tucker, J. L.; Urban, F. J.; Vazquez, E.; Wei, L. *Org. Process Res. Dev.* **2004**, *8*, 643–650.

[93]Frigoli, S.; Fuganti, C.; Malpezzi, L.; Serra, S. *Org. Process Res. Dev.* **2005**, *9*, 646–650.

In the example below, two successive Sonogashira couplings were conducted in a single pot. Upon completion of the first coupling, the alkyne was deprotected and coupled with a second aryl bromide.[94]

## 6.7.2   Preparation of Enynes

Enynes can be accessed through the coupling of a vinyl halide with a terminal alkyne under Sonogashira conditions. In the example below, a vinyl chloride was reacted under very mild conditions, which was necessary due to the known instability of the product.[95]

[94]Koenigsberger, K.; Chen, G.-P.; Wu, R. R.; Girgis, M. J.; Prasad, K.; Repic, O.; Blacklock, T. J. *Org. Process Res. Dev.* **2003**, *7*, 733–742.

[95]Alami, M.; Crousse, B.; Ferri, F. *J. Organomet. Chem.* **2001**, *624*, 114–123.

## 6.8  METAL-CATALYZED ALLYLIC SUBSTITUTION

The action of certain low valent transition-metal complexes on substrates containing electronegative leaving groups at an allylic position provides reactive, electrophilic, "π-allyl" intermediates. Although a range of nucleophiles may be employed in these reactions, the method has most generally been applied to the regio- and stereoselective construction of C−C, C−N, and C−O bonds, which will be discussed in this section. Pioneering work in the development of metal-catalyzed allylic substitution and application to organic synthesis was carried out by the research groups of Tsuji[96,97] and Trost[98,99] and has been a focus area for many others.

Palladium-catalyzed processes are undoubtedly the most well studied, and are typically an appropriate first choice for many substrates.[100,101] However, functionality-biased regioselectivity and erosion of stereochemistry in chiral π–allyl intermediates through facile π–σ–π isomerizations are limitations that must be addressed with most Pd-catalyzed allylation reactions. Thus, a number of alternative transition-metal complexes have been developed to address the restrictions of more traditional catalysts.[102–108]

Today, many catalyst systems are available for the generation of electrophilic π–allyl intermediates from a broad range of allylic alcohol derivatives. Each has advantages and disadvantages when cost, complexity, air or moisture sensitivity, and substrate compatibility are considered.

### 6.8.1  Carbon Nucleophiles

Soft, stabilized carbanions such as malonate, acetoacetate, or cyanoacetate salts represent the most common class of nucleophiles for metal-catalyzed allylic substitution reactions. The reaction conditions are often quite mild, and are generally tolerant of diverse functionality. In the example below, diethyl malonate is alkylated at the activated α carbon with an allylic carbonate under palladium(0) catalysis.[109] The racemic product was obtained in good

[96]Minami, I.; Shimizu, I.; Tsuji, J. *J. Organomet. Chem.* **1985**, *296*, 269–280.

[97]Tsuji, J. *Organic Synthesis with Palladium Compounds*; Springer-Verlag: Berlin, 1980.

[98]Trost, B. M.; Verhoeven, T. R. *J. Am. Chem. Soc.* **1980**, *102*, 4730–4743.

[99]Trost, B. M.; Van Vranken, D. L. *Chem. Rev.* **1996**, *96*, 395–422.

[100]Frost, C. G.; Howarth, J.; Williams, J. M. *J. Tetrahedron: Asymmetry* **1992**, *3*, 1089–1122.

[101]Trost, B. M. *Acc. Chem. Res.* **1980**, *13*, 385–393.

[102]Evans, P. A. *Modern Rhodium-Catalyzed Organic Reactions*; Wiley-VCH: Weinheim, Germany, 2005.

[103]Evans, P. A.; Nelson, J. D. *J. Am. Chem. Soc.* **1998**, *120*, 5581–5582.

[104]Takeuchi, R.; Kashio, M. *J. Am. Chem. Soc.* **1998**, *120*, 8647–8655.

[105]Trost, B. M.; Fraisse, P. L.; Ball, Z. T. *Angew. Chem., Int. Ed. Engl.* **2002**, *41*, 1059–1061.

[106]Ward, Y. D.; Villanueva, L. A.; Allred, G. D.; Liebeskind, L. S. *J. Am. Chem. Soc.* **1996**, *118*, 897–898.

[107]Lehmann, J.; Lloyd-Jones, G. C. *Tetrahedron* **1995**, *51*, 8863–8874.

[108]Bricout, H.; Carpentier, J.-F.; Mortreux, A. *J. Chem. Soc., Chem. Commun.* **1995**, 1863–1864.

[109]Krafft, M. E.; Kyne, G. M.; Hirosawa, C.; Schmidt, P.; Abboud, K. A.; L'Helias, N. *Tetrahedron* **2006**, *62*, 11782–11792.

yield and with a high degree of regioselectivity for the allylic terminus $\gamma$ (vs. $\alpha$) to the sulfone. The product of this reaction was also obtained with excellent selectivity for the (E)-double bond.

CH$_2$(CO$_2$Et)$_2$, 4Å MS
Pd$_2$dba$_3$, dppe

THF, reflux, 12 h

81%

In the following example from the process group at Merck & Co., palladium-catalyzed allylic alkylation of a protected dialkyl aminomalonate provided the (Z)-olefin preferentially. The reaction is tolerant of acetyl, Boc, or formyl protecting groups on nitrogen, and selectivity for the (Z)-olefin in the product is dependent upon the solvent, ligand, and palladium source utilized.[110]

[Pd(allyl)Cl]$_2$, dppe
NaH, DMF, 90°C

73%, 12:1 Z:E

Several asymmetric variations on this reaction have been reported. For racemic allylic electrophiles, asymmetry can be conveyed to alkylation products by using chiral organometallic complexes and taking advantage of facile $\pi$–$\sigma$–$\pi$ isomerization of the intermediate.[100] In the example below from Trost and coworkers, allylic alkylation of a geminal diacetate proceeds in high yield and with excellent stereoselectivity.[111] This particular example represents the synthetic equivalent of a challenging aldol condensation product.

[{$\eta^3$-C$_3$H$_5$PdCl}$_2$], L*

NaH, DME, 0–5°C

92%, >19 : 1 dr

99% ee

L* = Ph$_2$P      NH    HN      PPh$_2$

[110]Steinhuebel, D.; Palucki, M.; Davies, I. W. *J. Org. Chem.* **2006**, *71*, 3282–3284.
[111]Trost, B. M.; Ariza, X. *Angew. Chem., Int. Ed. Engl.* **1997**, *36*, 2635–2637.

Rhodium catalysis provides an attractive option for asymmetric C−C bond formation in cases where chirality is already present in the electrophilic component, since a high degree of regio- and stereochemical integrity is retained from starting material to product with these catalysts. Alkylation of stabilized carbon nucleophiles with chiral non-racemic allylic carbonates in the presence of a phosphite-modified Wilkinson's catalyst ($RhCl(PPh_3)_3$/$P(OMe)_3$) provided products in high yield and with impressive conservation of enantiomeric excess.[103]

Hartwig and coworkers have reported that silyl enol ethers of acetophenone derivatives serve as effective carbon nucleophiles for Ir-catalyzed allylic alkylation reactions.[112] Silyl enol ethers derived from aliphatic ketones also participate in the reaction to provide alkylation products with excellent regio- and enantioselectivities, although lower yields were reported. Evans and coworkers have demonstrated utility with Rh-phosphite catalysts in the allylic alkylation of copper enolates.[113,114]

Hard carbon nucleophiles (e.g., organozinc,[115] Grignard,[116] or organoaluminum[117] reagents) participate in transition-metal-catalyzed allylic alkylation reactions in a slightly different manner. Rather than attack metal allyl intermediates in an $S_N2'$ fashion, hard carbon-based nucleophiles are thought to add directly to the metal center. Subsequent

[112]Graening, T.; Hartwig, J. F. *J. Am. Chem. Soc.* **2005**, *127*, 17192–17193.

[113]Evans, P. A.; Leahy, D. K. *J. Am. Chem. Soc.* **2003**, *125*, 8974–8975.

[114]Evans, P. A.; Leahy, D. K.; Slieker, L. M. *Tetrahedron: Asymmetry* **2003**, *14*, 3613–3618.

[115]Evans, P. A.; Uraguchi, D. *J. Am. Chem. Soc.* **2003**, *125*, 7158–7159.

[116]Hayashi, T.; Konishi, M.; Yokota, K.; Humada, M. *J. Chem. Soc., Chem. Commun.* **1981**, 313–314.

[117]Matsushita, H.; Negishi, E. *J. Chem. Soc., Chem. Commun.* **1982**, 160–161.

reductive elimination typically provides carbon–carbon bonds with net inversion of stereochemistry.[100]

Several catalyst systems have shown effectiveness in allylic alkylation reactions with hard carbon nucleophiles. The representative example below from Kobayashi and coworkers utilizes a boronate ester nucleophile with a nickel catalyst to provide the alkylation product in high yield and with inversion of stereochemistry at carbon.[118]

## 6.8.2    Nitrogen Nucleophiles

The ability to form C−N bonds in a regio- and stereoselective manner via transition-metal-catalyzed allylic amination methodology provides a powerful tool to the practicing synthetic chemist. Trost and coworkers have provided several elegant examples of the use of chiral palladium catalysts for the desymmetrization of diastereomeric allylic carbonates. In the following example, the palladium catalyst was forced to approach the substrate from the face that was sterically hindered by the acetonide functionality. Despite this steric challenge, the reaction proceeded with excellent stereoselectivity to provide a single allylic amine product in high yield.[119]

[118]Kobayashi, Y.; Mizojiri, R.; Ikeda, E. *J. Org. Chem.* **1996**, *61*, 5391–5399.
[119]Trost, B. M.; Sorum, M. T. *Org. Process Res. Dev.* **2003**, *7*, 432–435.

In the example below from chemists at GlaxoSmithKline, a palladium-catalyzed allylic amination of an allylic 1,4-diacetate was followed by an intramolecular allylic etherification to yield a chiral, non-racemic morpholine derivative in high yield.[120] The Trost ligand proved optimal for enantioselectivity in this reaction, and a slow addition of the amino alcohol was found to minimize formation of a diamine impurity. For more on transition-metal-catalyzed allylic etherification, see Section 6.8.3.

Evans and coworkers reported complementary allylic amination methodology utilizing enantiomerically enriched allylic carbonates and a trimethylphosphite-modified Wilkinson's catalyst (RhCl(Ph$_3$)$_4$/P(OMe)$_3$). Under the conditions reported, allylation of N-tosyl N-benzylamine proceeds in high yield and with excellent conservation of regio- and stereochemistry.[121]

Hartwig and coworkers have reported that allylic amination reactions with chiral iridium catalysts provide branched products with high enantioselectivity.[122] In the example below, a linear achiral allylic carbonate is converted in good yield to a chiral-branched allylic amide via reaction with potassium trifluoroacetamide in the presence of a novel Ir catalyst.[123]

[120]Wilkinson, M. C. *Tetrahedron Lett.* **2005**, *46*, 4773–4775.

[121]Evans, P. A.; Robinson, J. E.; Nelson, J. D. *J. Am. Chem. Soc.* **1999**, *121*, 6761–6762.

[122]Ohmura, T.; Hartwig, J. F. *J. Am. Chem. Soc.* **2002**, *124*, 15164–15165.

[123]Pouy, M. J.; Leitner, A.; Weix, D. J.; Ueno, S.; Hartwig, J. F. *Org. Lett.* **2007**, *9*, 3949–3952.

### 6.8.3 Oxygen Nucleophiles

Ether formation may also be accomplished via metal-catalyzed allylic alkylation methodology, although the scope is somewhat limited in comparison to soft, stabilized carbon or
nitrogen nucleophiles. Evans and Leahy reported that sterically and electronically encumbered phenols serve as effective nucleophiles in Rh-catalyzed allylic etherification reactions.
In the following example, an optically enriched allylic carbonate is converted to a mixture of
allyl aryl ethers with excellent selectivity for the branched product.[124] This method is
especially attractive for target-oriented synthesis in that the stereochemistry of the carbonate
is translated to the product without the aid of expensive chiral ligands. This Rh catalyst has
also been applied to the regio- and stereoselective allylic etherification of aliphatic copper
alkoxides.[125]

Hartwig and coworkers have also applied their Ir-based catalyst system to enantioselective allylic etherification reactions.[126] In the representative example below, an achiral
allylic acetate is converted to a chiral allylic ether in good yield and enantioselectivity, and
with excellent preference for the branched regioisomer.

[124]Evans, P. A.; Leahy, D. K. *J. Am. Chem. Soc.* **2000**, *122*, 5012–5013.
[125]Evans, P. A.; Leahy, D. K. *J. Am. Chem. Soc.* **2002**, *124*, 7882–7883.
[126]Ueno, S.; Hartwig, J. F. *Angew. Chem., Int. Ed. Engl.* **2008**, *47*, 1928–1931.

In the following example from process chemists at Abbott Laboratories, a very impressive palladium-catalyzed allylic etherification reaction was carried out on an erythromycin A analog.[127] In this case, alkylation occurred at a highly sterically-congested tertiary alcohol to provide the allylic ether in nearly quantitative yield. Furthermore, the linear product with Z olefin geometry was formed exclusively from either the primary or secondary (shown) allylic carbonate.

## 6.9 METAL-CATALYZED COUPLING OF ALKENES (HECK COUPLING)

The Heck reaction is one of the most versatile methods for adding carbon nucleophiles to alkenes, and many reviews have been written addressing various aspects of it.[13,128–131] All oxidation states of carbon have been shown to add to the alkene under palladium catalysis, but the best reaction is achieved with $sp^2$ nucleophiles, such as aryl halides, and will be the only type covered herein. Almost any functionality is tolerated in the nucleophilic partner from a reactivity sense, although substituents can alter the

[127]Stoner, E. J.; Peterson, M. J.; Allen, M. S.; DeMattei, J. A.; Haight, A. R.; Leanna, M. R.; Patel, S. R.; Plata, D. J.; Premchandran, R. H.; Rasmussen, M. *J. Org. Chem.* **2003**, *68*, 8847–8852.

[128]Heck, R. F. *Acc. Chem. Res.* **1979**, *12*, 146–151.

[129]Beletskaya, I. P.; Cheprakov, A. V. *Chem. Rev.* **2000**, *100*, 3009–3066.

[130]Link, J. T. *Organic Reactions (New York)* **2002**, *60*, No pp given.

[131]Crisp, G. T. *Chem. Soc. Rev.* **1998**, *27*, 427–436.

regioselectivity. Substitution on the alkene partner is more limiting. Mono-substituted alkenes have the highest reactivity; increasing the substitution can greatly reduce the rate of reaction. Neutral, cationic, and anionic mechanisms have been proposed for the reaction; conditions dictate which is thought to be operational, and the substrates, especially the alkene, dictate which is preferred. As with almost all metal-mediated processes the optimal conditions (catalyst source, ligand, base, solvent) need to be identified by screening. However, for Heck reactions, it can be generalized that if both partners are reactive, such as aryl halides and electron-deficient alkenes, coupling occurs under the mildest conditions with monodentate ligands. Reactions of less-reactive substrates, such as electron-rich alkenes or vinyl halides may work better under halide-free conditions or with polydentate ligands. The selection of the base can also have a profound effect on the rate and product distribution. The stereochemical outcome of the reaction is generally predicted by consideration of the intermediate: the palladium nucleophile adds in a *syn* fashion to the alkene, generating a σ-palladium species that must rotate for the required *syn* Pd-H elimination. Substrates in which there is no strong conformational or electronic preference for the elimination can give regiochemical mixtures of isomers. Cyclic substrates containing bonds that cannot rotate to achieve the required conformation for *syn* elimination give products of β-hydride elimination, often as mixtures of regioisomers and stereoisomers.

### 6.9.1   Formation of Aryl Alkenes

The coupling of aryl halides or sulfonates, such as triflates, with alkenes is the most commonly reported variation of the Heck reaction. The order of reactivity of the aryl halide partner is I > Br >> Cl. Aryl bromides are most frequently used because of their availability and reactivity. A wide array of functional groups is tolerated on the aryl group, and many different alkene partners have been successfully employed. The example below, utilizing an electron-deficient alkene, demonstrates the selectivity of the palladium insertion for the aryl bromide over the chloride.[132]

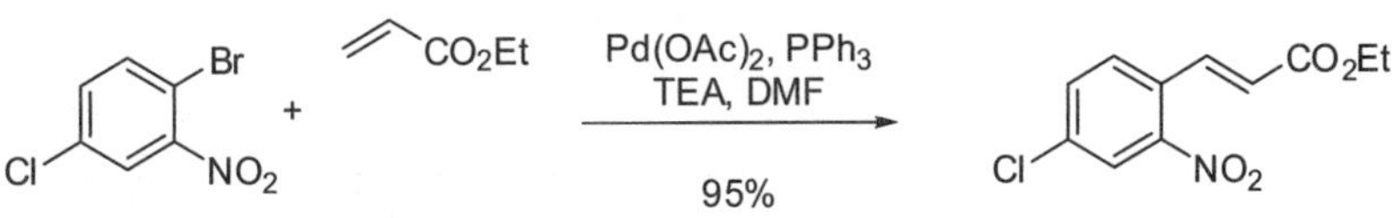

Aryl chlorides are attractive partners for coupling because of their generally lower cost than the corresponding bromides, but their lower reactivity precluded their use in Heck reactions for many years. More reactive catalysts have been developed that will couple even unactivated aryl chlorides, as shown below. Bulky, electron-rich ligands such as P(*t*-Bu)$_3$ form very reactive catalysts, especially when combined with methyldicyclohexylamine as the base.[133]

[132]Caron, S.; Vazquez, E.; Stevens, R. W.; Nakao, K.; Koike, H.; Murata, Y. *J. Org. Chem.* **2003**, *68*, 4104–4107.
[133]Littke, A. F.; Fu, G. C. *J. Am. Chem. Soc.* **2001**, *123*, 6989–7000.

84%

The regioselectivity of the addition is usually driven by steric considerations. In the example below, the nature of the protecting groups on the allyl amine had a strong influence on the regioisomeric purity of the product.[134] The bis(Boc) derivative was found to be optimal, and the reaction was run under ligand-free conditions.

The electronic nature of substituents on the alkene can enhance or conflict with the inherent steric bias. In the first example above, the ester substituent complemented the regioselectivity of addition based on sterics, resulting in a coupling that occurred at room temperature. Electron-rich alkenes were at one time problematic substrates because the electronics favor addition at the α rather than the β carbon, resulting in slow reaction rates and mixtures of regioisomers. The regioselectivity can be controlled by the appropriate choice of ligand and additives, however. In the example below, a chelating ligand was used to direct the addition to the α carbon.[135] The aryl ketone is isolated, resulting from hydrolysis of the initially formed enol ether.

[134]Ripin, D. H. B.; Bourassa, D. E.; Brandt, T.; Castaldi, M. J.; Frost, H. N.; Hawkins, J.; Johnson, P. J.; Massett, S. S.; Neumann, K.; Phillips, J.; Raggon, J. W.; Rose, P. R.; Rutherford, J. L.; Sitter, B.; Stewart, A. M., III; Vetelino, M. G.; Wei, L. *Org. Process Res. Dev.* **2005**, *9*, 440–450.
[135]Jiang, X.; Lee, G. T.; Prasad, K.; Repic, O. *Org. Process Res. Dev.* **2008**, *12*, 1137–1141.

Intramolecular Heck reactions have been used to assemble very complex substrates with tertiary or quaternary centers that would otherwise be quite challenging to manufacture.[136]

For substrates in which the inherent regioselectivity of the β-hydride elimination is poor, introduction of a temporary biasing agent, such as a silane, can be used to help control the outcome. In the example below, a trimethylsilyl group was used to control elimination away from the ring. A mixture of vinyl silanes was obtained, but was of no consequence after protodesilylation. In the absence of the silyl group, a mixture of regioisomers was obtained, and the trisubstituted olefin was also a mixture of stereoisomers.[137]

Asymmetric induction can be achieved with the use of chiral ligands.[138] The enantioselectivity achieved is generally modest,[139] but for some substrate and catalyst combinations, it can be quite high.[140]

[136]Campos, K. R.; Journet, M.; Lee, S.; Grabowski, E. J. J.; Tillyer, R. D. *J. Org. Chem.* **2005**, *70*, 268–274.

[137]Tietze, L. F.; Modi, A. *Eur. J. Org. Chem.* **2000**, 1959–1964.

[138]Shibasaki, M.; Boden, C. D. J.; Kojima, A. *Tetrahedron* **1997**, *53*, 7371–7395.

[139]Busacca, C. A.; Grossbach, D.; Campbell, S. J.; Dong, Y.; Eriksson, M. C.; Harris, R. E.; Jones, P.-J.; Kim, J.-Y.; Lorenz, J. C.; McKellop, K. B.; O'Brien, E. M.; Qiu, F.; Simpson, R. D.; Smith, L.; So, R. C.; Spinelli, E. M.; Vitous, J.; Zavattaro, C. *J. Org. Chem.* **2004**, *69*, 5187–5195.

[140]Minatti, A.; Zheng, X.; Buchwald, S. L. *J. Org. Chem.* **2007**, *72*, 9253–9258.

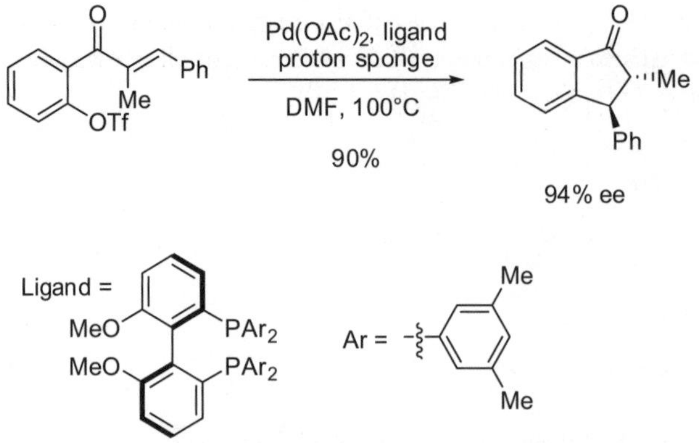

90%

94% ee

### 6.9.2 Reductive Heck

In most Heck reactions, the product of the reaction is a substituted alkene. However, if a reducing agent is present in the reaction, the alkene can be reduced. Formic acid or formate salts are the most commonly used reducing agents.[141]

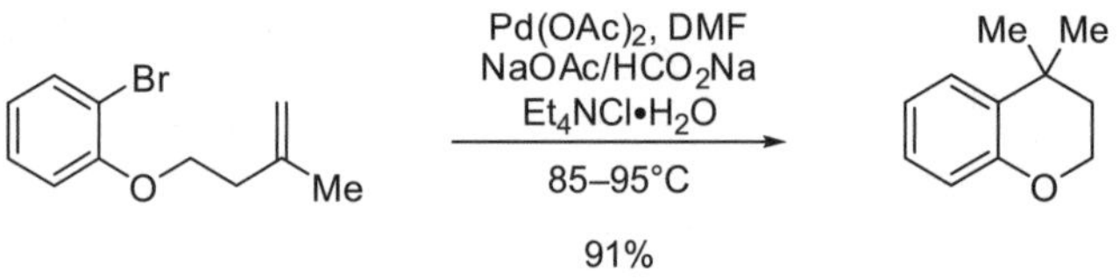

91%

The reduction can also be carried out as a separate step. Note that in the following example, the aryl triflate was used instead of the aryl bromide.[142] Triflates may require the use of chelating ligands in order to work well.

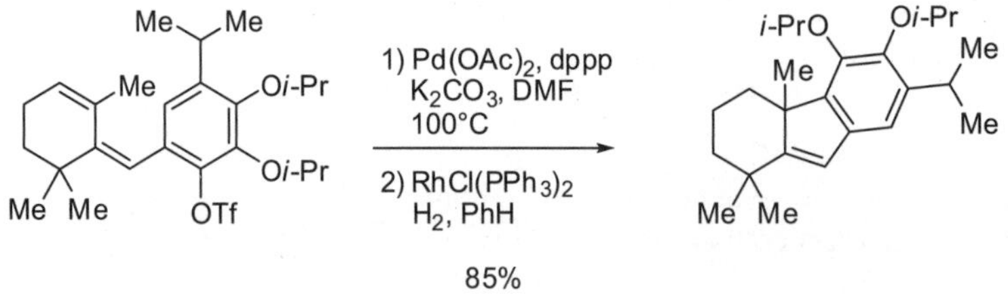

85%

### 6.9.3 Formation of Dienes

The second most common nucleophilic partner in the Heck reaction is a vinyl halide, which gives a diene as the product. The reaction is less robust than the aryl version, and mixtures of olefin isomers may be obtained. However, there are several examples of remarkable

[141]Liu, P.; Huang, L.; Lu, Y.; Dilmeghani, M.; Baum, J.; Xiang, T.; Adams, J.; Tasker, A.; Larsen, R.; Faul, M. M. *Tetrahedron Lett.* **2007**, *48*, 2307–2310.
[142]Planas, L.; Mogi, M.; Takita, H.; Kajimoto, T.; Node, M. *J. Org. Chem.* **2006**, *71*, 2896–2898.

selectivity, especially considering the ability of the palladium to isomerize alkenes. In the following example, the stereochemistry of the olefins in the alkene was not perturbed, and no products of π-allyl formation were observed.[143]

In other cases, special ligands have been developed to promote the reaction.[144]

Catalytic systems can also be tuned to give the expected product of insertion β to the activating group of the electrophilic partner or the product of 1,2-migration of the alkenyl palladium(II) species, resulting in an apparent α insertion.[145]

[143]Bienaymé, H.; Yezeguelian, C. Tetrahedron **1994**, 50, 3389–3396.
[144]Lemhadri, M.; Battace, A.; Berthiol, F.; Zair, T.; Doucet, H.; Santelli, M. Synthesis **2008**, 1142–1152.
[145]Ebran, J.-P.; Hansen, A. L.; Gogsig, T. M.; Skrydstrup, T. J. Am. Chem. Soc. **2007**, 129, 6931–6942.

## 6.10    Pd- AND Cu-CATALYZED ARYL C−N BOND FORMATION

Nitrogen heterocycles are key constituents of many natural products and pharmaceuticals. Synthesis of these compounds frequently starts from the appropriately functionalized aniline derivatives involving multiple chemical manipulations. In special cases where the corresponding aryl halides are highly activated, direct $S_N Ar$ substitutions have been reported (e.g., reaction of $n$-butylamine with 2,4-dinitrochlorobenzene proceeds in 83% yield in chloroform at ambient temperature in the presence of a phase-transfer catalyst).[146] Copper-catalyzed reactions are well established, and recent advances have been reviewed.[147] Although Pd-catalyzed aminations have received more recent attention, the Cu-catalyzed couplings can in many cases be competitive and potentially more economical.

The first preparation of aryl amines from an aryl bromide using $Bu_3SnNEt_2$ in the presence of $PdCl_2[P(o\text{-tol})_3]_2$ was reported by Kosugi and Migita in 1983.[148] A severe limitation of this process was the use of an alkyltin reagent that made it less attractive for industrial purposes. However, significant improvements have been reported that make this strategy more synthetically useful.[149–152] Buchwald and Hartwig independently demonstrated that under optimized conditions and with the appropriate choice of base (e.g, $NaOt$-Bu or $Cs_2CO_3$), and ligand (e.g., BINAP), tin-free aminations of aryl iodides and bromides can be performed under relatively mild conditions. Since then several improvements on palladium- and copper-catalyzed N-arylation have been reported from numerous laboratories for successful cross-couplings, in some cases utilizing the less expensive and more widely available aryl chlorides.[13] Numerous catalysts and conditions have been reported for these couplings, and selection of the optimal system will in most cases be substrate dependent and will involve some degree of optimization. The examples shown below are divided by catalyst and substrate type following the sequence of palladium-catalyzed (ArCl, HetArCl, ArBr, HetArBr), then copper-catalyzed reactions.

The following chloroarene–primary amine coupling was developed for the synthesis of torcetrapib.[153] Phenylboronic acid is used to activate the catalyst, and the phosphine ligand (first reported by Buchwald)[154] was selected from an extensive screening of both monodentate and chelating ligands. Buchwald has described the benefits of these biarylphosphine ligands,[155] which are also useful in heterocyclic cross-couplings.[156,157]

[146]Ross, S. D. *Tetrahedron* **1969**, *25*, 4427–4436.

[147]Ley, S. V.; Thomas, A. W. *Angew. Chem., Int. Ed. Engl.* **2003**, *42*, 5400–5449.

[148]Kosugi, M.; Kameyama, M.; Migita, T. *Chemistry Letters* **1983**, 927–928.

[149]Wolfe, J. P.; Wagaw, S.; Marcoux, J.-F.; Buchwald, S. L. *Acc. Chem. Res.* **1998**, *31*, 805–818.

[150]Hartwig, J. F. *Acc. Chem. Res.* **1998**, *31*, 852–860.

[151]Hartwig, J. F. *Pure Appl. Chem.* **1999**, *71*, 1417–1423.

[152]Surry, D. S.; Buchwald, S. L. *Angew. Chem., Int. Ed.* **2008**, *47*, 6338–6361.

[153]Damon, D. B.; Dugger, R. W.; Hubbs, S. E.; Scott, J. M.; Scott, R. W. *Org. Process Res. Dev.* **2006**, *10*, 472–480.

[154]Old, D. W.; Wolfe, J. P.; Buchwald, S. L. *J. Am. Chem. Soc.* **1998**, *120*, 9722–9723.

[155]Surry, D. S.; Buchwald, S. L. *Angew. Chem., Int. Ed. Engl.* **2008**, *47*, 6338–6361.

[156]Charles, M. D.; Schultz, P.; Buchwald Stephen, L. *Org. Lett.* **2005**, *7*, 3965–3968.

[157]Anderson, K. W.; Tundel, R. E.; Ikawa, T.; Altman, R. A.; Buchwald, S. L. *Angew. Chem. Int. E. Engl.* **2006**, *45*, 6523–6527.

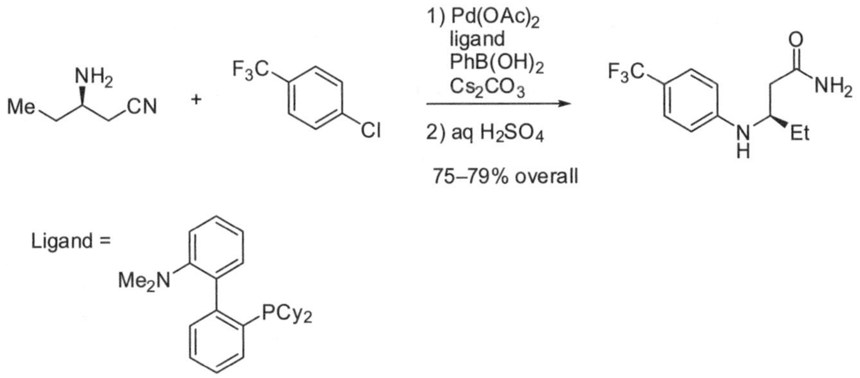

75–79% overall

Ligand =

Pd-catalyzed amination of a 3-chlorobenzimidazole was utilized in a synthesis of Norastemizole, a potent, non-sedating histamine $H_1$-receptor antagonist, as reported by Senanayake at Sepracor.[158]

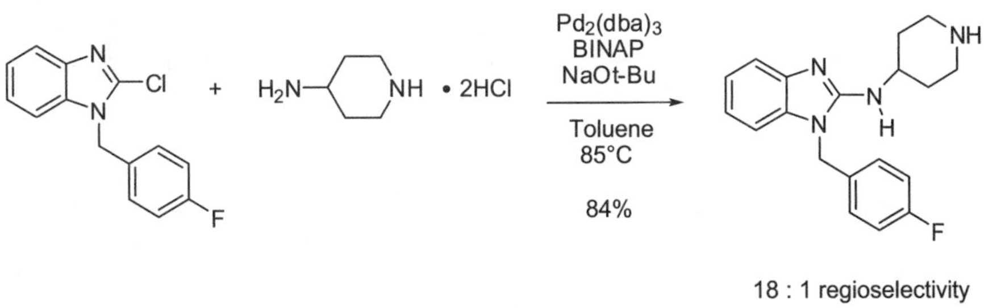

18 : 1 regioselectivity

Interestingly, while the Pd-catalyzed conditions preferentially provided the product of primary amine coupling, base-catalyzed displacement ($K_2CO_3$, glycol, 140°C) furnished the tertiary amine by preferential attack of the more nucleophilic secondary amine.

Hartwig has reported remarkable catalytic efficiency with Josiphos ligands with primary amine couplings.[159] With primary amines, chelating ligands are particularly important to avoid formation of two amines binding to palladium to give a stable, catalytically inactive species.[160]

Singer and coworkers have developed a non-proprietary bis-pyrazole phosphine ligand, bippyphos,[161] which is commercially available. Utilization of this ligand in a coupling with 3-chloropyridine is shown below.[162] Ureas[163] and hydroxylamines[164] have also been effectively N-arylated using bippyphos as the ligand.

[158]Hong, Y.; Tanoury, G. J.; Wilkinson, H. S.; Bakale, R. P.; Wald, S. A.; Senanayake, C. H. *Tetrahedron Lett.* **1997**, *38*, 5607–5610.

[159]Shen, Q.; Ogata, T.; Hartwig, J. F. *J. Am. Chem. Soc.* **2008**, *130*, 6586–6596.

[160]Widenhoefer, R. A.; Buchwald, S. L. *Organometallics* **1996**, *15*, 3534–3542.

[161]Singer, R. A.; Dore, M.; Sieser, J. E.; Berliner, M. A. *Tetrahedron Lett.* **2006**, *47*, 3727–3731.

[162]Withbroe, G. J.; Singer, R. A.; Sieser, J. E. *Org. Process Res. Dev.* **2008**, *12*, 480–489.

[163]Kotecki, B. J.; Fernando, D. P.; Haight, A. R.; Lukin, K. A. *Org. Lett.* **2009**, *11*, 947–950.

[164]Porzelle, A.; Woodrow, M. D.; Tomkinson, N. C. O. *Org. Lett.* **2009**, *11*, 233–236.

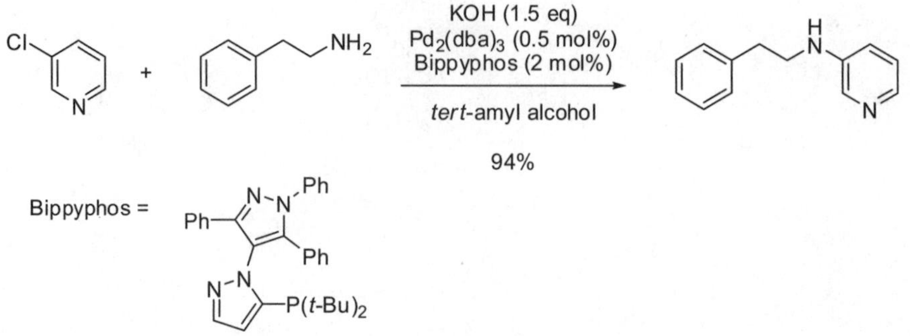

Cl
N
NH2
KOH (1.5 eq)
Pd2(dba)3 (0.5 mol%)
Bippyphos (2 mol%)
tert-amyl alcohol
94%
H
N
N
Bippyphos =
Ph
N-N
Ph
Ph
N-N
P(t-Bu)2


Bromoarenes have been widely utilized in Pd-catalyzed aminations, and frequently lead to more robust couplings than the corresponding (and less expensive) chloroarenes. The following two examples from *Organic Syntheses* represent a chelating diphosphine ligand (rac-BINAP)[165] as well as a monodentate biphenyl ligand ($Cy_2$P-2-naphthyl).[166]

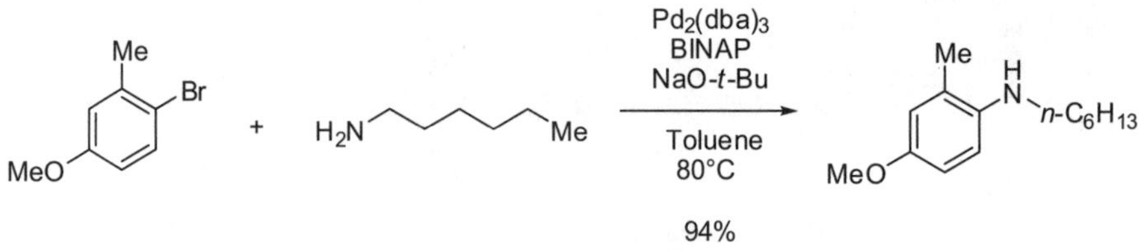

Me
Br
MeO
H2N
Me
Pd2(dba)3
BINAP
NaO-t-Bu
Toluene
80°C
94%
Me
H
N
n-C6H13
MeO


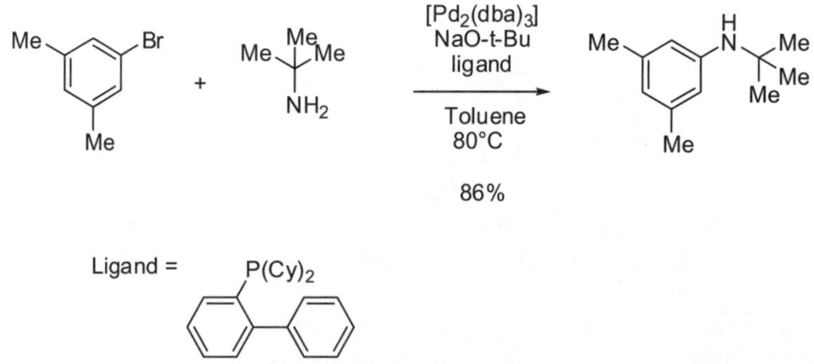

Me
Br
Me
Me
Me
Me
NH2
[Pd2(dba)3]
NaO-t-Bu
ligand
Toluene
80°C
86%
Me
H
N
Me
Me
Me
Me
Ligand =
P(Cy)2


Amides such as oxazolidinones can also serve as coupling partners with bromoarenes, as shown in the example below. The coupling was utilized in a synthesis of DuP-721, as developed by workers at Abbott.[167] Analogous couplings of oxazolidinones with aryl chlorides have also been reported.[168]

[165]Wolfe, J. P.; Buchwald, S. L. *Org. Synth.* **2002**, *78*, 23–35.
[166]Zhang, W.; Lu, Y.; Moore, J. S. *Org. Synth.* **2007**, *84*, 163–176.
[167]Madar, D. J.; Kopecka, H.; Pireh, D.; Pease, J.; Pliushchev, M.; Sciotti, R. J.; Wiedeman, P. E.; Djuric, S. W. *Tetrahedron Lett.* **2001**, *42*, 3681–3684.
[168]Ghosh, A.; Sieser, J. E.; Riou, M.; Cai, W.; Rivera-Ruiz, L. *Org. Lett.* **2003**, *5*, 2207–2210.

77%

N-arylation of amides has been a challenging problem, and Buchwald has reported that a hindered, biaryl monophosphine ligand is effective at promoting couplings with aryl chlorides, as shown in the example below.[169]

99%

Ligand =

2-Bromopyridines can also serve as useful coupling partners, as shown in the example below from Banyu and Merck in the synthesis of a muscarinic-receptor antagonist.[170] Following Buchwald and Hartwig's protocol, they used benzophenone imine as an ammonia surrogate that was subsequently hydrolyzed to the desired aminopyridine.[171,172] Lithium hexamethyl-disilazide (LiN(TMS)$_2$) has also been used as an ammonia surrogate.[173,174] In the example below, tri-n-butylphosphine was used to irreversibly coordinate the residual palladium and allow removal from the product, which was pulled into aqueous acid. Workers at Pfizer have described a screening method for identifying effective palladium purges by recrystallization or reslurry in the presence of various diamine ligands.[175]

[169]Ikawa, T.; Barder, T. E.; Biscoe, M. R.; Buchwald, S. L. *J. Am. Chem. Soc.* **2007**, *129*, 13001–13007.

[170]Mase, T.; Houpis, I. N.; Akao, A.; Dorziotis, I.; Emerson, K.; Hoang, T.; Iida, T.; Itoh, T.; Kamei, K.; Kato, S.; Kato, Y.; Kawasaki, M.; Lang, F.; Lee, J.; Lynch, J.; Maligres, P.; Molina, A.; Nemoto, T.; Okada, S.; Reamer, R.; Song, J. Z.; Tschaen, D.; Wada, T.; Zewge, D.; Volante, R. P.; Reider, P. J.; Tomimoto, K. *J. Org. Chem.* **2001**, *66*, 6775–6786.

[171]Mann, G.; Hartwig, J. F.; Driver, f. M. S.; Fernandez-Rivas, C. *J. Am. Chem. Soc.* **1998**, *120*, 827–828.

[172]Wolfe, J. P.; Ahman, J.; Sadighi, J. P.; Singer, R. A.; Buchwald, S. L. *Tetrahedron Lett.* **1997**, *38*, 6367–6370.

[173]Huang, X.; Buchwald, S. L. *Org. Lett.* **2001**, *3*, 3417–3419.

[174]Lee, S.; Jorgensen, M.; Hartwig, J. F. *Org. Lett.* **2001**, *3*, 2729–2732.

[175]Flahive, E. J.; Ewanicki, B. L.; Sach, N. W.; O'Neill-Slawecki, S. A.; Stankovic, N. S.; Yu, S.; Guinness, S. M.; Dunn, J. *Org. Process Res. Dev.* **2008**, *12*, 637–645.

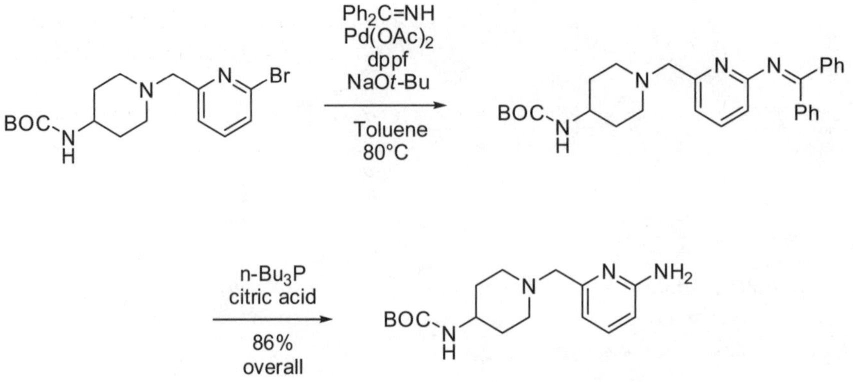

A fluorous-tagged BOC-protected amine ($C_8F_{17}CH_2CH_2C(Me_2)OCONH_2$) has also been used as an ammonia surrogate. The fluorous reagent allows for solid-liquid or liquid-liquid fluorous extraction techniques to be utilized prior to cleavage of the protecting group.[176]

Several copper catalysts have been developed for formation of C(aryl)−N bonds, and this area was recently reviewed.[147] Although less general than the Pd-catalyzed aminations, copper offers advantages in terms of cost, toxicity, and environmental impact. Selected examples are shown below.

Buchwald reported efficient coupling of aryl- and heteroarylbromides with primary amines catalyzed by copper iodide and diethylsalicylamide as a ligand.[177]

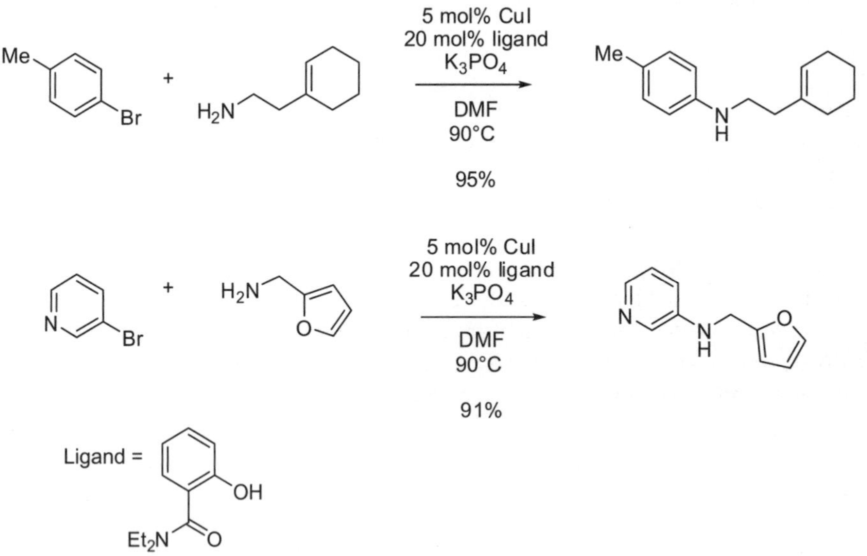

[176]Trabanco, A. A.; Vega, J. A.; Fernandez, M. A. *J. Org. Chem.* **2007**, *72*, 8146–8148.
[177]Kwong, F. Y.; Buchwald, S. L. *Org. Lett.* **2003**, *5*, 793–796.

Recent improvements to this method through use of β-diketones[178] and β-keto-esters[179] have been reported. Copper-catalyzed N-arylations have also been reported with β-amino acids[180] and BINOL[181] as ligands.

Buchwald first reported that diamines serve as effective ligands in Cu-catalyzed N-arylation of amides,[182,183] as shown in the example below.

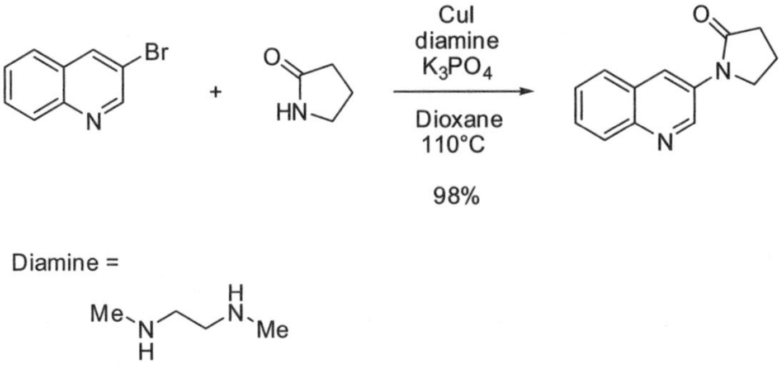

98%

Diamine =

N-Arylation of heterocycles (e.g., pyrazoles, pyrroles) has also been utilized with these ligands.[184] Note that in the example below, excellent regioselectivity is observed (>20 : 1 N-1 : N-2). The Buchwald group has also published mechanistic studies on these diamine-mediated couplings.[185]

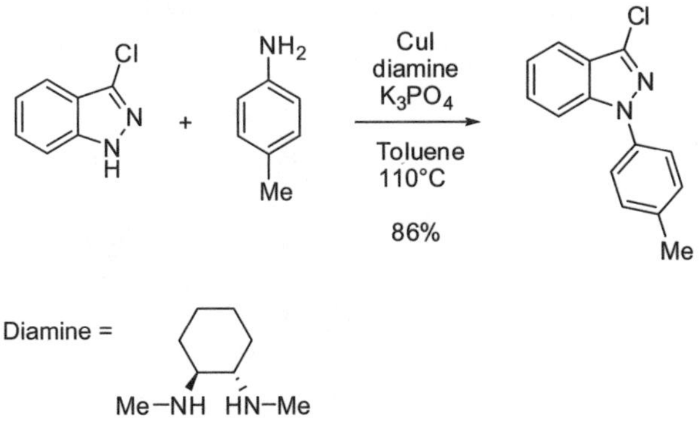

86%

Diamine =

Me−NH  HN−Me

[178]Shafir, A.; Buchwald, S. L. *J. Am. Chem. Soc.* **2006**, *128*, 8742–8743.

[179]Lv, X.; Bao, W. *J. Org. Chem.* **2007**, *72*, 3863–3867.

[180]Ma, D.; Cai, Q.; Zhang, H. *Org. Lett.* **2003**, *5*, 2453–2455.

[181]Jiang, D.; Fu, H.; Jiang, Y.; Zhao, Y. *J. Org. Chem.* **2007**, *72*, 672–674.

[182]Klapars, A.; Antilla, J. C.; Huang, X.; Buchwald, S. L. *J. Am. Chem. Soc.* **2001**, *123*, 7727–7729.

[183]Klapars, A.; Huang, X.; Buchwald, S. L. *J. Am. Chem. Soc.* **2002**, *124*, 7421–7428.

[184]Antilla, J. C.; Baskin, J. M.; Barder, T. E.; Buchwald, S. L. *J. Org. Chem.* **2004**, *69*, 5578–5587.

[185]Strieter, E. R.; Bhayana, B.; Buchwald Stephen, L. *J. Am. Chem. Soc.* **2009**, *131*, 78–88.

A copper-catalyzed N-arylation of an oxazolidinone was utilized in a synthesis of the κ-agonist CJ-15,161.[186,187]

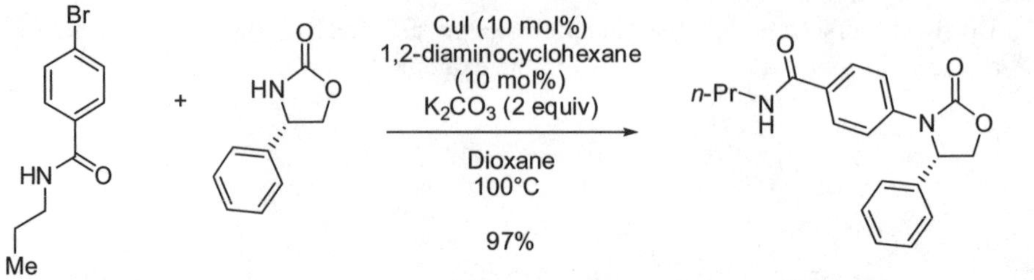

97%

3-Bromofuran was utilized in the example below from Padwa for his synthesis of 3-amidofurans, a useful class of dienes for Diels–Alder methodologies developed in his lab.[188]

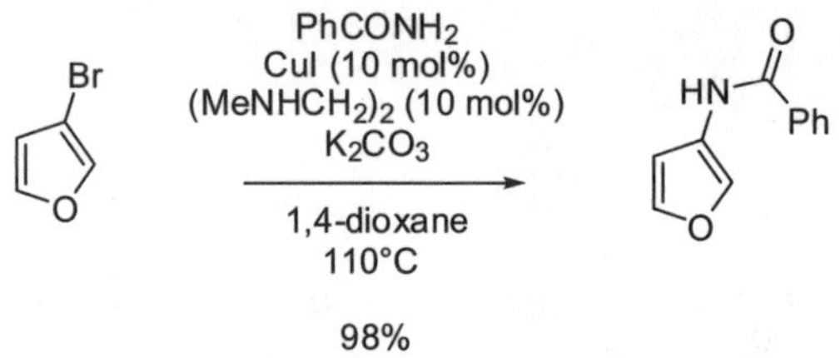

98%

Buchwald has demonstrated both N- and O-arylation of 1,2-aminoalcohols.[189] In the examples below, N-selective arylation is realized with strong base (NaOH) in aqueous DMSO; the presence of water was found to enhance the N- vs. O-selectivity. In the second example, use of a milder base (Cs$_2$CO$_3$) and a secondary amine provided selective O-arylation.

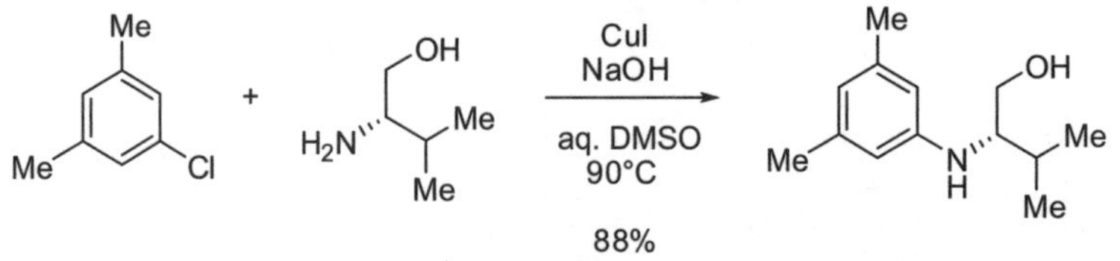

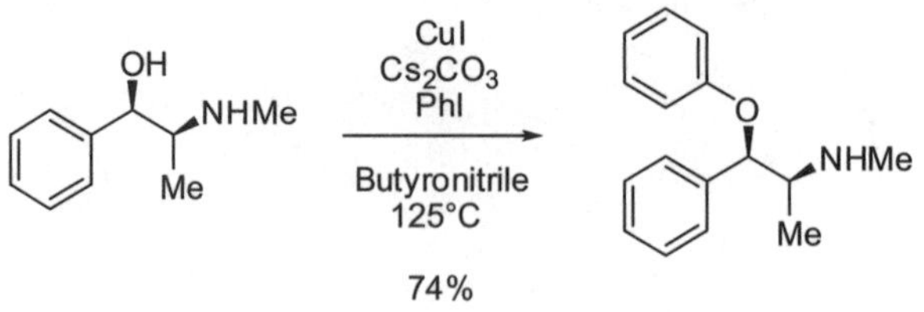

88%

74%

[186]Ghosh, A.; Sieser, J. E.; Caron, S.; Couturier, M.; Dupont-Gaudet, K.; Girardin, M. *J. Org. Chem.* **2006**, *71*, 1258–1261.

[187]Ghosh, A.; Sieser, J. E.; Caron, S.; Watson, T. J. N. *Chem. Commun.* **2002**, 1644–1645.

[188]Padwa, A.; Crawford, K. R.; Rashatasakhon, P.; Rose, M. *J. Org. Chem.* **2003**, *68*, 2609–2617.

[189]Job, G. E.; Buchwald, S. L. *Org. Lett.* **2002**, *4*, 3703–3706.

Subsequent to this initial report, the Buchwald group described that the choice of ligand could provide nearly complete control of N- vs. O-regioselectivity with iodoarenes.[190]

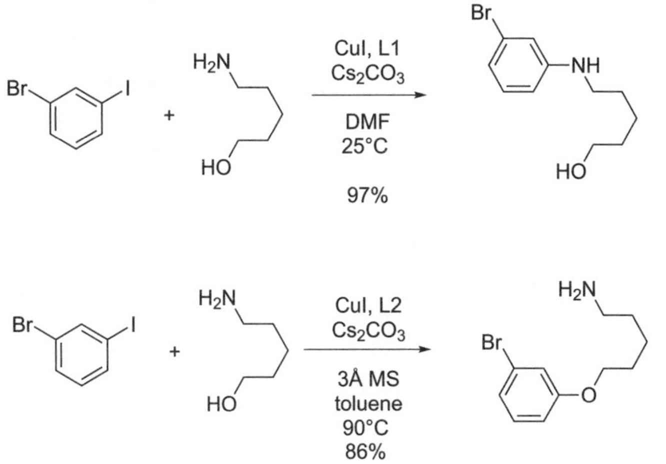

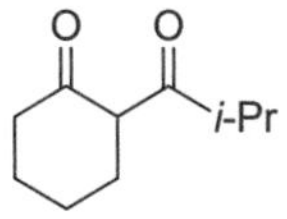

**L1**

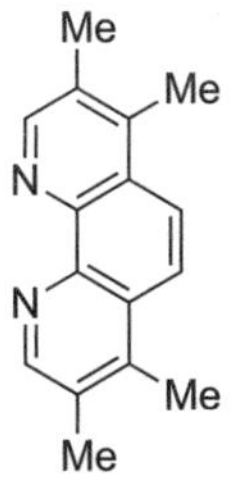

**L2**

Amino acids can also be effectively N-arylated, as shown below in work from Ma and coworkers in their synthesis of martinellic acid.[191]

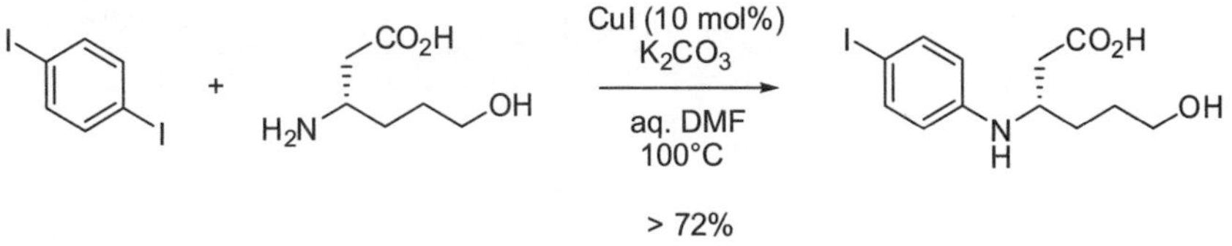

Copper-mediated couplings of amines and arylboronic acids represent an alternative to haloarene couplings. Buchwald has reported copper-catalyzed couplings in the presence of myristic acid ($n$-$C_{14}H_{29}COOH$); couplings with anilines are generally higher yielding (70–90%), but the system does tolerate alkylamines, as shown in the example below.[192] Chan has reported similar couplings with a wide variety of amine nucleophiles, including amides, imides, carbamates, and sulfonamides, using triethylamine or pyridine as an additive.[193]

[190]Shafir, A.; Lichtor, P. A.; Buchwald, S. L. *J. Am. Chem. Soc.* **2007**, *129*, 3490–3491.

[191]Ma, D.; Xia, C.; Jiang, J.; Zhang, J.; Tang, W. *J. Org. Chem.* **2003**, *68*, 442–451.

[192]Antilla, J. C.; Buchwald, S. L. *Org. Lett.* **2001**, *3*, 2077–2079.

[193]Chan, D. M. T.; Monaco, K. L.; Wang, R.-P.; Winters, M. P. *Tetrahedron Lett.* **1998**, *39*, 2933–2936.

N-arylation of imidazoles with arylboronic acids has been developed by Collman. These couplings are catalytic in copper and utilize tetramethylethylenediamine (TMEDA) as the ligand.[194] The couplings also work in water, as shown in the second example below.[195]

## 6.11 Pd- AND Cu-CATALYZED ARYL C−O BOND FORMATION

In general, metal-catalyzed C(aryl)−O bond formation is less well-developed than the analogous C−N bond forming reactions, particularly with primary or secondary alcohols, which are prone to β-hydride elimination of the PdAr(OR)$L_n$ intermediate. Buchwald has developed a useful class of ligands for effecting these transformations, even with electron-neutral aryl chlorides, a particularly challenging class of substrates.[196] The bulky phosphine ligand serves to suppress ß-hydride elimination while promoting reductive elimination.

[194]Collman, J. P.; Zhong, M.; Zhang, C.; Costanzo, S. *J. Org. Chem.* **2001**, *66*, 7892–7897.
[195]Collman, J. P.; Zhong, M.; Zeng, L.; Costanzo, S. *J. Org. Chem.* **2001**, *66*, 1528–1531.
[196]Torraca, K. E.; Huang, X.; Parrish, C. A.; Buchwald, S. L. *J. Am. Chem. Soc.* **2001**, *123*, 10770–10771.

Secondary alcohols are particularly demanding in these types of couplings (being more easily oxidized), as are aryl halides lacking ortho substitution. For these challenging substrates, a more-hindered ligand was developed, as shown below.[197]

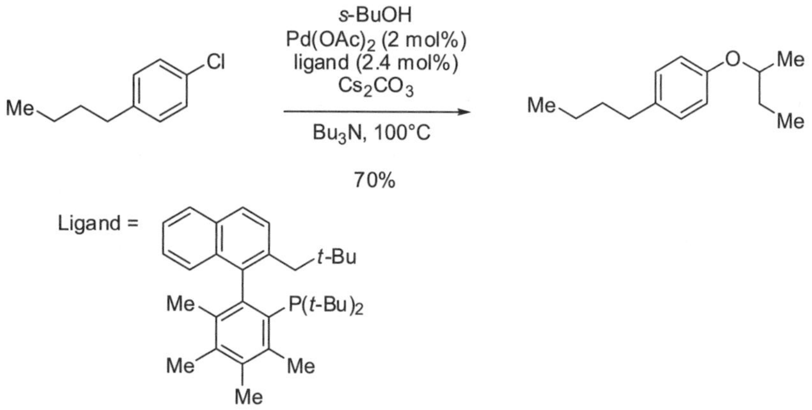

Phenols can be directly prepared from aryl halides with potassium hydroxide and a catalyst derived from $Pd_2(dba)_3$ and a hindered biphenyl phosphine ligand.[198] This transformation can also be done indirectly, by arylation of alcohols such as $t$-BuOH or $PhCH_2OH$ and subsequent dealkylation.

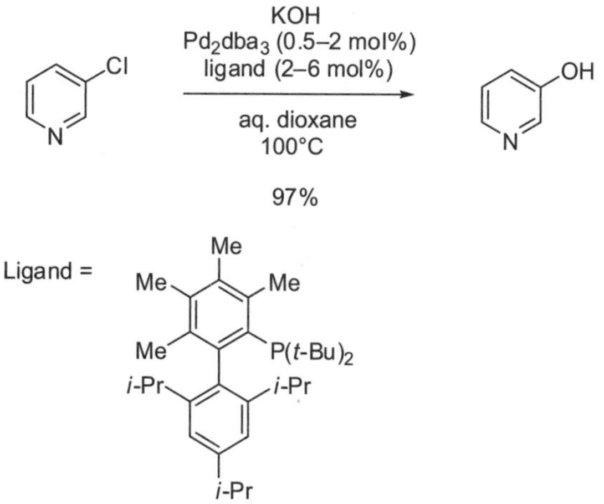

Copper-catalyzed variants of C(aryl)−O bond-forming reactions can also be useful, although they are generally narrower in scope. The classic Ullman diaryl ether formation is catalyzed by copper, and involves rather harsh conditions in its traditional variants (high temperatures, pyridine solvent). Song and coworkers at Merck have developed a particularly useful variant that uses CuCl and an inexpensive 1,3-diketone ligand.[199]

[197]Vorogushin, A. V.; Huang, X.; Buchwald, S. L. *J. Am. Chem. Soc.* **2005**, *127*, 8146–8149.

[198]Anderson, K. W.; Ikawa, T.; Tundel, R. E.; Buchwald, S. L. *J. Am. Chem. Soc.* **2006**, *128*, 10694–10695.

[199]Buck, E.; Song, Z. J.; Denmark, S. E.; Baird, J. D. *Org. Synth.* **2005**, *82*, 69–74.

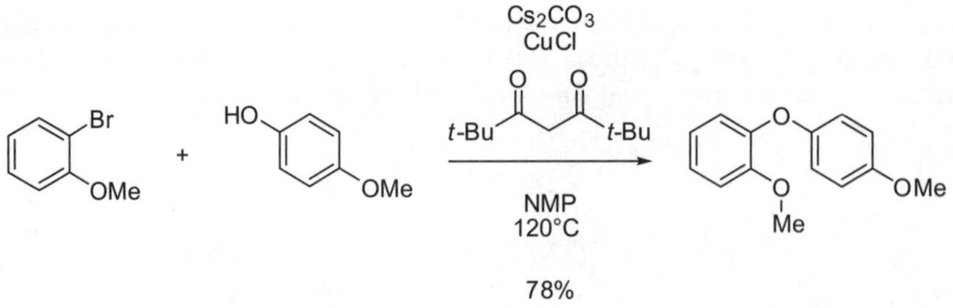

78%

Beller has reported N-methylimidazole as an effective ligand for Cu-catalyzed O-arylation of phenols.[200]

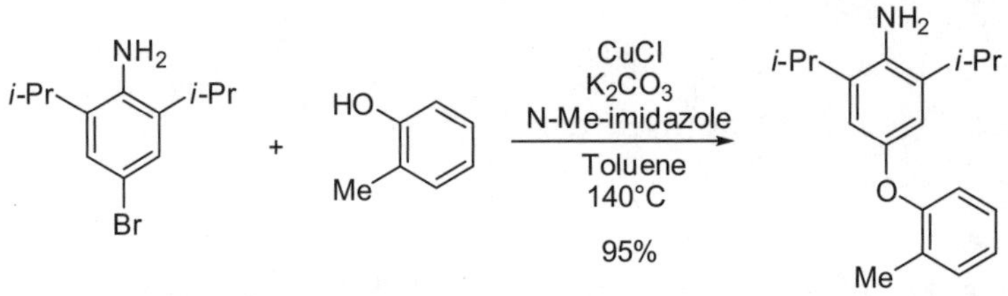

95%

Primary and secondary alcohols can be coupled with iodoarenes with a catalyst derived from CuI and 1,10-phenanthroline.[201]

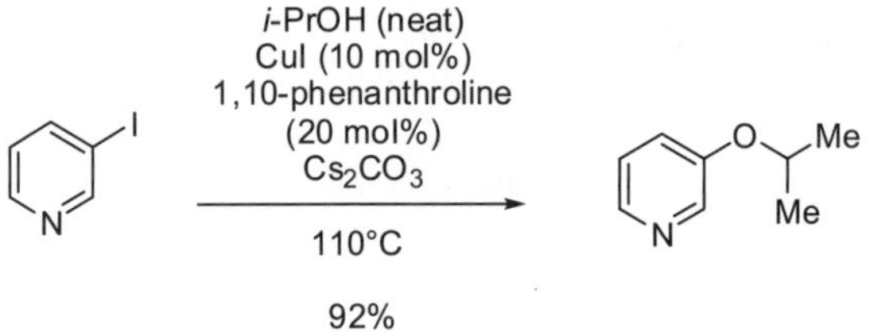

92%

In aniline-containing substrates, protection of the aniline as the 2,5-dimethylpyrrole gives more successful coupling.[202] Deprotection to the aniline is achieved by treatment with hydroxylamine hydrochloride and triethylamine in aqueous ethanol.

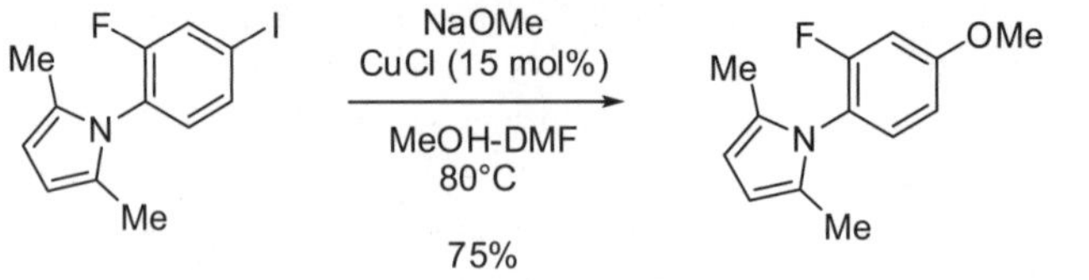

75%

[200]Schareina, T.; Zapf, A.; Cotte, A.; Mueller, N.; Beller, M. *Org. Process Res. Dev.* **2008**, *12*, 537–539.
[201]Wolter, M.; Nordmann, G.; Job, G. E.; Buchwald, S. L. *Org. Lett.* **2002**, *4*, 973–976.
[202]Ragan, J. A.; Jones, B. P.; Castaldi, M. J.; Hill, P. D.; Makowski, T. W. *Org. Synth.* **2002**, *78*, 63–72.

Heterocyclic electrophiles can be successfully coupled with benzyl alcohol, as shown in the example below.[203]

74%

Stoichiometric copper-promoted coupling of boronic acids and phenols has been developed by Evans and others as an effective route to the vancomycin class of antibiotics.[204]

76%

Aryl trifluoroborates can also be effectively coupled with alcohols.[205] This is a particularly convenient method if the trifluoroborate is readily available, as these compounds are frequently well-behaved, crystalline solids.

85%

## 6.12  Pd- AND Cu-CATALYZED ARYL C—S BOND FORMATION

The Pd-catalyzed cross-coupling of thiols and aryl halides has been known since 1980,[206] and later work has shown that several classes of ligands deliver catalysts with good substrate scope and activity. Copper-catalyzed couplings are also known, but are limited to aryl iodides.[207,208] The examples selected below all utilize commercially available ligands and allow for couplings with aryl bromides and/or chlorides.

[203]Nara, S. J.; Jha, M.; Brinkhorst, J.; Zemanek, T. J.; Pratt, D. A. *J. Org. Chem.* **2008**, *73*, 9326–9333.

[204]Evans, D. A.; Katz, J. L.; West, T. R. *Tetrahedron Lett.* **1998**, *39*, 2937–2940.

[205]Quach, T. D.; Batey, R. A. *Org. Lett.* **2003**, *5*, 1381–1384.

[206]Migita, T.; Shimizu, T.; Asami, Y.; Shiobara, J.; Kato, Y.; Kosugi, M. *Bull. Chem. Soc. Jpn.* **1980**, *53*, 1385–1389.

[207]Kwong, F. Y.; Buchwald, S. L. *Org. Lett.* **2002**, *4*, 3517–3520.

[208]Bates, C. G.; Gujadhur, R. K.; Venkataraman, D. *Org. Lett.* **2002**, *4*, 2803–2806.

Buchwald has reported the use of DiPPF as a ligand for the Pd-catalyzed coupling of thiols and aryl bromides and chlorides. This catalyst system offers good substrate scope with a reasonably priced ligand.[209]

Hartwig has reported the use of Josiphos as a ligand, leading to a catalyst with significantly higher turnover numbers, allowing for much lower catalyst loadings (as low as 0.001 mol% with aryl bromides). For larger scale applications, the benefit of these low catalyst loadings is likely to overcome the higher cost of this ligand.[210] In both of these examples, the sterically encumbered ligand serves to suppress catalyst inhibition by the thiol substrate.

The much less expensive Xantphos ligand has also been used with both alkyl- and aryl-thiols.[211] It does appear that this catalyst system operates best with aryl bromides and iodides; the only example of an aryl chloride is with 4-nitrophenylchloride.

[209]Murata, M.; Buchwald, S. L. *Tetrahedron* **2004**, *60*, 7397–7403.
[210]Fernandez-Rodriguez, M. A.; Shen, Q.; Hartwig, J. F. *J. Am. Chem. Soc.* **2006**, *128*, 2180–2181.
[211]Itoh, T.; Mase, T. *Org. Lett.* **2004**, *6*, 4587–4590.

Xantphos was also utilized by workers at Pfizer, albeit with an aryl iodide.[175] This paper describes investigation of several methods for removal of residual palladium, a problem frequently encountered with pharmaceutical applications of this type of chemistry. These workers also described studies on removal of residual sulfur contaminants that interfered with downstream metal-mediated couplings.[212]

Aryl triflates have also been used as electrophiles in Pd-catalyzed thiol couplings.[213] Although they must usually be prepared from the corresponding phenol, a wider range of phenols is typically available than aryl bromides or chlorides. Note that a higher catalyst loading (10 mol%) is required with Tol-BINAP relative to some of the ligands in the previous examples.

OTf

+ HS(CH$_2$)$_3$CH$_3$

Pd(OAc)$_2$ (10 mol%)
(R)-Tol-BINAP (11 mol%)
NaO-t-Bu

Toluene
80°C

S(CH$_2$)$_3$CH$_3$

95%

Boronic acids have also been successfully coupled with alkylthiols using a stoichiometric copper reagent.[214]

BnO$_2$C–N(H)–CH(SH)–CO$_2$H

PhB(OH)$_2$
Cu(OAc)$_2$ (1.5 equiv)
pyridine (3 equiv)
3A MS

DMF
110°C

BnO$_2$C–N(H)–CH(SPh)–CO$_2$H

79%

## 6.13 CARBONYLATION REACTIONS

The metal-mediated carbonylation of aryl halides offers an alternative to the addition of an organometallic reagent to carbonyl electrophiles such as carbon dioxide or dimethyl formamide. The advantages of the palladium-mediated process are its mild reaction conditions and compatibility with a number of functional groups that might not be compatible with an organomagnesium or organolithium reagent. The reaction usually proceeds with the aryl or vinyl halide in the presence of a palladium catalyst, a base, and carbon monoxide. The choice of reagent to intercept the acyl palladium intermediate dictates the nature of the final product.

[212]Xiang, Y.; Caron, P.-Y.; Lillie, B. M.; Vaidyanathan, R. *Org. Process Res. Dev.* **2008**, *12*, 116–119.

[213]Zheng, N.; McWilliams, J. C.; Fleitz, F. J.; Armstrong, J. D., III; Volante, R. P. *J. Org. Chem.* **1998**, *63*, 9606–9607.

[214]Herradura, P. S.; Pendola, K. A.; Guy, R. K. *Org. Lett.* **2000**, *2*, 2019–2022.

### 6.13.1    Formation of Aldehydes

In order to obtain an aldehyde, the acyl palladium species must be reduced with a hydride source. One of the most popular reagents is triethylsilane, a mild reducing agent that does not require a harsh workup. In some cases, reduction of the oxidative insertion product prior to carbonylation can be problematic. In the example below, reaction conditions were optimized to achieve 92% yield of the benzaldehyde derivative.[215]

### 6.13.2    Formation of Esters

Aryl and vinyl halides can be converted to the corresponding aryl and vinyl esters via transition-metal catalysis in the presence of CO and an alcohol. Depending on the nature of the substrate, the reaction might proceed under an atmosphere of carbon monoxide[216] or might require more forcing conditions such as elevated temperature or pressure.[217]

[215]Ashfield, L.; Barnard, C. F. *J. Org. Process Res. Dev.* **2007**, *11*, 39–43.

[216]Friesen, R. W.; Ducharme, Y.; Ball, R. G.; Blouin, M.; Boulet, L.; Cote, B.; Frenette, R.; Girard, M.; Guay, D.; Huang, Z.; Jones, T. R.; Laliberte, F.; Lynch, J. J.; Mancini, J.; Martins, E.; Masson, P.; Muise, E.; Pon, D. J.; Siegl, P. K. S.; Styhler, A.; Tsou, N. N.; Turner, M. J.; Young, R. N.; Girard, Y. *J. Med. Chem.* **2003**, *46*, 2413–2426.

[217]Daniewski, A. R.; Liu, W.; Puentener, K.; Scalone, M. *Org. Process Res. Dev.* **2002**, *6*, 220–224.

### 6.13.3   Formation of Amides

Because amines are excellent nucleophiles, they will react with acyl palladium intermediates to provide amides. In the first example below, only 1.2 equivalents of the amine were necessary to achieve reaction completion. Interestingly, the reaction also proceeded exclusively with the amine despite the presence of two equivalents of water.[218] In the second example below, a dibromopyridine was carbonylated with high regioselectivity (98:2).[219]

## 6.14   METAL-MEDIATED CYANATION

While cyanation of an arene can be achieved by nucleophilic aromatic substitution (see Section 4.6.1), one of the most reliable methods for the conversion of an aryl halide to the nitrile is under palladium-mediated catalysis. The reaction usually utilizes zinc cyanide as the cyanide source in a polar aprotic solvent such as DMF. In some protocols, the zinc cyanide is added slowly to minimize the cyanide concentration since it can act as an inhibitor of the palladium catalyst.[220] As shown in the first example below, an aminopyridine may not require protection to undergo the transformation.[221] In the second example, the catalyst was pre-formed using diethyl zinc as the reducing agent and the reaction proceeded to completion with only 0.6 equivalents of zinc cyanide. In this example, chemoselectivity between the bromide and the chloride was achieved and the chiral secondary alcohol did not require protection.[222]

[218]Etridge, S. K.; Hayes, J. F.; Walsgrove, T. C.; Wells, A. S. *Org. Process Res. Dev.* **1999**, *3*, 60–63.

[219]Wu, G. G.; Wong, Y.; Poirier, M. *Org. Lett.* **1999**, *1*, 745–747.

[220]Ryberg, P. *Org. Process Res. Dev.* **2008**, *12*, 540–543.

[221]Beaudin, J.; Bourassa, D. E.; Bowles, P.; Castaldi, M. J.; Clay, R.; Couturier, M. A.; Karrick, G.; Makowski, T. W.; McDermott, R. E.; Meltz, C. N.; Meltz, M.; Phillips, J. E.; Ragan, J. A.; Ripin, D. H. B.; Singer, R. A.; Tucker, J. L.; Wei, L. *Org. Process Res. Dev.* **2003**, *7*, 873–878.

[222]Chen, C.-y.; Frey, L. F.; Shultz, S.; Wallace, D. J.; Marcantonio, K.; Payack, J. F.; Vazquez, E.; Springfield, S. A.; Zhou, G.; Liu, P.; Kieczykowski, G. R.; Chen, A. M.; Phenix, B. D.; Singh, U.; Strine, J.; Izzo, B.; Krska, S. W. *Org. Process Res. Dev.* **2007**, *11*, 616–623.

There are some examples of aryl chlorides participating in this reaction, although they usually require slightly more forcing conditions.[223]

Buchwald has introduced a copper-catalyzed process whereby an aryl bromide is first converted to the aryl iodide that then undergoes the cyanation with sodium cyanide. A 1,2-diamine is used to accelerate the Cu-catalyzed cyanation process.[224]

## 6.15   METATHESIS REACTIONS

### 6.15.1   Introduction

Catalytic metathesis reactions have been used in the petrochemical and polymer industries since the 1950s, but it was not until the 1990s that catalytic metathesis found

[223]Wang, X.; Zhi, B.; Baum, J.; Chen, Y.; Crockett, R.; Huang, L.; Eisenberg, S.; Ng, J.; Larsen, R.; Martinelli, M.; Reider, P. *J. Org. Chem.* **2006**, *71*, 4021–4023.
[224]Zanon, J.; Klapars, A.; Buchwald, S. L. *J. Am. Chem. Soc.* **2003**, *125*, 2890–2891.

significant use in the synthesis of complex molecules.[225] Since then, catalytic metathesis has become a reliable, efficient, and functional-group-tolerant reaction that chemists can count on as a key reaction in the later stages of complex molecule synthesis.[226–228] In recognition of the significance of this reaction to the field of organic chemistry, the 2005 Nobel Prize in Chemistry was awarded to Yves Chauvin, Robert Grubbs, and Richard Schrock.[229]

### 6.15.2 Cross-Metathesis

Intermolecular metathesis is often referred to as cross-metathesis. This process is widely used in polymerization reactions (see Section 6.15.4), but is less common in organic synthesis, where most applications of metathesis reactions are intramolecular ring-closing reactions (see Section 6.15.3).

#### 6.15.2.1  Alkene Cross-Metathesis

$$R_1\text{---CH=CH}_2 \; + \; M\text{=CH-}R_2 \; \rightleftharpoons \; \underset{R_1 \quad R_2}{\overset{\text{--M}}{\square}} \; \rightleftharpoons \; \underset{R_2}{\overset{R_1}{||}} \; + \; M\text{=}$$

Mechanistically, the cross-metathesis of alkenes proceeds by formal [2 + 2] reaction of a metal alkylidene with an alkene.[230,231] The resulting metallacyclobutane can then undergo a cycloreversion reaction to provide starting materials (via degenerate metathesis) or a new alkene and metal alkylidene pair. Metathesis reactions are reversible, unless a volatile alkene product like ethylene is expelled from solution, and self-metathesis (metathesis of two identical olefins) is often in competition with the desired cross-metathesis reactions. In addition to challenges associated with reversibility and self-metathesis, mixtures of E- and Z-olefin products can also be problematic. For these reasons, alkene cross-metathesis is a challenge with many classes of alkene substrates. Nonetheless, cross-metathesis reactions can be effective, especially if it is practical to use an excess of one alkene component. This was shown by Morimoto and coworkers during the course of their synthesis of ( + )-omaezakianol.[232] In this case, the use of 4.7 equivalents of the triepoxy alkene component is unfortunate, but the 87% yield of E-alkene and the functional group compatibility of the reaction are impressive.

[225]Astruc, D. *New J. Chem.* **2005**, *29*, 42–56.
[226]Nicolaou, K. C.; Ortiz, A.; Denton, R. M. *Chim. Oggi* **2007**, *25*, 70–72, 74–76.
[227]Grubbs, R. H. *Tetrahedron* **2004**, *60*, 7117–7140.
[228]Grubbs, R. H. *Handbook of Metathesis*; Wiley-VCH: Weinheim, Germany, 2003.
[229]Rouhi, M. *Chem. Eng. News* **2005**, *83*, 8.
[230]Waetzig, J. D.; Hanson, P. R. *Chemtracts* **2006**, *19*, 157–167.
[231]Connon, S. J.; Blechert, S. *Angew. Chem., Int. Ed. Engl.* **2003**, *42*, 1900–1923.
[232]Morimoto, Y.; Okita, T.; Kambara, H. *Angew. Chem., Int. Ed. Engl.* **2009**, *48*, 2538–2541.

(1.0 equiv)    (4.7 equiv)

MesN⤵NMes
Cl‚Ph
Cl‷Ru⟋
PCy₃
(0.1 equiv)

CH₂Cl₂ (0.05 M)
40°C
87%

Electron-deficient olefins are one of the best substrate classes for alkene cross-metathesis, because they typically undergo very slow self-metathesis and often allow for good *E:Z* selectivity.[231] List and Michrowska were able to take advantage of this reactivity in their synthesis of ricciocarpin A.[233] In this case, the relatively high reaction concentration, the relatively low catalyst loading, and the low cost of the alkene used in excess (crotonalde-hyde), make this a very practical way to functionalize the terminal alkene. It is noteworthy that the sterically hindered enone is not reactive in this metathesis reaction. Sterically hindered alkenes react much slower than electronically similar sterically unhindered alkenes in metathesis reactions.[230]

### 6.15.2.2 Alkane Cross-Metathesis

Alkane cross-metathesis is a valuable reaction that allows for the modification of alkane chain lengths at industrial scales, but this process will not be discussed in detail here, because the mechanism of this process (dehydrogenation/metathesis/hydrogenation) is identical to alkene cross-metathesis during the key carbon–carbon bond forming/breaking steps.[234]

---

[233]Michrowska, A.; List, B. *Nat. Chem.* **2009**, *1*, 225–228.

[234]Basset, J.-M.; Coperet, C.; Soulivong, D.; Taoufik, M.; Thivolle-Cazat, J. *Angew. Chem., Int. Ed. Engl.* **2006**, *45*, 6082–6085.

### 6.15.2.3 Alkyne Cross-Metathesis

While alkyne cross-metathesis has not yet reached a level of practicality that would allow it to be counted on in organic synthesis, the oligomerization of diynes can be quite effective. Mechanistically, alkyne metathesis is analogous to alkene metathesis (see Section 6.15.2.1), except that the reaction proceeds via a metallacyclobutadiene intermediate instead of a metallacyclobutane intermediate. It is also worth noting that terminal alkynes, unlike terminal alkenes, are not typically suitable substrates.[235] Like alkene metathesis, alkyne metathesis reactions suffer from reversibility and self-metathesis pathways, but E:Z selectivity is not an issue with alkyne metathesis. While many alkene and alkyne metathesis strategies limit reversibility by expelling a volatile alkene/alkyne byproduct from solution, another practical strategy in alkyne metathesis is the precipitation of an insoluble alkyne byproduct from solution.[236] This strategy was developed by Moore and Zhang for the synthesis of shape-persistent arylene ethynylene macrocycles and was successfully used to generate 3.8 grams of a carbazole-based macrocycle.[237]

### 6.15.3 Ring-Closing Metathesis

Ring-closing metathesis (RCM) is the most commonly applied metathesis strategy in organic synthesis, but this reaction does present some challenges. The reaction is typically dilute, in order to avoid competing intermolecular processes, and the necessary catalyst

[235]Schrock, R. R. *Polyhedron* **1995**, *14*, 3177–3195.
[236]Zhang, W.; Moore, J. S. *Adv. Synth. Cat.* **2007**, *349*, 93–120.
[237]Zhang, W.; Moore, J. S. *J. Am. Chem. Soc.* **2004**, *126*, 12796.

loadings are sometimes high.[238] Despite these challenges, ring-closing metathesis is a valuable process that can typically be counted on to be efficient, chemoselective, and tolerant of various functional groups.

### 6.15.3.1  Alkene Ring-Closing Metathesis

Alkene RCM is mechanistically identical to alkene cross-metathesis (see Section 6.15.2.1), except that the two alkenes undergoing metathesis are tethered to each other. As with alkene cross-metathesis, it can be difficult with larger rings to obtain the desired *E:Z* selectivity in alkene ring-closing metathesis.[238] While there are many impressive and elegant examples of ring-closing metathesis in the literature,[226] the large scale synthesis of hepatitis C protease inhibitor BILN 2061 by scientists at Boehringer-Ingelheim stands out as a landmark achievement in metathesis chemistry.[239–244]

[238]Hoveyda, A. H.; Zhugralin, A. R. *Nature* **2007**, *450*, 243–251.

[239]Farina, V.; Shu, C.; Zeng, X.; Wei, X.; Han, Z.; Yee, N. K.; Senanayake, C. H. *Org. Process Res. Dev.* **2009**, *13*, 250–254.

[240]Randolph, J. T.; Zhang, X.; Huang, P. P.; Klein, L. L.; Kurtz, K. A.; Konstantinidis, A. K.; He, W.; Kati, W. M.; Kempf, D. J. *Bioorg. Med. Chem. Lett.* **2008**, *18*, 2745–2750.

[241]Shu, C.; Zeng, X.; Hao, M.-H.; Wei, X.; Yee, N. K.; Busacca, C. A.; Han, Z.; Farina, V.; Senanayake, C. H. *Org. Lett.* **2008**, *10*, 1303–1306.

[242]Yee, N. K.; Farina, V.; Houpis, I. N.; Haddad, N.; Frutos, R. P.; Gallou, F.; Wang, X.-J.; Wei, X.; Simpson, R. D.; Feng, X.; Fuchs, V.; Xu, Y.; Tan, J.; Zhang, L.; Xu, J.; Smith-Keenan, L. L.; Vitous, J.; Ridges, M. D.; Spinelli, E. M.; Johnson, M.; Donsbach, K.; Nicola, T.; Brenner, M.; Winter, E.; Kreye, P.; Samstag, W. *J. Org. Chem.* **2006**, *71*, 7133–7145.

[243]Tsantrizos, Y. S.; Ferland, J.-M.; McClory, A.; Poirier, M.; Farina, V.; Yee, N. K.; Wang, X.-j.; Haddad, N.; Wei, X.; Xu, J.; Zhang, L. *J. Organomet. Chem.* **2006**, *691*, 5163–5171.

[244]Faucher, A.-M.; Bailey, M. D.; Beaulieu, P. L.; Brochu, C.; Duceppe, J.-S.; Ferland, J.-M.; Ghiro, E.; Gorys, V.; Halmos, T.; Kawai, S. H.; Poirier, M.; Simoneau, B.; Tsantrizos, Y. S.; Llinas-Brunet, M. *Org. Lett.* **2004**, *6*, 2901–2904.

This process is impressive for several key reasons: (1) The reaction concentration is relatively high for an RCM reaction, which allows for reasonable throughput in standard manufacturing equipment;[239] (2) the reaction is performed in toluene, which is much "greener" (see Section 16.3) than dichloromethane, the typical solvent for RCM reactions; (3) the reaction is complete in 30 minutes; and (4) the reaction proceeds with a very low catalyst loading (0.1 mol%). The low catalyst loading is important for two reasons. First, the catalyst is expensive, and therefore needs to be minimized for an industrial process. Second, the removal of ruthenium impurities from the product required "heroic efforts" using earlier technology,[242] which utilized higher catalyst loadings, while no special efforts were required for ruthenium removal using this optimized approach.[239]

### 6.15.3.2   Alkyne Ring-Closing Metathesis

Alkyne ring-closing metathesis (RCAM) is mechanistically identical to alkyne cross-metathesis (see Section 6.15.2.3), except that the two alkynes undergoing metathesis are tethered to each other. Typically, RCAM is not efficient for the synthesis of ring sizes smaller than 12, because the strain due to alkyne incorporation is too significant for the reaction to be favorable.[245] However, RCAM does have a significant advantage over RCM for the formation of larger rings, because there are no *E:Z* isomer issues, and methods have been developed for the synthesis of both *E* and *Z* alkenes via RCAM and partial reduction.[236] For this reason, RCAM is complementary to RCM for the synthesis of large rings.

The preparation of *Z*-alkenes via alkyne metathesis can be accomplished in a straightforward fashion by Lindlar-type reduction of the cyclic alkyne to the corresponding *Z*-alkene. This strategy was used by Fürstner and Grela for the synthesis of prostaglandin $E_2$-1,15-lactone.[246]

The synthesis of *E*-alkenes by RCAM and partial reduction is a more involved process, but both Trost[247] and Fürstner[248] have developed hydrosilylation/desilylation approaches for this purpose. In the example below, Fürstner and coworkers utilize a tungsten catalyst for RCAM and a ruthenium catalyst for hydrosilylation, followed by desilylation using silver fluoride. The end result is selective formation of the macrocylic *E*-alkene.[248] It is worth noting that, as this example shows, it is possible to perform alkyne metathesis in the presence of alkenes, however, it is difficult to perform alkene metathesis in the presence of alkynes.[236]

[245]Furstner, A.; Seidel, G. *Angew. Chem., Int. Ed. Engl.* **1998**, *37*, 1734–1736.
[246]Furstner, A.; Grela, K. *Angew. Chem., Int. Ed. Engl.* **2000**, *39*, 1234–1236.
[247]Trost, B. M.; Ball, Z. T.; Joege, T. *J. Am. Chem. Soc.* **2002**, *124*, 7922–7923.
[248]Lacombe, F.; Radkowski, K.; Seidel, G.; Furstner, A. *Tetrahedron* **2004**, *60*, 7315–7324.

### 6.15.3.3  Enyne Ring-Closing Metathesis

Enyne ring-closing metathesis (RCEYM) is a metal-mediated ring-closing reaction that converts a tethered alkene and alkyne into a cyclic diene.[249,250] While not as commonly employed as RCM, the fact that this process generates a ring and a readily functionalized diene makes this a valuable reaction. In the key step of their synthesis of (+)-anatoxin-A, Martin and coworkers generated the bridged bicyclic core via RCEYM.[251]

[249]Mori, M. *Adv. Synth. Cat.* **2007**, *349*, 121–135.
[250]Villar, H.; Frings, M.; Bolm, C. *Chem. Soc. Rev.* **2007**, *36*, 55–66.
[251]Brenneman, J. B.; Machauer, R.; Martin, S. F. *Tetrahedron* **2004**, *60*, 7301–7314.

### 6.15.4  Metathesis Polymerization

While a thorough discussion of metathesis polymerization reactions is well beyond the scope of this book, it is worth noting that acyclic diene metathesis (ADMET), acyclic diyne metathesis (ADIMET), ring-opening alkyne metathesis, and ring-opening alkene metathesis polymerization (ROMP) have all proven to be valuable methods for the synthesis of polymers with unique physical and/or biological properties.[236,252–254]

[252]Bielawski, C. W.; Grubbs, R. H. *Prog. Polym. Sci.* **2007**, *32*, 1–29.
[253]Nomura, K.; Yamamoto, N.; Ito, R.; Fujiki, M.; Geerts, Y. *Macromolecules* **2008**, *41*, 4245–4249.
[254]Weychardt, H.; Plenio, H. *Organometallics* **2008**, *27*, 1479–1485.

# 7

# REARRANGEMENTS

David H.B. Ripin

## 7.1 INTRODUCTION

Rearrangement reactions can be powerful methods for the relay of stereochemistry, functional group interconversion, and altering the atomic connectivity. Simple and easily prepared substrates can, in many cases, be transformed into significantly more complex molecules. The distinction of a rearrangement from other varieties of chemical transformations can be difficult; for the purposes of this chapter, rearrangements are defined as a change in atomic connectivity of a molecule in a concerted or stepwise manner. Some of the reactions shown in this chapter are not synthetically useful, but are worth being aware of as mechanisms for impurity formation in other reactions.

## 7.2 [1,2]-REARRANGEMENTS

### 7.2.1 Carbon to Carbon Migrations of Carbon and Hydrogen

*7.2.1.1 Wagner–Meerwein and Related Reactions* The Wagner–Meerwein rearrangement is a term used to describe the cationic [1,2]-shift of carbon or hydrogen. These reactions can be initiated with an acid catalyst ionizing an alcohol or halide, or through the generation of halonium ions from the treatment of olefins with a halogenating agent. In cases where there is a driving force favoring a single product, the reaction can be synthetically useful.[1,2] Aluminum trichloride is typically employed in generating cations from alkyl chlorides, while protic acids are often employed for the ionization of alcohols. Solvents for the chloride processes must be aprotic and unreactive with the carbocation intermediates. In the case of alcohol ionization, a solvent that reversibly traps the cations can be employed.

[1]Rieke, R. D.; Bales, S. E.; Hudnall, P. M.; Poindexter, G. S. *Org. Synth.* **1980**, *59*, 85–94.
[2]Salaun, J.; Fadel, A. *Org. Synth.* **1986**, *64*, 50–56.

*Practical Synthetic Organic Chemistry: Reactions, Principles, and Techniques*, First Edition.
Edited by Stéphane Caron.
© 2011 John Wiley & Sons, Inc. Published 2011 by John Wiley & Sons, Inc.

On more complex substrates, selectivity for a single product can be problematic. One system in which the Wagner–Meerwein rearrangement is frequently employed is in the functionalization of camphor and related bicyclic compounds. In the terpene literature, Wagner–Meerwein rearrangement is used to refer to carbon substituents other than methyl and the reactions with a [1,2]-shift of a methyl group are considered Nametkin rearrangement.[3]

Cationic [1,2]-shifts are frequently employed in the rearrangement of pyrroles and thiophenes; these rearrangements are followed by loss of a proton to produce the aromatic system.[4,5]

[3]de la Moya Cerero, S.; Martinez, A. G.; Vilar, E. T.; Fraile, A. G.; Maroto, B. L. *J. Org. Chem.* **2003**, *68*, 1451–1458.

[4]Kinugawa, M.; Nagamura, S.; Sakaguchi, A.; Masuda, Y.; Saito, H.; Ogasa, T.; Kasai, M. *Org. Process Res. Dev.* **1998**, *2*, 344–350.

[5]Vicenzi, J. T.; Zhang, T. Y.; Robey, R. L.; Alt, C. A. *Org. Process Res. Dev.* **1999**, *3*, 56–59.

*7.2.1.2    Pinacol Rearrangement: Vicinal Diols to Ketones or Aldehydes*    The pinacol rearrangement of vicinal diols is an acid-catalyzed cationic rearrangement process. The more highly substituted alcohol departs, leaving a carbocation. A neighboring hydrogen or carbon migrates to the cation, resulting in the formation of a ketone or aldehyde.[6,7] Protic acids are most frequently employed, although other ionization methods have been used. A number of Lewis acids have been utilized to effect the semi-pinacol rearrangement including the easily handled $BF_3 \cdot OEt_2$.

Epoxides or β-haloalcohols can also undergo a similar reaction called the semi-pinacol rearrangement. The ready availability of diols and epoxides from olefins provides rapid access to pinacol substrates, allowing preparation of many potentially challenging compounds.[8]

In the case of an asymmetric diol, the more highly substituted bridgehead carbon can ionize first, followed by [1,2]-hydride shift. The unusual method for the generation of the cation was preferred in this case to avoid decomposition of the acid-sensitive substrate.[9]

[6]Martinelli, M. J.; Khau, V. V.; Horcher, L. M. *J. Org. Chem.* **1993**, *58*, 5546–5547.

[7]Barnier, J. P.; Champion, J.; Conia, J. M. *Org. Synth.* **1981**, *60*, 25–29.

[8]Ooi, T.; Maruoka, K.; Yamamoto, H. *Org. Synth.* **1995**, *72*, 95–103.

[9]DeCamp, A. E.; Mills, S. G.; Kawaguchi, A. T.; Desmond, R.; Reamer, R. A.; DiMichele, L.; Volante, R. P. *J. Org. Chem.* **1991**, *56*, 3564–3571.

In some substrates, the migration can occur with stereocontrol.[10] The inversion below, resulting in product of >99% ee, demonstrates in certain substrates the stereochemistry at the ionizing center is retained, possibly through a concerted process.

In the example below, the semi-pinacol rearrangement of the epoxide shown proceeds with a [2,3]-sigmatropic shift of the hydride in an allylic variant of the reaction.[11]

### 7.2.1.3    Expansion and Contraction of Rings

Cationic [1,2]-shifts including Wagner–Meerwein rearrangements and pinacol rearrangements are frequently employed in the expansion or contraction of ring systems. The pinacol transformation below goes through an expanded ring intermediate.[12]

[10]Shionhara, T.; Suzuki, K. *Synthesis* **2003**, *141*–146.
[11]Dolle, R. E.; Kruse, L. I. *J. Org. Chem.* **1986**, *51*, 4047–4053.
[12]Overman, L. E.; Rishton, G. M. *Org. Synth.* **1993**, *71*, 63–71.

The following are two pinacol rearrangements that result in ring expansion.[13,14]

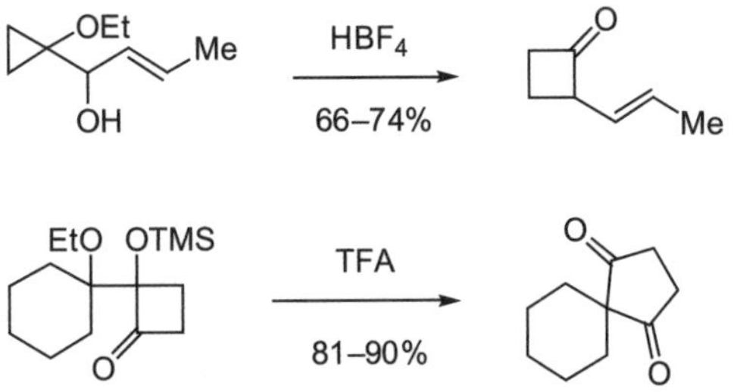

The Tiffeneau–Demyanov ring expansion is similar to a pinacol rearrangement, but the carbocation is generated from a diazo intermediate that is in turn generated from an α-hydroxy amine.[15]

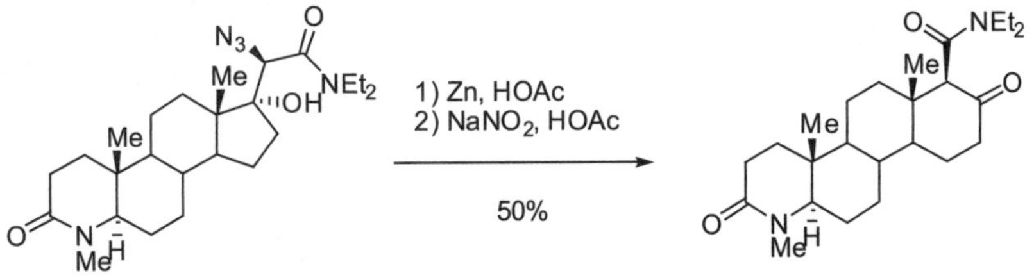

The Ciamician–Dennstedt reaction is the cyclopropanation of a pyrrole or indole, followed by ring-expanding rearrangement to the 3-halopyridine or quinoline.[16]

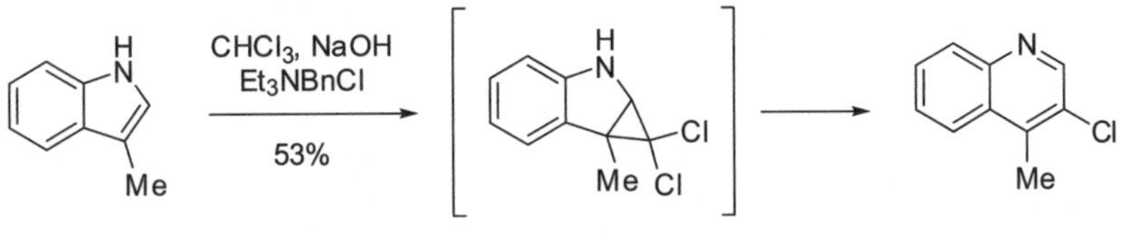

### 7.2.1.4 Rearrangements of Ketones and Aldehydes

The α-ketol rearrangement is a rearrangement of an α-hydroxy ketone, where the alcohol and ketone are transposed. In the case shown below, hemiketal opening was followed by enolization of the ester and aldol reaction with the C-3 carbonyl.[17] The resulting tertiary alcohol underwent a [1,2]-migration of the carbon subsistent to produce the tertiary alcohol shown on multi-kilogram scale.

[13]Miller, S. A.; Gadwood, R. C. *Org. Synth.* **1989**, *67*, 210–221.

[14]Nakamura, E.; Kuwajima, I. *Org. Synth.* **1987**, *65*, 17–25.

[15]Steinberg, N. G.; Rasmusson, G. H.; Reynolds, G. F.; Hirshfield, J. H.; Arison, B. H. *J. Org. Chem.* **1984**, *49*, 4731–4733.

[16]Kwon, S.; Nishimura, Y.; Ikeda, M.; Tamura, Y. *Synthesis* **1976**, 249.

[17]Koch, G.; Jeck, R.; Hartmann, O.; Kuesters, E. *Org. Process Res. Dev.* **2001**, *5*, 211–215.

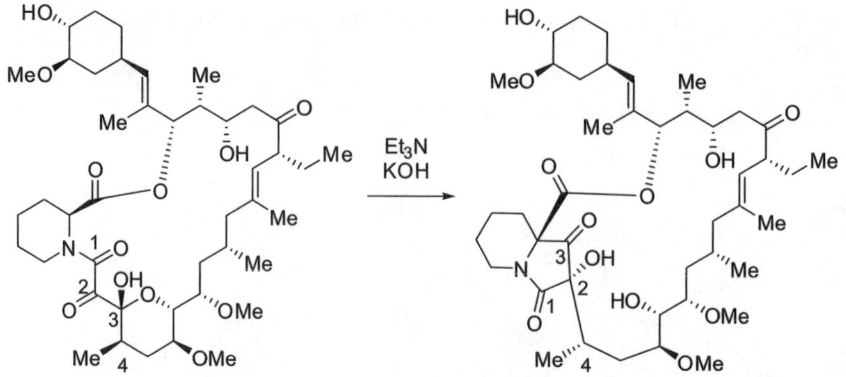

The α-ketol rearrangement can be run under acidic conditions as well.[18] The product to starting material distribution in the rearrangement is under thermodynamic control; therefore, a driving force for the reaction such as relief of ring strain (as above) is a requirement for a complete reaction.

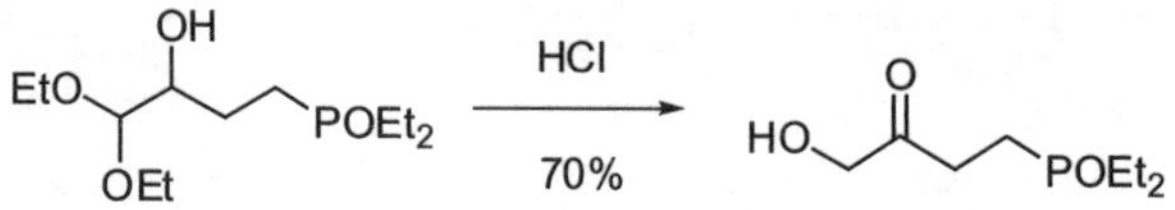

**7.2.1.5   Dienone to Phenol Rearrangement**   When a 4,4-disubstituted cyclohexadienone is treated with acid, one of the C-4 substituents migrates to the 3- or 5-position with concomitant aromatization of the 6-membered ring to a phenol. The direction of migration is dominated by electronic considerations.[19]

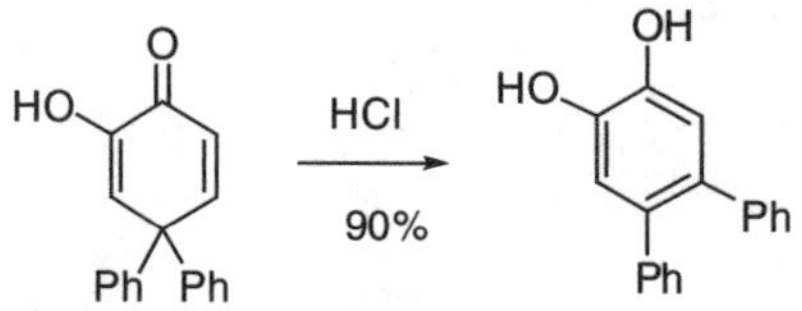

**7.2.1.6   Benzil to Benzilic Acid Rearrangement**   The benzil to benzilic acid rearrangement is formally the reaction shown below. Hydroxide bases are most commonly employed, but other nucleophilic bases will work as well.

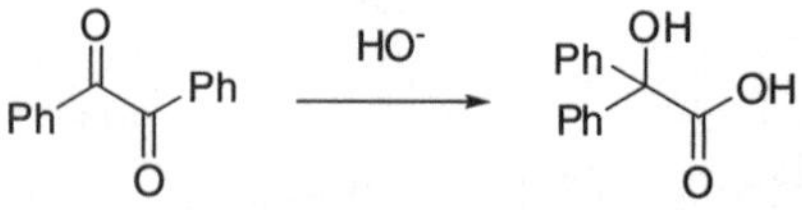

[18]Page, P.; Blonski, C.; Perie, J. *Tetrahedron Lett.* **1995**, *36*, 8027–8030.
[19]Frimer, A. A.; Marks, V.; Sprecher, M.; Gilinsky-Sharon, P. *J. Org. Chem.* **1994**, *59*, 1831–1834.

This rearrangement has been utilized in a synthesis of dihydropyrroles in a reaction with acetonitrile.[20]

### 7.2.1.7 Favorskii Rearrangement: Anionic Rearrangement of α-Haloketones

The Favorskii rearrangement occurs when an α-haloketone is treated with a strong, nucleophilic base. The mechanism has been the subject of extensive study, with surprising results. The rearrangement is not a simple migration of the carbon substituent that does not bear the halide (bottom pathway below), *because either the $R_1$-bearing carbon or the $R_2$-bearing carbon can migrate regardless of the position of the halogen.* This finding, in conjunction with labeling experiments, points to the intermediacy of a cyclopropanone, which ring opens on addition of alkoxide to result in the more stable anion that is protonated by the alcoholic solvent. It is important to take this into consideration when planning a synthesis utilizing this rearrangement, which is most useful for symmetrical substrates.

The Favorskii rearrangement is also useful for producing strained rings. In the case below, treating the bromoketone below with NaOMe results in a ring contraction of the bicyclo [3.2.1]octane ring system to a bicyclo[2.2.1]heptane.[21] Alkoxide bases are most commonly employed for this transformation.

[20]Akabori, S.; Ohtomi, M.; Takahashi, K.; Sakamoto, Y.; Ichinohe, Y. *Synthesis* **1980**, 900–901.
[21]Bai, D.; Xu, R.; Chu, G.; Zhu, X. *J. Org. Chem.* **1996**, *61*, 4600–4606.

In the case where the α-carbon of the ketone not appended to the halide is not enolizable, the reaction is called the quasi-Favorskii rearrangement.[22] This rearrangement obviously does not proceed via the cyclopropanone intermediate but rather by the [1,2]-shift mechanism above. In the case shown, an aniline base was used resulting in an amide product rather than an ester.

Amines with a more nucleophilic lone-pair of electrons can also effect the quasi-Favorskii rearrangement.[23]

Quasi-Favorskii rearrangements of ketones and ketals can be induced on treatment of appropriate substrates with a Lewis acid.[24]

### 7.2.1.8 Arndt–Eistert Synthesis: Homologation of Carboxylic Acids via Wolff Rearrangement of α-Diazoketones

The Ardnt–Eistert synthesis is an effective synthetic method for the homologation of carboxylic acids by one methylene. In the first step of the process, diazomethane is added to the activated carboxylic acid to form the α-diazoketone intermediate, which then undergoes Wolff rearrangement on treatment with a silver salt to

[22] Ares, J. J.; Outt, P. E.; Kakodkar, S. V.; Buss, R. C.; Geiger, J. C. *J. Org. Chem.* **1993**, *58*, 7903–7905.
[23] Sanchez, J. P.; Parcell, R. F. *J. Heterocycl. Chem.* **1990**, *27*, 1601–1607.
[24] Maiti, S. B.; Chaudhuri, S. R. R.; Chatterjee, A. *Synthesis* **1987**, 806–809.

produce the desired homologated product. Two examples are shown below. The use of diazomethane is highly hazardous, and is a significant drawback to this process.[25,26] Diazomethane can be generated in a continuous process to reduce the risks on larger scale.[27]

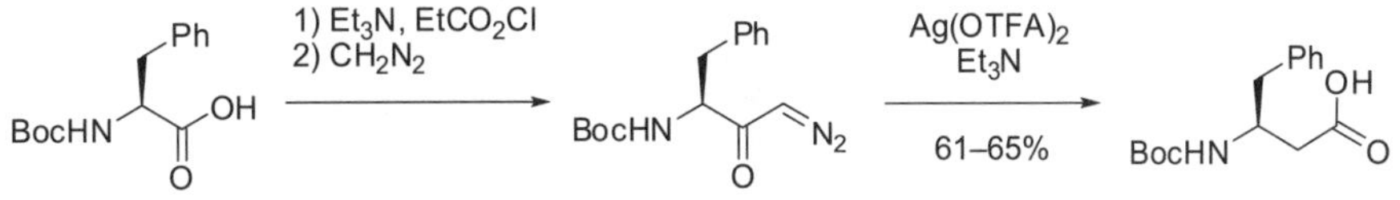

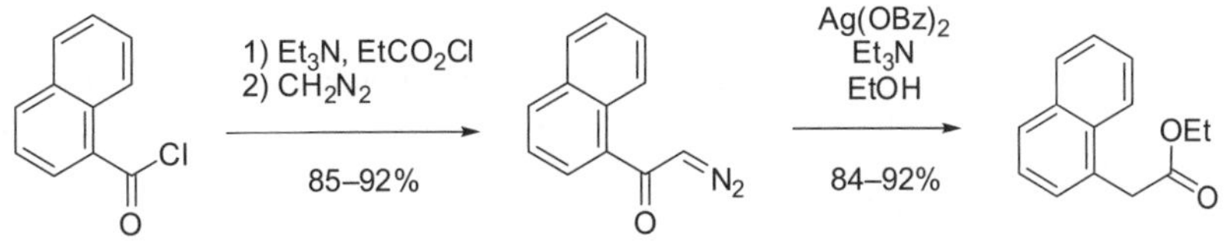

An alternative to Ardnt–Eistert synthesis is shown below, and does not require the use of diazomethane.[28]

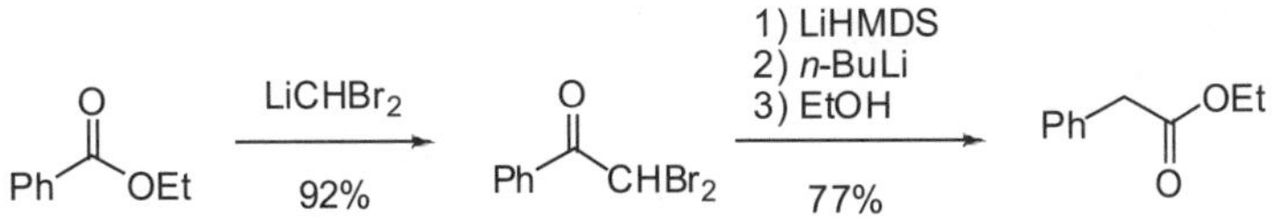

### 7.2.1.9 Homologation of Aldehydes and Ketones

The homologation of ketones and the conversion of aldehydes to ketones can be accomplished through the use of a variety of methods involving rearrangements.

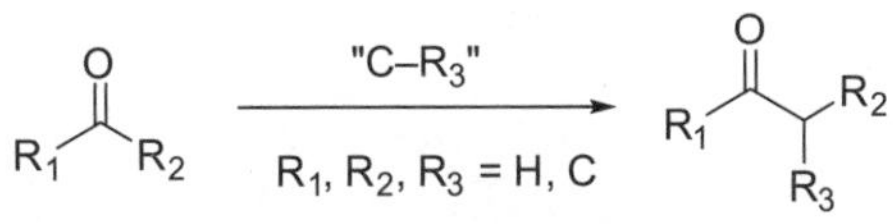

A traditional method for effecting this transformation is through the reaction of a diazo compound with a ketone or aldehyde in the presence of a Lewis acid.[29] In the second example shown, the use of the large Lewis acid MAD ((2,6-di-$t$-Bu-4-MePhO)$_2$AlMe) improved selectivity to 97:3 from 60:40 with Me$_3$Al.[30] The aryldiazomethanes can be generated *in situ* from the corresponding tosyl hydrozones.[31]

[25]Linder, M. R.; Steurer, S.; Podlech, J. *Org. Synth.* **2003**, *79*, 154–164.

[26]Lee, V.; Newman, M. S. *Org. Synth.* **1970**, *50*, 77–80.

[27]Proctor, L. D.; Warr, A. J. *Org. Process Res. Dev.* **2002**, *6*, 884–892.

[28]Kowalski, C. J.; Reddy, R. E. *J. Org. Chem.* **1992**, *57*, 7194–7208.

[29]Loeschorn, C. A.; Nakajima, M.; McCloskey, P. J.; Anselme, J. P. *J. Org. Chem.* **1983**, *48*, 4407–4410.

[30]Maruoka, K.; Concepcion, A. B.; Yamamoto, H. *Synthesis* **1994**, 1283–1290.

[31]Angle, S. R.; Neitzel, M. L. *J. Org. Chem.* **2000**, *65*, 6458–6461.

For the insertion of a $CH_2$ group, diazomethane can be used in the above procedures. A number of safer alternatives to diazomethane have also been developed for use in this reaction including $TMSCHN_2$[32] and ethyl diazoacetate.[33]

Some methods involving a similar rearrangement without the use of diazo compounds have also been reported.[34-36]

[32]Aoyama, T.; Shioiri, T. *Synthesis* **1988**, 228–229.
[33]Dave, V.; Warnhoff, E. W. *J. Org. Chem.* **1983**, *48*, 2590–2598.
[34]Villieras, J.; Perriot, P.; Normant, J. F. *Synthesis* **1979**, 968–970.
[35]Katritzky, A. R.; Toader, D.; Xie, L. *J. Org. Chem.* **1996**, *61*, 7571–7577.
[36]Satoh, T.; Mizu, Y.; Kawashima, T.; Yamakawa, K. *Tetrahedron* **1995**, *51*, 703–710.

### 7.2.1.10  Fritsch–Buttenberg–Wiechell  Rearrangement:  Acetylenes  from  1,1-Disubstituted Olefins

The Fritsch–Buttenberg–Wiechell rearrangement is a synthesis of acetylenes from 1,1-disubstituted-2-haloolefins. The reaction is useful for converting asymmetrically substituted ketones, which can be readily accessed, into asymmetrically substituted acetylenes.

In general, the R groups are aryl groups, olefins, or acetylenes. The example below depicts the synthesis of diynes using this method.[37]

The rearrangement can be combined with the formation of the 1,1-disubstituted olefin, as shown in the example below.[38]

### 7.2.1.11  Other Carbon to Carbon Migrations of Carbon

An oxidative ring contraction using hydrogen peroxide and sulfuric acid has been accomplished in the ring contraction below.[39]

[37]Shi Shun, A. L. K.; Chernick, E., T.; Eisler, S.; Tykwinski, R. R. *J. Org. Chem.* **2003**, *68*, 1339–1347.
[38]Mouries, V.; Waschbuesch, R.; Carran, J.; Savignac, P. *Synthesis* **1998**, 271–274.
[39]Challenger, S.; Derrick, A.; Silk, T. V. *Synth. Commun.* **2002**, *32*, 2911–2918.

***7.2.1.12    Other Carbon to Carbon Migrations of Hydrogen***    The anion migration shown below was run on a 50 g scale.[40]

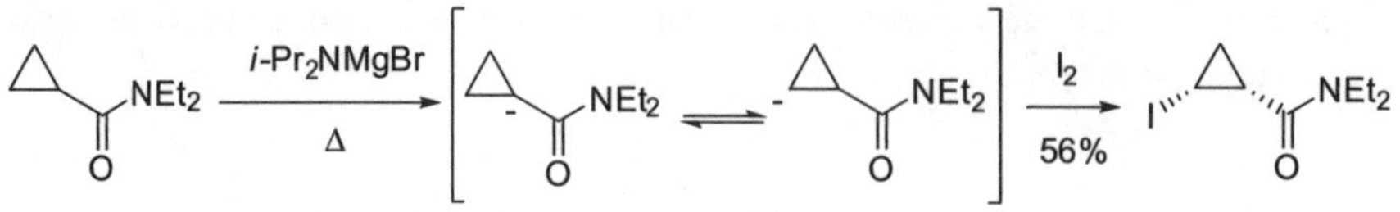

## 7.2.2    Carbon to Carbon Migrations of Other Groups

***7.2.2.1    Migration of Halogen, Hydroxy, Amino, and Other Groups***    The migration of a heteroatom from carbon to carbon is not uncommon. A variety of migrations is shown below, but it is not all encompassing. Heteroatoms can migrate to a radical, cationic, or anionic center to create a more stable intermediate, which then goes on to further reactions.[41,42]

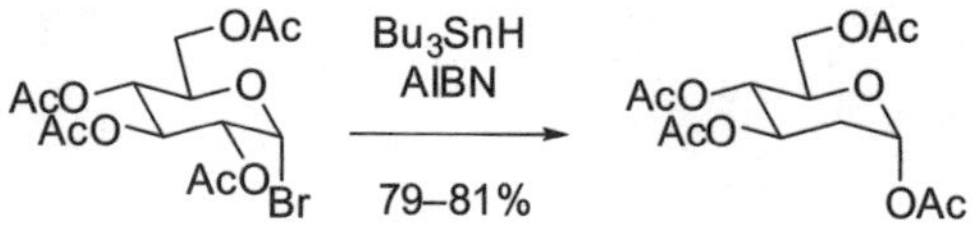

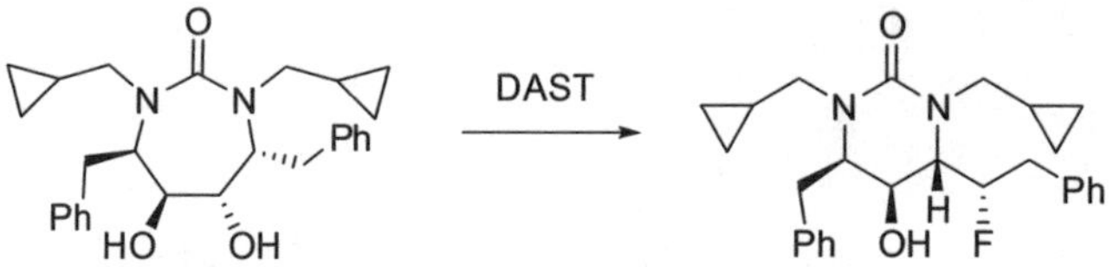

The Willgerodt reaction is a useful method for the migration of a carbonyl oxygen to the end of a carbon chain, resulting in a carboxylic acid. This rearrangement was utilized early in the development of naproxen on 500 kg scale.[43,44]

[40]Cai, S.; Dimitroff, M.; McKennon, T.; Reider, M.; Robarge, L.; Ryckman, D.; Shang, X.; Therrien, J. *Org. Process Res. Dev.* **2004**, *8*, 353–359.

[41]Giese, B.; Groeninger, K. S. *Org. Synth.* **1990**, *69*, 66–71.

[42]Lucca, G. V. D. *J. Org. Chem.* **1998**, *63*, 4755–4766.

[43]Harrington, P. J.; Lodewijk, E. *Org. Process Res. Dev.* **1997**, *1*, 72–76.

[44]Harrison, I. T.; Lewis, B.; Nelson, P.; Rooks, W.; Roszkowski, A.; Tomolonis, A.; Fried, J. H. *J. Med. Chem.* **1970**, *13*, 203–205.

The Rupe rearrangement also involves the migration of oxygen, in this case a propargylic alcohol.[45]

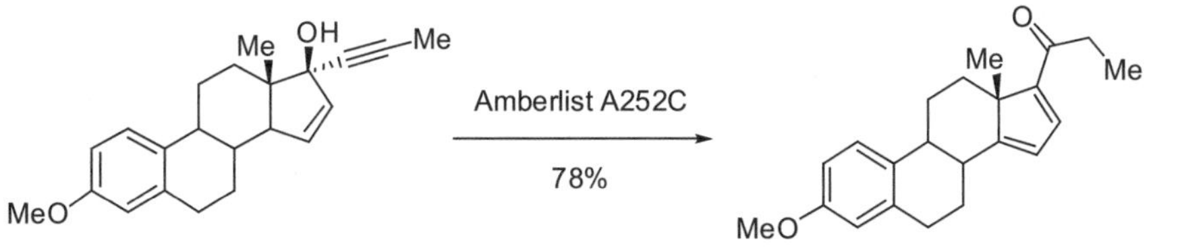

### 7.2.2.2 Neber Rearrangement: Carbon to Carbon Migration of Nitrogen of Activated Oximes

The Neber rearrangement involves the enolization of an activated oxime, displacement of the leaving group by the enolate to form an azirine, and hydrolysis or ketalization of the azirine to the α-aminoketone or α-aminoketal. A Neber rearrangement has been utilized to synthesize a 3-pyridylaminomethyl ketal.[46–48] While this was an effective method for the synthesis of kilogram quantities of the material, it has been noted that *the tosyloxime intermediate in the rearrangement reaction is shock sensitive*[46] and decomposes at a low temperature via a Beckmann rearrangement (Section 7.2.3.5).[49] One solution to the instability of this intermediate is to keep the intermediate in solution and bring it directly into the rearrangement reaction.

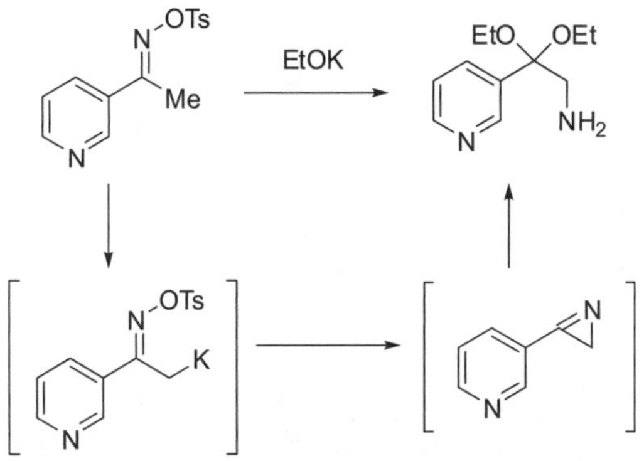

The Neber rearrangement can also be executed on hydrazonium salts as shown below.[50]

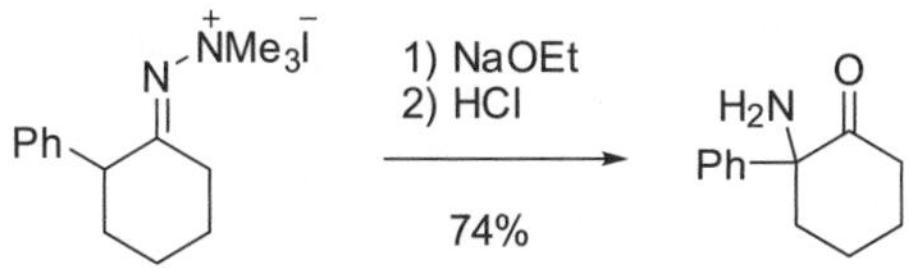

[45]Tilstam, U.; Weinmann, H. *Org. Process Res. Dev.* **2002**, *6*, 384–393.

[46]Chung, J. Y. L.; Ho, G.-J.; Chartrain, M.; Roberge, C.; Zhao, D.; Leazer, J.; Farr, R.; Robbins, M.; Emerson, K.; Mathre, D. J.; McNamara, J. M.; Hughes, D. L.; Grabowski, E. J. J.; Reider, P. J. *Tetrahedron Lett.* **1999**, *40*, 6739–6743.

[47]LaMattina, J. L.; Suleske, R. T. *Org. Synth.* **1986**, *64*, 19–26.

[48]LaMattina, J. L.; Suleske, R. T. *Synthesis* **1980**, 329–330.

[49]am Ende, D. J.; Brown Ripin, D. H.; Weston, N. P. *Thermochim. Acta* **2004**, *419*, 83–88.

[50]Parcell, R. F.; Sanchez, J. P. *J. Org. Chem.* **1981**, *46*, 5229–5231.

An interesting variant on the procedure was reported directly from the primary amine. The authors note that the dichloroamine intermediate explodes at 100°C, so caution must be used when utilizing this procedure.[51]

### 7.2.2.3 Payne Rearrangement: Rearrangement of α-Hydroxyepoxides

The Payne rearrangement is the transformation of a 1-hydroxy-2,3-epoxide to a 3-hydroxy-1,2-epoxide. The rearrangement is under thermodynamic control, but the rearrangement can be synthetically useful when the rearranged epoxide reacts preferentially to the starting material.[52]

In an analogous rearrangement, an intermediate sulfate ester rendered the reaction irreversible and provided a good yield of product.[22]

### 7.2.3  Carbon to Nitrogen Migrations of Carbon

### 7.2.3.1 Hofmann Rearrangement: Primary Amides to Amines or Carbamates

The Hofmann rearrangement transforms a primary amide into a carbamate, essentially transposing the carbon and nitrogen of the amide functionality. The mechanism of the reaction proceeds via oxidation of the amide nitrogen to install a leaving group, typically a halide, followed by [1,2]-migratory displacement of the leaving group by the carbon appended to the carbonyl, resulting in an isocyanate, which is then quenched with a nucleophile, typically an alcohol.

Two preparations demonstrate the utility of the Hofmann rearrangement utilizing different oxidants: PhI(OTFA)$_2$[53] and NBS, DBU.[54]

[51]Coffen, D. L.; Hengartner, U.; Katonak, D. A.; Mulligan, M. E.; Burdick, D. C.; Olson, G. L.; Todaro, L. J. *J. Org. Chem.* **1984**, *49*, 5109–5113.
[52]Behrens, C. H.; Ko, S. Y.; Sharpless, K. B.; Walker, F. J. *J. Org. Chem.* **1985**, *50*, 5687–5696.
[53]White, E. *Org. Synth.* **1973**, *Coll. Vol. V*, 336–339.
[54]Keillor, J. W.; Huang, X. *Org. Synth.* **2002**, *78*, 234–238.

Hofmann rearrangements are used on large scale to afford a transformation similar to that of the Curtius (Section 7.2.3.2), without the intermediacy of an acyl azide. Common oxidants used on commercial scale are the inexpensive and easy to handle NaOCl and NaOBr.[55] The second transformation shown is noteworthy as it produces an *N*-aminomethyl amide.[56]

A Hofmann rearrangement was utilized to convert 5-cyanovaleramide to methyl *N*-cyanobutyl carbamate using bromine as the oxidant.[57] In this procedure, the brominated amide intermediate was added to refluxing methanol to effect rearrangement.

The reaction below was run on over 23 kg material;[58] in this case iodosobenzene was found to provide superior results to hypochlorites and other common oxidants.

[55]Hoekstra, M. S.; Sobieray, D. M.; Schwindt, M. A.; Mulhern, T. A.; Grote, T. M.; Huckabee, B. K.; Hendrickson, V. S.; Franklin, L. C.; Granger, E. J.; Karrick, G. L. *Org. Process Res. Dev.* **1997**, *1*, 26–38.

[56]Ogasa, T.; Ikeda, S.; Sato, M.; Tamaoki, K. *Preparation of N-(2,2,5,5-tetramethylcyclopentanecarbonyl)-(S)-1,1-diaminoethane p-toluenesulfonate as a sweetener intermediate.* JP02233651 1990, 5 pp.

[57]Faul, M. M.; Ratz, A. M.; Sullivan, K. A.; Trankle, W. G.; Winneroski, L. L. *J. Org. Chem.* **2001**, *66*, 5772–5782.

[58]Zhang, L.-h.; Chung, J. C.; Costello, T. D.; Valvis, I.; Ma, P.; Kauffman, S.; Ward, R. *J. Org. Chem.* **1997**, *62*, 2466–2470.

***7.2.3.2   Curtius Rearrangement: Acyl Azides to Amines or Carbamates***   The Curtius rearrangement effects essentially the same transformation as the Hofmann rearrangement. In this process, an acyl azide, generally formed either from the acid chloride and sodium azide, or through oxidation of an acyl hydrazide, rearranges to an isocyanate, which is then trapped with a nucleophile, typically an alcohol. The transformation below depicts the rearrangement of a dienyl substrate that would not be compatible with the conditions of the Hofmann rearrangement.[59]

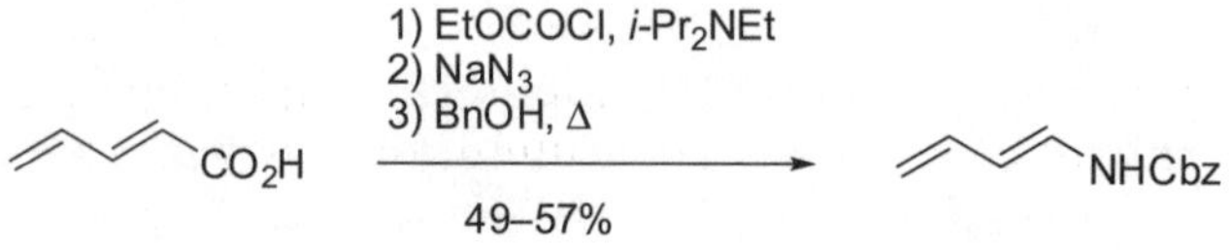

$$\text{1) EtOCOCl, } i\text{-Pr}_2\text{NEt}$$
$$\text{2) NaN}_3$$
$$\text{3) BnOH, } \Delta$$

49–57%

A number of examples of Curtius rearrangements are reported in the literature. A well-studied Curtius rearrangement in which the issues critical to running the process safely were identified is depicted below.[60,61]

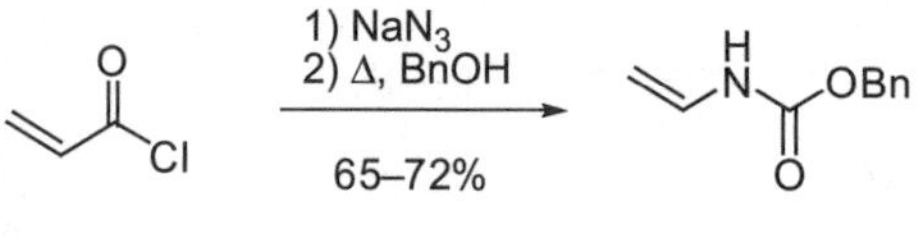

1) NaN$_3$
2) $\Delta$, BnOH

65–72%

As the Curtius rearrangement involves the intermediacy of an acyl azide that must be heated to promote rearrangement, extensive safety testing was undertaken in order to run this reaction safely on a large scale. The key to safe operation in this case was the controlled addition of a solution of acyl azide to a heated solution of benzyl alcohol in toluene in order to keep the quantity of azide being heated at any one time to a minimum. The decomposition of azide to isocyanate and reaction with benzyl alcohol to produce product is rapid, and thus the addition rate controls the rate of decomposition of the azide intermediate. Addition of potassium carbonate to the benzyl alcohol solution minimized an acid-catalyzed addition of benzyl alcohol to the product that produced an aminal side product. A modification was made to the procedure in which the acyl azide is added to hot toluene and the resultant isocyanate was distilled into a receiving vessel containing benzyl alcohol at lower (0°C) temperature. This process also reduced the amount of side product produced and provided the product in high enough purity to crystallize directly from the reaction mixture.

There are also some large scale examples of Curtius rearrangements wherein the acyl azide was generated by oxidation of an acyl hydrazine. 3-Phenoxypropionyl hydrazide was oxidized to the acyl azide in solution using NaNO$_2$ and HCl; this acyl azide was then rearranged by slowly adding the acyl azide to a hot solvent to effect rearrangement.[62] An ergoline derivative was synthesized using a similar procedure.[63]

[59]Jessup, P. J.; Petty, C. B.; Roos, J.; Overman, L. E. *Org. Synth.* **1980**, *59*, 1–9.

[60]am Ende, D. J.; DeVries, K. M.; Clifford, P. J.; Brenek, S. J. *Org. Process Res. Dev.* **1998**, *2*, 382–392.

[61]Govindan, C. K. *Org. Process Res. Dev.* **2002**, *6*, 74–77.

[62]Madding, G. D.; Smith, D. W.; Sheldon, R. I.; Lee, B. *J. Heterocycl. Chem.* **1985**, *22*, 1121–1126.

[63]Baenziger, M.; Mak, C. P.; Muehle, H.; Nobs, F.; Prikoszovich, W.; Reber, J. L.; Sunay, U. *Org. Process Res. Dev.* **1997**, *1*, 395–406.

### 7.2.3.3  Lossen Rearrangement: Hydroxamic Imides to Amines or Carbamates

O-Activated hydroxamic acids undergo Lossen rearrangement to produce carbamates or ureas after reaction of the isocyanate intermediate with an alcohol or amine.

Imides of hydroxylamine also undergo Lossen rearrangement to carbamates in a transformation similar to the Curtius or Hofmann rearrangement. In the reaction below, the utility of N-Boc-O-mesyl hydroxylamine for effecting the Lossen rearrangement of activated carboxylic acids is demonstrated.[64] N,O-Bis(ethoxycarbonyl)hydroxylamine has also been demonstrated to be a useful reagent for this transformation.[65]

NMM = N-methyl morpholine

A Lossen rearrangement was employed in the synthesis of benz[cd]indol-2(1H)-one and a derivative thereof.[66] Safety concerns drove the workers to employ o,p-dinitrophenol rather than chloride as a leaving group.

[64]Stafford, J. A.; Gonzales, S. S.; Barrett, D. G.; Suh, E. M.; Feldman, P. L. *J. Org. Chem.* **1998**, *63*, 10040–10044.
[65]Anilkumar, R.; Chandrasekhar, S.; Sridhar, M. *Tetrahedron Lett.* **2000**, *41*, 5291–5293.
[66]Marzoni, G.; Varney, M. D. *Org. Process Res. Dev.* **1997**, *1*, 81–84.

***7.2.3.4   Schmidt Rearrangement: Ketones to Amides***   The ketone to amide transformation achieved in the Beckmann rearrangement (Section 7.2.3.5) can also be affected by treating a ketone with azide under acidic conditions, although the Beckmann rearrangement is preferred. The azidoalcohol initially formed rearranges to an amide with displacement of nitrogen.[67]

A modified procedure for this transformation has been published; in this case, the rearrangement is photochemically induced.[68]

Interestingly, the reaction can be carried out in an intramolecular fashion with an azidoketone substrate.[69,70] This is a nice method for taking relatively simple aliphatic azides and converting them into complex nitrogen-containing ring systems.

Trapping the azide in an intramolecular fashion with a carbocation instead of a ketone results in the formation of an amine after reduction of the intermediate imine.[71]

[67]Galvez, N.; Moreno-Manas, M.; Sebastian, R. M.; Vallribera, A. *Tetrahedron* **1996**, *52*, 1609–1616.

[68]Brands, K. M. J.; Payack, J. F.; Rosen, J. D.; Nelson, T. D.; Candelario, A.; Huffman, M. A.; Zhao, M. M.; Li, J.; Craig, B.; Song, Z. J.; Tschaen, D. M.; Hansen, K.; Devine, P. N.; Pye, P. J.; Rossen, K.; Dormer, P. G.; Reamer, R. A.; Welch, C. J.; Mathre, D. J.; Tsou, N. N.; McNamara, J. M.; Reider, P. J. *J. Am. Chem. Soc.* **2003**, *125*, 2129–2135.

[69]Iyengar, R.; Schildknegt, K.; Morton, M.; Aube, J. *J. Org. Chem.* **2005**, *70*, 10645–10652.

[70]Mossman, C. J.; Aube, J. *Tetrahedron* **1996**, *52*, 3403–3408.

[71]Pearson, W. H.; Walavalkar, R. *Tetrahedron* **2001**, *57*, 5081–5089.

*7.2.3.5  Beckmann Rearrangement: Oximes to Amides*   The Beckmann rearrangement involves the rearrangement of an oxime to an amide through the migration of an alkyl group from the oxime carbon to nitrogen with displacement of the oxygen. Quenching the iminium ion with water produces an amide. This transformation is frequently used to expand carbocycles. The oxygen of the oxime must be activated for displacement, and this is typically accomplished through the use of an acid catalyst. These reactions tend to be very exothermic. A useful preparation using an aluminum catalyst is shown below.[72]

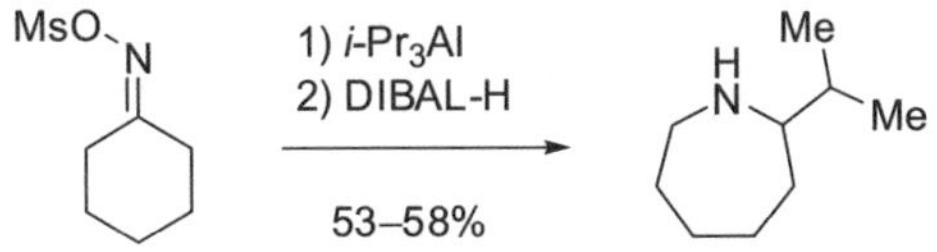

A Beckmann rearrangement was utilized in the synthesis of azithromycin. In this case, the alcohol is activated for displacement by sulfonylation. The facile Beckmann rearrangement of O-tosyloximes leads to low levels of stability in the intermediates used in Neber rearrangements (Section 7.2.2.2).[73,74]

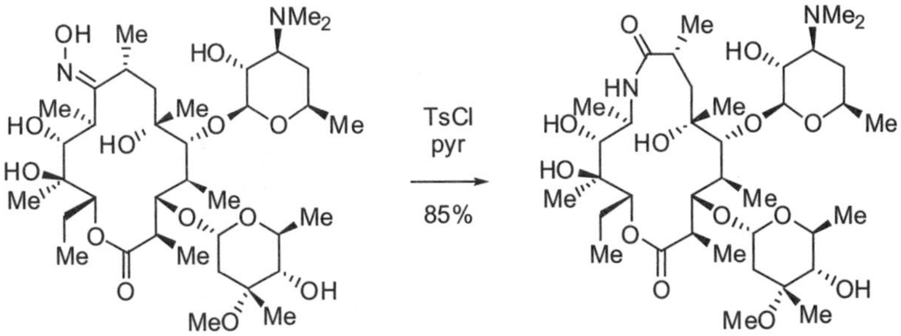

A Beckmann fragmentation results in some cases if there is an oxygen substituent $\alpha$ to the oxime. This fragmentation was used to transform milbemycin VM-44866 into SB-201561.[75] The five step sequence was carried out in 30% overall yield.

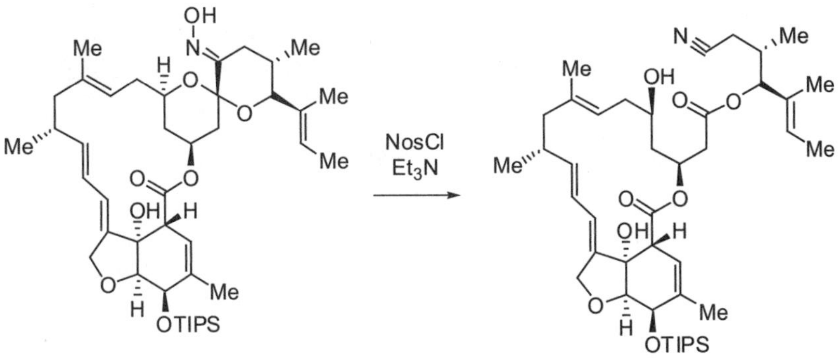

[72]Maruoka, K.; Nakai, S.; Yamamoto, H. *Org. Synth.* **1988**, *66*, 185–193.

[73]Kerdesky, F. A. J.; Premchandran, R.; Wayne, G. S.; Chang, S.-J.; Pease, J. P.; Bhagavatula, L.; Lallaman, J. E.; Arnold, W. H.; Morton, H. E.; King, S. A. *Org. Process Res. Dev.* **2002**, *6*, 869–875.

[74]Yang, B. V.; Goldsmith, M.; Rizzi, J. P. *Tetrahedron Lett.* **1994**, *35*, 3025–3028.

[75]Andrews, I. P.; Dorgan, R. J. J.; Harvey, T.; Hudner, J. F.; Hussain, N.; Lathbury, D. C.; Lewis, N. J.; Macaulay, G. S.; Morgan, D. O.; et al. *Tetrahedron Lett.* **1996**, *37*, 4811–4814.

The Tiemann rearrangement is a variant of the Beckmann rearrangement; in this case, the oxime of an amide is rearranged to produce a diimide product. A large excess of base is required for this transformation, limiting its utility.[76] This methodology is useful for synthesizing strained or asymmetrically substituted carbodiimides.

81%

### 7.2.3.6 Stieglitz Rearrangements and Related Reactions: Cationic Carbon to Nitrogen Migration of Carbon

The Stieglitz rearrangement is similar to the Wagner–Meerwein rearrangement, but with a nitrogen-centered cation. The cation is traditionally generated from the N-chloride with a silver salt. Activation of a hydroxylamine as the sulfonate ester is also sufficient to initiate rearrangement and has obvious advantages over the silver method.[77]

48%

The sequence below depicts a Diels–Alder reaction followed by Stieglitz rearrangement.[78]

42%

### 7.2.4 Carbon to Oxygen Migrations of Carbon

### 7.2.4.1 Baeyer–Villiger Oxidation: Ketones or Aldehydes to Esters

The Baeyer–Villiger oxidation is widely used and selection of reagents and reaction conditions can often have an effect on the selectivity of rearrangement. Peroxides such as peracids, hydrogen peroxide, and inorganic peroxides can be used in this reaction, and a variety of oxidation compatible solvents can be utilized.

Isatin can be converted selectively to either isatoic anhydride or 2,3-dioxo-1,4-benzoxazine.[79] The product obtained is dependent on the oxidant used; hydrogen peroxide in acetic

[76]Richter, R.; Tucker, B.; Ulrich, H. *J. Org. Chem.* **1983**, *48*, 1694–1700.
[77]Hoffman, R. V.; Kumar, A.; Buntain, G. A. *J. Am. Chem. Soc.* **1985**, *107*, 4731–4736.
[78]Renslo, A. R.; Danheiser, R. L. *J. Org. Chem.* **1998**, *63*, 7840–7850.
[79]Reissenweber, G.; Mangold, D. *Angew. Chem., Int. Ed. Engl.* **1980**, *92*, 196–197.

acid leads to formation of anhydride while the use of $K_2S_2O_8$ in sulfuric acid results in production of the benzoxazine.

79%                    95%

In general, there is selectivity for more highly substituted carbons to migrate. The compatibility of functionalities such as amines[80] and epoxides[81] are demonstrated by the examples below. In the case of the synthesis of a prostaglandin analog, the generation of *N*-oxides as intermediates in the reaction appears to improve selectivity for the desired lactone over the undesired, and careful control of conditions resulted in the hydrolysis of the undesired lactone in the presence of desired one, thereby facilitating purification.

60–63%

R = p-PhPh

*m*-CPBA

87%

Finally, a notable example of subtle regiochemical control by selection of reagents is presented.[82] Using trifluoroperacetic acid as the oxidant, generated *in situ* from urea–hydrogen peroxide complex (UHP) and trifluoroacetic anhydride (TFAA), very high (98:2) selectivity for the desired product was observed at 80% conversion. Using other oxidant systems, the selectivity was notably lower.

TFAA, UHP

[80]Coleman, M. J.; Crookes, D. L.; Hill, M. L.; Singh, H.; Marshall, D. R.; Wallis, C. J. *Org. Process Res. Dev.* **1997**, *1*, 20–25.

[81]Flisak, J. R.; Gombatz, K. J.; Holmes, M. M.; Jarmas, A. A.; Lantos, I.; Mendelson, W. L.; Novack, V. J.; Remich, J. J.; Snyder, L. *J. Org. Chem.* **1993**, *58*, 6247–6254.

[82]Varie, D. L.; Brennan, J.; Briggs, B.; Cronin, J. S.; Hay, D. A.; Rieck, J. A., III; Zmijewski, M. J. *Tetrahedron Lett.* **1998**, *39*, 8405–8408.

### 7.2.5  Heteroatom to Carbon Migrations

#### 7.2.5.1  Stevens Rearrangement: [1,2]-Rearrangement of Ammonium, Oxonium, or Sulfur Ylides with Migration of Carbon

*Sulfur Ylides with Migration of Carbon*    The Stevens rearrangement is the [1,2]-migration of a carbon substituent from a tetrasubstituted ammonium salt, a sulfur ylide, or an oxonium ylide. The rearrangement requires a strong base to deprotonate a carbon adjacent to the heteroatom. As more than one substituent of the ammonium salt frequently has a proton that can be abstracted, mixtures of products are not unusual in this reaction. Further, if one of the substituents is allylic or benzylic, the potential for a [2,3]-rearrangement (the Sommelet–Hauser rearrangement in the case of an aryl group, Section 7.3.2.10) to compete exists.

There are examples of Stevens rearrangements being used in the literature. The ammonium ylide is generated by any of a number of methods, including deprotonation with a strong base such as NaNH$_2$,[83] treating a methyltrimethylsilane group with fluoride,[5] or the reaction of a tertiary amine with a diazonium.[84]

Oxonium ylides generated from the diazo compound can also undergo a Stevens rearrangement, although this reaction is also plagued by the formation of side products.[85]

[83]Elmasmodi, A.; Cotelle, P.; Barbry, D.; Hasiak, B.; Couturier, D. *Synthesis* **1989**, 327–329.
[84]West, F. G.; Naidu, B. N. *J. Org. Chem.* **1994**, *59*, 6051–6056.
[85]Eberlein, T. H.; West, F. G.; Tester, R. W. *J. Org. Chem.* **1992**, *57*, 3479–3482.

Interestingly, it was found that small variations in the reaction conditions for the rearrangement of a benzylic ammonium system could drive the reaction via a Stevens or a Sommelet–Hauser pathway.[86]

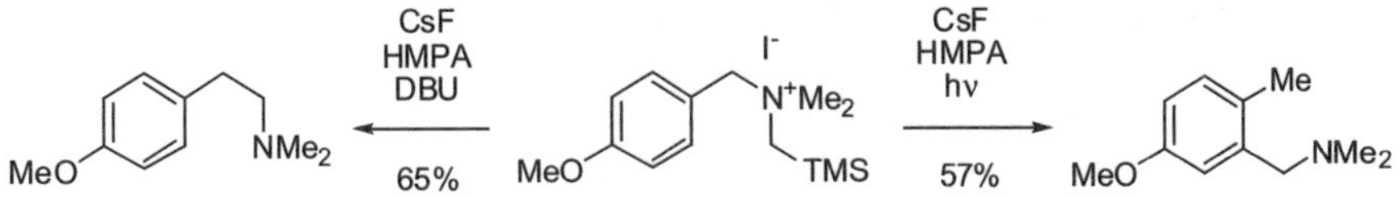

Ring contraction has been achieved on a sulfur ylide system as shown below.[87]

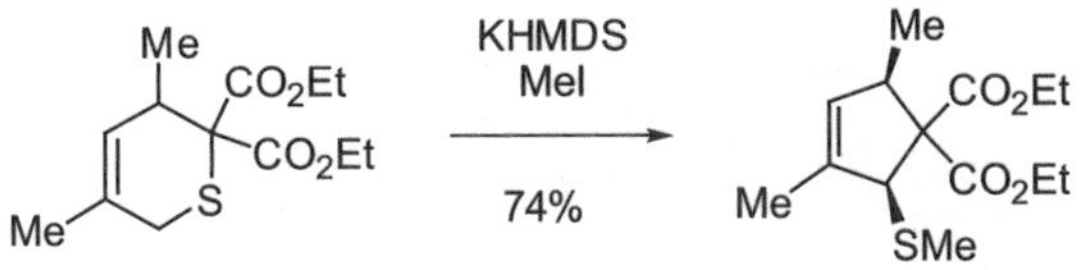

### 7.2.5.2 *[1,2]-Meisenheimer Rearrangement*: $R_3NO$ to $R_2NOR$

The Meisenheimer rearrangement of *N*-oxides proceeds via a similar mechanism as the Stevens rearrangement (Section 7.2.5.1). This reaction can be synthetically useful for converting an amine into an aminoalcohol as shown below, or can be a side reaction if another pathway of reaction is desired for the *N*-oxide.[88]

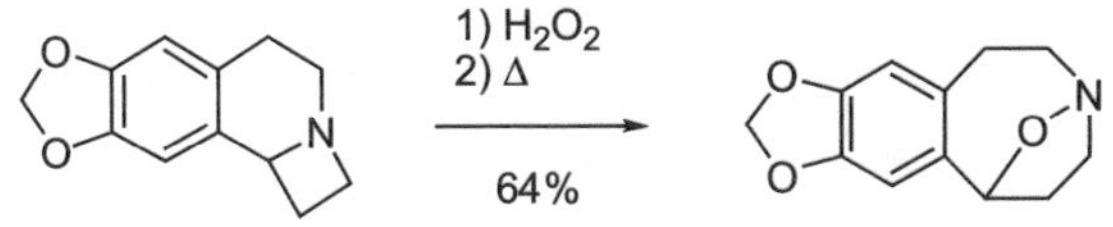

In another example, rearrangement occurs at low temperature once the substrate is oxidized.[89]

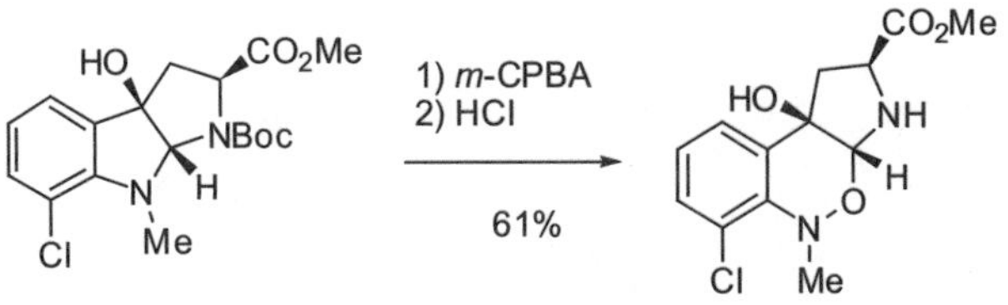

### 7.2.5.3 *Other Heteroatom to Carbon Rearrangements*

The sulfenyl transfer reaction below was useful in the preparation of a geminal aminosulfide in the synthesis of an antibiotic.[90]

[86]Tanaka, T.; Shirai, N.; Sugimori, J.; Sato, Y. *J. Org. Chem.* **1992**, *57*, 5034–5036.

[87]Larsen, R. D.; Corley, E. G.; King, A. O.; Carroll, J. D.; Davis, P.; Verhoeven, T. R.; Reider, P. J.; Labelle, M.; Gauthier, J. Y.; et al. *J. Org. Chem.* **1996**, *61*, 3398–3405.

[88]Yoneda, R.; Sakamoto, Y.; Oketo, Y.; Harusawa, S.; Kurihara, T. *Tetrahedron* **1996**, *52*, 14563–14576.

[89]Didier, C.; Critcher, D. J.; Walshe, N. D.; Kojima, Y.; Yamauchi, Y.; Barrett, A. G. M. *J. Org. Chem.* **2004**, *69*, 7875–7879.

[90]Gordon, E. M.; Chang, H. W.; Cimarusti, C. M.; Toeplitz, B.; Gougoutas, J. Z. *J. Am. Chem. Soc.* **1980**, *102*, 1690–1702.

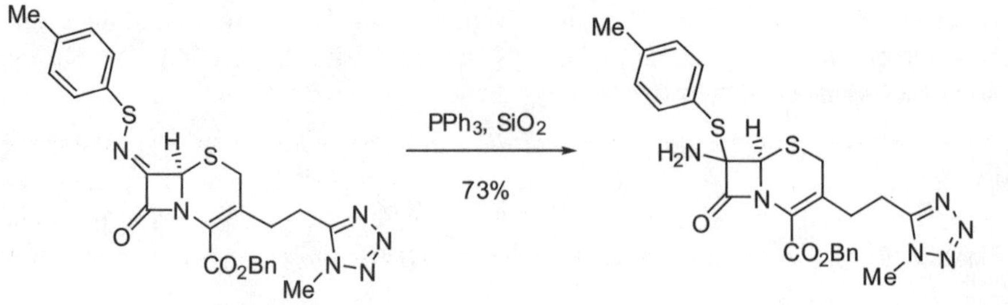

## 7.2.6    Carbon to Heteroatom Rearrangements

### 7.2.6.1    *Brook Rearrangement: Carbon to Oxygen Migration of Silicon*    The Brook rearrangement is the anionic [1,2]-shift of silicon from carbon to oxygen. The retro-Brook rearrangement is sometimes called the silicon-Wittig rearrangement and requires the use of strong bases such as alkyl lithium reagents or metal amides. The most common Brook rearrangement is the [1,2]-shift, but longer range shifts are also reported.[91] Some examples of Brook rearrangements are shown below.[92,93]

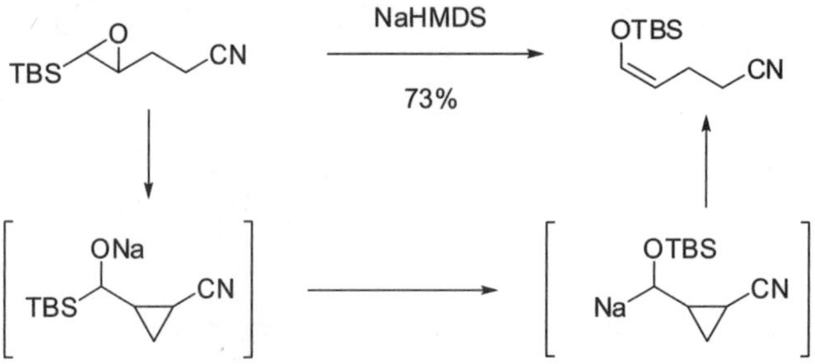

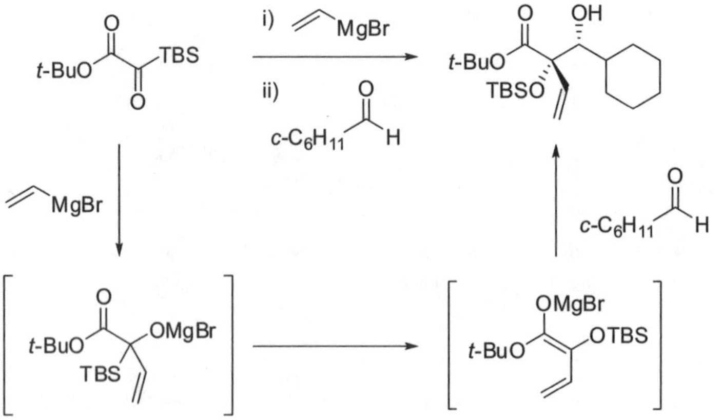

[91]Moser, W. H. *Tetrahedron* **2001**, *57*, 2065–2084.

[92]Okugawa, S.; Masu, H.; Yamaguchi, K.; Takeda, K. *J. Org. Chem.* **2005**, *70*, 10515–10523.

[93]Nicewicz, D. A.; Johnson, J. S. *J. Am. Chem. Soc.* **2005**, *127*, 6170–6171.

## 7.3  OTHER REARRANGEMENTS

### 7.3.1  Electrocyclic Rearrangements

***7.3.1.1  Rearrangements of Cyclobutenes and 1,3-Cyclohexadienes***    The thermal conversion of cyclobutenes to butadienes is a conrotatory process. In general the ring-opening process is of more utility than the reverse photochemically induced reaction of butadienes to the cyclobutene. The reaction is useful in the preparation of quinines as shown below[94]

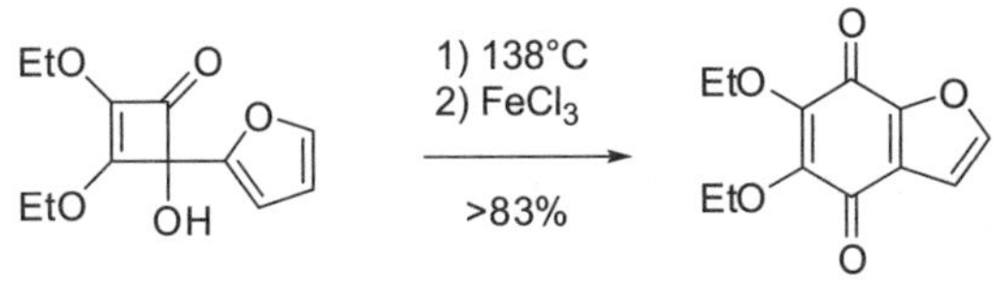

Cyclohexadienes can be converted to 1,3,5-trienes photochemically in a disrotatory process, the reverse reaction can be achieved thermally and is also useful. A useful application of this chemistry is to the synthesis of vitamin D analogs from cholesterol precursors.[95] The process shown involves photochemical ring opening, photochemical olefin inversion, and a thermally induced [1,7] sigmatropic hydride transfer reaction in the conversion of ergosterol to a vitamin D prodrug.

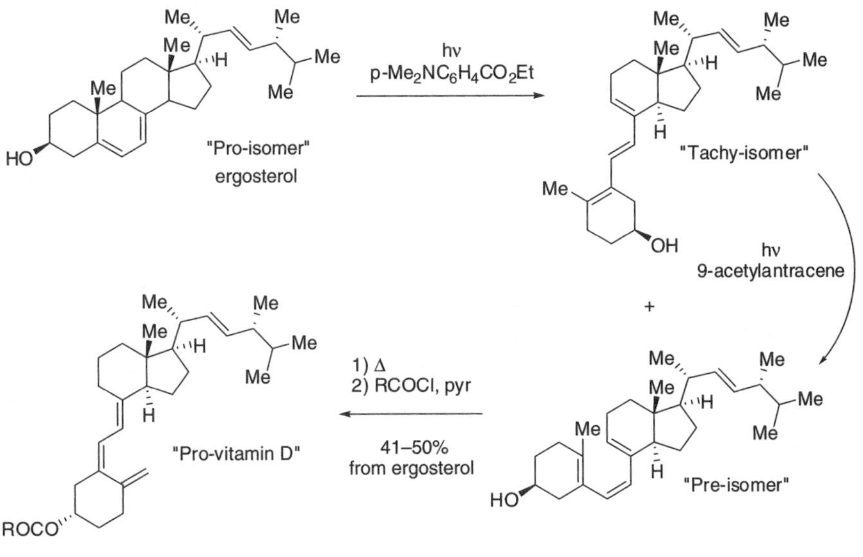

***7.3.1.2  Stilbenes to Phenanthrenes***    Stilbenes can be converted to phenanthrenes in a photochemically induced process followed by air oxidation. This reaction was utilized in a synthesis of steganone on 3 gram scale.[96]

---

[94]Perri, S. T.; Rice, P.; Moore, H. W. *Org. Synth.* **1990**, *69*, 220–225.

[95]Okabe, M. *Org. Synth.* **1999**, *76*, 275–286.

[96]Krow, G. R.; Damodaran, K. M.; Michener, E.; Wolf, R.; Guare, J. *J. Org. Chem.* **1978**, *43*, 3950–3953.

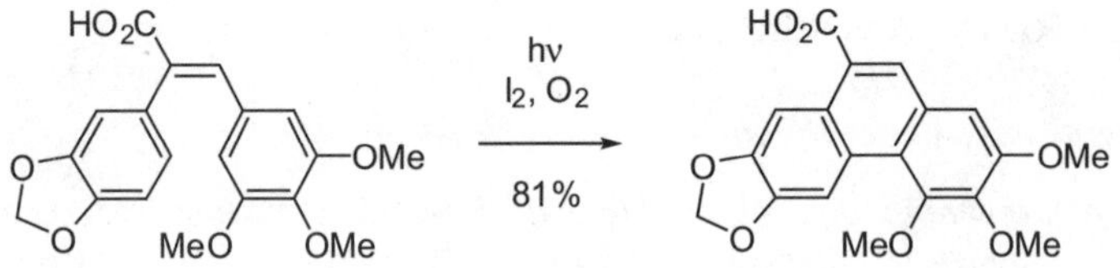

## 7.3.2   Sigmatropic Rearrangements

### 7.3.2.1   Cyclopropylimine Rearrangement   The cyclopropylimine rearrangement is a [1,3]-carbon shift from carbon to nitrogen. The rearrangement is a cationic process and proceeds under conditions where the imine is protonated.[97]

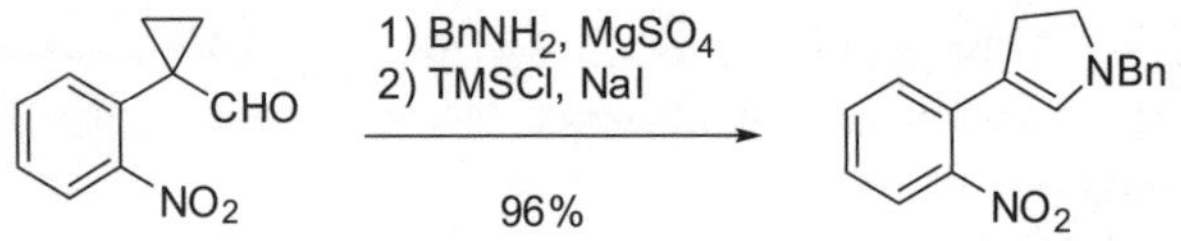

The reverse of this reaction can be accomplished photochemically and has been utilized in the preparation below.[98]

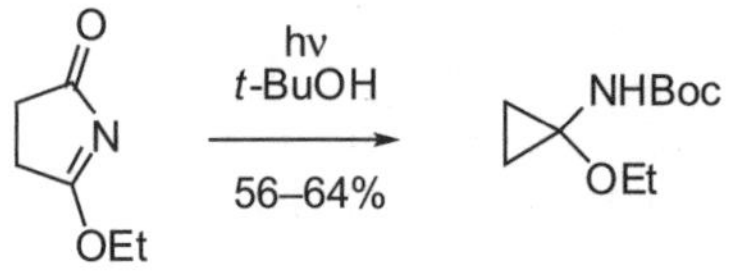

Nitrogen variants of the rearrangement can be used to construct heterocycles.[99] The route depicted below uses the Neber rearrangement (Section 7.2.2.2) to synthesize the azirine intermediate.

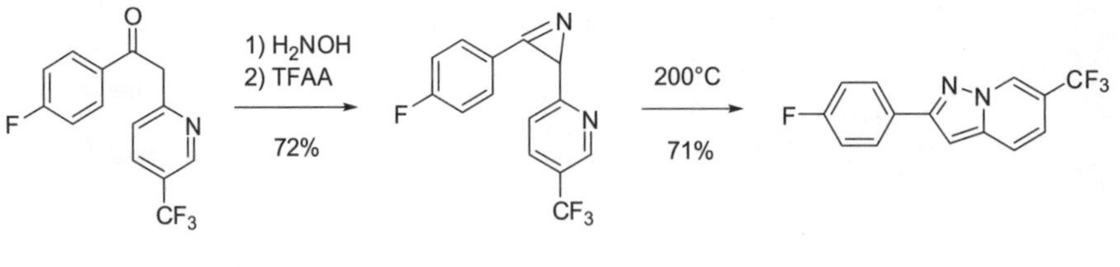

### 7.3.2.2   Cyclopropanes to Allenes: The Skattebøl and Related Rearrangements
1,1-Dihalocyclopropanes undergo rearrangement to allenes when treated with an alkyl lithium reagent at low temperature in a process known as the Doering–Moore–Skattebøl

[97]Rawal, V. H.; Michoud, C.; Monestel, R. *J. Am. Chem. Soc.* **1993**, *115*, 3030–3031.

[98]Crockett, G. C.; Koch, T. H. *Org. Synth.* **1980**, *59*, 132–140.

[99]Stevens, K. L.; Jung, D. K.; Alberti, M. J.; Badiang, J. G.; Peckham, G. E.; Veal, J. M.; Cheung, M.; Harris, P. A.; Chamberlain, S. D.; Peel, M. R. *Org. Lett.* **2005**, *7*, 4753–4756.

rearrangement. This process, which involves the intermediacy of a carbene, has been well reviewed,[100] and an example is below.[101]

When the cyclopropane is substituted with a vinyl group, cyclopentadienes are formed in a variant of the rearrangement known as the Skattebøl rearrangement.[102]

The same transformation can be accomplished thermally. In the case of a cyclopropene, the vinyl carbene can be trapped with an electron-deficient olefin as shown in the preparation below.[103] This process has also been well reviewed.[104]

### 7.3.2.3 Cope Rearrangement

The Cope rearrangement is a [3,3]-sigmatropic rearrangement of a 1,5 hexadiene under thermal conditions. The reaction is under thermodynamic control, and requires a driving force to go to completion. Two driving forces commonly used are ring strain relief[105] and re-aromatization. In the second example below, Claisen rearrangement (Section 7.3.2.4) is followed by Cope rearrangement to provide the desired product[106]

[100]Sydnes, L. K. *Chem. Rev.* **2003**, *103*, 1133–1150.
[101]Patterson, J. W. *J. Org. Chem.* **1990**, *55*, 5528–5531.
[102]Paquette, L. A.; McLaughlin, M. L. *Org. Synth.* **1990**, *68*, 220–226.
[103]Boger, D. L.; Brotherton, C. E.; Georg, G. I. *Org. Synth.* **1987**, *65*, 32–41.
[104]Baird, M. S. *Chem. Rev.* **2003**, *103*, 1271–1294.
[105]Limanto, J.; Snapper, M. L. *J. Am. Chem. Soc.* **2000**, *122*, 8071–8072.
[106]Newhouse, B. J.; Bordner, J.; Augeri, D. J.; Litts, C. S.; Kleinman, E. F. *J. Org. Chem.* **1992**, *57*, 6991–6995.

Much more frequently employed is the oxy-Cope rearrangement, where the driving force is the formation of an enol, which tautomerizes to the ketone. The rearrangement is greatly accelerated if the alcohol is deprotonated, resulting in an enolate product.[107]

An analog to the Cope rearrangement is the divinylcyclopropane rearrangement, which can be effectively utilized to prepare 7-membered rings.[108]

1 : 2.2

### 7.3.2.4   Claisen Rearrangement

The Claisen rearrangement is in close analogy to the Cope rearrangement, but in this case the 1,5-diene contains heteroatoms. A key difference from the Cope rearrangement is the thermodynamic stability of the newly formed products, thus leading to complete reaction. A number of variants of the Claisen rearrangement have been reported.[109–111] The example below is an example of a classical Claisen rearrangement.[112]

The rearrangement can occur with multiple heteroatoms in the system.[113]

[107]Paquette, L. A.; Poupart, M. A. *J. Org. Chem.* **1993**, *58*, 4245–4253.
[108]Wallock, N. J.; Bennett, D. W.; Siddiquee, T.; Haworth, D. T.; Donaldson, W. A. *Synthesis* **2006**, 3639–3646.
[109]Lutz, R. P. *Chem. Rev.* **1984**, *84*, 205–247.
[110]Castro, A. M. M. *Chem. Rev.* **2004**, *104*, 2939–3002.
[111]Ziegler, F. E. *Chem. Rev.* **1988**, *88*, 1423–1452.
[112]Vogel, D. E.; Buechi, G. H. *Org. Synth.* **1988**, *66*, 29–36.
[113]Pilgram, K. H.; Skiles, R. D.; Kleier, D. A. *J. Org. Chem.* **1988**, *53*, 38–41.

There are a number of variations on the classical Claisen rearrangement, some of which are detailed here.

The Eschenmoser–Claisen reaction is the reaction of an allylic alcohol with dimethyl acetamide dimethylacetal (DMA-DMA) to generate an enamine intermediate, which undergoes rearrangement and hydrolysis to result in the amide product.[114]

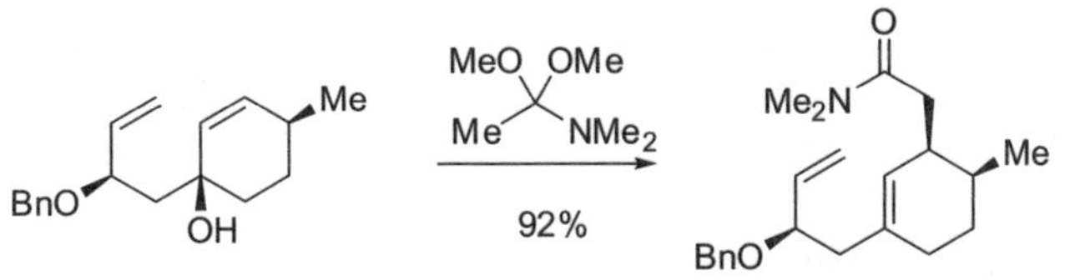

The Johnson–Claisen is analogous to the Eschenmoser–Claisen with an orthoester replacing the DMA-DMA. Two cases where this reaction was used to synthesize allenes are shown below.[115,116] The stereochemistry of the allene is determined by the stereochemistry of the alcohol.

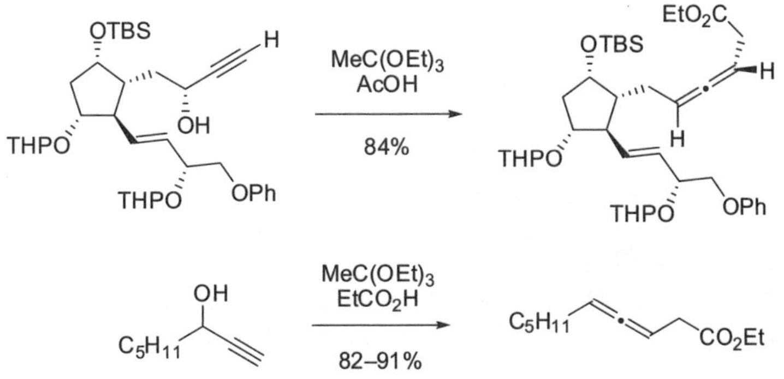

The Ireland–Claisen variant is the rearrangement of a silylketene acetal, as shown below.[117]

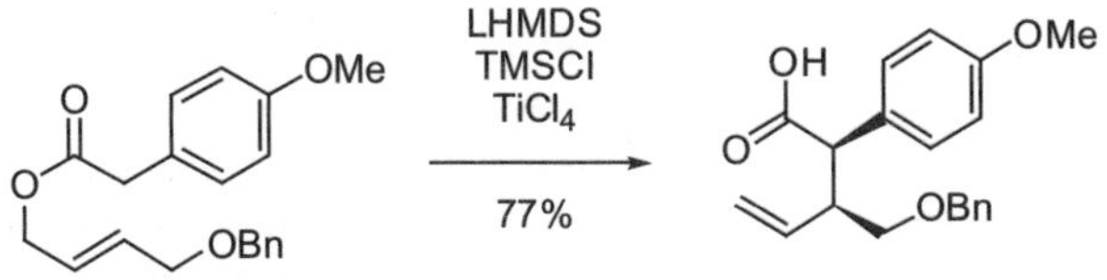

The Carroll rearrangement entails the rearrangement of the enolate of a β-ketoester.[118]

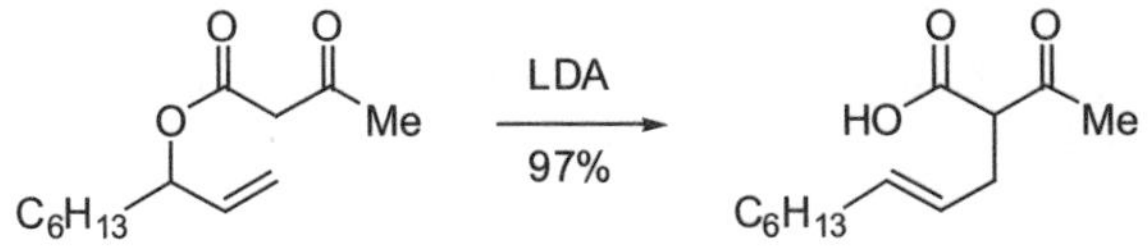

[114]Daniewski, A. R.; Wovkulich, P. M.; Uskokovic, M. R. *J. Org. Chem.* **1992**, *57*, 7133–7139.

[115]Cooper, G. F.; Wren, D. L.; Jackson, D. Y.; Beard, C. C.; Galeazzi, E.; Van Horn, A. R.; Li, T. T. *J. Org. Chem.* **1993**, *58*, 4280–4286.

[116]Tsuboi, S.; Masuda, T.; Mimura, S.; Takeda, A. *Org. Synth.* **1988**, *66*, 22–28.

[117]Koch, G.; Kottirsch, G.; Wietfeld, B.; Kuesters, E. *Org. Process Res. Dev.* **2002**, *6*, 652–659.

[118]Wilson, S. R.; Augelli, C. E. *Org. Synth.* **1990**, *68*, 210–219.

Claisen rearrangements on aromatic systems are usually followed by re-aromatization. In the case below, a propargylic phenol undergoes rearrangement despite the distance between the reacting centers.[119]

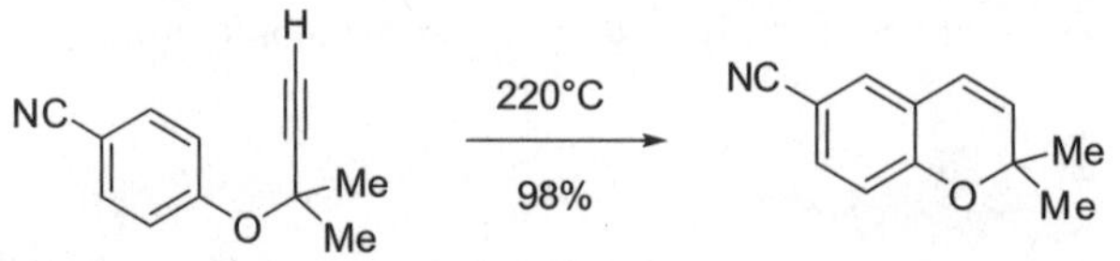

A few methods for the asymmetric catalysis of the Claisen reaction have been reported, such as the preparation below to produce chiral allylamines from trichloroacetimidates.[120]

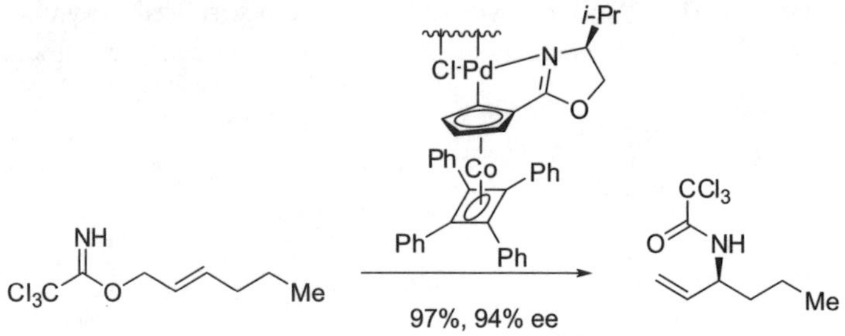

Asymmetric catalysis is possible due to an acceleration of the rate of rearrangement when the allylic oxygen is cationic. An example below demonstrates the facile rearrangements of allylic oxonium ions.[121]

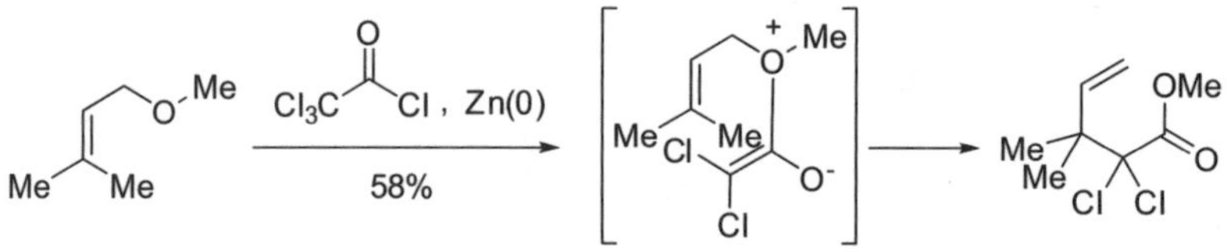

The stereochemical outcome of the Claisen and Cope rearrangements is determined by olefin and enolate geometry, stereocenters on the cyclic transition state, and exocyclic stereocenters including ring geometries. Some simplistic generalizations are discussed below, but do not take into consideration stereocenters outside the transition state, such as ring geometries that can affect the stereochemical outcome of the reaction. When possible, the reaction proceeds through a chair transition state. Which chair is accessed is dependent on the substitution patterns of the carbons in the 6-membered ring transition state. For the *trans/trans* case depicted below, $R_2$ and/or $R_4$ groups with a larger A values lead to higher selectivity for the indicated product.

[119]Bogaert-Alvarez, R. J.; Demena, P.; Kodersha, G.; Polomski, R. E.; Soundararajan, N.; Wang, S. S. Y. *Org. Process Res. Dev.* **2001**, *5*, 636–645.
[120]Anderson, C. E.; Overman, L. E.; Watson, M. P.; Maddess, M. L.; Lautens, M. *Org. Synth.* **2005**, *82*, 134–139.
[121]Malherbe, R.; Rist, G.; Bellus, D. *J. Org. Chem.* **1983**, *48*, 860–869.

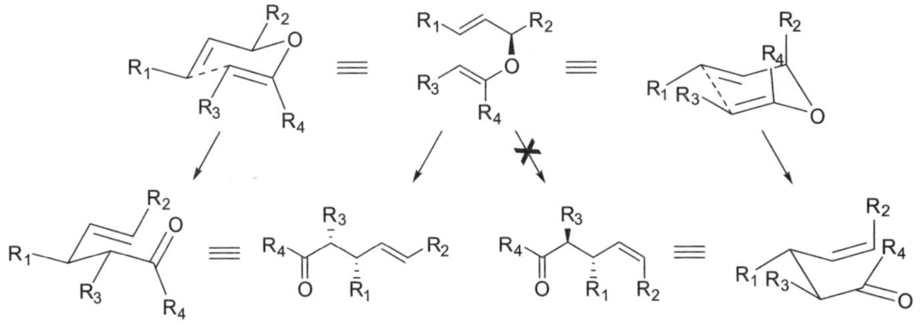

Variants of the Claisen rearrangement with nitrogen, sulfur, and other heteroatoms are also known. An example of the aza-Claisen rearrangement used to make quinolines is shown below.[122]

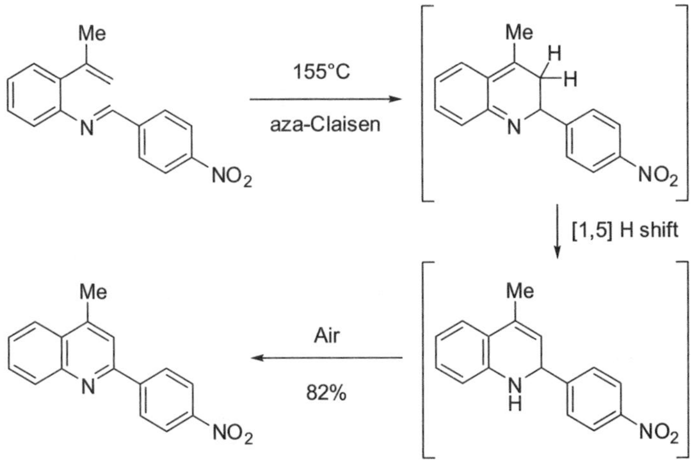

In the example below, the cationic aza-Claisen rearrangement is followed by a Mannich reaction to form the product shown.[123]

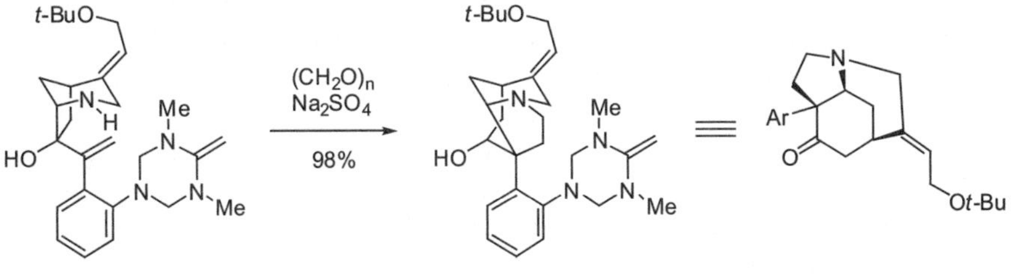

**7.3.2.5 Fischer Indole Synthesis**   The Fischer indole synthesis involves the [3,3]-sigmatropic rearrangement of an aryl hydrazone under acid catalysis and when combined with the hydrazone formation provides a convergent method for coupling an aldehyde or

[122]Qiang, L. G.; Baine, N. H. *J. Org. Chem.* **1988**, *53*, 4218–4222.
[123]Knight, S. D.; Overman, L. E.; Pairaudeau, G. *J. Am. Chem. Soc.* **1995**, *117*, 5776–5788.

ketone partner with an aryl hydrazine. Two examples of this process are depicted below.[124,125] For meta-substituted aryl rings, regioselectivity can be an issue in this process. Protic acids are commonly used in this reaction.

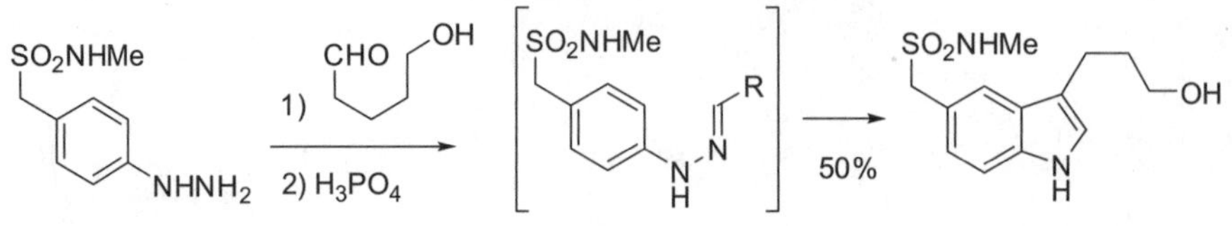

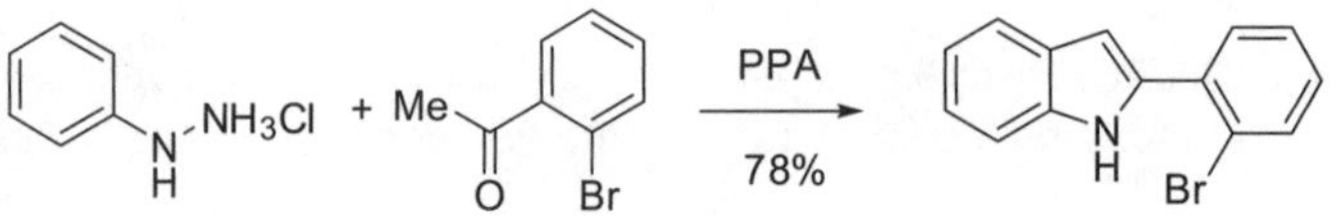

An oxo variant of the Fischer process is depicted below where an O-aryl oxime undergoes [3,3] sigmatropic rearrangement to produce the benzofuran shown.[126]

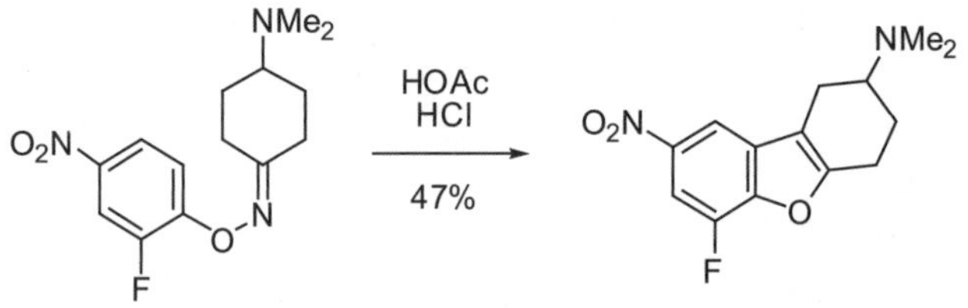

### 7.3.2.6    Boekelheide Rearrangement: 2-Alkylpyridine N-Oxide Rearrangements    The Boekelheide rearrangement is commonly used to functionalize the carbon substituent at the C-2 position of a pyridine. The N-oxide is acylated, and loss of a proton from the C-2 position sets up a [3,3]-sigmatropic shift to produce the product shown. Acetic anhydride is most frequently used for this transformation. This reaction is used frequently in industry to exploit the relatively facile N-oxide formation to functionalize the C-2 position under relatively mild conditions.[127]

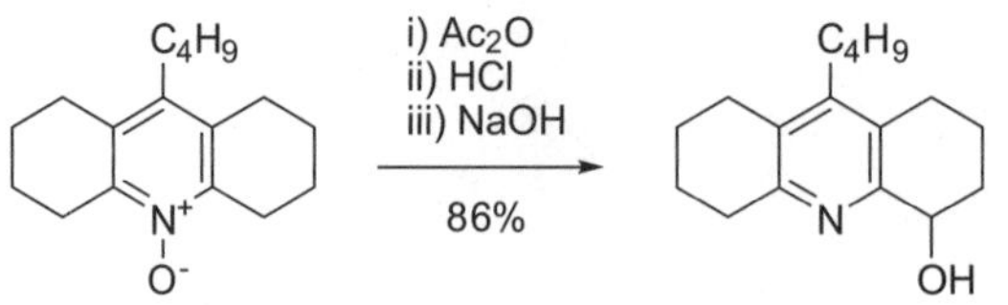

[124]Brodfuehrer, P. R.; Chen, B.-C.; Sattelberg, T. R., Sr., ; Smith, P. R.; Reddy, J. P.; Stark, D. R.; Quinlan, S. L.; Reid, J. G.; Thottathil, J. K.; Wang, S. *J. Org. Chem.* **1997**, *62*, 9192–9202.

[125]Ikemoto, N.; Liu, J.; Brands, K. M. J.; McNamara, J. M.; Reider, P. J. *Tetrahedron* **2003**, *59*, 1317–1325.

[126]Guzzo, P. R.; Buckle, R. N.; Chou, M.; Dinn, S. R.; Flaugh, M. E.; Kiefer, A. D., Jr., ; Ryter, K. T.; Sampognaro, A. J.; Tregay, S. W.; Xu, Y.-C. *J. Org. Chem.* **2003**, *68*, 770–778.

[127]Bell, T. W.; Cho, Y. M.; Firestone, A.; Healy, K.; Liu, J.; Ludwig, R.; Rothenberger, S. D. *Org. Synth.* **1990**, *69*, 226–237.

**7.3.2.7  [2,3]-Wittig Rearrangement: Oxygen to Carbon Shift of Carbon**    The [2,3]-Wittig rearrangement,[128] sometimes called the [2,3]-Wittig–Still rearrangement, involves the anionic rearrangement of allylic ethers as shown below. The reaction utilizes the relatively facile alcohol alkylation reaction, relaying that into a carbon to carbon bond-forming step. The requirement for strong bases to achieve this transformation limits functional group compatibility and solvent choice.[129,130]

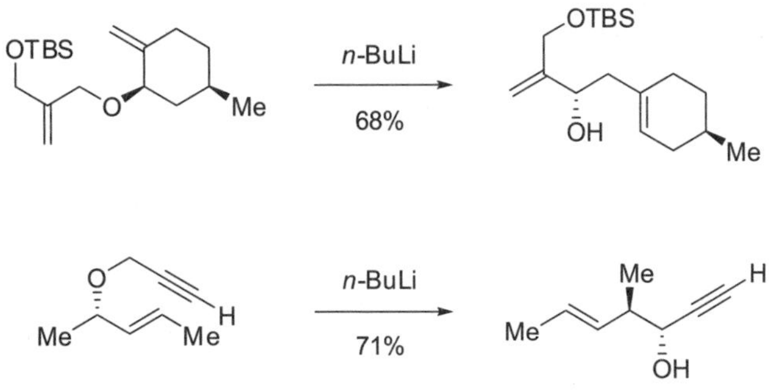

Electron-withdrawing groups or metal exchangeable groups such as a stannane are frequently appended to the ether to increase the acidity of the proton to be abstracted.[131,132]

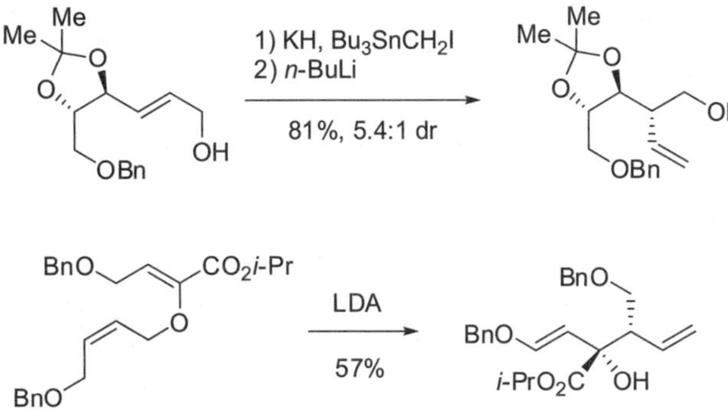

The stereochemical outcome of this reaction has been well studied, and predictive models now exist.[133,134]

[128]Nakai, T.; Mikami, K. *Chem. Rev.* **1986**, *86*, 885–902.
[129]Liang, J.; Hoard, D. W.; Khau, V. V.; Martinelli, M. J.; Moher, E. D.; Moore, R. E.; Tius, M. A. *J. Org. Chem.* **1999**, *64*, 1459–1463.
[130]Wovkulich, P. M.; Shankaran, K.; Kiegiel, J.; Uskokovic, M. R. *J. Org. Chem.* **1993**, *58*, 832–839.
[131]Ghosh, A. K.; Wang, Y. *Tetrahedron* **1999**, *55*, 13369–13376.
[132]Pollex, A.; Millet, A.; Mueller, J.; Hiersemann, M.; Abraham, L. *J. Org. Chem.* **2005**, *70*, 5579–5591.
[133]Mikami, K.; Nakai, T. *Synthesis* **1991**, 594–604.
[134]Mikami, K.; Azuma, K.; Nakai, T. *Tetrahedron* **1984**, *40*, 2303–2308.

The use of chiral bases has been shown to induce asymmetry in some systems.[135]

The nitrogen analog of the Wittig rearrangement is also useful. In the example below, the silyl group on the olefin profoundly effects selectivity; on the substrate lacking a TMS group the diastereoselectivity was 1:1.[136]

### 7.3.2.8 [2,3]-Meisenheimer Rearrangement: Nitrogen to Oxygen Migration of Carbon in Tertiary N-Oxides

The [2,3]-Meisenheimer rearrangement is the allylic version of the [1,2]-variant discussed earlier (Section 7.2.5.2). This rearrangement can occur under very mild conditions.[137]

[135]Marshall, J. A.; Wang, X. J. *J. Org. Chem.* **1992**, *57*, 2747–2750.
[136]Anderson, J. C.; Siddons, D. C.; Smith, S. C.; Swarbrick, M. E. *J. Org. Chem.* **1996**, *61*, 4820–4823.
[137]Majumdar, K. C.; Jana, G. H. *J. Org. Chem.* **1997**, *62*, 1506–1508.

***7.3.2.9   [2,3]-Sulfoxide, Selenoxide, and Sulfilimine Rearrangements***   The [2,3]-rearrangement of sulfoxides, selenoxides, and their analogs provide a method for stereochemical relay of an established stereocenter bearing S or Se into an alcohol or amine. In the case of sulfur, the reverse reaction is also synthetically useful. When this method is used following the ene reaction of selenium dioxide or a reagent of the structure $RN=Se=NR$ or $RN=S=NR$ (Section 10.3.8.2), the allylic position of an olefin can be oxidized to the alcohol or amine.

Allylic sulfoxides undergo [2,3]-rearrangement to form an allylic sulfinate, which can be reduced to the allylic alcohol using mild reducing agents such as a phosphate; this reaction is known as the Mislow–Evans rearrangement. The rearrangement is under thermodynamic control. As a result, the allylic sulfoxide is typically favored unless a driving force is present, such as a phosphine that acts as a reducing agent. Allylic alcohols can be converted to the rearranged sulfoxide by derivatization of the alcohol with a sulfinyl chloride.[138,139]

The nitrogen analog to this rearrangement is the sulfilimine rearrangement. Again, the use of a reducing agent to drive the reaction and free the amine is necessary.[140]

Selenoxides undergo a more facile [2,3]-rearrangement to the selenate, which is readily reduced to the allylic alcohol. Hydroxide can be used instead of a reducing agent to displace the selenium from the newly formed alcohol.[141,142]

[138]Koprowski, M.; Krawczyk, E.; Skowronska, A.; McPartlin, M.; Choi, N.; Radojevic, S. *Tetrahedron* **2001**, *57*, 1105–1118.
[139]Lu, T. J.; Liu, S. W.; Wang, S. H. *J. Org. Chem.* **1993**, *58*, 7945–7947.
[140]Dolle, R. E.; Li, C. S.; Novelli, R.; Kruse, L. I.; Eggleston, D. *J. Org. Chem.* **1992**, *57*, 128–132.
[141]Kshirsagar, T. A.; Moe, S. T.; Portoghese, P. S. *J. Org. Chem.* **1998**, *63*, 1704–1705.
[142]Zanoni, G.; Porta, A.; Castronovo, F.; Vidari, G. *J. Org. Chem.* **2003**, *68*, 6005–6010.

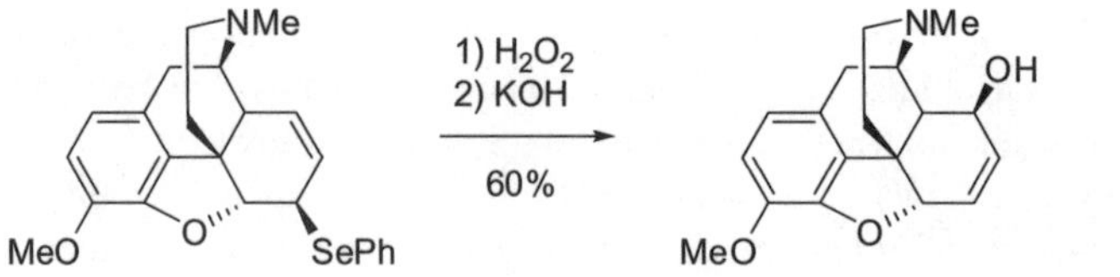

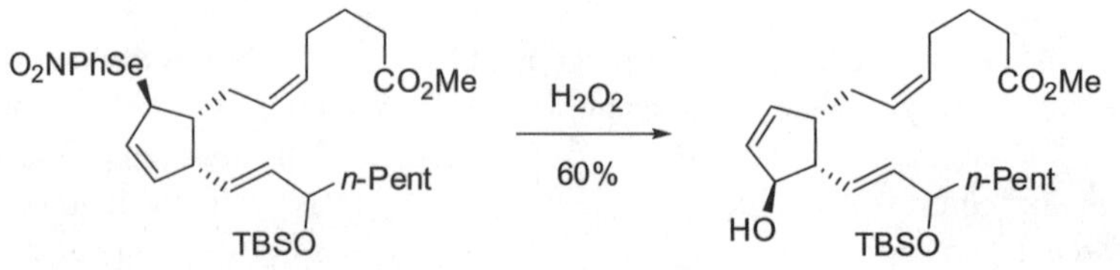

This rearrangement also has a nitrogen analog.[143]

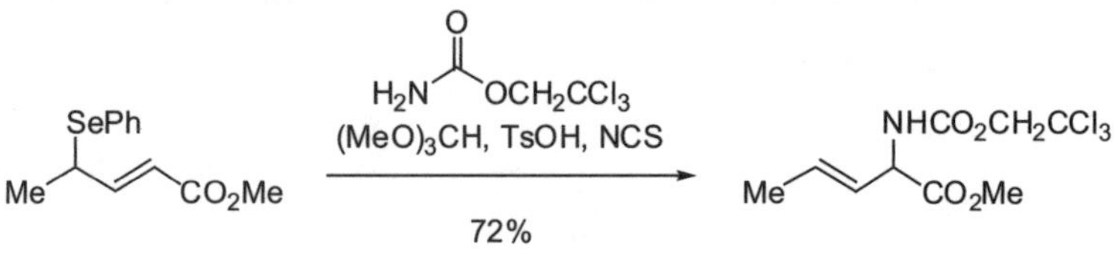

### 7.3.2.10 Sommelet–Hauser Rearrangement and Related Reactions: [2,3]-Sigmatropic Rearrangements of Ylides

The Sommelet–Hauser rearrangement is the [2,3]-migration of a dialkylamino group from a benzylic ammonium salt. The reaction can be categorized as a [2,3]-sigmatropic rearrangement of a nitrogen ylide, and examples of a sulfur ylide or an oxonium ylide undergoing the same transformation can be found. This is the [2,3]-variant of the [1,2]-Stevens rearrangement discussed in Section 7.2.5.1. The rearrangement requires a strong base to deprotonate one of the hydrogens on a carbon adjacent to the nitrogen. As more than one substituent of the ammonium salt frequently has a proton that can be abstracted, mixtures of products are not unusual in this reaction. Further, the potential for a [1,2]-rearrangement (the Stevens rearrangement) to compete exists.

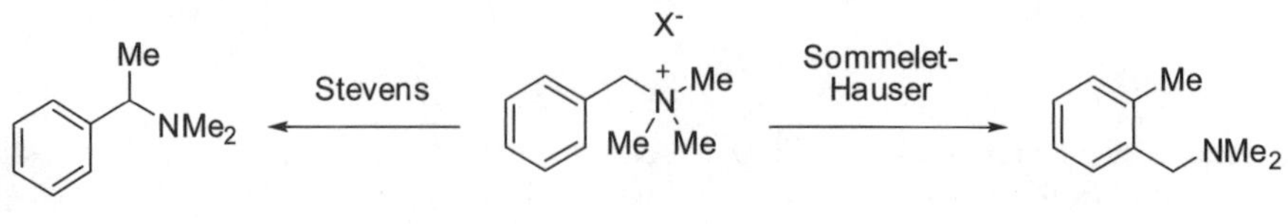

An example of a sulfur variant of the Sommelet–Hauser rearrangement is shown below. In this case, a weak base, triethylamine, is sufficient to deprotonate the ylide to initiate the rearrangement.[144]

[143]Shea, R. G.; Fitzner, J. N.; Fankhauser, J. E.; Spaltenstein, A.; Carpino, P. A.; Peevey, R. M.; Pratt, D. V.; Tenge, B. J.; Hopkins, P. B. *J. Org. Chem.* **1986**, *51*, 5243–5252.
[144]Inoue, S.; Ikeda, H.; Sato, S.; Horie, K.; Ota, T.; Miyamoto, O.; Sato, K. *J. Org. Chem.* **1987**, *52*, 5495–5497.

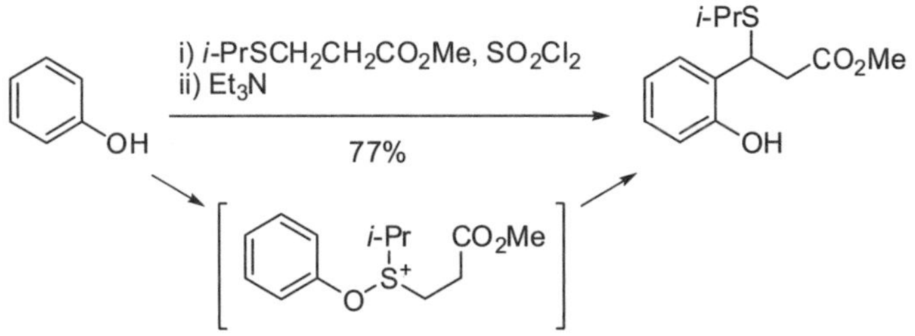

### 7.3.2.11   Benzidine Rearrangement

**7.3.2.11   Benzidine Rearrangement**   The benzidine rearrangement is formally the reaction shown below. The major product is shown, but others with a different substitution pattern are also formed. The synthetic utility of this reaction is limited, but chemists should be aware of this rearrangement when handling diarylhydrazines.

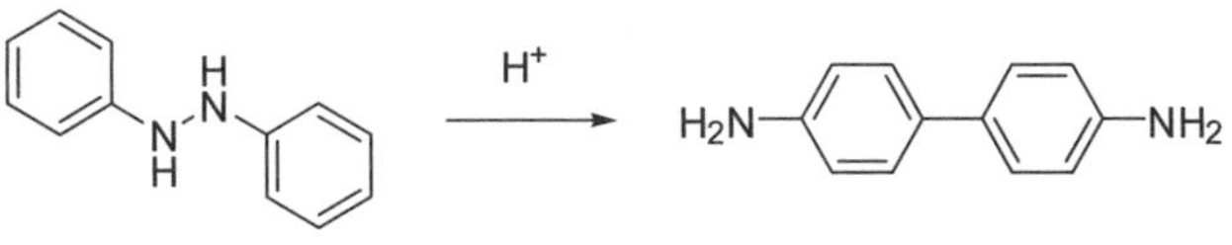

While not particularly useful synthetically, a variant of the reaction was used in the example below.[145] On treatment with polyphosphoric acid (PPA), the substrate below underwent a Fischer indole synthesis with a second molecule of substrate, leading to the benzidine rearrangement substrate. After rearrangement and Friedel–Crafts acylation, the product shown was formed in low yield.

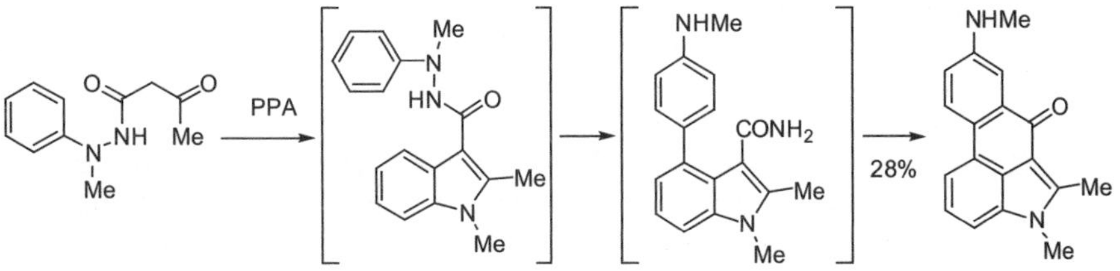

### 7.3.3   Other Cyclic Rearrangements

**7.3.3.1   Di-π-methane and Related Rearrangements**   The di-π-methane rearrangement has been well reviewed.[146] The reaction is a photochemically induced diradical rearrangement that results in a dramatic transformation. Use of this method is limited by the requirement for a photochemical apparatus. Most commonly used are the aza-di-π-methane rearrangement and the oxa-di-π-methane rearrangement, which can generate some highly

[145]Kornet, M. J.; Thio, A. P.; Tolbert, L. M. *J. Org. Chem.* **1980**, *45*, 30–32.
[146]Zimmerman, H. E.; Armesto, D. *Chem. Rev.* **1996**, *96*, 3065–3112.

complex and strained ring systems. Some examples of two manifolds of the aza-di-π-methane rearrangement are shown below.[147,148]

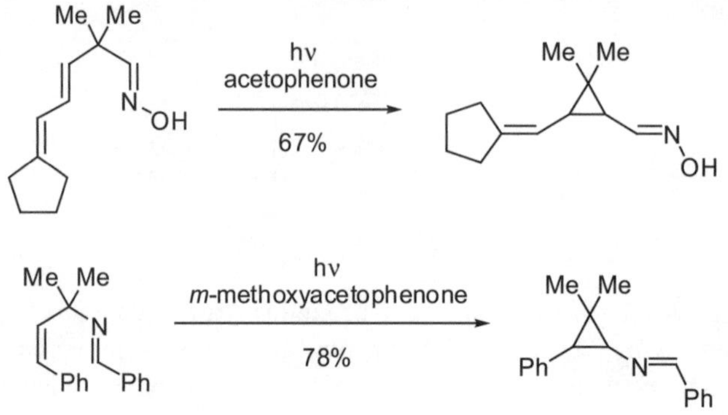

Some examples of the oxa-di-π-methane rearrangement are shown below.[149,150] The first example was demonstrated on 2 gram scale.

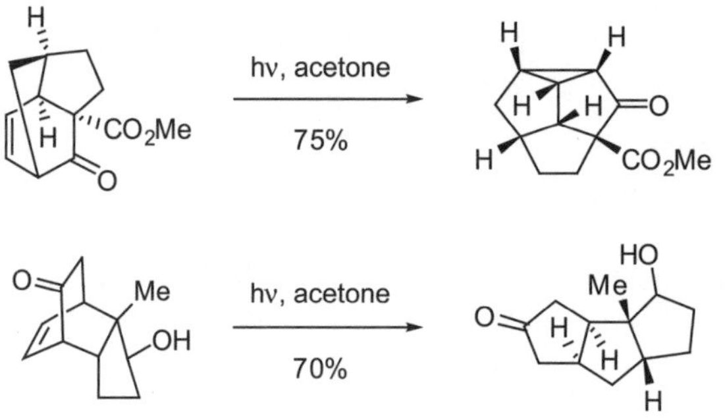

### 7.3.4 Acyclic Rearrangements

***7.3.4.1 Migration of Double Bonds***    Olefin migrations as a result of electrocyclic or sigmatropic rearrangements (Sections 7.3.1 and 7.3.2) and allylic displacement reactions are covered elsewhere in this book (Chapter 1 and Section 6.8). Olefins and acetylenes can be migrated into conjugation with other olefins or acetylenes with strong bases such as hydroxide.[151,152]

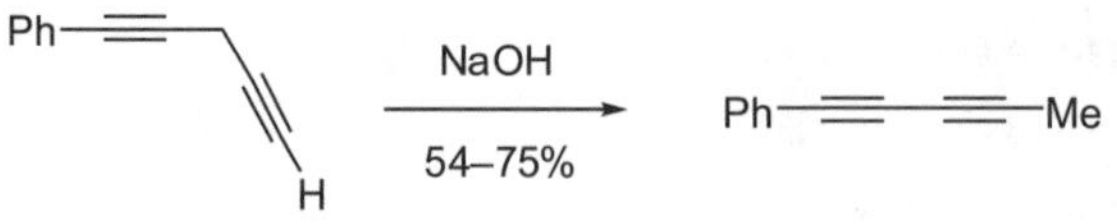

[147]Armesto, D.; Gallego, M. G.; Horspool, W. M.; Agarrabeitia, A. R. *Tetrahedron* **1995**, *51*, 9223–9240.

[148]Armesto, D.; Caballero, O.; Ortiz, M. J.; Agarrabeitia, A. R.; Martin-Fontecha, M.; Torres, M. R. *J. Org. Chem.* **2003**, *68*, 6661–6671.

[149]Schultz, A. G.; Lavieri, F. P.; Snead, T. E. *J. Org. Chem.* **1985**, *50*, 3086–3091.

[150]Singh, V.; Vedantham, P.; Sahu, P. K. *Tetrahedron* **2004**, *60*, 8161–8169.

[151]Taniguchi, H.; Mathai, I.; Miller, S. I. *Org. Synth.* **1970**, *50*, 97–101.

[152]Stoeckel, K.; Sondheimer, F. *Org. Synth.* **1974**, *54*, 1–10.

Acetylenes can be migrated to the terminal position in the zipper reaction.[153]

To migrate an olefin into conjugation with a carbonyl, a weak acid or base is needed.[154]

Allylic amines can be isomerized to the enamines using a rhodium catalyst. The use of chiral ligands allows for the introduction of stereochemistry in some cases.[155]

Allyl ethers can be isomerized to the enol ether using a variety of reagents including Pd/C,[156] KO$t$-Bu, and Rh(PPh$_3$)$_3$Cl. The enol ether can then be hydrolyzed to remove the allyl group.

[153]Hoye, R. C.; Baigorria, A. S.; Danielson, M. E.; Pragman, A. A.; Rajapakse, H. A. *J. Org. Chem.* **1999**, *64*, 2450–2453.

[154]Fieser, L. F. *Org. Synth.* **1955**, *35*, 43–49.

[155]Tani, K.; Yamagata, T.; Otsuka, S.; Kumobayashi, H.; Akutagawa, S. *Org. Synth.* **1989**, *67*, 33–43.

[156]Boss, R.; Scheffold, R. *Angew. Chem., Int. Ed. Engl.* **1976**, *15*, 558–559.

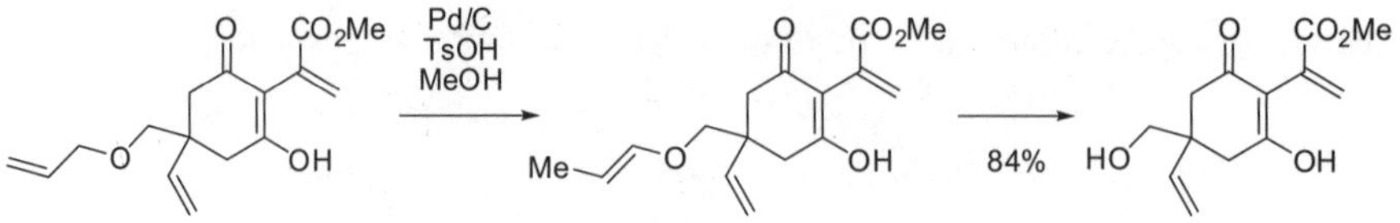

### 7.3.4.2 Hydride Shifts

*7.3.4.2 Hydride Shifts*   Long- or short-range hydride shifts can occur when a radical or cationic intermediate rearranges to a more stable radical or cation. This process does not have to be a sigmatropic rearrangement. A preparation utilizing a hydride shift is shown below.[157] These rearrangements are possible any time a radical or cationic intermediate is present in a reaction and the shift of a proximal hydride would result in a more stable intermediate or an intermediate more easily trapped than the original.

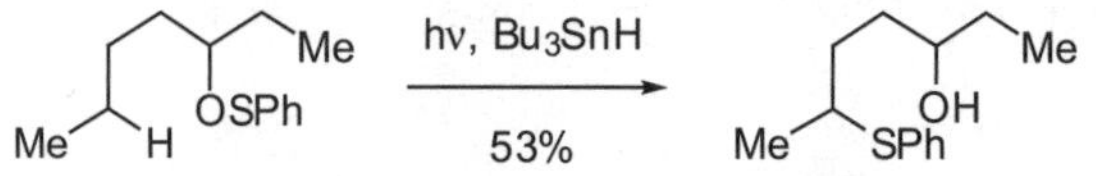

An example of a [1,5]-hydride shift is shown below. In this case, the transfer of hydride results in an iminium ion that can be trapped with an electron-rich aromatic ring to form a diazocine.[158]

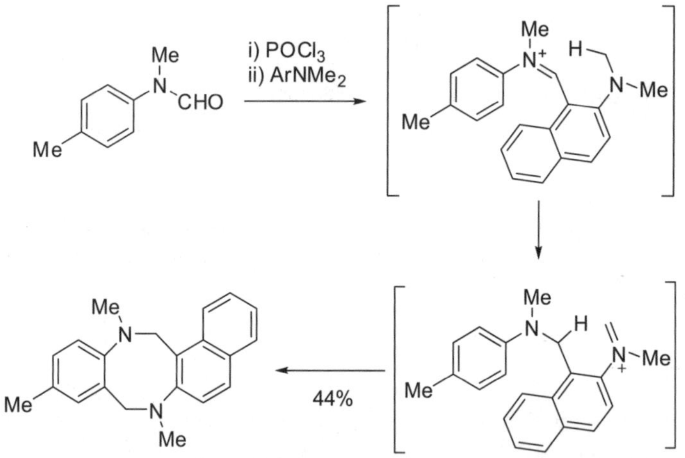

*7.3.4.3 Newman–Kwart Rearrangement: O-Phenylthiocarbanate to S-Phenylcarbamate*   The Newman–Kwart rearrangement is a high-temperature procedure for the conversion of a phenol to a thiophenol. Temperatures in excess of 200°C are required for the rearrangement to proceed, limiting solvent choice for the transformation.[159]

[157]Patrovic, G.; Saicic, R. N.; Cekovic, Z. *Org. Synth.* **2005**, *81*, 244–253.
[158]Cheng, Y.; Yang, H.-B.; Liu, B.; Meth-Cohn, O.; Watkin, D.; Humphries, S. *Synthesis* **2002**, 906–910.
[159]Bowden, S. A.; Burke, J. N.; Gray, F.; McKown, S.; Moseley, J. D.; Moss, W. O.; Murray, P. M.; Welham, M. J.; Young, M. *J. Org. Process Res. Dev.* **2004**, *8*, 33–44.

#### 7.3.4.4 Nitrosamide Decomposition

*7.3.4.4 Nitrosamide Decomposition* Nitrosation of an amide is sometimes employed to activate the amide to hydrolysis. If this strategy is employed, there are a couple of rearrangements that need to be considered to prevent byproduct formation. When a nitrosamide is heated gently, nitrogen is extruded, resulting in the ester product. This is a nice procedure for converting an amine to an alcohol.[53]

In other cases, the *N*-nitrosylcarbamate decomposes to the diazo compound.[160]

#### 7.3.4.5 The Achmatowitz Reaction: Oxidative Furan Rearrangement

*7.3.4.5 The Achmatowitz Reaction: Oxidative Furan Rearrangement* Under oxidative conditions, appropriately substituted furans can be rearranged to the pyridinium salt or a pyrone.[161,162]

[160]Van Leusen, A. M.; Strating, J. *Org. Synth.* **1977**, *57*, 95–102.
[161]Guo, H.; O'Doherty, G. A. *Org. Lett.* **2006**, *8*, 1609–1612.
[162]Peese, K. M.; Gin, D. Y. *Org. Lett.* **2005**, *7*, 3323–3325.

# 8

# ELIMINATIONS

Sally Gut Ruggeri

## 8.1 INTRODUCTION

Elimination chemistry is one of the classical approaches for the formation of multiple bonds (C=C, C=X, etc.). Many of the oldest methods for carrying out eliminations are still synthetically practical, although some modern variations have increased the scope of sensitive functional groups that can be tolerated. The discussion below describes useful methods for carrying out eliminations, where the elimination is the focus of the reaction. Other transformations where the elimination is not the primary focus, such as the elimination of water in the formation of an imine from a carbonyl, or amide formation from carboxylic acids, are captured in Section 2.7.1 for imine and Section 2.26 for amide formation.

## 8.2 FORMATION OF ALKENES

Alkenes are readily formed by the elimination of a wide variety of functional groups. The four most general and widely applicable types of conditions to effect this transformation, depending on the substituent being eliminated, are treatment with acid, base, dehydrating reagents, and/or heat. The references listed below cite some of the most mild and/or generally useful conditions to effect these transformations. While the reactivity is usually straightforward, the selectivity of the elimination can be complicated, and competitive formation of regioisomers and geometrical isomers can be an issue. In some cases the control is excellent, due either to substrate or reagent bias. The section is roughly divided by elimination of H-X or X-X′.

*Practical Synthetic Organic Chemistry: Reactions, Principles, and Techniques*, First Edition.
Edited by Stéphane Caron.
© 2011 John Wiley & Sons, Inc. Published 2011 by John Wiley & Sons, Inc.

### 8.2.1 Elimination of Alcohols

The elimination of water under acidic conditions is one of the classical reactions of organic synthesis. Mineral acids are good reagents for this transformation,[1] if the rest of the substrate is stable to their use.

If mineral acids are not tolerated by the substrate, organic sulfonic acids are good substitutes.[2]

This method works best for tertiary or secondary alcohols; primary alcohols require very forcing conditions to lose water, and should be functionalized for elimination. The classic method of water removal is preferred: use of toluene or another suitable water-immiscible solvent that forms an azeotrope with water attached to a Dean–Stark trap or equivalent.

### 8.2.2 Elimination of Ethers

The elimination of an unactivated alkyl ether is not a facile reaction, and usually requires relatively harsh conditions.[3]

For some substrates, the elimination will occur under milder conditions, such as treatment with a sulfonic acid, but usually only when the ether is tertiary.[4]

[1] Norris, J. F. *Org. Synth.* **1927**, *7*, 76–77.
[2] Barnett, C. J.; Copley-Merriman, C. R.; Maki, J. *J. Org. Chem.* **1989**, *54*, 4795–4800.
[3] Weizmann, C.; Sulzbacher, M.; Bergmann, E. *J. Am. Chem. Soc.* **1948**, *70*, 1153–1158.
[4] Sebahar, P. R.; Williams, R. M. *J. Am. Chem. Soc.* **2000**, *122*, 5666–5667.

There are some special cases in which an ether will eliminate more easily. One example is the elimination of an epoxide to form an allylic alcohol, which is accomplished by treatment with a hindered alkyl amide base.[5]

LCIA = lithium cyclohexylisopropylamide

Ethers activated by neighboring groups can be readily induced to eliminate under relatively mild conditions. Elimination of acetals or ketals is usually accomplished by treatment with a Brønsted or Lewis acid,[6] and elimination will occur if no nucleophile is present to trap the resulting oxocarbenium ion.

The elimination of an allylic ether to a diene is also relatively facile (see Section 8.3.1). If the ether is β to an electron-withdrawing group, the elimination will occur under mild or neutral conditions (see Section 8.2.12).

### 8.2.3   Elimination of Esters

Some of the mildest and most commonly used conditions for elimination of an alcohol utilize an ester to facilitate the elimination. The ester can take the form of an acetate,[7] sulfate,[8] sulfonate,[9] or phosphate,[10] and is often generated *in situ*. An interesting variation of the

[5]Mander, L. N.; Thomson, R. J. *J. Org. Chem.* **2005**, *70*, 1654–1670.
[6]Castro, S.; Peczuh, M. W. *J. Org. Chem.* **2005**, *70*, 3312–3315.
[7]Kuethe, J. T.; Wong, A.; Qu, C.; Smitrovich, J.; Davies, I. W.; Hughes, D. L. *J. Org. Chem.* **2005**, *70*, 2555–2567.
[8]Hauske, J. R.; Guadliana, M.; Kostek, G.; Schulte, G. *J. Org. Chem.* **1987**, *52*, 4622–4625.
[9]Subramanyam, C.; Noguchi, M.; Weinreb, S. M. *J. Org. Chem.* **1989**, *54*, 5580–5585.
[10]McGuire, M. A.; Sorenson, E.; Owings, F. W.; Resnick, T. M.; Fox, M.; Baine, N. H. *J. Org. Chem.* **1994**, *59*, 6683–6686.

sulfonate elimination is the use of Furukawa's reagent (mesyl chloride and DMAP), where addition of water as a co-solvent is necessary for efficient conversion.[11] Primary, secondary, and tertiary alcohols will eliminate when activated in this way.

## 8.2.4  Elimination of Xanthates

Xanthates are typically used to effect the reduction of alcohols, but in the absence of a radical initiator or hydride source, will eliminate instead (Chugaev elimination). For unactivated xanthates, high temperatures are generally required.[12]

[11]Comins, D. L.; Al-Awar, R. S. *J. Org. Chem.* **1992**, *57*, 4098–4103.
[12]Paquette, L. A.; Tsui, H.-C. *J. Org. Chem.* **1996**, *61*, 142–145.

Where a neighboring leaving group is present, radical initiation of the xanthate cleavage leads to an alkene. When the leaving group is nitro, this methodology presents an alternative to alkene formation from carbonyls via the Henry reaction.[13] The product of the elimination is usually the thermodynamically favored isomer.

DLP = lauroyl peroxide

### 8.2.5    Elimination of Ammonium or Sulfonium Salts

Amines do not readily eliminate under most reaction conditions. However, elimination of the corresponding quaternary ammonium salts, also known as the Hofmann elimination, can be achieved under basic conditions. If the quaternary ammonium salt is not activated for elimination by neighboring groups, high temperatures are required.[14] This methodology has been exemplified on many cyclic ammonium salts as an indirect method to control the stereochemistry or olefin geometry of acyclic systems.[15]

[13]Ouvry, G.; Quiclet-Sire, B.; Zard, S. Z. *Org. Lett.* **2003**, *5*, 2907–2909.
[14]Matsubara, S.; Matsuda, H.; Hamatani, T.; Schlosser, M. *Tetrahedron* **1988**, *44*, 2855–2863.
[15]Decodts, G.; Dressaire, G.; Langlois, Y. *Synthesis* **1979**, 510–513.

Substrates in which the leaving group is activated by an electron-withdrawing group occur at lower temperatures and with milder bases.[16]

The variation in which the leaving group is sulfonium is also known, and eliminates under similar conditions.[17]

## 8.2.6 Elimination of *N*-Oxides

The elimination of *N*-oxides to produce alkenes (Cope elimination) is most easily performed by heating the substrate; elimination occurs at relatively low temperatures.[18,19] There can be an inherent safety issue with these reactions, as *N*-oxides are often high-energy intermediates, and the thermal decomposition of the starting material should be assessed before carrying out the reaction. Interestingly, the corresponding Hofmann elimination in the first example below gave the regioisomeric alkene (i.e., elimination toward the less-substituted carbon).

[16]Tarzia, G.; Balsamini, C.; Spadoni, G.; Duranti, E. *Synthesis* **1988**, 514–517.

[17]Yamato, T.; Kobayashi, K.; Arimura, T.; Tashiro, M.; Yoshihira, K.; Kawazoe, K.; Sato, S.; Tamura, C. *J. Org. Chem.* **1986**, *51*, 2214–2218.

[18]Woolhouse, A. D.; Gainsford, G. J.; Crump, D. R. *J. Heterocycl. Chem.* **1993**, *30*, 873–880.

[19]Langlois, N.; Rakotondradany, F. *Tetrahedron* **2000**, *56*, 2437–2448.

### 8.2.7  Elimination of Diazonium Salts

Treatment of diazo species with rhodium catalysts gives the formal elimination product.[20] These reactions actually occur via β-hydride transfer, and transfer of a non-hydride substituent is often a competing, if not preferential process. One interesting aspect of these reactions is that they often give exclusively the *cis* olefin, which can be difficult to produce from most elimination processes.

### 8.2.8  Elimination of Hydrazones

The elimination of hydrazones with base is well precedented. When sterically undemanding bases[21] or thermolytic conditions are used, the reaction usually gives rise to the thermodynamically favored product (Bamford–Stevens reaction), while bulkier bases usually yield the kinetically favored product (Shapiro reaction) (see Section 9.3.4). For cases in which only one mode of elimination is possible, the conditions of the Shapiro reaction are generally preferred, as they proceed at lower temperatures. A variety of bases have been used, but alkyl lithium[22] and lithium amide bases[23] are generally preferred.

[20]Taber, D. F.; Herr, R. J.; Pack, S. K.; Geremia, J. M. *J. Org. Chem.* **1996**, *61*, 2908–2910.
[21]Loev, B.; Kormendy, M. F.; Snader, K. M. *J. Org. Chem.* **1966**, *31*, 3531–3534.
[22]Tsantali, G. G.; Takakis, I. M. *J. Org. Chem.* **2003**, *68*, 6455–6458.
[23]Siemeling, U.; Neumann, B.; Stammler, H.-G. *J. Org. Chem.* **1997**, *62*, 3407–3408.

### 8.2.9   Elimination of Sulfoxides and Selenoxides

Sulfides and selenides can be oxidized to their corresponding sulfoxides and selenoxides, and will eliminate to produce an alkene. The elimination of sulfoxides usually requires more forcing conditions (i.e., higher temperatures) while the selenoxides eliminate more easily.[24]

X = SOMe    Mesitylene, Δ, 62%

X = SePh    30% $H_2O_2$, rt, 83%

The key to selecting the best conditions for a given elimination is the proper choice of oxidant that is compatible with other functional groups (see Chapter 10).

### 8.2.10   Elimination of Halides

Halides are readily eliminated under basic conditions to form alkenes. A wide variety of bases and substrates are accommodated by the reaction, and many substitution patterns can be achieved, including the formation of enol ethers.[25–27] Use of DBU or DBN as the base is preferred where competitive nucleophilic displacement is an issue.

[24]Bartley, D. M.; Coward, J. K. *J. Org. Chem.* **2005**, *70*, 6757–6774.
[25]Baker, S. R.; Harris, J. R. *Synth. Commun.* **1991**, *21*, 2015–2023.
[26]Frey, L. F.; Marcantonio, K. M.; Chen, C.-y.; Wallace, D. J.; Murry, J. A.; Tan, L.; Chen, W.; Dolling, U. H.; Grabowski, E. J. J. *Tetrahedron* **2003**, *59*, 6363–6373.
[27]Price, C. C.; Judge, J. M. *Org. Synth.* **1965**, *45*, 22–24.

If the substrate has 1,1- or 1,2-dihalo substitution, a vinyl halide is produced.[28] The vinyl halide can be further eliminated to the corresponding alkyne with excess strong base (see Section 8.4.4).

## 8.2.11   Elimination of Nitriles

Unactivated cyano groups are difficult to eliminate, but the reaction can be accomplished under strongly basic conditions.[29]

Nitriles with an α-amino group will eliminate when treated with TMSOTf.[30]

The reverse Strecker reaction or cyanohydrin formation also formally involve the elimination of a nitrile, but will be addressed in the chapter on their formation and/or reaction (see Section 2.22).

## 8.2.12   Elimination β to an Electron-Withdrawing Group

The elimination of a functional group β to an electron-withdrawing group (retro conjugate addition) is extremely facile under mild conditions. The elimination can occur under acidic, basic, or neutral conditions. Carbonyl groups,[31,32] sulfones,[33] nitriles, and nitro groups are among the many that facilitate the elimination.

[28]Champness, N. R.; Khlobystov, A. N.; Majuga, A. G.; Schroder, M.; Zyk, N. V. *Tetrahedron Lett.* **1999**, *40*, 5413–5416.

[29]Palecek, J.; Paleta, O. *Synthesis* **2004**, 521–524.

[30]Ahlbrecht, H.; Dueber, E. O. *Synthesis* **1983**, 56–57.

[31]Maragni, P.; Mattioli, M.; Pachera, R.; Perboni, A.; Tamburini, B. *Org. Process Res. Dev.* **2002**, *6*, 597–605.

[32]Ito, Y.; Fujii, S.; Nakatsuka, M.; Kawamoto, F.; Saegusa, T. *Org. Synth.* **1980**, *59*, 113–122.

[33]Chang, S.-J.; Fernando, D.; Fickes, M.; Gupta, A. K.; Hill, D. R.; McDermott, T.; Parekh, S.; Tian, Z.; Wittenberger, S. J. *Org. Process Res. Dev.* **2002**, *6*, 329–335.

A less common elimination of this category arises when the group at the β carbon is a silyl group. The preferred conditions to obtain the unsaturated product use copper salts to effect the oxidative elimination.[34]

## 8.2.13   Elimination of Diol Derivatives

Diols can be eliminated to form alkenes, but require some form of functionalization to do so. An additional complication is that depending on the substitution pattern, elimination can occur to form a diene instead (see Section 8.3.3) The conditions for the elimination of bis (sulfonates) are among the mildest to achieve reductive elimination.[35]

Phosphorus-based derivatives will also eliminate, but the byproducts make these conditions somewhat less preferable.[36]

[34]Schmoldt, P.; Mattay, J. *Synthesis* **2003**, 1071–1078.
[35]Nishiyama, S.; Toshima, H.; Kanai, H.; Yamamura, S. *Tetrahedron* **1988**, *44*, 6315–6324.
[36]Mereyala, H. B.; Gadikota, R. R.; Sunder, K. S.; Shailaja, S. *Tetrahedron* **2000**, *56*, 3021–3026.

Diols can be converted to cyclic orthoformates or orthoformamides by reaction with a trialkyl orthoformate[37] or *N,N*-dimethylformamide dimethyl acetal, respectively. These derivatives will eliminate to the alkene on heating. While the amino analogue requires acylation prior to elimination, it is generally preferred because of the lower temperatures[38] for the elimination.

Some of the mildest conditions for elimination of diols involve bis(xanthates);[39] the drawback to this method are the conditions required to form the xanthates, since $CS_2$ is toxic and the alkylating agents present worker safety issues.

Cyclic thiocarbonates can be eliminated to alkenes by treatment with phosphites at elevated temperatures;[40] the reaction temperature is lowered when reagents such as diazaphospholidines are used.[41]

[37]Trost, B. M.; Romero, A. G. *J. Org. Chem.* **1986**, *51*, 2332–2342.
[38]Henegar, K. E.; Ashford, S. W.; Baughman, T. A.; Sih, J. C.; Gu, R.-L. *J. Org. Chem.* **1997**, *62*, 6588–6597.
[39]Park, H. S.; Lee, H. Y.; Kim, Y. H. *Org. Lett.* **2005**, *7*, 3187–3190.
[40]Kaneko, S.; Nakajima, N.; Shikano, M.; Katoh, T.; Terashima, S. *Tetrahedron* **1998**, *54*, 5485–5506.
[41]Chu, C. K.; Bhadti, V. S.; Doboszewski, B.; Gu, Z. P.; Kosugi, Y.; Pullaiah, K. C.; Van Roey, P. *J. Org. Chem.* **1989**, *54*, 2217–2225.

## 8.2.14   Elimination of Epoxides and Episulfides

Epoxides can be eliminated by treatment with iodide. If no neighboring group participation is possible, further activation of the resulting iodohydrin is necessary for good conversion. Reagents that have commonly been used include strong acids,[42] anhydrides,[43] and silylating reagents.[44] If the epoxide is activated by an electron-withdrawing group, the elimination occurs under milder conditions.[45]

Epoxides and episulfides can be directly converted to their corresponding alkene derivatives by treatment with phosphines[46] or phosphites.[47]

[42]Sarma, D. N.; Sharma, R. P. *Chem. Ind.* **1984**, 712–713.

[43]Sonnet, P. E. *J. Org. Chem.* **1978**, *43*, 1841–1842.

[44]Caputo, R.; Mangoni, L.; Neri, O.; Palumbo, G. *Tetrahedron Lett.* **1981**, *22*, 3551–3552.

[45]Righi, G.; Bovicelli, P.; Sperandio, A. *Tetrahedron* **2000**, *56*, 1733–1737.

[46]Denney, D. B.; Boskin, M. J. *J. Am. Chem. Soc.* **1960**, *82*, 4736–4738.

[47]Neureiter, N. P.; Bordwell, F. G. *J. Am. Chem. Soc.* **1959**, *81*, 578–580.

The epoxides usually give inversion while the episulfides generally proceed with retention of configuration. Because the elimination of epoxides with these reagents only occurs at very high temperatures, an alternative method to form the betaine is to open the epoxide with a metal phosphide and quaternize *in situ*.[48]

Alternatively, the epoxide can be converted to the episulfide to lower the temperature for elimination; this method is less attractive because of the need for extra steps and the conditions required for the conversion.

### 8.2.15  Elimination of α-Halosulfones

Alkenes can be formed by elimination of α-halosulfones (Ramberg–Bäcklund reaction) under strongly basic conditions. Substrates containing sensitive functionality have been reacted successfully, despite the relatively harsh conditions.[49] The halide may be an existing functional group in the substrate or can be generated *in situ*.[50]

[48]Vedejs, E.; Fuchs, P. L. *J. Am. Chem. Soc.* **1971**, *93*, 4070–4072.

[49]Choi, B. S.; Chang, J. H.; Choi, H.-w.; Kim, Y. K.; Lee, K. K.; Lee, K. W.; Lee, J. H.; Heo, T.; Nam, D. H.; Shin, H. *Org. Process Res. Dev.* **2005**, *9*, 311–313.

[50]Pasetto, P.; Franck, R. W. *J. Org. Chem.* **2003**, *68*, 8042–8060.

The intermediate episulfone can also be eliminated under thermolytic conditions.[51]

### 8.2.16 Elimination of Aziridines

Aziridines are not easily eliminated under reaction conditions that tolerate a wide range of functional groups. The *N*-nitroso derivative eliminates at low temperatures, but is likely to have safety issues with regard to thermal decomposition.[52]

Iodonium salts can also give the eliminated product, but the reaction often results in multiple byproducts.[53]

### 8.2.17 Elimination of Dihalides

Dihalides can be eliminated to form alkenes under relatively mild conditions, such as treatment with iodide[54] or metallic samarium.[55]

[51]Muccioli, A. B.; Simpkins, N. S.; Mortlock, A. *J. Org. Chem.* **1994**, *59*, 5141–5143.
[52]Lee, K.; Kim, Y. H. *Synth. Commun.* **1999**, *29*, 1241–1248.
[53]Padwa, A.; Eastman, D.; Hamilton, L. *J. Org. Chem.* **1968**, *33*, 1317–1322.
[54]Yang, J.; Bauld, N. L. *J. Org. Chem.* **1999**, *64*, 9251–9253.
[55]Wang, L.; Zhang, Y. *Tetrahedron* **1999**, *55*, 10695–10712.

Dehalogenation in dry DMF at elevated temperatures has also been reported.[56]

Depending on the relative stereochemistry of the dihalide and the conditions used, either *cis* or *trans* alkenes can be produced. If the first elimination can occur away from the vicinal halide, a diene may be produced instead (see Section 8.3.1).

## 8.2.18  Elimination of Haloethers

The elimination of haloethers can be accomplished via metal–halogen exchange.[57] The resulting alkene is the product of *syn* elimination, although some *trans* product may be observed.

More classically, the elimination can be carried out using zinc dust (Boord reaction).[58]

The corresponding halothioethers can also be eliminated, generally under milder conditions. The reaction is most easily carried out with iodide,[59] and gives the product of *anti* elimination.

[56]Khurana, J. M.; Maikap, G. C. *J. Org. Chem.* **1991**, *56*, 2582–2584.
[57]Maeda, K.; Shinokubo, H.; Oshima, K. *J. Org. Chem.* **1996**, *61*, 6770–6771.
[58]Beusker, P. H.; Aben, R. W. M.; Seerden, J.-P. G.; Smits, J. M. M.; Scheeren, H. W. *Eur. J. Org. Chem.* **1998**, 2483–2492.
[59]Helmkamp, G. K.; Pettitt, D. J. *J. Org. Chem.* **1964**, *29*, 3258–3262.

### 8.2.19  Elimination of Hydroxy- or Haloacids

Several methods exist for simultaneous dehydration and decarboxylation of hydroxyacids. One of the mildest is treatment of the starting material with dimethylformamide dimethyl acetal and moderate heating, which gives the product of *anti* elimination.[60]

The elimination can also be effected by treatment with acetic anhydride and heat.[61]

Several groups have used Mitsunobu conditions for this type of elimination, but these conditions are generally less preferable because of the difficulty in purging the byproducts, such as triphenylphosphine oxide and substituted hydrazines.

Haloacids give rise to alkenes with good stereospecificity by treatment with a mild base.[62]

### 8.2.20  Elimination of Hydroxysulfones

The elimination of β-hydroxyphenylsulfones (Julia–Lythgoe olefination, see Section 2.19) has classically been accomplished by reduction with sodium amalgam. The stereochemistry of the resulting alkene is not usually controlled, unless the substrate possesses an inherent bias in the elimination.[63]

R = TBDPS

[60]Frater, G.; Mueller, U.; Guenther, W. *Tetrahedron* **1984**, *40*, 1269–1277.
[61]Alexander, B. H.; Barthel, W. F. *J. Org. Chem.* **1958**, *23*, 389–391.
[62]Fuller, C. E.; Walker, D. G. *J. Org. Chem.* **1991**, *56*, 4066–4067.
[63]Wang, Q.; Sasaki, N. A. *J. Org. Chem.* **2004**, *69*, 4767–4773.

The elimination is better controlled, and the use of amalgam can be avoided, by using heterocycle-substituted sulfones.[64] After addition of the corresponding anion to an aldehyde, the intermediate eliminates *in situ*, and the stereocontrol can be high with the correct choice of base and solvent.

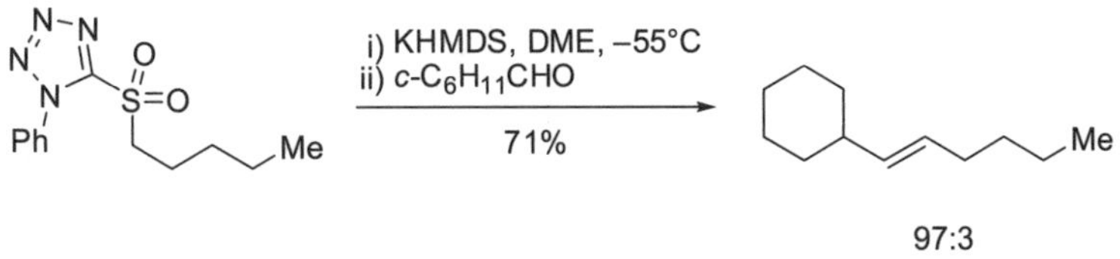

i) KHMDS, DME, −55°C
ii) *c*-C$_6$H$_{11}$CHO

71%

97:3

### 8.2.21  Elimination of β-Silyl Alcohols

The elimination of a β-silyl alcohol (Peterson reaction) is well documented. The reaction is stereospecific, and gives the product of *syn* elimination under basic conditions.[65] Under acidic conditions, the reaction results in *anti* elimination.[66]

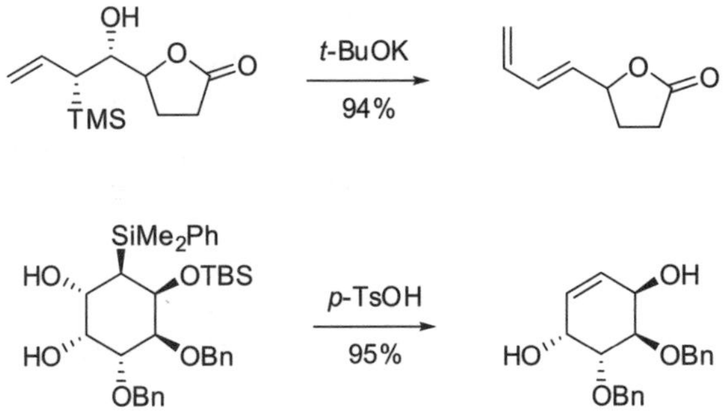

*t*-BuOK

94%

*p*-TsOH

95%

### 8.2.22  Elimination of β-Silyl Esters, Sulfides, and Sulfones

The elimination of groups other than an alcohol β to a silyl group is less well known than the Peterson reaction, but has also been demonstrated. Esters[67] and sulfones[68] readily eliminate on treatment with fluoride ion.

[64]Blakemore, P. R.; Cole, W. J.; Kocienski, P. J.; Morley, A. *Synlett* **1998**, 26–28.

[65]Keck, G. E.; Romer, D. R. *J. Org. Chem.* **1993**, *58*, 6083–6089.

[66]Heo, J.-N.; Holson, E. B.; Roush, W. R. *Org. Lett.* **2003**, *5*, 1697–1700.

[67]Shibuya, A.; Okada, M.; Nakamura, Y.; Kibashi, M.; Horikawa, H.; Taguchi, T. *Tetrahedron* **1999**, *55*, 10325–10340.

[68]Najera, C.; Sansano, J. M. *Tetrahedron* **1994**, *50*, 5829–5844.

Enol silanes can be generated from addition of nucleophiles to acyl silanes with a β leaving group.[69] The intermediate alkoxide undergoes a Brook rearrangement (see Section 7.2.6.1) followed by elimination.

## 8.3 FORMATION OF DIENES

Dienes can usually be formed under very similar conditions to those used for generation of alkenes. In this section, substrates and conditions are described that would not eliminate in the analogous "monoene" system, or that present issues that would not be otherwise encountered.

### 8.3.1 Formation of Dienes from Allylic Systems

Allylic X systems, where X can be alcohol, ether, ester, halide, sulfone, or other leaving groups can yield dienes. Allylic alcohols eliminate under acidic conditions at temperatures much lower than the saturated analogues.[70,71]

[69]Reich, H. J.; Holtan, R. C.; Bolm, C. *J. Am. Chem. Soc.* **1990**, *112*, 5609–5617.
[70]Hikawa, H.; Yokoyama, Y.; Murakami, Y. *Synthesis* **2000**, 214–216.
[71]Cuadrado, P.; Gonzalez-Nogal, A. M.; Sanchez, A.; Sarmentero, M. A. *Tetrahedron* **2003**, *59*, 5855–5859.

Allylic ethers also eliminate under neutral or mildly acidic conditions with heat.[72]

Allylic esters, halides, and sulfones eliminate by treatment with Pd(0); if a nucleophile is not present to trap the π-allyl intermediate, hydride elimination may occur.[73]

Allylic sulfones and halides will also eliminate under basic conditions[74–76]

Since allylic groups eliminate so readily, the choice of reaction conditions should be selected based on consideration of the other functional groups in the substrate, and what they will tolerate.

[72]Hajos, Z. G.; Doebel, K. J.; Goldberg, M. W. *J. Org. Chem.* **1964**, *29*, 2527–2533.

[73]Shim, S.-B.; Ko, Y.-J.; Yoo, B.-W.; Lim, C.-K.; Shin, J.-H. *J. Org. Chem.* **2004**, *69*, 8154–8156.

[74]Hill, K. W.; Taunton-Rigby, J.; Carter, J. D.; Kropp, E.; Vagle, K.; Pieken, W.; McGee, D. P. C.; Husar, G. M.; Leuck, M.; Anziano, D. J.; Sebesta, D. P. *J. Org. Chem.* **2001**, *66*, 5352–5358.

[75]Sellen, M.; Baeckvall, J. E.; Helquist, P. *J. Org. Chem.* **1991**, *56*, 835–839.

[76]Bridges, A. J.; Fischer, J. W. *J. Org. Chem.* **1984**, *49*, 2954–2961.

### 8.3.2  Formation of Dienes from 1,4-Dihalo-2-Butenes

Treatment of 1,4-dihalo-2-butene derivatives with reducing agents such as $Zn^0$ results in reductive elimination to produce the corresponding diene.[77]

If these substrates are exposed to basic conditions, simple allylic elimination occurs, as noted in Section 8.3.1. However, if a nucleophile capable of displacing one of the halides under basic conditions is present, a substituted diene is produced.[78]

3 : 1 E : Z

### 8.3.3  Formation of Dienes from Diols

Vicinal diols can be eliminated to form dienes, most easily under the same kinds of conditions that convert alcohols to alkenes, such as via conversion to and elimination of a diester derivative.[79]

If the diols are inherently susceptible (i.e., tertiary, benzylic, or allylic), they will eliminate under solvolytic conditions.[80] These eliminations are complicated by the potential to undergo a pinacol rearrangement, which is often a competitive, or even dominant, process (see Section 7.2.1.2).

[77]Kamabuchi, A.; Miyaura, N.; Suzuki, A. *Tetrahedron Lett.* **1993**, *34*, 4827–4828.
[78]Burke, S. D.; Cobb, J. E.; Takeuchi, K. *J. Org. Chem.* **1990**, *55*, 2138–2151.
[79]Niederl, J. B.; Silverstein, R. M. *J. Org. Chem.* **1949**, *14*, 10–13.
[80]Wagner, R. A.; Brinker, U. H. *Synthesis* **2001**, 376–378.

### 8.3.4    Formation of Dienes from δ-Elimination

Analogous to the elimination of a β leaving group, activated groups δ to an α,β-unsaturated system will eliminate to form a diene. The reaction is readily accomplished under basic conditions.[81]

### 8.3.5    Formation of Dienes from Sulfolenes

Dienes can be synthesized by extrusion of $SO_2$ from sulfolenes. The preferred conditions employ a mild base at elevated temperatures. Activated substrates will eliminate in refluxing ethanol,[82] while less activated substrates require the higher temperatures that can be achieved with toluene.[83]

[81] Areces, P.; Carrasco, E.; Mancha, A.; Plumet, J. *Synthesis* **2006**, 946–948.

[82] Manchand, P. S.; Yiannikouros, G. P.; Belica, P. S.; Madan, P. *J. Org. Chem.* **1995**, *60*, 6574–6581.

[83] Winkler, J. D.; Quinn, K. J.; MacKinnon, C. H.; Hiscock, S. D.; McLaughlin, E. C. *Org. Lett.* **2003**, *5*, 1805–1808.

### 8.3.6 Formation of Dienes Via Retro Diels–Alder Reactions

A pericyclic rearrangement can be used to build dienes. The substrate for the fragmentation often arises from a Diels–Alder reaction, with the reverse reaction eliminating an alternate dienophile. Acetylenes have been used as the eliminating group, but usually require high temperatures. Other groups, such as nitriles[84] or $CO_2$,[85] eliminate under milder conditions.

### 8.3.7 Formation of Dienes from Extrusion of CO

Fragmentation of molecules containing a cyclopentenone in which CO can be extruded are synthetically useful, and occur thermolytically at reasonable temperatures to give dienes.[86]

[84]Selnick, H. G.; Brookes, L. M. *Tetrahedron Lett.* **1989**, *30*, 6607–6610.
[85]Noguchi, M.; Kakimoto, S.; Kawakami, H.; Kajigaeshi, S. *Bull. Chem. Soc. Jpn.* **1986**, *59*, 1355–1362.
[86]Lin, C. T.; Chou, T. C. *J. Org. Chem.* **1990**, *55*, 2252–2254.

## 8.4   FORMATION OF ALKYNES

A variety of functional groups can be eliminated to form alkynes, and many of the conditions are similar to those used for formation of alkenes. Substituted alkynes are generally formed by introduction of an intact acetylene unit, but can be synthesized by elimination. The most useful methods for disubstituted alkynes rely on formation of a terminal alkyne followed by functionalization of the terminus, often in the same pot. The methods available for directly generating disubstituted alkynes are often limited to substrates containing at least one aromatic substituent, where the direction of elimination is fixed.

### 8.4.1   Formation of Alkynes from Ketones

Conversion of ketones to alkynes is most easily carried out via the intermediacy of phosphoranyl enolates.[87]

An analogous reaction has also been carried out on β-phosphoranyl ketones, but these substrates generally require very high temperatures[88] or microwave assistance[89] to eliminate, making them less useful. Alkyl-substituted alkynes can be generated by a two-step procedure in which benzotriazole (Bt) serves as a dummy ligand during the alkyne formation, then can be displaced by an alkyl- or arylmetal species.[90]

Pyridinium salts have been shown to effect the dehydration of α-ketosulfonamides.[91]

[87]Negishi, E.; King, A. O.; Tour, J. M. *Org. Synth.* **1986**, *64*, 44–49.
[88]Heard, N. E.; Turner, J. *J. Org. Chem.* **1995**, *60*, 4302–4304.
[89]RamaRao, V. V. V. N. S.; Reddy, G. V.; Maitraie, D.; Ravikanth, S.; Yadla, R.; Narsaiah, B.; Rao, P. S. *Tetrahedron* **2004**, *60*, 12231–12237.
[90]Katritzky, A. R.; Abdel-Fattah, A. A. A.; Wang, M. *J. Org. Chem.* **2002**, *67*, 7526–7529.
[91]Leclercq, M.; Brienne, M. J. *Tetrahedron Lett.* **1990**, *31*, 3875–3878.

### 8.4.2   Formation of Alkynes from Bis(Hydrazones)

Alkynes are readily formed from bis(hydrazones) by treatment with a copper salt[92] or iodine in the presence of guanidine bases.[93]

BTMG = *t*-butyltetramethylguanidine

### 8.4.3   Formation of Alkynes from Dihalides

Alkynes may be synthesized from 1,1- or 1,2-dihalides by treatment with a strong base.[94,95] The reaction can also be catalyzed by the use of phase-transfer catalysts.[96]

### 8.4.4   Formation of Alkynes from Vinyl Halides

Alkynes can also be formed from vinyl halides, which are often intermediates in the elimination from the dihalides. The elimination is successful with mono-halo,[97]

[92]Ito, S.; Nomura, A.; Morita, N.; Kabuto, C.; Kobayashi, H.; Maejima, S.; Fujimori, K.; Yasunami, M. *J. Org. Chem.* **2002**, *67*, 7295–7302.

[93]Barton, D. H. R.; Bashiardes, G.; Fourrey, J. L. *Tetrahedron* **1988**, *44*, 147–162.

[94]Smith, L. I.; Falkof, M. M. *Org. Synth.* **1942**, *22*, 50–51.

[95]Marshall, J. A.; Yanik, M. M. *J. Org. Chem.* **2001**, *66*, 1373–1379.

[96]Dehmlow, E. V.; Lissel, M. *Tetrahedron* **1981**, *37*, 1653–1658.

[97]Toussaint, D.; Suffert, J. *Org. Synth.* **1999**, *76*, 214–220.

1,1-dihalo,[98] or 1,2-dihalo alkenes.[99] Of these, the most commonly used in recent times is the Corey–Fuchs reaction. As with many syntheses in which a terminal alkyne is generated, the product is often trapped under the basic reaction conditions with an electrophile.

### 8.4.5  Formation of Alkynes from Elimination of Sulfones

Elimination of a sulfur-containing moiety has been used to generate alkynes. The most preparatively useful reactions involve elimination of a vinyl sulfone, either as an isolated intermediate,[100] or generated *in situ* in the course of the reaction.[101] The latter reaction can be particularly useful, since the intermediate vinyl metal species can be used to introduce more functionality prior to the elimination.

[98]Mori, M.; Tonogaki, K.; Kinoshita, A. *Org. Synth.* **2005**, *81*, 1–13.
[99]Kende, A. S.; Fludzinski, P. *Org. Synth.* **1986**, *64*, 73–79.
[100]Otera, J.; Misawa, H.; Sugimoto, K. *J. Org. Chem.* **1986**, *51*, 3830–3833.
[101]Yoshimatsu, M.; Kawahigashi, M.; Shimizu, H.; Kataoka, T. *J. Chem. Soc., Chem. Commun.* **1995**, 583–584.

Sulfide[102] and sulfoxide[103] eliminations are also known, but are generally only feasible on small scale due to the conditions required.

## 8.5  FORMATION OF C=N BONDS

Imines are readily formed by the addition of reasonably nucleophilic amines to aldehydes or ketones followed by dehydration (see Section 2.7.1). Other C=N functional groups may be formed by standard dehydrating conditions from a suitable precursor.

### 8.5.1  Carbodiimides from Ureas

Ureas are readily dehydrated by activation with sulfonyl chlorides to form the corresponding carbodiimides.[104] They may also be formed by treatment with a phosphine and halogen.[105]

## 8.6  FORMATION OF NITRILES

Nitriles are readily formed by dehydration of amides, oximes, and similar functional groups. The standard conditions for dehydration of an alcohol are most commonly used. Newer methods tend to be milder than the older conditions and, therefore, more compatible with sensitive functional groups.

### 8.6.1  Nitriles from Amides

Several relatively mild conditions have been reported for the dehydration of amides, preferably by activation as a sulfonate,[106] acetate,[107] or iminoyl chloride.[108]

[102]Sato, T.; Tsuchiya, H.; Otera, J. *Synlett* **1995**, 628–630.

[103]Nakamura, S.; Kusuda, S.; Kawamura, K.; Toru, T. *J. Org. Chem.* **2002**, *67*, 640–647.

[104]Zhang, M.; Vedantham, P.; Flynn, D. L.; Hanson, P. R. *J. Org. Chem.* **2004**, *69*, 8340–8344.

[105]Gibson, F. S.; Park, M. S.; Rapoport, H. *J. Org. Chem.* **1994**, *59*, 7503–7507.

[106]McLaughlin, M.; Mohareb, R. M.; Rapoport, H. *J. Org. Chem.* **2003**, *68*, 50–54.

[107]Ashwood, M. S.; Alabaster, R. J.; Cottrell, I. F.; Cowden, C. J.; Davies, A. J.; Dolling, U. H.; Emerson, K. M.; Gibb, A. D.; Hands, D.; Wallace, D. J.; Wilson, R. D. *Org. Process Res. Dev.* **2004**, *8*, 192–200.

[108]Krynitsky, J. A.; Carhart, H. W. *Org. Synth.* **1952**, *32*, 65–67.

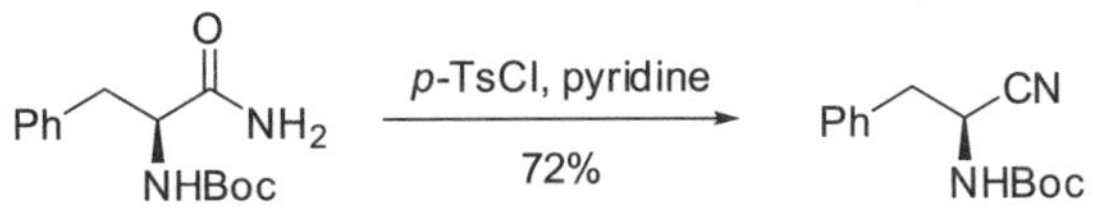

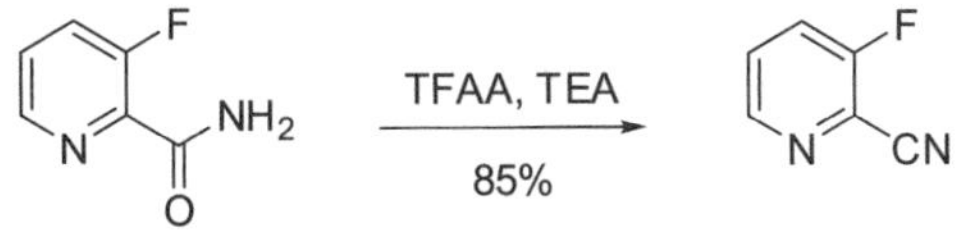

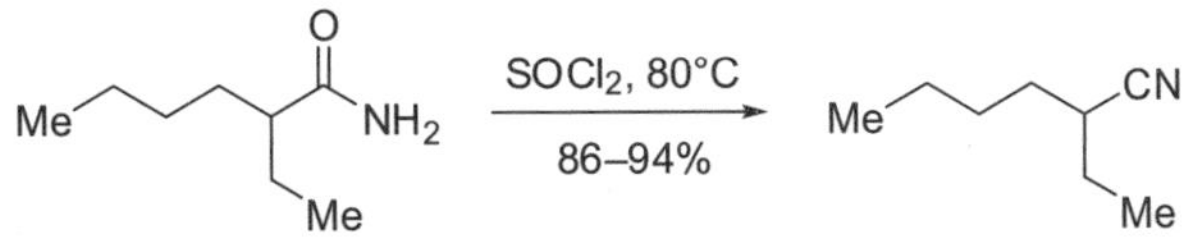

A more unusual, but exceedingly mild elimination has been reported for the sulfonamide derivative, which eliminates by treatment with a mild base.[109] The limitation to this method is the need to make the acyl sulfonamide, which can be generated from the acid[110] or installed directly, as shown below.

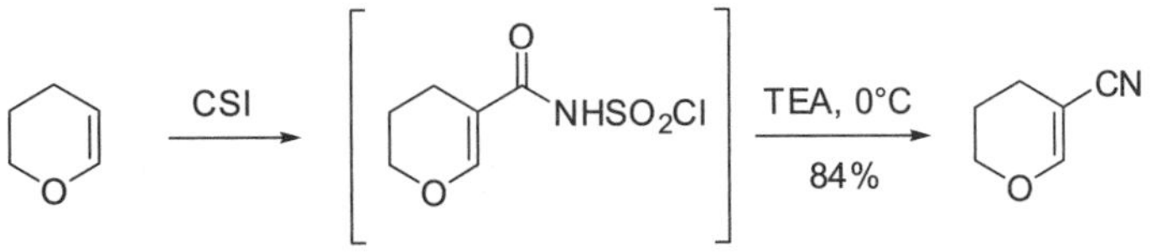

CSI = chlorosulfonylisocyanate

### 8.6.2  Nitriles from Cleavage of N–O Bonds

The cleavage of N–O bonds has been used to generate nitriles, particularly the dehydration of oximes. A wide range of dehydrating conditions has been employed. Some of the mildest involve formation of the silyl aldoximine followed by base decomposition[111] or dehydration with thionyl chloride.[112]

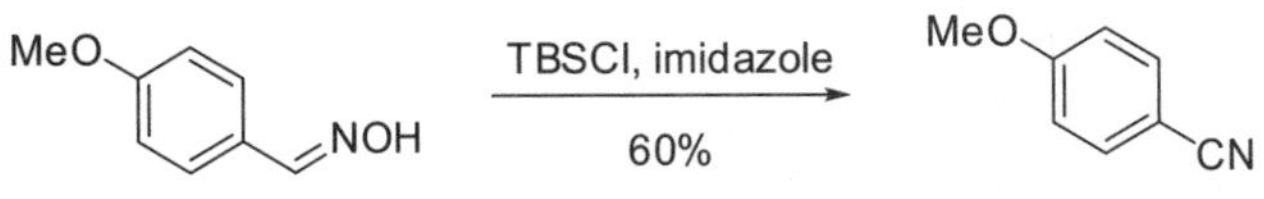

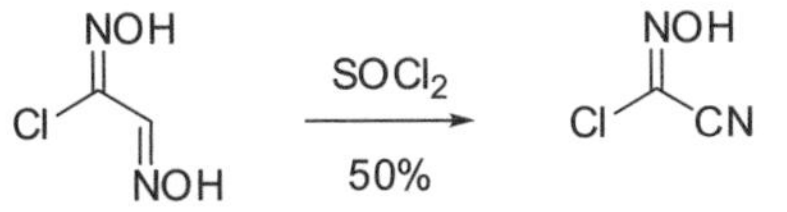

[109]Vorbruggen, H.; Krolikiewicz, K. *Tetrahedron* **1994**, *50*, 6549–6558.
[110]Lohaus, G. *Org. Synth.* **1970**, *50*, 18–21.
[111]Ortiz-Marciales, M.; Pinero, L.; Ufret, L.; Algarin, W.; Morales, J. *Synth. Commun.* **1998**, *28*, 2807–2811.
[112]Kozikowski, A. P.; Adamcz, M. *J. Org. Chem.* **1983**, *48*, 366–372.

Oximes substituted with a carboxylic acid can spontaneously decarboxylate during the dehydration. This is most conveniently accomplished by treatment with an anhydride.[113]

Isoxazoles can also serve as latent cyanoaldehydes, but must usually be trapped *in situ*. The ring opening occurs readily with almost any base.[114]

### 8.6.3  Isonitriles from Formamides

Isonitriles are readily formed from the corresponding *N*-formyl precursors. Most of the standard dehydrating conditions work well for this transformation, including the elimination of sulfonate,[115] sulfate,[116] and phosphate derivatives.[117]

[113]Divald, S.; Chun, M. C.; Joullie, M. M. *J. Org. Chem.* **1976**, *41*, 2835–2846.
[114]Tarsio, P. J.; Nicholl, L. *J. Org. Chem.* **1957**, *22*, 192–193.
[115]Swindell, C. S.; Patel, B. P.; DeSolms, S. J.; Springer, J. P. *J. Org. Chem.* **1987**, *52*, 2346–2355.
[116]Meyers, A. I.; Bailey, T. R. *J. Org. Chem.* **1986**, *51*, 872–875.
[117]Cristau, P.; Vors, J.-P.; Zhu, J. *Tetrahedron* **2003**, *59*, 7859–7870.

## 8.7  FORMATION OF KETENES AND RELATED COMPOUNDS

Ketenes, isocyanates, and isothiocyanates are all readily formed by elimination from a suitable precursor, such as an acid, amide or carbamate derivative.

### 8.7.1  Ketenes

Ketenes are easily formed by treatment of the corresponding activated acid precursor with mild base or heating. Acid chlorides are the preferred precursor to ketenes, as they eliminate with very mild bases,[118] or spontaneously on standing.

A ketene may also be formed by treatment of an α-halo acid halide with zinc.[119]

Esters with α-activating groups have been demonstrated to eliminate to the corresponding ketene on heating, and can be trapped *in situ* with an alcohol or amine to form the ester or amide product.[120] In this approach, hindered esters are preferred over less-substituted esters, since the ketene formation occurs at lower temperatures.

[118]Schmittel, M.; Von Seggern, H. *J. Am. Chem. Soc.* **1993**, *115*, 2165–2177.
[119]Baigrie, L. M.; Seiklay, H. R.; Tidwell, T. T. *J. Am. Chem. Soc.* **1985**, *107*, 5391–5396.
[120]Witzeman, J. S.; Nottingham, W. D. *J. Org. Chem.* **1991**, *56*, 1713–1718.

## 8.7.2 Isocyanates and Isothiocyanates

Isocyanates form readily from the corresponding carbamoyl chloride by treatment with a mild base.[121]

Isocyanates are also formed during the Curtius rearrangement (see Section 7.2.3.2), but are not usually isolated during that reaction. The potential for high-energy decomposition of the intermediate acyl azide makes it a less than desirable method.

Isothiocyanates may be formed from amines by a two-step procedure involving formation of the dithiocarbamate followed by carboxylation and elimination with base. Caution should be used with this procedure as it liberates carbonyl sulfide (COS).[122]

## 8.7.3 Ketimines

Substituted amides can be dehydrated to form ketimines. The elimination requires relatively strong dehydrating conditions to be successful. The most commonly employed conditions utilize triphenylphosine/bromine to effect the transformation.[123] Other dehydrating conditions, such as $PCl_5$ have also been demonstrated.[124]

[121] Sitzmann, M. E.; Gilligan, W. H. *J. Org. Chem.* **1985**, *50*, 5879–5881.

[122] Hodgkins, J. E.; Reeves, W. P. *J. Org. Chem.* **1964**, *29*, 3098–3099.

[123] Motoyoshiya, J.; Teranishi, A.; Mikoshiba, R.; Yamamoto, I.; Gotoh, H.; Enda, J.; Ohshiro, Y.; Agawa, T. *J. Org. Chem.* **1980**, *45*, 5385–5387.

[124] Hiroi, K.; Sato, S. *Chem. Pharm. Bull. (Tokyo).* **1985**, *33*, 2331–2338.

## 8.8  FRAGMENTATIONS

### 8.8.1  Grob Fragmentations

Carbon frameworks containing an alcohol at one terminus and a leaving group at the other, of the general structure $HO-C-C-C-X$, can be fragmented to give a carbonyl and an alkene (Grob fragmentation) if the correct orbital overlap is present. A base strong enough to deprotonate the alcohol is all that is needed to effect this transformation.[125] The fragmentation is frequently employed to synthesize medium-sized rings.[126]

The product aldehyde or ketone is not always stable to isolation, so sodium borohydride or another reducing agent is often added to reduce the carbonyl produced *in situ*.[127]

The 3-aza variant of this fragmentation gives rise to an amino alcohol after reduction of the resulting imine and aldehyde.[55]

[125]Mehta, G.; Karmakar, S.; Chattopadhyay, S. K. *Tetrahedron* **2004**, *60*, 5013–5017.
[126]Villagomez-Ibarra, R.; Alvarez-Cisneros, C.; Joseph-Nathan, P. *Tetrahedron* **1995**, *51*, 9285–9300.
[127]Lamers, Y. M. A. W.; Rusu, G.; Wijnberg, J. B. P. A.; De Groot, A. *Tetrahedron* **2003**, *59*, 9361–9369.

## 8.8.2   Eschenmoser Fragmentations

The elimination and subsequent fragmentation of an epoxy-hydrazone to a keto-alkyne (Eschenmoser fragmentation) is readily accomplished under very mild conditions. The hydrazone is usually formed *in situ*, and the reaction can be carried out at low temperatures in acetic acid[128] or with mild heating.[129]

## 8.8.3   Formation of Benzenes

Bicyclo[3.1.0]hexane rings that are appropriately substituted with leaving groups, usually halogens, will spontaneously or under basic conditions fragment to give benzene rings.[130]

[128]Kocienski, P. J.; Ostrow, R. W. *J. Org. Chem.* **1976**, *41*, 398–400.
[129]Dai, W.; Katzenellenbogen, J. A. *J. Org. Chem.* **1993**, *58*, 1900–1908.
[130]Jenneskens, L. W.; De Wolf, W. H.; Bickelhaupt, F. *Synthesis* **1985**, 647–649.

This methodology has been used to synthesize rings with substitution patterns that would be difficult to achieve by nucleophilic or electrophilic aromatic substitution reactions (see Chapters 4 and 5).

### 8.8.4   Extrusion of $N_2$

Cyclopropanes can be synthesized in good yield by the extrusion of nitrogen from 1-pyrazolines. The reaction is most easily carried out by refluxing the starting material in a medium- to high-boiling solvent.[131] The isomeric 2-pyrazolines will also eliminate, but only under conditions that isomerize it to the 1-isomer first.

Triazolines will also extrude nitrogen to form aziridines. The temperature of the extrusion can vary depending on the nitrogen substituent; the carbamate derivative eliminates at moderate temperatures,[132] while alkyl or aryl substituents can require much higher temperatures.

### 8.8.5   Extrusion of Sulfur

As noted earlier, sulfur extrusion from episulfides is a well-established method for formation of alkenes (see Section 8.2.15). A special case arises when the sulfur is part of a β-dicarbonyl

---

[131]Srivastava, V. P.; Roberts, M.; Holmes, T.; Stammer, C. H. *J. Org. Chem.* **1989**, *54*, 5866–5870.
[132]Tanida, H.; Tsuji, T.; Irie, T. *J. Org. Chem.* **1966**, *31*, 3941–3947.

system or some functional equivalent. Treatment of these systems with a phosphine also results in extrusion of the sulfide, in this case to form the β-dicarbonyl or a tautomer, but at much lower temperatures.[133]

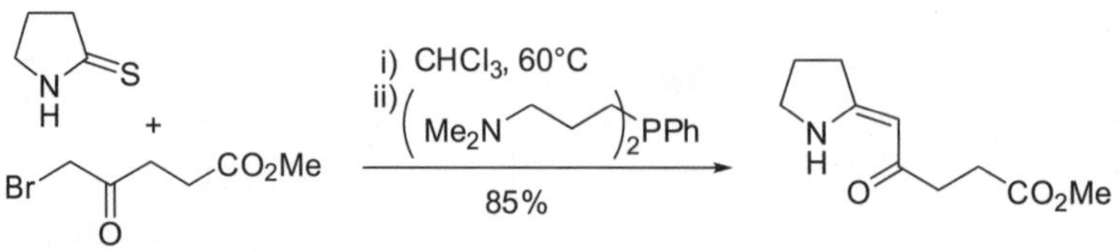

### 8.8.6  Multi-Component Extrusions

Several different multi-component extrusions, of the general formula shown below, have been carried out to form alkenes. While these eliminations are generally not practical as truly preparative processes, they can be carried out on the laboratory scale. The most useful usually involve a sulfide as one of the components, removed with a phosphine, and another group that eliminates under the high temperatures of the reaction, such as nitrogen.[134]

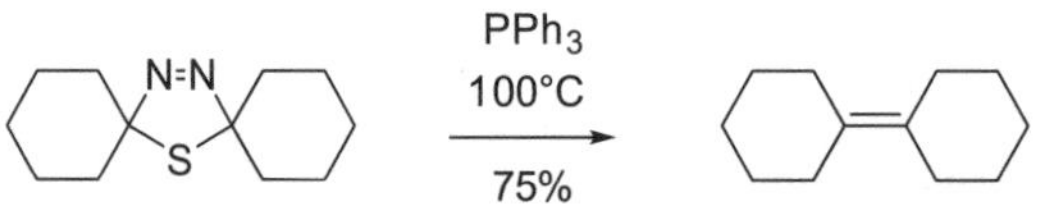

## 8.9  DEHYDRATING REAGENTS

Table 8.1 summarizes some of the more commonly used reagents for dehydration. Specific references may be found in the relevant sections above for the reagents that have the widest applicability or are preferred options. References given below for these reagents are for review articles on their use. The final three entries in the table are not generally suitable for large scale use because of availability, stability, safety, or toxicological concerns. However, they may be well suited to small scale use, especially in the academic environment. Finally, the series *Encyclopedia of Reagents for Organic Synthesis*[135] is a good reference source on the physical properties, handling recommendations, and reactivity of these and other reagents.

[133]Roth, M.; Dubs, P.; Goetschi, E.; Eschenmoser, A. *Helv. Chim. Acta* **1971**, *54*, 710–734.

[134]Barton, D. H. R.; Willis, B. J. *J. Chem. Soc., Perkin Trans. 1* **1972**, 305–310.

[135]Paquette, L. A., (editor-in-chief) *Encyclopedia of Reagents for Organic Synthesis*; John Wiley & Sons Ltd: West Sussex, U.K., 1995.

**TABLE 8.1  Reagents for Dehydration**

| Reagent | Structure | Uses | Limitations |
|---|---|---|---|
| Sulfuric acid ($H_2SO_4$) | $HO-\overset{\displaystyle O}{\underset{\displaystyle O}{S}}-OH$ | Alkenes from 2°, 3° OH, β-silyl OH | Not compatible with acid-sensitive functionality and some solvents |
| p-Toluenesulfonic acid (tosic acid, p-TsOH))[136] | tolyl–SO₃H (Me-substituted) | Alkenes from 1°, 2°, 3° OH<br>Nitriles from oximes<br>N-oxides from nitro | May not be compatible with acid-sensitive functionality |
| Acetic anhydride ($Ac_2O$) | (acetic anhydride structure) Me, O | Alkenes from 1°, 2°, 3° OH<br>Nitriles from aldoximes | Competitive acylation of other nucleophilic functional groups<br>Oximes give N-acyl enamines |
| Trifluoroacetic anhydride (TFAA) | $F_3C$, O, $CF_3$ | Nitriles from amides, oximes | Competitive acylation of other nucleophilic functional groups |
| Methanesulfonyl chloride (MsCl) | $Me-\overset{\displaystyle O}{\underset{\displaystyle O}{S}}-Cl$ | Alkenes from 1°, 2°, 3° OH, β-silyl OH, iodohydrins<br>Isonitriles from formamides | Competitive sulfonylation and/or chlorination without elimination |
| Toluenesulfonyl chloride (TsCl) | tolyl–SO₂Cl (Me-substituted) | Nitriles from amides<br>Isonitriles from formamides<br>Carbodiimides from ureas | Sulfonylation of alcohols or amines may be competitive |
| Thionyl chloride ($SOCl_2$) | $Cl-\overset{\displaystyle O}{S}-Cl$ | Nitriles from amides, oximes<br>Isonitriles from formamides | May not be compatible with acid-sensitive functionality |

(continued)

TABLE 8.1 (*Continued*)

| Reagent | Structure | Uses | Limitations |
|---|---|---|---|
| Phosphorus oxychloride (POCl₃)[137] | $Cl-\underset{\underset{Cl}{\|}}{\overset{\overset{O}{\|}}{P}}-Cl$ | Alkenes from<br>Nitriles from amides<br>Isonitriles from formamides<br>Carbodiimides from ureas | May not be compatible with acid-sensitive functionality; competitive chloro-dehydration |
| Cyanuric chloride (Trichloroisocyanuric acid, 2,4,6-trichloro-1,3,5-triazine)[138] | | Nitriles from amides, aldoximes | Competitively converts alcohols to chlorides |
| Phosphorus pentoxide (P₂O₅) (Eaton's reagent when in MsOH) | | Nitriles from amides | |
| Burgess reagent[139] | $Et_3N^+-\overset{O^-}{\underset{O}{S}}-N^--\overset{O}{\underset{}{C}}-OMe$ | Alkenes from 2°, 3°, allylic alcohols<br>Nitriles from 1° amides<br>Isonitriles from formamides | Does not work for 1° alcohols<br>Competitive $S_N2$ pathway<br>Cost, availability |
| Martin sulfurane[140,141] | | Alkenes from 2°, 3° alcohols or activated 1° alcohols | Cost, availability |
| DCC | | Alkenes from activated alcohols<br>Nitriles from amides, oximes<br>Ketenes from carboxylic acids | Severe irritant, potential sensitizer<br>Difficult to remove urea byproducts |

[136] D'Onofrio, F.; Scettri, A. *Synthesis* **1985**, 1159–1161.
[137] Sharma, S. D.; Kanwar, S. *Indian J. Chem. B Org* **1998**, *37B*, 965–978.
[138] Giacomelli, G.; Porcheddu, A.; de Luca, L. *Curr. Org. Chem.* **2004**, *8*, 1497–1519.
[139] Khapli, S.; Dey, S.; Mal, D. *J. Indian Inst. Sci.* **2001**, *81*, 461–476.
[140] Martin, J. C.; Arhart, R. J. *J. Am. Chem. Soc.* **1971**, *93*, 4327–4329.
[141] Brain, C. T.; Brunton, S. A. *Synlett* **2001**, 382–384.

# 9

# REDUCTIONS

SALLY GUT RUGGERI, STÉPHANE CARON, PASCAL DUBÉ, NATHAN D. IDE,
KRISTIN E. PRICE, JOHN A. RAGAN, AND SHU YU

## 9.1  INTRODUCTION

The reduction of functional groups is one of the fundamental reactions of organic chemistry. Many bond-forming transformations are carried out on highly oxidized molecules because they possess greater reactivity. For example, the deprotonation of a methylene alpha to a carbonyl is trivial compared to an analogous substrate where the carbonyl is reduced. After facilitating the desired chemistry, the oxidized functionality often needs to be reduced to a lower oxidation state to synthesize the compound of interest. Many methods have been developed to effect the reduction of a wide variety of functional groups; the two most commonly used are hydride-based reagents or conditions that employ hydrogen gas with a catalyst. The following chapter describes the most successfully utilized conditions and is arranged by the type of bond being reduced.

## 9.2  REDUCTION OF C–C BONDS

### 9.2.1  Reduction of Alkynes

*9.2.1.1  Reduction of Alkynes to Alkanes*    The reduction of alkynes to alkanes is well precedented for both alkyl- and aryl-substituted alkynes, and can be carried out routinely by catalytic hydrogenation in high yield. The reactions are often carried out in alcohols, ethanol, and methanol in particular. The most frequently used catalysts involve carbon-supported palladium, which is readily available (see Section 20.3.4.3). For most reactions, the addition of hydrogen is diffusion-controlled; sufficient hydrogen pressure (45 psi) as well as good agitation is required for fast conversion. For less-reactive substrates, mild heating is

*Practical Synthetic Organic Chemistry: Reactions, Principles, and Techniques*, First Edition.
Edited by Stéphane Caron.
© 2011 John Wiley & Sons, Inc. Published 2011 by John Wiley & Sons, Inc.

sometimes required to effect reaction completion. The catalyst is generally safe to store and handle, but can ignite the vapor in the headspace of a reaction vessel, particularly when methanol is used as solvent. Water-wet catalysts should be used whenever possible to reduce this potential. Another way to charge the catalyst is to make a toluene or 2-propanol slurry of the catalyst; the transfer of the slurry rarely causes any problems. The catalyst is easily removed (and potentially recycled as well) after the reaction by simple filtration through diatomaceous earth. The alcohol-wet catalyst can also ignite when left dry in air, and should be treated with water before disposal. Under these reaction conditions, a few other functional groups, such as olefins, benzyl ethers, and esters, are also reduced. The aforementioned principles are exemplified in Novartis' practical synthesis of 6-[2-(2,5-dimethoxyphenyl) ethyl]-4-ethylquinazoline.[1]

A similar strategy was used in synthesis of racemic muscone on high-tonnage scale.[2] No scrambling of the chiral center was observed, indicating fast hydrogenation and little migration of the multiple bonds.

### 9.2.1.2   Reduction of Alkynes to Alkenes

The reduction of alkynes to *cis* alkenes is most frequently achieved by hydrogenation using a poisoned palladium catalyst, such as a Lindlar catalyst (palladium on calcium carbonate, lead-poisoned). A Lindlar catalyst is specific to the reduction of alkynes to *cis* alkenes and is rarely used to mediate other hydrogenations. The efficiency of this transformation is often improved when pyridine or quinoline is used as an additive. Under such conditions, over-reduction and migration of the multiple bond(s) are rare. The utility of this transformation is exemplified in Loreau's synthesis of (9Z,12E)-[1-13C]-octadeca-9,12-dienoic acid;[3] note that chemoselectivity was achieved and no migration of the C−C double bond occurred.

[1] Koenigsberger, K.; Chen, G.-P.; Wu, R. R.; Girgis, M. J.; Prasad, K.; Repic, O.; Blacklock, T. J. *Org. Process Res. Dev.* **2003**, *7*, 733–742.

[2] Fehr, C.; Galindo, J.; Farris, I.; Cuenca, A. *Helv. Chim. Acta* **2004**, *87*, 1737–1747.

[3] Loreau, O.; Maret, A.; Poullain, D.; Chardigny, J. M.; Sebedio, J. L.; Beaufrere, B.; Noel, J. P. *Chem. Phys. Lipids* **2000**, *106*, 65–78.

$$n\text{-}C_5H_{11} \quad \text{(alkyne-ol)} \quad \xrightarrow[\substack{\text{Quinoline} \\ \text{EtOAc} \\ 97\%}]{\text{H}_2, \text{Lindlar}} \quad n\text{-}C_5H_{11} \quad \text{(cis-alkene-ol)}$$

While alkynes are efficiently reduced to *cis* alkenes by catalytic hydrogenation, they cannot be easily reduced to *trans* alkenes under similar conditions. Instead, hydride-based reducing agents are employed. A limitation to this methodology is poor chemoselectivity due to the reactivity of the hydride. This transformation is more often used early in the synthesis of simpler substrates, as illustrated in Novartis' process for a dual MMP/TNF inhibitor.[4]

$$\text{HO} \quad \text{(alkyne-OBn)} \quad \xrightarrow[\substack{\text{THF, 0°C} \\ 80\%}]{\text{Red-Al}} \quad \text{HO} \quad \text{(trans-alkene-OBn)}$$

Red-Al = sodium bis(2-methoxyethoxy) aluminum hydride

It should be pointed out that alkynyl ethers can also be reduced to *cis* or *trans* enol ethers using the method described above. For example, both (*Z*)- and (*E*)-1-menthoxy-1-butene are successfully produced from 1-menthoxy-1-butyne, as reported by Greene and coworkers.[5]

$$\text{(trans enol ether)} \quad \xleftarrow[\substack{\text{THF} \\ 94\%}]{\text{LiAlH}_4} \quad \text{(menthoxy-butyne)} \quad \xrightarrow[\substack{\text{H}_2 (1 \text{ atm}) \\ 90\%}]{\substack{\text{Pd/BaSO}_4 \\ \text{pyridine}}} \quad \text{(cis enol ether)}$$

## 9.2.2  Reduction of Alkenes

For the reduction of alkenes that are not in conjugation with an electron-withdrawing group(s), the most commonly used method is again hydrogenation, which will be discussed in this section. For conjugate reduction, see Section 9.2.4.

### 9.2.2.1  Reduction of Alkenes to Alkanes Without Facial Selectivity   The reduction of alkenes to alkanes can be achieved using heterogeneous catalysts under similar conditions to the hydrogenation of alkynes (*vide supra*). This is a technology that has been utilized extensively, especially when no stereocontrol is necessary. Suitable substrates include terminal alkenes, 1,2-disubstituted alkenes, and other alkenes with certain symmetry in the product. 1,2-Alkenes with aryl substitutents are more challenging to reduce, as conjugation

[4]Koch, G.; Kottirsch, G.; Wietfeld, B.; Kuesters, E. *Org. Process Res. Dev.* **2002**, *6*, 652–659.
[5]Kann, N.; Bernardes, V.; Greene, A. E. *Org. Synth.* **1997**, *74*, 13–22.

is lost in the process. In these cases, mild heating can increase the likelihood of reduction, as demonstrated by Novartis in their synthesis of methyl 5-[2-(2,5-dimethoxyphenyl)ethyl]-2-hydroxybenzoate.[6]

Catalytic hydrogenation works equally well on enamides. In Pfizer's synthesis of a thromboxane receptor,[7] an enamide function was reduced using 5% Pd/C as catalyst. Note that the alkene in conjugation with the ethyl ester was also reduced in this example. For more on the reduction of $\alpha,\beta$-unsaturated carbonyl compounds, see Section 9.2.4.

In the reduction of 1,1-disubstitued alkenes where the two substituents are different, the product will be racemic. If the racemates can be easily resolved, this is a viable route to a chiral target (see Chapter 14). This strategy was used by Neurocrine Biosciences in the synthesis of NBI-75043.[8] Note that a platinum catalyst was employed in this case, an approach often used for sulfur-containing substrates, where poisoning of palladium catalyst can be severe.

[6]Kucerovy, A.; Li, T.; Prasad, K.; Repic, O.; Blacklock, T. J. *Org. Process Res. Dev.* **1997**, *1*, 287–293.
[7]Waite, D. C.; Mason, C. P. *Org. Process Res. Dev.* **1998**, *2*, 116–120.
[8]Gross, T. D.; Chou, S.; Bonneville, D.; Gross, R. S.; Wang, P.; Campopiano, O.; Ouellette, M. A.; Zook, S. E.; Reddy, J. P.; Moree, W. J.; Jovic, F.; Chopade, S. *Org. Process Res. Dev.*, **2008**, *12*, 929–939.

When multiple olefin moieties are present, it is possible to effect selective reduction for some substrates. One general rule is that the difficulty of hydrogenation increases as substitution increases (i.e., in terms of reactivity: terminal alkene > disubstituted olefin > trisubstituted olefin > tetrasubstituted olefin). The selectivity is amplified when a soluble palladium catalyst is used, such as Wilkinson's catalyst [chlorotris(triphenylphosphine) rhodium(I)], as illustrated by Ireland in the following example; note that the enone function remains intact.[9]

For highly substituted olefins, especially tetrasubstituted olefins, heterogeneous hydrogenation is sometimes difficult, and undesired olefin isomerization can occur prior to hydrogenation, which will afford undesired product. For instance, in the synthesis of (*R*)-muscone, Firmenich scientists noticed that Pd/C was efficient for reduction of the olefin shown,[10] but partial racemization occurred in the process due to olefin migration. On the other hand, when Crabtree's catalyst [(tricyclohexylphosphine)(1,5-cyclooctadiene)(pyridine)iridium(I) hexafluorophosphate] was used,[11] (*R*)-muscone formed cleanly.

***9.2.2.2 Reduction of Alkenes to Alkanes with Facial Selectivity*** When 1,1-disubstituted, trisubstituted and tetrasubstituted olefins are hydrogenated, the delivery of dihydrogen to different faces of the olefin will potentially afford a pair of isomers. A high degree of facial selection can be achieved based on the presence of pre-existing chiral centers in the substrate (diastereotopic), which makes one face easier to access than the other (substrate controlled, diastereoselective). In this regard, homogeneous hydrogenation using Crabtree's catalyst is particular effective, as illustrated in Padwa's total synthesis of (+)-stenine.[12]

[9]Ireland, R. E.; Bey, P. *Org. Synth.* **1973**, *53*, 63–65.
[10]Branca, Q.; Fischli, A. *Helv. Chim. Acta* **1977**, *60*, 925–944.
[11]Saudan, L. A. *Acc. Chem. Res.* **2007**, *40*, 1309–1319.
[12]Ginn, J. D.; Padwa, A. *Org. Lett.* **2002**, *4*, 1515–1517.

When the substrate is achiral and facial selection is desired, a chiral reagent can be used (reagent controlled, enantioselective). For instance, if the metal center is modified by a chiral ligand(s), the delivery of dihydrogen can occur preferentially at one face, giving high enantio excess in the product. This possibility was first demonstrated on a special family of olefins–enamides by Knowles in the now-famous Monsanto process for the commercial manufacture of L-DOPA (Rh/CAMP, see Section 9.2.4.2). In the last 40 years, great improvements have been made in the efficiency of the catalysts for the hydrogenation of enamides, the most remarkable being the one made by Burk at DuPont, who introduced DuPhos in 1991. DuPhos is to date the most general ligand for the hydrogenation of enamides, often with >98% selectivity.

Unfortunately, for the hydrogenation of other families of C–C double bonds, the outcomes are much less predictable. This brings about a sad but true fact with following: In general there is no single "best" catalyst for any class of enantioselective reductions. It is not typical to just pick a chiral catalyst/ligand off the shelf and get good results on the first try; for each specific substrate, some screening of reaction conditions will likely be required, including metal, ligand, additive, solvent, temperature, and hydrogen pressure among others. In addition, screening of known ligands may not provide satisfactory results. Under such circumstances, one needs to systematically synthesize chiral ligands and study the ligand structure–activity relationship, using empirical results to guide the design of more efficient ligands. Homogeneous hydrogenations often require elevated temperature and much higher hydrogen pressures (20–50 bar are commonplace), an aspect that should be considered before investing a large amount of resources.

In part for these reasons, there are fewer successful homogeneous hydrogenation precedents, and most of them are reported by industry. However, some trends have been observed since 2000 that warrant further discussion. Most of the successfully scaled-up homogeneous hydrogenations were reported on olefins with Lewis basic heteroatom(s) in the vicinity, such as allyl alcohols or allyl amines. These heteroatoms may serve as ligands (anchoring points) in the transition state that brings the olefin, hydride, and chiral ligand together on the metal center. Such tight transition states are required for high-level facial differentiation.

The classic example of a ruthenium-catalyzed enantioselective hydrogenation of allylic alcohols is from Noyori's work on (S)-(−)-citronellol.[13] Note that the distal trisubstituted olefin is intact, indicating again the role of coordination of the allylic hydroxyl group to the catalyst.

[13]Takaya, H.; Ohta, T.; Inoue, S.-i.; Tokunaga, M.; Kitamura, M.; Noyori, R. *Org. Synth.* **1995**, *72*, 74–85.

Merck has reported[14] the rhodium-catalyzed hydrogenation of a trisubstituted olefin with an amide function at the allylic position. Despite extensive screening of reaction conditions, the maximum enantiomeric excess (ee) achieved was only 88%, demonstrating the challenge of enantioselective hydrogenation. The chemistry has been scaled up on pilot plant scale.

The most difficult class of substrates for enantioselective hydrogenation is an alkene without proximal heteroatoms that can bind the catalyst. In this arena, Pfaltz's P, N-ligand-based iridium catalyst is one of the better systems to try. Although these transformations have not been demonstrated on scale, they hold the promise to become powerful tools in the future. At least on lab scale, this class of catalyst has been reported to catalyze the hydrogenation of trisubstituted vinyl ethers, trisubstituted allylic alcohols, and trisubstituted alkyl/alkyl olefins. In the latter case, remarkable chemoselectivity was achieved as well.

[14]Limanto, J.; Shultz, C. S.; Dorner, B.; Desmond, R. A.; Devine, P. N.; Krska, S. W. *J. Org. Chem.* **2008**, *73*, 1639–1642.

BАrF⁻ = tetrakis[3,5-bis (trifluoromethyl) phenyl] borate

> 99% conv., > 94% ee

Pfaltz reported that the counter ion is critical for the performance of the catalyst. Considering its importance, a few extra remarks are warranted. Pfaltz and coworkers observed that although hydrogenation is fast at the initial stage for a number of iridium catalysts, the reaction slows down rapidly as the reaction progresses, probably due to product inhibition of the catalyst.[15] Catalyst deactivation is markedly reduced when a non-coordinating counter ion, such as tetraarylborate, is employed. Caution should be taken not to introduce extra counter ions that could potentially poison the catalyst.

This catalytic system tolerates diverse substitution patterns and the presence of hetero-atoms in the substrates.

[15]Bell, S.; Wuestenberg, B.; Kaiser, S.; Menges, F.; Netscher, T.; Pfaltz, A. *Science* **2006**, *311*, 642–644.

50 bar H₂
CH₂Cl₂, rt, 2 h

100% conv., 98% ee

Reduction of unactivated tetrasubstituted olefins is the most challenging hydrogenation. In this field, iridium is again the metal that has performed the best. Earlier pioneering work by Crabtree, who showed that his reagent can catalyze hydrogenation of tetrasubstituted olefins, laid the foundation for asymmetric hydrogenation. Pfaltz's P,N ligands are the most efficient.[16] Note that the hydrogenation of tetrasubstituted olefins generates two adjacent chiral centers in a single step.

$[Ir(L)(COD)]^+ BArF^-$

50 bar H₂

CH₂Cl₂, rt

>99% conversion
94% ee

L =

### 9.2.3 Reduction of Aromatic Rings and Heterocycles

The reduction of aromatic rings generally requires forceful conditions, as aromaticity is lost in the process. For this reason, benzene rings are difficult to reduce, while partial reduction of naphthalene rings is easier than benzene rings. Similar trends have been observed for 5-, and 6-membered heterocycles and their benzo-fused counter parts. There are two general methods for aromatic ring reduction: dissolving metal reduction (Birch reduction) and hydrogenation. The former is an older technology that requires little experimentation (typical conditions are lithium or sodium in refluxing liquid ammonia). Since many functional groups are reduced under these conditions, the utility of this transformation is

[16]Roseblade, S. J.; Pfaltz, A. *Acc. Chem. Res.* **2007**, *40*, 1402–1411.

limited. Hydrogenation is more amenable to a broader class of substrates and can be enantioselective. Screening of the catalyst and ligands is to be expected for this type of substrate and the required hydrogen pressure is often high.

### 9.2.3.1   Reduction of Benzene and Naphthalene Rings

The classic method to reduce benzene or naphthalene rings to partially reduced aromatic rings is a dissolving metal reduction, commonly known as a Birch reduction. This method works for both electron-rich[17] and electron-deficient[18] benzene rings to yield dihydrobenzene with isolated double bonds as justified by the *principle of least motion*.[19] Alkyl and alkoxy benzenes yield 2,5-dihydro products; benzoates afford 1,4-dihydro products.[20] The yields are normally high.

77–92%

89–95%

The benzene ring can also be reduced to cyclohexane by hydrogenation under forcing conditions.[21] Hoffmann-La Roche reported an efficient synthesis of oseltamivir phosphate (Tamiflu), where they fully reduced a pentasubstituted benzene ring.[22] Among the metal catalysts investigated, platinum and nickel were inactive, while rhodium and ruthenium both gave desired product. The best conditions were Ru/Al$_2$O$_3$ in ethyl acetate at 60°C, but 100 bar H$_2$ was required to effect the desired transformation. The high level of stereocontrol is very impressive.

82%

[17]Paquette, L. A.; Barrett, J. H. *Org. Synth.* **1969**, *49*, 62–65.

[18]Kuehne, M. E.; Lambert, B. L. *Org. Synth.* **1963**, *43*, 22–24.

[19]Hine, J. *Advances in Physical Organic Chemistry* **1977**, *15*, 1–61.

[20]Birch, A. J.; Hinde, A. L.; Radom, L. *J. Am. Chem. Soc.* **1980**, *102*, 3370–3376.

[21]Rylander, P. *Catalytic Hydrogenation in Organic Syntheses*; Academic Press: New York, N. Y., **1979**.

[22]Zutter, U.; Iding, H.; Spurr, P.; Wirz, B. *J. Org. Chem.* **2008**, *73*, 4895–4902.

Novartis reported a reduction of a naphthalene moiety using lithium in butanol at 95°C to afford 1,4-dihydronaphthalene.[23] The reduction occurs at the more-substituted side of the naphthalene. The transformation was scaled up to 8.2 kg without incident. The fact that the naphthalene ring was only partially reduced, and the benzene ring is intact in the product, once again demonstrates the striking difference in the reactivity of these ring systems.

### 9.2.3.2 Reduction of Pyridines and Quinolines

*9.2.3.2 Reduction of Pyridines and Quinolines*   Compared to a phenyl ring, a pyridine ring is easier to reduce by hydrogenation (not in absolute terms, but relative to all-carbon arenes).[24] This is especially true when the nitrogen atom on the substrate is activated with an alkyl or acyl group. Pfizer reported the synthesis of *cis-N*-protected-3-methylamino-4-methylpiperidine via the hydrogenation of the corresponding pyridine using a 5% Rh/C catalyst.[25] Screening of solvents revealed that acetic acid is optimal, likely because the pyridine ring is hydrogenated more easily when protonated. Under such conditions, only a moderate 70–80 psi hydrogen pressure is required. The chemistry has been successfully demonstrated on >50 kg scale.

Pfaltz has demonstrated the feasibility of enantioselective hydrogenation of substituted pyridines, as in the example below.

[23]Baenziger, M.; Cercus, J.; Stampfer, W.; Sunay, U. *Org. Process Res. Dev.* **2000**, *4*, 460–466.

[24]Freifelder, M. *Advan. Catalysis* **1963**, *14*, 203–253.

[25]Cai, W.; Colony, J. L.; Frost, H.; Hudspeth, J. P.; Kendall, P. M.; Krishnan, A. M.; Makowski, T.; Mazur, D. J.; Phillips, J.; Ripin, D. H. B.; Ruggeri, S. G.; Stearns, J. F.; White, T. D. *Org. Process Res. Dev.* **2005**, *9*, 51–56.

98%, 90% ee

For partial reduction of pyridines, the classical approach is to treat a pyridinium salt with sodium borohydride. This approach was also employed by Pfizer scientists in the large scale synthesis of *cis-N*-benzyl-3-methylamino-4-methylpiperidine.[26]

73%

Benzofused heterocycles are easier to hydrogenate, as only part of the aromaticity is lost during the process. For this family of compounds, even enantioselective hydrogenation has been achieved on reasonable scale. Often times, the strategy is to simultaneously activate the catalyst (by addition of iodine), as well as the substrate (by addition of acid). Iridium-based catalysts are more successful than others, in part because Ir(I) can be activated with iodine, presumably due to the oxidation of Ir(I) to more active Ir(III). Although this transformation has only been demonstrated on lab scale, as illustrated in the synthesis of (−)-augustrueine, its application on large scale is expected in the future.

>94%, 94% ee

In addition to hydrogen gas, Hantzsch esters have also been employed as the hydrogen source for the enantioselective reduction of quinolines, in a fashion similar to the Meerwein–Ponndorf–Verley reaction, (see Section 9.4.1.2). The advantage of using Hantzsch esters as the hydrogen source is that no special equipment is required. The transformation below

[26]Ripin, D. H. B.; Abele, S.; Cai, W.; Blumenkopf, T.; Casavant, J. M.; Doty, J. L.; Flanagan, M.; Koecher, C.; Laue, K. W.; McCarthy, K.; Meltz, C.; Munchhoff, M.; Pouwer, K.; Shah, B.; Sun, J.; Teixeira, J.; Vries, T.; Whipple, D. A.; Wilcox, G. *Org. Process Res. Dev.* **2003**, *7*, 115–120.

provides yet another example for the empirical rules of arene reduction: Partial reduction is easier than full reduction, and pyridine rings are easier to reduce than phenyl rings.

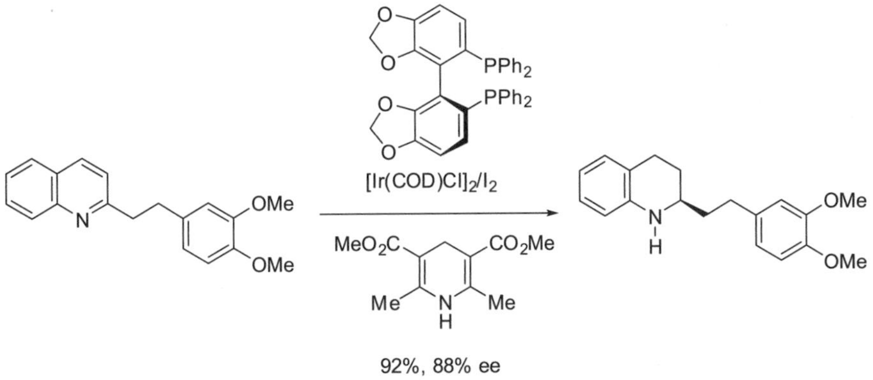

92%, 88% ee

### 9.2.3.3 *Reduction of Furans*

Pfaltz has reported[27] the enantioselective hydrogenation of furans with respectable chiral induction. The potential and utility of this transformation has yet to be explored.

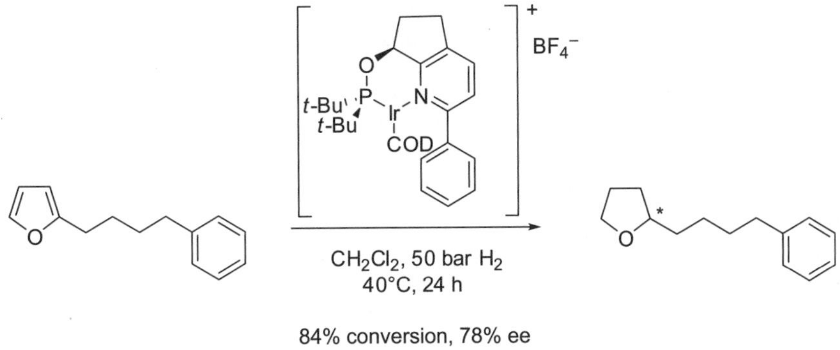

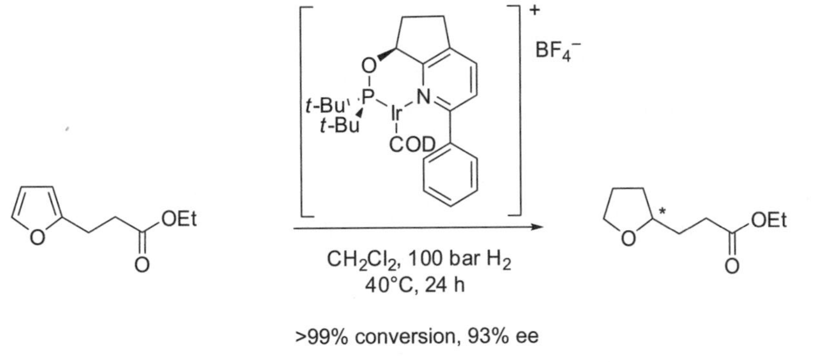

84% conversion, 78% ee

>99% conversion, 93% ee

[27]Saiser, S.; Smidt, S. P.; Pfaltz, A. *Angew. Chem., Int. Ed. Engl.* **2006**, *45*, 5194–5197.

*9.2.3.4  Reduction of Pyrroles and Indoles*    Pyrroles can be partially reduced via dissolving metal reduction. The transformation works well on those pyrroles that bear electron-withdrawing groups, as described in the following example by Cowley,[28] which provided the *trans* isomer as the major product. The authors also demonstrated the scalability of the reaction and the utility of the partially reduced pyrrole in the total synthesis of hyacinthacine A1 and 1-epiaustraline.[29]

Pyrroles are not normally reduced by hydrogenation, as the pyrrolidine products typically poison the catalyst.[30] Pyrrolidines can be efficiently prepared via alkylation of amines. The pyrrole moiety of an indole, on the other hand, can be easily reduced with borohydride, hydrogen or silanes. BMS reported that $Et_3SiH/TFA$ is particularly efficient in their process for a 5HT2C-receptor agonist.[31] The authors did not report the yield of the transformation, because the crude product showed satisfactory purity and was used directly in the next step, but the chemistry was demonstrated on 15 kg scale.

## 9.2.4  Conjugate Reductions

This section will focus on reductions of alkenes and alkynes that are conjugated to an electron-withdrawing group such as a ketone or ester.

*9.2.4.1  Reduction of Enones*    Conjugated enones have most frequently been reduced by transition metal-catalyzed hydrogenation or by dissolving metal reduction. The stereochemical outcome can be different for the two methods, with the former method typically delivering hydrogen from the less-hindered face of the olefin, while the latter method delivers the more thermodynamically stable isomer (via protonation of the intermediate radical anion). Examples of both a hydrogenation[32] and a dissolving metal reduction[33] are given below; they exemplify the divergent stereochemical outcomes described above.

[28]Donohoe, T. J.; Headley, C. E.; Cousins, R. P. C.; Cowley, A. *Org. Lett.* **2003**, *5*, 999–1002.

[29]Donohoe, T. J.; Sintim, H. O.; Hollinshead, J. *J. Org. Chem.* **2005**, *70*, 7297–7304.

[30]Hegedus, L.; Mathe, T. *Applied Catalysis, A: General* **2002**, *226*, 319–322.

[31]Hobson, L. A.; Nugent, W. A.; Anderson, S. R.; Deshmukh, S. S.; Haley, J. J., III; Liu, P.; Magnus, N. A.; Sheeran, P.; Sherbine, J. P.; Stone, B. R. P.; Zhu, J. *Org. Process Res. Dev.* **2007**, *11*, 985–995.

[32]McMurry, J. E. *J. Am. Chem. Soc.* **1968**, *90*, 6821–6825.

[33]Paquette, L. A.; Sauer, D. R.; Cleary, D. G.; Kinsella, M. A.; Blackwell, C. M.; Anderson, L. G. *J. Am. Chem. Soc.* **1992**, *114*, 7375–7387.

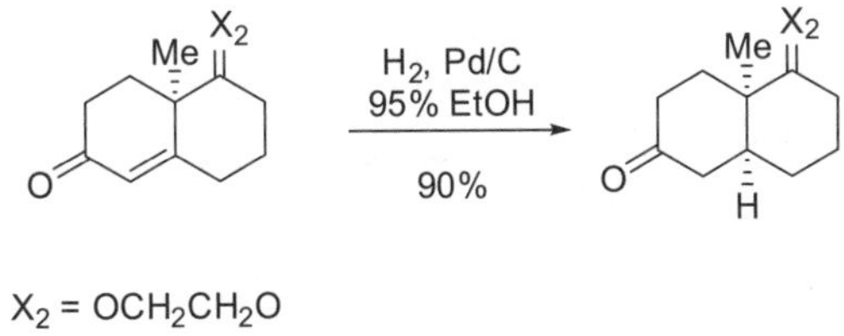

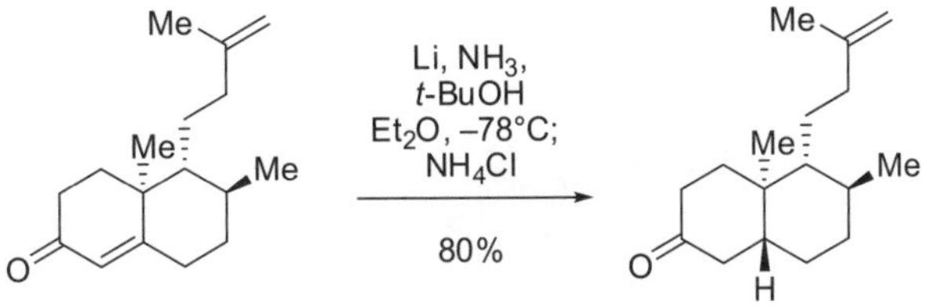

X_2 = OCH_2CH_2O

Asymmetric enone reductions have been reported by Buchwald,[34,35] Lipshutz,[36,37] Mashima,[38] and others. An interesting dynamic kinetic resolution from Buchwald is shown below, which utilizes a copper-catalyzed hydrosilylation with poly(methylhydrosiloxane) (PMHS).[35]

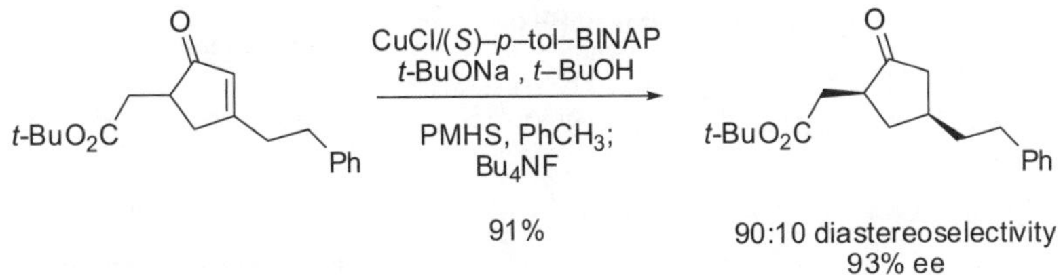

91%    90:10 diastereoselectivity
93% ee

Lipshutz has also developed a copper-catalyzed hydrosilylation catalyst with wide substrate generality, reducing ketones, imines, and α,β-unsaturated esters in addition to enones. An enone reduction is shown below.[37]

[34]Moritani, Y.; Appella, D. H.; Jurkauskas, V.; Buchwald, S. L. *J. Am. Chem. Soc.* **2000**, *122*, 6797–6798.

[35]Jurkauskas, V.; Buchwald Stephen, L. *J. Am. Chem. Soc.* **2002**, *124*, 2892–2893.

[36]Lipshutz, B. H.; Frieman, B. A. *Angewandte Chemie International Edition in English* **2005**, *44*, 6345–6348.

[37]Lipshutz, B. H.; Frieman, B. A.; Tomaso Jr., A. E. *Angewandte Chemie International Edition in English* **2006**, *45*, 1259–1264.

[38]Ohshima, T.; Tadaoka, H.; Hori, K.; Sayo, N.; Mashima, K. *Chem.-Eur. J.* **2008**, *14*, 2060–2066.

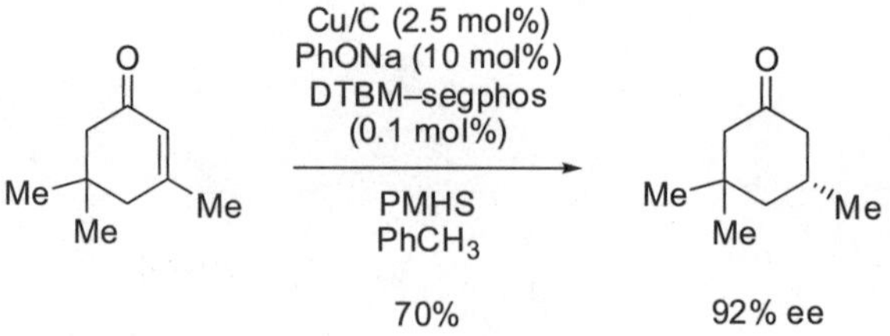

Cu/C (2.5 mol%)
PhONa (10 mol%)
DTBM–segphos
(0.1 mol%)

PMHS
PhCH$_3$

70%                                        92% ee

DTBM–segphos =

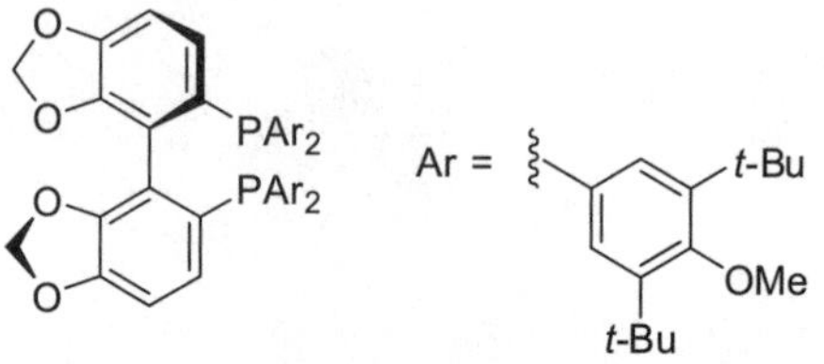

Ar =

An asymmetric hydrogenation using a Ru–Me–DuPHOS catalyst was used by workers at Firmenich to generate (+)-*cis*-methyl dihydrojasmonate,[39] the active component in the perfume ingredient methyl dihydrojasmonate. Interestingly, this perfume agent was originally used as a racemate containing a 90:10 mixture of *trans/cis* isomers, such that the commercial mixture contained just 5% of the active olfactory component.

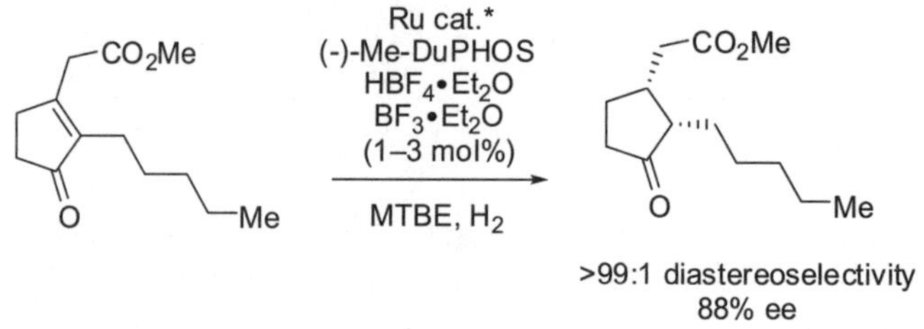

Ru cat.*
(-)-Me-DuPHOS
HBF$_4$•Et$_2$O
BF$_3$•Et$_2$O
(1–3 mol%)

MTBE, H$_2$

>99:1 diastereoselectivity
88% ee

Ru cat.* =
Ru(n$^4$-1,5-COD)(n$^3$-2-propenyl)$_2$=

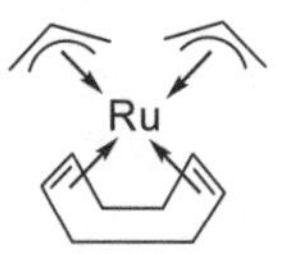

(-)-Me-DuPHOS =

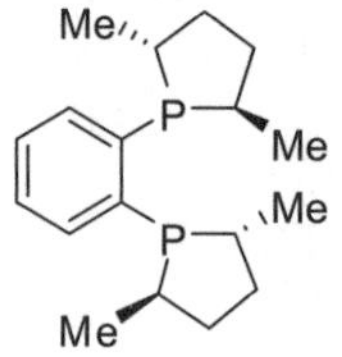

### 9.2.4.2 *Reduction of α,β-Unsaturated Acids and Derivatives*

Olefins conjugated with esters, amides, and nitriles can frequently be selectively reduced by metal-catalyzed hydrogenation. Two examples are shown below for an *N*-acyloxazolidinone[40] and an

[39]Dobbs, D. A.; Vanhessche, K. P. M.; Brazi, E.; Rautenstrauch, V.; Lenoir, J.-Y.; Genet, J.-P.; Wiles, J.; Bergens, S. H. *Angew. Chem., Int. Ed. Engl.* **2000**, *39*, 1992–1995.
[40]Burke, T. R.; Liu, D.-G.; Gao, Y. *J. Org. Chem.* **2000**, *65*, 6288–6291.

ester-nitrile.[41]

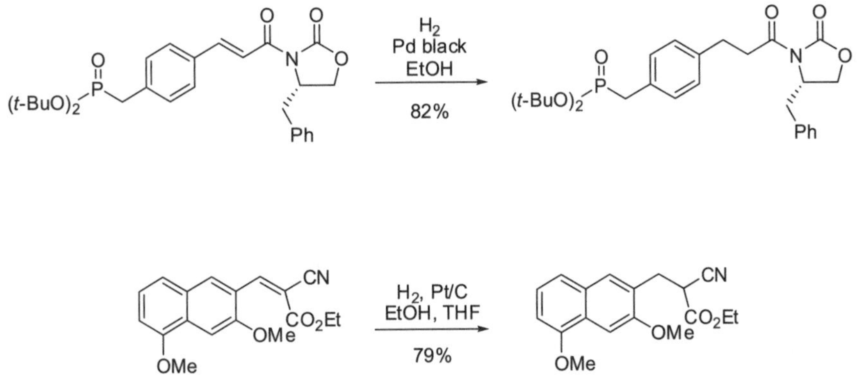

An example in which a pyridine ring is simultaneously hydrogenated over a Rh/Al$_2$O$_3$ catalyst is shown below.[42]

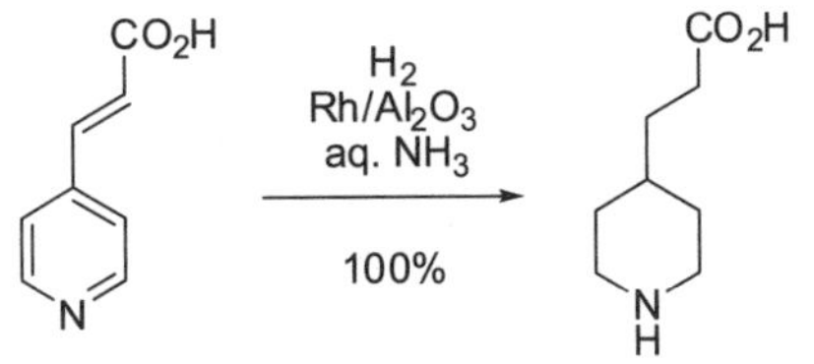

A particularly useful subset of this reaction class is the asymmetric hydrogenation of α,β-unsaturated amides and esters containing a β-amino (or N-acyl) moiety. While working at Monsanto in the 1970's, Knowles pioneered this class of reaction.[43,44] This chemistry plays an important role in the production of many fine chemicals and pharmaceuticals. Noyori has also been a major contributor in this area,[44] and developed the widely used BINAP ligand.[45] Numerous other chiral disphosphine ligands have also been developed.

Feringa has developed a BINOL-derived monodentate ligand (MonoPhos), which is also useful in the asymmetric hydrogenation of enamides.[46] This ligand is particularly attractive because of its low cost.

[41]Baenziger, M.; Kuesters, E.; La Vecchia, L.; Marterer, W.; Nozulak, J. *Org. Process Res. Dev.* **2003**, *7*, 904–912.

[42]Cohen, J. H.; Bos, M. E.; Cesco-Cancian, S.; Harris, B. D.; Hortenstine, J. T.; Justus, M.; Maryanoff, C. A.; Mills, J.; Muller, S.; Roessler, A.; Scott, L.; Sorgi, K. L.; Villani, F. J., Jr.; Webster, R. R. H.; Weh, C. *Org. Process Res. Dev.* **2003**, *7*, 866–872.

[43]Knowles, W. S. *Acc. Chem. Res.* **1983**, *16*, 106–112.

[44]Knowles, W. S.; Noyori, R. *Acc. Chem. Res.* **2007**, *40*, 1238–1239.

[45]Miyashita, A.; Yasuda, A.; Takaya, H.; Toriumi, K.; Ito, T.; Souchi, T.; Noyori, R. *J. Am. Chem. Soc.* **1980**, *102*, 7932–7934.

[46]van den Berg, M.; Minnaard, A. J.; Schudde, E. P.; van Esch, J.; de Vries, A. H. M.; de Vries, J. G.; Feringa, B. L. *J. Am. Chem. Soc.* **2000**, *122*, 11539–11540.

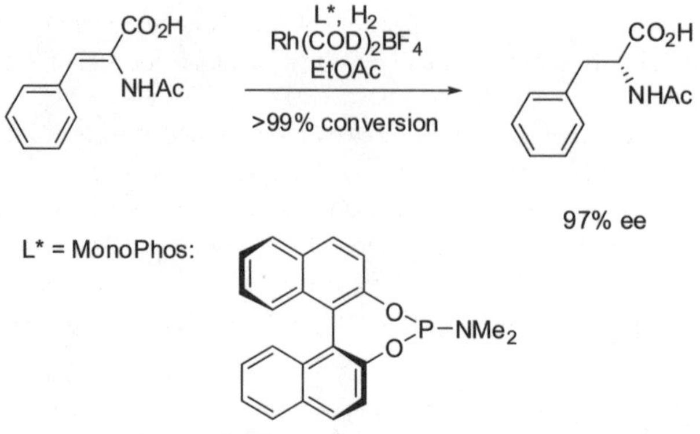

L* = MonoPhos:

A useful example from Burk's group is shown below, using the commercially available Et-DuPHOS ligand.[47]

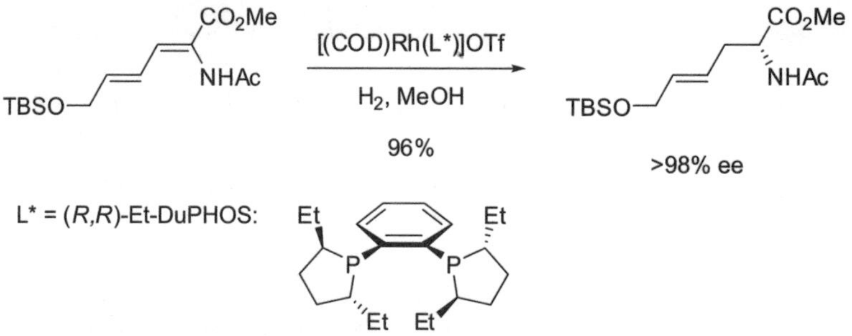

L* = (R,R)-Et-DuPHOS:

An example from the synthesis of sitagliptin phosphate is shown below.[48] The chiral phosphine ligand JOSIPHOS was used in this reaction. A dramatic pH dependence for this substrate (possibly related to the presence of the unprotected $NH_2$ moiety) was noted, and residual ammonium chloride was identified as serving the role of enhancing both reaction rate and enantioselectivity.

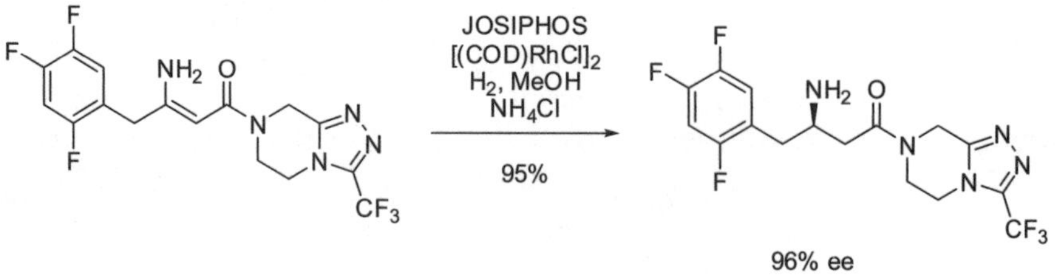

JOSIPHOS =

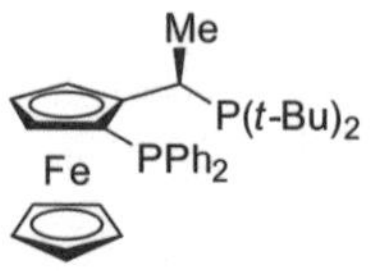

[47]Burk, M. J.; Allen, J. G.; Kiesman, W. F. *J. Am. Chem. Soc.* **1998**, *120*, 657–663.
[48]Clausen, A. M.; Dziadul, B.; Cappuccio, K. L.; Kaba, M.; Starbuck, C.; Hsiao, Y.; Dowling, T. M. *Org. Process Res. Dev.* **2006**, *10*, 723–726.

As the above examples demonstrate, a variety of chiral phosphine ligands have been used for these asymmetric hydrogenations, and identification of the optimal ligand will frequently be substrate specific.

An interesting reduction of a β-amino enone was developed at Abbott for the synthesis of ritonavir. Unlike the previous examples, this reduction utilizes an achiral reagent, and the resulting diastereoselectivity arises due to induction from the pre-existing stereocenter in the substrate. With careful choice of acid and solvent, useful levels of diastereoselectivity were realized on 30 kg scale (83:6:4:2).[49] The two-stage reduction is necessary to optimize diastereoselectivity; the first charge of reagent reduces the enamine, and the subsequent charge reduces the ketone.

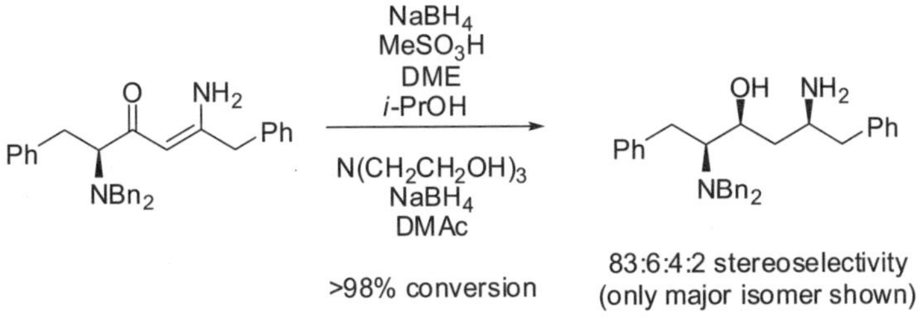

>98% conversion

83:6:4:2 stereoselectivity
(only major isomer shown)

Chemoselective reduction of a β-amino acrylate is shown below, which also utilizes a sodium borohydride–acid reagent.[42]

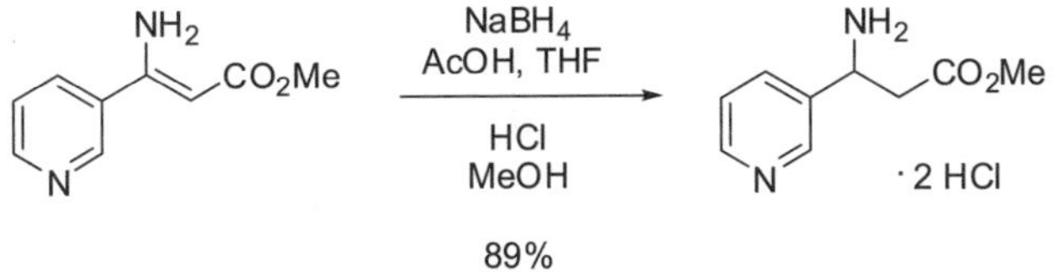

89%

A cobalt-catalyzed reduction of an α,β-unsaturated thiazolidine-2,4-dione is shown below. A series of Design of Experiments (DOE) studies led to the optimal conditions for this reduction.[50]

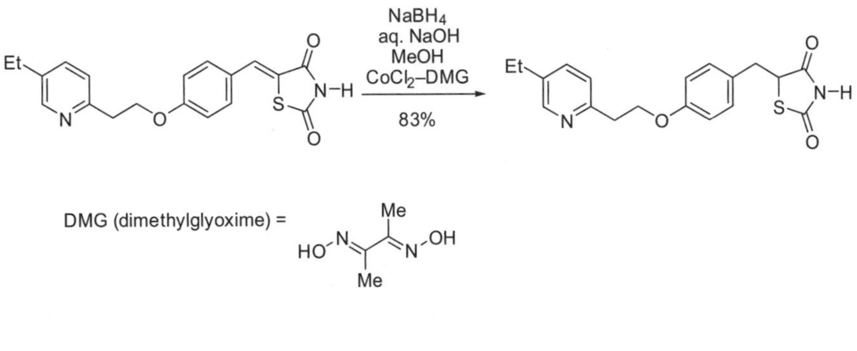

83%

DMG (dimethylglyoxime) =

[49]Haight, A. R.; Stuk, T. L.; Allen, M. S.; Bhagavatula, L.; Fitzgerald, M.; Hannick, S. M.; Kerdesky, F. A. J.; Menzia, J. A.; Parekh, S. I.; Robbins, T. A.; Scarpetti, D.; Tien, J.-H. J. *Org. Process Res. Dev.* **1999**, *3*, 94–100.
[50]Les, A.; Pucko, W.; Szelejewski, W. *Org. Process Res. Dev.* **2004**, *8*, 157–162.

*9.2.4.3  Reduction of Conjugated Alkynes*   Reduction of conjugated alkynes can be challenging in terms of chemoselectivity and avoiding over-reduction to the alkane, but with the proper choice of catalyst and reaction conditions useful levels of selectivity can be realized. For conversion to the *Z*-olefin, conjugated alkynes can be hydrogenated over a poisoned catalyst, also known as a Lindlar reduction. The poisoned catalyst is necessary to prevent over-reduction to the saturated alkane. An example of reduction to a *Z*-enoate is shown below.[51]

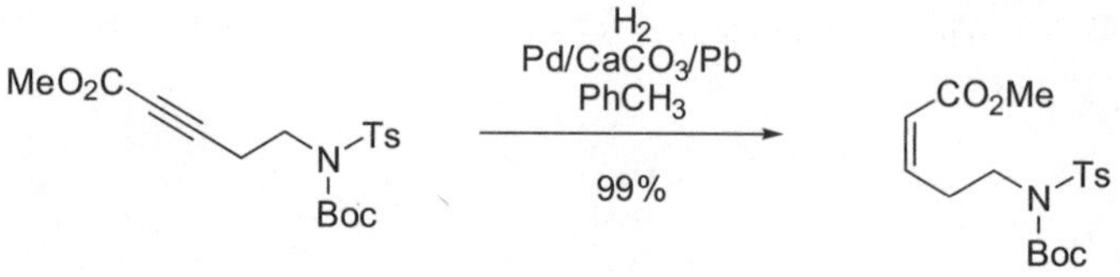

Reduction to an *E*-olefin can be achieved with Red-Al, as shown in the example below.[52,53] This reaction requires an adjacent free hydroxyl, and is not specific to conjugated alkynes.[54] Remarkably, the ester and epoxide functionalities are not affected by the reaction conditions.

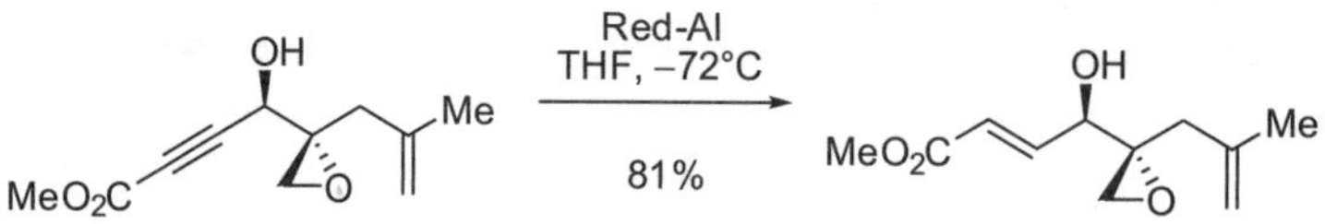

Red-Al = NaAlH$_2$(OCH$_2$CH$_2$OMe)$_2$

If full reduction to the saturated ester is desired, this can be readily effected with standard hydrogenation catalysts. An example is shown below.[55]

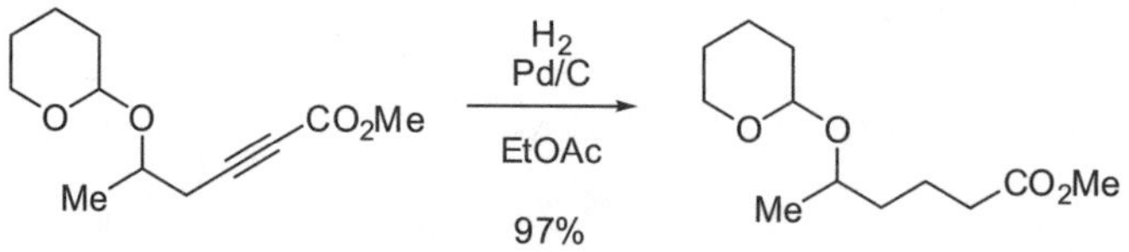

## 9.3   REDUCTION OF C–N BONDS

This section will review practical methods for the reduction of various carbon–nitrogen bonds. These include popular reactions such as the cleavage of benzyl amines and reductive

[51]Becker, M. H.; Chua, P.; Downham, R.; Douglas, C. J.; Garg, N. K.; Hiebert, S.; Jaroch, S.; Matsuoka, R. T.; Middleton, J. A.; Ng, F. W.; Overman, L. E. *J. Am. Chem. Soc.* **2007**, *129*, 11987–12002.
[52]Albert, B. J.; Sivaramakrishnan, A.; Naka, T.; Czaicki, N. L.; Koide, K. *J. Am. Chem. Soc.* **2007**, *129*, 2648–2659.
[53]Meta, C. T.; Koide, K. *Org. Lett.* **2004**, *6*, 1785–1787.
[54]Jones, T. K.; Denmark, S. E. *Org. Synth.* **1986**, *64*, 182–188.
[55]White, J. D.; Somers, T. C.; Reddy, G. N. *J. Org. Chem.* **1992**, *57*, 4991–4998.

amination of carbonyl species. With regard to the latter, emphasis will be on the reduction itself as opposed to the overall transformation from the carbonyl species (see Section 2.7.1).

### 9.3.1  Reduction of Carbon–Nitrogen Single Bonds

*9.3.1.1  Reduction of Aromatic Carbon–Nitrogen Single Bonds*    The reduction of aromatic amines is typically carried out via the deamination of diazonium salts. Hypophosphorous acid has been used by Fu and coworkers to effect the clean reduction of a diazonium salt that was generated *in situ*.[56]

Hydrogen peroxide has also been employed as the reducing agent. A rare example of such reactivity is exemplified by the work of Barbero.[57] High yields were obtained for a variety of arene diazonium sulfonimidates. It is noteworthy that molecular oxygen is formed as a byproduct, raising flammability concerns.

The reduction can also be carried under metal catalysis, where the hydrogen is abstracted from the solvent. Two of the preferred catalysts are iron(II) sulfate[58] and copper metal.[59]

[56]Fu, X.; McAllister, T. L.; Shi, X.; Thiruvengadam, T. K.; Tann, C.-H.; Lee, J. *Org. Process Res. Dev.* **2003**, *7*, 692–695.

[57]Barbero, M.; Degani, I.; Dughera, S.; Fochi, R. *Synthesis* **2004**, 2386–2390.

[58]Wassmundt, F. W.; Kiesman, W. F. *J. Org. Chem.* **1995**, *60*, 1713–1719.

[59]Hartman, G. D.; Biffar, S. E. *J. Org. Chem.* **1977**, *42*, 1468–1469.

***9.3.1.2  Reduction of Aliphatic Nitro Groups***    Aliphatic nitro groups, can be cleaved under a variety of conditions for which the synthetic limitations have been reviewed.[60] Lithium aluminum hydride can effect the reduction of tertiary nitro groups adjacent to tosyl hydrazones.[61]

Tributyltin hydride is a standard reagent for this transformation, and is the most versatile, as it can also be applied to the reduction of secondary nitro groups. The latter substrates require more forcing conditions and an excess of tin hydride is typically used.[62]

***9.3.1.3  Reduction of Benzylic Amines***    The most common cleavage of aliphatic carbon–nitrogen bonds involves the reduction of benzylic amines. Palladium-catalyzed hydrogenolysis constitutes the preferred method for this transformation, but each substrate requires screening to identify the optimal conditions. Benzylamines are generally less susceptible to catalytic hydrogenolysis than benzyl ethers (see Section 9.4.3.3), but their cleavage can be facilitated by the addition of acid. Through careful control of the conditions, Merck scientists were able to selectively cleave an exocyclic benzylamine bond in the presence of an endocylic benzylamine and a benzyloxy function.[63]

[60]Ono, N.; Kaji, A. *Synthesis* **1986**, 693–704.
[61]Rosini, G.; Ballini, R.; Zanotti, V. *Synthesis* **1983**, 137–139.
[62]Ono, N.; Miyake, H.; Kaji, A. *J. Org. Chem.* **1984**, *49*, 4997–4999.
[63]Zhao, M. M.; McNamara, J. M.; Ho, G.-J.; Emerson, K. M.; Song, Z. J.; Tschaen, D. M.; Brands, K. M. J.; Dolling, U.-H.; Grabowski, E. J. J.; Reider, P. J.; Cottrell, I. F.; Ashwood, M. S.; Bishop, B. C. *J. Org. Chem.* **2002**, *67*, 6743–6747.

Alternative hydrogen sources, such as cyclohexene,[64] ammonium formate,[65] and borane–amines[66] have also been used successfully. In addition to palladium on carbon, Pearlman's catalyst can also be relied on for benzylamine hydrogenolysis.[67]

Complementary to these palladium-mediated reactions, dissolved metals can also promote the cleavage of benzylic C−N bonds. Although few examples have been reported in the literature, Joshi and coworkers have published a detailed evaluation and successful application of a lithium/ammonia reduction on multi-kilogram scale.[68]

## 9.3.2   Reduction of Imines

### 9.3.2.1   *Achiral Reductions*   Metal hydrides are the reagents of choice for the reduction of imine derivatives. Among these, sodium triacetoxyborohydride (STAB) is one of the most commonly employed reagents to effect reductive aminations due to its mildness and stability

[64]Patel, M. K.; Fox, R.; Taylor, P. D. *Tetrahedron* **1996**, *52*, 1835–1840.

[65]Slade, J.; Bajwa, J.; Liu, H.; Parker, D.; Vivelo, J.; Chen, G.-P.; Calienni, J.; Villhauer, E.; Prasad, K.; Repic, O.; Blacklock, T. J. *Org. Process Res. Dev.* **2007**, *11*, 825–835.

[66]Couturier, M.; Andresen, B. M.; Jorgensen, J. B.; Tucker, J. L.; Busch, F. R.; Brenek, S. J.; Dube, P.; Ende, D. J.; Negri, J. T. *Org. Process Res. Dev.* **2002**, *6*, 42–48.

[67]Yasuda, N.; Hsiao, Y.; Jensen, M. S.; Rivera, N. R.; Yang, C.; Wells, K. M.; Yau, J.; Palucki, M.; Tan, L.; Dormer, P. G.; Volante, R. P.; Hughes, D. L.; Reider, P. J. *J. Org. Chem.* **2004**, *69*, 1959–1966.

[68]Joshi, D. K.; Sutton, J. W.; Carver, S.; Blanchard, J. P. *Org. Process Res. Dev.* **2005**, *9*, 997–1002.

over a wide pH range. A review of its chemical properties and synthetic scope, including large scale applications, has been published,[69] and a representative example is presented.[70]

Borane–amine complexes are often used as alternatives to borohydrides. These compounds are often solids, facilitating handling and storage. Borane can be released using acids or other protic additives. Connolly has reported the reduction of a nitro-containing imine using *tert*-butylamine-borane activated with methanesulfonic acid.[71]

In the absence of sensitive functional groups, lithium aluminium hydride can be used to reduce carbon–nitrogen bonds. This is the case for the synthesis of (−)-MIB as reported by Chen.[72]

Metal-catalyzed hydrogenations of imines have been extensively utilized, with impressive selectivity and tolerance of a wide variety of functional groups. Palladium on carbon under an atmosphere of hydrogen is the standard system for such reductions.[63,73]

[69]Abdel-Magid, A. F.; Mehrman, S. J. *Org. Process Res. Dev.* **2006**, *10*, 971–1031.

[70]Conlon, D. A.; Jensen, M. S.; Palucki, M.; Yasuda, N.; Um, J. M.; Yang, C.; Hartner, F. W.; Tsay, F.-R.; Hsiao, Y.; Pye, P.; Rivera, N. R.; Hughes, D. L. *Chirality* **2005**, *17*, S149–S158.

[71]Connolly, T. J.; Constantinescu, A.; Lane, T. S.; Matchett, M.; McGarry, P.; Paperna, M. *Org. Process Res. Dev.* **2005**, *9*, 837–842.

[72]Chen, Y. K.; Jeon, S.-J.; Walsh, P. J.; Nugent, W. A.; Kendall, C.; Wipf, P. *Org. Synth.* **2005**, *82*, 87–92.

[73]Brands, K. M. J.; Krska, S. W.; Rosner, T.; Conrad, K. M.; Corley, E. G.; Kaba, M.; Larsen, R. D.; Reamer, R. A.; Sun, Y.; Tsay, F.-R. *Org. Process Res. Dev.* **2006**, *10*, 109–117.

i) Pd/C, H$_2$
EtOH
ii) TsOH•H$_2$O

94%

In cases where chemoselectivity is hard to obtain by other methods, rhodium may be a useful alternative. Its use can also alleviate issues arising from catalyst deactivation through amine coordination or reduction of carbon–halide bonds.[74]

Rh/C, H$_2$
NH$_4$OAc, EtOH

70%

### 9.3.2.2 Substrate Control in Stereoselective Reductions

Imines bearing chiral non-racemic substituents can be reduced with high stereoselectivity by various methods. Additionally, various temporary chiral auxiliaries have been developed to access enantio-enriched primary amines. Colyer and coworkers,[75] along with the Ellman group,[76] have studied the reagent-controlled reduction of sulfinyl imines. The reduction of such imines with either sodium borohydride or L-Selectride provides complementary diastereomers with high selectivity.

NaBH$_4$, MeOH

84%, >99% de

L-Selectride
THF

90%, >99% de

This reactivity of sulfinylimines has been exploited by Han and coworkers using 9-BBN as the reducing agent. Following the reduction, the sulfinyl group is smoothly cleaved using methanolic HCl.[77]

[74]Thompson, W. J.; Jones, J. H.; Lyle, P. A.; Thies, J. E. *J. Org. Chem.* **1988**, *53*, 2052–2055.

[75]Colyer, J. T.; Andersen, N. G.; Tedrow, J. S.; Soukup, T. S.; Faul, M. M. *J. Org. Chem.* **2006**, *71*, 6859–6862.

[76]Tanuwidjaja, J.; Peltier, H. M.; Ellman, J. A. *J. Org. Chem.* **2007**, *72*, 626–629.

[77]Han, Z.; Koenig, S. G.; Zhao, H.; Su, X.; Singh, S. P.; Bakale, R. P. *Org. Process Res. Dev.* **2007**, *11*, 726–730.

Phenethyl imines have also been used to effect stereoinduction. Reduction of the imine has traditionally been performed with Raney nickel or palladium on carbon, since the reductive cleavage of the auxiliary can be performed *in situ* following the reduction.[78]

**9.3.2.3 Enantioselective Reductions** Access to enantioenriched secondary amines through racemic reduction followed by resolution is often more economical and practical than the enantioselective reduction with chiral non-racemic reagents and catalysts (see Chapter 14). A few enantioselective reductions of imines have been performed on gram scale, and some of the most promising methods will be discussed herein. One of the most practical methods is the reduction of *N*-benzoyl hydrazones using a cationic rhodium catalyst, as per the work of Burk.[79,80] Although many enantioselective hydrogenation procedures require over 1000 psi of hydrogen, Burk's system proceeds with pressures of 60 psi while still giving high yields and enantioselectivities for a range of aromatic and glyoxylic hydrazones.

[78]Storace, L.; Anzalone, L.; Confalone, P. N.; Davis, W. P.; Fortunak, J. M.; Giangiordano, M.; Haley, J. J., Jr.; Kamholz, K.; Li, H.-Y.; Ma, P.; Nugent, W. A.; Parsons, R. L., Jr.; Sheeran, P. J.; Silverman, C. E.; Waltermire, R. E.; Wood, C. C. *Org. Process Res. Dev.* **2002**, *6*, 54–63.

[79]Burk, M. J.; Feaster, J. E. *J. Am. Chem. Soc.* **1992**, *114*, 6266–6267.

[80]Burk, M. J.; Martinez, J. P.; Feaster, J. E.; Cosford, N. *Tetrahedron* **1994**, *50*, 4399–4428.

Interesting labeling experiments shed light on the mechanism of the reaction, supporting the direct reduction of the hydrazone vs. the enamine tautomer (see Section 9.2.2.2).

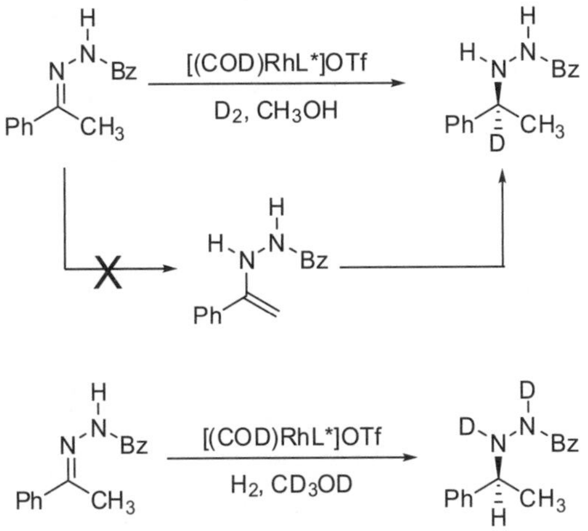

L* = (R,R)-Et-DuPHOS

Ruthenium-based catalytic transfer hydrogenation can also be employed in the enantio-selective reduction of imines.[81] The hydrogenation of dihydroisoquinolines has been extensively studied and typically provides the desired tetrahydroisoquinolines with high selectivities.[82]

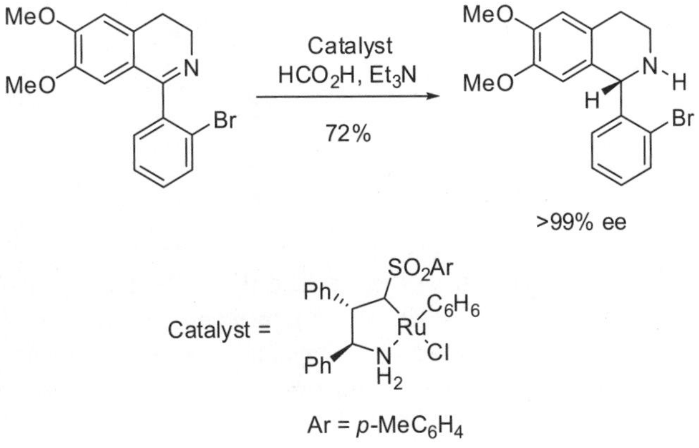

>99% ee

Catalyst =

Ar = p-MeC$_6$H$_4$

A novel approach features air-stable rhenium catalysts with dimethylphenylsilane (DMPS-H) as the stoichiometric-reducing agent. Taking advantage of the umpolung reactivity of these high oxidation state catalysts enabled the mild reduction of a variety of aryl, heteroaryl, and alkyl ketimines.[83]

[81]Noyori, R.; Hashiguchi, S. *Acc. Chem. Res.* **1997**, *30*, 97–102.
[82]Vedejs, E.; Trapencieris, P.; Suna, E. *J. Org. Chem.* **1999**, *64*, 6724–6729.
[83]Nolin, K. A.; Ahn, R. W.; Toste, F. D. *J. Am. Chem. Soc.* **2005**, *127*, 12462–12463.

Several organocatalysts have been developed, and are complementary to the metal-catalyzed hydrogenation. With these catalysts, the use of high pressures of hydrogen gas can be circumvented by alternative hydride sources. An enantioselective reduction of N-aryl ketimines can be induced by sulfinamide[84] or amino acid-derived catalysts with trichlorosilane.[85,86]

Alternatively, List,[87] and MacMillan[88] have used BINOL-derived phosphonic acids in conjunction with Hantzsch esters as the reducing agent to perform the enantioselective reduction of a variety of ketimines. For some substrates, the reduction can also be highly chemoselective, as shown by the reductive amination of a methyl ketone in the presence of an ethyl ketone.

[84]Pei, D.; Wang, Z.; Wei, S.; Zhang, Y.; Sun, J. *Org. Lett.* **2006**, *8*, 5913–5915.

[85]Malkov, A. V.; Mariani, A.; MacDougall, K. N.; Kocovsky, P. *Org. Lett.* **2004**, *6*, 2253–2256.

[86]Wang, Z.; Ye, X.; Wei, S.; Wu, P.; Zhang, A.; Sun, J. *Org. Lett.* **2006**, *8*, 999–1001.

[87]Hoffmann, S.; Seayad, A. M.; List, B. *Angew. Chem., Int. Ed. Engl.* **2005**, *44*, 7424–7427.

[88]Storer, R. I.; Carrera, D. E.; Ni, Y.; MacMillan, D. W. C. *J. Am. Chem. Soc.* **2006**, *128*, 84–86.

Additionally, this catalytic system has provided the best results for aliphatic ketones, being able to stereodifferenciate nearly symmetrical ketones.

### 9.3.3  Reduction of C=N to CH$_2$

The reduction of imines to methylenes, as originally developed by Wolff and Kishner, has seen very sparse attention in the literature. Typical conditions to perform such transformations involve derivatization of a ketone into a hydrazone followed by treatment with a base at elevated temperature.[89] Myers has recently reported the synthesis of N-*tert*-butlydimethyl-silylhydrazones and their use in Wolff–Kishner reactions under scandium triflate catalysis.[90] Using their conditions, the reduction of hydrazones to methylenes can even be performed at ambient temperature.

[89]Confalonieri, G.; Marotta, E.; Rama, F.; Righi, P.; Rosini, G.; Serra, R.; Venturelli, F. *Tetrahedron* **1994**, *50*, 3235–3250.
[90]Furrow, M. E.; Myers, A. G. *J. Am. Chem. Soc.* **2004**, *126*, 5436–5445.

Alternate conditions involve the treatment of a tosyl hydrazone with an *in situ* generated borohydride.[91] This procedure also allows for the reaction to occur at ambient temperature.

## 9.3.4   Reduction of C=N to Alkenes

Reduction of a hydrazone to an alkene typically occurs by treatment with alkyl lithium reagents. The use of methyl lithium complexed with lithium bromide in MTBE allowed Faul and coworkers to avoid reduction of the aryl bromide through metal–halide exchange.[92] Interestingly, the reaction could also be performed at ambient temperature.

α,α'-Tetrasubstituted ketones, although traditionally poor substrates, can be reduced via their corresponding hydrazones by the use of LDA as the base. The desired trisubstituted olefin can thus be obtained in high yield.[93]

Additionally, Yamamoto reported that the use of a cyclopropyl-derived hydrazone allows for the use of only catalytic amounts of LDA.[94]

[91] Kabalka, G. W.; Summers, S. T. *J. Org. Chem.* **1981**, *46*, 1217–1218.

[92] Faul, M. M.; Ratz, A. M.; Sullivan, K. A.; Trankle, W. G.; Winneroski, L. L. *J. Org. Chem.* **2001**, *66*, 5772–5782.

[93] Siemeling, U.; Neumann, B.; Stammler, H.-G. *J. Org. Chem.* **1997**, *62*, 3407–3408.

[94] Maruoka, K.; Oishi, M.; Yamamoto, H. *J. Am. Chem. Soc.* **1996**, *118*, 2289–2290.

It is noteworthy that this reaction can also be carried out with sodium methoxide as the base at higher temperatures.[95]

### 9.3.5   Reduction of C≡N to C=N/CHO

The chemoselective reduction of nitriles to imines, which often yields aldehydes after hydrolysis, is traditionally performed using diisobutylaluminum hydride (DIBAL-H) as the reducing agent.

Substrates sensitive to hydride can be reduced under hydrogenation conditions; palladium on carbon or Raney nickel are the most common metals for this transformation. Performing the reaction in the presence of an acid additive and water enables a facile reduction to the primary imine, which is then kinetically suited for hydrolysis, thus avoiding formation of the undesired primary amine. Bell and coworkers have demonstrated this method in the reduction of an aminocyano pyrimidine.[96]

Raney nickel is also a common metal for this type of catalysis, and alternative sources of hydrogen can be used. Moss and coworkers disclosed an example of formic acid serving a dual purpose: hydrogen precursor and acid additive.[97]

[95]Koot, W.-J.; Ley, S. V. *Tetrahedron* **1995**, *51*, 2077–2090.

[96]Bell, T. W.; Beckles, D. L.; Debetta, M.; Glover, B. R.; Hou, Z.; Hung, K.-Y.; Khasanov, A. B. *Org. Prep. Proced. Int.* **2002**, *34*, 321–325.

[97]Moss, N.; Ferland, J.-M.; Goulet, S.; Guse, I.; Malenfant, E.; Plamondon, L.; Plante, R.; Deziel, R. *Synthesis* **1997**, 32–34.

### 9.3.6 Reduction of C≡N to Primary Amines

The reduction of nitriles to primary amines can be accomplished with various borohydrides. *In situ* generated sodium trifluoroacetoxyborohydride was used by Denyer to effect the chemioselective reduction of a benzylnitrile in the presence of a nitro group.[98]

Merck scientists have used zinc borohydride to fully reduce an ester and a nitrile. The reducing agent was pre-formed from zinc chloride and lithium borohydride.[99]

Raney nickel has commonly been used for the reduction of nitriles. The use of ammonia was found to be essential for high conversion on larger scale.[100,101]

[98]Denyer, C. V.; Bunyan, H.; Loakes, D. M.; Tucker, J.; Gillam, J. *Tetrahedron* **1995**, *51*, 5057–5066.

[99]Nelson, T. D.; LeBlond, C. R.; Frantz, D. E.; Matty, L.; Mitten, J. V.; Weaver, D. G.; Moore, J. C.; Kim, J. M.; Boyd, R.; Kim, P.-Y.; Gbewonyo, K.; Brower, M.; Sturr, M.; McLaughlin, K.; McMasters, D. R.; Kress, M. H.; McNamara, J. M.; Dolling, U. H. *J. Org. Chem.* **2004**, *69*, 3620–3627.

[100]Haight, A. R.; Bailey, A. E.; Baker, W. S.; Cain, M. H.; Copp, R. R.; DeMattei, J. A.; Ford, K. L.; Henry, R. F.; Hsu, M. C.; Keyes, R. F.; King, S. A.; McLaughlin, M. A.; Melcher, L. M.; Nadler, W. R.; Oliver, P. A.; Parekh, S. I.; Patel, H. H.; Seif, L. S.; Staeger, M. A.; Wayne, G. S.; Wittenberger, S. J.; Zhang, W. *Org. Process Res. Dev.* **2004**, *8*, 897–902.

[101]Watson, T. J.; Ayers, T. A.; Shah, N.; Wenstrup, D.; Webster, M.; Freund, D.; Horgan, S.; Carey, J. P. *Org. Process Res. Dev.* **2003**, *7*, 521–532.

## 9.4   REDUCTION OF C–O BONDS

### 9.4.1   Reduction of Aldehydes and Ketones

#### 9.4.1.1   *Reduction with Hydride Donors*   One of the most common methods to reduce aldehydes and ketones is with a hydride donor, typically sodium borohydride or a variation thereof. Sodium borohydride offers selectivity toward aldehydes and ketones over acids/amides/esters.[102]

Sodium borohydride also can function in a range of solvents and acidities, as seen in the example of a two-step asymmetric Mannich reaction, reduction sequence from Merck.[103]

When an appropriate neighboring group is present, it is possible to obtain diastereoselectivity using standard hydride reagents. Chemists at SmithKline Beecham have demonstrated that with an appropriately hindered chiral primary amine, derived from L-phenylalanine, sodium borohydride can be used to provide a 27:1 ratio of desired *S,S*-diastereomer relative to the undesired *S,R*.[104] This ratio could be further improved through crystallization.

This selectivity is largely dependent on the steric bulk of the neighboring groups in the reduction, as demonstrated in the reduction of a similar compound with the smaller Boc-protected amine. In this example, N-Selectride (sodium tri-*sec*-butylborohydride solution) was the most selective for the desired isomer.[105] The authors continued screening and found

[102]Belecki, K.; Berliner, M.; Bibart, R. T.; Meltz, C.; Ng, K.; Phillips, J.; Brown Ripin, D. H.; Vetelino, M. *Org. Process Res. Dev.* **2007**. *11*, 754–761.

[103]Janey, J. M.; Hsiao, Y.; Armstrong, J. D., III *J. Org. Chem.* **2006**, *71*, 390–392.

[104]Diederich, A. M.; Ryckman, D. M. *Tetrahedron Lett.* **1993**, *34*, 6169–6172.

[105]Urban, F. J.; Jasys, V. J. *Org. Process Res. Dev.* **2004**, *8*, 169–175.

that standard Meerwein–Pondorff–Verley conditions gave even greater selectivity (see Section 9.4.1.2.).

Chiral hydroxy or alkoxy ketones can in many cases be reduced diastereoselectivity with the appropriate choice of conditions that take into account sterics and any conformational effects.[106] In some cases, hindered borohydrides can give high selectivity. In the example below, K-Selectride (potassium tri-*sec*-butylborohydride solution) was used to effect reduction in 85% yield with a 97:3 ratio of diastereomers favoring the desired *anti* product.[107]

By modifying the reducing agents and conditions, the *syn* products can also be produced, as seen in the reduction of a β-hydroxyketone.[106] It is important to note, in these cases the reducing agents selected are substrate dependent. In some cases, the Selectride reagents provide formation of the *syn* isomer as in the case where a bulky hydride was needed to force equatorial attack on the 6-membered ring allowing isolation of the kinetic product.[106]

[106]Romeyke, Y.; Keller, M.; Kluge, H.; Grabley, S.; Hammann, P. *Tetrahedron* **1991**, *47*, 3335–3346.
[107]Mickel, S. J.; Niederer, D.; Daeffler, R.; Osmani, A.; Kuesters, E.; Schmid, E.; Schaer, K.; Gamboni, R.; Chen, W.; Loeser, E.; Kinder, F. R., Jr.; Konigsberger, K.; Prasad, K.; Ramsey, T. M.; Repic, O.; Wang, R.-M.; Florence, G.; Lyothier, I.; Paterson, I. *Org. Process Res. Dev.* **2004**. *8*, 122–130.

Lithium aluminum hydride can also be used to reduce aldehydes[108] and ketones but is far less chemoselective toward other carbonyl functionalities.

In some cases, stabilized borane complexes can be used in place of borohydride salts or other hydride donors for the reduction of ketones. This method has been most commonly used for the asymmetric reduction of ketones. The Corey group has developed a series of asymmetric oxazaborolidine catalysts, known as CBS catalysts, derived from the reaction of chiral amino alcohols and borane complexes that can be used for the selective reduction of ketones.[109] Only a few select examples will be included here because this method has been reviewed elsewhere.[109,110]

As CBS catalysts are moisture and air sensitive, a number of *in situ* generated oxazaborolidine catalysts have been developed.[111] In these cases, chiral amino alcohols are reacted with a borane–methyl sulfide complex to generate the oxazaborolidines prior to reaction. The exact conditions needed for catalyst formation are ligand dependent. Quallich illustrated that (1*S*,2*R*)-2-amino-1,2-diphenylethanol, a commercially available amino alcohol, would complex with borane at room temperature in THF or toluene. The catalyst generated *in situ* could then be used to reduce prochiral ketones selectively in high yield.

Researchers at Bayer have scaled a similar *in situ* generated oxazaborolidine catalyst that relies on the traditional Corey CBS ligand to synthesize kilograms of a chiral chlorohydrin.[112] They found that controlling the ketone addition time and the temperature were crucial for maintaining high enantioselectivity.

### 9.4.1.2 Meerwein–Ponndorf–Verley Reaction
The Meerwein–Ponndorf–Verley (MPV) reaction, originally described in 1925, involves the reduction of a ketone or aldehyde with an

[108]Aycock, D. F. *Org. Process Res. Dev.* **2007**, *11*, 156–159.

[109]Corey, E. J.; Helal, C. J. *Angew. Chem., Int. Ed. Engl.* **1998**, *37*, 1986–2012.

[110]Cho, B. T. *Tetrahedron* **2006**, *62*, 7621–7643.

[111]Quallich, G. J.; Woodall, T. M. *Synlett* **1993**, 929–930.

[112]Duquette, J.; Zhang, M.; Zhu, L.; Reeves, R. S. *Org. Process Res. Dev.* **2003**, *7*, 285–288.

aluminum trialkoxide and has been extended to include a number of boron variations.[113] When using aluminum triisopropoxide as the reductant, acetone is eliminated as a byproduct, making the reaction reversible unless the acetone is removed throughout the reaction. As a result, the aluminum variation of the MPV reaction is typically run at elevated temperatures. Much like the hydride donors above, these reductants are required in equimolar ratios to the molecule being reduced, creating extensive metal waste.

Although not as common as the use of hydride donors, the MPV reaction provides unique selectivity in some cases. In the case of the Boc-protected phenylalanine below, reduction with aluminum triisopropoxide provided a ratio of 24:1 between the *S,R*- and *S,S*-diastereomers.[105] Although the *S,S* diastereomer was originally desired, the enhanced selectivity caused the researchers to redesign the route to include the MPV conditions followed by an alcohol inversion.

The most common variation of the standard MPV reaction is the use of trialkyl and dialkyl-halo boron reagents. In these cases, the hydride is abstracted from the alkyl groups, generating alkenes incapable of reversing the reaction.[113] Dialkylboranes such as 9-borabicyclo[3.3.1]nonane (9-BBN) can be applied to aldehydes and ketones but are relatively expensive and are pyrophoric, limiting their utility on large scale.[114]

One of the most common alkyl borane reagents for the asymmetric reduction of ketones is diisopinocampheylchloroborane (DIP-Cl). DIP-Cl was developed by H.C. Brown and coworkers in the 1980's and can reduce a wide range of prochiral ketones with stereoselectivity at or below room temperature.[115] The use of DIP-Cl and related species has been extensively reviewed.[113,116]

The (−)-DIP-Cl method has also been applied to the synthesis of chiral chlorohydrins.[117] Both high purity commercial DIP-Cl and DIP-Cl made from lower purity pinene (80% ee) gave reasonable enantioselectivities providing evidence for non-linear effects (see Chapter 14 on chirality).

[113]Cha, J. S. *Org. Process Res. Dev.* **2006**, *10*, 1032–1053.

[114]Chidambaram, R.; Kant, J.; Zhu, J.; Lajeunesse, J.; Sirard, P.; Ermann, P.; Schierling, P.; Lee, P.; Kronenthal, D. *Org. Process Res. Dev.* **2002**, *6*, 632–636.

[115]Chandrasekharan, J.; Ramachandran, P. V.; Brown, H. C. *J. Org. Chem.* **1985**, *50*, 5446–5448.

[116]Brown, H. C.; Ramachandran, P. V.; Chandrasekharan, J. *Heteroat. Chem.* **1995**, *6*, 117–131.

[117]Scott, R. W.; Fox, D. E.; Wong, J. W.; Burns, M. P. *Org. Process Res. Dev.* **2004**, *8*, 587–592.

Complete conversion
93% ee from commercial DIP-Cl
90% ee from 80% ee pinene DIP-Cl

No yield provided,
contaminated with pinene

(–)-DIP-Cl

One complication when using DIP-Cl or similar boron species is the need to safely oxidize the boron carbon bonds after reaction. In this case, the basic oxidation conditions caused partial cyclization to the epoxide; as a result, sodium hydroxide was added to drive the epoxidation to completion. A further complication is the challenge of separating the product and the pinene-related byproducts when chromatography is not an option. In the case of this particular chlorohydrin, the authors also explored Quallich's modified oxazaborolidine catalyst and obtained 95% yield with 84% ee.[117] Although CBS-like catalysis provided slightly lower ee, it allowed the authors to isolate the chlorohydrin and crystallize to 95% ee.

### 9.4.1.3 Reduction via Catalytic Hydrogenation (or Transfer Hydrogenation) Many aldehydes and ketones can be reduced with molecular hydrogen (or under transfer hydrogenation conditions) in the presence of an appropriate catalyst. In the example below, Zaidi and coworkers used Pd/C to reduce only one of the two aldehydes of terephthalaldehyde in high yield.[118]

Most of the examples that use catalytic hydrogenation (or transfer hydrogenation) for the reduction of carbonyls demonstrate asymmetric reduction of ketones. Many of these asymmetric reductions grew from the work of Noyori and coworkers who have shown that chiral ruthenium, rhodium, and iridium complexes can be used to successfully reduce both ketones and imines.[119] The ligands for these metals can be chiral diamines, amino alcohols, or amino phosphines. In these cases, it is often possible to use molecular hydrogen, formic acid/triethylamine mixtures, or isopropanol as the stoichiometric reductant.[119,120] Further examples on the reduction of imines and conjugated systems are included in Section 9.3.2.3.

---

[118]Zaidi, S. H. H.; Loewe, R. S.; Clark, B. A.; Jacob, M. J.; Lindsey, J. S. *Org. Process Res. Dev.* **2006**, *10*, 304–314.
[119]Ikariya, T.; Murata, K.; Noyori, R. *Org. Biomol. Chem.* **2006**, *4*, 393–406.
[119]Ikariya, T.; Murata, K.; Noyori, R. *Org. Biomol. Chem.* **2006**, *4*, 393–406.
[120]Miyagi, M.; Takehara, J.; Collet, S.; Okano, K. *Org. Process Res. Dev.* **2000**, *4*, 346–348.

Researchers at Lilly carried out an asymmetric transfer hydrogenation in lieu of a classical resolution because of the poorly crystalline intermediates.[121] For this particular case, the ruthenium-catalyzed reduction provided 84% diastereomeric excess (de) and 95% yield. This chemistry was compared directly against an oxazaborolidine catalyst derived from (S)-(−)-diphenylprolinol that provides 93% yield and 76% de with the opposite selectivity.

As with many enantioselective processes, development of a highly selective asymmetric ketone reduction requires extensive screening of metal–ligand combinations, along with detailed optimization of the reaction conditions. Researchers from Solvias have detailed this development process as well as final conditions for their reduction of 3,5-bistrifluoromethyl acetophenone.[122]

### 9.4.1.4 Biocatalytic Routes to Reduction of Aldehydes and Ketones

Asymmetric reduction of ketones can in some cases be carried out in the presence of an appropriate enzyme. Similar to the chiral organometallic catalysts, biocatalytic routes often require

[121]Merschaert, A.; Boquel, P.; Van Hoeck, J.-P.; Gorissen, H.; Borghese, A.; Bonnier, B.; Mockel, A.; Napora, F. *Org. Process Res. Dev.* **2006**, *10*, 776–783.

[122]Naud, F.; Spindler, F.; Rueggeberg, C. J.; Schmidt, A. T.; Blaser, H.-U. *Org. Process Res. Dev.* **2007**, *11*, 519–523.

extensive screening and optimization to identify the best enzyme, host, medium, loading, and workup conditions for the reaction. Once developed, biocatalysis benefits can include: high enantioselectivity, low catalyst cost, and mild reaction conditions.[117] Disadvantages can include relatively large volumes needed for these reactions and the potential for poor product recovery. Further discussion and examples of biocatalysis can be found in Chapter 14.

***9.4.1.5    Aldol–Tishchenko Reaction***    The Tishchenko reaction involves the combination of two equivalents of aldehyde in the presence of a Lewis acid catalyst, such as aluminum, to form an ester. One equivalent of aldehyde is reduced while one equivalent is oxidized.[123]

Often times, the Tishchenko reaction is not run as a stand-alone reaction but in combination with an aldol condensation providing a diastereoselective route to a mono-protected diol. This can be carried out as a one-[124] or two-step process.[125] To carry out this reaction, especially as a one-pot process, the catalyst needs to be sufficiently basic to catalyze the aldol condensation as well as providing a sufficiently Lewis acidic cation for the reduction reaction. Woerpel and coworkers developed a one-pot approach that relies on lithium diisopropylamide (LDA) to form the desired enolate; the lithium cations catalyze the reduction step.[124] This method provides good yields and very high diastereoselectivity arising from the cyclic transition states formed during the reduction portion of the reaction.

[123]Ooi, T.; Miura, T.; Takaya, K.; Maruoka, K. *Tetrahedron Lett.* **1999**, *40*, 7695–7698.
[124]Bodnar, P. M.; Shaw, J. T.; Woerpel, K. A. *J. Org. Chem.* **1997**, *62*, 5674–5675.
[125]Toermaekangas, O. P.; Koskinen, A. M. P. *Org. Process Res. Dev.* **2001**, *5*, 421–425.

***9.4.1.6   Reduction of Acetals and Ketals***   Acetals and ketals are commonly used as protecting groups for aldehydes, ketones, and diols. Reductive cleavage of acetals can be used to prepare alcohols, diols, or ethers depending on the conditions chosen.

Most commonly, ketals and acetals are reduced using hydride reagents in the presence of a Brønsted or Lewis acid. When reducing a cyclic ketal to a hydroxyether, the less-hindered ether is typically favored.[126] This is demonstrated in the deprotection of a chiral diol by scientists at Gilead.[127] In this case, trimethylsilyl triflate was used as the activating acid with borane–methyl sulfide complex acting as the reductant. When rapidly quenched, this reaction was found to provide the desired 3-pentyl ether in a 10:1:1 ratio relative to the undesired ether and the fully deprotected diol.

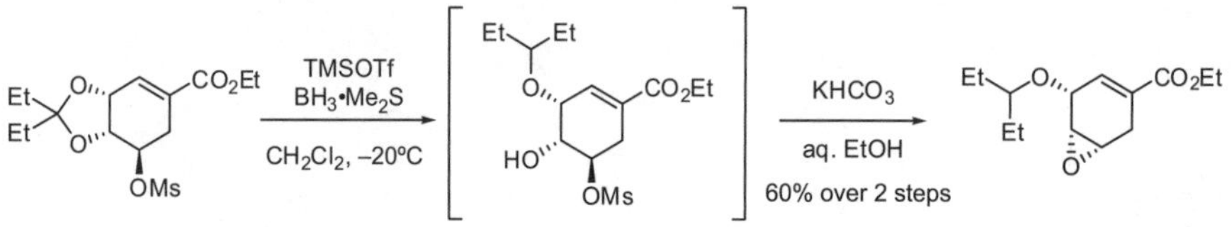

There are cases where the more-hindered ether can be formed selectively through formation of cyclic intermediates.[126] Soderquist and coworkers have shown by using 9-BBN, a Lewis acidic, reducing agent, they could selectively form hindered benzyl ethers when deprotecting unsymmetrical diols.[126]

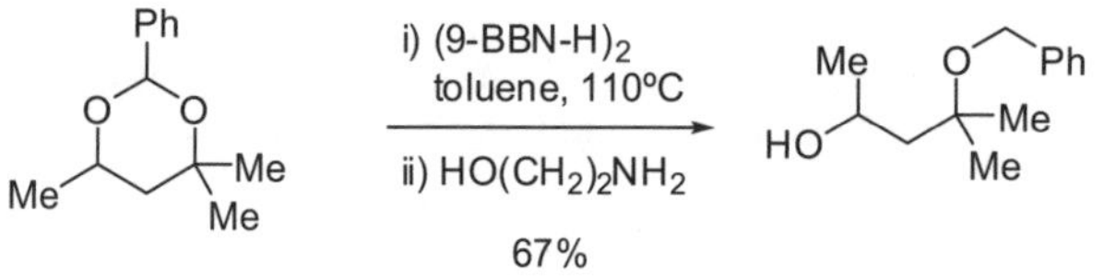

These methods can also be applied to the reduction of hemi-acetals, as seen below in the reduction of a lactol to the cyclic ether using triethylsilane under acidic conditions.[128]

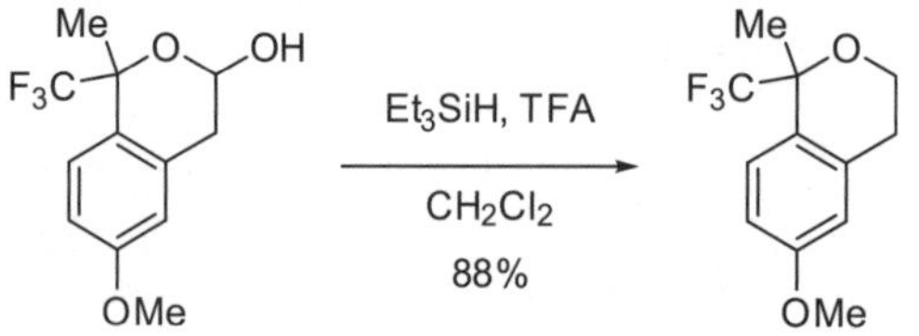

### 9.4.2   Reduction of Ketones to Alkanes

***9.4.2.1   Wolff–Kishner Reduction***   One of the oldest and still most common methods for the reduction of an aldehyde or ketone to the corresponding alkane is with the Wolff–Kishner

[126]Soderquist, J. A.; Kock, I.; Estrella, M. E. *Org. Process Res. Dev.* **2006**, *10*, 1076–1079.

[127]Rohloff, J. C.; Kent, K. M.; Postich, M. J.; Becker, M. W.; Chapman, H. H.; Kelly, D. E.; Lew, W.; Louie, M. S.; McGee, L. R.; Prisbe, E. J.; Schultze, L. M.; Yu, R. H.; Zhang, L. *J. Org. Chem.* **1998**, *63*, 4545–4550.

[128]Caron, S.; Do, N. M.; Sieser, J. E.; Arpin, P.; Vazquez, E. *Org. Process Res. Dev.* **2007**, *11*, 1015–1024.

reduction. In this reaction, the aldehyde or ketone is reacted with hydrazine to form a hydrazone that can then be reduced, leaving behind the alkane. This reaction is discussed in more detail in the section on the reduction of C—N bonds (see Section 9.3.3).

***9.4.2.2 Clemmensen Reduction*** The Clemmensen reduction uses activated zinc and acid to reduce an aldehyde or ketone to the corresponding alkane. As the original conditions required zinc amalgam in refluxing aqueous hydrochloric acid, gentler variations have been developed that use powdered zinc under anhydrous conditions.[129] These anhydrous conditions are exemplified in the reduction of a tricyclic ketone to an alkane in the preparation of an anti-cancer agent.[130]

***9.4.2.3 Silane-Mediated Reduction*** A number of other methods have been developed that use silanes, both small molecule and solid supported, to reduce specific classes of ketone or aldehyde. Benzylic keto-acids can be effectively reduced using a combination of triethylsilane and trifluoroacetic acid.[131]

### 9.4.3 Reduction of Alcohols to Alkanes

***9.4.3.1 Reduction of Aliphatic Alcohols*** The most common method for the reduction of unhindered alcohols is the Barton deoxygenation reaction (see Section 11.1.1), which is a two-step process involving the activation of the alcohol as a xanthate, followed by treatment with a tin or silyl hydride to complete the reduction (see Section 9.5.1). Although originally developed with stoichiometric tin hydride, it has been modified to require only stoichiometric tin with a more innocuous stoichiometric reductant.[132] The Fu group has successfully reduced a number of xanthates using a tributyltin hydride polymethylhydrosiloxane (PMHS) combination that compared favorably with the standard Barton conditions.[132]

[129]Vedejs, E. *Organic Reactions (New York)* **1975**, *22*, 401–422.

[130]Kuo, S.-C.; Chen, F.; Hou, D.; Kim-Meade, A.; Bernard, C.; Liu, J.; Levy, S.; Wu, G. G. *J. Org. Chem.* **2003**, *68*, 4984–4987.

[131]Ashcroft, C. P.; Challenger, S.; Derrick, A. M.; Storey, R.; Thomson, N. M. *Org. Process Res. Dev.* **2003**, *7*, 362–368.

[132]Lopez, R. M.; Hays, D. S.; Fu, G. C. *J. Am. Chem. Soc.* **1997**, *119*, 6949–6950.

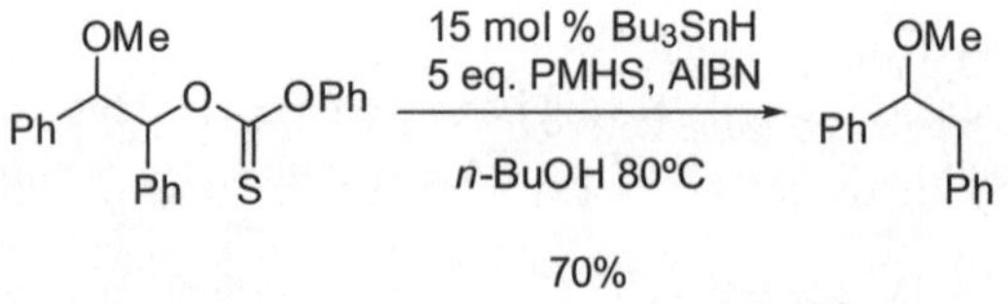

70%

In a similar fashion, unhindered alcohols can be reduced in a single step via Mitsunobu replacement of the hydroxyl with nitrobenzene–sulfinic acid (NBSH), which may be reduced *in situ*.[133]

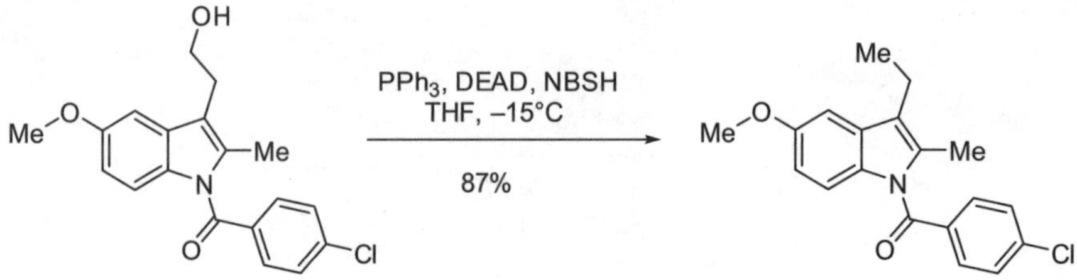

87%

NBSH = nitrobenzene–sulfinic acid

### 9.4.3.2 *Reduction of Benzylic Alcohols*

Unlike aliphatic hydroxyl groups and phenols,[134] benzylic hydroxyl groups can in some cases be reduced to the alkane under hydrogenation conditions. The selectivity of this method is illustrated in the example from Storz and coworkers where the benzylic hydroxyl could be removed in the presence of both an ester and an aliphatic alcohol.[135]

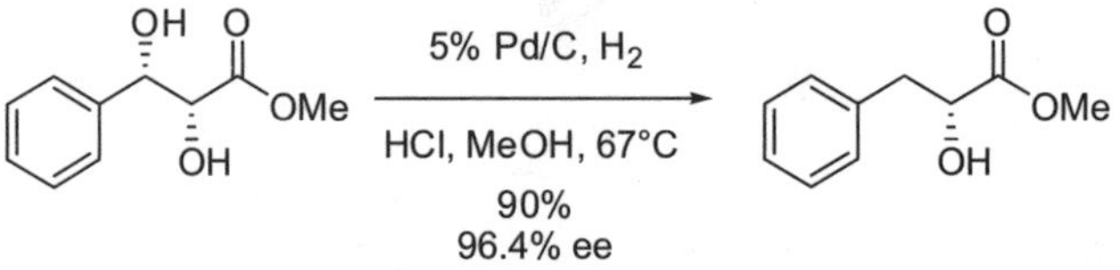

90%
96.4% ee

In some cases, the benzylic hydroxyl needs to be further activated to prevent reduction of other functional groups. In the following example, the benzylic alcohol is acetylated and then reduced, along with the nitro functionality, but without reduction of the ketone.[136]

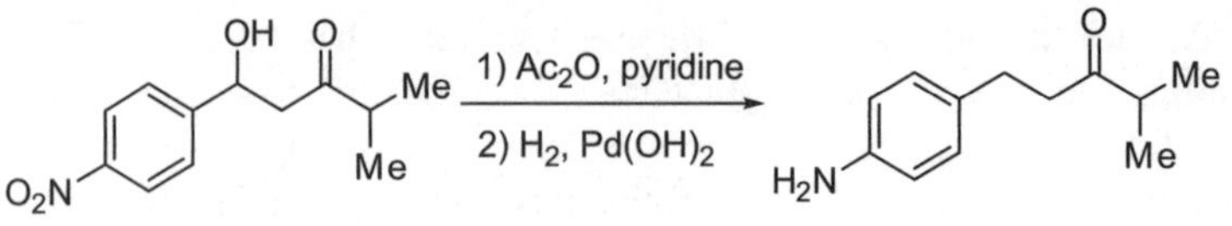

[133]Myers, A. G.; Movassaghi, M.; Zheng, B. *J. Am. Chem. Soc.* **1997**, *119*, 8572–8573.

[134]Ujvary, I.; Mikite, G. *Org. Process Res. Dev.* **2003**, *7*, 585–587.

[135]Storz, T.; Dittmar, P.; Fauquex, P. F.; Marschal, P.; Lottenbach, W. U.; Steiner, H. *Org. Process Res. Dev.* **2003**, *7*, 559–570.

[136]Deering, C. F.; Huckabee, B. K.; Lin, S.; Porter, K. T.; Rossman, C. A.; Wemple, J. *Org. Process Res. Dev.* **2000**, *4*, 596–600.

***9.4.3.3  Reduction of Benzylic Ethers***    Benzylic ethers, like their alcohol counterparts, can be reduced to the alkane and free alcohol under hydrogenation conditions. Because of their susceptibility to reductive cleavage, benzylic functionalities are commonly used to protect free alcohols. They are stable under many organic reaction conditions, but may be removed easily in the presence of a catalyst and hydrogen (see Section 9.3.1.3 for cleavage of benzylic amines).[137] Many catalysts can be used but palladium on carbon is the most common.

Like many other organometallic-catalyzed transformations, some catalyst screening is needed when a specific solvent is required or there are other reducible groups in the molecule. In the following case, the authors chose to use palladium hydroxide on carbon because it allowed for hydrogenolysis of the benzyl group without reducing the indazole ring.[138]

### 9.4.4  Reduction of Carboxylic Acid Derivatives

***9.4.4.1  Reduction of Carboxylic Acids to Primary Alcohols***    Lithium aluminum hydride is an effective reagent for the reduction of carboxylic acids to primary alcohols. In the example below, both a carboxylic acid and a carbamate are fully reduced to a $N$-methylamino alcohol in high yield.[139] Borane is also a very effective, and exhibits chemoselectivity for carboxylic acids. In the example shown, the acid is reduced selectively in the presence of an $\alpha,\beta$-unsaturated ester.[140]

[137]Greene, T. W.; Wuts, P. G. M. *Protective Groups in Organic Synthesis. 3rd ed*; Wiley: New York, N. Y., 1999.

[138]Saenz, J.; Mitchell, M.; Bahmanyar, S.; Stankovic, N.; Perry, M.; Craig-Woods, B.; Kline, B.; Yu, S.; Albizati, K. *Org. Process Res. Dev.* **2007**, *11*, 30–38.

[139]Becker, C. W.; Dembofsky, B. T.; Hall, J. E.; Jacobs, R. T.; Pivonka, D. E.; Ohnmacht, C. J. *Synthesis* **2005**, 2549–2561.

[140]Lobben, P. C.; Leung, S. S.-W.; Tummala, S. *Org. Process Res. Dev.* **2004**, *8*, 1072–1075.

***9.4.4.2 Reduction of Esters to Aldehydes*** The partial reduction of esters to aldehydes is generally accomplished by addition of diisobutylaluminum hydride at low temperature. These reaction conditions are often difficult to control as the scale of the reaction is increased, and over-reduction is always observed to a certain extent. The example below was successfuly conducted on 17.5 kg of starting material and the aldehyde was purified by crystallization of the bisulfite adduct.[141]

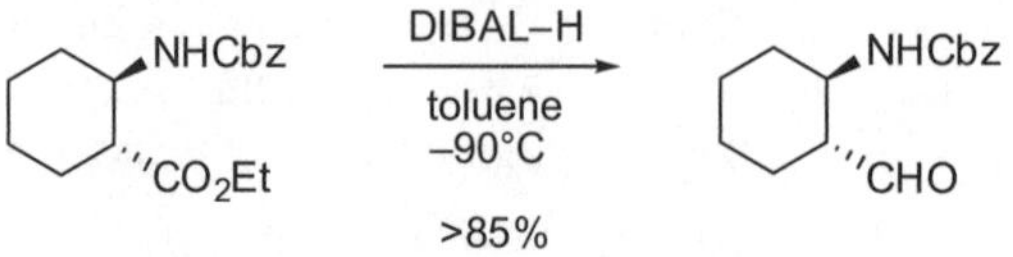

Diisobutylaluminum hydride is also the reagent of choice for the reduction of lactones to lactols without further reduction.[142]

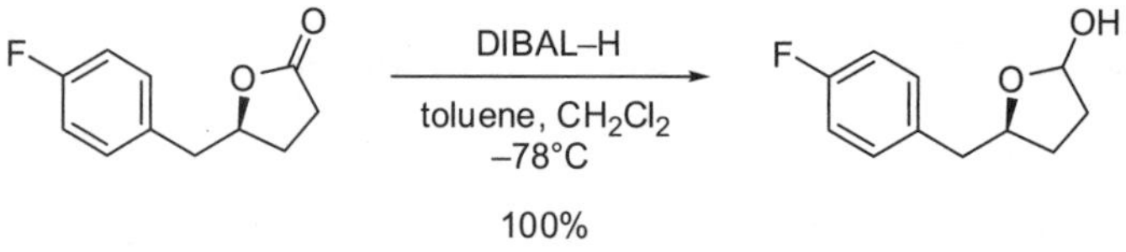

***9.4.4.3 Reduction of Esters to Primary Alcohols*** The reduction of esters to primary alcohols is a very common transformation that can be accomplished by a variety of reducing agents. Lithium borohydride[143] and sodium borohydride, with or without a Lewis acid additive such as calcium chloride,[144] are often utilized and can offer the advantages of chemoselectivity and ease of workup. In the last example shown below, sodium borohydride in toluene offered the lowest level of epimerization of the alkyl side chain, and rapidly reduced the aldehyde intermediate.[100]

[141]Yue, T.-Y.; McLeod, D. D.; Albertson, K. B.; Beck, S. R.; Deerberg, J.; Fortunak, J. M.; Nugent, W. A.; Radesca, L. A.; Tang, L.; Xiang, C. D. *Org. Process Res. Dev.* **2006**, *10*, 262–271.
[142]Cai, X.; Chorghade, M. S.; Fura, A.; Grewal, G. S.; Jauregui, K. A.; Lounsbury, H. A.; Scannell, R. T.; Yeh, C. G.; Young, M. A.; Yu, S.; Guo, L.; Moriarty, R. M.; Penmasta, R.; Rao, M. S.; Singhal, R. K.; Song, Z.; Staszewski, J. P.; Tuladhar, S. M.; Yang, S. *Org. Process Res. Dev.* **1999**, *3*, 73–76.
[143]Hu, B.; Prashad, M.; Har, D.; Prasad, K.; Repic, O.; Blacklock, T. J. *Org. Process Res. Dev.* **2007**, *11*, 90–93.
[144]Kato, T.; Ozaki, T.; Tsuzuki, K.; Ohi, N. *Org. Process Res. Dev.* **2001**, *5*, 122–126.

In cases where chemoselectivity is not required, lithium aluminum hydride[145] and Red-Al[146] are very effective reagents for the reduction of esters to alcohols.

### 9.4.4.4 Reduction of Anhydride and Mixed Anhydrides to Aldehydes

The reduction of anhydrides or mixed anhydrides to aldehydes is rare, and the full reduction to a primary alcohol followed by an oxidation is usually employed. One exception is the partial reduction of cyclic anhydrides that yield an aldehyde, usually residing in the hydroxylactone form. Lithium tri-*tert*-butoxyaluminohydride is generally the reagent of choice for this transformation and has proven to be superior to diisobutylaluminum hydride.[147]

[145]Caille, J.-C.; Govindan, C. K.; Junga, H.; Lalonde, J.; Yao, Y. *Org. Process Res. Dev.* **2002**, *6*, 471–476.

[146]Srinivas, K.; Srinivasan, N.; Reddy, K. S.; Ramakrishna, M.; Reddy, C. R.; Arunagiri, M.; Kumari, R. L.; Venkataraman, S.; Mathad, V. T. *Org. Process Res. Dev.* **2005**, *9*, 314–318.

[147]Ocain, T. D.; Deininger, D. D.; Russo, R.; Senko, N. A.; Katz, A.; Kitzen, J. M.; Mitchell, R.; Oshiro, G.; Russo, A.; et al. *J. Med. Chem.* **1992**, *35*, 823–832.

#### 9.4.4.5 Reduction of Anhydride and Mixed Anhydrides to Primary Alcohols

Anhydrides can easily be reduced to diols using a reducing agent such as lithium aluminum hydride. In the example below, a cyclic anhydride is reduced to the diol in high yield.[148]

Carboxylic acids can also be modified *in situ* to be rendered more reactive toward a milder reducing agent. For example, a mixed anhydride can be easily generated and reacted with sodium borohydride to provide the primary alcohol.[149] Similarly, in the second example shown, the acylimidazole can be prepared and selectively reduced in the presence of an *N*-sulfonylamide.[150]

[148]Spiegel, D. A.; Njardarson, J. T.; Wood, J. L. *Tetrahedron* **2002**, *58*, 6545–6554.

[149]Nishino, Y.; Komurasaki, T.; Yuasa, T.; Kakinuma, M.; Izumi, K.; Kobayashi, M.; Fujiie, S.; Gotoh, T.; Masui, Y.; Hajima, M.; Takahira, M.; Okuyama, A.; Kataoka, T. *Org. Process Res. Dev.* **2003**, *7*, 649–654.

[150]Ashcroft, C. P.; Challenger, S.; Clifford, D.; Derrick, A. M.; Hajikarimian, Y.; Slucock, K.; Silk, T. V.; Thomson, N. M.; Williams, J. R. *Org. Process Res. Dev.* **2005**, *9*, 663–669.

***9.4.4.6 Reduction of Amides and Imides to Aldehydes*** Amides can be reduced to the corresponding aldehydes with a number of reducing agents that usually require cryogenic conditions to minimize further reduction. Weinreb amides are often selected as a suitable protected form of a carboxylate that can easily provide an aldehyde after a reduction. A tetrahydrofuran solution of lithium aluminum hydride is a convenient reagent choice,[151] and Red-Al (bis(2-methoxyethoxy)aluminum hydride) can also be utilized.[152] Pseudoephedrine amides that have been used as chiral auxiliaries can be reduced to the aldehyde upon treatment with lithium triethoxyaluminum hydride generated *in situ* from reduction of ethyl acetate with lithium aluminum hydride.[153]

Lactams, especially imides, can be reduced to hemiaminals, which can easily be converted to aminoaldehydes under mild conditions.[154] In cases where the nitrogen is protected as a carbamate, chemoselectivity is observed in the reduction.[155] In some instances the equilibrium favors the aldehyde.[156]

[151]Mickel, S. J.; Sedelmeier, G. H.; Niederer, D.; Schuerch, F.; Koch, G.; Kuesters, E.; Daeffler, R.; Osmani, A.; Seeger-Weibel, M.; Schmid, E.; Hirni, A.; Schaer, K.; Gamboni, R.; Bach, A.; Chen, S.; Chen, W.; Geng, P.; Jagoe, C. T.; Kinder, F. R., Jr.; Lee, G. T.; McKenna, J.; Ramsey, T. M.; Repic, O.; Rogers, L.; Shieh, W.-C.; Wang, R.-M.; Waykole, L. *Org. Process Res. Dev.* **2004**, *8*, 107–112.

[152]Loiseleur, O.; Koch, G.; Cercus, J.; Schuerch, F. *Org. Process Res. Dev.* **2005**, *9*, 259–271.

[153]Myers, A. G.; Yang, B. H.; Chen, H. *Org. Synth.* **2000**, *77*, 29–44.

[154]Stuk, T. L.; Assink, B. K.; Bates, R. C., Jr.; Erdman, D. T.; Fedij, V.; Jennings, S. M.; Lassig, J. A.; Smith, R. J.; Smith, T. L. *Org. Process Res. Dev.* **2003**, *7*, 851–855.

[155]DeGoey, D. A.; Chen, H.-J.; Flosi, W. J.; Grampovnik, D. J.; Yeung, C. M.; Klein, L. L.; Kempf, D. J. *J. Org. Chem.* **2002**, *67*, 5445–5453.

[156]Shuman, R. T.; Rothenberger, R. B.; Campbell, C. S.; Smith, G. F.; Gifford-Moore, D. S.; Gesellchen, P. D. *J. Med. Chem.* **1993**, *36*, 314–319.

**9.4.4.7  *Reduction of Amides to Alkylamines***   While partial reduction of amides to aldehydes can be achieved by careful control of stoichiometry of the reducing agent and temperature, full reduction to the alkylamine will proceed if the temperature is raised. Lithium aluminum hydride is often the reducing agent of choice for this reaction either in toluene,[141] tetrahydrofuran,[157] or in toluene using the bis-THF complex.[101] In these cases, lithium aluminum hydride has proved to be superior to Red-Al, which also resulted in the formation of the isoindole along with significant defluorination. However, in cases where substrate sensitivity might not be an issue, Red-Al represents a good choice of reagent.[158]

[157]Shieh, W.-C.; Chen, G.-P.; Xue, S.; McKenna, J.; Jiang, X.; Prasad, K.; Repic, O.; Straub, C.; Sharma, S. K. *Org. Process Res. Dev.* **2007**, *11*, 711–715.

[158]Alimardanov, A. R.; Barrila, M. T.; Busch, F. R.; Carey, J. J.; Couturier, M. A.; Cui, C. *Org. Process Res. Dev.* **2004**, *8*, 834–837.

Borane is also an effective reducing agent, but often results in a borane–amine complex that requires additional manipulation in order to be cleaved.[66]

### 9.4.4.8 Reduction of Carbamates to N-Methylamines 
Carbamates can be fully reduced to a methyl group upon reaction with an excess of a reducing agent. The reducing agent of choice for this transformation is lithium aluminum hydride. The substrate and hydride are combined, generally with controlled cooling, and reaction occurs by increasing the temperature. Methyl and ethyl carbamates both perform equally well.[25,159] In the third example below, both a Cbz carbamate and a benzylic ketone are fully reduced.[160]

[159]Webb, R. R., II; Venuti, M. C.; Eigenbrot, C. *J. Org. Chem.* **1991**, *56*, 4706–4713.
[160]Macor, J. E.; Chenard, B. L.; Post, R. J. *J. Org. Chem.* **1994**, *59*, 7496–7498.

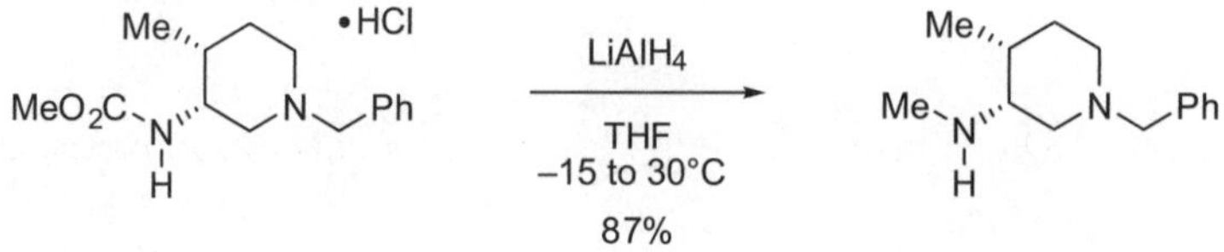

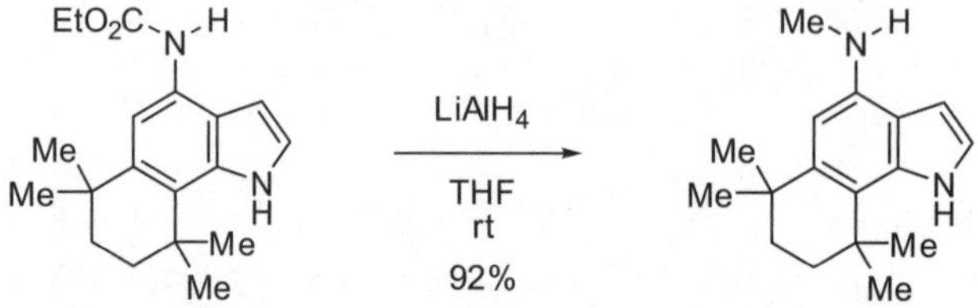

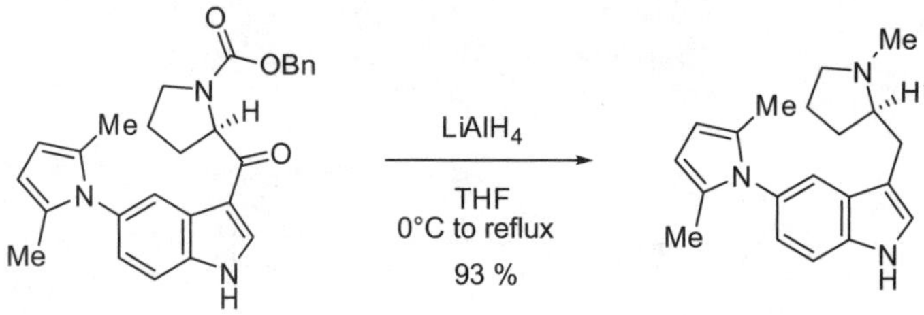

Red-Al in toluene is also an acceptable reagent for this transformation. In the case shown below, it proved to be a superior reducing agent to lithium aluminum hydride.[161] In the last example below, borane is utilized to cleave an N—O bond, reduce an amide, and also reduce a carbamate to the *N*-methyl derivative in a single operation.[162]

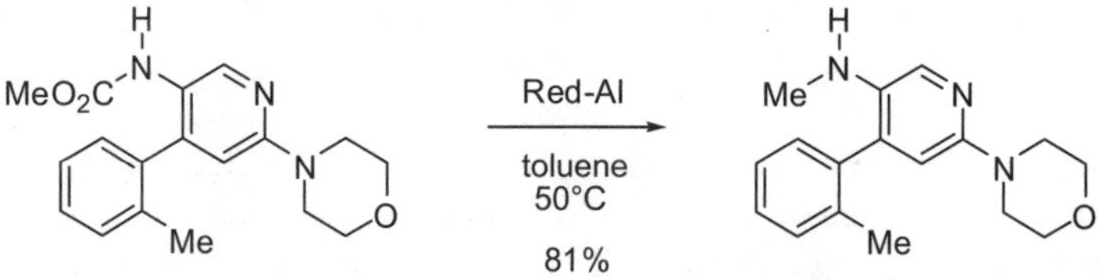

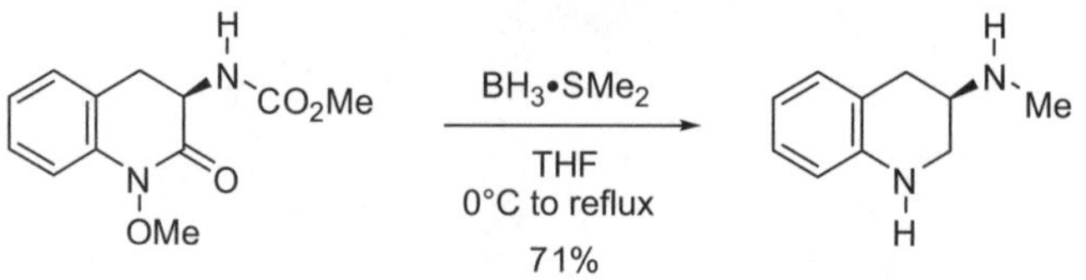

[161]Hoffmann-Emery, F.; Hilpert, H.; Scalone, M.; Waldmeier, P. *J. Org. Chem.* **2006**, *71*, 2000–2008.
[162]Romero, A. G.; Darlington, W. H.; McMillan, M. W. *J. Org. Chem.* **1997**, *62*, 6582–6587.

## 9.5   REDUCTION OF C–S BONDS

The reduction of carbon–sulfur bonds is well precedented for sulfides, sulfoxides, sulfones and other sulfur functional groups, and can be carried out with a variety of reagents. Although other reagents can be very useful for specific functional groups, nickel-based reagents have unique reactivity in reducing almost all carbon–sulfur bonds. The classic reagent is Raney nickel, and more recently, nickel boride has been used. Nickel complex reducing agents (NiCRAs) have also been used, but are not generally recommended as they use NaH in their preparation. Nickel metal is carcinogenic, and reagents based on it are usually assumed to be the same. Raney nickel is pyrophoric, has widely varying reactivity depending on how it is prepared, and is deactivated with aging; large excesses are usually required to get good conversion. It can also reduce other types of functional groups, such as amides or olefins, but can be deactivated with reagents like acetone to increase selectivity. Nickel boride,[163] prepared from a nickel salt and sodium borohydride, is more reproducible and its stoichiometry is more easily controlled, but large excesses of it are still required for reduction since it has reduced reactivity. Despite these drawbacks, these reagents are a staple for reducing carbon–sulfur bonds, and must be considered when such a transformation is desired. Another classic reagent is a metal amalgam, but it is not recommended because of the issues associated with mercury.

Few of the examples discussed below have been executed on very large scale. This is due in part to the perceived inelegance in utilizing functional groups that are not part of the final molecule, and in part because of the stench associated with lower oxidation state analogues. These groups can have a strong directing effect, however, either in reactivity or selectivity, and have found use as control elements in synthesis. Another general issue arises in substrates where a carbon–sulfur double bond is reduced without cleavage: the primary thiol produced is extremely prone to oxidative dimerization, and prevention of disulfide formation can be very challenging. For this reason, thioesters are often cleaved under reducing, rather than hydrolytic conditions.

### 9.5.1   Reduction of Alkyl Sulfides, Sulfoxides, and Sulfones

Alkyl sulfides are reduced under dissolving metal conditions, most mildly by zinc in acid.[164] In the example below, ammonium chloride was sufficiently acidic to effect the desired reduction without hydrolysis of the acetonide.

Other electron-transfer conditions are also effective, and a wide variety of conditions have been employed. The most commonly used reagent is sodium amalgam, but less offensive alternatives, such as magnesium in methanol[165] or samarium diiodide[166] can be used to achieve the desulfurization.

[163]Back, T. G.; Baron, D. L.; Yang, K. *J. Org. Chem.* **1993**, *58*, 2407–2413.

[164]Liu, G.; Wang, Z. *Synthesis* **2001**, 119–127.

[165]Porta, A.; Vidari, G.; Zanoni, G. *J. Org. Chem.* **2005**, *70*, 4876–4878.

[166]Blakemore, P. R.; Browder, C. C.; Hong, J.; Lincoln, C. M.; Nagornyy, P. A.; Robarge, L. A.; Wardrop, D. J.; White, J. D. *J. Org. Chem.* **2005**, *70*, 5449–5460.

These reductions can also be carried out by Raney nickel, which is effective for alkyl sulfides,[167] sulfoxides,[168] and sulfones.[169]

Nickel boride can be used in place of Raney nickel, and usually reacts in a very similar fashion.[170]

[167]Inomata, K.; Barrague, M.; Paquette, L. A. *J. Org. Chem.* **2005**, *70*, 533–539.

[168]Garcia Ruano, J. L.; Rodriguez-Fernandez, M. M.; Maestro, M. C. *Tetrahedron* **2004**, *60*, 5701–5710.

[169]Sadanandan, E. V.; Srinivasan, P. C. *Synthesis* **1992**, 648–650.

[170]Alcaide, B.; Casarrubios, L.; Dominguez, G.; Sierra, M. A. *J. Org. Chem.* **1994**, *59*, 7934–7936.

Tin hydrides are also effective reducing agents, although less preferred, but may be used in situations where the other functional groups present are sensitive to the conditions above.[171]

Tin hydrides have been the reagents of choice for most Barton reductions. Although this is usually carried out as a deoxygenation procedure (see Section 9.4.3.1), it is also effective in achieving desulfurization.[172] The xanthate reduction has also been carried out using lauroyl peroxide to initiate the radical mechanism.[173] Although this reagent has environmental advantages over tin-containing reagents, reductions using it are typically slower than those using tin hydride, and the peroxide has to be added in small portions at reflux, creating additional handling problems. If reaction pathways other than quenching are possible once the carbon radical has been generated, they occur more frequently under these conditions, and have been purposefully used to create more complex molecules.

A special case exists where the sulfoxide is part of a thioaminal. In this instance, sodium borohydride in pyridine has been found to be very effective for selective cleavage of the C–S bond.[174]

[171]Liang, Q.; Zhang, J.; Quan, W.; Sun, Y.; She, X.; Pan, X. *J. Org. Chem.* **2007**, *72*, 2694–2697.

[172]Alameda-Angulo, C.; Quiclet-Sire, B.; Schmidt, E.; Zard, S. Z. *Org. Lett.* **2005**, *7*, 3489–3492.

[173]Gagosz, F.; Zard, S. Z. *Org. Lett.* **2003**, *5*, 2655–2657.

[174]Volonterio, A.; Vergani, B.; Crucianelli, M.; Zanda, M.; Bravo, P. *J. Org. Chem.* **1998**, *63*, 7236–7243.

### 9.5.2 Reduction of Vinyl Sulfides, Sulfoxides, and Sulfones

Not many practical examples of the reduction of vinyl sulfides exist that do not also result in the reduction of the alkene. The reduction has been achieved with a catalytic amount of a nickel complex in the presence of a stoichiometric Grignard.[175]

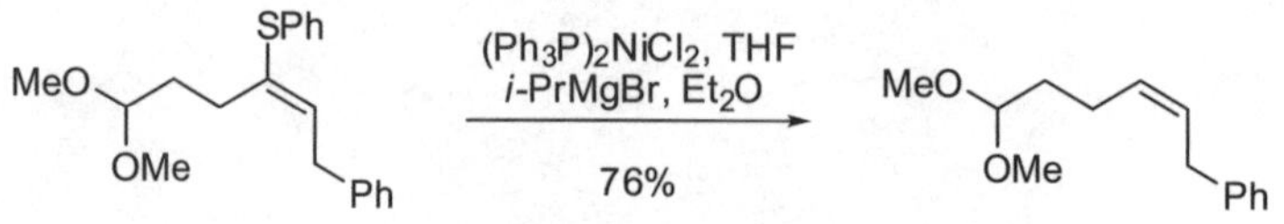

One special case is the example shown below, in which a dithioketene acetal was mono-reduced under electron-transfer conditions.[176]

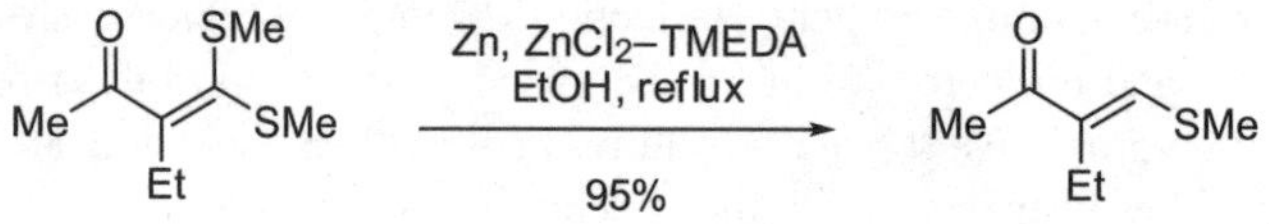

Vinylic sulfoxides can be reduced with samarium iodide.[177] The drawbacks to its use are cost and relative unavailability on large scale.

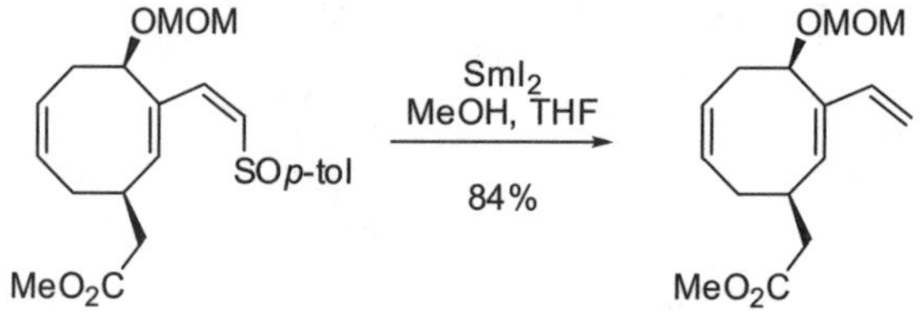

The cleavage of vinylic sulfone bonds is fairly well precedented in the literature. While sodium amalgam has classically been used for the reduction, it is not preferred due to the toxic nature of the reagent and the occasional weak control of the double-bond geometry. A preferred method employs dithionite as the reducing agent under very mild conditions.[178] It has been touted as an alternative to the usual Julia olefination procedure, since the vinyl sulfone can be generated by base-induced cleavage of the precursor acetate.

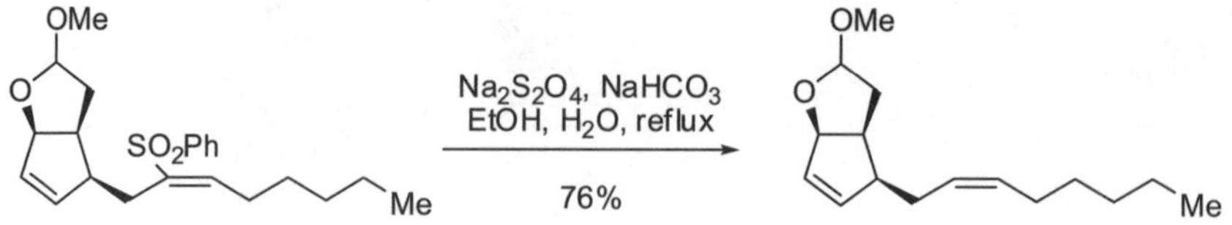

[175]Trost, B. M.; Ornstein, P. L. *Tetrahedron Lett.* **1981**, *22*, 3463–3466.
[176]Yadav, K. M.; Suresh, J. R.; Patro, B.; Ila, H.; Junjappa, H. *Tetrahedron* **1996**, *52*, 4679–4686.
[177]Paley, R. S.; Estroff, L. A.; Gauguet, J.-M.; Hunt, D. K.; Newlin, R. C. *Org. Lett.* **2000**, *2*, 365–368.
[178]Porta, A.; Re, S.; Zanoni, G.; Vidari, G. *Tetrahedron* **2007**, *63*, 3989–3994.

A secondary, much less preferred, method that may be used involves metal exchange with the sulfone followed by protonation.[179]

### 9.5.3    Reduction of Aryl Sulfides, Sulfoxides, and Sulfones

Aryl sulfides[180] and sulfoxides[181] are readily reduced with Raney nickel. Note that in the second example below, the thioester was also reduced. Aryl sulfones are relatively inert to these conditions.

Aryl sulfoxides can also be reductively cleaved with Grignard reagents, as impressively demonstrated in the example below, where the aryl chloride bond remained untouched.[182]

Few examples for the reduction of aryl sulfones exist. One option is the use of LAH,[183] but it will not be compatible with other easily reduced functional groups.

[179]Chinkov, N.; Majumdar, S.; Marek, I. *Synthesis* **2004**, 2411–2417.
[180]Gerster, J. F.; Hinshaw, B. C.; Robins, R. K.; Townsend, L. B. *J. Heterocycl. Chem.* **1969**, *6*, 207–213.
[181]Garcia Ruano, J. L.; Fernandez-Ibanez, M. A.; Maestro, M. C. *Tetrahedron* **2006**, *62*, 12297–12305.
[182]Ogawa, S.; Furukawa, N. *J. Org. Chem.* **1991**, *56*, 5723–5726.
[183]Ghera, E.; Ben-David, Y.; Rapoport, H. *J. Org. Chem.* **1983**, *48*, 774–779.

### 9.5.4 Reduction of Thioketones

Thioketones are readily reduced by mild agents such as sodium borohydride.[184]

### 9.5.5 Reduction of Thioesters and Thioamides

The reduction of mono-thioesters is accomplished with many of the same reducing agents as the all-oxygen analogues. LAH is one of the most frequently used and successful reagents for the transformation.[185]

This reaction has been used many times for the conversion of an alcohol to a thiol, since the intermediate thioester can be readily formed by a Mitsunobu reaction of the starting alcohol with thioacetic acid.[186]

A special case of thioester reduction occurs when a palladium catalyst is used with a silane. In these reactions, an aldehyde is obtained as the product, and other potentially reactive functional groups are not touched.[187]

[184]Barrett, A. G. M.; Braddock, D. C.; Christian, P. W. N.; Pilipauskas, D.; White, A. J. P.; Williams, D. J. *J. Org. Chem.* **1998**, *63*, 5818–5823.

[185]Costa, M. d. C.; Teixeira, S. G.; Rodrigues, C. B.; Ryberg Figueiredo, P.; Marcelo Curto, M. J. *Tetrahedron* **2005**, *61*, 4403–4407.

[186]Yang, X.-F.; Mague, J. T.; Li, C.-J. *J. Org. Chem.* **2001**, *66*, 739–747.

[187]Kimura, M.; Seki, M. *Tetrahedron Lett.* **2004**, *45*, 3219–3223.

The reduction of dithioesters can be accomplished with hydride reagents such as LAH. As with the corresponding oxygen analogs, it is extremely difficult to stop the reduction at the thioaldehyde stage, and the primary sulfide is almost inevitably the product.[188]

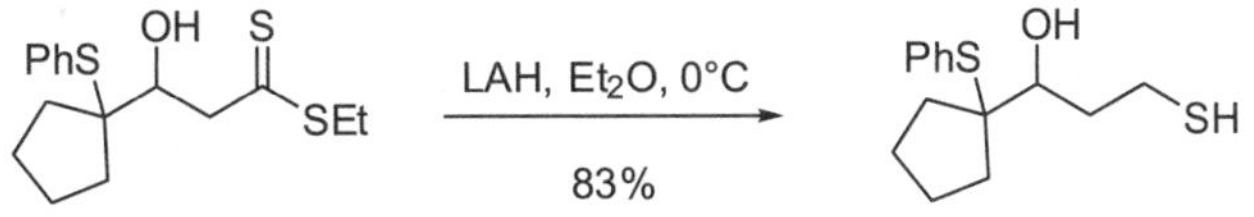

83%

Thioamides are reduced under a variety of conditions. They can be directly reduced by Raney nickel[189] or nickel boride,[190] as shown below.

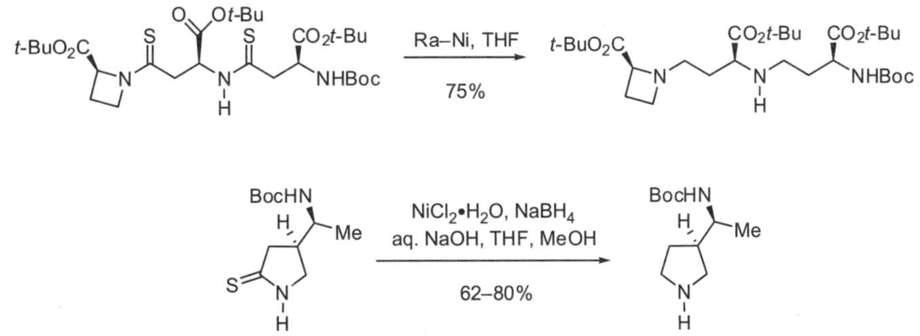

75%

62–80%

Thioamides are also frequently reduced by formation of a thioimidate derivative and treatment with a reducing agent such as nickel boride. The thioimidate can be formed discretely[191] or generated *in situ*[192] prior to reduction.

[188]Eames, J.; Jones, R. V. H.; Warren, S. *Tetrahedron Lett.* **1996**, *37*, 707–710.

[189]Miyakoshi, K.; Oshita, J.; Kitahara, T. *Tetrahedron* **2001**, *57*, 3355–3360.

[190]Beylin, V.; Boyles, D. C.; Curran, T. T.; Macikenas, D.; Parlett, R. V. I. V.; Vrieze, D. *Org. Process Res. Dev.* **2007**, *11*, 441–449.

[191]Jean, M.; Le Roch, M.; Renault, J.; Uriac, P. *Org. Lett.* **2005**, *7*, 2663–2665.

[192]Raucher, S.; Klein, P. *J. Org. Chem.* **1986**, *51*, 123–130.

## 9.5.6 Reduction of Miscellaneous Thiocarbonyls

Thiocarbonyls that are part of complex moieties can be selectively reduced under many of the same conditions under which other carbon–sulfur bonds are cleaved. In the example below, a 2-thioxo-4-thiazolidinone was reduced with zinc in acetic acid to cleanly give the des-thio product.[193]

Similarly, 2-thioxo-4-quinazolinones can be reduced with nickel boride. The reduction can be stopped at the mono-reduced product or more fully reduced depending on the ratio of hydride to nickel. The authors propose the mono-reduced product as an intermediate.[194]

[193]Hansen, M. M.; Grutsch, J. L., Jr. *Org. Process Res. Dev.* **1997**, *1*, 168–171.
[194]Khurana, J. M.; Kukreja, G. *J. Heterocycl. Chem.* **2003**, *40*, 677–679.

## 9.6  REDUCTION OF C–X BONDS

The reduction of carbon–halogen bonds is a fundamental reaction of organic chemistry,[195] and is relatively straightforward to carry out when the halide is iodide, bromide, or chloride (I > Br > Cl). A wide variety of reagents have been used for these tranformations, most typically hydrogen gas or hydride sources. The choice of reagent is often dependent on a consideration of other functional groups in the molecule. When the halide is fluoride, the reduction is much more difficult to achieve, and is not synthetically practical at an $sp^3$ carbon without some form of neighboring activation. Vinyl or aryl fluorides are more easily reduced, but are still more challenging than the other halogen analogues. For all the halides, neighboring group activation, such as an electron-withdrawing group or $\pi$ system, facilitates the reduction.

### 9.6.1  Alkyl Halide Reductions

The reduction of alkyl halides (I, Br, Cl) can be carried out under catalytic hydrogenation conditions. The reactivity generally follows the order of tertiary halides > secondary > primary. A base is often added to the reaction to neutralize the acid generated, which may act as a catalyst poison. The reduction is usually stereospecific, giving the product consistent with retention of configuration of the starting halide. As shown by the penem example below, hydrogenolysis under catalytic palladium gave a good yield of the desired product with reasonable control of the stereoisomers.[196] The use of germanium hydride (posited to be a less toxic alternative to tin hydride[197]) was shown to be stereoselective,[198]

[195]Pinder, A. R. *Synthesis* **1980**, 425–452.
[196]Norris, T.; Ripin, D. H. B.; Ahlijanian, P.; Andresen, B. M.; Barrila, M. T.; Colon-Cruz, R.; Couturier, M.; Hawkins, J. M.; Loubkina, I. V.; Rutherford, J.; Stickley, K.; Wei, L.; Vollinga, R.; de Pater, R.; Maas, P.; de Lang, B.; Callant, D.; Konings, J.; Andrien, J.; Versleijen, J.; Hulshof, J.; Daia, E.; Johnson, N.; Sung, D. W. L. *Org. Process Res. Dev.* **2005**, *9*, 432–439.
[197]Bowman, W. R.; Krintel, S. L.; Schilling, M. B. *Org. Biomol. Chem.* **2004**, *2*, 585–592.
[198]Norris, T.; Dowdeswell, C.; Johnson, N.; Daia, D. *Org. Process Res. Dev.* **2005**, *9*, 792–799.

and presumably arises from hydrogen atom transfer to the sterically less-hindered face of the molecule.

$$\beta{:}\alpha = 90{:}10$$

Either diastereomer $\xrightarrow[\substack{\text{CH}_3\text{CN, 80°C} \\ \\ 53\text{–}72\%}]{\substack{n\text{-Bu}_3\text{GeH} \\ \text{AIBN}}}$ $\beta$ only

The hydrogenolysis of alkyl halide bonds can also be carried out with hydride reagents. Not surprisingly, the order of reactivity is inverse to the palladium-mediated reduction, and follows the pattern primary halide > secondary > tertiary. Sodium borohydride in aprotic polar solvents[199] is the mildest reagent of this type that has been employed, and tolerates a wide variety of functional groups, including other halides. Modifiers such as $ZnCl_2$[200] or other Lewis acids may also be added to the reaction. Stronger hydride donors such as LAH[201] or its modified versions may be needed for less-reactive substrates.

[199]Gonzalez, J.; Foti, M. J.; Elsheimer, S. *Org. Synth.* **1995**, *72*, 225–231.

[200]Jacks, T. E.; Belmont, D. T.; Briggs, C. A.; Horne, N. M.; Kanter, G. D.; Karrick, G. L.; Krikke, J. J.; McCabe, R. J.; Mustakis, J. G.; Nanninga, T. N.; Risedorph, G. S.; Seamans, R. E.; Skeean, R.; Winkle, D. D.; Zennie, T. M. *Org. Process Res. Dev.* **2004**, *8*, 201–212.

[201]Jefford, C. W.; Gunsher, J.; Hill, D. T.; Brun, P.; Le Gras, J.; Waegell, B. *Org. Synth.* **1971**, *51*, 60–65.

Tributyltin hydride is a standard reagent for these transformations, and is usually the reagent of choice for tertiary halides.[202] Its use should be avoided whenever possible on large scale because of its toxicity. If it must be used, methods have been developed to minimize the amount of tin required, such as employing catalytic n-Bu$_3$SnH with NaBH$_4$ as a stoichiometric reductant.[203] A variety of tricks can be employed to separate the tin-containing residues from the product while avoiding chromatography, including selective extractions[204] and immobilization of the stannane. Samarium iodide can also be used for tertiary halide reductions, and in the presence of a chiral proton source, has been shown to proceed with reasonable stereoinduction.[205]

As mentioned previously, alkyl fluorides are not reduced easily, and on a practical level only in substrates in which the fluoride is activated, such as allylic fluorides[206] or α-fluoroketones.[207] For some substrates a net dehydrofluorination can be effected via an elimination mechanism followed by reduction.[208] Although not mentioned by the authors, it seems likely that in the final example below, the fluoride is displaced by the neighboring alcohol and the resulting epoxide is reduced by LAH.[209]

[202]Vijgen, S.; Nauwelaerts, K.; Wang, J.; Van Aerschot, A.; Lagoja, I.; Herdewijn, P. *J. Org. Chem.* **2005**, *70*, 4591–4597.

[203]Attrill, R. P.; Blower, M. A.; Mulholland, K. R.; Roberts, J. K.; Richardson, J. E.; Teasdale, M. J.; Wanders, A. *Org. Process Res. Dev.* **2000**, *4*, 98–101.

[204]Berge, J. M.; Roberts, S. M. *Synthesis* **1979**, 471–472.

[205]Nakamura, Y.; Takeuchi, S.; Ohgo, Y.; Yamaoka, M.; Yoshida, A.; Mikami, K. *Tetrahedron* **1999**, *55*, 4595–4620.

[206]Yamazaki, T.; Hiraoka, S.; Sakamoto, J.; Kitazume, T. *Org. Lett.* **2001**, *3*, 743–746.

[207]Chikashita, H.; Ide, H.; Itoh, K. *J. Org. Chem.* **1986**, *51*, 5400–5405.

[208]Qiu, X.-L.; Meng, W.-D.; Qing, F.-L. *Tetrahedron* **2004**, *60*, 5201–5206.

[209]Yamazaki, T.; Asai, M.; Onogi, T.; Lin, J. T.; Kitazume, T. *J. Fluorine Chem.* **1987**, *35*, 537–555.

## 9.6.2  Acid Halides to Aldehydes

The reduction of acid halides, usually acid chlorides, to their corresponding aldehydes is known as the Rosenmund reduction, and is a classical method for converting carboxylic acids to aldehydes without having to reoxidize the over-reduction product. It is best carried out under hydrogenolysis conditions. In most cases, a catalyst poison or a deactivated catalyst may be needed to prevent over-reduction.[210]

Hydride reagents have also been used for the reduction, and should be considered when the substrate possesses functionality incompatible with hydrogenation conditions.[211]

[210]Maligres, P. E.; Houpis, I.; Rossen, K.; Molina, A.; Sager, J.; Upadhyay, V.; Wells, K. M.; Reamer, R. A.; Lynch, J. E.; Askin, D.; Volante, R. P.; Reider, P. J.; Houghton, P. *Tetrahedron* **1997**, *53*, 10983–10992.
[211]Siggins, J. E.; Larsen, A. A.; Ackerman, J. H.; Carabateas, C. D. *Org. Synth.* **1973**, *53*, 52–55.

In an analogous fashion, iminoyl halides can also be reduced to their corresponding imines; hydrolysis then provides the aldehyde,[212] or the imine derivative can be isolated.[213]

93–95%

85%

### 9.6.3    Vinyl Halide Reductions

Vinyl halides are most easily reduced with zinc in acid.[214–216] For many substrates, the control of double bond geometry is good to excellent, and occurs with retention.

80%

86%

The reduction can also be carried out with palladium catalysis, and for gem-dibromides, the regiochemical control can be high. The drawback to use of palladium is the need for strong hydride sources to reduce the Pd–C bond; tributyltin hydride is the reagent most commonly used for this purpose.[217] Hydrogen gas can be used to effect the reduction, but over-reduction to the saturated product can be difficult to stop, especially for vinylic chlorides and fluorides. An interesting complementary method for the mono-reduction of gem-dibromides is that of Hirao,[218] that uses basic dialkylphosphite for the debromination, and gives the opposite regioisomer to the palladium process. Unfortunately, this procedure seems to be limited to this type of substrate, and has not been shown to have wider applicability.

[212]Williams, J. W. *Org. Synth.* **1943**, *23*, 63–65.
[213]Sakamoto, T.; Okamoto, K.; Kikugawa, Y. *J. Org. Chem.* **1992**, *57*, 3245–3248.
[214]Yang, X.; Wang, Z.; Fang, X.; Yang, X.; Wu, F.; Shen, Y. *Synthesis* **2007**, 1768–1778.
[215]Yamamoto, I.; Sakai, T.; Yamamoto, S.; Ohta, K.; Matsuzaki, K. *Synthesis* **1985**, 676–677.
[216]Viger, A.; Coustal, S.; Schambel, P.; Marquet, A. *Tetrahedron* **1991**, *47*, 7309–7322.
[217]Rahman, S. M. A.; Sonoda, M.; Ono, M.; Miki, K.; Tobe, Y. *Org. Lett.* **2006**, *8*, 1197–1200.
[218]Hirao, T.; Masunaga, T.; Ohshiro, Y.; Agawa, T. *J. Org. Chem.* **1981**, *46*, 3745–3747.

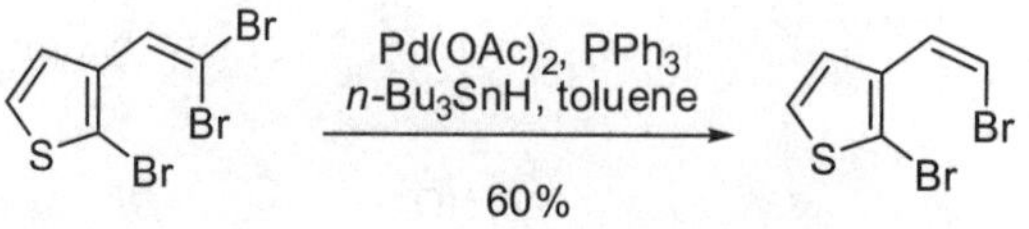

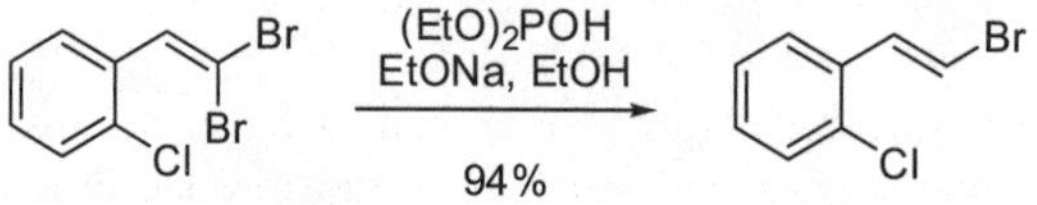

60%

94%

Vinyl halides can also be reduced by metal–halogen exchange followed by quenching,[219] and is often a side process when trapping with more complex electrophiles. This method is less preferred because of the high reactivity of reagents such as *n*-BuLi, which often react with other functional groups, and the instability of the lithiated species.

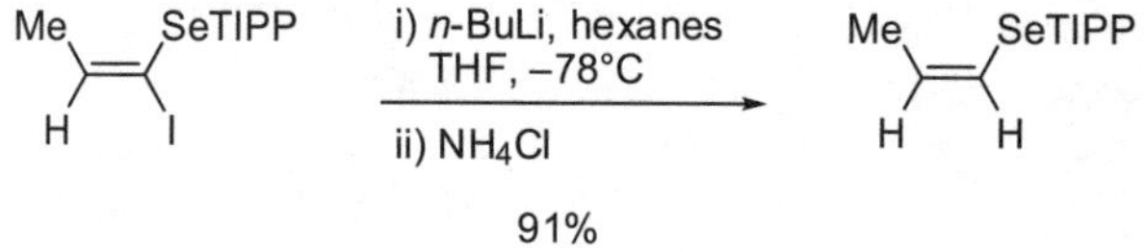

91%

TIPP = 2,4,6-triisopropylphenyl

### 9.6.4    Aryl Halide Reductions

Aryl halides can be readily reduced under hydrogenolysis conditions,[220] often at very low pressures of hydrogen.[221] Hydrogen sources other than $H_2$ gas, such as DMF, have also been shown to be effective for these reactions.[222]

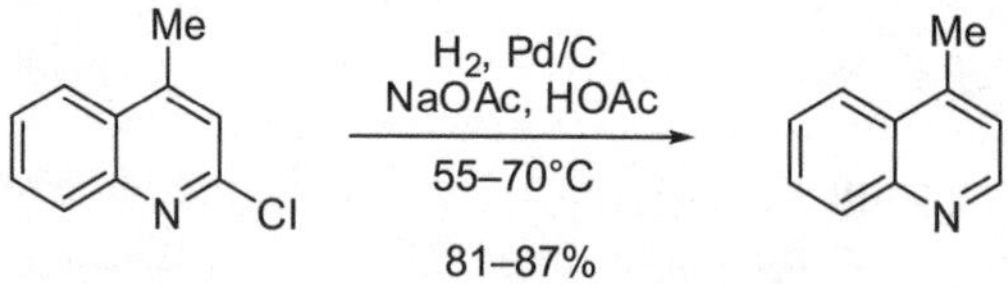

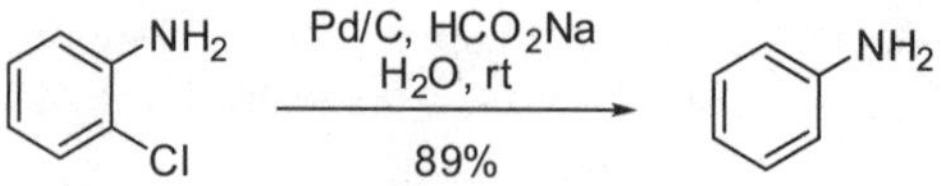

81–87%

89%

[219]Perez-Balado, C.; Marko, I. E. *Tetrahedron* **2006**, *62*, 2331–2349.

[220]Neumann, F. W.; Sommer, N. B.; Kaslow, C. E.; Shriner, R. L. *Org. Synth.* **1946**, *26*, 45–49.

[221]Arcadi, A.; Cerichelli, G.; Chiarini, M.; Vico, R.; Zorzan, D. *Eur. J. Org. Chem.* **2004**, 3404–3407.

[222]Zawisza, A. M.; Muzart, J. *Tetrahedron Lett.* **2007**, *48*, 6738–6742.

The reduction can also be carried out under dissolving metal conditions, such as zinc in acetic acid.[223]

89–90%

As with the vinyl halides, metal–halogen exchange and quench may be a viable option for some substrates,[224] although less preferred for the same reasons. In addition, the potential for a "hydrogen dance" to occur can be a complicating factor.[225]

97%

R =

As with the alkyl fluorides, aryl fluorides are less readily reduced, and in many cases the apparent reduction occurs by a prior elimination, such as benzyne or quinide-type formation, followed by reduction.[226] It is not practical to design these elimination–reduction reactions into a synthetic scheme, since it can be difficult to carry out cleanly, but these are side processes that should be considered when carrying out reductions.

90%

## 9.7  REDUCTION OF HETEROATOM–HETEROATOM BONDS

### 9.7.1  Reduction of Nitrogen–Nitrogen Bonds

The reduction of nitrogen–nitrogen bonds is often an important step in synthetic sequences that utilize azides, hydrazones, and diazo compounds. Aminations and cycloadditions that involve diazodicarboxylates are often followed by nitrogen–nitrogen bond reductions, in

[223]Gronowitz, S.; Raznikiewicz, T. *Org. Synth.* **1964**, *44*, 9–11.
[224]Li, Z.; Liang, X.; Wan, B.; Wu, F. *Synthesis* **2004**, 2805–2808.
[225]Sammakia, T.; Stangeland, E. L.; Whitcomb, M. C. *Org. Lett.* **2002**, *4*, 2385–2388.
[226]Holland, D. G.; Moore, G. J.; Tamborski, C. *J. Org. Chem.* **1964**, *29*, 3042–3046.

order to reveal the desired amine or diamine functionality. Hydrogenolysis is typically the most practical approach for these types of reductions.[227,228]

### 9.7.1.1  Reduction of Azides

Azides are a versatile functional group and they can serve as protected amines. Conversion of azides to the corresponding amines can be readily accomplished by reduction. The most practical azide reductions for larger scale reactions and/or simple substrates typically involve hydrogenolysis.[229]

Reactions requiring the selective reduction of an azide in the presence of other reducible functionality will probably best be accomplished via Staudinger reduction with a phosphine as the reducing agent. This approach is also particularly convenient for small scale reactions. Using triphenylphosphine, Mapp and coworkers were able to selectively reduce an azide in the presence of a benzylamine, an olefin, and a nitrogen–oxygen bond.[230] All of these groups would likely be reduced under hydrogenation conditions. While this azide reduction was performed on a small scale (15 mg), it showcases the outstanding chemoselectivity that phosphines can provide for azide reductions.

[227]Bournaud, C.; Bonin, M.; Micouin, L. *Org. Lett.* **2006**, *8*, 3041–3043.
[228]Guanti, G.; Banfi, L.; Narisano, E. *Tetrahedron* **1988**, *44*, 5553–5562.
[229]Rowland, E. B.; Rowland, G. B.; Rivera-Otero, E.; Antilla, J. C. *J. Am. Chem. Soc.* **2007**, *129*, 12084–12085.
[230]Rowe, S. P.; Casey, R. J.; Brennan, B. B.; Buhrlage, S. J.; Mapp, A. K. *J. Am. Chem. Soc.* **2007**, *129*, 10654–10655.

## 9.7.2    Reduction of Nitrogen–Oxygen Bonds

The efficient reduction of nitrogen–oxygen bonds is a critical transformation in many synthetic sequences that involve oximes, nitrones, nitroso, and nitro compounds. In most cases, the reduction method of choice would be catalytic hydrogenation, but functional group compatibility might force an alternative reagent choice. Additionally, many of these functional groups can undergo multiple stages of reduction, and effective partial reduction can call for more specialized conditions. Various cycloaddition strategies, including those that involve nitrones and nitroalkenes, require reduction of the cycloadduct nitrogen–oxygen bonds. Gallos and coworkers developed an intramolecular nitrone cycloaddition and reduction sequence during their development of synthetic strategies for the total synthesis of pentenomycin I and naplanocin A.[231] The product of the nitrone [3 + 2] cycloaddition could be reduced with zinc and acetic acid or via catalytic hydrogenolysis with palladium on carbon. The benzylamine was untouched in the reaction with zinc and acetic acid, while hydrogenolysis resulted in complete debenzylation.

Another synthetic sequence that often requires the reduction of a nitrogen–oxygen bond is the organocatalytic asymmetric α-oxidation of ketones and aldehydes. This approach can be an effective method for the enantioselective installation of a hydroxyl group, but these reactions typically utilize nitroso compounds as oxidants, and thus a reduction must be performed in order to reveal the desired hydroxyl group. Hayashi and coworkers utilized an

[231]Gallos, J. K.; Stathakis, C. I.; Kotoulas, S. S.; Koumbis, A. E. *J. Org. Chem.* **2005**, *70*, 6884–6890.

L-proline catalyzed α-aminoxylation/hydrogenolysis sequence in their synthesis of fumagillol and several related angiogenesis inhibitors.[232]

### 9.7.2.1 Reduction of Nitro Groups to Amines

*9.7.2.1  Reduction of Nitro Groups to Amines*    The reduction of nitro compounds to the corresponding amines is typically straightforward, and it can be accomplished with a variety of reagents. Catalytic hydrogenation, either under hydrogen atmosphere[233] or via transfer hydrogenation,[234] is often the most practical method for this transformation. When a nitro group must be selectively reduced in the presence of other reducible functional groups, sodium dithionite[235] and iron with acetic acid[236] can be highly effective reagents.

[232]Yamaguchi, J.; Toyoshima, M.; Shoji, M.; Kakeya, H.; Osada, H.; Hayashi, Y. *Angew. Chem., Int. Ed. Engl.* **2006**, *45*, 789–793.

[233]Dorow, R. L.; Herrinton, P. M.; Hohler, R. A.; Maloney, M. T.; Mauragis, M. A.; McGhee, W. E.; Moeslein, J. A.; Strohbach, J. W.; Veley, M. F. *Org. Process Res. Dev.* **2006**, *10*, 493–499.

[234]Walz, A. J.; Sundberg, R. J. *J. Org. Chem.* **2000**, *65*, 8001–8010.

[235]Chandregowda, V.; Rao, G. V.; Reddy, G. C. *Org. Process Res. Dev.* **2007**, *11*, 813–816.

[236]Huelgas, G.; Bernes, S.; Sanchez, M.; Quintero, L.; Juaristi, E.; de Parrodi, C. A.; Walsh, P. J. *Tetrahedron* **2007**, *63*, 12655–12664.

### 9.7.2.2  *Partial Reduction of Aromatic Nitro Compounds*

The nitrogen–oxygen bonds of aromatic nitro compounds can be partially or fully reduced, depending on the reaction conditions. Reduction of nitro compounds to the corresponding hydroxylamines is relatively straightforward with zinc and ammonium chloride.[237] Zinc and ammonium chloride can also be used, under carefully controlled conditions, to generate the nitroso product.[238] However, it is typically more practical to reduce the nitro derivative to the hydroxylamine with zinc and ammonium chloride, followed by partial oxidation up to the nitroso derivative using iron trichloride.[239]

### 9.7.2.3  *Partial Reduction of Aliphatic Nitro Compounds*

The partial reduction of aliphatic nitro compounds is also possible, but the initial reduction generates an oxime instead of the tautomeric nitroso derivative, unless tautomerization is impossible due to substitution $\alpha$ to the nitro group. The preferred method for the partial reduction of aliphatic nitro compounds to the corresponding oxime is the benzylation/elimination protocol developed by Czekelius and Carreira.[240] In this approach, the nitro group is alkylated with benzyl bromide, followed by a base-induced elimination to generate the oxime and benzaldehyde.

Analogous to the transformation with aromatic nitro compounds, the partial reduction of aliphatic nitro compounds to the corresponding hydroxylamines can be accomplished with zinc and ammonium chloride.[241]

[237]Evans, D. A.; Song, H.-J.; Fandrick, K. R. *Org. Lett.* **2006**, *8*, 3351–3354.

[238]Kimbaris, A.; Cobb, J.; Tsakonas, G.; Varvounis, G. *Tetrahedron* **2004**, *60*, 8807–8815.

[239]Standaert, R. F.; Park, S. B. *J. Org. Chem.* **2006**, *71*, 7952–7966.

[240]Czekelius, C.; Carreira, E. M. *Angew. Chem., Int. Ed. Engl.* **2005**, *44*, 612–615, S612/611–S612/620.

[241]Jin, C.; Burgess, J. P.; Kepler, J. A.; Cook, C. E. *Org. Lett.* **2007**, *9*, 1887–1890.

### 9.7.3  Reduction of Oxygen–Oxygen Bonds

The reduction of oxygen–oxygen bonds is not a particularly common process in organic synthesis, but it is a critical reaction in many sequences that involve singlet oxygen, ozone, or peroxides.

***9.7.3.1  Reduction of Peroxides***   Reactions with singlet oxygen, including ene reactions and cycloadditions, generate peroxide products that often require reduction. Additionally, nucleophilic additions of peroxides can be useful for the installation of hydroxyl groups, but a reduction is required to convert the peroxide intermediates to the corresponding hydroxyl products. The preferred methods for the reduction of peroxides are hydrogenolysis or zinc and acetic acid.[242] Wood and coworkers utilized a Diels–Alder reaction with singlet oxygen and a reduction of the resulting peroxide in their synthesis of the BCE ring system of ryanodine.[243]

***9.7.3.2  Reduction of Ozonides***   The reaction of ozone with alkenes results in the formation of ozonides. The oxygen–oxygen bonds of these ozonides can be reduced

[242]Robinson, T. V.; Taylor, D. K.; Tiekink, E. R. T. *J. Org. Chem.* **2006**, *71*, 7236–7244.
[243]Wood, J. L.; Graeber, J. K.; Njardarson, J. T. *Tetrahedron* **2003**, *59*, 8855–8858.

with various reagents, such as trimethyl phosphite,[244] to provide carbonyl compounds. Alternatively, when ozonolysis reactions are performed in methanol, the ozonides react with methanol to generate peroxyaldehydes, which can be converted to the corresponding diols with sodium borohydride.[145] Many other reducing agents can be used to reduce ozonides (or methoxy-hydroperoxides if executed in methanol), such as dimethylsulfide, triphenylphosphine, or zinc in acetic acid. Dimethylsulfide is convenient at a laboratory scale, but is not preferred on larger scale due to the stench associated with this reagent. Sodium bisulfite has also been used, which delivers the bisulfite adduct of the aldehyde.[245]

DPM = diphenylmethyl

### 9.7.4 Reduction of Oxygen–Sulfur Bonds

While the reduction of sulfoxides and sulfones to the corresponding sulfides is not a particularly common reaction in organic synthesis, these transformations can be successfully completed. The partial reduction of sulfones to sulfoxides, however, is not usually possible.

[244]Bernasconi, E.; Lee, J.; Sogli, L.; Walker, D. *Org. Process Res. Dev.* **2002**, *6*, 169–177.

[245]Ragan, J. A.; am Ende, D. J.; Brenek, S. J.; Eisenbeis, S. A.; Singer, R. A.; Tickner, D. L.; Teixeira, J. J., Jr.; Vanderplas, B. C.; Weston, N. *Org. Process Res. Dev.* **2003**, *7*, 155–160.

***9.7.4.1   Reduction of Sulfones to Sulfides***   The two most practical approaches for the reduction of sulfones to sulfides are magnesium/methanol[246] and aluminum hydride reagents.[247]

***9.7.4.2   Reduction of Sulfoxides to Sulfides***   There are many methods for the reduction of sulfoxides to sulfides, and the best method will be highly dependent upon the scale of the reaction and compatibility with other functional groups on the molecule. One of the preferred methods for this transformation, developed by Drabowicz and Oae, utilizes trifluoroacetic anhydride (TFAA) and sodium iodide.[248,249] A green approach for the reduction of sulfoxides to sulfides using a molybdenum catalyst and polymethylhydrosiloxane (PMHS) in water has been reported by Fernandes and Romão.[250] Borane,[251] phosphorus pentasulfide,[252] and phosphorus trichloride[253] are also effective reagents for converting sulfoxides to sulfides.

[246]Khurana, J. M.; Sharma, V.; Chacko, S. A. *Tetrahedron* **2006**, *63*, 966–969.

[247]Harpp, D. N.; Heitner, C. *J. Org. Chem.* **1970**, *35*, 3256–3259.

[248]Drabowicz, J.; Oae, S. *Synthesis* **1977**, 404–405.

[249]Cho, B. T.; Shin, S. H. *Tetrahedron* **2005**, *61*, 6959–6966.

[250]Fernandes, A. C.; Romao, C. C. *Tetrahedron* **2006**, *62*, 9650–9654.

[251]Madec, D.; Mingoia, F.; Macovei, C.; Maitro, G.; Giambastiani, G.; Poli, G. *Eur. J. Org. Chem.* **2005**, 552–557.

[252]Ma, S.; Hao, X.; Meng, X.; Huang, X. *J. Org. Chem.* **2004**, *69*, 5720–5724.

[253]Oh, K. *Org. Lett.* **2007**, *9*, 2973–2975.

## 9.7.5  Reduction of Disulfides to Thiols

Disulfides are often used as a protected form of the more oxidatively unstable thiol group, and can be readily reduced to the corresponding thiols under a variety of conditions. While disulfide reductions are often accomplished in biochemical research by the addition of thiols, sodium borohydride is typically used for this transformation in organic synthesis.[254]

[254]Wheelhouse, R. T.; Jennings, S. A.; Phillips, V. A.; Pletsas, D.; Murphy, P. M.; Garbett, N. C.; Chaires, J. B.; Jenkins, T. C. *J. Med. Chem.* **2006**, *49*, 5187–5198.

## 9.7.6 Reduction of Phosphine Oxides to Phosphines

The reduction of phosphine oxides to phosphines is an important step in the synthesis of many chiral and achiral phosphine ligands. There are multiple reagents that can reduce phosphine oxides, but trichlorosilane is typically used. In this reaction, the trichlorosilane deoxygenates the phosphine oxide to provide trichlorosilanol and the phosphine. This method is particularly useful because the reduction occurs with retention of configuration at phosphorus.[255–257] P-Chirogenic phosphine oxides can also be reduced to the corresponding phosphines with inversion of configuration at phosphorus. This is accomplished by alkylation with methyl triflate, followed by reduction with lithium aluminum hydride.[258]

[255]Wu, H.-C.; Yu, J.-Q.; Spencer, J. B. *Org. Lett.* **2004**, *6*, 4675–4678.

[256]Piras, E.; Lang, F.; Ruegger, H.; Stein, D.; Worle, M.; Grutzmacher, H. *Chem.-Eur. J.* **2006**, *12*, 5849–5858.

[257]Odinets, I. L.; Vinogradova, N. M.; Matveeva, E. V.; Golovanov, D. D.; Lyssenko, K. A.; Keglevich, G.; Kollar, L.; Roeschenthaler, G.-V.; Mastryukova, T. A. *J. Organomet. Chem.* **2005**, *690*, 2559–2570.

[258]Imamoto, T.; Kikuchi, S.-i.; Miura, T.; Wada, Y. *Org. Lett.* **2001**, *3*, 87–90.

# 10

# OXIDATIONS

David H. Brown Ripin

## 10.1 INTRODUCTION

The oxidation of organic compounds generally refers to the replacement of a functional group (or atom) with a more electronegative functional group, or the conversion of a lone pair of electrons to the $N-O$, $S-O$, $N-X$ or $S-X$. Most commonly oxidized are $C-H$, $C-C$, $C-O$, and $C-N$ bonds, and lone pairs on nitrogen and sulfur. Oxidation reactions are powerful tools to convert a position that is protected in a lower oxidation state to the desired functionality, and for the activation of otherwise unfunctionalized positions. A wide variety of reagents have been developed over the years for the oxidation of various functional groups, with varying reactivity, stability, ease of preparation and use, and waste products. Yet despite their power as a synthetic tool and abundant use in academic research, oxidation reactions as a whole tend to be avoided in commercial processes.[1-3] This disparity is likely due to a mixture of factors including the exothermicity of the reactions, stability of reagents, toxic waste products, and general preference not to adjust oxidation state. While the use of stable, easily prepared reagents that generate minimal amounts of waste is obviously preferred, this is frequently not an option in the course of a synthesis. This said, there are a number of oxidation reactions that are safe, produce innocuous byproducts, and are amenable to large scale execution.

---

[1]Oxidation reactions as a whole comprise as little as 3% of the reactions used on a preparative scale in the pharmaceutical industry.

[2]Schmid, C. R.; Bryant, J. D.; Dowlatzedah, M.; Phillips, J. L.; Prather, D. E.; Schantz, R. D.; Sear, N. L.; Vianco, C. S. *J. Org. Chem.* **1991**, *56*, 4056–4058.

[3]Clark, J. D.; Weisenburger, G. A.; Anderson, D. K.; Colson, P.-J.; Edney, A. D.; Gallagher, D. J.; Kleine, H. P.; Knable, C. M.; Lantz, M. K.; Moore, C. M. V.; Murphy, J. B.; Rogers, T. E.; Ruminski, P. G.; Shah, A. S.; Storer, N.; Wise, B. E. *Org. Process Res. Dev.* **2004**, *8*, 51–61.

---

*Practical Synthetic Organic Chemistry: Reactions, Principles, and Techniques*, First Edition.
Edited by Stéphane Caron.
© 2011 John Wiley & Sons, Inc. Published 2011 by John Wiley & Sons, Inc.

The examples selected provide practical experimental details for the oxidations depicted. Examples of the use of a variety of reagents are provided when available; the selection of the appropriate method requires consideration of the reactivity of the substrate, the scale of the reaction, and the resultant safety and waste disposal issues. Excluded from discussion are biotransformations and oxidative aromatic substitution reactions (see Chapter 5). This chapter is organized by the oxidized functionality for ease in looking up specific transformations.

*Caution: Many of the oxidants, oxidation reactions, and products described herein have the potential to release large amounts of energy in an uncontrolled fashion. Investigators considering running a large scale oxidation reaction should consult the literature, run appropriate safety tests, and take proper precautions when running the reaction.*

## 10.2   OXIDATION OF C–C SINGLE AND DOUBLE BONDS

The oxidation of carbon–carbon bonds encompasses a number of transformations. Many of the reactions of olefins are formally oxidations, but are covered in Chapter 3, on the electrophilic substitution of olefins. This section will primarily focus on the oxidative cleavage of C–C double and single bonds, and on the oxidation of double and triple bonds to the corresponding ketones or diketones. Not included in this section are rearrangement reactions that result in the oxidative cleavage of a C–C bond such as the Baeyer–Villiger, Beckmann, Hoffman and other rearrangements (see Chapter 7).

### 10.2.1   Oxidative Cleavage of Glycols β-Aminoalcohols, α-Hydroxyaldehydes and Ketones, and Related Compounds

Periodates are the reagent of choice to achieve the oxidative cleavage of vicinal diols to produce aldehydes and ketones, as exemplified in the preparation of 2,3-*O*-isopropylidene-*D*-glyceraldehyde from a *D*-mannitol derivative.[2]

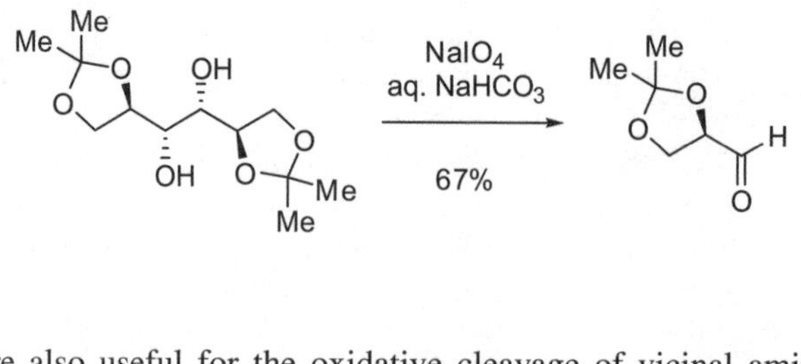

Periodates are also useful for the oxidative cleavage of vicinal aminoalcohols. The sequence shown below,[3] in which oxidative cleavage of the aminoalcohol to the imine, was executed on 150 kg scale. Sodium periodate was preferred over lead tetraacetate for this oxidation due to the less-hazardous waste products and significantly lower toxicity of the reagent.

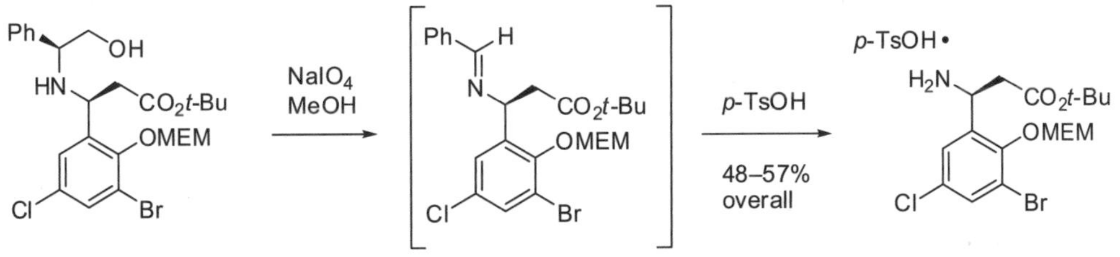

Hydrogen peroxide-mediated oxidation can also be utilized to effect oxidative cleavage in some cases.[4]

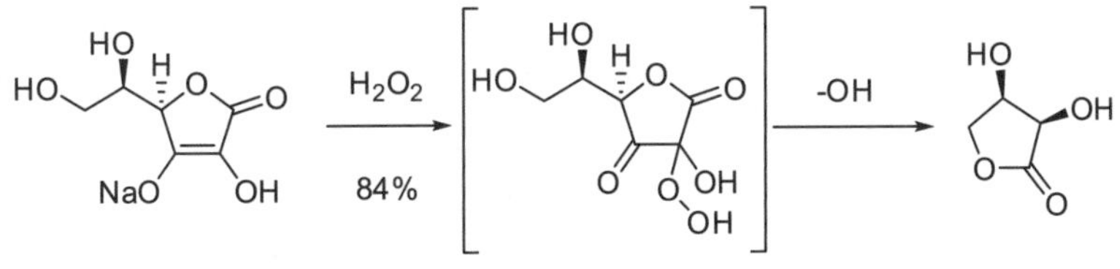

## 10.2.2    Ozonolysis

The ozonolytic cleavage of olefins[5] is followed by a variety of workups to convert the intermediate ozonide into alcohols, amines, esters, aldehydes, and other functionalities. While ozonolysis is a very powerful process for the oxidative cleavage of olefins, the intermediate ozonides, and ozone itself dissolved in organic solvents, are highly hazardous and should be handled with extreme care, even on research scale. Complete decomposition of intermediate ozonides should be confirmed by a test for residual peroxides (such as fresh peroxide test strips), especially prior to heating the material. The reaction is very general, and selectivity for less-substituted olefins over more highly substituted olefins can be achieved through the use of the appropriate indicator. Indicators can also be used to determine reaction completion before the concentration of dissolved ozone in solution rises to a hazardous level. On large scale, a number of factors need to be considered in designing a process, including the flammability of the solvent being used, whether to dilute the ozone stream with nitrogen to lower the flammability of the gas phase in the reactor, and the nature of the intermediates. Use of a protic solvent leads to the formation of alkoxy or acyloxy hydroperoxides, rather than the primary ozonide, which can improve the safety factor of the reaction.

*10.2.2.1   Ozonolysis Followed by Reduction to the Alcohol*    Alcohols can be prepared by ozonolysis followed by workup with a reducing agent such as NaBH$_4$. The 100 g preparation depicted below details the care that should be employed in the sodium borohydride quench of the ozonide; in this case the addition of the ozonide to the reducing agent was carried out over 45 minutes.[6]

[4]Dunigan, J.; Weigel, L. O. *J. Org. Chem.* **1991**, *56*, 6225–6227.
[5]Van Ornum, S. G.; Champeau, R. M.; Pariza, R. *Chem. Rev.* **2006**, *106*, 2990–3001.
[6]Faul, M. M.; Winneroski, L. L.; Krumrich, C. A.; Sullivan, K. A.; Gillig, J. R.; Neel, D. A.; Rito, C. J.; Jirousek, M. R. *J. Org. Chem.* **1998**, *63*, 1961–1973.

Although typically carried out in alcohols or $CH_2Cl_2$, ozonolysis has been reported using water as the solvent.[7] Other reducing agents that will convert an aldehyde to an alcohol can be used in the workup in place of sodium borohydride.

***10.2.2.2  Ozonolysis to the Ketone or Aldehyde Oxidation State***    Ozonolysis of the bicyclic carbamate shown to generate the bis-aldehyde was followed by treatment with benzylamine and sodium cyanoborohydride to generate the bicyclic amine.[8]Dimethylsulfide (DMS) was utilized to reduce the intermediate ozonide to the corresponding bis-aldehyde. Triphenylphosphine and trimethylphosphite are other reagents commonly used to reduce ozonides to the corresponding aldehyde or ketone.

An alternative ozonolysis sequence on the above substrate was described in which the methoxyhydroperoxide is reduced by hydrogenation over Pt/C to generate the bis-aldehyde equivalent, and benzylamine and $HCO_2H$ are added to affect reductive amination after further hydrogenation over the same catalyst.[9]

The ozonolytic cleavage of a cyclopenene to a keto-aldehyde, and its subsequent aldol condensation has been described, using DMS to reduce the intermediate ozonide.[10]

A commonly used method for the ozonolysis of cyclic olefins reported in *Organic Syntheses* results in a dialdehyde product with one end differentiated from the other as an acetal.[11] It should be noted that this procedure works well for acyclic olefins and 6-membered rings, but other ring sizes do not behave well in this reaction due to the formation of cyclic acetals.

[7]Fleck, T. J.; McWhorter, W. W., Jr.; DeKam, R. N.; Pearlman, B. A. *J. Org. Chem.* **2003**, *68*, 9612–9617.
[8]Fray, A. H.; Augeri, D. J.; Kleinman, E. F. *J. Org. Chem.* **1988**, *53*, 896–899.
[9]Brooks, P. R.; Caron, S.; Coe, J. W.; Ng, K. K.; Singer, R. A.; Vazquez, E.; Vetelino, M. G.; Watson, H. H., Jr.; Whritenour, D. C.; Wirtz, M. C. *Synthesis* **2004**, 1755–1758.
[10]Fleming, F. F.; Shook, B. C. *Org. Synth.* **2002**, *78*, 254–264.
[11]Claus, R. E.; Schreiber, S. L. *Org. Synth.* **1986**, *64*, 150–156.

The scale-up of the ozonolysis of the olefin below has been described.[12] The primary ozonide was trapped by methanol to generate the methoxy-hydroperoxide, which was treated with aqueous sodium bisulfite to effect simultaneous peroxide reduction and bisulfite adduct formation in 57% yield on 2.3 kg scale.

### 10.2.2.3 Ozonolysis Resulting in Carboxylic Acids or Esters

The electron-rich furan ring is readily oxidized, and can serve as a masked carboxylic acid or as a precursor to other heterocycles. A 2-substituted furan was utilized as a masked carboxylic acid in the synthesis of a cephalosporin.[13] Ozonolysis of the furan below and oxidative workup provided the acid in 77% yield.

The ozonolysis of simple olefins can be followed by an oxidative workup to produce the carboxylic acid directly. The most common methods for oxidative workup are hydrogen peroxide in formic acid (performic acid is most likely the oxidant)[14] or oxygen in the presence of a carbocylic acid.[15]

Methods for the ozonolysis of cyclic olefins to the aldehyde/acid oxidation state have also been published and widely used.[11]

[12]Ragan, J. A.; am Ende, D. J.; Brenek, S. J.; Eisenbeis, S. A.; Singer, R. A.; Tickner, D. L.; Teixeira, J. J., Jr.; Vanderplas, B. C.; Weston, N. *Org. Process Res. Dev.* **2003**, *7*, 155–160.

[13]Bodurow, C. C.; Boyer, B. D.; Brennan, J.; Bunnell, C. A.; Burks, J. E.; Carr, M. A.; Doecke, C. W.; Eckrich, T. M.; Fisher, J. W.; et al. *Tetrahedron Lett.* **1989**, *30*, 2321–2324.

[14]Baraldi, P. G.; Pollini, G. P.; Simoni, D.; Barco, A.; Benetti, S. *Tetrahedron* **1984**, *40*, 761–764.

[15]Habib, R. M.; Chiang, C. Y.; Bailey, P. S. *J. Org. Chem.* **1984**, *49*, 2780–2784.

***10.2.2.4   Ozonolysis Followed by Criegee Rearrangement***   The Criegee rearrangement is reminiscent of the Baeyer–Villiger oxidation and proceeds through a similar mechanism. The propenyl side chain below was cleaved by ozonolysis followed by Criegee rearrangement of the intermediate methoxy-hydroperoxide to generate the acetate of the desired alcohol.[16,17] This procedure allows the use of an olefin as a masked alcohol.

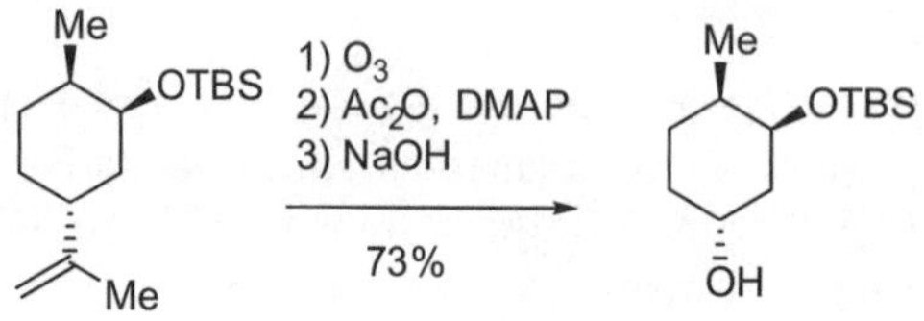

Another Criegee rearrangement on a sensitive substrate, is shown below.[17]

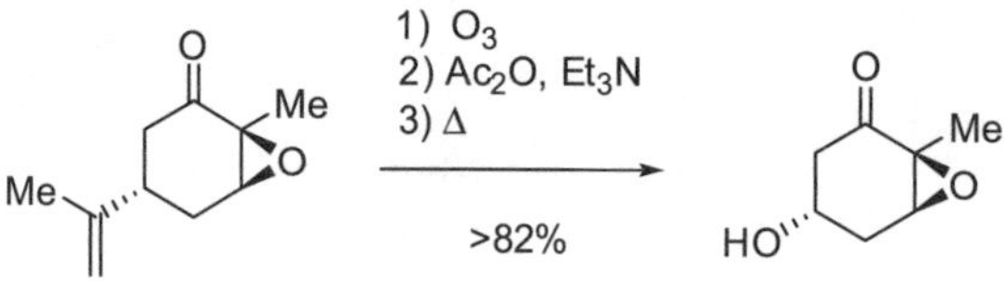

### 10.2.3   Oxidative Cleavage of Double Bonds and Aromatic Rings

Double bonds can be cleaved with a variety of inorganic reagents, with potassium permanganate being preferred.[18] If the intermediacy of a diol is employed, periodate reagents are also preferred reagents.[9] Permanganate is preferred over catalytic $OsO_4$ with either *N*-methylmorpholine-*N*-oxide or sodium chlorite as stoichiometric oxidants (for dihydroxylation methods, see Section 3.9.1). Oxidative cleavage with $NaIO_4$ in aqueous dichloromethane generated a solution of the bis-aldehyde, which was condensed with benzylamine and reduced with $NaBH(OAc)_3$ directly.[19]

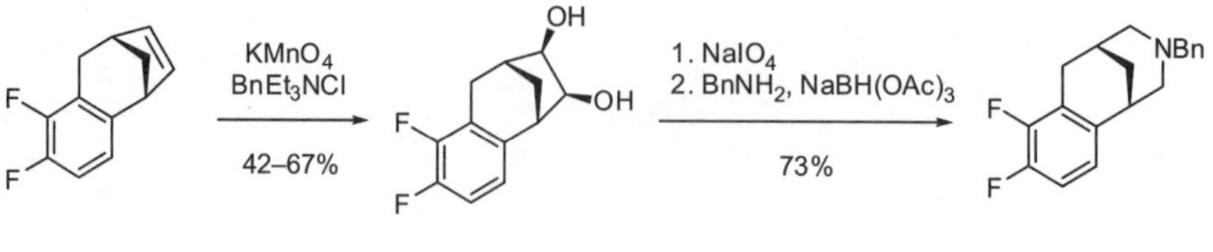

[16]Varie, D. L.; Brennan, J.; Briggs, B.; Cronin, J. S.; Hay, D. A.; Rieck, J. A., III; Zmijewski, M. J. *Tetrahedron Lett.* **1998**, *39*, 8405–8408.

[17]Schreiber, S. L.; Liew, W. F. *Tetrahedron Lett.* **1983**, *24*, 2363–2366.

[18]Frost, H. N. Synthesis of 5,6-difluoro-11-aza-tricyclo[7.3.1.0$^{2,7}$]trideca-2(7),3,5-triene Presented at the Northeast Regional Meeting of the American Chemical Society, Fairfield, CT, July, 2005.

[19]Bashore, C. G.; Vetelino, M. G.; Wirtz, M. C.; Brooks, P. R.; Frost, H. N.; McDermott, R. E.; Whritenour, D. C.; Ragan, J. A.; Rutherford, J. L.; Makowski, T. W.; Brenek, S. J.; Coe, J. W. *Org. Lett.* **2006**, *8*, 5947–5950.

Electron-rich olefins such as enamines can be oxidatively cleaved directly by treatment with periodate reagents. A strategy based on this oxidation was developed for oxidation of activated aromatic methyl groups.[20]

Noyori and coworkers have reported the oxidative cleavage of cyclohexene to adipic acid ($HO_2C(CH_2)_4CO_2H$) with 30% hydrogen peroxide and catalytic $Na_2WO_4 \cdot 2H_2O$ and a phase-transfer catalyst ($Me(n\text{-octyl})_3NHSO_4$), both at 1 mol% loading.[21]

Oxidative cleavage of a highly conjugated aromatic system was accomplished with hydrogen peroxide and catalytic $WO_4$ to provide phenanthrenedicarboxyxlic acid.[22]

## 10.2.4   Oxidative Cleavage of Alkyl Groups from Rings

Oxidative loss of alkyl groups from phenyl ethers can be accomplished via oxidation to the quinone oxidation state, followed by loss of an alkyl cation. For alkyl groups that will not form stable cations, the dienone–phenol rearrangement is an alternative reaction pathway for the rearomatization of the ring system (Section 7.2.1.5). The di-*t*-butylphenol below was oxidized using $MnO_2$ and the resulting intermediate was trapped *para* to the phenol. Following loss of the *t*-butyl group affected with $TiCl_4$, the depicted product was obtained in 55% yield.[23]

---

[20]Vetelino, M. G.; Coe, J. W. *Tetrahedron Lett.* **1994**, *35*, 219–222.

[21]Sato, K.; Aoki, M.; Noyori, R. *Science* **1998**, *281*, 1646–1647.

[22]Young, E. R. R.; Funk, R. L. *J. Org. Chem.* **1998**, *63*, 9995–9996.

[23]Hickey, D. M. B.; Leeson, P. D.; Carter, S. D.; Goodyear, M. D.; Jones, S. J.; Lewis, N. J.; Morgan, I. T.; Mullane, M. V.; Tricker, J. Y. *J. Chem. Soc., Perkin Trans.* **1988**, *1*, 3097–3102.

55%

The Baeyer–Villiger reaction (Section 7.2.4.1) is a common method for the cleavage of aryl–carbon bonds.

### 10.2.5   Oxidative Decarboxylation

Dimethyl-1,3-acetonedicarboxylate has been prepared by oxidative decarboxylation of citric acid.[24] This procedure is a modification of an *Organic Syntheses* procedure, which utilized fuming $H_2SO_4$.[25]

The use of catalytic amounts of $AgNO_3$ with stoichiometric $Na_2S_2O_8$, and the use of $CuCl_2$ with oxygen have also been reported for oxidative decarboxylations.[26]

Decarboxylations of unactivated carboxylic acids (not adjacent to an oxygen substituent) generally proceed through radical intermediates, limiting their synthetic utility. In one process, a carboxylic acid is decarboxylated with elimination to produce the olefin. Although usually accomplished with $Pb(OAc)_4$ and a copper catalyst,[27] more appealing conditions have been published, including $PdCl_2$ and pivaloyl anhydride[28] or iodobenzene diacetate and a copper catalyst.[29]

[24]Kotha, S.; Joseph, A.; Sivakumar, R.; Manivannan, E. *Indian J. Chem. B Org* **1998**, *37B*, 397–398.

[25]Adams, R.; Chiles, H. M. *Org. Synth.* **1925**, *V*, 53–54.

[26]Bjorsvik, H.-R.; Liguori, L.; Minisci, F. *Org. Process Res. Dev.* **2000**, *4*, 534–543.

[27]Hanessian, S.; Sahoo, S. P. *Tetrahedron Lett.* **1984**, *25*, 1425–1428.

[28]Goossen, L. J.; Rodriguez, N. *Chem. Commun.* **2004**, 724–725.

[29]Concepcion, J. I.; Francisco, C. G.; Freire, R.; Hernandez, R.; Salazar, J. A.; Suarez, E. *J. Org. Chem.* **1986**, *51*, 402–404.

Oxidative replacement of a carboxylic acid with a halide, known as the Hunsdiecker reaction, can be accomplished under fairly mild conditions.[30]

More commonly, this reaction is run using NCS, NBS, or NIS as the oxidant and halide source, along with a catalyst such as PhIO[31] or Bu$_4$NOCOCF$_3$.[32]

Oxidative decarboxylation can also be achieved electrochemically in a process known as the Kolbe decarboxylation, as demonstrated by the example below from *Organic Syntheses*.[33]

### 10.2.6 Oxidative Decyanation

The oxidative conversion of arylacetonitriles into the corresponding acetophenone is accomplished by deprotonation followed by trapping with oxygen and expulsion of cyanide anion. As such, the acidity of the benzylic proton dictates the strength of base needed to accomplish the transformation. In the case of benzophenones, aqueous potassium carbonate is a strong enough base.[34] Care must be used when working with oxygen in the presence of a flammable solvent, particularly on large scale, to avoid undesired combustion.

For substrates with a lower acidity, such as acetophenone, sodium hydroxide is a suitable base;[35] with hydrolysis-prone substrates LDA can be employed.[36]

[30]You, H.-W.; Lee, K.-J. *Synlett* **2001**, 105–107.
[31]Graven, A.; Joergensen, K. A.; Dahl, S.; Stanczak, A. *J. Org. Chem.* **1994**, *59*, 3543–3546.
[32]Naskar, D.; Roy, S. *Tetrahedron* **2000**, *56*, 1369–1377.
[33]Lakner, F. J.; Chu, K. S.; Negrete, G. R.; Konopelski, J. P. *Org. Synth.* **1996**, *73*, 201–214.
[34]Kulp, S. S.; McGee, M. J. *J. Org. Chem.* **1983**, *48*, 4097–4098.
[35]Donetti, A.; Boniardi, O.; Ezhaya, A. *Synthesis* **1980**, 1009–1011.
[36]Parker, K. A.; Kallmerten, J. *J. Org. Chem.* **1980**, *45*, 2614–2620.

### 10.2.7   Oxidation of Olefins to Aldehydes and Ketones

The Wacker process converts olefins to the corresponding aldehyde or ketone. The carbonyl is selectively installed at the internal position of a terminal olefin to make the ketone rather than at the primary position to make the aldehyde. The process uses a palladium catalyst, a copper co-oxidant, and oxygen as the stoichiometric oxidant.[37]

$$C_8H_{17}\diagdown\diagup \quad \xrightarrow[\substack{65-73\%}]{\substack{PdCl_2,\ CuCl,\ O_2 \\ DMF,\ H_2O}} \quad C_8H_{17}\diagup\overset{O}{\diagdown}Me$$

## 10.3   OXIDATION OF C–H BONDS

### 10.3.1   Aromatization of 6-Membered Rings

The aromatization of 6-membered rings can be accomplished in a number of ways. The most practical method is dehydrogenation using a metal catalyst, although other reagents can be useful for this transformation as well. Generally, some level of unsaturation must exist in the ring as a handle for aromatization.

Aromatization of cyclic enamines can be accomplished using Pd/C, with nitrobenzene as the hydrogen scavenger.[38] This procedure has been applied to cyclohexanone oximes[39] and to dihydropyridines.[40]

$$\xrightarrow[\substack{86\%}]{\substack{5\%\ Pd/C,\ 4\text{\AA}\ MS \\ PhNO_2,\ PhCH_3}}$$

Another common reagent used for the aromatization of 6-membered rings is DDQ.[41]

$$\xrightarrow[\substack{88\%}]{\substack{DDQ,\ CH_2Cl_2}}$$

Aromatization of the dihydropyridine below with sulfur was exploited in a regiospecific synthesis of the product depicted.[42]

[37]Tsuji, J.; Nagashima, H.; Nemoto, H. *Org. Synth.* **1984**, *62*, 9–13.

[38]Cossy, J.; Belotti, D. *Org. Lett.* **2002**, *4*, 2557–2559.

[39]Matsumoto, M.; Tomizuka, J.; Suzuki, M. *Synth. Commun.* **1994**, *24*, 1441–1446.

[40]Nakamichi, N.; Kawashita, Y.; Hayashi, M. *Org. Lett.* **2002**, *4*, 3955–3957.

[41]Adams, J.; Belley, M. *J. Org. Chem.* **1986**, *51*, 3878–3881.

[42]Lantos, I.; Gombatz, K.; McGuire, M.; Pridgen, L.; Remich, J.; Shilcrat, S. *J. Org. Chem.* **1988**, *53*, 4223–4227.

In the synthesis of a serotonin receptor agonist, aromatization was accomplished by treatment with $MnO_2$.[43] An interesting solvent effect was noted: HOAc greatly accelerated the reaction such that only two equivalents of $MnO_2$ were required.

In the synthesis of voriconazole, a key ethylpyrimidine was prepared by addition of ethyl Grignard to a pyrimidine followed by *in situ* oxidation with iodine.[44]

Another interesting aromatization is found in the synthesis of 10-hydroxycamptothe-cin.[45] The reaction actually proceeds by oxidation *para* to the aniline nitrogen followed by a rapid aromatization. Oxygenation of the fully aromatized substrate does not work, lending evidence that the oxygenation reaction proceeds first. The desired product can also be further oxidized, but the authors found that by careful selection of solvent, the product would precipitate from the reaction mixture as it was formed, protecting it from further oxidation.

[43]Martinelli, M. J. *J. Org. Chem.* **1990**, *55*, 5065–5073.

[44]Butters, M.; Ebbs, J.; Green, S. P.; MacRae, J.; Morland, M. C.; Murtiashaw, C. W.; Pettman, A. J. *Org. Process Res. Dev.* **2001**, *5*, 28–36.

[45]Lipshutz, B. H.; Wood, M. R. *J. Am. Chem. Soc.* **1994**, *116*, 11689–11702.

### 10.3.2    Dehydrogenations Yielding Carbon–Carbon Bonds

A few methods have been developed to directly introduce α,β-unsaturation from the corresponding saturated carbonyl. Saegusa's method is to treat a silyl enol ether with palladium to effect dehydrogenation.[46]

Using DDQ in conjunction with bis(trimethylsilyl)trifluoroacetamide (BSTFA) accomplishes the direct oxidation of the amide below to produce finasteride in high yield.[47] This reagent mixture provided a nice alternative to the more commonly employed reagent phenylselenic anhydride, which is not attractive due to the toxicity of the reagent and waste streams produced.

Another reagent for introducing unsaturation proximal to a ketone is IBX (*o*-iodoxybenzoic acid).[52] As an alternative, $HIO_3$ and $I_2O_5$ have also been reported to work well for this transformation.[48]

### 10.3.3    Halogenation α to a Ketone, Aldehyde, or Carboxylic Acid

The halogenation of enolizable carbons α to a carbonyl is a widely utilized reaction, and a wide variety of reagents can be used for this transformation. Acidic, basic, and radical reaction mechanisms often achieve different levels of selectivity and reactivity.

[46]Torneiro, M.; Fall, Y.; Castedo, L.; Mourino, A. *Tetrahedron* **1997**, *53*, 10851–10870.

[47]Bhattacharya, A.; DiMichele, L. M.; Dolling, U. H.; Douglas, A. W.; Grabowski, E. J. J. *J. Am. Chem. Soc.* **1988**, *110*, 3318–3319.

[48]Nicolaou, K. C.; Montagnon, T.; Baran, P. S. *Angew. Chem., Int. Ed. Engl.* **2002**, *41*, 993–996.

*10.3.3.1   Chlorination*   Chlorination is most often achieved using NCS or sulfuryl chloride, although chlorine and chlorophosphorus reagents can also be used. Other useful reagents include trichloroisocyanuric acid (TCCA)[49] and *N,N'*-dichlorodimethyl hydantoin (NDDH). TCCA is worth considering as a replacement for NCS due to its relatively low cost, high organic solubility, and the active chlorine content is 92%, because all three chlorines are active.

If the chlorinated product contains additional hydrogen atoms, the higher acidity of the product frequently results in polyhalogenated products. Avoiding over-reaction is the greatest challenge in this type of reaction.

In the example below, 3-acetylpyridine can be chlorinated with NCS under acidic conditions that protect the pyridine nitrogen from oxidation on large scale in 83% yield.[50]

Chlorination of ketones can also be accomplished with sulfuryl chloride as exemplified below.[51–53] In the second example, chlorination of the more highly substituted position, resulting from the more stable enol, is noteworthy.

The chlorination of acids and esters requires more forcing conditions. A strong oxidant that can be generated *in situ* is $OCl_2$, but this requires not only the use of chlorine gas, but mixing in a second gas as well.[54] The reaction can also be accomplished using $PCl_5$,[55] or $Cl_2$ and $PCl_3$ (known as the Hell–Volhard–Zelinskii reaction).

[49]Tilstam, U.; Weinmann, H. *Org. Process Res. Dev.* **2002**, *6*, 384–393.

[50]Duquette, J.; Zhang, M.; Zhu, L.; Reeves, R. S. *Org. Process Res. Dev.* **2003**, *7*, 285–288.

[51]Masilamani, D.; Rogic, M. M. *J. Org. Chem.* **1981**, *46*, 4486–4489.

[52]Ikemoto, N.; Liu, J.; Brands, K. M. J.; McNamara, J. M.; Reider, P. J. *Tetrahedron* **2003**, *59*, 1317–1325.

[53]Warnhoff, E. W.; Martin, D. G.; Johnson, W. S. *Org. Synth.* **1957**, *37*, 8–12.

[54]Ogata, Y.; Sugimoto, T.; Inaishi, M. *Org. Synth.* **1980**, *59*, 20–26.

[55]Gerlach, H.; Kappes, D.; Boeckman, R. K., Jr.; Maw, G. N. *Org. Synth.* **1993**, *71*, 48–55.

In the case of an ester, enolate formation with a strong base can be followed by chlorination with a wide variety of chlorinating agents, including hexachloroethane.[56]

**10.3.3.2 Bromination**   Bromination is more common than chlorination, and a wider variety of reagents are useful for this transformation. Particularly useful brominating agents are Br$_2$, NBS, dibromodimethylhydantoin (DBDMH), pyridinium hydrobromide perbromide (PHP), and phenyltrimethylammonium perbromide (PTAB)

The simplest and least expensive brominating agent is Br$_2$, which is more easily handled than chlorine.[57] In some cases, an acid such as HBr[58] or PBr$_3$[59] is added to catalyze the reaction.

[56]Boeckman, R. K., Jr.; Perni, R. B.; Macdonald, J. E.; Thomas, A. J. *Org. Synth.* **1988**, *66*, 194–202.
[57]Rappe, C. *Org. Synth.* **1973**, *53*, 123–127.
[58]Gaudry, M.; Marquet, A. *Org. Synth.* **1976**, *55*, 24–27.
[59]Ashcroft, M. R.; Hoffmann, H. M. R. *Org. Synth.* **1978**, *58*, 17–24.

In cases where selectivity is an issue, bromine may not be the optimal choice of reagent. Bromine in methanol produced a 63 : 37 ratio of bromoketones below.[60] The authors took advantage of the large rate difference of the subsequent reaction with PPh$_3$. The primary bromoketone reacts much faster with PPh$_3$, allowing the isolation of the primary phosphonium salt in 57% overall yield (115 kg).

57% overall

Cupric bromide (CuBr$_2$) is known to be a mild brominating agent for ketones.[61] Treatment of the ketone with CuBr$_2$ in EtOAc results in 85% conversion to the bromoketone.[62] Although CuBr$_2$ often demonstrates useful selectivity differences over Br$_2$, a major detriment is that it requires the use of two equivalents of CuBr$_2$, generating a large amount of metal-containing waste.

Although rare, aldehydes can also be α-brominated. The aldehyde shown was brominated with dibromobarbituic acid.[63] The bromide was not isolated but carried into the next step as a crude solution.

[60]Stuk, T. L.; Assink, B. K.; Bates, R. C., Jr.; Erdman, D. T.; Fedij, V.; Jennings, S. M.; Lassig, J. A.; Smith, R. J.; Smith, T. L. *Org. Process Res. Dev.* **2003**, *7*, 851–855.

[61]Kochi, J. K. *J. Am. Chem. Soc.* **1955**, *77*, 5274–5278.

[62]Bai, D.; Xu, R.; Chu, G.; Zhu, X. *J. Org. Chem.* **1996**, *61*, 4600–4606.

[63]Barnett, C. J.; Wilson, T. M.; Kobierski, M. E. *Org. Process Res. Dev.* **1999**, *3*, 184–188.

Other reagents that are often used to brominate ketones are pyridinium hydrobromide perbromide (PHP) and phenyltrimethylammonium perbromide (PTAB). Unlike $Br_2$, PHP and PTAB are stable crystalline solids, which increase the ease of handling. They act like $Br_2$ and in some cases offer superior results; for example, the methylketones below were brominated with PHP[64] and PTAB,[65] respectively.

The bromination of acids and esters require the use of more powerful oxidants. Bromine with phosphorus[66] or $PCl_3$[67] (Hell–Volhard–Zelinskii reaction) can be used. In the case of an acid chloride, NBS alone can be employed,[68] but in the case of the less acidic ester, NBS requires conditions which initiate radical formation.[69]

[64]Conrow, R. E.; Dean, W. D.; Zinke, P. W.; Deason, M. E.; Sproull, S. J.; Dantanarayana, A. P.; DuPriest, M. T. *Org. Process Res. Dev.* **1999**, *3*, 114–120.

[65]Jacques, J.; Marquet, A. *Org. Synth.* **1973**, *53*, 111–115.

[66]Price, C. C.; Judge, J. M. *Org. Synth.* **1965**, *45*, 22–24.

[67]Carpino, L. A.; McAdams, L. V., III *Org. Synth.* **1970**, *50*, 31–35.

[68]Harpp, D. N.; Bao, L. Q.; Coyle, C.; Gleason, J. G.; Horovitch, S. *Org. Synth.* **1976**, *55*, 27–31.

[69]Muehlemann, C.; Hartmann, P.; Obrecht, J. P. *Org. Synth.* **1993**, *71*, 200–206.

***10.3.3.3  Iodination***  Iodination is employed less frequently than chlorination and bromination, and is most commonly accomplished with $I_2$ or NIS.[70]

α-Halogenation of amides is not as common as the corresponding reaction of ketones. A general procedure for α-iodination or α-bromination of secondary amides was applied to the steroid below.[71] Treatment with TMSI and $I_2$ produces the α-iodoamide in 98% yield.

***10.3.3.4  Fluorination***  Electrophilic introduction of fluorine is the most difficult halogenation to run, and *the reagents employed generally pose safety hazards.* Nevertheless, the importance of fluorine in biologically active compounds has driven the development of a number of methods for its introduction. The most widely known reagent for this task is 1-(chloromethyl)-4-fluoro-1,4-diazabicyclo[2.2.2]octane bis(tetrafluoroborate) (Selectfluor); and in the example below, treatment of the *in situ* generated enol ether of the steroidal ketone produces the fluoroketone in 93% yield.[72]

A comparison of the reaction of several fluorinating agents (*N*-fluorobenzenesulfonimide [NFSI], *N*-fluoropyridinium pyridine heptafluorodiborate [NFPy], and Selectfluor) with three key steroidal 3,5-dienol acetates to produce the fluorinated products has been reported.[73] In general, Selectfluor had the best combination of reactivity and minimal byproduct formation, but procedures for using all three are provided.

[70]Colon, I.; Griffin, G. W.; O'Connell, E. J., Jr. *Org. Synth.* **1972**, *52*, 33–35.

[71]King, A. O.; Anderson, R. K.; Shuman, R. F.; Karady, S.; Abramson, N. L.; Douglas, A. W. *J. Org. Chem.* **1993**, *58*, 3384–3386.

[72]Koenigsberger, K.; Chen, G.-P.; Vivelo, J.; Lee, G.; Fitt, J.; McKenna, J.; Jenson, T.; Prasad, K.; Repic, O. *Org. Process Res. Dev.* **2002**, *6*, 665–669.

[73]Reydellet-Casey, V.; Knoechel, D. J.; Herrinton, P. M. *Org. Process Res. Dev.* **1997**, *1*, 217–221.

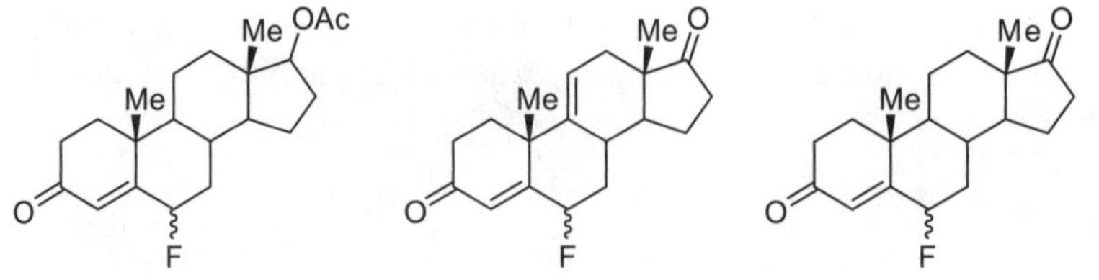

Another reagent reported in an *Organic Syntheses* preparation is *N*-fluoropyridinium triflate.[74]

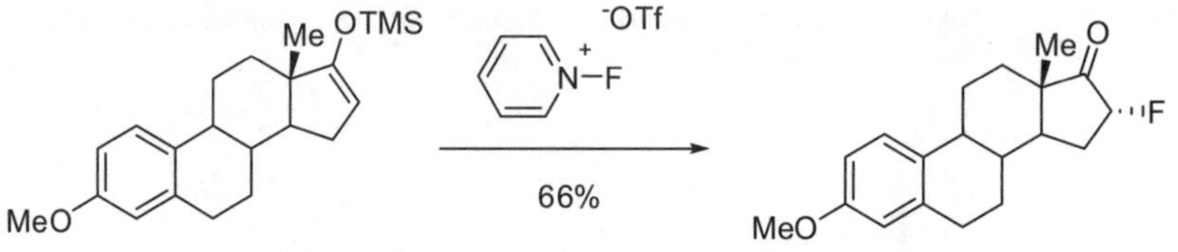

66%

**10.3.3.5 Haloform Reaction** The haloform reaction, the solvolysis of a trihalomethyl ketone to the ester or acid, is a useful procedure for the cleavage of methyl ketones.[75–77] Sodium hydroxide in water is most commonly used for this transformation, in the presence of a hypochlorite or hypobromite if the halogenation runs concurrently with the cleavage reaction. The cleavage is facile and does not require heat to proceed.

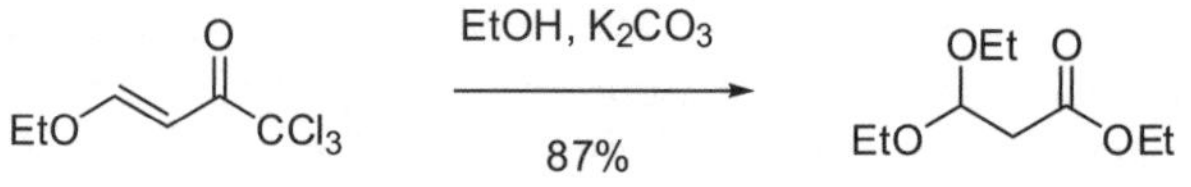

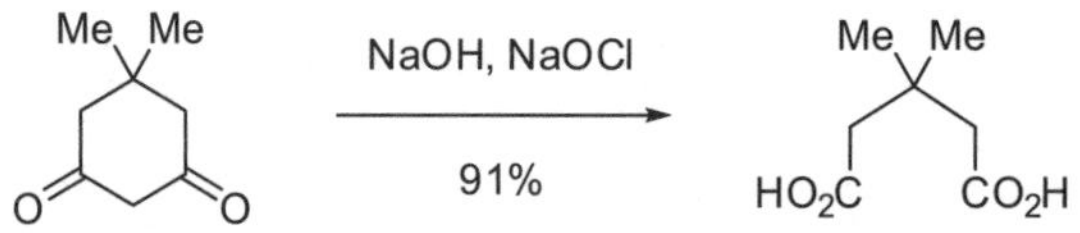

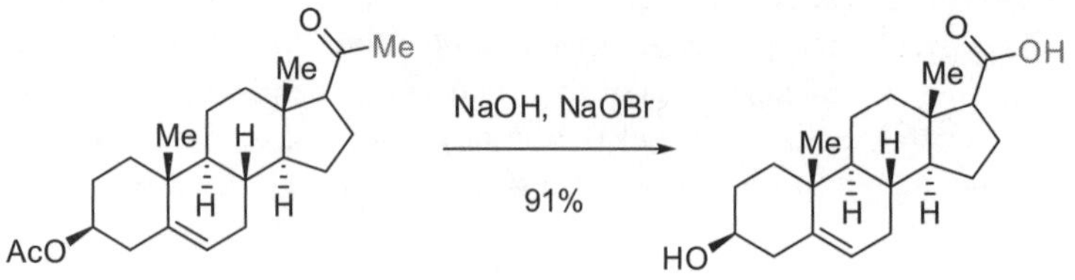

EtOH, $K_2CO_3$

87%

NaOH, NaOCl

91%

NaOH, NaOBr

91%

[74]Umemoto, T.; Fukami, S.; Tomizawa, G.; Harasawa, K.; Kawada, K.; Tomita, K. *J. Am. Chem. Soc.* **1990**, *112*, 8563–8575.

[75]Tietze, L. F.; Voss, E.; Hartfiel, U. *Org. Synth.* **1990**, *69*, 238–244.

[76]Smith, W. T., Jr.; McLeod, G. L. *Org. Synth.* **1951**, *31*, 40–42.

[77]Staunton, J.; Eisenbraun, E. J. *Org. Synth.* **1962**, *42*, 4–7.

When following a Friedel–Crafts reaction, the haloform reaction essentially allows the electrophilic introduction of a carboxylic acid or ester to an aromatic system.[78]

*10.3.3.6  Cleavage of Ketones with MNH2*    Nitrogen nucleophiles can participate in the haloform reaction.[79] When following a Friedel–Crafts acylation (Section 5.5.1), the amino variant of the haloform reaction essentially allows the electrophilic introduction of an amide to an aromatic system. The cleavage on an unenolizable ketone with ammonia is known as the Haller–Bauer reaction.

### 10.3.4  Oxygenation α to a Ketone, Aldehyde, or Carboxylic Acid

Via the enolate, ketones, aldehydes, and carboxylic acid derivatives can be oxygenated at the α-position. Useful reagents for the introduction of oxygen include oxygen, peracids, dimethyldioxirane (DMDO), and oxaziridines.

Introduction of oxygen α to a carbonyl is much less common on large scale than halogenation. This is likely due to the more hazardous nature of common oxygenating agents. One interesting example of a large scale (10 kg) oxygenation that illustrates some of the safety concerns is the synthesis of 6-hydroxybuspirone from buspirone.[9] Treatment with NaHMDS generated the enolate, which was then oxygenated with oxygen in 71% yield. Triethylphosphite had to be present in the reaction mixture before the introduction of $O_2$ so that the intermediate peroxide was reduced and did not build up in the reaction mixture. Additionally, the concentration of oxygen in the headspace of the reactor had to be carefully controlled so as to not reach levels where it would form a combustible mixture with the solvent vapors.

[78]Bailey, D. M.; Johnson, R. E.; Albertson, N. F. *Org. Synth.* **1971**, *51*, 100–102.
[79]Sukornick, B. *Org. Synth.* **1960**, *40*, 103–104.

The enolate oxygenation with oxygen can be run to produce the alcohol or ketone, depending on the use of a reducing agent under specific conditions.[80]

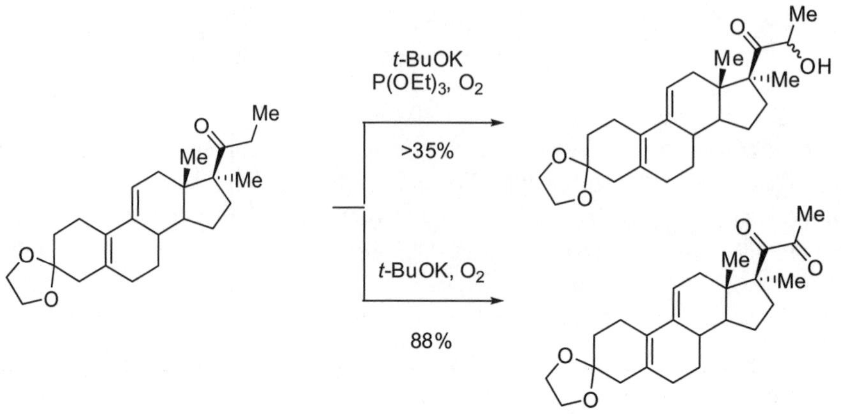

The use of DMDO has also been reported to be effective.[81]

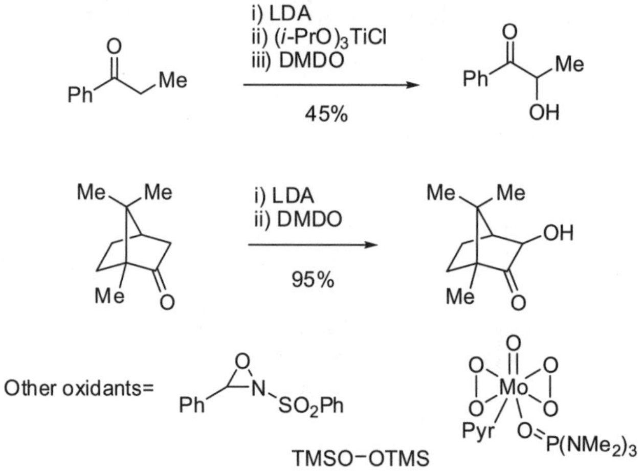

As shown above, other reagents are also useful for the latter transformation, and provide varying levels of diastereoselectivity.[82] The use of oxaziridines has been described in a diastereoselective and enantioselective process.[83] For enantioselective oxygenations of achiral enolates, the most common oxaziridines used are derivatives of camphor.[84]

[80]Crocq, V.; Masson, C.; Winter, J.; Richard, C.; Lemaitre, G.; Lenay, J.; Vivat, M.; Buendia, J.; Prat, D. *Org. Process Res. Dev.* **1997**, *1*, 2–13.

[81]Adam, W.; Mueller, M.; Prechtl, F. *J. Org. Chem.* **1994**, *59*, 2358–2364.

[82]Adam, W.; Korb, M. N. *Tetrahedron* **1996**, *52*, 5487–5494.

[83]Davis, F. A.; Chen, B. C. *Chem. Rev.* **1992**, *92*, 919–934.

[84]Davis, F. A.; Sheppard, A. C.; Chen, B. C.; Haque, M. S. *J. Am. Chem. Soc.* **1990**, *112*, 6679–6690.

84%, 95%ee

An alternative strategy is to make an intermediate enol acetate or enol silane and epoxidize or dihydroxylate the olefin (see Section 3.9).

### 10.3.5    Introduction of Nitrogen α to a Ketone, Aldehyde, or Carboxylic Acid

*10.3.5.1  Aliphatic Diazonium Coupling*    The enolates of ethyl 2-methylacetoacetate and methyl 2-methylmalonate were utilized to trap *p*-methoxyphenyl azide as the diazo intermediate, which after deacetylation and decarboxylation, respectively, resulted in the α-iminoester.[85] The overall process is known as the Japp–Klingermann reaction.

Another example can be found in *Organic Syntheses.*[86]

*10.3.5.2  Nitrosation of Activated Carbon–Hydrogen Bonds*    Nitrous acid adds to acidic position of organic compounds to make the corresponding oxime.[87,88] The same transformation can be accomplished with RONO reagents such as methyl nitrite[89] or isoamyl nitrite.[90] On compounds with only one acidic proton, the nitrosated product is the final product.

[85]Bessard, Y. *Org. Process Res. Dev.* **1998**, *2*, 214–220.
[86]Reynolds, G. A.; Van Allan, J. A. *Org. Synth.* **1952**, *32*, 84–86.
[87]Ferris, J. P.; Sanchez, R. A.; Mancuso, R. W. *Org. Synth.* **1973**, *Coll.* Vol. V, 32–35.
[88]Zambito, A. J.; Howe, E. E. *Org. Synth.* **1960**, *40*, 21–23.
[89]Itoh, M.; Hagiwara, D.; Kamiya, T. *Org. Synth.* **1980**, *59*, 95–101.
[90]Wheeler, T. N.; Meinwald, J. *Org. Synth.* **1972**, *52*, 53–58.

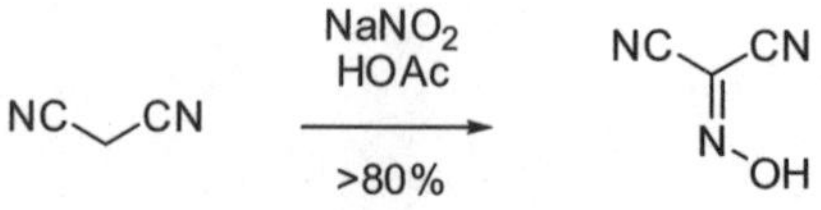

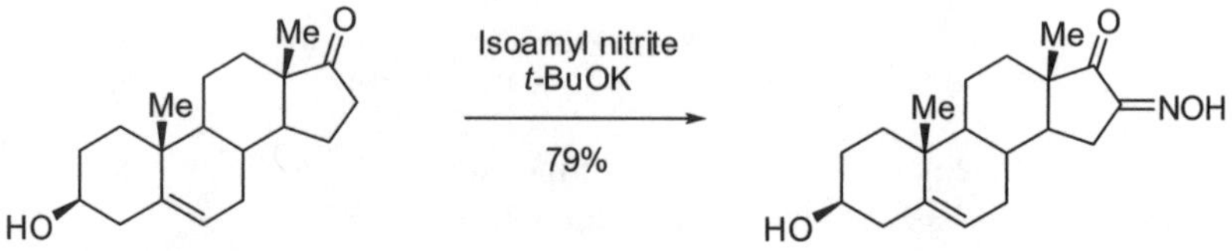

### 10.3.5.3 Formation of Diazo Compounds

One classic example of the direct introduction of nitrogen can be found in the synthesis of thienamycin.[91] The ketoester was converted to the diazo compound in 90% yield.

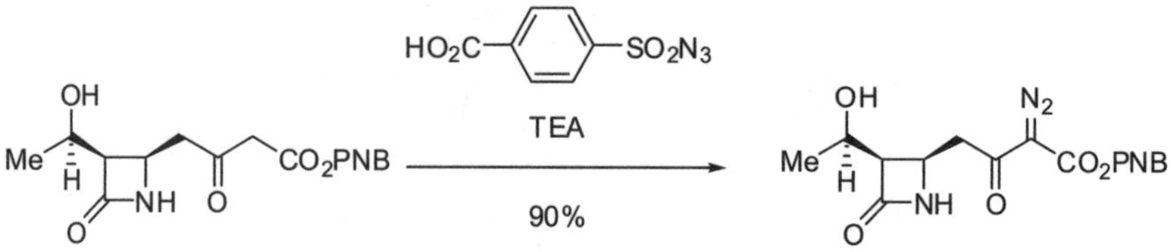

Likewise, loracarbef was prepared by a similar strategy.[13] Diazotization was accomplished in 85% yield with $p$-dodecylbenzenesulfonyl azide, a reagent that has been shown to have a better safety profile than other sulfonyl azides.[92]

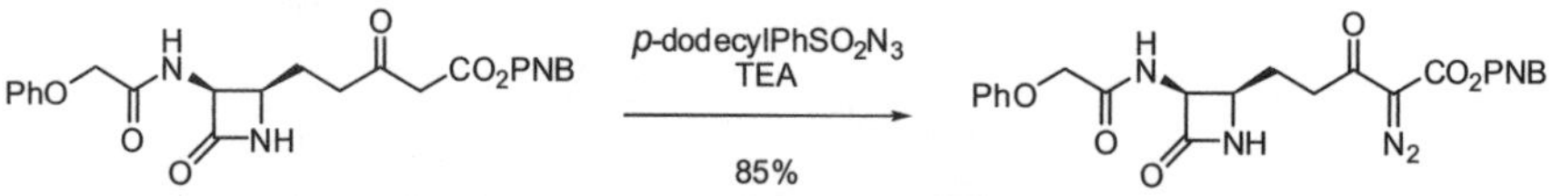

Two preparations for the "diazo transfer reaction" are shown below. The first reaction can be achieved using TEA[93] or with NaOH and a phase-transfer catalyst.[94] The second example demonstrates a diazo transfer with deacylation.[95]

[91]Salzmann, T. N.; Ratcliffe, R. W.; Christensen, B. G.; Bouffard, F. A. *J. Am. Chem. Soc.* **1980**, *102*, 6161–6163.
[92]Hazen, G. G.; Weinstock, L. M.; Connell, R.; Bollinger, F. W. *Synth. Commun.* **1981**, *11*, 947–956.
[93]Regitz, M.; Hocker, J.; Liedhegener, A. *Org. Synth.* **1973**, *Coll.* Vol. V, 179–183.
[94]Ledon, H. J. *Org. Synth.* **1980**, *59*, 66–71.
[95]Regitz, M.; Rueter, J.; Liedhegener, A. *Org. Synth.* **1971**, *51*, 86–89.

### 10.3.5.4 *Amination α to a Carbonyl*

The Neber rearrangement (Section 7.2.2.2) is an effective method for the introduction of an amine next to a ketone.[96–98]

Common reagents for the trapping of an enolate to introduce nitrogen are diazocarboxylates. These lead to the carboxyhydrazino derivatives, which can be difficult to reduce to the amine or carbamate.[99,100]

DBAD=BocN=NBoc

[96]Chung, J. Y. L.; Ho, G.-J.; Chartrain, M.; Roberge, C.; Zhao, D.; Leazer, J.; Farr, R.; Robbins, M.; Emerson, K.; Mathre, D. J.; McNamara, J. M.; Hughes, D. L.; Grabowski, E. J. J.; Reider, P. J. *Tetrahedron Lett.* **1999**, *40*, 6739–6743.

[97]LaMattina, J. L.; Suleske, R. T. *Synthesis* **1980**, 329–330.

[98]LaMattina, J. L.; Suleske, R. T. *Org. Synth.* **1986**, *64*, 19–26.

[99]Evans, D. A.; Britton, T. C.; Dellaria, J. F., Jr. *Tetrahedron* **1988**, *44*, 5525–5540.

[100]Evans, D. A.; Nelson, S. G. *J. Am. Chem. Soc.* **1997**, *119*, 6452–6453.

Oxaziridines can also be used to transfer nitrogen to an enolate, as shown below. This methodology introduces a single nitrogen to the substrate, obviating the need for a hydrazine reduction, but suffers from the disadvantage that the reagents are more expensive and less readily available.[101]

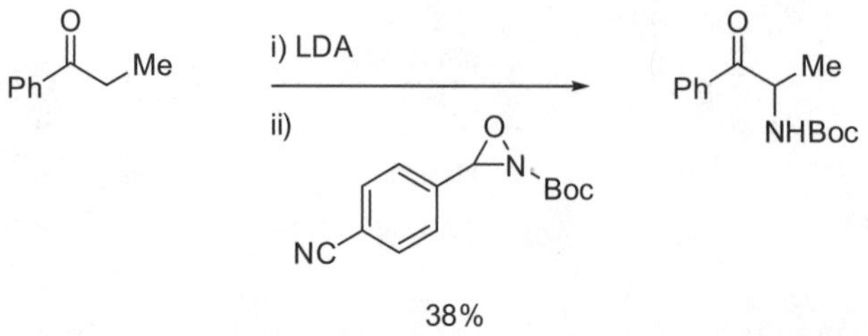

38%

*O*-Nitroarylhydroxylamines can also be used as an electrophilic nitrogen source with enolates.[102]

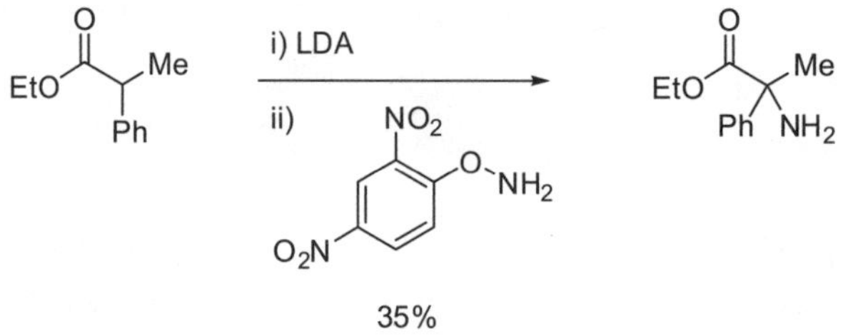

35%

### 10.3.6    Sulfenation and Selenylation of Ketones, Aldehydes, and Esters

Enolates can be sulfenated and selenated using electrophilic sulfur or selenium reagents. The selenylation is frequently followed by selenoxide elimination to form the $\alpha,\beta$-unsaturated compound. Phenylselenic anhydride accomplishes both of these transformations.[47,103]

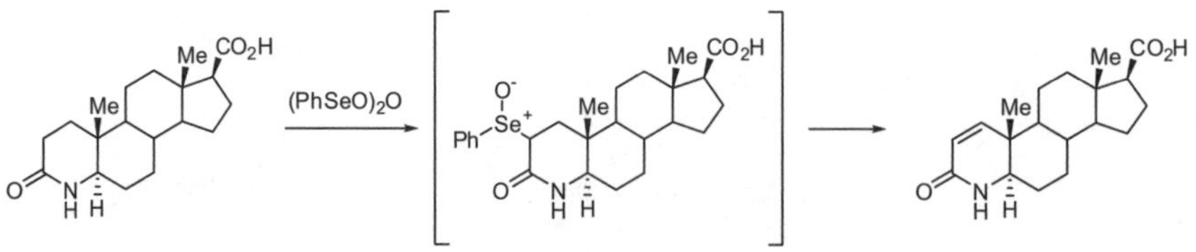

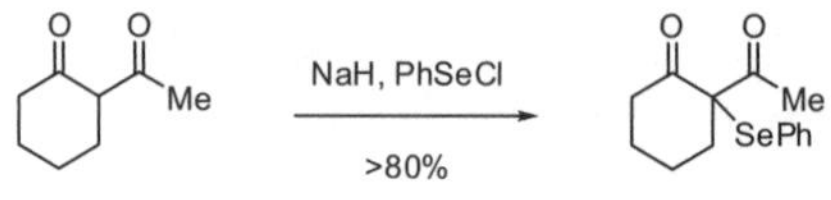

>80%

[101]Vidal, J.; Guy, L.; Sterin, S.; Collet, A. *J. Org. Chem.* **1993**, *58*, 4791–4793.
[102]Radhakrishna, A. S.; Loudon, G. M.; Miller, M. J. *J. Org. Chem.* **1979**, *44*, 4836–4841.
[103]Renga, J. M.; Reich, H. J. *Org. Synth.* **1980**, *59*, 58–65.

Disulfides and PhSCl are the most frequently employed reagents for electrophilic sulfination.[104]

### 10.3.7  Sulfonylation of Aldehydes, Ketones, and Acids

Sulfur trioxide reacts α to carbonyl compounds containing an acidic proton to produce the sulfonic acid.[105] Solvents other than carbon tetrachloride should be selected if utilizing this transformation.

### 10.3.8  Allylic and Benzylic Halogenation

For the halogenation of allylic or benzylic positions, different halogen sources can be used including $Cl_2$, $Br_2$, or N-halo amides or imides (N-bromosuccinimide [NBS], 1,3-dibromo-5,5-dimethylhydantoin [DBDMH], etc).[106–112] Free radical initiation is generally accomplished by azo derivatives or photochemically. DuPont has developed a number of azo derivatives (including VAZO-52) that offer different stabilities so one can tailor the reaction to different solvent and temperature combinations.[113]

[104]Stuetz, P.; Stadler, P. A. *Org. Synth.* **1977**, *56*, 8–14.

[105]Weil, J. K.; Bistline, R. G., Jr.; Stirton, A. J. *Org. Synth.* **1956**, *36*, 83–86.

[106]Yasuda, N.; Huffman, M. A.; Ho, G.-J.; Xavier, L. C.; Yang, C.; Emerson, K. M.; Tsay, F.-R.; Li, Y.; Kress, M. H.; Rieger, D. L.; Karady, S.; Sohar, P.; Abramson, N. L.; DeCamp, A. E.; Mathre, D. J.; Douglas, A. W.; Dolling, U.-H.; Grabowski, E. J. J.; Reider, P. J. *J. Org. Chem.* **1998**, *63*, 5438–5446.

[107]Srinivas, K.; Srinivasan, N.; Krishna, M. R.; Reddy, C. R.; Arunagiri, M.; Lalitha, R.; Reddy, K. S. R.; Reddy, B. S.; Reddy, G. M.; Reddy, P. P.; Kumar, M. K.; Reddy, M. S. *Org. Process Res. Dev.* **2004**, *8*, 952–954.

[108]Shimizu, H.; Shimizu, K.; Kubodera, N.; Mikami, T.; Tsuzaki, K.; Suwa, H.; Harada, K.; Hiraide, A.; Shimizu, M.; Koyama, K.; Ichikawa, Y.; Hirasawa, D.; Kito, Y.; Kobayashi, M.; Kigawa, M.; Kato, M.; Kozono, T.; Tanaka, H.; Tanabe, M.; Iguchi, M.; Yoshida, M. *Org. Process Res. Dev.* **2005**, *9*, 278–287.

[109]Hayler, J. D.; Howie, S. L. B.; Giles, R. G.; Negus, A.; Oxley, P. W.; Walsgrove, T. C.; Whiter, M. *Org. Process Res. Dev.* **1998**, *2*, 3–9.

[110]Larsen, R. D.; Corley, E. G.; King, A. O.; Carroll, J. D.; Davis, P.; Verhoeven, T. R.; Reider, P. J.; Labelle, M.; Gauthier, J. Y.; et al. *J. Org. Chem.* **1996**, *61*, 3398–3405.

[111]Aeilts, S. L.; Cefalo, D. R.; Bonitatebus, P. J., Jr.; Houser, J. H.; Hoveyda, A. H.; Schrock, R. R. *Angew. Chem., Int. Ed. Engl.* **2001**, *40*, 1452–1456.

[112]Conlon, D. A.; Drahus-Paone, A.; Ho, G.-J.; Pipik, B.; Helmy, R.; McNamara, J. M.; Shi, Y.-J.; Williams, J. M.; Macdonald, D.; Deschenes, D.; Gallant, M.; Mastracchio, A.; Roy, B.; Scheigetz, J. *Org. Process Res. Dev.* **2006**, *10*, 36–45.

[113]Anderson, A. G.; Gridnev, A.; Moad, G.; Rizzardo, E.; Thang, S. H. *Polymerization initiators for controlling polymer molecular weight and structure.* WO 9830601 **1998**, 40 pp. http://www2.dupont.com/vazo/en_US, last visited May 8, 2011.

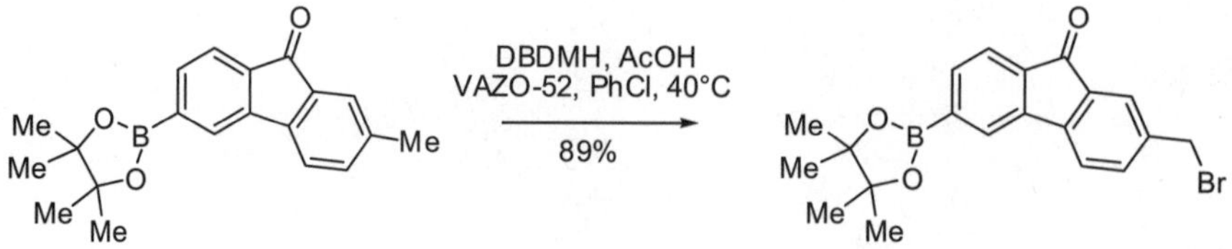

DBDMH, AcOH
VAZO-52, PhCl, 40°C
89%

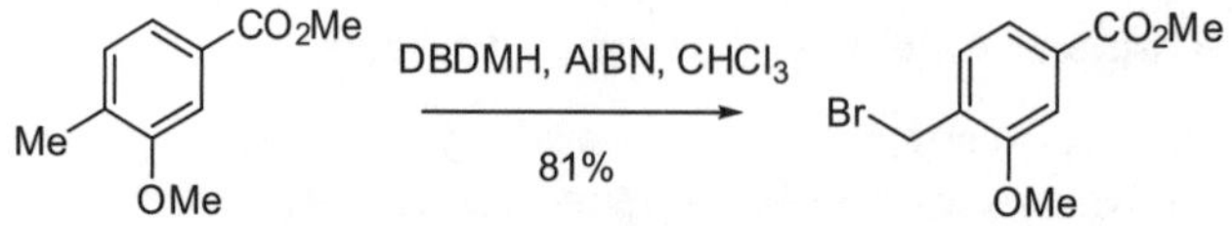

DBDMH, AIBN, CHCl₃
81%

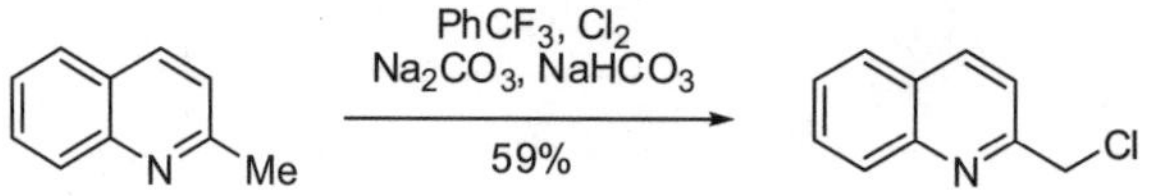

PhCF₃, Cl₂
Na₂CO₃, NaHCO₃
59%

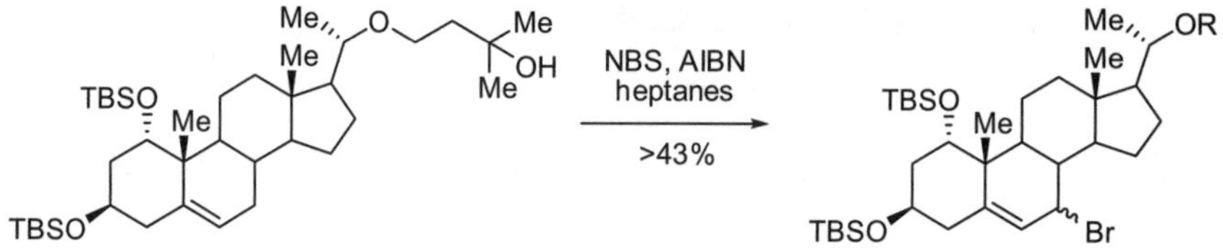

NBS, AIBN
heptanes
>43%

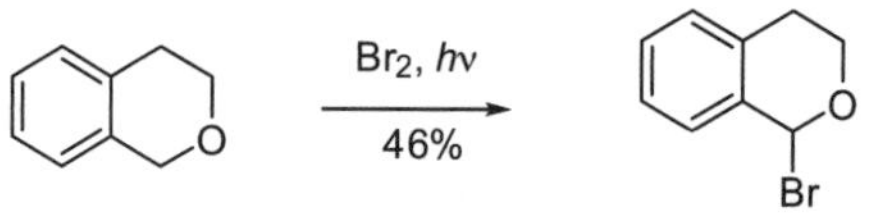

Br₂, hν
46%

In one noteworthy example, free radical bromination of a benzylic position required the photochemical initiation of an ordinary light bulb.[114]

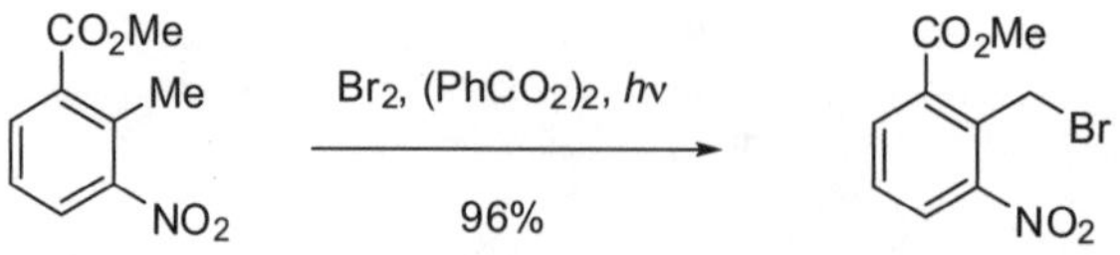

Br₂, (PhCO₂)₂, hν
96%

***10.3.8.1  Oxygenations***   Most allylic oxygenations in the literature are chromium based and therefore environmentally unfriendly. A ruthenium-catalyzed *t*-butylhydroperoxide oxidation that produced the enone below in 75% yield was developed and could be performed on kilogram scale.[115]

[114]Soederberg, B. C.; Shriver, J. A.; Wallace, J. M. *Org. Synth.* **2003**, *80*, 75–84.
[115]Miller, R. A.; Li, W.; Humphrey, G. R. *Tetrahedron Lett.* **1996**, *37*, 3429–3432.

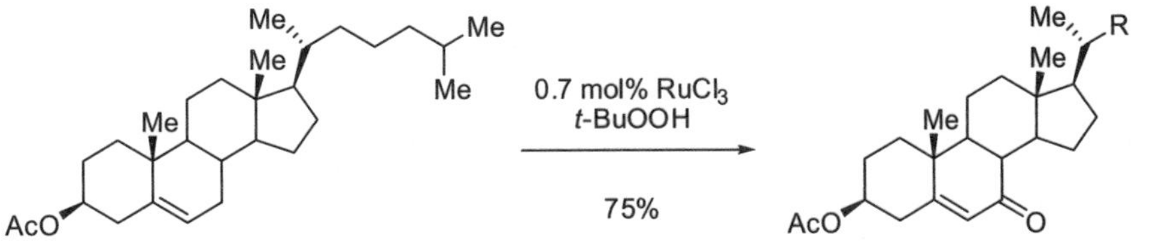

75%

The relatively high acidity of the benzylic proton below allowed for its easy ionization. Oxidation with the $O_2$ in air yielded the hemiketal.[116]

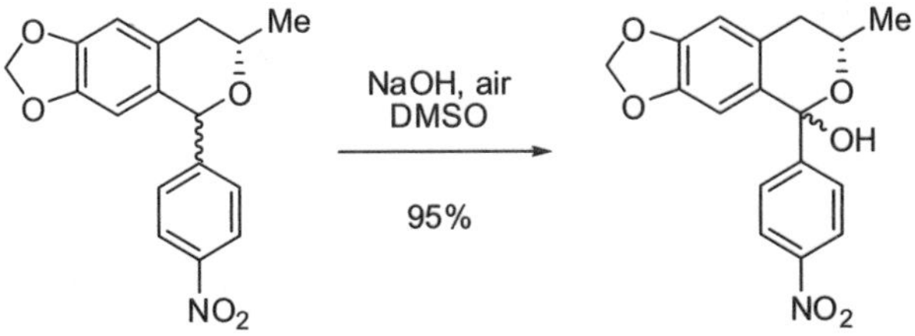

95%

A classic example of benzylic oxidation is the removal of *p*-methoxybenzyl (PMB) ethers by oxidation, often with DDQ. Two interesting large-scale examples of this reaction can be found in the synthesis of discodermolide by Novartis. A DDQ oxidation that was performed under anhydrous conditions resulted in the adjacent alcohol adding to the intermediate benzylic cation to form a PMP acetal in 50% yield. In a later stage of the synthesis, two PMB-protecting groups were removed simultaneously.[117]

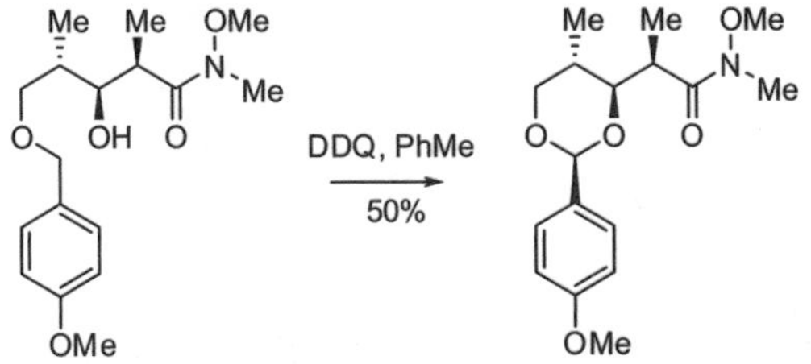

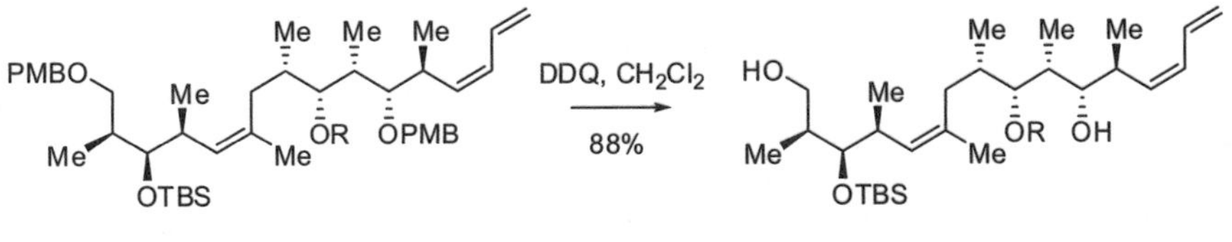

[116]Anderson, B. A.; Hansen, M. M.; Harkness, A. R.; Henry, C. L.; Vicenzi, J. T.; Zmijewski, M. J. *J. Am. Chem. Soc.* **1995**, *117*, 12358–12359.

[117]Mickel, S. J.; Sedelmeier, G. H.; Niederer, D.; Schuerch, F.; Koch, G.; Kuesters, E.; Daeffler, R.; Osmani, A.; Seeger-Weibel, M.; Schmid, E.; Hirni, A.; Schaer, K.; Gamboni, R.; Bach, A.; Chen, S.; Chen, W.; Geng, P.; Jagoe, C. T.; Kinder, F. R., Jr.; Lee, G. T.; McKenna, J.; Ramsey, T. M.; Repic, O.; Rogers, L.; Shieh, W.-C.; Wang, R.-M.; Waykole, L. *Org. Process Res. Dev.* **2004**, *8*, 107–112.

The oxidation of ethylbenzene to benzoic acid using oxygen and a manganese catalyst was reported.[118,119]

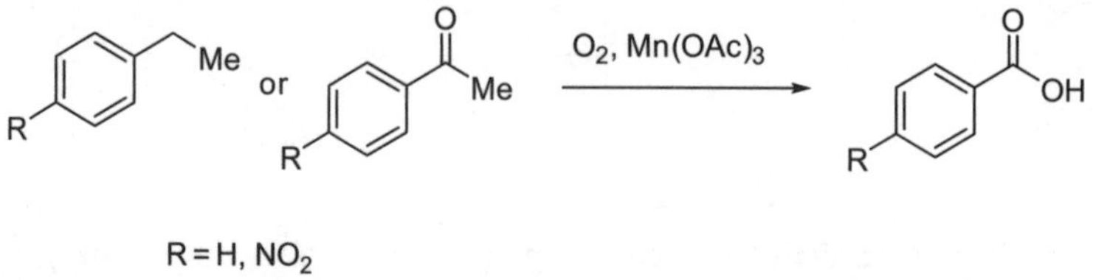

R = H, NO$_2$

Potassium permanganate was utilized to produce the triacid below following complete oxidation of all carbon substituents on the aryl ring.[120] In cases where overoxidation is not going to be an issue, this is the reagent of choice for this transformation due to its effectiveness in this transformation and low cost.

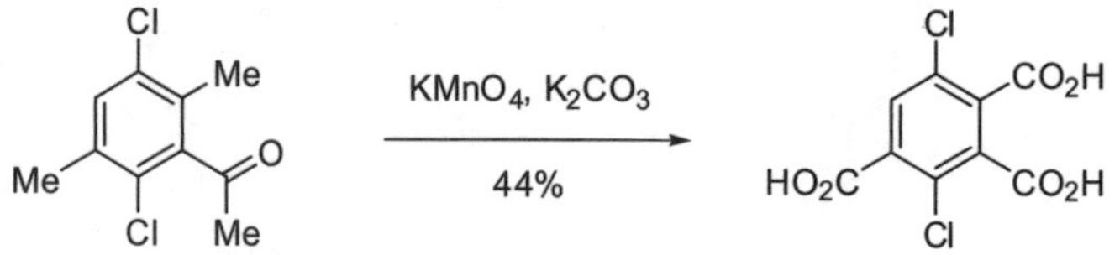

The classic method for the hydroxylation of allylic positions is through the use of SeO$_2$ and hydrogen peroxide.[121] The requirement for stoichiometric quantities of selenium in this process is a drawback to the method.

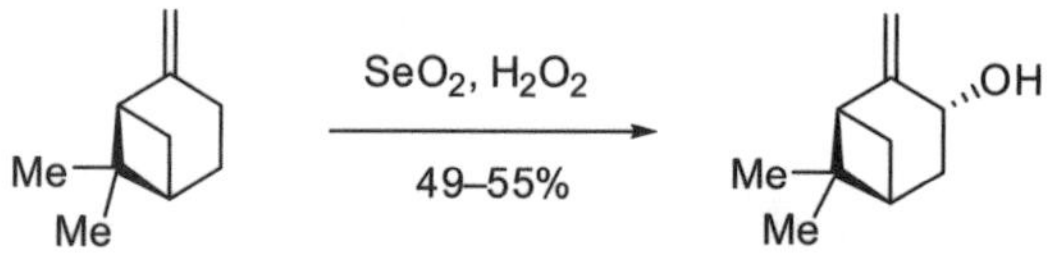

### 10.3.8.2  *Allylic Amination*

The allylic position of olefins can be aminated with reagents of the structure RN = Se = NR or RN = S = NR in direct analogy to the allylic oxidation using selenium dioxide.[122,123]

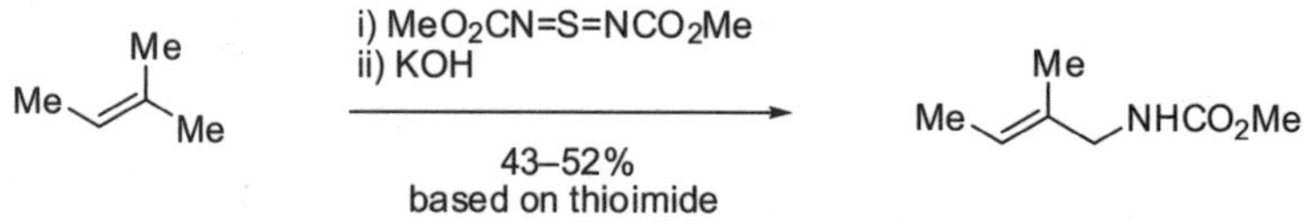

This reaction proceeds through an ene reaction followed by [2,3] sigmatropic rearrangement (Section 7.3.2) and reduction of the N—S or N—Se bond. This mechanism, which involves the olefin moving back and forth, results in an overall amination of the allylic position. Less preferable, but also useful, is the selenium analog of the above reaction.[124]

[118]Bukharkina, T. V.; Digurov, N. G.; Mil'ko, S. B.; Shelud'ko, A. B. *Org. Process Res. Dev.* **1999**, *3*, 404–408.

[119]Bukharkina, T. V.; Grechishkina, O. S.; Digurov, N. G.; Krukovskaya, N. V. *Org. Process Res. Dev.* **2003**, *7*, 148–154.

[120]Lyttle, M. H.; Carter, T. G.; Cook, R. M. *Org. Process Res. Dev.* **2001**, *5*, 45–49.

[121]Coxon, J. M.; Dansted, E.; Hartshorn, M. P. *Org. Synth.* **1977**, *56*, 25–27.

[122]Kresze, G.; Braxmeier, H.; Muensterer, H. *Org. Synth.* **1987**, *65*, 159–165.

[123]Katz, T. J.; Shi, S. *J. Org. Chem.* **1994**, *59*, 8297–8298.

[124]Bruncko, M.; Khuong, T.-A. V.; Sharpless, K. B. *Angew. Chem., Int. Ed. Engl.* **1996**, *35*, 454–456.

Reaction with an azodicarboxylate and a Lewis acid results in an ene reaction with the olefin, in this case the final product has the olefin transposed one carbon from its origination.[125]

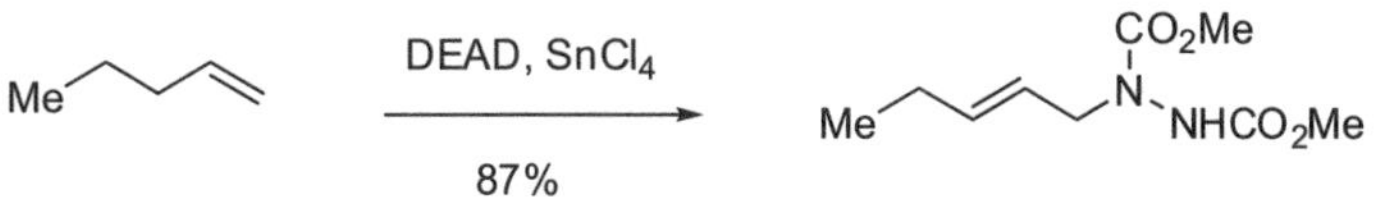

Allylic amination with transposition of the olefin can also be achieved using a hydroxylamine and an iron catalyst.[126]

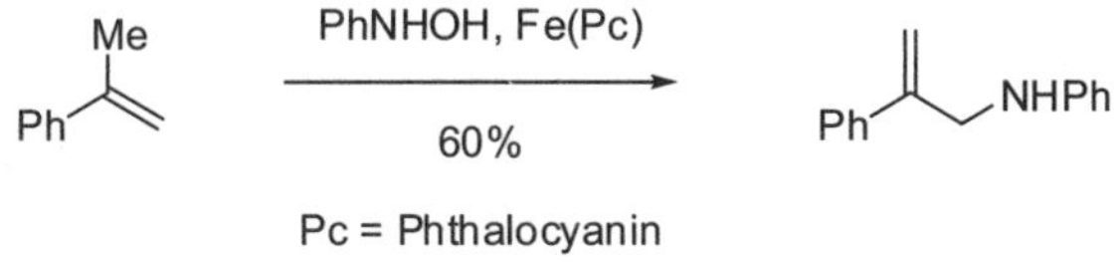

Pc = Phthalocyanin

Another methodology for the introduction of an allylic nitrogen is through the use of π-allyl chemistry.[127]

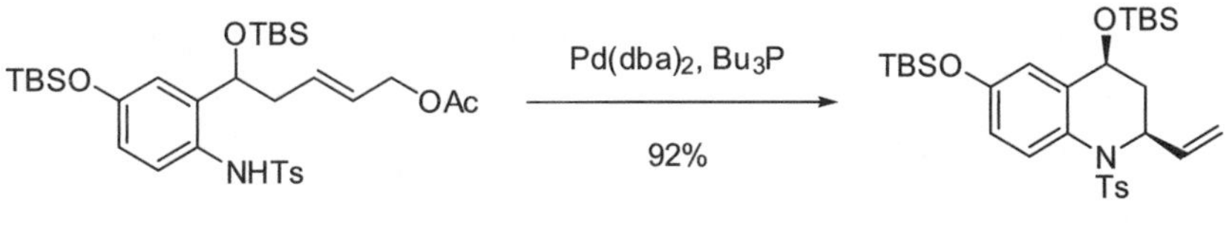

## 10.3.9   Nitrene Insertion into Carbon–Hydrogen Bonds

Nitrenes are known to insert into unactivated C–H bonds. The nitrene is typically generated from the thermolysis of an azide and is most frequently trapped intramolecularly. Two examples are shown below. *Caution should be exercised when considering thermolyzing an azide.*[128,129]

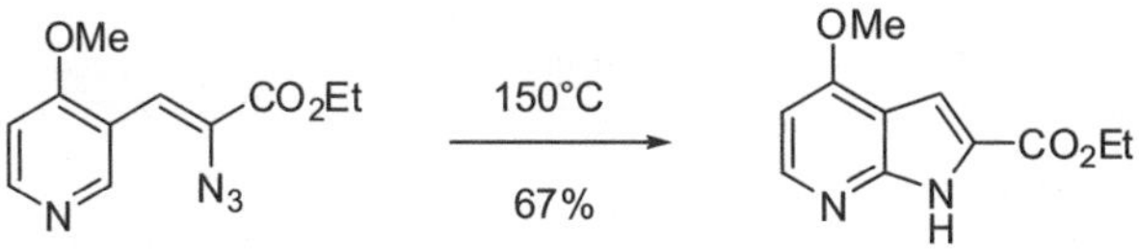

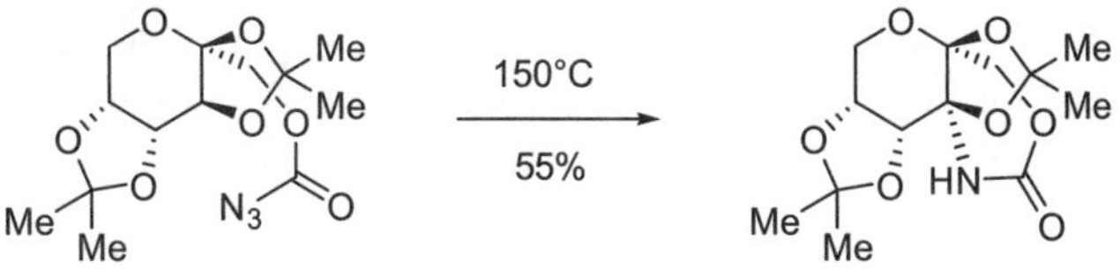

[125]Brimble, M. A.; Heathcock, C. H. *J. Org. Chem.* **1993**, *58*, 5261–5263.

[126]Johannsen, M.; Joergensen, K. A. *J. Org. Chem.* **1994**, *59*, 214–216.

[127]Hara, O.; Sugimoto, K.; Hamada, Y. *Tetrahedron* **2004**, *60*, 9381–9390.

[128]Molina, P.; Fresneda, P. M.; Delgado, S. *J. Org. Chem.* **2003**, *68*, 489–499.

[129]Banks, M. R.; C., J. I. G.; Gosney, I.; Gould, R. O.; Hodgson, P. K. G.; McDougall, D. *Tetrahedron* **1998**, *54*, 9765–9784.

Nitrenes can also be generated by the oxidation of amide and sulfonamide nitrogens with $PhI(OAc)_2$ and a metal catalyst.[130,131]

## 10.4   OXIDATION OF CARBON–OXYGEN BONDS, AND AT CARBON BEARING AN OXYGEN SUBSTITUENT

### 10.4.1   Oxidation of Alcohols to Aldehydes and Ketones

The oxidation of alcohols to the corresponding aldehydes, ketones, or carboxylic acid derivatives is one of the most commonly utilized chemical transformations in organic synthesis and remains an active research area for the identification of more effective and practical methods.[49,132–134] This is due in part to the development and availability of a plethora of orthogonal-protecting groups for alcohols, which often allow for chemoselective deprotection of the desired alcohol prior to its oxidation to the required derivative.[135] For the oxidation of primary alcohols, TEMPO has become the reagent of choice in industry. Moffatt and modified Moffatt processes are also very useful for the oxidation of primary alcohols, and are the method of choice for secondary alcohols. It is rarely necessary to resort to stoichiometric metal oxidations with the wide range of oxidants available, but there are a number of effective catalytic metal reagents available for alcohol oxidation if necessary.

**TABLE 10.1   Preferred Methods for Oxidation of Alcohols**

| Preference | Primary Alcohol to Aldehyde | Primary Alcohol to Acid | Secondary Alcohol to Ketone |
|---|---|---|---|
| 1 | TEMPO | TEMPO/NaClO$_2$ | Moffat |
| 2 | SO$_3$·pyridine | Catalytic metal mediated | TEMPO |
| 3 | Moffatt | Stoichiometric metal | SO$_3$·pyridine |
| 4 | Catalytic metal mediated | | Catalytic metal mediated |
| 5 | Stoichiometric metal | | Stoichiometric metal |

*10.4.1.1   TEMPO-Mediated Processes*   2,2,6,6-Tetramethylpiperidin-1-oxyl (TEMPO)[136] is a hydroxyl radical catalyst that is successful under mild conditions, generally at room

[130]Espino, C. G.; Wehn, P. M.; Chow, J.; Du Bois, J. *J. Am. Chem. Soc.* **2001**, *123*, 6935–6936.

[131]Davies, H. M. L.; Long, M. S. *Angew. Chem., Int. Ed. Engl.* **2005**, *44*, 3518–3520.

[132]Kwon, M. S.; Kim, N.; Park, C. M.; Lee, J. S.; Kang, K. Y.; Park, J. *Org. Lett.* **2005**, *7*, 1077–1079.

[133]Matano, Y.; Hisanaga, T.; Yamada, H.; Kusakabe, S.; Nomura, H.; Imahori, H. *J. Org. Chem.* **2004**, *69*, 8676–8680.

[134]Mori, K.; Hara, T.; Mizugaki, T.; Ebitani, K.; Kaneda, K. *J. Am. Chem. Soc.* **2004**, *126*, 10657–10666.

[135]Greene, T. W.; Wuts, P. G. M. **1991**, 10–175.

[136]De Nooy, A. E. J.; Besemer, A. C.; Van Bekkum, H. *Synthesis* **1996**, 1153–1174.

temperature, using inexpensive co-oxidants such as bleach (NaOCl). It can be chemoselective for a primary alcohols[137] and is not prone to over-oxidation under the appropriate conditions. TEMPO is particularly effective for the preparation of chiral α-amino and α-alkoxy aldehydes.[138]

Because the oxidation is significantly slower on secondary alcohols, TEMPO-mediated oxidations do not have the same frequency of use for large scale preparation of ketones from secondary alcohols as do Moffatt-type oxidations.[139]

While NaOCl has been the most utilized co-oxidant in the industry, other alternatives such as CuCl in the presence of oxygen[140] and iodine[141] have also been employed.

***10.4.1.2 Moffatt and Modified-Moffatt Processes*** The Moffatt oxidation, originally introduced in 1965,[142] has proven to be one of the methods of choice for the preparation of ketones from secondary alcohols. Conditions are listed in Table 10.2. This procedure has the advantage of generating an oxosulfenium ion that is deprotonated under mild conditions. A unique feature to these procedures is that the oxidation occurs in a stepwise manner, and the aldehyde or ketone products are not produced until the second stage of the process when excess oxidant is not present to further oxidize the substrate. For primary alcohols particularly, the $SO_3$·pyridine activating agent is the easiest and most practical to use, as cryogenic temperatures are not needed and additional reagent can be used to push the reaction to completion. In cases where a more reactive oxidant is needed, such as secondary alcohols, the most practical method experimentally is to premix the alcohol and DMSO in solvent, and add the activating agent to the cooled solution: a "reverse addition order" Moffatt. TFAA is the best activating agent in this case because gaseous byproducts are not produced in the reaction. If off-gassing is not a concern, oxalyl chloride and thionyl chloride also work very well. In some cases, substrates prone to side reactions require a screen of activating agents. Some useful ones are shown in Table 10.2.

The Swern modification,[143] utilizing oxalyl chloride as the activating agent, has been used extensively in the pharmaceutical industry.[43]

[137]Semmelhack, M. F.; Chou, C. S.; Cortes, D. A. *J. Am. Chem. Soc.* **1983**, *105*, 4492–4494.

[138]Leanna, M. R.; Sowin, T. J.; Morton, H. E. *Tetrahedron Lett.* **1992**, *33*, 5029–5032.

[139]Urban, F. J.; Anderson, B. G.; Orrill, S. L.; Daniels, P. J. *Org. Process Res. Dev.* **2001**, *5*, 575–580.

[140]Ernst, H. *Pure Appl. Chem.* **2002**, *74*, 2213–2226.

[141]Miller, R. A.; Hoerrner, R. S. *Org. Lett.* **2003**, *5*, 285–287.

[142]Pfitzner, K. E.; Moffatt, J. G. *J. Am. Chem. Soc.* **1965**, *87*, 5661–5670.

[143]Mancuso, A. J.; Huang, S.-L.; Swern, D. *J. Org. Chem.* **1978**, *43*, 2480–2482.

**TABLE 10.2  Activating Agents Used in Moffat Processes**

| Activating Agent | Advantage | Product | Ref. | Yield |
|---|---|---|---|---|
| DCC | Chemoselective oxidation of an alcohol in the presence of a thioether | | 144 | 60% |
| PhPCl$_2$ | Oxidation of a very sensitive substrate | | 145 | >82% |

| Reagent | Structure | Notes | | Yield |
|---|---|---|---|---|
| Acetic anhydride | | α-Chloroketone was the sole product when using oxalyl chloride | 152 | 78% |
| P$_2$O$_5$ | | Inexpensive reagent, cryogenic conditions not required, MTM ether formation minimized | 146 | 75% |
| Trifluoroacetic anhydride | | Byproducts observed with oxalyl chloride | 147 | 52% |

[144] Confalone, P. N.; Baggiolini, E.; Hennessy, B.; Pizzolato, G.; Uskokovic. M. R. *J. Org. Chem.* **1981**, *46*, 4923–4927.
[145] Cvetovich, R. J.; Leonard, W. R.; Amato, J. S.; DiMichele, L. M.; Reamer, R. A.; Shuman, R. F.; Grabowski, E. J. J. *J. Org. Chem.* **1994**, *59*, 5838–5840.
[146] Carpenter, D. E.; Imbordino, R. J.; Maloney, M. T.; Moeslein, J. A.; Reeder, M. R.; Scott, A. *Org. Process Res. Dev.* **2002**, *6*, 721–728.
[147] Izumi, H.; Futamura, S. *J. Org. Chem.* **1999**, *64*, 4502–4505.

1) $(COCl)_2$, DMSO
2) $Et_3N$

HO～～OBn  $\xrightarrow{\hspace{2cm}}$  OHC～～OBn

90%

A number of activating agents have been used in place of oxalyl chloride to prevent problems such as CO and $CO_2$ formation, and the formation of methylthiomethyl ethers (MTM ethers) (see Table 10.2). Operationally, a process in which the activating agent is added to a solution of all of the other reactants at reaction temperature is the simplest to execute. The rate of reaction of the activating agents with DMSO is so much higher than with the alcohols that acylated byproducts are not seen.

All of the procedures in Table 10.2 require cryogenic temperatures with the exception of the $P_2O_5$ procedure, suggesting that a different mechanistic pathway is operative with this activating agent. This makes the $P_2O_5$ protocol comparable to the Parikh–Doering process described below.

One process which has gained popularity in industry is the Parikh–Doering oxidation, which is the use of DMSO activated with the $SO_3$·pyridine complex.[148] This reagent has the practicality of being a solid and offers the advantage that the reaction can be carried out at or near room temperature. This also provides the flexibility to charge additional reagent in the case of an incomplete reaction due to the compatibility of the reactive intermediate with amine bases. For example, phenyl alaninol was oxidized to the corresponding aldehyde on a 190 kg scale without any loss of the chiral purity as part of the synthesis of an HIV protease inhibitor[149]

$Bn_2N$〜(Ph)(OH)  $\xrightarrow[\text{99.9\% ee}]{\substack{SO_3\text{•pyr, DMSO}\\Et_3N\\\\>95\% \text{ yield}}}$  $Bn_2N$〜(Ph)(CHO)

This reagent is significantly less reactive than intermediates generated in the examples in Table 10.2, and requires extended reaction times at elevated temperature for the oxidation of secondary alcohols. Its use has been documented by Boehringer Ingelheim as a convenient procedure for the preparation of 2-hydroxy-3-pinanone in two steps from α-pinene.[150]

Another modification of the Moffatt oxidation is the CoreyKim protocol[151] where the chlorosulfenium ion is generated by oxidation of dimethyl sulfide by either chlorine or N-chlorosuccinimide.[152]

[148]Parikh, J. R.; Doering, W. v. E. *J. Am. Chem. Soc.* **1967**, *89*, 5505–5507.

[149]Liu, C.; Ng, J. S.; Behling, J. R.; Yen, C. H.; Campbell, A. L.; Fuzail, K. S.; Yonan, E. E.; Mehrotra, D. V. *Org. Process Res. Dev.* **1997**, *1*, 45–54.

[150]Krishnamurthy, V.; Landi, J., Jr.; Roth, G. P. *Synth. Commun.* **1997**, *27*, 853–860.

[151]Corey, E. J.; Kim, C. U. *J. Am. Chem. Soc.* **1972**, *94*, 7586–7587.

[152]Danheiser, R. L.; Fink, D. M.; Okano, K.; Tsai, Y. M.; Szczepanski, S. W. *J. Org. Chem.* **1985**, *50*, 5393–5396.

**10.4.1.3 Metal-Mediated Processes**   Metal-mediated oxidations were of primary importance before 1980, prior to the introduction of more environmentally friendly methods.

In recent years, the use of a metal catalyst to promote oxidation has gained in popularity, especially in the case of secondary alcohols that cannot over-oxidize. One such example is tetra-*n*-propyl ammonium perruthenate (TPAP),[153] which is capable of selective oxidation of a very sensitive macrolide using NMO as the co-oxidant.[154]

Another efficient catalytic reagent is $RuO_4$, usually generated from $RuCl_3$ and a co-oxidant. It is only suitable for oxidation of secondary alcohols, since primary alcohols produce a carboxylic acid. Approximately 1 mol% of ruthenium is employed, usually in aqueous acetonitrile, and the preferred co-oxidant is sodium bromate ($NaBrO_3$) due to its reactivity, cost, and innocuous side products.[155] An acidic buffer such as acetic acid can be employed if the substrate or product is sensitive to the high pH resulting from the co-oxidant.[156] In another metal-mediated oxidation, catalytic amounts of $Na_2WO_4$ in the presence of $H_2O_2$ has been demonstrated as an effective oxidant of secondary alcohols in the presence of a phase-transfer catalyst. It also proved to be chemoselective for secondary over primary alcohols as demonstrated in the oxidation below.[157]

[153] Ares, J. J.; Outt, P. E.; Kakodkar, S. V.; Buss, R. C.; Geiger, J. C. *J. Org. Chem.* **1993**, *58*, 7903–7905.

[154] Jones, A. B. *J. Org. Chem.* **1992**, *57*, 4361–4367.

[155] Fleitz, F. J.; Lyle, T. A.; Zheng, N.; Armstrong, J. D., III; Volante, R. P. *Synth. Commun.* **2000**, *30*, 3171–3180.

[156] Belyk, K. M.; Leonard, W. R., Jr.; Bender, D. R.; Hughes, D. L. *J. Org. Chem.* **2000**, *65*, 2588–2590.

[157] Sato, K.; Aoki, M.; Takagi, J.; Noyori, R. *J. Am. Chem. Soc.* **1997**, *119*, 12386–12387.

Reagents such as chromium trioxide ($CrO_3$) in pyridine,[158–160] pyridinum chlorochromate (PCC),[161] and pyridinum dichromate (PDC),[162] have been used extensively in academia but sparsely in industry. One example of such a process has been demonstrated in the preparation of an α-amino aldehyde using $CrO_3$ in pyridine without loss of chiral purity.[163]

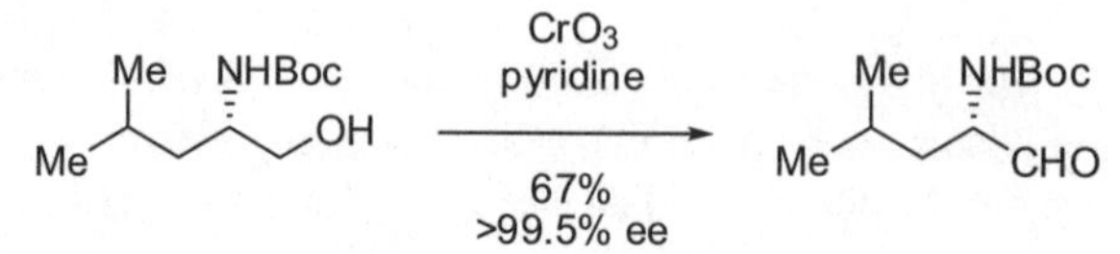

There are few stoichiometric metal-mediated oxidations of secondary alcohols reported from process groups since 1980. A classic example is the oxidation below using $CrO_3$ in the Merck synthesis of cortisone.[164] These processes are not recommended.

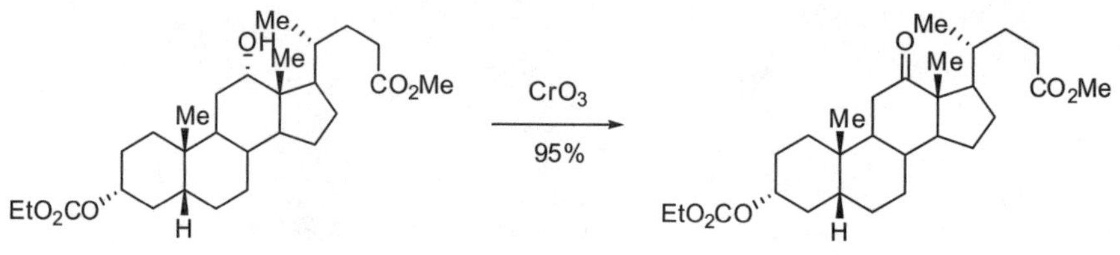

**10.4.1.4  Alternative Methods**   The Dess–Martin reagent[165] is rarely used on large scale for a variety of reasons, but an example is shown below.[166]

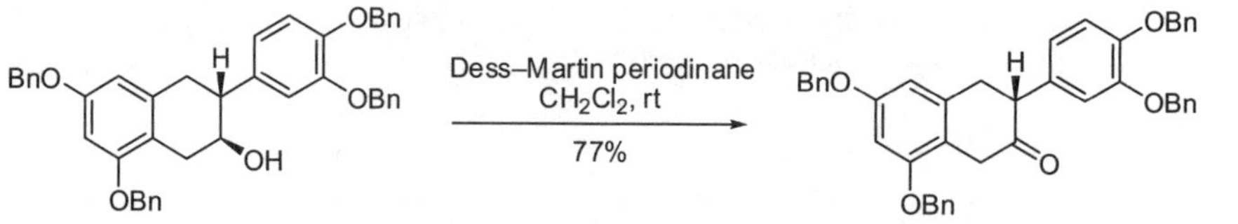

An interesting oxidation of the cholic acid derivative below was accomplished on 17 kg scale by simply using aqueous NaOCl[167] in the presence of KBr in a mixture of EtOAc and water. This procedure afforded a 92% yield of the desired ketone.[158]

[158]Arosio, R.; Rossetti, V.; Beratto, S.; Talamona, A.; Crisafulli, E. *Process for the production of chenodeoxycholic and ursodeoxycholic acids.* EP 424232, 1991, 19 pp.

[159]Horak, V.; Moezie, F.; Klein, R. F. X.; Giordano, C. *Synthesis* **1984**, 839–840.

[160]Salman, M.; Babu, S. J.; Kaul, V. K.; Ray, P. C.; Kumar, N. *Org. Process Res. Dev.* **2005**, *9*, 302–305.

[161]Corey, E. J.; Suggs, J. W. *Tetrahedron Lett.* **1975**, 2647–2650.

[162]Corey, E. J.; Schmidt, G. *Tetrahedron Lett.* **1979**, 399–402.

[163]Rittle, K. E.; Homnick, C. F.; Ponticello, G. S.; Evans, B. E. *J. Org. Chem.* **1982**, *47*, 3016–3018.

[164]Pines, S. H. *Org. Process Res. Dev.* **2004**, *8*, 708–724.

[165]Dess, D. B.; Martin, J. C. *J. Org. Chem.* **1983**, *48*, 4155–4156.

[166]Sarma, D. N.; Sharma, R. P. *Chem. Ind.* **1984**, 712–713.

[167]Stevens, R. V.; Chapman, K. T.; Weller, H. N. *J. Org. Chem.* **1980**, *45*, 2030–2032.

Another oxidation that is seldom used on large scale is the Oppenauer oxidation. In general, this procedure suffers from the fact that a large excess of a sacrificial ketone must be employed in order to drive the equilibrium towards the substrate oxidation, and that it is difficult to drive the reaction to completion despite long reaction times. A large number of catalysts for the Oppenauer oxidation have been reported; below is an example using potassium *t*-butoxide as the catalyst and benzophenone as the hydride acceptor. In this case, the reaction is driven to completion following oxidation by retro-aldol reaction to form the relatively stable enolate that does not undergo reduction under the reaction conditions.[159]

### 10.4.1.5 Oxidation of Benzylic and Allylic Alcohols

*MnO2 Oxidation* Manganese dioxide has been used for the preparation of a key intermediate in the synthesis of isotretinoin at 1 kg scale in >95% yield.[160]

*DDQ Oxidation* Oxidations using DDQ for the preparation of aldehydes and ketones are rare. One reported example is the preparation of a HMG-CoA reductase inhibitor side chain through the chemoselective oxidation of an allylic alcohol in the presence of a secondary alcohol in a very sensitive product.[168]

[168]Tempkin, O.; Abel, S.; Chen, C.-P.; Underwood, R.; Prasad, K.; Chen, K.-M.; Repic, O.; Blacklock, T. J. *Tetrahedron* **1997**, *53*, 10659–10670.

### 10.4.1.6   Oxidation of Diols to Lactones: Selective Oxidation of Primary or Secondary Alcohols

In some cases, diols can be oxidized to the lactone with selectivity for oxidation at primary versus secondary alcohols. The reaction proceeds via oxidation to the aldehyde, hemiacetal formation, and oxidation of the hemiacetal to the lactone. This transformation can be achieved with a variety of reagents. An example using TEMPO as the oxidant was used in a complex total synthesis.[169] Sodium bromite has also been reported as a co-oxidant with TEMPO.[170]

Another oxidant that has been reported to accomplish this type of transformation is TCCA.[171] Sodium bromite alone has been reported to oxidize diols to the lactol without further oxidation to the lactone.[100]

### 10.4.2   Oxidation of Primary Alcohols to Carboxylic Acids

### 10.4.2.1   TEMPO/Sodium Chlorite Oxidation of Alcohols to Carboxylic Acids and Derivatives

Primary alcohols can be directly oxidized to carboxylic acids in a single operation by tandem oxidation to the aldehyde with TEMPO followed by a second oxidation with $NaClO_2$.[172]

[169]Hansen, T. M.; Florence, G. J.; Lugo-Mas, P.; Chen, J.; Abrams, J. N.; Forsyth, C. J. *Tetrahedron Lett.* **2002**, *44*, 57–59.

[170]Inokuchi, T.; Matsumoto, S.; Nishiyama, T.; Torii, S. *J. Org. Chem.* **1990**, *55*, 462–466.

[171]Hiegel, G. A.; Gilley, C. B. *Synth. Commun.* **2003**, *33*, 2003–2009.

[172]Song, Z. J.; Zhao, M.; Desmond, R.; Devine, P.; Tschaen, D. M.; Tillyer, R.; Frey, L.; Heid, R.; Xu, F.; Foster, B.; Li, J.; Reamer, R.; Volante, R.; Grabowski, E. J.; Dolling, U. H.; Reider, P. J.; Okada, S.; Kato, Y.; Mano, E. *J. Org. Chem.* **1999**, *64*, 9658–9667.

*10.4.2.2 Metal-Mediated Oxidation of Alcohols to Carboxylic Acids and Derivatives* As stated in previous sections, non-catalytic metal-mediated oxidations have the disadvantage of generating a large waste effluent and often lead to difficult workups that are cumbersome at scale. $KMnO_4$ in the presence of a phase-transfer catalyst proved to be efficient in the preparation of pentafluoropentanoic acid.[173] A procedure for the oxidation of primary and secondary alcohols using a catalytic amount of chromium trioxide and 2.5 equivalents of periodic acid has also been developed.[174] As a last ditch effort, a Jones oxidation protocol can be employed, but this process is not recommended.[175]

## 10.5  OXIDATION OF ALDEHYDES TO CARBOXYLIC ACIDS AND DERIVATIVES

### 10.5.1  Sodium Chlorite Oxidation of Aldehydes to Carboxylic Acids and Derivatives

Probably the most practical method for the oxidation of an aldehyde to the carboxylic acid is sodium chlorite in the presence of a hypochlorite and chlorine scavenger such as sulfamic acid or an electron-rich olefin or arene. In the example shown, $H_2O_2$ is used as the hypochlorite and chlorine scavenger.[176] While it seems counterintuitive to use an oxidant to eliminate hypochlorite and chlorine, $H_2O_2$ reacts with HOCl to produce HCl, $H_2O$, and $O_2$, which are innocuous side products. For substrates not prone to chlorination, no chlorine scavenger is necessary.[177]

[173]Mahmood, A.; Robinson, G. E.; Powell, L. *Org. Process Res. Dev.* **1999**, *3*, 363–364.

[174]Zhao, M.; Li, J.; Song, Z.; Desmond, R.; Tschaen, D. M.; Grabowski, E. J. J.; Reider, P. J. *Tetrahedron Lett.* **1998**, *39*, 5323–5326.

[175]Thottathil, J. K.; Moniot, J. L.; Mueller, R. H.; Wong, M. K. Y.; Kissick, T. P. *J. Org. Chem.* **1986**, *51*, 3140–3143.

[176]Daniewski, A. R.; Garofalo, L. M.; Hutchings, S. D.; Kabat, M. M.; Liu, W.; Okabe, M.; Radinov, R.; Yiannikouros, G. P. *J. Org. Chem.* **2002**, *67*, 1580–1587.

[177]Ruggeri, S. G.; Bill, D. R.; Bourassa, D. E.; Castaldi, M. J.; Houck, T. L.; Ripin, D. H. B.; Wei, L.; Weston, N. *Org. Process Res. Dev.* **2003**, *7*, 1043–1047.

### 10.5.1.1 Hydrogen Peroxide Oxidation of Aldehydes to Carboxylic Acids and Derivatives

Hydrogen peroxide has been reported as a safe and effective reagent for the preparation of a benzoic acid intermediate. The benzaldehyde below could be oxidized under a variety of conditions, and $NaClO_2$ proved to be acceptable in the presence of sulfamic acid ($NH_2SO_3H$). However, chlorination of the aromatic ring was observed and could not be eliminated. In order to circumvent this problem, $H_2O_2$ under basic conditions was identified as an inexpensive alternative for the preparation.[178]

### 10.5.1.2 Metal-Mediated Oxidations of Aldehydes to Carboxylic Acids and Derivatives

For reasons discussed previously, non-catalytic metal-mediated processes are now used infrequently in oxidations performed on large scale. These processes are not recommended. Some exceptions are shown.[179,180]

[178]Cook, D. C.; Jones, R. H.; Kabir, H.; Lythgoe, D. J.; McFarlane, I. M.; Pemberton, C.; Thatcher, A. A.; Thompson, D. M.; Walton, J. B. *Org. Process Res. Dev.* **1998**, *2*, 157–168.

[179]Wuts, P. G. M.; Ritter, A. R. *J. Org. Chem.* **1989**, *54*, 5180–5182.

[180]Raggon, J. W.; Welborn, J. M.; Godlewski, J. E.; Kelly, S. E.; LaCour, T. G. *Org. Prep. Proced. Int.* **1995**, *27*, 233–236.

*10.5.1.3  Oxidation of Bisulfite Adducts*    The bisulfite addition adducts of aldehydes can be oxidized to the carboxylic acid under modified Moffat conditions.[181]

## 10.5.2   Oxidation of Carboxylic Acids to Peroxyacids

The oxidation of carboxylic acids to the peroxyacid is generally accomplished *in situ* in the course of running oxidation reactions, as the peroxyacids tend to be thermally unstable. Formic acid reacts almost instantly with hydrogen peroxide to generate performic acid; acetic acid requires prolonged heating with hydrogen peroxide to accomplish the same transformation. The reaction of acid chlorides or anhydrides with hydrogen peroxide very quickly generates the peroxyacid. Examples of *in situ* peroxy acid formation can be found in Sections 10.7.10 (Table 10.4, examples 2 and 3), 10.8.5.1.2, and 10.8.5.2.1 on the oxidation of nitrogen and sulfur.

## 10.5.3   Oxidation of Phenols and Anilines to Quinones

Galanthamine was synthesized from a fairly simple substrate via an intramolecular oxidative aromatic coupling reaction on 12 kg scale; the oxidation was achieved using $K_2[Fe(CN)_6]$.[182]

An oxidation followed by reduction was used to demethylate a methyl phenyl ether.[183] Following Cbz deprotection, a second phenol to quinone oxidation was effected with Fremy's salt [$NO(SO_3Na)$ or $NO(SO_3K)$] to afford the benzodiazepine.

[181]Wuts, P. G. M.; Bergh, C. L. *Tetrahedron Lett.* **1986**, *27*, 3995–3998.
[182]Kueenburg, B.; Czollner, L.; Froehlich, J.; Jordis, U. *Org. Process Res. Dev.* **1999**, *3*, 425–431.
[183]Hayes, J. F. *Synlett* **1999**, 865–866.

## 10.5.4   Oxidation α to Oxygen

Although it is known that simple ethers will oxidize via a radical mechanism to form hazardous peroxides, there are few synthetically useful examples. One such example is the oxidation of an acetal to an ester using ozone.[184] Interestingly, if the ozonolysis is prolonged, the primary alcohol will slowly oxidize to the carboxylic acid.

Ethers can be oxidized to the corresponding ester using ruthenium tetroxide.[185]

## 10.6   OXIDATION OF CARBON–NITROGEN BONDS, AND AT CARBON BEARING A NITROGEN SUBSTITUENT

### 10.6.1   Dehydrogenation of Amines to Imines and Nitriles

Oxidation of a hydrazide to the corresponding hydrazone under Swern conditions has been reported.[186] Oxidation of a cephalosporin to the 7α-formamido cephalosporin via oxidation of the imine to a quinone[187] has been carried out on 1 kg scale.

[184]Urban, F. J.; Jasys, V. J. *Org. Process Res. Dev.* **2004**, *8*, 169–175.

[185]Carlsen, P. H. J.; Katsuki, T.; Martin, V. S.; Sharpless, K. B. *J. Org. Chem.* **1981**, *46*, 3936–3938.

[186]Mancuso, A. J.; Swern, D. *Synthesis* **1981**, 165–185.

[187]Berry, P. D.; Brown, A. C.; Hanson, J. C.; Kaura, A. C.; Milner, P. H.; Moores, C. J.; Quick, J. K.; Saunders, R. N.; Southgate, R.; Whittall, N. *Tetrahedron Lett.* **1991**, *32*, 2683–2686.

Barium permanganate can also be used.[188]

Oxidation of primary amines to the corresponding nitrile has been reported under a variety of mild conditions. In the example shown, trichlorocyanuric acid (TCCA) with catalytic TEMPO was used to dehydrogenate benzylamine to benzonitrile in 90% yield.[189]

Other systems reported to effect the same transformation include catalytic Ru/Al$_2$O$_3$[190] or Cu[191] with O$_2$ as the oxidant, the use of iodosobenzene,[192] and electrochemical methods.[135,193,194]

***10.6.1.1  Oxidation α to Nitrogen***   The Nef reaction is the conversion of a primary or secondary alkylnitro compounds to the corresponding aldehyde or ketone.[195] The transformation can be accomplished under oxidative, hydrolytic, or reductive conditions, but regardless of the method, a net oxidation at carbon is achieved.

[188]Firouzabadi, H.; Seddighi, M.; Mottaghinejad, E.; Bolourchian, M. *Tetrahedron* **1990**, *46*, 6869–6878.

[189]Chen, F.-e.; Kuang, Y.-y.; Dai, H.-f.; Lu, L.; Huo, M. *Synthesis* **2003**, 2629–2631.

[190]Yamaguchi, K.; Mizuno, N. *Angew. Chem., Int. Ed. Engl.* **2003**, *42*, 1480–1483.

[191]Capdevielle, P.; Lavigne, A.; Maumy, M. *Synthesis* **1989**, 453–454.

[192]Moriarty, R. M.; Vaid, R. K.; Duncan, M. P.; Ochiai, M.; Inenaga, M.; Nagao, Y. *Tetrahedron Lett.* **1988**, *29*, 6913–6916.

[193]Feldhues, U.; Schaefer, H. J. *Synthesis* **1982**, 145–146.

[194]Shono, T.; Matsumura, Y.; Inoue, K. *J. Am. Chem. Soc.* **1984**, *106*, 6075–6076.

[195]Ballini, R.; Petrini, M. *Tetrahedron* **2004**, *60*, 1017–1047.

The method of choice will depend on functional group compatibility with the various reaction conditions. Hydrolytic conditions are reported with hydroxide base followed by sulfuric acid as the hydrolyzing agent.[196] Although significantly slower, the transformation can be accomplished in wet acetonitrile with DBU as the base.[197] Reductive conditions typically involve alkoxide base and $TiCl_3$.[198] Oxidative conditions also use strong base, hydroxide or alkoxide, and an oxidant such as $KMnO_4$,[199] DMDO, Oxone, $H_2O_2$, and TPAP.

Carbamates can be oxidized to form the $N$-acylimminium ion electrochemically,[200] or through the use of strong oxidants. The product is the $\alpha$-alkoxy carbamate in the case of electrochemical oxidation, and the imide in the case of Dess–Martin periodinane.[201]

[196]Grethe, G.; Mitt, T.; Williams, T. H.; Uskokovic, M. R. *J. Org. Chem.* **1983**, *48*, 5309–5315.
[197]Ballini, R.; Bosica, G.; Fiorini, D.; Petrini, M. *Tetrahedron Lett.* **2002**, *43*, 5233–5235.
[198]Hauser, F. M.; Baghdanov, V. M. *J. Org. Chem.* **1988**, *53*, 4676–4681.
[199]Steliou, K.; Poupart, M. A. *J. Org. Chem.* **1985**, *50*, 4971–4973.
[200]Shono, T.; Matsumura, Y.; Tsubata, K. *Org. Synth.* **1985**, *63*, 206–213.
[201]Nicolaou, K. C.; Mathison, C. J. N. *Angew. Chem., Int. Ed. Engl.* **2005**, *44*, 5992–5997.

1,2,4-Triazenes can be oxidized to 1,2,4-triazoles using hypochlorites, Dess–Martin periodinane, and TPAP.[202]

| Reagent | Yield |
| --- | --- |
| NaOCl | 45% |
| Ca(OCl)$_2$ | 63% |
| Dess–Martin | 52% |
| TPAP | 45% |

***10.6.1.2 Oxidation of Aldoximes and Hydrazones of Aldehydes***   Aldoximes can be oxidized to the corresponding nitrile oxide by chlorination followed by elimination with base. Although the nitrile oxides are unstable, they can be trapped with alkenes or alkynes to form isoxazolines and isoxazoles respectively. To synthesize isoxazole, the acetaldoxime was oxidized to the corresponding nitrile oxide, which then reacted with methyl propiolate.[203]

Hydrazones of aldehydes can be oxidized using NCS.[204]

## 10.7   OXIDATION OF NITROGEN FUNCTIONALITIES

### 10.7.1   Diazotization of Amines

The most common nitrogen oxidation run on large scale in the literature is diazotization of an aniline. In general, the diazo compound is not isolated, and reacted *in situ*, although in a few rare cases isolations were reported. In almost every case, the diazotization is run using NaNO$_2$ in mineral acid as the oxidant. TFA can be used as the acid as well. A rarely used alternative to NaNO$_2$ is *t*-amyl nitrite, but stability of the reagent is a concern. Several examples are provided in Table 10.3, categorized by the manner in which the intermediate is trapped.

[202]Paulvannan, K.; Hale, R.; Sedehi, D.; Chen, T. *Tetrahedron* **2001**, *57*, 9677–9682.
[203]Bell, D.; Crowe, E. A.; Dixon, N. J.; Geen, G. R.; Mann, I. S.; Shipton, M. R. *Tetrahedron* **1994**, *50*, 6643–6652.
[204]Paulvannan, K.; Chen, T.; Hale, R. *Tetrahedron* **2000**, *56*, 8071–8076.

**TABLE 10.3  Amine diazotization and trapping**

| Scheme | Reaction Following Diazotization | Reagents Used | Ref. |
|---|---|---|---|
| [structure, 51%] | Reduction | i) NaNO$_2$, HCl; ii) H$_3$PO$_2$, CuSO$_4$ or H$_2$, Ni, HCl | 205 |
| [structure] | Halogenation | NaNO$_2$, HBr, KBr | 206 |
| [structure, 50%] | Cyanation | i) NaNO$_2$, H$_2$SO$_4$; ii) CuCN, NaCN | 207 |
| [structure, 73%] | Electrophilic aromatic substitution (Pschorr ring closure) | NaNO$_2$  H$_3$PO$_4$ | 106 |
| [structure, 24%] | Beech reaction | i) NaNO$_2$, HCl; ii) NH$_2$OH•HCl, (CH$_2$O)$_m$, CuSO$_4$, Na$_2$SO$_3$, NaOAc | 208 |

538

| | Reaction | Conditions | Ref |
|---|---|---|---|
| | Carbonylation | i) $NaNO_2$, HCl; ii) $H_2NSO_3H$, $PdCl_2$, CO, $H_2O$ | 209 |
| | Sulfonylation | i) $NaNO_2$, HCl, HOAc; ii) $SO_2$, CuCl | 52 |
| | Hydroxylation | $NaNO_2$, $H_2SO_4$ | 210 |
| | Reduction | i) $NaNO_2$, HCl; ii) $Na_2SO_3$ | 211 |

205 De Jong, R. L.; Davidson, J. G.; Dozeman, G. J.; Fiore, P. J.; Giri, P.; Kelly, M. E.; Puls, T. P.; Seamans, R. E. *Org. Process Res. Dev.* **2001**, *5*, 216–225.

206 Bunegar, M. J.; Dyer, U. C.; Green, A. P.; Gott, G. G.; Jaggs, C. M.; Lock, C. J.; Mead, B. J. V.; Spearing, W. R.; Tiffin, P. D.; Tremayne, N.; Woods, M. *Org. Process Res. Dev.* **1998**, *2*, 334–336.

207 Nielsen, M. A.; Nielsen, M. K.; Pittelkow, T. *Org. Process Res. Dev.* **2004**, *8*, 1059–1064.

208 Herr, R. J.; Fairfax, D. J.; Meckler, H.; Wilson, J. D. *Org. Process Res. Dev.* **2002**, *6*, 677–681.

209 Siegrist, U.; Rapold, T.; Blaser, H.-U. *Org. Process Res. Dev.* **2003**, *7*, 429–431.

210 Storz, T.; Dittmar, P.; Fauquex, P. F.; Marschal, P.; Lottenbach, W. U.; Steiner, H. *Org. Process Res. Dev.* **2003**, *7*, 559–570.

211 Faul, M. M.; Ratz, A. M.; Sullivan, K. A.; Trankle, W. G.; Winneroski, L. L. *J. Org. Chem.* **2001**, *66*, 5772–5782.

### 10.7.2 Oxidations of Hydrazines and Hydrazones

The majority of examples of hydrazine oxidation on large scale are run in the context of Curtius rearrangements (see Section 7.2.3.2) wherein the acyl azide is generated by oxidation of an acyl hydrazine. The ergoline derivative below was synthesized via hydrazine oxidation with sodium nitrite to form the intermediate acyl azide, followed by heating to effect rearrangement.[212]

Hydrazones can be oxidized to the corresponding azo compound using a variety of oxidants including NaOCl, $I_2$, $H_2O_2$, and AcOOH. However, there are few examples of these reagents being used.

An oxidation of an N−N bond leading to aromatization was accomplished using bromine or sodium nitrite.[213]

Oximes can be oxidized to the diazo compound with the use of $H_2NCl$.[90]

### 10.7.3 Amination of Nitrogen

Although diazotization is the most common method in the literature for conversion of N–H to N–N bonds, electrophilic sources of nitrogen have also been reported. The O-nitroarylhydroxylamine shown was found to be a useful source of electrophilic nitrogen.[214] Several related hydroxylamines were evaluated from a safety and yield perspective.

[212]Baenziger, M.; Mak, C. P.; Muehle, H.; Nobs, F.; Prikoszovich, W.; Reber, J. L.; Sunay, U. *Org. Process Res. Dev.* **1997**, *1*, 395–406.

[213]Fields, S. C.; Parker, M. H.; Erickson, W. R. *J. Org. Chem.* **1994**, *59*, 8284–8287.

[214]Boyles, D. C.; Curran, T. T.; Parlett, R. V. I. V.; Davis, M.; Mauro, F. *Org. Process Res. Dev.* **2002**, *6*, 230–233.

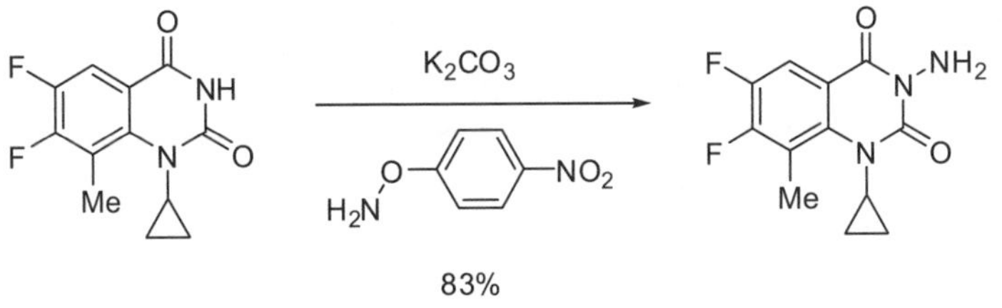

83%

Oxaziridines can also be used to transfer nitrogen to an amine.[101]

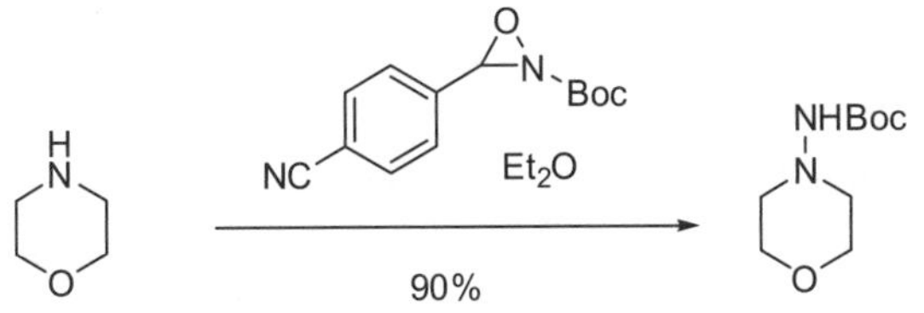

90%

Another reagent that can be used for this transformation is hydroxylamine-*O*-sulfonic acid.[215]

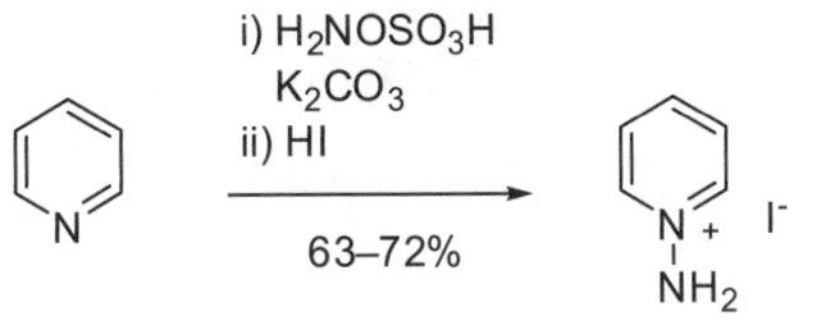

63–72%

## 10.7.4   Oxidation of Amines to Azo or Azoxy Compounds

Diazo compounds can be trapped by nucleophiles to form azo dyes. The enolate of ethyl 2-methylacetoacetate was utilized to trap *p*-methoxyphenyl azide as the azo intermediate in the example below.[85]

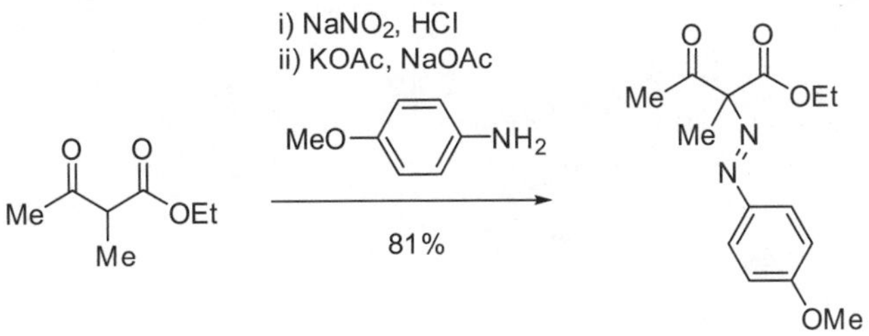

81%

[215]Goesl, R.; Meuwsen, A. *Org. Synth.* **1963**, *43*, 1–3.

Benzotriazoles can be synthesized from *o*-diaminobenzenes through generation of the diazo compound followed by intramolecular trapping of the intermediate.[216]

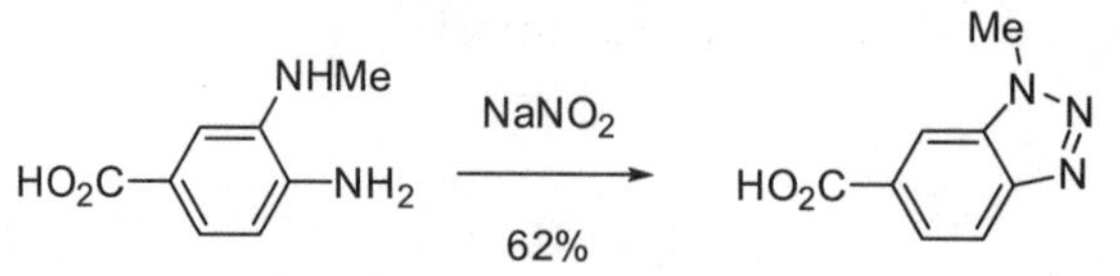

62%

Anilines can be converted to the corresponding azo compounds on treatment with oxidants. The dimeric azo compounds are produced when permanganate[188] or hydrogen peroxide in acetic acid is used as the oxidant.

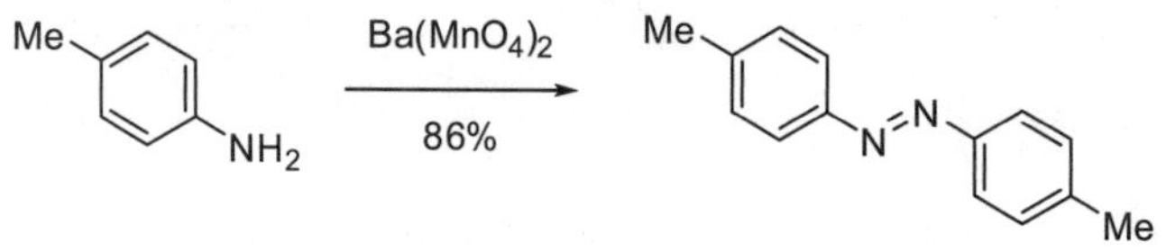

86%

Mixed azo compounds can be produced by trapping a diazotized aniline with a second aniline.

### 10.7.5   Oxidation of Primary Amines to Hydroxylamines

When a primary amine is alkylated with bromoacetonitrile, oxidized to the nitrone (as in the previous section), and reacted with hydroxylamine, the primary hydroxylamine is formed.[217]

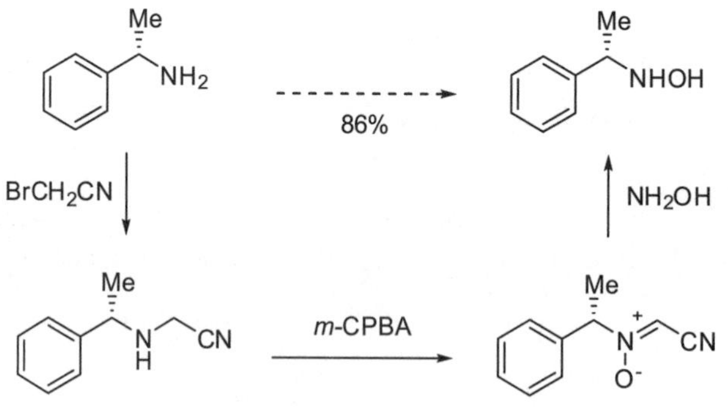

86%

BrCH₂CN

NH₂OH

*m*-CPBA

[216]De Knaep, A. G. M.; Vandendriessche, A. M. J.; Daemen, D. J. E.; Dingenen, J. J.; Laenen, K. D.; Nijs, R. L.; Pauwels, F. L. J.; Van den Heuvel, D. F.; Van der Eycken, F. J.; Vanierschot, R. W. E.; Van Laar, G. M. L. W.; Verstappen, W. L. A.; Willemsens, B. L. A. *Org. Process Res. Dev.* **2000**, *4*, 162–166.
[217]Tokuyama, H.; Kuboyama, T.; Fukuyama, T. *Org. Synth.* **2003**, *80*, 207–218.

### 10.7.6    Oxidation of Nitrogen to Nitroso Compounds

***10.7.6.1    Nitrone Formation***    Secondary amines can be converted to the corresponding nitrone using hydrogen peroxide catalyzed by $Na_2WO_4$[218,219] or $MeReO_3$.[220] The same transformation has been reported using *m*-CPBA.[217]

Anilines can be oxidized to the corresponding nitroso derivative in a two-stage process using DMS/NBS followed by *m*-CPBA.[221]

***10.7.6.2    Oxidation of Hydroxylamine to Nitroso Compounds***    Tertiary or aryl hydroxylamines can be oxidized to the nitroso compound using NaOBr.[222]

PCC[223] and silver carbonate[224] have also been utilized for this oxidation.

Benzofurazans can be synthesized by oxidative cyclization from an *o*-nitroaniline using sodium hypochlorite as the oxidant.[177]

[218]Stappers, F.; Broeckx, R.; Leurs, S.; Van den Bergh, L.; Agten, J.; Lambrechts, A.; Van den Heuvel, D.; De Smaele, D. *Org. Process Res. Dev.* **2002**, *6*, 911–914.

[219]Murahashi, S.; Shiota, T.; Imada, Y. *Org. Synth.* **1992**, *70*, 265–271.

[220]Goti, A.; Cardona, F.; Soldaini, G. *Org. Synth.* **2005**, *81*, 204–212.

[221]Taylor, E. C.; Tseng, C. P.; Rampal, J. B. *J. Org. Chem.* **1982**, *47*, 552–555.

[222]Calder, A.; Forrester, A. R.; Hepburn, S. P. *Org. Synth.* **1972**, *52*, 77–82.

[223]Wood, W. W.; Wilkin, J. A. *Synth. Commun.* **1992**, *22*, 1683–1686.

[224]Demeunynck, M.; Tohme, N.; Lhomme, M. F.; Lhomme, J. *J. Heterocycl. Chem.* **1984**, *21*, 501–503.

### 10.7.7 Nitrosation of Secondary Amines and Amides

Amides can be nitrosated using $N_2O_4$[225], ClNO,[226] or $N_2O_3$[227] (generated from $H_2NO_3$ and $NaNO_2$) to make the N-nitrosoamide. These nitrosamides are useful in the nitrosamide decomposition (Section 7.3.4.4) to make the corresponding ester, or can be hydrolyzed much more easily than an amide.

Secondary anilines can be converted to the corresponding N-nitrosoaniline using sodium nitrite.[228]

Secondary amines can be nitrosated with alkylnitrites as well. The N-nitrosoamines acidify the hydrogen's α to the amine and allow anionic chemistry to be performed.[229]

### 10.7.8 Oxidation of Primary Amines, Oximes, or Nitroso Compounds to Nitro Compounds

Tertiary or aryl amines can be oxidized to the nitro compound using $KMnO_4$.[222]

[225]White, E. *Org. Synth.* **1973**, *Coll. Vol. V*, 336–339.
[226]Van Leusen, A. M.; Strating, J. *Org. Synth.* **1977**, *57*, 95–102.
[227]Huisgen, R.; Bast, K. *Org. Synth.* **1962**, *42*, 69–72.
[228]Thoman, C. J.; Voaden, D. J. *Org. Synth.* **1965**, *45*, 96–99.
[229]Enders, D.; Pieter, R.; Renger, B.; Seebach, D. *Org. Synth.* **1978**, *58*, 113–122.

Anilines can be oxidized to the nitroaryl compound using dimethyldioxirane that is generated *in situ* from acetone and oxone under phase-transfer conditions.[230]

### 10.7.9   Oxidation of Tertiary Amines to Amine Oxides and Elimination to Form Imines

Oxidation of a tertiary amine to the *N*-oxide followed by elimination of water results in the formation of an imine that can be tautomerized to the enamine, or hydrolyzed to the corresponding carbonyl compound and the *N*-dealkylated product.[154]

Oxidation can also be accomplished with *m*-CPBA or other peracids, and iron catalysts can be used to facilitate the elimination reaction.[231]

The Polonovski reaction involves the reaction of an *N*-oxide with an acid anhydride to result in elimination, followed by tautomerization to form an enamine, hydrolytic dealkylation, or addition of nucleophiles such as cyanide to the iminium ion.[232]

[230]Zabrowski, D. L.; Moormann, A. E.; Beck, K. R., Jr. *Tetrahedron Lett.* **1988**, *29*, 4501–4504.
[231]Monkovic, I.; Wong, H.; Bachand, C. *Synthesis* **1985**, 770–773.
[232]Grieco, P. A.; Inana, J.; Lin, N. H. *J. Org. Chem.* **1983**, *48*, 892–895.

## 10.7.10    Oxidation of Pyridines to Pyridine *N*-Oxides

Pyridine *N*-oxides are useful intermediates in the synthesis of a number of pharmaceutically active compounds. Table 10.4 summarizes some of the more common oxidants used in this transformation.

## 10.7.11    Halogenation or Sulfination of Amines and Amides

Aromatization of a 3-carboxytetrahydro-β-carbolines using trichlorocyanuric acid (TCCA)[49] via chlorination of the nitrogen followed by elimination has been reported.[236,237]

**TABLE 10.4    Oxidation of pyridines to pyridine *N*-oxides**

| Reaction | Reagents | Reference |
|---|---|---|
| | i) MeReO$_3$, H$_2$O$_2$ <br> ii) TsCl, K$_2$CO$_3$ | 233 |
| | H$_2$O$_2$, TFA | 234 |
| | Urea-H$_2$O$_2$, TFAA | 235 |

[233]Payack, J. F.; Vazquez, E.; Matty, L.; Kress, M. H.; McNamara, J. *J. Org. Chem.* **2005**, *70*, 175–178.
[234]Nettekoven, M.; Jenny, C. *Org. Process Res. Dev.* **2003**, *7*, 38–43.
[235]Caron, S.; Do, N. M.; Sieser, J. E. *Tetrahedron Lett.* **2000**, *41*, 2299–2302.
[236]Haffer, G.; Nickisch, K.; Tilstam, U. *Heterocycles* **1998**, *48*, 993–998.
[237]Haffer, G.; Nickisch, K. *reparation of β-carbolines*. DE 4240672, 1994, 5 pp.

Morpholine can be chlorinated with bleach to make a useful chlorinating agent.[238]

Methylcarbamate can be chlorinated with chlorine gas in acetic acid.[122]

Piperizine can be converted to the *N*-thiophenyl derivative on treatment with disulfides.[239]

## 10.8   OXIDATION OF SULFUR, AND AT CARBON ADJACENT TO SULFUR

### 10.8.1   Pummerer Rerrangement

The Pummerer rearrangement is the rearrangement of a sulfoxide to the α-acetoxysulfide. Although not a net oxidation, the carbon atom appended to sulfur undergoes oxidation. This reaction can be employed in the dealkylation of mercaptans.[240]

[238]Girard, G. R.; Bondinell, W. E.; Hillegass, L. M.; Holden, K. G.; Pendleton, R. G.; Uzinskas, I. *J. Med. Chem.* **1989**, *32*, 1566–1571.
[239]Walinsky, S. W.; Fox, D. E.; Lambert, J. F.; Sinay, T. G. *Org. Process Res. Dev.* **1999**, *3*, 126–130.
[240]Young, R. N.; Gauthier, J. Y.; Coombs, W. *Tetrahedron Lett.* **1984**, *25*, 1753–1756.

1) *m*-CPBA
2) TFAA
3) MeOH

97%

When base is added to the reaction, the intermediate sulfur ylide is deprotonated to produce the vinylsulfide.[241]

1) *m*-CPBA
2) TFAA, Et$_3$N

97%

## 10.8.2 Formation of α-Halosulfides

When a sulfoxide is treated with an activating agent in the presence of halide, a Pummerer-like reaction occurs to form the α-halosulfide. Treating a sulfoxide with DAST results in the formation of α-fluorosulfides.[242] Treating a sulfide with NCS[243] or thionyl chloride[244] directly results in oxidation of sulfur, elimination, and addition of chloride at the α-carbon.

DAST
SbCl$_3$
>80%

NCS, CCl$_4$
100%

Mono- or di-chlorination or bromination can be achieved using sulfuryl chloride or NBS, respectively.[245]

SO$_2$Cl$_2$
73%

NBS
58%

## 10.8.3 Halogenation of Sulfoxides, Sulfones, and Phosphine Oxides

Sulfoxides and sulfones can be halogenated at the α-position in direct analogy to the halogenation α to ketones, aldehydes, and carboxylic acids (Section 10.3.3).

[241]Bakuzis, P.; Bakuzis, M. L. F. *J. Org. Chem.* **1985**, *50*, 2569–2573.

[242]McCarthy, J. R.; Matthews, D. P.; Paolini, J. P. *Org. Synth.* **1995**, *72*, 209–215.

[243]Abbaspour Tehrani, K.; Boeykens, M.; Tyvorskii, V. I.; Kulinkovich, O.; De Kimpe, N. *Tetrahedron* **2000**, *56*, 6541–6548.

[244]Van der Veen, J. M.; Bari, S. S.; Krishnan, L.; Manhas, M. S.; Bose, A. K. *J. Org. Chem.* **1989**. *54*, 5758–5762.

[245]Ogura, K.; Kiuchi, S.; Takahashi, K.; Iida, H. *Synthesis* **1985**, 524–525.

Sulfones that are further acidified at the α-position by the presence of an additional electron-withdrawing group can be halogenated under mild conditions.[246]

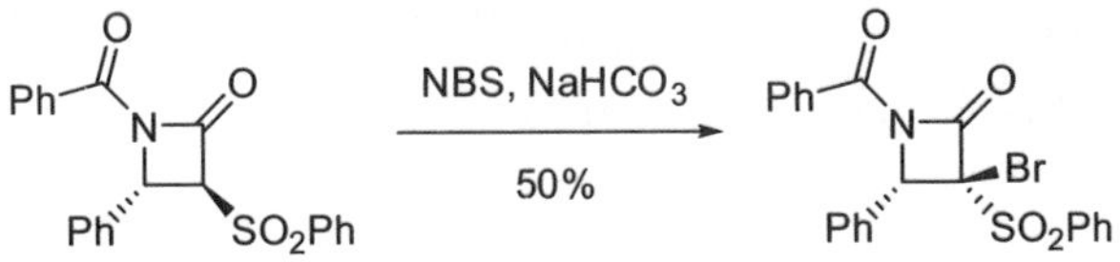

Deprotonation of a sulfone, a sulfoximine, or a phosphine oxide followed by quenching with iodine results in the iodo derivative.[247]

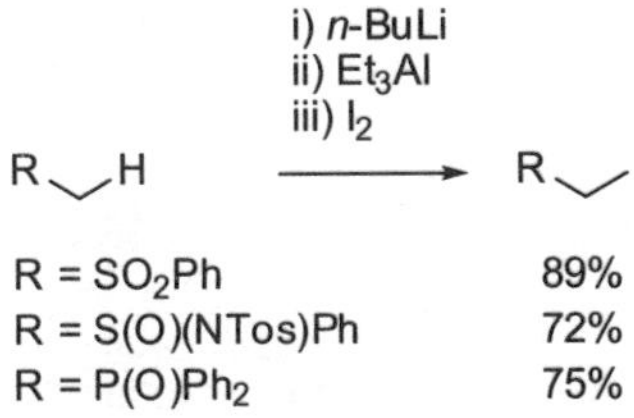

R = SO$_2$Ph          89%
R = S(O)(NTos)Ph      72%
R = P(O)Ph$_2$        75%

### 10.8.4 Oxidation of Mercaptans and Other Sulfur Compounds to Sulfonic Acids or Sulfonyl Chlorides

#### 10.8.4.1 Peroxide-Based Oxidations
*N*-Phenylthiourea was oxidized to its corresponding amidine sulfonic acid using hydrogen peroxide as the stoichiometric oxidant with a molybdenum catalyst.[248]

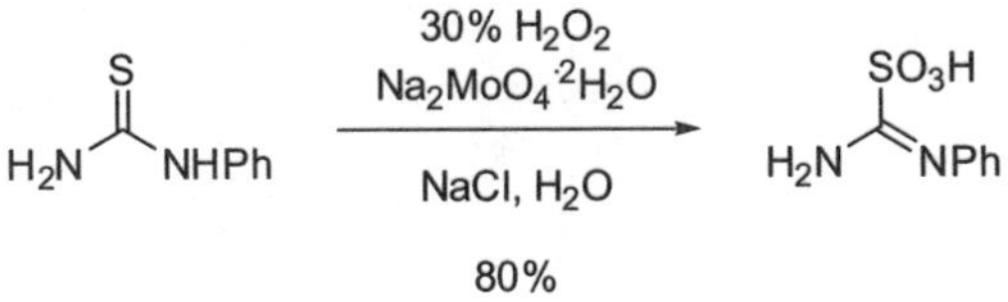

#### 10.8.4.2 Chlorine Oxidations
Oxidation to a sulfonic acid has also been achieved using chlorine as the oxidant. In many cases, the intermediate sulfonyl chloride is trapped with an amine to form the sulfonamide derivative, as shown in the example below.[249]

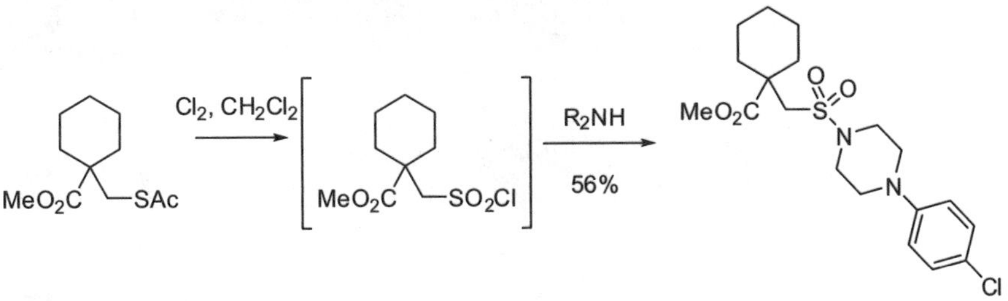

[246]Freihammer, P. M.; Detty, M. R. *J. Org. Chem.* **2000**, *65*, 7203–7207.

[247]Imamoto, T.; Koto, H. *Synthesis* **1985**, 982–983.

[248]Maryanoff, C. A.; Stanzione, R. C.; Plampin, J. N.; Mills, J. E. *J. Org. Chem.* **1986**, *51*, 1882–1884.

[249]Atkins, R. J.; Banks, A.; Bellingham, R. K.; Breen, G. F.; Carey, J. S.; Etridge, S. K.; Hayes, J. F.; Hussain, N.; Morgan, D. O.; Oxley, P.; Passey, S. C.; Walsgrove, T. C.; Wells, A. S. *Org. Process Res. Dev.* **2003**, *7*, 663–675.

One unusual case of a sulfide oxidation to a sulfonic acid equivalent is shown below.[64] Direct chlorine oxidation of the thioether to its corresponding sulfonyl chloride was successful on small scale but erratic during scale-up. Therefore, the two-step procedure to the desired sulfonamide was further broken down into three steps in order to better control the chemistry: oxidation to the sulfenyl chloride, amination, and oxidation to sulfonamide.

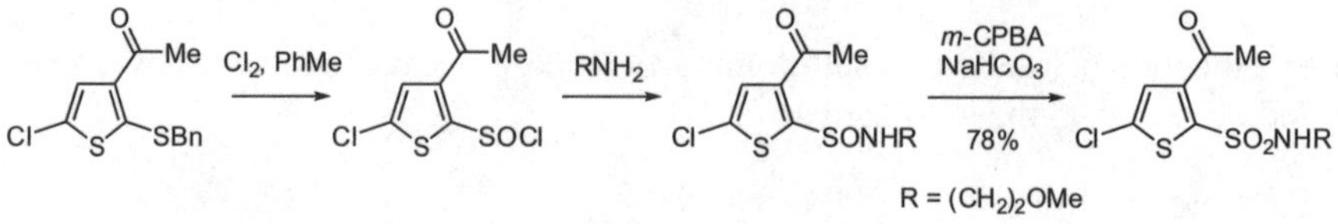

R = (CH$_2$)$_2$OMe

### 10.8.5   Oxidation of Sulfides to Sulfoxides and Sulfones

#### 10.8.5.1   Oxidation of a Sulfide to a Sulfoxide

*Peroxide-Based Reagents*   One of the most common sulfur oxidations found in pharmaceutical research and production is the oxidation of a sulfide to a sulfoxide. The oxidation occurs with a very wide variety of reagents, with the main issue for the reaction being limiting the amount of over-oxidation to the sulfone, usually controlled by the stoichiometry of the oxidant. Historically, hydrogen peroxide has been the most commonly used stoichiometric oxidant to achieve the desired transformation. This transformation can be catalyzed by a large number of catalysts.[250,251] Using trifluoroethanol as the solvent can reduce the level of over-oxidation seen in this process, but is not required.

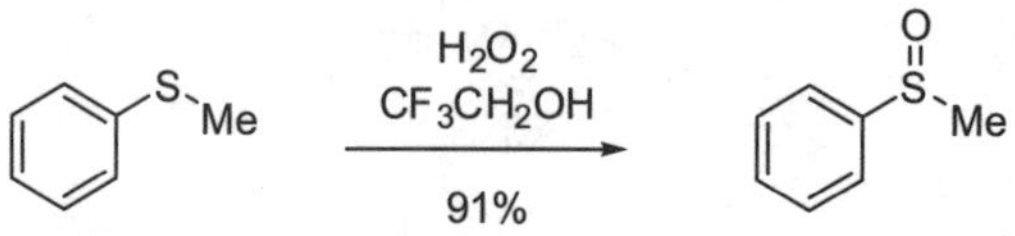

91%

*Peracid Oxidations*   Another common class of reagents for the oxidation of sulfide to sulfoxide is organic peracids, such as peracetic acid, -phthalimidohexanoic peracid, or *m*-CPBA. Pantoprazole was synthesized (50 kg) by oxidation with peracetic acid in a mixture of dichloromethane, water and methanol to allow the reaction temperature to be lowered, minimizing the production of sulfone.[252]

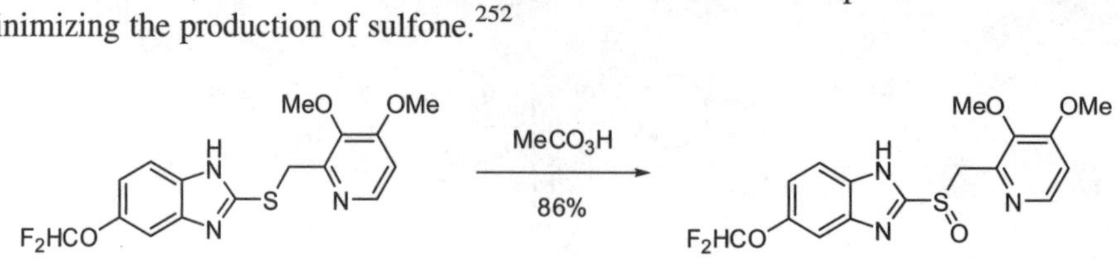

86%

[250]Kaczorowska, K.; Kolarska, Z.; Mitka, K.; Kowalski, P. *Tetrahedron* **2005**, *61*, 8315–8327.

[251]Ravikumar, K. S.; Kesavan, V.; Crousse, B.; Bonnet-Delpon, D.; Begue, J.-P. *Org. Synth.* **2003**, *80*, 184–189.

[252]Mathad, V. T.; Govindan, S.; Kolla, N. K.; Maddipatla, M.; Sajja, E.; Sundaram, V. *Org. Process Res. Dev.* **2004**, *8*, 266–270.

The use of peracids has also been demonstrated for other structural classes, such as cephalosporins, as shown in the synthesis below (37 kg).[253] The peracid reagents are usually inexpensive and readily available. The major detraction from their use, aside from the usual safety issues, is the need to purge the resulting organic acid, which can be a much more difficult task than with inorganic reagents.

In some cases, the exact nature of the oxidizing species is not clear, especially when hydrogen peroxide is used as the oxidant in acidic media.[254] For example, the conversion of acetic acid to peracetic acid with hydrogen peroxide is reported to be slow in the absence of a stronger acid catalyst,[255] but some of the oxidations carried out in such a system occur at elevated temperatures for prolonged reaction times. In these instances, the reaction rate may be dependent on the conversion of the acid to the peracid, but this issue is not usually discussed in the publications reviewed. For example, the reaction below required refluxing temperatures for 17 hours.[256]

*Inorganic Oxidants*  Inorganic oxidants have also been used to oxidize sulfides to sulfoxides. These reagents are relatively inexpensive and generate byproducts that are often more readily purged from the product than organic-based oxidants; some are also non-hazardous and/or environmentally benign. Oxone is effective at oxidizing sulfides to the sulfoxide or sulfone depending on reaction temperature and time.[257] Other useful oxidants include sodium perborate[258] and sodium hypochlorite.[250]

[253]Bernasconi, E.; Lee, J.; Roletto, J.; Sogli, L.; Walker, D. *Org. Process Res. Dev.* **2002**, *6*, 152–157.
[254]Fieser, L. F.; Fieser, M. *Reagents for Organic Synthesis*; Wiley: New York, N. Y. **1967**.
[255]Sawaki, Y.; Ogata, Y. *Bull. Chem. Soc. Jpn.* **1965**, *38*, 2103–2106.
[256]Anderson, E. L.; Post, A.; Staiger, D. S.; Warren, R. *J. Heterocycl. Chem.* **1980**, *17*, 597–598.
[257]Webb, K. S. *Tetrahedron Lett.* **1994**, *35*, 3457–3460.
[258]McKillop, A.; Tarbin, J. A. *Tetrahedron* **1987**, *43*, 1753–1758.

*Stereoselective Oxidations* In systems where the oxidation of a sulfide to a sulfoxide can lead to the formation of a new chiral center, stereoselective oxidation of the sulfur may be achieved by either substrate or reagent control. Oxone has been successfully utilized to control the relative stereochemistry of a sulfide oxidation, presumably due to the steric bulk of the inorganic complex, in the synthesis of a penem side chain.[259]

>10:1 *trans:cis*

The Kagan modification[260] of the Sharpless reagent has been successfully scaled up for a number of substrates. In these cases, alkyl peroxides give the best stereoselectivity. Conveniently, they can be purchased in anhydrous form or with low water content, since the water level is often critical to the success of the asymmetric induction. The enantioselectivity is highest for rigid substrates or those in which there is a large disparity in size between the two substituents on sulfur, such as the substrate below.[261,262]

D-DET = D-diethyl tartrate

### 10.8.5.2 Oxidation of a Sulfide to a Sulfone

*Peroxide-Based Reagents* Sulfones are another form of oxidized sulfur commonly found in pharmaceuticals. As mentioned previously, in many cases the sulfone can be installed at the correct oxidation state by direct sulfonylation, but in some cases it has been formed by oxidation of the corresponding sulfide. Once again, hydrogen peroxide is the most commonly utilized stoichiometric oxidant, and has been demonstrated for a wide range of substrates. In the past 15 years, its use with catalytic sodium tungstate has been particularly exploited, since the reaction is usually carried out under phase-transfer conditions, and the byproducts are water soluble. These conditions have been used in the reaction below.[263]

[259]Quallich, G. J.; Lackey, J. W. *Tetrahedron Lett.* **1990**, *31*, 3685–3686.

[260]Zhao, S. H.; Samuel, O.; Kagan, H. B. *Tetrahedron* **1987**, *43*, 5135–5144.

[261]Bowden, S. A.; Burke, J. N.; Gray, F.; McKown, S.; Moseley, J. D.; Moss, W. O.; Murray, P. M.; Welham, M. J.; Young, M. J. *Org. Process Res. Dev.* **2004**, *8*, 33–44.

[262]Zhao, S. H.; Samuel, O.; Kagan, H. B. *Org. Synth.* **1990**, *68*, 49–55.

[263]Giles, M. E.; Thomson, C.; Eyley, S. C.; Cole, A. J.; Goodwin, C. J.; Hurved, P. A.; Morlin, A. J. G.; Tornos, J.; Atkinson, S.; Just, C.; Dean, J. C.; Singleton, J. T.; Longton, A. J.; Woodland, I.; Teasdale, A.; Gregertsen, B.; Else, H.; Athwal, M. S.; Tatterton, S.; Knott, J. M.; Thompson, N.; Smith, S. J. *Org. Process Res. Dev.* **2004**, *8*, 628–642.

As with the sulfoxides, there are some cases where the actual oxidizing species is ambiguous, since hydrogen peroxide in an organic acid is a commonly used system. For example, in the conversion below, hydrogen peroxide is reported to be the oxidant, but residual TFA/TFAA from the previous step is not removed prior to addition of the peroxide, making it unclear whether peroxide or trifluoroperacetic acid is the oxidant.[264]

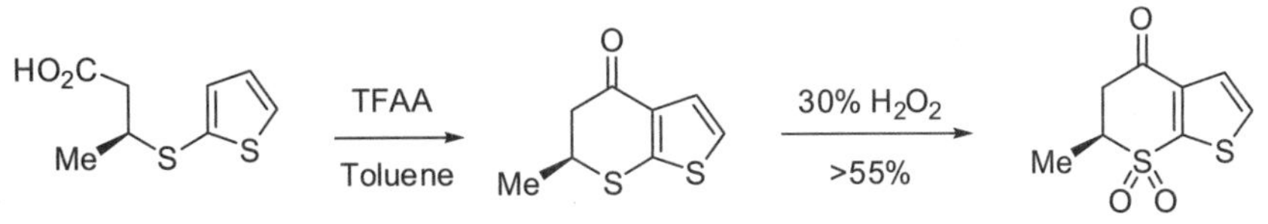

*Peracid Oxidations*    Peracids have also been employed to achieve the oxidation to sulfones. The reaction below was run on 100 kg scale using peracetic acid oxidation of the protected amino-alcohol.[265] Other simple sulfone building blocks have been synthesized by oxidation using both *m*-CPBA[266] and magnesium monoperoxyphthalate (MMPP).[267]

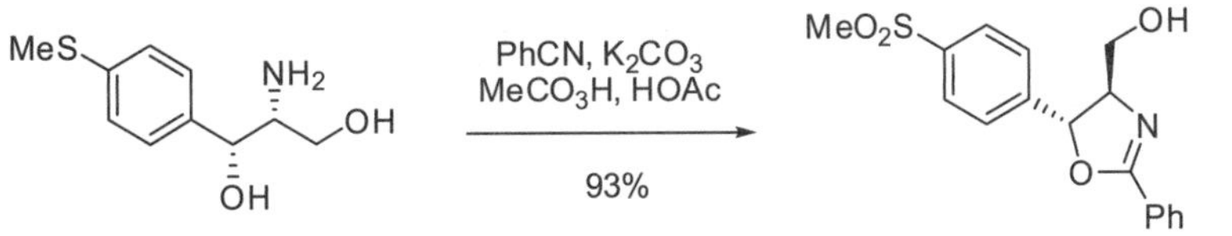

*Inorganic Oxidants*    Inorganic oxidants have been used to effect large scale oxidations to sulfones, as shown below.[268]

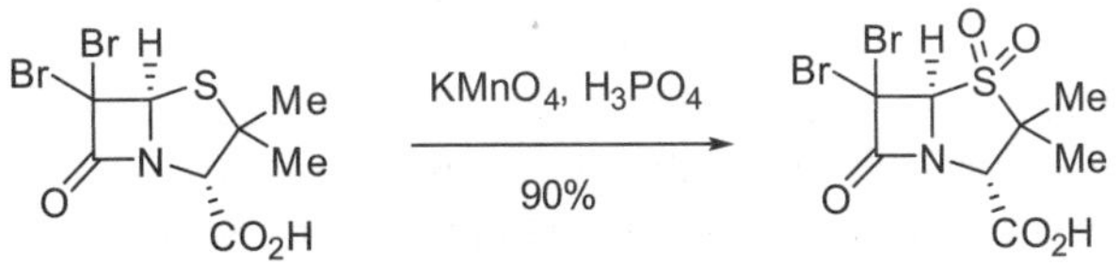

One special inorganic oxidant that is more commonly used in this situation than in other sulfide oxidations is Oxone, since it readily gives the sulfone oxidation state with little contamination from the corresponding sulfoxide, and tolerates a wide variety of functional groups. In the synthesis of a COX-2 inhibitor, the sulfone was cleanly formed in high yield, and the residual palladium from a previous step was also purged during the oxidation.[96] Oxone is inexpensive and the resulting salts are easily separated from most products;

[264]Tempkin, O.; Blacklock, T. J.; Burke, J. A.; Anastasia, M. *Tetrahedron: Asymmetry* **1996**, *7*, 2721–2724.

[265]Schumacher, D. P.; Clark, J. E.; Murphy, B. L.; Fischer, P. A. *J. Org. Chem.* **1990**, *55*, 5291–5294.

[266]Kaptein, B.; van Dooren, T. J. G. M.; Boesten, W. H. J.; Sonke, T.; Duchateau, A. L. L.; Broxterman, Q. B.; Kamphuis, J. *Org. Process Res. Dev.* **1998**, *2*, 10–17.

[267]Therien, M.; Gauthier, J. Y.; Leblanc, Y.; Leger, S.; Per*rier, H.; Prasit, P.; Wang, Z. *Synthesis* **2001**, 1778–1779.

[268]Volkmann, R. A.; Carroll, R. D.; Drolet, R. B.; Elliott, M. L.; Moore, B. S. *J. Org. Chem.* **1982**, *47*, 3344–3345.

however, its greatest drawback is its high molecular weight relative to the amount of oxygen it delivers, which requires large mass charges relative to most substrates.

### 10.8.5.3  Oxidation of Selenides

The oxidation of a selenide is usually followed by elimination or rearrangement.[269,270]

### 10.8.6  Oxidation of Mercaptans to Disulfides

The oxidation of a mercaptan to a disulfide can be accomplished with a wide array of oxidants. The most common choices are halides,[271] peroxides,[251] hypochlorites,[272] and even sulfoxides.[239]

[269]Kshirsagar, T. A.; Moe, S. T.; Portoghese, P. S. *J. Org. Chem.* **1998**, *63*, 1704–1705.
[270]Zanoni, G.; Porta, A.; Castronovo, F.; Vidari, G. *J. Org. Chem.* **2003**, *68*, 6005–6010.
[271]Basha, A.; Brooks, D. W. *J. Org. Chem.* **1993**, *58*, 1293–1294.
[272]Ramadas, K.; Srinivasan, N. *Synth. Commun.* **1995**, *25*, 227–234.

## 10.9   OXIDATION OF OTHER FUNCTIONALITY

### 10.9.1   Oxidation of Primary Halides

The Sommelet reaction is the oxidation of an alkyl chloride with hexamethylenetetramine (HMTA). The reaction proceeds through alkylation at nitrogen, elimination of the quaterinary amine salt to form an imine, and intramolecular hydride transfer.[109]

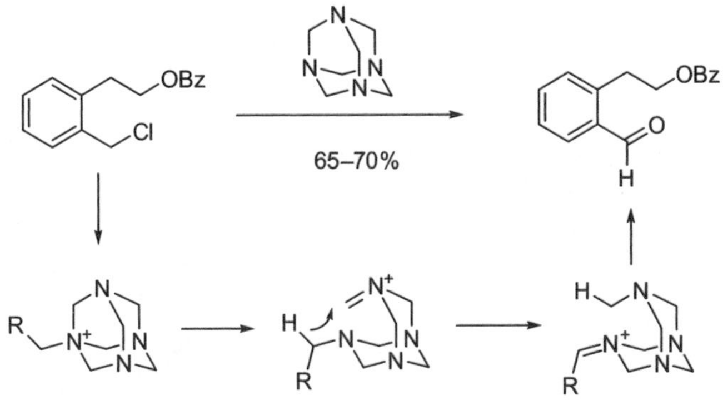

Barium permanganate can also be used to achieve this transformation, although over-oxidation to the carboxylic acid is a problem with this reagent.[188]

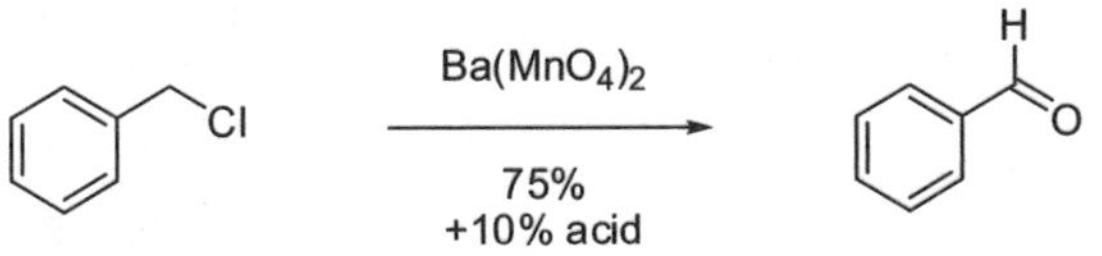

Oxidation of primary halides by displacement with DMSO and base-mediated elimination of dimethylsulfide (similar to a Swern oxidation) can also be effective for this transformation. Silver-mediated variants can be quite efficient,[273] but metal-free examples are also known, as shown below for preparation of a bis-aldehyde.[274]

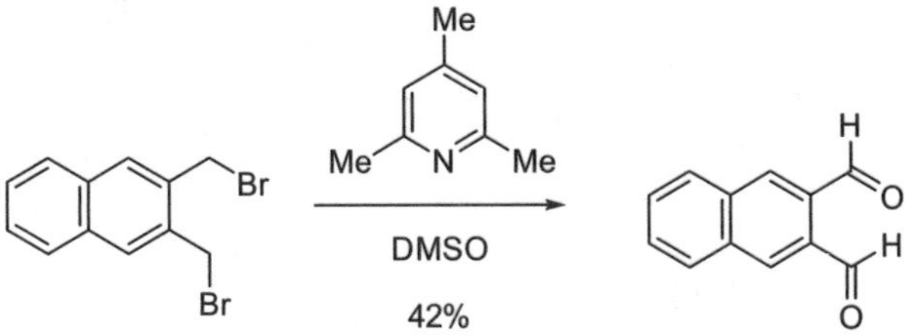

[273]Ganem, B.; Boeckman, R. K., Jr. *Tetrahedron Lett.* **1974**, 917–920.
[274]Wilcox, C. F., Jr.; Weber, K. A. *J. Org. Chem.* **1986**, *51*, 1088–1094.

## 10.9.2 Oxidation of C−Si Bonds: The Tamao Oxidation

The Tamao oxidation is the formation of an alcohol via the oxidative cleavage of a C−Si bond promoted by fluoride.[275,276]

[275]Ogasa, T.; Ikeda, S.; Sato, M.; Tamaoki, K. *Preparation of N-(2,2,5,5-tetramethylcyclopentanecarbonyl)-(S)-1,1-diaminoethane p-toluenesulfonate as a sweetener intermediate* JP02233651, **1990**, 5 pp
[276]Itami, K.; Mitsudo, K.; Yoshida, J.-I. *J. Org. Chem.* **1999**, *64*, 8709–8714.

# 11

# SELECTED FREE RADICAL REACTIONS

Nathan D. Ide

## 11.1 INTRODUCTION

Many useful bond-forming reactions and functional-group manipulations involve free radicals. Because this class of reactions includes oxidations, reductions, and bond-forming reactions, a number of free radical reactions are best discussed in other chapters. For that reason, this chapter was envisioned as a place to discuss some selected radical reactions that do not fit elsewhere in this book. This chapter is not, nor is it intended to be, a comprehensive discussion of radical chemistry.

### 11.1.1 Radical Reactions Discussed in Other Chapters

Radical reductions (dehalogenations, decarboxylations, deoxygenations, reactions involving radical anions, etc.) will be discussed in the context of reductions (Chapter 9). Radical oxidations (halogenations, hydroxylations, reactions involving radical cations, etc.) will be discussed in the context of oxidations (Chapter 10). Reactions involving aromatic diazonium salts (Sandmeyer reaction, Meerwein arylation, etc.) will be discussed in the context of nucleophilic aromatic substitution (Chapter 4). Radical reactions that involve addition across double bonds will be discussed in the context of the corresponding unsaturated functional groups. There are radical reactions in Chapter 3.

## 11.2 RADICAL CYCLIZATIONS

Radical cyclizations are a broad class of reactions that are useful in organic synthesis, largely due to their functional group tolerance. These reactions typically involve a radical and a carbon–carbon multiple bond or a carbon–heteroatom multiple bond. Because of the huge variety of reactions in this class, a full discussion is beyond the scope of this book.

*Practical Synthetic Organic Chemistry: Reactions, Principles, and Techniques*, First Edition.
Edited by Stéphane Caron.
© 2011 John Wiley & Sons, Inc. Published 2011 by John Wiley & Sons, Inc.

### 11.2.1   Atom Transfer Radical Cyclizations

While many radical cyclizations depend upon tributyltin hydride for the reduction of the cyclized radical intermediate (see Section 3.5.8), the avoidance of tin is highly desirable from a practical standpoint. For this reason, atom transfer radical cyclizations stand out as one of the most practical radical cyclization protocols.

During the course of their formal synthesis of stemoamide, Cossy and coworkers converted an iodoester to the corresponding lactone (1 : 1 mixture of iodide epimers) by treatment with 0.30 equivalents of dilauroyl peroxide (DLP) in refluxing benzene.[1] The DLP presumably homolyzes to initiate the radical reaction, followed by abstraction of iodine from the starting material and 5-*exo*-trig radical cyclization. The cyclized radical intermediate could then abstract the iodine from a molecule of starting material to generate product and propagate the chain reaction.

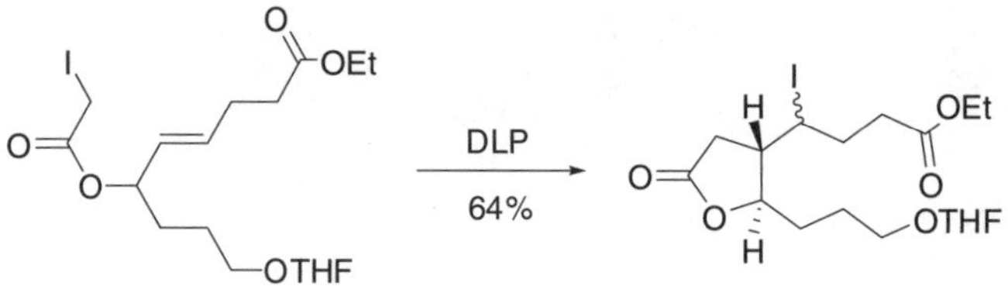

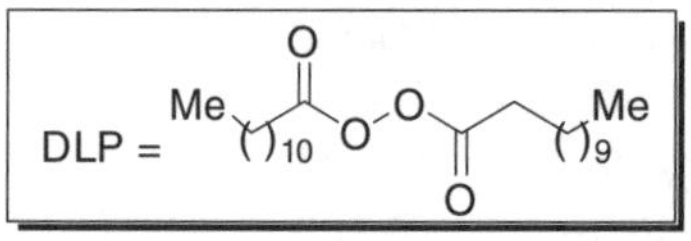

In the absence of overwhelming steric bias, reactions of this type will proceed via 5-*exo*-trig cyclization rather than 6-*endo*-trig cyclization.[2]

Atom transfer radical cyclizations can also be promoted by transition-metal catalysts, such as copper(I) complexes.[3] Speckamp and coworkers have utilized catalytic amounts of copper(I) chloride 2,2'-bipyridine (bpy) complexes as promoters for chlorine transfer radical cyclizations.[4] In these reactions, the chlorine in the starting material is transferred to the copper(I) complex, to generate a radical and a copper(II) complex. The radical intermediate undergoes cyclization, and the copper(II) complex transfers a chlorine back to the cyclized radical intermediate, thus generating product and regenerating the copper(I) complex. The reaction shown below was slow (two days in refluxing 1,2-dichloroethane), but the yield is impressive.

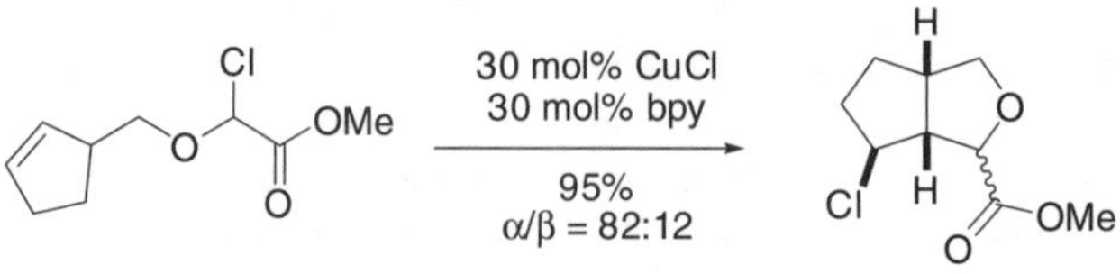

[1]Bogliotti, N.; Dalko, P. I.; Cossy, J. *J. Org. Chem.* **2006**, *71*, 9528–9531.
[2]Baldwin, J. E. *J. Chem. Soc., Chem. Commun.* **1976**, 734–736.
[3]Clark, A. J. *Chem. Soc. Rev.* **2002**, *31*, 1–11.
[4]Udding, J. H.; Tuijp, K. C. J. M.; van Zanden, M. N. A.; Hiemstra, H.; Speckamp, W. N. *J. Org. Chem.* **1994**, *59*, 1993–2003.

## 11.3   RADICAL ALLYLATION

Radical allylation is a powerful method for forming carbon–carbon bonds and has been successfully utilized in the synthesis of several complex molecules.[5] This reaction is particularly useful because it not only forms a carbon–carbon bond, but also allows for additional functionalization via manipulation of the alkene portion of the allyl unit.

### 11.3.1   Keck Radical Allylation

The traditional method for radical allylation, sometimes referred to as the Keck radical allylation,[6] involves allyltributyltin, a radical initiator and an alkyl halide starting material.[7] This process proceeds via a radical chain reaction. The reaction scope is quite broad, as the transformation can be accomplished with various alkyl halides in the presence of a wide variety of functional groups. The use of organotin compounds is a concern from an environmental and safety perspective, so the most practical radical allylations use the smallest possible amount of allyltributyltin. While many examples in the literature use a large excess of this reagent, the reaction can be accomplished in high yield with only a slight excess of allyltributyltin. This was shown to be the case by Raman and coworkers during their synthesis of carba-sugars.[8] They were able to conduct a radical allylation with a secondary alkyl chloride, allyltributyltin (1.1 equivalents), and azobisisobutyronitrile (AIBN) as an initiator. This process allowed for the isolation of a 92% yield of the allylated product, as a single diastereomer, on multigram scale.

### 11.3.2   Tin-Free Radical Allylations

While radical allylations with allyltributyltin have proven to be robust and effective reactions for the synthesis of complex molecules, the practicality of this methodology is significantly diminished by the use of allyltributyltin. Tin reagents not only represent a hazard to workers and the environment, but they can also cause problems when they contaminate compounds that are being evaluated for toxicity and/or biological activity.[9] In an effort to obviate the need for organotin reagents, several tin-free radical allylation methods have been developed. While none of these methods have yet been shown to have the substrate scope of the Keck

---

[5]Jarosz, S.; Kozlowska, E. *Pol. J. Chem.* **1998**, *72*, 815–831.

[6]Keck, G. E.; Yates, J. B. *J. Am. Chem. Soc.* **1982**, *104*, 5829–5831.

[7]Kurti, L.; Czako, B.; editors, *Strategic Applications of Named Reactions in Organic Synthesis*; Academic Press: Burlington, MA, 2005.

[8]Ramana, C. V.; Chaudhuri, S. R.; Gurjar, M. K. *Synthesis* **2007**, 523–528.

[9]Le Guyader, F.; Quiclet-Sire, B.; Seguin, S.; Zard, S. Z. *J. Am. Chem. Soc.* **1997**, *119*, 7410–7411.

radical allylation, it is likely that these methods could be advantageous in certain circumstances.

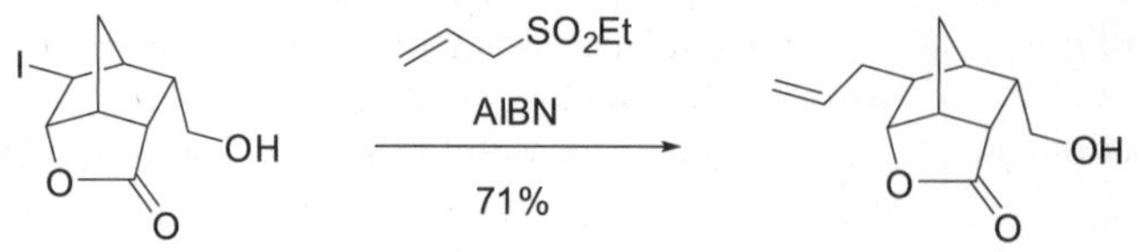

Zard and coworkers have developed a strategy that allows for the replacement of allyltin reagents with allyl sulfones.[9] This methodology allows for the conversion of a variety of alkyl iodides to the corresponding allyl-substituted compounds. In the example shown above, a secondary alkyl iodide is converted (71% yield) to the corresponding allyl compound in the presence of allylethylsulfone and AIBN. Analogous to the Keck radical allylation, the reaction is broad in scope and tolerant of many functional groups. Unfortunately, the requisite allylsulfones are not as readily available, from commercial sources, as the corresponding allyltin reagents. Despite this drawback, the benefits of performing radical allylations without tin are significant and would suggest that this methodology could indeed provide the most practical radical allylation for applications where tin byproducts would be undesirable.

While the approach of Zard and coworkers utilizes alkyliodides and allyl sulfones, an alternative strategy developed by Renaud and coworkers involves boronates and allyl sulfones.[10] In this approach, an alkene undergoes dimethylacetamide-promoted hydroboration and the resulting boronate is subjected to oxygen-initiated radical allylation. This reaction provides products that have undergone a net hydroallylation. This approach is highlighted in the figure below, which depicts the hydroallylation of α-pinene, a process that proceeds in 80% yield. This strategy is desirable because it allows for the use of alkenes as starting materials and obviates the need for tin reagents. In situations where an alkene is a more logical intermediate than an alkyl halide, this technique could be the most practical approach for radical allylation.

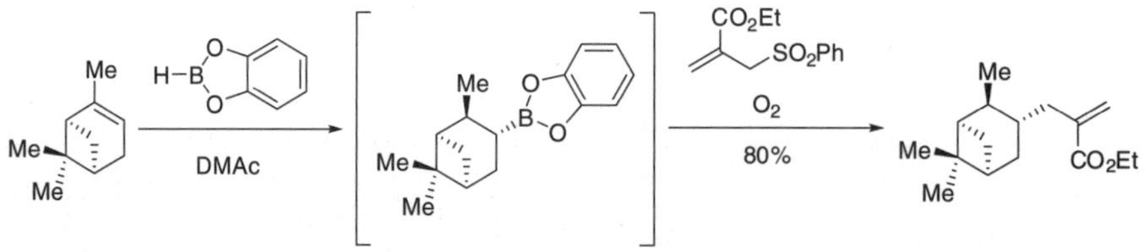

## 11.4   REMOTE FUNCTIONALIZATION REACTIONS

Radical remote functionalization reactions are a powerful set of methods for the functionalization of complex molecules.[11,12] These reactions typically involve the generation of a heteroatom-centered radical, which proceeds to abstract a pendant hydrogen atom, typically

[10]Darmency, V.; Scanlan, E. M.; Schaffner, A. P.; Renaud, P.; Sui, B.; Curran, D. P. *Org. Synth.* **2006**, *83*, 24–30.
[11]Majetich, G.; Wheless, K. *Tetrahedron* **1995**, *51*, 7095–7129.
[12]Reese, P. B. *Steroids* **2001**, *66*, 481–497.

via a 6-membered ring that incorporates the abstracted hydrogen atom (hydrogen abstraction at the δ-position). The generation of the carbon-centered radical allows for functionalization at that position. Depending on many factors, including the number of positions with accessible hydrogen atoms and their relative reactivities, these remote functionalization reactions can vary from poorly selective and low yielding to exquisitely selective and high yielding. While the conditions and reagents required for remote radical functionalization may not be the most practical, the transformations accomplished via these methods would be difficult or even impossible to accomplish using alternative methods.

## 11.4.1   Barton Nitrite Ester Reaction

The Barton nitrite ester reaction involves the photolytic homolysis of a nitrite ester.[7,13,14] This homolysis results in the formation of an alkoxyl radical, which abstracts a remote hydrogen via a 6-membered ring transition state. The resulting carbon-centered radical then recombines with the NO radical that was generated by photolysis. Tautomerization then leads to the desired oxime product.

This reaction was utilized by Hakimelahi and coworkers in the preparation of a carbacephem antibiotic.[15] In this case, which is shown below, the captodative carbon-centered radical is stabilized by the adjacent nitrogen atom and ester group. Stabilization of the carbon-centered radical by an adjacent heteroatom is not a requirement for a successful remote radical functionalization, but it is often highly beneficial. In this reaction, the oxime was formed in a 7 : 5 ratio favoring the *anti* product.

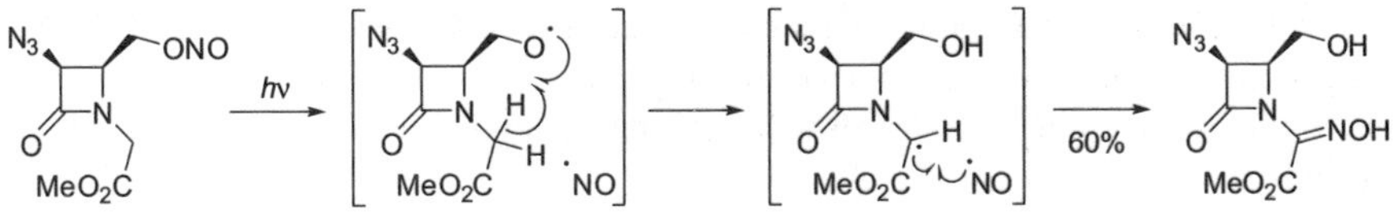

Konoike and coworkers were able to utilize a Barton nitrite ester reaction for the preparation of an oxime intermediate in 85% yield.[16] This transformation, which is depicted below, required exclusion of oxygen to prevent formation of nitrate products. While the concentration of the photolytic reaction was low by industrial standards (67 mM, 30 L/kg), the reaction was still practical enough to allow for the production of 237 g of product from 14 batches in the photoreactor. This reaction has also been carried out using a flow microreactor.[17]

[13]Barton, D. H. R.; Beaton, J. M.; Geller, L. E.; Pechet, M. M. *J. Am. Chem. Soc.* **1960**, *82*, 2640–2641.

[14]Robinson, C. H.; Gnoj, O.; Mitchell, A.; Wayne, R.; Townley, E.; Kabasakalian, P.; Oliveto, E. P.; Barton, D. H. R. *J. Am. Chem. Soc.* **1961**, *83*, 1771–1772.

[15]Hakimelahi, G. H.; Li, P.-C.; Moosavi-Movahedi, A. A.; Chamani, J.; Khodarahmi, G. A.; Ly, T. W.; Valiyev, F.; Leong, M. K.; Hakimelahi, S.; Shia, K.-S.; Chao, I. *Org. Biomol. Chem.* **2003**, *1*, 2461–2467.

[16]Konoike, T.; Takahashi, K.; Araki, Y.; Horibe, I. *J. Org. Chem.* **1997**, *62*, 960–966.

[17]Sugimoto, A.; Sumino, Y.; Takagi, M.; Fukuyama, T.; Ryu, I. *Tetrahedron Lett.* **2006**, *47*, 6197–6200.

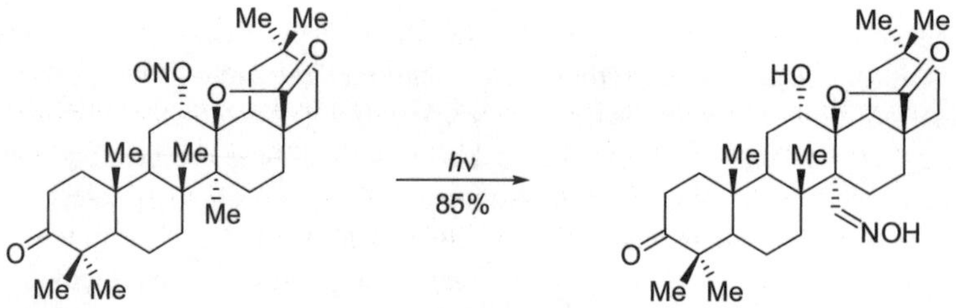

## 11.4.2   Hofmann–Löffler–Freytag Reaction

The Hofmann–Löffler–Freytag reaction is a method for remote radical functionalization that is related to the Barton nitrite ester reaction, but involves the homolytic cleavage of a nitrogen–halogen bond, rather than a nitrite ester.[7]

This transformation was used by Ban and coworkers in their synthesis of various 1,3-diaza heterocycles.[18] Their method, which is shown below, involves treatment of a diamine starting material with N-chlorosuccinimide and triethylamine. The resulting chloroamine is directly irradiated with ultraviolet light, which results in homolysis of the nitrogen–chlorine bond. The nitrogen-centered radical then abstracts the adjacent hydrogen atom via a 7-membered ring transition state. Recombination of the carbon-centered radical with the chlorine radical and displacement of the resulting chloride by the secondary nitrogen gives the tricyclic product in high yield. The tertiary nitrogen of the starting material likely plays a key role in this reaction. It stabilizes the carbon-centered radical and also assists with the displacement of the chloride, by allowing the displacement to proceed via an iminium intermediate. While this example showcases a reaction that is highly efficient, generating product in quantitative yield, the Hofmann–Löffler–Freytag reaction is clearly not a widely applicable reaction. As reported by Ban and coworkers,[18] seemingly minor changes in substrate structure can result in dramatic changes in the efficiency of this reaction. This is not surprising, when one considers the complex sequence of events that must occur in order to generate product via this transformation.

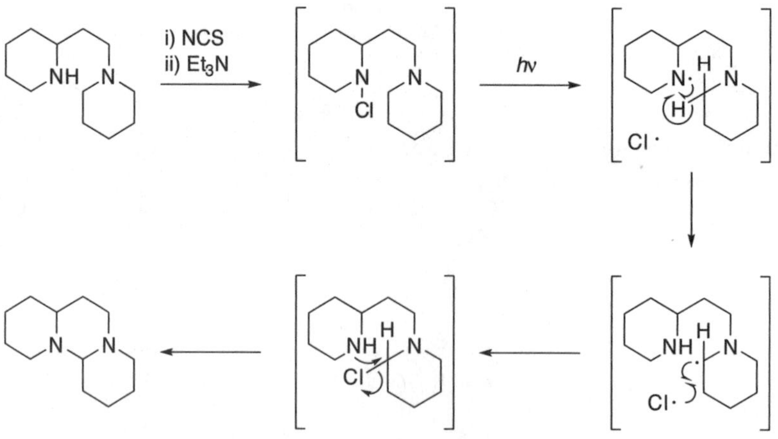

[18]Kimura, M.; Ban, Y. *Synthesis* **1976**, 201–202.

As shown below, the Hofmann–Löffler–Freytag reaction can also be used to generate alkyl chlorides if the reaction halts prior to intramolecular alkylation of the secondary nitrogen. This variation of the reaction was used by Uskoković and coworkers during the course of their synthesis of meroquinene.[19] In this case, the alkylation step is disfavored because the alkylation would have to proceed via an unactivated (no adjacent heteroatoms) primary alkyl chloride to generate a bridged bicyclic product.

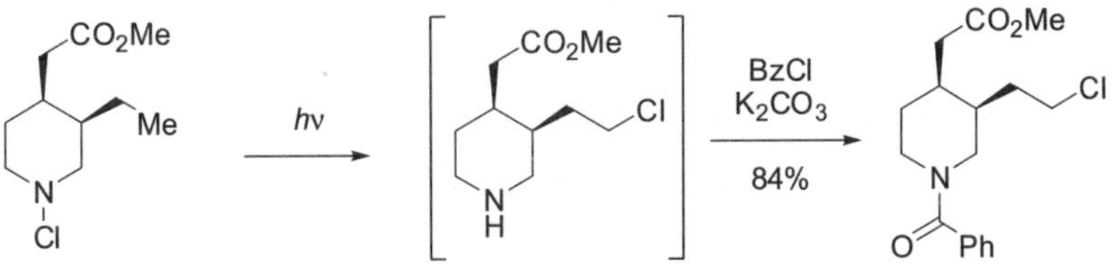

A useful variation of the Hofmann–Löffler–Freytag reaction, which uses iodine and either (diacetoxyiodo)benzene (DIB) or iodosylbenzene, has been developed by Suárez and coworkers.[20] While most Hofmann–Löffler–Freytag reactions are initiated by ultraviolet light, the Suárez modification proceeds under irradiation with visible light.

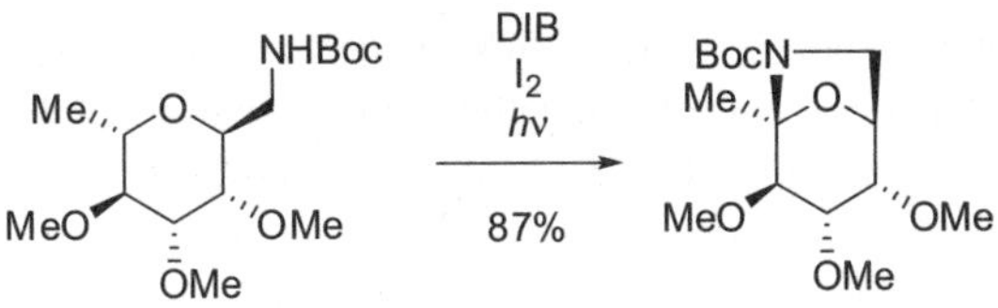

### 11.4.3   Hypohalite Reaction

The hypohalite reaction is analogous to the Hofmann–Löffler–Freytag reaction, except for the involvement of an oxygen-centered radical rather than a nitrogen-centered radical. This reaction has traditionally been accomplished using lead tetraacetate and iodine under photolytic conditions. The reaction typically requires a large excess of lead tetraacetate, which is undesirable. The reaction can also be promoted by mercury- and selenium-derived reagents,[11] but these are also undesirable due to their toxicity. Suárez and coworkers have developed a convenient protocol for the hypoiodite reaction, which utilizes hypervalent iodine reagents.[21]

When dihydrotigogenin 3-acetate was irradiated in the presence of DIB and iodine, a 92% yield of tigogenin acetate was obtained. This reaction is particularly interesting because

[19]Uskokovic, M. R.; Henderson, T.; Reese, C.; Lee, H. L.; Grethe, G.; Gutzwiller, J. *J. Am. Chem. Soc.* **1978**, *100*, 571–576.

[20]Francisco, C. G.; Herrera, A. J.; Suarez, E. *J. Org. Chem.* **2003**, *68*, 1012–1017.

[21]De Armas, P.; Concepcion, J. I.; Francisco, C. G.; Hernandez, R.; Salazar, J. A.; Suarez, E. *J. Chem. Soc., Perkin Trans. 1* **1989**, 405–411.

it forms the 6-membered ring preferentially over a 5-membered ring, which suggests a 7-membered ring transition state for the abstraction of hydrogen by the alkoxyl radical. This preferential reactivity is likely due to the tendency to form carbon-centered radicals adjacent to oxygen and other radical-stabilizing heteroatoms.

During their efforts toward the synthesis of calcitriol analogues, Mouriño and coworkers investigated conditions and reagents for a key hypoiodite reaction.[22] They found that DIB was at least as effective as lead tetraacetate and that reactions promoted by DIB proceeded under more practical conditions. While the lead tetraacetate reactions required high dilution in benzene, the DIB-promoted reactions proceeded efficiently at much higher concentrations in cyclohexane. Additionally, the DIB-promoted reactions required less reagent (1.5 equivalents DIB vs. 4.5 equivalents lead tetraacetate) and could be initiated by either irradiation or sonication. When using DIB, sonication allowed for higher reaction concentrations than were possible with the photo-initiated reactions.

| Conditions | Yield |
|---|---|
| DIB, $I_2$, $h\nu$ | 93% |
| DIB, $I_2$, )))) | 96% |
| Pb(OAc)$_4$, $h\nu$ | 92% |

)))) = sonication

## 11.5   THE HUNSDIECKER REACTION

The Hunsdiecker reaction is a method for the conversion of carboxylic acids to the corresponding decarboxylated aryl, alkyl, or vinyl halides.[7] The original conditions for this transformation, as developed by Borodin and Hunsdiecker, utilized silver carboxylates

[22]Moman, E.; Nicoletti, D.; Mourino, A. *J. Org. Chem.* **2004**, *69*, 4615–4625.

and elemental bromine.[23] The mechanism of this reaction is shown below. The silver carboxylate is converted to the corresponding acyl hypobromite, which can undergo homolysis either thermally or photolytically. The resulting carboxyl radical can decarboxylate to reveal a carbon-centered radical, which then abstracts a bromine atom from elemental bromine. The carbon-centered radical could, alternatively, recombine with the previously generated bromine radical or abstract bromine from another acyl hypobromite molecule.

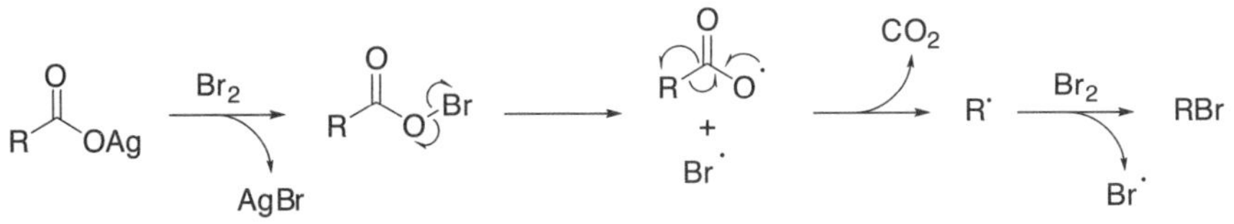

The original Hunsdiecker reaction is not very practical, because it requires isolation and drying of the thermally sensitive silver carboxylates. For this reason, a number of modifications have been developed, including approaches with other metals (Hg, Tl, Mn, Pb, etc.),[23] but most of these variations lack the functional group compatibility observed for the more practical Barton[24] and Suárez[25] modifications.

### 11.5.1    Barton Modification of the Hunsdiecker Reaction

An example of the Barton modification of the Hunsdiecker reaction is shown below.[26] In this reaction, the carboxylic acid is converted to the acid chloride, which is treated with sodium 2-thioxopyridin-1(2H)-olate, 4-dimethylaminopyridine (DMAP) and iodoform in refluxing cyclohexane. This procedure results in the formation and thermolysis of the corresponding thiohydroxamic ester. Homolysis of the nitrogen–oxygen bond of the thiohydroxamic ester results in a carboxylate radical, which decarboxylates to provide a carbon-centered radical, followed by abstraction of iodine from iodoform. This process converts the carboxylic acid to the corresponding iodide in 65% yield. Even better yields have been observed with simpler starting materials, but this example showcases the most significant feature of this approach, which is the observed functional group tolerance.

[23]Li, J. J. *Name Reactions for Functional Group Transformations*; Wiley: Hoboken, N. J., 2007, p. 623–629.

[24]Barton, D. H. R.; Crich, D.; Motherwell, W. B. *Tetrahedron Lett.* **1983**, *24*, 4979–4982.

[25]Concepcion, J. I.; Francisco, C. G.; Freire, R.; Hernandez, R.; Salazar, J. A.; Suarez, E. *J. Org. Chem.* **1986**, *51*, 402–404.

[26]Barton, D. H. R.; Crich, D.; Motherwell, W. B. *Tetrahedron* **1985**, *41*, 3901–3924.

## 11.5.2   Suráez Modification of the Hunsdiecker Reaction

The Suárez modification of the Hunsdiecker reaction relies upon irradiation of a carboxylic acid with (diacetoxyiodo)benzene (DIB) and iodine in refluxing cyclohexane.[25] Mechanistically, it has been proposed that the carboxylic acid displaces an acetate from the DIB, and reaction of the resulting adduct with iodine results in acyl hypoiodite formation. The intermediate acyl hypoiodite then undergoes a decarboxylative iodination to provide the alkyl iodide, as a 3 : 2 mixture of epimers. It is significant that the reaction proceeds in high yield with only 1.1 equivalents of DIB and 1.0 equivalent of $I_2$. Similar conditions have also been reported for decarboxylative bromination.[27] It should also be noted that this reaction proceeds efficiently for primary and secondary carboxylic acids, but tertiary carboxylic acids tend to form olefins via oxidation of the tertiary radical intermediate to the corresponding carbocation, followed by elimination.

[27]Camps, P.; Lukach, A. E.; Pujol, X.; Vazquez, S. *Tetrahedron* **2000**, *56*, 2703–2707.

Interestingly, the oxidative olefin formation can be encouraged with primary and secondary carboxylic acids, if excess DIB and catalytic cupric acetate (without iodine) are utilized, but this procedure is not particularly practical, as it requires 5.0 equivalents of DIB.[25]

### 11.5.3  Catalytic Hunsdiecker Reaction with α,β-Unsaturated Carboxylic Acids

A catalytic variant of the Hunsdiecker reaction has been developed by Roy and coworkers for use with α,β-unsaturated carboxylic acids.[28] This reaction utilizes catalytic lithium acetate and stoichiometric *N*-bromosuccinimide (NBS).

This reaction works best for electron-rich α,β-unsaturated carboxylic acids and it typically provides the (*E*)-bromoalkenes with good stereoselectivity. Useful modifications

[28]Chowdhury, S.; Roy, S. *J. Org. Chem.* **1997**, *62*, 199–200.

include the use of microwave heating[29] and alternative bases, such as triethylamine.[30] It is important to note that this reaction most likely does not proceed via a radical pathway. An ionic mechanism has been proposed,[31] and is shown below.

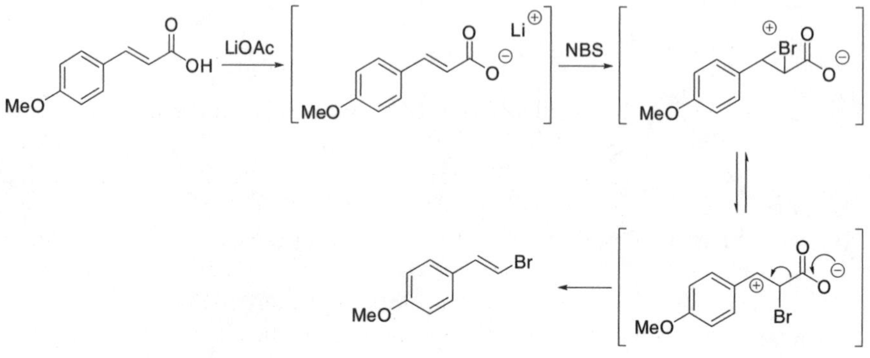

In this proposed mechanism, the base deprotonates the $\alpha,\beta$-unsaturated carboxylic acid, and the olefin then reacts with NBS to generate a bromonium carboxylate (which can also be drawn as a benzylic carbocation), followed by decarboxylative elimination to generate the bromoalkene.

### 11.5.4  Nitro-Hunsdiecker Reaction

Roy and coworkers have also developed a nitro-Hunsdiecker reaction, which uses nitric acid and substoichiometric AIBN to convert electron-rich $\alpha,\beta$-unsaturated carboxylic acids and aromatic acids to the corresponding nitro derivatives.[32]

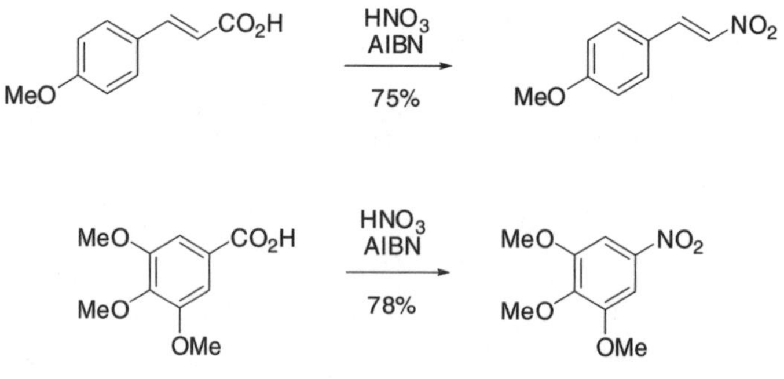

The conversion of electron-rich benzoic acid derivatives to the mono-nitro products is significant because other approaches to these products, via nitration of the appropriate

[29]Kuang, C.; Yang, Q.; Senboku, H.; Tokuda, M. *Synthesis* **2005**, 1319–1325.
[30]Das, J. P.; Roy, S. *J. Org. Chem.* **2002**, *67*, 7861–7864.
[31]Naskar, D.; Roy, S. *J. Chem. Soc., Perkin Trans. 1* **1999**, 2435–2436.
[32]Das, J. P.; Sinha, P.; Roy, S. *Org. Lett.* **2002**, *4*, 3055–3058.

benzene derivative, often suffer from competitive oxidation reactions.[33] The mechanism of this reaction is not well understood at this time, but the requirement of a substoichiometric amount of AIBN strongly suggests a radical mechanism.

## 11.6  THE MINISCI REACTION

The Minisci reaction involves the reaction of nucleophilic carbon-centered radicals with heteroaromatic systems, typically under acidic conditions.[7,34,35] The acidic conditions, although not a strict requirement, result in protonation of the heteroaromatic system and thereby activate the system for reaction with nucleophilic radicals. The power of the Minisci reaction, and the reason that it has seen use in organic synthesis, is that the selectivity of the reaction is complementary to that observed with electrophilic aromatic functionalization reactions, such as Friedel–Crafts acylations/alkylations.

Minisci and coworkers developed the reaction shown below, which allows for the conversion of iodo sugars to the corresponding heterocycle-bound derivatives.[36] In this case, a nucleophilic carbon-centered radical is generated from the iodo sugar using benzoyl peroxide. The radical adds to the quinolinium ring, and the resulting stabilized radical is oxidized to the corresponding cation. Deprotonation provides the product in high yield.

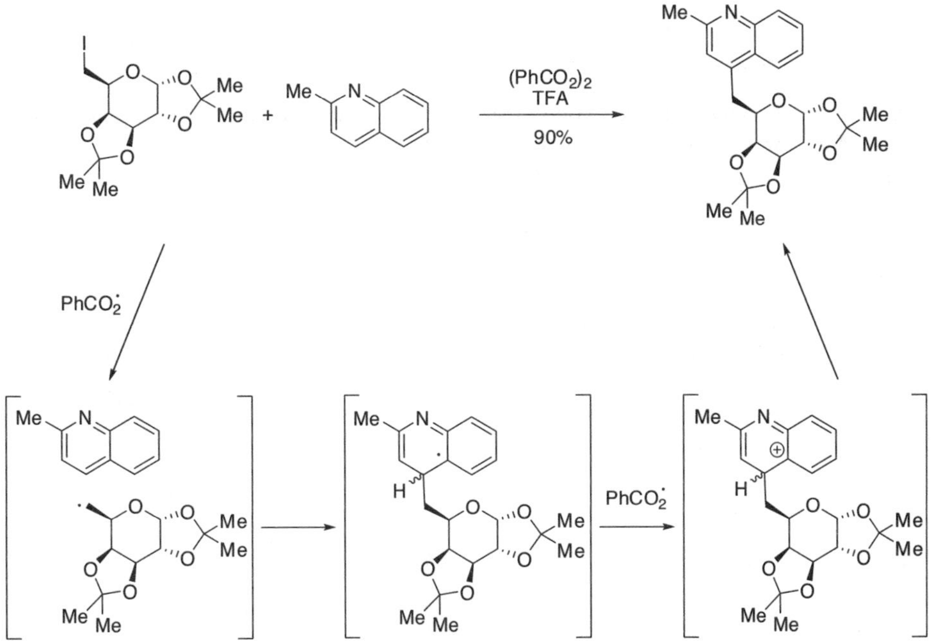

[33]Dwyer, C. L.; Holzapfel, C. W. *Tetrahedron* **1998**, *54*, 7843–7848.
[34]Minisci, F.; Fontana, F.; Vismara, E. *J. Heterocycl. Chem.* **1990**, *27*, 79–96.
[35]Punta, C.; Minisci, F. *Trends in Heterocyclic Chemistry* **2008**, *13*, 1–68.
[36]Vismara, E.; Donna, A.; Minisci, F.; Naggi, A.; Pastori, N.; Torri, G. *J. Org. Chem.* **1993**, *58*, 959–963.

As is the case for many radical reactions, the Minisci reaction can be performed using radicals generated under a wide variety of conditions. Cowden has developed a Minisci reaction that relies upon radical decarboxylation of amino acids and subsequent reaction with dihalopyridazines.[37] This reaction utilizes catalytic silver with stoichiometric ammonium persulfate and provides a variety of useful products in moderate to high yields. It is worth noting that the dihalopyridazine products of this methodology are functionalized appropriately to undergo a variety of nucleophilic aromatic substitutions and/or transition-metal-promoted cross-coupling reactions.

Baran and coworkers have developed a variation of the Minisci reaction that couples arylboronic acids to electron-deficient heterocycles using catalytic silver nitrate and excess potassium persulfate.[38] As with most Minisci reaction variants, the transformations are of high synthetic utility, but the yields are substrate dependant.

## 11.7   RADICAL CONJUGATE ADDITIONS

Radical conjugate additions involve the reaction of a nucleophilic radical with an electron-deficient olefin ($\alpha,\beta$-unsaturated amides, esters, ketones, sulfones, sulfoxides, etc.). Unlike ionic conjugate additions, which are readily reversible and often suffer from unfavorable equilibria, most radical conjugate additions are essentially irreversible. Additionally, radical conjugate additions typically react exclusively via 1,4-addition, without competitive 1,2-addition. Finally, radical reaction conditions are typically mild, which results in good functional group tolerance and minimal need for protecting groups.[39,40]

### 11.7.1   Intramolecular Radical Conjugate Additions

In this book, intramolecular radical conjugate additions will be considered a subset of radical cyclizations (see Sections 3.5.8 and 11.2).

[37]Cowden, C. J. *Org. Lett.* **2003**, *5*, 4497–4499.
[38]Seiple, I. B.; Su, S.; Rodriguez, R. A.; Gianatassio, R.; Fujiwara, Y.; Sobel, A. L.; Baran, P. S. *J. Am. Chem. Soc.* **2010**, *132*, 13194–13196.
[39]Srikanth, G. S. C.; Castle, S. L. *Tetrahedron* **2005**, *61*, 10377–10441.
[40]Zhang, W. *Tetrahedron* **2001**, *57*, 7237–7262.

## 11.7.2    Intermolecular Radical Conjugate Additions

Intermolecular radical conjugate additions have proven to be a useful method for carbon–carbon bond formation and for the establishment of chiral centers.[39–43] Sibi and co-workers have developed a catalytic asymmetric variant of this reaction, which is capable of generating two new stereocenters.[44] Radical initiation by triethylborane and oxygen results in formation of the *t*-butyl radical, which undergoes a conjugate addition to the alkene. The resulting radical is then trapped by the allyltributyltin, providing the second stereocenter and a tin radical for chain propagation. The reaction is enantioselective because the radical conjugate addition is accelerated by complexation of the acylox-azolidinone to the chiral Lewis acid. High enantioselectivities require bulky radicals in the addition step and the radical trapping occurs with *anti* diastereoselectivity. It is interesting to note that the use of copper(II) triflate, instead of magnesium(II) iodide, resulted in enantiomeric products, even though the same ligand was used. This reversal of enantioselectivity is likely due to different coordination geometries with copper and magnesium.

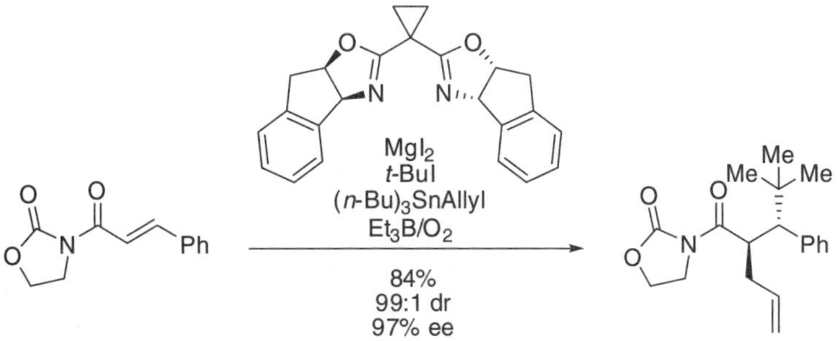

Renaud and Ollivier have developed a tin-free radical conjugate addition, which relies upon the hydroboration of olefins with catecholborane, followed by oxygen-initiated radical conjugate addition into unsaturated ketones and aldehydes.[45] As shown below, an alkene can be treated with catecholborane to provide the hydroboration product. Treatment of this intermediate with oxygen (air) generates the corresponding alkyl radical, which can add in a conjugate fashion to the enone. The authors have proposed that the conjugate addition results in the formation of a boron enolate intermediate, which could form via radical recombination or via a chain-propagating reaction with another equivalent of alkylcatecholborane. Quenching of the boron enolate results in formation of the conjugate addition product in good yield.

[41]Sibi, M. P.; Manyem, S. *Tetrahedron* **2000**, *56*, 8033–8061.
[42]Sibi, M. P.; Manyem, S.; Zimmerman, J. *Chem. Rev.* **2003**, *103*, 3263–3295.
[43]Sibi, M. P.; Porter, N. A. *Acc. Chem. Res.* **1999**, *32*, 163–171.
[44]Sibi, M. P.; Chen, J. *J. Am. Chem. Soc.* **2001**, *123*, 9472–9473.
[45]Ollivier, C.; Renaud, P. *Chem.-Eur. J.* **1999**, *5*, 1468–1473.

## 11.8 β-SCISSION REACTIONS

In addition to their utility for the formation of carbon-carbon bonds, radical reactions can be effective for the selective fragmentation of carbon-carbon bonds, typically via β-scission reactions. One of the most commonly employed β-scission approaches involves the generation of an alkoxy radical, which then undergoes fragmentation. Macdonald and O'Dell used this approach for the β-scission of 9-decalinol.[46] Treatment with mercuric oxide and iodine results in the formation of the corresponding hypoiodite, followed by homolysis to give the alkoxy radical and scission of the adjacent carbon-carbon bond. Finally, the primary radical is trapped with iodine to give a 68% yield of 2-(4-iodobutyl)cyclohexanone. The iodine source for the trapping of the primary radical could be elemental iodine, an iodine radical, or another hypoiodite molecule.

A similar approach was used by Cairns and Englund for the synthesis of ω-haloketones.[47,48] Conversion of 1-methylcyclopentanol to the corresponding hypochlorite was completed using chlorine and aqueous sodium hydroxide. Gentle heating (40°C)

[46]Macdonald, T. L.; O'Dell, D. E. *J. Org. Chem.* **1981**, *46*, 1501–1503.

[47]Cairns, T. L.; Englund, B. E. *J. Org. Chem.* **1956**, *21*, 140.

[48]Englund, B. E. *Preparing w-halo ketones by rearrangement of tertiary cycloaliphatic hypohalites* US2691682 1954, (4 pp).

resulted in rearrangement to the ω-chloroketone. This reaction probably proceeds in a similar fashion to the above reaction of 9-decalinol.

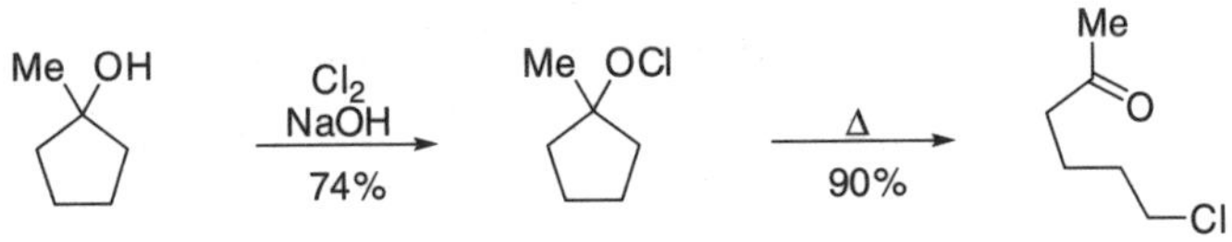

Samarium diiodide can also be used to promote β-scission reactions, and although samarium diiodide-mediated β-scissions are reduction reactions, they will be discussed in the context of radical reactions. One of the main advantages of samarium diiodide is its tendency to react in a highly chemoselective fashion. The resulting functional group compatibility has made it a useful reagent for the synthesis of complex molecules. Baran and coworkers utilized a key samarium diiodide-induced radical cleavage during their synthesis of ( + )-cortistatin A.[49] In this reaction, samarium diiodide reacts with the ketone to generate a ketyl radical, which results in cyclopropyl ring opening and radical debromination. The resulting samarium enolate is selectively trapped with 2,4,4,6-tetrabromo-2,5-cyclohexadienone (TBCHD), an electrophilic bromination reagent, to provide the desired α-bromoketone. The α-bromoketone was not isolated, but the isolation of an intermediate in 58% yield after two additional steps is indicative of a highly efficient radical cascade sequence.

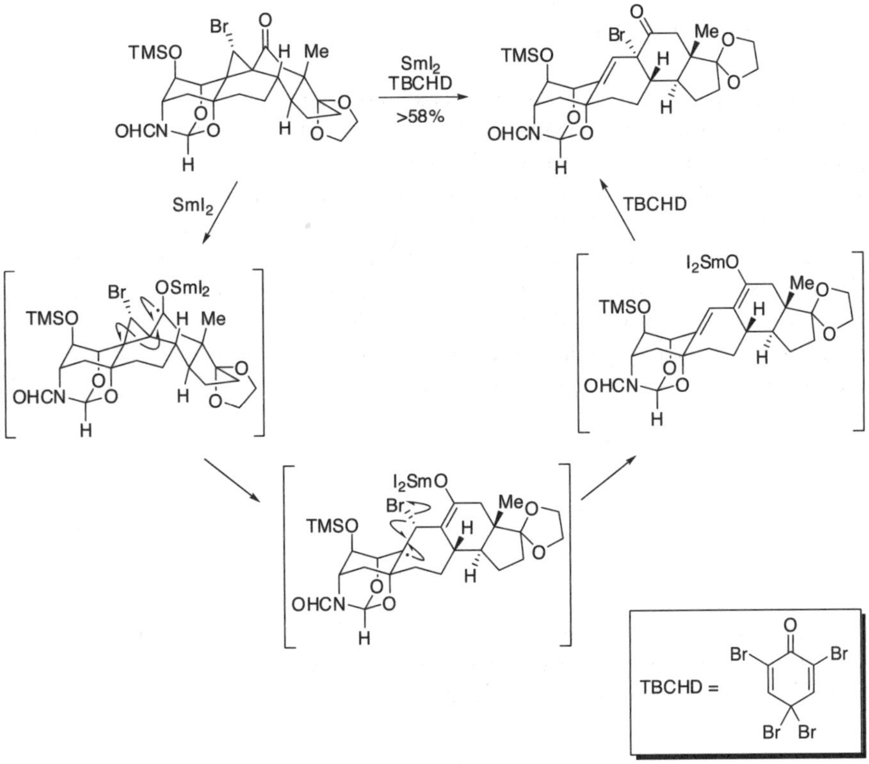

[49]Shenvi, R. A.; Guerrero, C. A.; Shi, J.; Li, C.-C.; Baran, P. S. *J. Am. Chem. Soc.* **2008**, *130*, 7241–7243.

## 11.9 FREE-RADICAL POLYMERIZATION

While a discussion of free-radical polymerization is well beyond the scope of this book, it is worth noting that radical chemistry finds far more use in the industrial manufacturing of polymers than it does in the manufacturing of small molecules.[50–52] In fact, radical polymerization is the most commonly employed method for synthesis of polymers on an industrial scale.[52]

[50]Kamigaito, M.; Ando, T.; Sawamoto, M. *Chem. Rev.* **2001**, *101*, 3689–3745.
[51]Matyjaszewski, K.; Braunecker, W. A. *Radical polymerization*; Wiley-VCH: Weinheim, Germany, 2007; Vol. 1.
[52]Matyjaszewski, K.; Xia, J. *Chem. Rev.* **2001**, *101*, 2921–2990.

# 12

# SYNTHESIS OF "NUCLEOPHILIC" ORGANOMETALLIC REAGENTS

David H. Brown Ripin

## 12.1  INTRODUCTION

The synthesis of organometallic reagents is a large topic that is the subject of many books and reviews.[1–4] This chapter covers useful methods for the synthesis of nucleophilic reagents commonly used in synthesis and is in no way meant to be comprehensive. For a very thorough treatment of this subject, see Schlosser's treatise.[1] Table 12.1 lists the metals discussed in this chapter, the common methods of synthesis of the active species and their uses. Included in the category of "nucleophilic" organometallic reagents are the metalated reagents used in cross-coupling reactions. The subset of metalated heterocycles has been reviewed.[2] Not included are transition-metal catalysts; these are discussed in the appropriate chapters on transformations.

[1]Schlosser, M.;Editor, *Organometallics in Synthesis: A Manual; 2nd ed*; Wiley: West Sussex, UK, 2002.
[2]Chinchilla, R.; Najera, C.; Yus, M. *Chem. Rev.* **2004**, *104*, 2667–2722.
[3]Crabtree, R.; Mongos, M. *Comprehensive Organometallic Chemistry III, 13 Volume Set*; Elsevier: Oxford, UK, 2006.
[4]Abel, E. W.; Stone, F. G. A.; Wilkinson, G.; Editors *Comprehensive Organometallic Chemistry II: A Review of the Literature 1982–1994, 14 Volume Set*; Pergamon: Oxford, UK, 1995.

*Practical Synthetic Organic Chemistry: Reactions, Principles, and Techniques*, First Edition.
Edited by Stéphane Caron.
© 2011 John Wiley & Sons, Inc. Published 2011 by John Wiley & Sons, Inc.

**TABLE 12.1   Common "Nucelophillic" Reagents and Their Uses**

| Metal | Methods of synthesis | Transmetalation precursors | Common Uses |
|-------|----------------------|----------------------------|-------------|
| Lithium | M−X exchange<br>Deprotonation<br>M−M exchange | Sn, Te | Deprotonations, metal–halogen exchange, addition to electrophiles, 1,2 addition to conjugated systems, transmetalation precursor |
| Boron | Hydrometalation<br>M−M exchange<br>M−X exchange | Li, Mg, Zr | Cross-couplings, oxidation of C−B bond to alcohol |
| Magnesium | M−X exchange<br>M−M exchange<br>Carbometalation<br>Hydrometalation | Li, Zr | Deprotonation of kinetically acidic protons, addition to electrophiles, 1,2 addition to conjugated systems |
| Aluminum | Carbometalation<br>Hydrometalation | Li, Zr | Reaction with acid chlorides, protonation or halogenation, cross-couplings |
| Silicon | M−M exchange<br>M−X exchange<br>Hydrosilation | Li, Mg | Allylations, protonation, cross-coupling |
| Titanium | M−M exchange<br>Homoenolate form | Li, Mg, Sn | Addition to carbonyls |
| Chromium | M−M exchange<br>M−X exchange | Li, Mg | Selective additions to carbonyls |
| Manganese | M−M exchange | Li, Mg | Conjugate additions |
| Iron | M−X exchange | | Carbonylations |
| Copper | M−M exchange | Li, Mg, Zn, Zr, Mn | Conjugate additions |
| Zinc | M−X exchange<br>M−M exchange<br>Homoenolate form | Li, Mg, Zr | Carbonyl additions |
| Zirconium | Hydrometalation | | Transmetalation precursor, cross-couplings |
| Indium | M−M exchange<br>M−X exchange | Sn | Allylations, cross-coupling |
| Tin | M−M exchange<br>$S_N2$<br>Hydrometalation<br>M−X exchange | Li, Mg | Cross-coupling, allylations, transmetalation precursor |
| Cerium | M−M exchange | Li, Mg | Addition to carbonyl with minimization of enolate formation |
| Bismuth | M−M exchange | Li, Mg | Cross-coupling |

## 12.2  SYNTHESIS OF "NUCLEOPHILIC" ORGANOMETALLIC REAGENTS

### 12.2.1  Lithium

***12.2.1.1  Deprotonation***  See Chapter 18 for a discussion of pKa and deprotonation, including examples of directed and enantioselective deprotonations. See Section 12.3 for a discussion of strategies for metalating heterocyclic compounds.

***12.2.1.2  Metal–Halogen Exchange***  Perhaps the most common method for generating aryllithium reagents is through a metal–halogen exchange reaction of an iodide or bromide with an alkyllithium reagent (e.g., *n*-BuLi or *t*-BuLi).[5] This method allows for the generation of an aryllithium with predictable regiochemistry. Useful solvents for the generation of organolithium reagents include aliphatic hydrocarbons, aromatic hydrocarbons, and ethers. THF can undergo ring opening if exposed to an organolithium for a prolonged period of time at ambient temperature, and toluene can be deprotonated at the benzylic position in the absence of more facile reaction pathways. When *t*-BuLi is employed, two equivalents of the reagent must be utilized: the first for metal–halogen exchange and the second for elimination of the resulting *t*-BuX.[6] The use of *n*-BuLi is highly preferred over *s*-BuLi and *t*-BuLi due to the relative ease and safety of handling. If the generation of butane is a concern in a large scale process, *n*-HexLi is a very useful replacement.

Metal–halogen exchange with lithium is a diffusion-controlled reaction that occurs faster than nucleophilic addition and at a competitive rate with deprotonation. This rate allows for the Barbier reaction wherein the organolithium reagent is generated in the presence of an electrophile.[7]

[5]Lipton, M. F.; Mauragis, M. A.; Maloney, M. T.; Veley, M. F.; VanderBor, D. W.; Newby, J. J.; Appell, R. B.; Daugs, E. D. *Org. Process Res. Dev.* **2003**, 7, 385–392.
[6]Bryce-Smith, D.; Blues, E. T. *Org. Synth.* **1973**, Coll. Vol. V, 1141–1145.
[7]Ennis, D. S.; Lathbury, D. C.; Wanders, A.; Watts, D. *Org. Process Res. Dev.* **1998**, 2, 287–289.

A less common method for the generation of organolithium reagents is the treatment of a halide with Li(0).[8] Phenyllithium can also be prepared by this method, although reflux is required.[9]

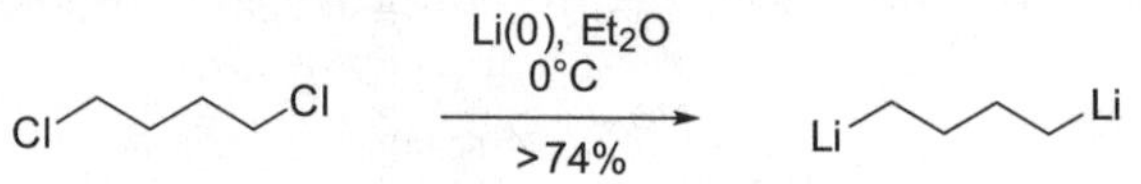

### 12.2.1.3 Metal–Metal Exchange
Organostannanes can be transmetalated to organolithium reagents on treatment with *n*-BuLi. This is not a desirable method as toxic stannanes need to be prepared and used in the process.[10]

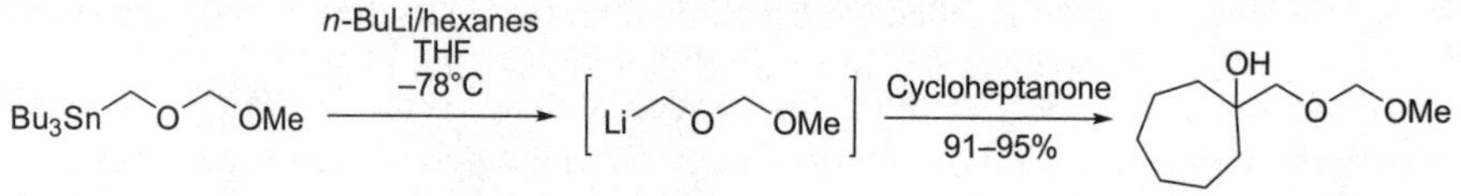

## 12.2.2 Boron

### 12.2.2.1 Hydroboration
One of the most common methods for the preparation of alkyl and vinyl boranes is through the hydroboration reaction (see Section 3.3.1.1.1 for a detailed discussion). An example is shown below.[11]

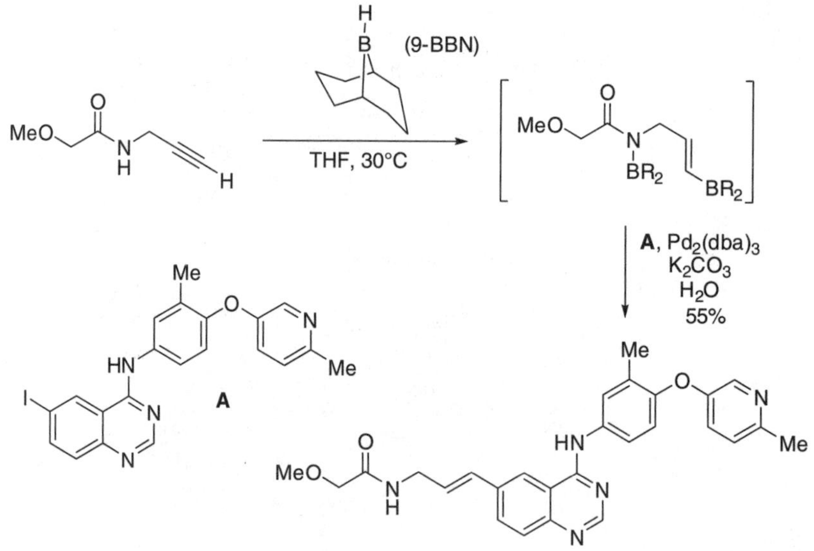

[8]Wender, P. A.; White, A. W.; McDonald, F. E. *Org. Synth.* **1992**, *70*, 204–214.
[9]Woodward, R. B.; Kornfeld, E. C. *Org. Synth.* **1949**, *29*, 44–46.
[10]Greco, M. N.; Rasmussen, C. R. *J. Org. Chem.* **1992**, *57*, 5532–5535.
[11]Dugger, R. W.; Ragan, J. A.; Brown Ripin, D. H. *Org. Process Res. Dev.* **2005**, *9*, 253–258.

This method can be used to synthesize allyl boranes as well.[12]

### 12.2.2.2  Metal–Metal Exchange

Boronic acid derivatives can be generated by treating an organolithium or Grignard reagent with a trialkylborate. This is the most common method for the preparation of boronic acids to be used in a Suzuki coupling.[13–15]

[12]Flamme, E. M.; Roush, W. R. *J. Am. Chem. Soc.* **2002**, *124*, 13644–13645.

[13]Caron, S.; Hawkins, J. M. *J. Org. Chem.* **1998**, *63*, 2054–2055.

[14]Winkle, D. D.; Schaab, K. M. *Org. Process Res. Dev.* **2001**, *5*, 450–451.

[15]Jacks, T. E.; Belmont, D. T.; Briggs, C. A.; Horne, N. M.; Kanter, G. D.; Karrick, G. L.; Krikke, J. J.; McCabe, R. J.; Mustakis, J. G.; Nanninga, T. N.; Risedorph, G. S.; Seamans, R. E.; Skeean, R.; Winkle, D. D.; Zennie, T. M. *Org. Process Res. Dev.* **2004**, *8*, 201–212.

Heterocycles containing nitrogen can be derivatized to stable boranes such as the pyridylborane shown below. A $200\,\text{kg}$[5] preparation using the organolithium reagent is depicted; the organolithium will "walk," resulting in a mixture of products if the temperature rises much above $-78°C$. The same transformation can be accomplished through the Grignard reagent generated by metalation of the bromide with $i$-PrMgCl at $0°C$; the Grignard reagent is stable up to $15°C$.[16]

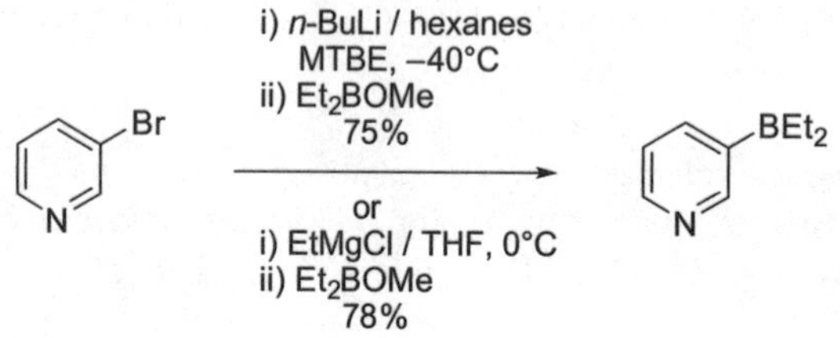

Vinylboranes can also be prepared using a transmetalation, as demonstrated by the $2.3\,\text{kg}$ preparation below.[17] The vinylborane was used directly in a Suzuki coupling reaction.

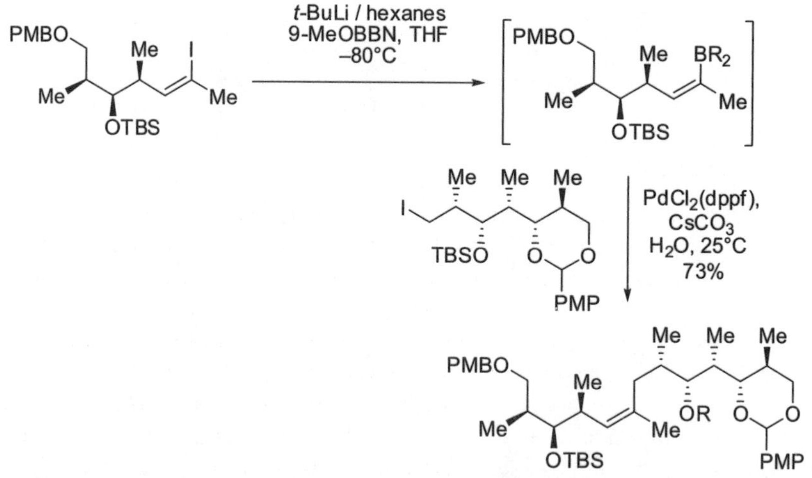

The zirconium-catalyzed hydroboration of acetylenes with pinacol borane can arguably be considered a transmetalation from Zr to B following hydrozirconation, and is an effective method for the hydroboration of acetylenes.[18]

[16]Cai, W.; Ripin, D. H. B. *Synlett* **2002**, 273–274.

[17]Mickel, S. J.; Sedelmeier, G. H.; Niederer, D.; Schuerch, F.; Koch, G.; Kuesters, E.; Daeffler, R.; Osmani, A.; Seeger-Weibel, M.; Schmid, E.; Hirni, A.; Schaer, K.; Gamboni, R.; Bach, A.; Chen, S.; Chen, W.; Geng, P.; Jagoe, C. T.; Kinder, F. R., Jr.; Lee, G. T.; McKenna, J.; Ramsey, T. M.; Repic, O.; Rogers, L.; Shieh, W.-C.; Wang, R.-M.; Waykole, L. *Org. Process Res. Dev.* **2004**, *8*, 107–112.

[18]Batt, D. G.; Houghton, G. C.; Daneker, W. F.; Jadhav, P. K. *J. Org. Chem.* **2000**, *65*, 8100–8104.

#### 12.2.2.3 Cross-Coupling with $R_2B$-$BR_2$

A mild method for the formation of boronic acids is via the Pd-catalyzed coupling of an aryl halide and bis(pinicolato)diborane. The cost of the reagent and potential for dimerization (through a Suzuki coupling) are issues to consider when selecting this methodology.[19] Only one of the two borons in the reagent is transferred.

#### 12.2.2.4 Other

Potassium trifluoroborates can be generated either from boronic acids[20] or from some hydroboration products[21] using $KHF_2$.

### 12.2.3 Magnesium

#### 12.2.3.1 Metal–Halogen Exchange

The most common method for generating a Grignard reagent is the reaction of Mg(0) with a halide. The reaction is highly exothermic, and requires elevated temperatures to proceed. The major difficulty with this process is initiating the reaction, with a hazardous situation arising if too much halide is added prior to initiation. Some useful procedures are referenced here, with a variety of initiators including phenyl Grignard formation with no initiation,[22] with iodine,[23] and with DIBAL-H activation.[24] Other useful initiators are TMSCl,[25] bromine, diiodoethane, dibromoethane,[26] or a Grignard reagent itself such as MeMgCl. A procedure for preparing an unsolvated Grignard reagent, n-BuMgCl, in methylcyclohexane with Mg(0) powder has also been reported.[6] Typically, 10% or so of

[19]Zembower, D. E.; Zhang, H. *J. Org. Chem.* **1998**, *63*, 9300–9305.

[20]Vedejs, E.; Chapman, R. W.; Fields, S. C.; Lin, S.; Schrimpf, M. R. *J. Org. Chem.* **1995**, *60*, 3020–3027.

[21]Clay, J. M.; Vedejs, E. *J. Am. Chem. Soc.* **2005**, *127*, 5766–5767.

[22]Brenner, M.; la Vecchia, L.; Leutert, T.; Seebach, D. *Org. Synth.* **2003**, *80*, 57–65.

[23]Braun, M.; Graf, S.; Herzog, S. *Org. Synth.* **1995**, *72*, 32–37.

[24]Tilstam, U.; Weinmann, H. *Org. Process Res. Dev.* **2002**, *6*, 384–393.

[25]Ace, K. W.; Armitage, M. A.; Bellingham, R. K.; Blackler, P. D.; Ennis, D. S.; Hussain, N.; Lathbury, D. C.; Morgan, D. O.; O'Connor, N.; Oakes, G. H.; Passey, S. C.; Powling, L. C. *Org. Process Res. Dev.* **2001**, *5*, 479–490.

[26]Aki, S.; Haraguchi, Y.; Sakikawa, H.; Ishigami, M.; Fujioka, T.; Furuta, T.; Minamikawa, J.-i. *Org. Process Res. Dev.* **2001**, *5*, 535–538.

the halide is premixed with the Mg(0) and the reaction is initiated with heat and an added initiator if needed, followed by slow addition of the remaining halide. Ethereal solvents such as THF are most commonly employed in the process and will not react with the Grignard reagent; unlike the case of organolithium reagents.

A less-appreciated method for the generation of aryl Grignard reagents is metal–halogen exchange with an alkyl Grignard.[27] This method has the advantages of proceeding at a low temperature (0–25°C) and circumventing the initiation problems of the reaction with Mg(0).

### 12.2.3.2 Metal–Metal Exchange

A vinyl Grignard reagent can be generated by hydrozirconation followed by transmetalation with an alkyl Grignard reagent.[28]

Organolithium reagents can also be transmetalated to the Grignard reagent on treatment with a Mg(II) salt, most commonly $MgCl_2$ or $MgBr_2$.[29]

[27]Leazer, J. L., Jr.; Cvetovich, R.; Maloney, K. M.; Danheiser, R. L. *Org. Synth.* **2005**, *82*, 115–119.

[28]Boulton, L. T.; Brick, D.; Fox, M. E.; Jackson, M.; Lennon, I. C.; McCague, R.; Parkin, N.; Rhodes, D.; Ruecroft, G. *Org. Process Res. Dev.* **2002**, *6*, 138–145.

[29]Gawley, R. E.; Zhang, P. *J. Org. Chem.* **1996**, *61*, 8103–8112.

***12.2.3.3  Hydromagnesiation***  Although rare, propargylic alcohols can be directly hydromagnesiated in the presence of a catalyst, rather than hydrozirconating followed by transmetalation.[30]

***12.2.3.4  Carbomagnesiation***  Propargylic alcohols can be carbomagnesiated in a process that can be used as a fragment coupling reaction.[31]

Carbomagnesiation reaction can also be accomplished in the presence of a Ti catalyst.[32]

### 12.2.4  Aluminum

***12.2.4.1  Carboalumination***  The carboalumination of acetylenes to vinylalanes is a common method for the preparation of complex metalated olefin species to be used in cross-coupling reactions and other processes. The process is generally conducted using a zirconium catalyst.[33] Interestingly, this process can be accelerated by the addition of water.[34] The reaction is highly regioselective in the case of terminal acetylenes, and can show useful

[30]Ogasa, T.; Ikeda, S.; Sato, M.; Tamaoki, K. *Preparation of N-(2,2,5,5-tetramethylcyclopentanecarbonyl)-(S)-1,1-diaminoethane p-toluenesulfonate as a sweetener intermediate* JP02233651, 1990, 5 pp.

[31]Bury, P.; Hareau, G.; Kocienski, P.; Dhanak, D. *Tetrahedron* **1994**, *50*, 8793–8808.

[32]Nii, S.; Terao, J.; Kambe, N. *J. Org. Chem.* **2004**, *69*, 573–576.

[33]Rand, C. L.; Van Horn, D. E.; Moore, M. W.; Negishi, E. *J. Org. Chem.* **1981**, *46*, 4093–4096.

[34]Wipf, P.; Waller, D. L.; Reeves, J. T. *J. Org. Chem.* **2005**, *70*, 8096–8102.

levels of regioselectivity with sterically unsymmetrical internal alkynes (e.g., methyl vs. α-branched alkyl).

Treating the vinylalane with an organolithium reagent produces the more nucleophilic alanate.[31]

### 12.2.4.2 *Hydroalumination*

Acetylenes can be hydroaluminated using DIBAL-H and LAH in a process analogous to the carboalumination reaction. Selectivity is seen in the case of terminal olefins or olefins with directing groups such as propargylic alcohols.[35,36]

[35]Negishi, E.; Takahashi, T.; Baba, S. *Org. Synth.* **1988**, *66*, 60–66.
[36]Havranek, M.; Dvorak, D. *J. Org. Chem.* **2002**, *67*, 2125–2130.

Red-Al is generally employed in the *trans* reduction of acetylenes to olefins and proceeds via a vinyl aluminate.[37,38]

The premixing of DIBAL-H with triethylamine results in deprotonation of the acetylenic proton to generate the alkynyl alane.[39]

### *12.2.4.3 Transmetalation* Organoaluminum reagents can be produced by transmetalation of a hydrozirconation product with $AlCl_3$.[40]

### 12.2.5  Silicon

Organosilicon reagents have found extensive use in organic synthesis. This is likely due to the large supply of silicon-containing starting materials, the relatively high stability of the organosilanes, and the diversity of reactions in which they can participate. Due to the high stability of these reagents, the derivitization of an organosilicon reagent into another organosilicon reagent is a common preparation in the literature. For the sake of brevity, this chapter will only focus on methods that generate the C−Si bond directly.

[37]Jones, A. B. *J. Org. Chem.* **1992**, *57*, 4361–4367.

[38]Ripin, D. H. B.; Bourassa, D. E.; Brandt, T.; Castaldi, M. J.; Frost, H. N.; Hawkins, J.; Johnson, P. J.; Massett, S. S.; Neumann, K.; Phillips, J.; Raggon, J. W.; Rose, P. R.; Rutherford, J. L.; Sitter, B.; Stewart, A. M., III; Vetelino, M. G.; Wei, L. *Org. Process Res. Dev.* **2005**, *9*, 440–450.

[39]Blanchet, J.; Bonin, M.; Micouin, L.; Husson, H. P. *J. Org. Chem.* **2000**, *65*, 6423–6426.

[40]Carr, D. B.; Schwartz, J. *J. Am. Chem. Soc.* **1979**, *101*, 3521–3531.

**12.2.5.1  Metal–Metal Exchange**   Arylsilanes can be generated from aryllithium or arylmagnesium reagents and a silylchloride.[41,42] This represents the most common way to generate aryl,[43] allyl, vinyl,[44] and alkyl silanes. Silyl imidazoles, siloxanes, and silyl triflates can also be used as the electrophilic silicon source in these processes.

**12.2.5.2  Hydrosilylation**   Acetylenes and olefins can be hydrosilylated in the presence of a catalyst, often with high levels of selectivity, as in the example below.[45] In this case, the vinylsilane intermediate is utilized in a cross-coupling reaction *in situ*. [RhCl$_2$($p$-cymene)]$_2$ has also been reported for this transformation.[46]

**12.2.5.3  Metal–Halogen Exchange**   Arylsiloxanes can be generated from the aryl halide by treatment with a palladium catalyst in the presence of HSi(OMe)$_3$.[47] This process, coupled with a copper-mediated amination of the siloxane, offers a mild, room-temperature alternative to the widely used Pd-catalyzed amination processes.

[41] Dondoni, A.; Merino, P. *Org. Synth.* **1995**, *72*, 21–31.
[42] Beak, P.; Kerrick, S. T.; Wu, S.; Chu, J. *J. Am. Chem. Soc.* **1994**, *116*, 3231–3239.
[43] Haebich, D.; Effenberger, F. *Synthesis* **1979**, 841–876.
[44] Denmark, S. E.; Ober, M. H. *Aldrichimica Acta* **2003**, *36*, 75–85.
[45] Denmark, S. E.; Wang, Z. *Org. Synth.* **2005**, *81*, 54–62.
[46] Angle, S. R.; Neitzel, M. L. *J. Org. Chem.* **2000**, *65*, 6458–6461.
[47] Anilkumar, R.; Chandrasekhar, S.; Sridhar, M. *Tetrahedron Lett.* **2000**, *41*, 5291–5293.

### 12.2.5.4   Use of Nucleophilic Silicon Reagents

Reagents featuring a metalated carbon substituted with silicon or featuring metalated silicon, have been demonstrated to be useful for the synthesis of organosilanes. The example below shows a readily available nucleophilic silicon-containing reagent being utilized in a cross-coupling reaction.[48]

### 12.2.6   Titanium

### 12.2.6.1   Metal–Metal Exchange

Metal–metal exchange is the primary method for the synthesis of alkyl or aryl titanium reagents. Some organolithium and organomagnesium reagents can be reacted directly with $TiCl_4$, although in many cases this results in the reduction of titanium. The use of an organozinc reagent avoids this reduction. An example is the preparation of $MeTiCl_3$. When synthesizing the etherate of the titanium reagent, MeLi can be employed.[49] In the absence of an ethereal solvent, $Me_2Zn$ must be used.[50] Chlorotitanium alkoxides can be used in place of $TiCl_4$.

A common reagent used in the preparation of olefins is dimethyl titanocene. This reagent is prepared by the treatment of $Cp_2TiCl_2$ with 2 equivalents of MeMgCl.[51] This is an alternative to the Tebbe reagent.[52]

[48]Hayashi, T.; Fujiwa, T.; Okamoto, Y.; Katsuro, Y.; Kumada, M. *Synthesis* **1981**, 1001–1003.

[49]Reetz, M. T.; Kyung, S. H.; Huellmann, M. *Tetrahedron* **1986**, *42*, 2931–2935.

[50]Reetz, M. T.; Westermann, J.; Steinbach, R. *Angew. Chem., Int. Ed. Engl.* **1980**, *19*, 900.

[51]Brands, K. M. J.; Payack, J. F.; Rosen, J. D.; Nelson, T. D.; Candelario, A.; Huffman, M. A.; Zhao, M. M.; Li, J.; Craig, B.; Song, Z. J.; Tschaen, D. M.; Hansen, K.; Devine, P. N.; Pye, P. J.; Rossen, K.; Dormer, P. G.; Reamer, R. A.; Welch, C. J.; Mathre, D. J.; Tsou, N. N.; McNamara, J. M.; Reider, P. J. *J. Am. Chem. Soc.* **2003**, *125*, 2129–2135.

[52]Paquette, L. A.; McLaughlin, M. L. *Org. Synth.* **1990**, *68*, 220–226.

Titanium tetrachloride can also exchange with tin in allylstannanes. The exchange occurs with retention of stereochemistry and transfer of allylic regiochemistry.[53]

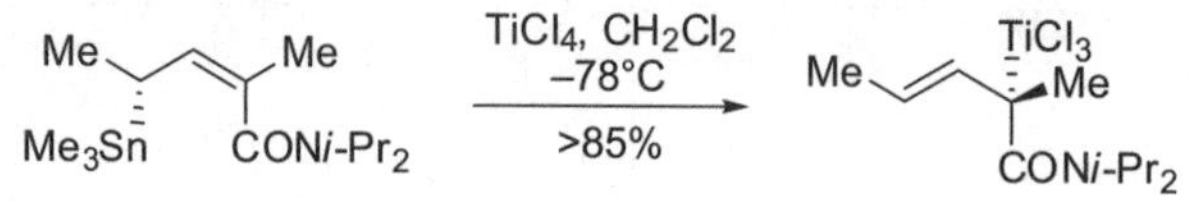

### 12.2.6.2 Other

A useful method for the generation of a titanium homoenolate is shown below.[54]

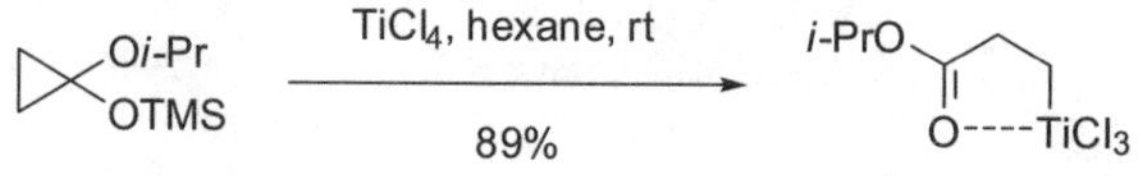

## 12.2.7 Chromium

### 12.2.7.1 Metal–Halogen and Metal–Metal Exchange

In the presence of a nickel(II) catalyst, vinyl or aryl halides and triflates can be converted to the organochromium species via the organonickel intermediate. Two examples are shown below.[55,56] Catalysts that are useful include Ni(II) salts such as NiCl$_2$ and Ni(acac)$_2$; palladium catalysts are less effective in this transformation, particularly in the case of the triflate. Nickel is converted to Ni(0) as the chromium goes from Cr(II) to Cr(III).

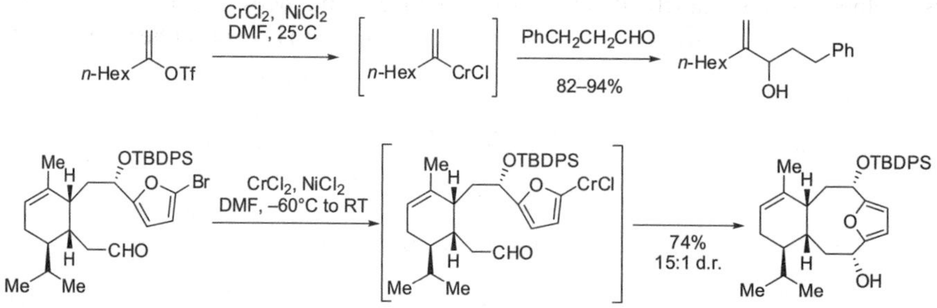

Allylic chromium reagents, generated *in situ* with an excess of Cr(II) and an allylic bromide, can be used in allylic addition reactions with aldehydes. The process suffers from the drawback that a large amount of chromium is required. Allylic chromium reagents can equilibrate olefin geometry and typically result in *anti*-addition products in cases where selectivity is an issue.[57]

[53]Kraemer, T.; Schwark, J. R.; Hoppe, D. *Tetrahedron Lett.* **1989**, *30*, 7037–7040.

[54]Nakamura, E.; Oshino, H.; Kuwajima, I. *J. Am. Chem. Soc.* **1986**, *108*, 3745–3755.

[55]Takai, K.; Sakogawa, K.; Kataoka, Y.; Oshima, K.; Utimoto, K. *Org. Synth.* **1995**, *72*, 180–188.

[56]Chen, X.-T.; Bhattacharya, S. K.; Zhou, B.; Gutteridge, C. E.; Pettus, T. R. R.; Danishefsky, S. J. *J. Am. Chem. Soc.* **1999**, *121*, 6563–6579.

[57]Nowotny, S.; Tucker, C. E.; Jubert, C.; Knochel, P. *J. Org. Chem.* **1995**, *60*, 2762–2772.

The Fischer-type carbenes of chromium can be generated by the reaction of an organolithium or Grignard reagent with $Cr(CO)_6$ followed by alkylation.[58]

$$Cr(CO)_6 \xrightarrow[\text{83\%}]{\substack{\text{i) MeLi, Et}_2\text{O, reflux} \\ \text{ii) Me}_3\text{O}^+ \text{BF}_4^-, \text{H}_2\text{O, RT}}} (OC)_5Cr= \overset{OMe}{\underset{Me}{\big\langle}}$$

## 12.2.8    Manganese

Organomanganese reagents are rarely used in synthesis, but do provide interesting selectivity in some additions. These relatively unreactive organometallic reagents can add in a 1,4-manner to alkylidene malonates, and can even add to an acid chloride in the presence of an aldehyde. The large quantities of manganese required are a significant drawback to these reagents.

One example of a useful organomanganese reagent is shown in the example below. This method is very practical, catalytic in copper, and does not require cryogenic conditions. The process is described as a copper-catalyzed addition rather than addition of a cuprate.[59]

*12.2.8.1  Metal–Metal Exchange*  The primary method for the synthesis of organomanganese reagents is via the addition of an organolithium or Grignard reagent to $MnX_2$, where $MnCl_2$ is the most convenient due to its commercial availability.[60]

[58]Hegedus, L. S.; McGuire, M. A.; Schultze, L. M. *Org. Synth.* **1987**, *65*, 140–145.

[59]Alami, M.; Marquais, S.; Cahiez, G. *Org. Synth.* **1995**, *72*, 135–146.

[60]Friour, G.; Cahiez, G.; Normant, J. F. *Synthesis* **1984**, 37–40.

By increasing the ratio of organolithium to $MnX_2$, dialkyl organomanganese reagents $R_2Mn$, and organomanganates $R_3MnLi$ can be generated. The reactivity of these reagents varies.[61]

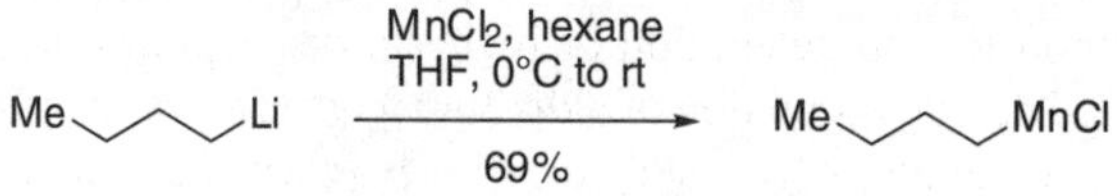
MnCl2, hexane
THF, 0°C to rt
Me
Li
69%
Me
MnCl

### 12.2.9 Iron

*12.2.9.1 Metal–Halogen Exchange* The addition of $Na_2Fe(CO)_4$ to an alkyl halide or acid chloride can be followed by alkylation, protonation, or oxidation of the acyl-iron intermediate to make ketones, aldehydes, or esters, respectively. A representative procedure published in *Organic Syntheses* is depicted below.[62] The same acyl-iron intermediate can be generated through the addition of an alkyllithium to $Fe(CO)_5$.

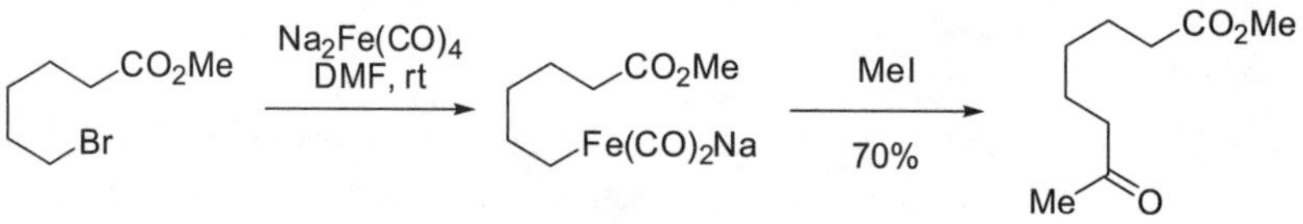
CO2Me
Br
Na2Fe(CO)4
DMF, rt
CO2Me
Fe(CO)2Na
MeI
70%
CO2Me
Me
O

### 12.2.10 Copper

Cuprates are unique organometallic reagents that are useful in 1,4-addition reactions and displacement reactions. The primary method of synthesis of cuprates is via a transmetalation reaction.

*12.2.10.1 Metal–Metal Exchange* A method for the generation of a catalytically active cuprate reagent from an aryl Grignard at 0°C is depicted below and was demonstrated on 11 kg scale.[63]

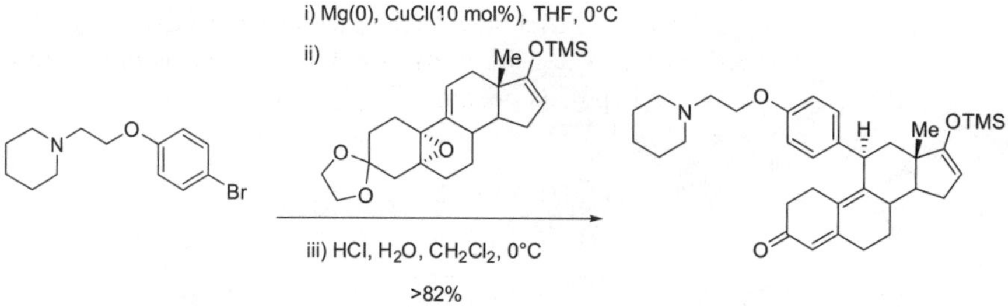
i) Mg(0), CuCl(10 mol%), THF, 0°C
ii)
Me
OTMS
N
O
Br
O
O
O
iii) HCl, H2O, CH2Cl2, 0°C
>82%
N
O
H
Me
OTMS
O

[61]Cahiez, G.; Alami, M. *Tetrahedron* **1989**, *45*, 4163–4176.
[62]Finke, R. G.; Sorrell, T. N. *Org. Synth.* **1980**, *59*, 102–112.
[63]Larkin, J. P.; Wehrey, C.; Boffelli, P.; Lagraulet, H.; Lemaitre, G.; Nedelec, A.; Prat, D. *Org. Process Res. Dev.* **2002**, *6*, 20–27.

One example of a useful cuprate generated from an organomanganese reagent is shown in Section 12.2.8. This method is very practical, catalytic in copper, and does not require cryogenic conditions. The process is described as a copper-catalyzed addition rather than addition of a cuprate.[59]

A complex method for the generation of higher-order cuprates described in *Organic Syntheses* is depicted below. Formation of a dialkylzincate followed by addition of $Me_2Cu(CN)Li$ generates the cuprate, which adds in a 1,4-manner to an enone, with the enolate trapped as the silyl enol ether to free copper to act in the catalytic cycle.[64]

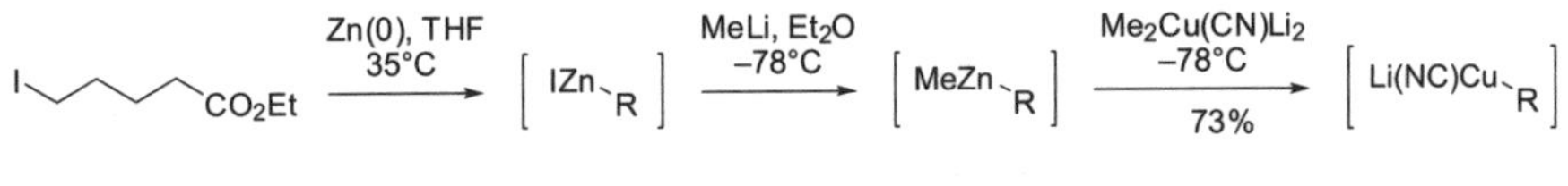

$$I\diagdown\diagup\diagdown CO_2Et \xrightarrow[35°C]{Zn(0),\ THF} \left[ IZn{\diagdown}R \right] \xrightarrow[-78°C]{MeLi,\ Et_2O} \left[ MeZn{\diagdown}R \right] \xrightarrow[73\%]{Me_2Cu(CN)Li_2\ \ -78°C} \left[ Li(NC)Cu{\diagdown}R \right]$$

A powerful method for the generation of vinyl cuprates is hydrozirconation followed by transmetalation with copper. This method has proven very useful in the synthesis of prostaglandins.[65]

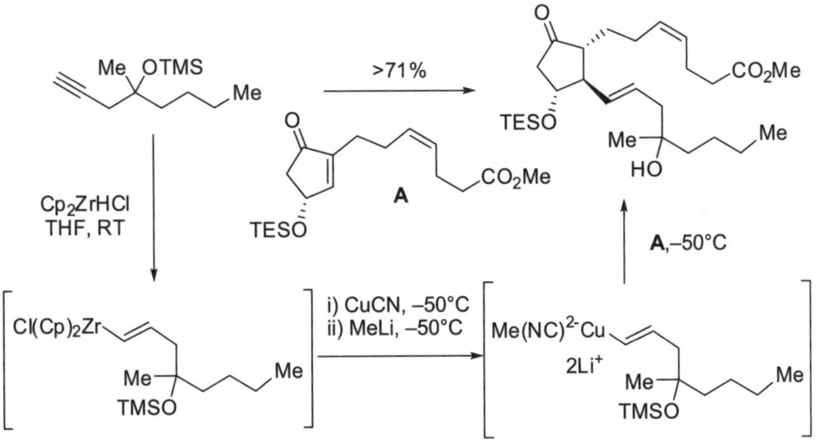

A method for the transmetalation of zirconates in the presence of catalytic quantities of copper has been developed. The cuprates are useful for 1,4-additions, additions to acid chlorides, allylic substitution, and epoxide openings.[66]

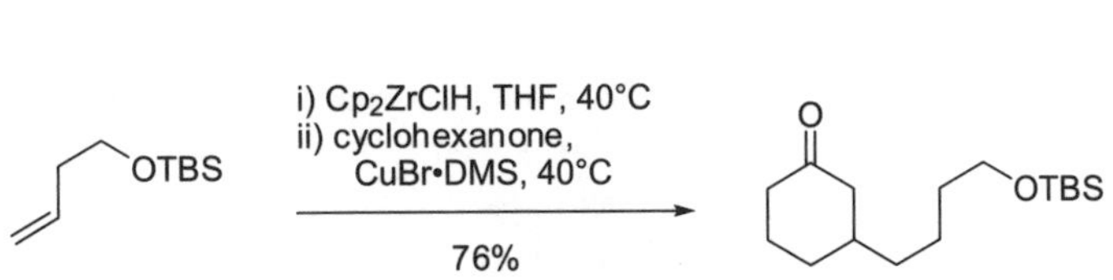

i) Cp2ZrClH, THF, 40°C
ii) cyclohexanone, CuBr·DMS, 40°C

76%

[64]Lipshutz, B. H.; Wood, M. R.; Tirado, R. *Org. Synth.* **1999**, *76*, 252–262.

[65]Dygos, J. H.; Adamek, J. P.; Babiak, K. A.; Behling, J. R.; Medich, J. R.; Ng, J. S.; Wieczorek, J. J. *J. Org. Chem.* **1991**, *56*, 2549–2552.

[66]Wipf, P.; Xu, W.; Smitrovich, J. H.; Lehmann, R.; Venanzi, L. M. *Tetrahedron* **1994**, *50*, 1935–1954.

Finally, in a transmetalation bonanza, the addition of $Me_3ZnLi$ allows for the transmetalation of the copper enolate that results from 1,4-addition. This frees the copper to participate in the catalytic cycle, and generates a more reactive enolate that can be alkylated more readily than the copper enolate.[67]

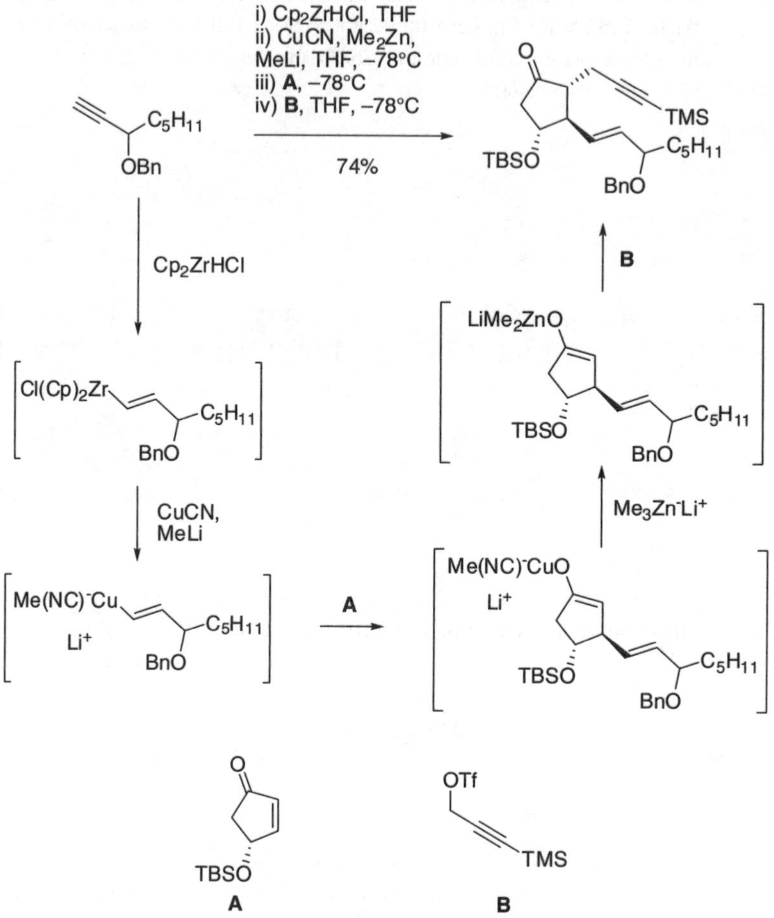
i) Cp2ZrHCl, THF
ii) CuCN, Me2Zn, MeLi, THF, –78°C
iii) A, –78°C
iv) B, THF, –78°C
74%
C5H11
OBn
O
TMS
TBSO
BnO
Cp2ZrHCl
B
LiMe2ZnO
Cl(Cp)2Zr
CuCN, MeLi
Me3Zn-Li+
Me(NC)-Cu
Li+
A
Me(NC)-CuO
OTf
TMS
TBSO
A
B

## 12.2.11 Zinc

The most practical methods for the production of alkylzinc reagents are through metal–halogen exchange using Zn(0) or metal–metal exchange from the reaction of organolithium, magnesium, or zirconium regents. A number of particularly mild but less practical methods are available for the generation of organozinc reagents of varying reactivity in the presence of a wide variety of functional groups.

Zincates are very useful reagents for use in catalyzed cross-coupling reactions and mild additions to carbonyls and are particularly attractive due to the wide functional group tolerance of these compounds.

[67]Lipshutz, B. H.; Wood, M. R. *J. Am. Chem. Soc.* **1994**, *116*, 11689–11702.

***12.2.11.1  Metal–Halogen Exchange***    The simplest method for the generation of alkylzincs is from Zn(0) and an iodide. These reactions are exothermic and suffer from the same initiation difficulties as Grignard reagents. Activated zinc, such as zinc–copper couple (Zn(Cu)), or the use of activating agents such as dibromoethane, TMSCl, and MsOH are useful to avoid these difficulties.[68,69] The middle reaction is an example of a Blaise reaction, the nitrile equivalent of a Reformatsky reaction, and demonstrates the ability to generate a zincate in the presence of an electrophile.[70]

The Simmons–Smith reagent, utilized in cyclopropanation reactions, can be generated by the reaction of diiodomethane with either zinc(0)[71] or diethyl zinc.[72]

***12.2.11.2  Metal–Metal Exchange***    Organolithium reagents will transmetalate to the zincate on treatment with ZnCl$_2$ or ZnBr$_2$. This method is not tolerant of reactive functional groups.[73,74] The use of an alkylzinc chloride in the third example is noteworthy, as the 2-lithiooxazole is known to ring open at temperatures much higher than –78°C, but the zinc reagent is stable at the 60°C reaction temperature.[75]

[68]Tamaru, Y.; Ochiai, H.; Nakamura, T.; Yoshida, Z. *Org. Synth.* **1989**, *67*, 98–104.

[69]Yeh, M. C. P.; Chen, H. G.; Knochel, P. *Org. Synth.* **1992**, *70*, 195–203.

[70]Choi, B. S.; Chang, J. H.; Choi, H.-w.; Kim, Y. K.; Lee, K. K.; Lee, K. W.; Lee, J. H.; Heo, T.; Nam, D. H.; Shin, H. *Org. Process Res. Dev.* **2005**, *9*, 311–313.

[71]Rieke, R. D.; Bales, S. E.; Hudnall, P. M.; Poindexter, G. S. *Org. Synth.* **1980**, *59*, 85–94.

[72]Charette, A. B.; Lebel, H. *Org. Synth.* **1999**, *76*, 86–100.

[73]Jensen, A. E.; Kneisel, F.; Knochel, P. *Org. Synth.* **2003**, *79*, 35–42.

[74]Smith, A. P.; Savage, S. A.; Love, J. C.; Fraser, C. L. *Org. Synth.* **2002**, *78*, 51–62.

[75]Reeder, M. R.; Gleaves, H. E.; Hoover, S. A.; Imbordino, R. J.; Pangborn, J. J. *Org. Process Res. Dev.* **2003**, *7*, 696–699.

Acetylenes can be hydrozirconated and transmetalated to zinc to generate the vinyl zinc species.[76]

**12.2.11.3   Other**   A method for the generation of the zinc homoenolate below is described in *Organic Syntheses*.[77]

## 12.2.12   Zirconium

Alkylzirconium reagents can be transmetalated to aluminum, boron, copper, nickel, palladium, tin, and zinc. When combined with the powerful hydrozirconation reaction, this is a very effective method for the synthesis of a variety of substituted olefins.

**12.2.12.1   Hydrozirconation**   By far the most common method for the creation of a carbon–zirconium bond is via the hydrozirconation process.[78] Acetylenes can be hydrozirconated followed by transmetalation, direct cross-coupling, or conversion to

[76]Wipf, P.; Xu, W. *Org. Synth.* **1997**, *74*, 205–211.
[77]Nakamura, E.; Kuwajima, I. *Org. Synth.* **1988**, *66*, 43–51.
[78]Wipf, P.; Jahn, H. *Tetrahedron* **1996**, *52*, 12853–12910.

oxidized products.[28,76] The reaction conditions are strongly reducing, therefore ketones, aldehydes, esters, nitriles, and epoxides are generally not tolerated.

The hydrozirconation of internal alkynes results in metalation at the less sterically hindered position. In the presence of a sub-stoichiometric amount of zirconium hydride, the product ratio is a result of the kinetic selectivity between the two positions. In the presence of excess reagent, the products will equilibrate to the thermodynamically dictated ratio, generally with higher selectivity for the less-hindered position. The exception to this generalization occurs when isomerization to the end of the carbon chain is possible (*see below*).

Hydrozirconation can also be accomplished using $Cp_2Zr(Cl)i$-Bu, generated *in situ* by the reaction of $t$-BuMgCl with $Cp_2ZrCl_2$.[79]

The hydrozirconation of olefins results in an alkylzirconium intermediate in which the zirconium can freely migrate and will isomerizes to the end of a chain, limiting its synthetic utility.[80]

[79] Makabe, H.; Negishi, E. *Eur. J. Org. Chem.* **1999**, 969–971.
[80] Gibson, T.; Tulich, L. *J. Org. Chem.* **1981**, *46*, 1821–1823.

### 12.2.13   Indium

Allylindium reagents, generated from an allylic mesylate, can be added to aldehydes in the presence of a palladium catalyst. The example below is representative.[81]

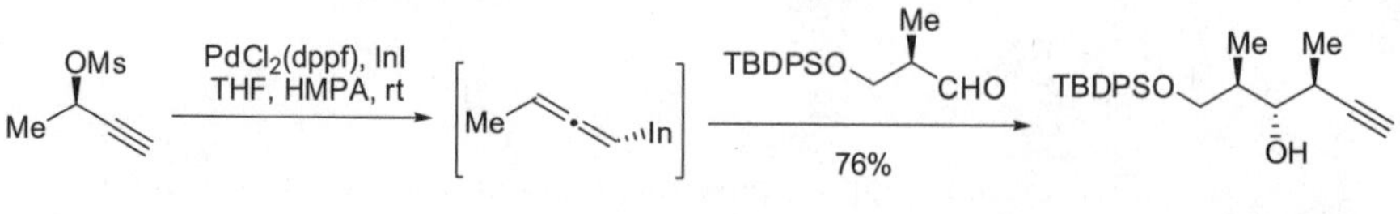

Allyl stannanes can be transmetalated using $InCl_3$. This can occur in allylic systems via an anti-$S_E2'$ mechanism as shown below, leading to a regioselectivity and diastereoselectivity unique from the originating allylstannane.[82]

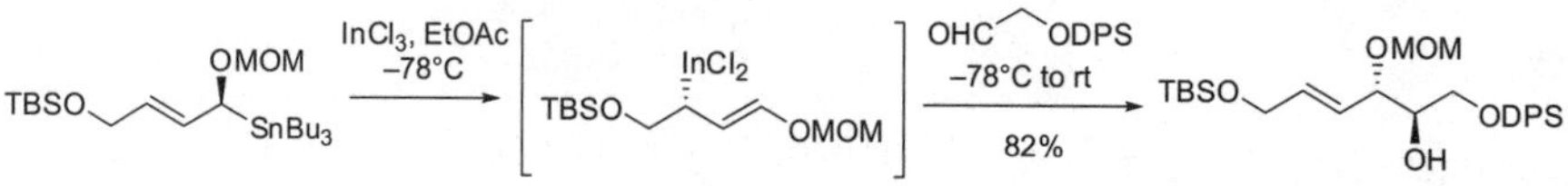

### 12.2.14   Tin

*12.2.14.1   Metal–Metal Exchange*    Grignard reagents and organolithium reagents react with tin chlorides to make organostannanes.[83,84]

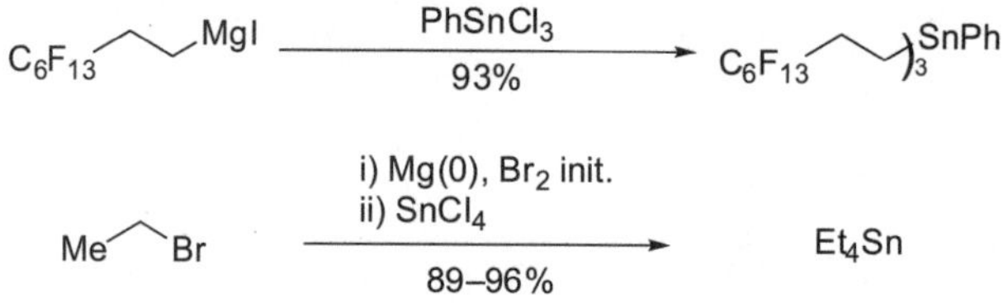

*12.2.14.2   Nucleophilic Sn*    Metalated stannanes (e.g., $Me_3SnLi$, $Bu_3SnLi$) can be prepared from the tin hydride and LDA, LiHMDS, NaHMDS, or KHMDS or by treating the tin chloride or bromide with lithium or sodium. They can displace leaving groups or add to electrophilic functionality to generate a variety of organostannanes. Some displacement reactions are shown below.[85,86] The procedure in the second example is a simpler preparation of the metalated tin species. In the case of the chiral propargylic mesylate, the stereochemistry of the starting material dictates the stereochemistry of the product, with the mesylate leaving anti-periplanar to the incoming tin nucleophiles.

[81]Johns, B. A.; Grant, C. M.; Marshall, J. A. *Org. Synth.* **2003**, *79*, 59–71.

[82]Marshall, J. A.; Garofalo, A. W. *J. Org. Chem.* **1996**, *61*, 8732–8738.

[83]Crombie, A.; Kim, S.-Y.; Hadida, S.; Curran, D. P. *Org. Synth.* **2003**, *79*, 1–10.

[84]Van Der Kerk, G. J. M.; Luijten, J. G. A. *Org. Synth.* **1963**, Coll. Voll. V, 881–883.

[85]Newcomb, M.; Courtney, A. R. *J. Org. Chem.* **1980**, *45*, 1707–1708.

[86]Fields, S. C.; Parker, M. H.; Erickson, W. R. *J. Org. Chem.* **1994**, *59*, 8284–8287.

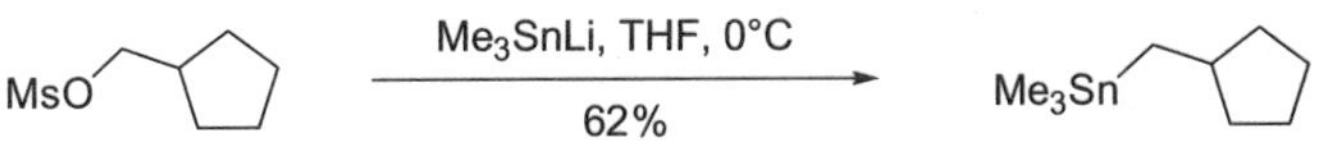

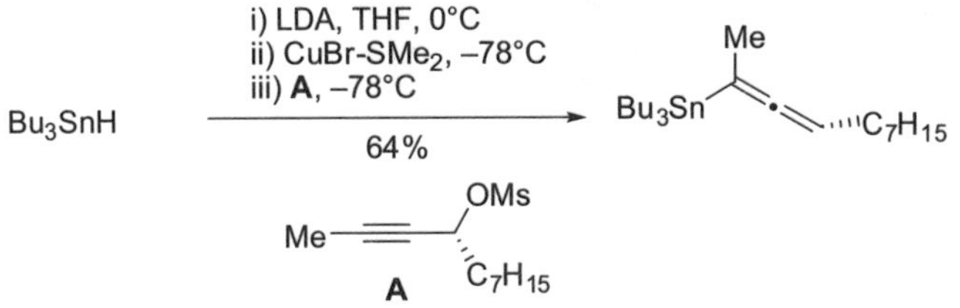

Metalated stannanes can participate in a 1,4-addition to an unsaturated ketone, as shown below.[87] In the case of aldehydes, 1,2-addition is observed.[82]

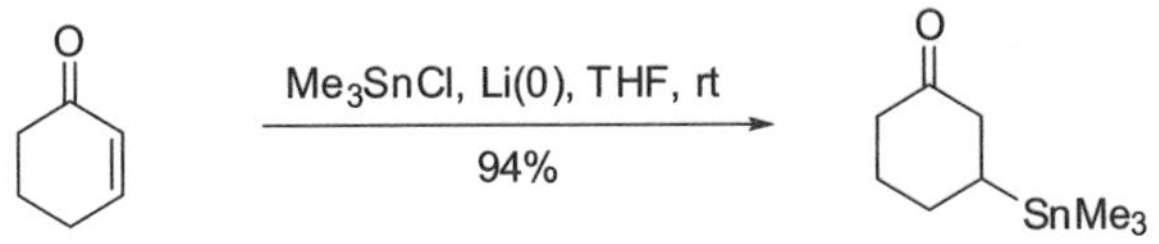

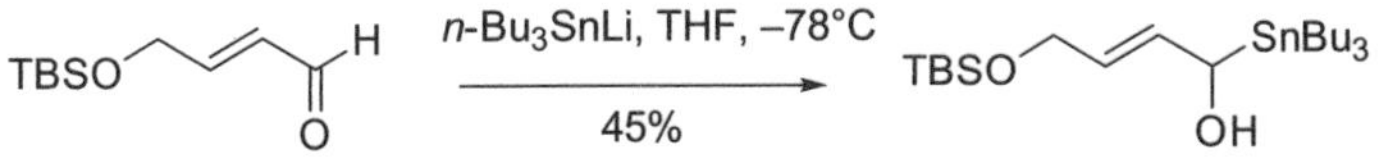

Acyl stannanes can be prepared via the addition of a metalated stannane to an ester or thioester.[88]

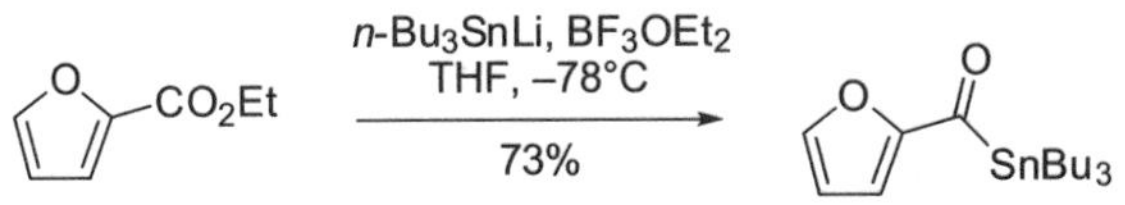

***12.2.14.3 Cross-Coupling with R₃Sn-SnR₃*** A mild method for the formation of stannanes is via the Pd-catalyzed coupling of an aryl halide and distannane. This transformation may be better achieved via a Suzuki coupling (Section 6.2). The cost of the reagent and potential for dimerization (through a Stille coupling) are issues to consider when selecting this methodology.[19]

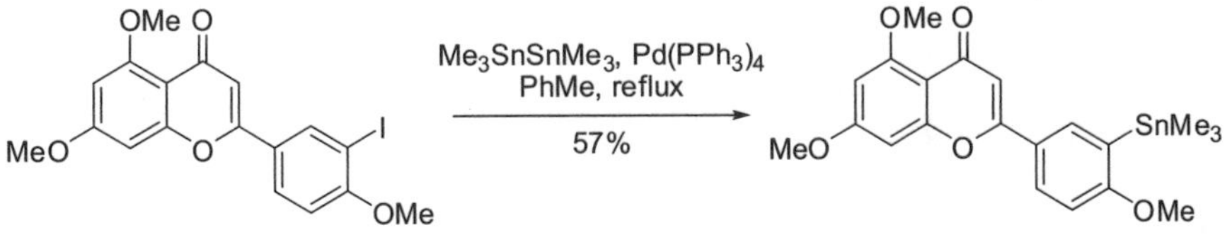

[87]Wickham, G.; Olszowy, H. A.; Kitching, W. *J. Org. Chem.* **1982**, *47*, 3788–3793.
[88]Capperucci, A.; Degl'Innocenti, A.; Faggi, C.; Reginato, G.; Ricci, A.; Dembech, P.; Seconi, G. *J. Org. Chem.* **1989**, *54*, 2966–2968.

**12.2.14.4 *Hydrostannation*** Hydrostannation is an effective method for the synthesis of cross-coupling components and has been well reviewed.[89] The uncatalyzed free-radical hydrostannation of acetylenes frequently results in a mixture of olefin isomers due to the isomerization of olefins under the reaction conditions. In some cases, a directing group can lead to useful selectivity in the reaction. The use of a Pd catalyst in the process can also lead to clean production of the *E*-olefin.[90,91]

Employing a molybdenum catalyst gives access to the α-stannylated product.[92]

Hydrostannylation of terminal acetylenes in the presence of a Lewis acid results in the Z-olefin.[93]

The hydrostannation reaction can be combined with a Pd-mediated cross-coupling that is catalytic in tin. The tin chloride that is generated as a byproduct of the cross-coupling reaction is reduced with PMHS to regenerate the tin hydride for the hydrostannylation process.[94]

[89]Smith, M. B.; March, J. *March's Advanced Organic Chemistry: Reactions, Mechanisms, and Structure, 5th ed*; Wiley: Chichester, UK, 2000.

[90]Sai, H.; Ogiku, T.; Nishitani, T.; Hiramatsu, H.; Horikawa, H.; Iwasaki, T. *Synthesis* **1995**, 582–586.

[91]Bordwell, F. G.; Fried, H. E.; Hughes, D. L.; Lynch, T. Y.; Satish, A. V.; Whang, Y. E. *J. Org. Chem.* **1990**, *55*, 3330–3336.

[92]Kazmaier, U.; Schauss, D.; Pohlman, M.; Raddatz, S. *Synthesis* **2000**, 914–916.

[93]Asao, N.; Liu, J.-X.; Sudoh, T.; Yamamoto, Y. *J. Org. Chem.* **1996**, *61*, 4568–4571.

[94]Gallagher, W. P.; Maleczka, R. E., Jr., *J. Org. Chem.* **2005**, *70*, 841–846.

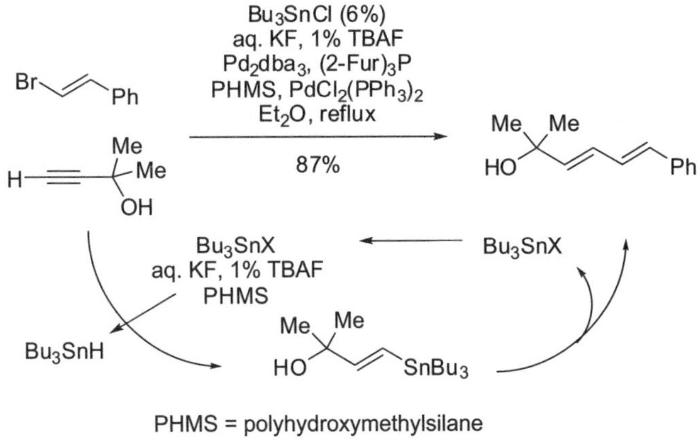

Olefins can also be hydrostannylated in the presence of a Pd catalyst.[95]

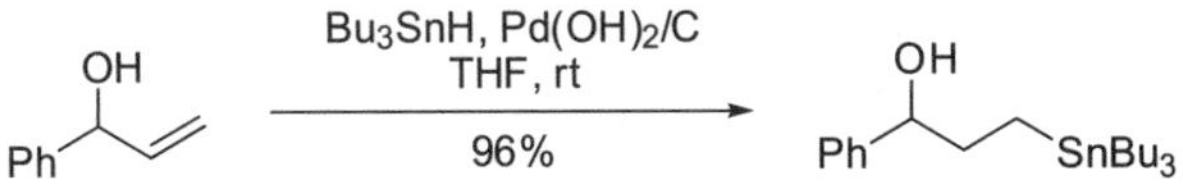

**12.2.14.5  Electrophilic Tin**  Allylic stannanes and silanes can be converted to allylic stannanes via an $S_E2'$ mechanism. In this process, a relatively simple stannane or silane can be converted to a significantly more complex one.[96]

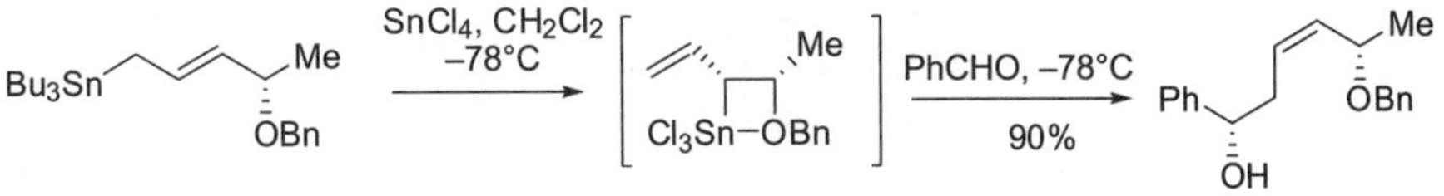

## 12.2.15  Cerium

Organocerium reagents are soft nucleophiles that are frequently employed in the addition of nucleophiles to enolizable carbonyls. The reagents are prepared by transmetalation from magnesium or lithium and cerium trichloride. The cerium salt is available as the hydrate and must be rigorously dried prior to use. Two large scale preparations are shown below.[63,97]

[95]Lautens, M.; Kumanovic, S.; Meyer, C. *Angew. Chem., Int. Ed. Engl.* **1996**, *35*, 1329–1330.
[96]McNeill, A. H.; Thomas, E. J. *Synthesis* **1994**, 322–334.
[97]Takeda, N.; Imamoto, T. *Org. Synth.* **1999**, *76*, 228–238.

## 12.2.16    Bismuth

Organobismuth reagents can be produced by the transmetalation of an organolithium reagent with $BiCl_3$.[98] The organobismuth reagents can be stable and isolable.

## 12.3    STRATEGIES FOR METALATING HETEROCYCLES

Strategies for the metalation of heterocycles have been reviewed.[2] A variety of strategies exist for the metalation of heterocycles at different positions; selection of the appropriate method is frequently dictated by the availability of starting materials. There are strategies, pitfalls, and common methods for the metalation of some heterocycles with specific substitution patterns.

Deprotonation or metal–halogen exchange to make the aryllithium or aryl Grignard reagent is frequently the first or only step used in the metalation of a heterocycle. Tables with pKa values for the various positions of heterocycles are not generally available, but the order in which protons will be abstracted by base is well known for many ring systems. In many cases, the lithiated heterocycle can undergo a lithium migration to form a more stable aryllithium, so careful control of temperature may be necessary to prevent this unwanted side reaction. The use of a metal–halogen exchange reaction with an alkyl Grignard reagent to form a positionally stable anion is often a more practical method than lithium–halogen exchange to prevent anion "walking." This method does require higher temperatures and times than the very fast lithium–halogen exchange reaction. Metal–halogen exchange

[98]Brands, K. M. J.; Dolling, U.-H.; Jobson, R. B.; Marchesini, G.; Reamer, R. A.; Williams, J. M. *J. Org. Chem.* **1998**, *63*, 6721–6726.

processes catalyzed by palladium have also allowed for the direct substitution of a halogen with boron or tin.

### 12.3.1    Strategies for Metalating 5-Membered Heterocycles

For pyrroles, furans, and thiophenes, the C-2 position can be deprotonated preferentially over the C-3 position.[2] Metalation at C-3 is usually accomplished with a metal–halogen exchange reaction.[5,16,99] Bases employed for C-2 deprotonation are LDA[100,101] or, more commonly, $n$-, $s$-, and $t$-BuLi.[102] The presence of directing substituents can change the selectivity of deprotonation.[103]

In the case of indoles, benzo($b$)furans, and benzo[$b$]thiophenes, again the C-2 position can be deprotonated;[2,100,102] the remaining positions are best selectively metalated using metal–halogen exchange.[5,16,99] The presence of a large group on nitrogen, such as a TIPS group, can drive the deprotonation to C-3.[104]

In the case of imidazoles, the C-2 proton is the first to deprotonate,[105] followed by the C-5 position if C-2 does not bear a proton.[2] A common strategy is to deprotonate C-2, protect it with a silyl group, and then deprotonate at C-5.[106] C-4 is generally metalated with a metal–halogen exchange reaction. If the C-2 position bears a substituent with protons, competitive deprotonation of the C-2 substituent[105] can compete with deprotonation at C-5, and the choice of base can shift the selectivity.[107]

[99]Amat, M.; Hadida, S.; Sathyanarayana, S.; Bosch, J. *Org. Synth.* **1997**, *74*, 248–256.

[100]Alvarez-Ibarra, C.; Quiroga, M. L.; Toledano, E. *Tetrahedron* **1996**, *52*, 4065–4078.

[101]Pomerantz, M.; Amarasekara, A. S.; Dias, H. V. R. *J. Org. Chem.* **2002**, *67*, 6931–6937.

[102]Rewcastle, G. W.; Janosik, T.; Bergman, J. *Tetrahedron* **2001**, *57*, 7185–7189.

[103]Grimaldi, T.; Romero, M.; Pujol, M. D. *Synlett* **2000**, 1788–1792.

[104]Matsuzono, M.; Fukuda, T.; Iwao, M. *Tetrahedron Lett.* **2001**, *42*, 7621–7623.

[105]Pettersen, D.; Amedjkouh, M.; Ahlberg, P. *Tetrahedron* **2002**, *58*, 4669–4673.

[106]Carpenter, A. J.; Chadwick, D. J. *Tetrahedron* **1986**, *42*, 2351–2358.

[107]Evans, D. A.; Cee, V. J.; Smith, T. E.; Santiago, K. J. *Org. Lett.* **1999**, *1*, 87–90.

Deprotonate if C-2 substituent is not H → [imidazole structure, R on N1, positions 1, 3]

First to deprotonate: frequently protected with silyl to access C-5 anion

Metal–halogen exchange →

The strategy for metalation of oxazoles and thiazoles is similar to that for imidazoles.[2] In the case of oxazoles, deprotonation at the 2-position generates an anion that will exist in equilibrium with a ring-opened isonitrile anion at low temperatures.[108] This anion will react at C-4, resulting in low selectivity. Thiazoles do not suffer from the same problem.[41] C-5 can be directly deprotonated if C-2 is substituted.[109] C-4 is generally metalated with a metal–halogen exchange reaction. If the C-2 position bears a substituent with protons, competitive deprotonation of the C-2 substituent[105] can compete with deprotonation at C-5, and the choice of base can shift the selectivity.[107]

Deprotonate if C-2 substituent is not H → [oxazole/thiazole structure, X at 1, N at 3]

First to deprotonate: frequently protected with silyl to access C-5 anion

Metal–halogen exchange →

X=O will ring-open to enolate and alkylate at C-4

X=O, S

Pyrazoles can be deprotonated at the C-5 position using $n$-BuLi[110] or LDA in the case of $N$-alkyl or $N$-aryl pyrazoles, and at the C-3 position in the case of $N$-alkoxy pyrazoles using $n$-BuLi[111] or LDA.[112] C-4 can be metalated through metal–halogen exchange. A selective electrophilic halogenation is often possible at C-4 to access the substrate for metal–halogen exchange.[112]

When R = alkyl or aryl, first to deprotonate $n$-BuLi → [pyrazole structure, R on N1, positions 1, 2]

Metal–halogen exchange →

When R = OR, first to deprotonate $n$-BuLi

Isoxazoles and isothiazoles can also be deprotonated at C-3, but in these cases, rapid ring opening through cleavage of the N−X bond occurs.[2] Metalation at C-4 and C-5 can be accomplished through metal–halogen exchange.

Deprotonate when C-3 blocked $n$-BuLi → [isoxazole/isothiazole structure, X at 1, N at 2]

First to deprotonate: ring opens rapidly $n$-BuLi

X=O, S

[108]Vedejs, E.; Monahan, S. D. *J. Org. Chem.* **1996**, *61*, 5192–5193.

[109]Marcantonio, K. M.; Frey, L. F.; Murry, J. A.; Chen, C.-Y. *Tetrahedron Lett.* **2002**, *43*, 8845–8848.

[110]Singer, R. A.; Dore, M.; Sieser, J. E.; Berliner, M. A. *Tetrahedron Lett.* **2006**. *47*, 3727–3731.

[111]Cali, P.; Begtrup, M. *Tetrahedron* **2002**, *58*, 1595–1605.

[112]Balle, T.; Vedso, P.; Begtrup, M. *J. Org. Chem.* **1999**, *64*, 5366–5370.

## 12.3.2   Strategies for Metalating 6-Membered Heterocycles

Electron-deficient aromatic rings such as pyridines can be deprotonated, although the choice of base becomes critical in these processes. Uncomplexed alkyllithium reagents tend to add to the ring rather than deprotonating; the addition of a complexing agent such as lithium diethylamino ethoxide can change the balance in favor of deprotonation. Lithium amides such as LDA and LTMP can also be used for deprotonation. The aryllithiums are prone to migration of the anion ("walking") and self-reaction to give coupled products. Different bases may be affected differently by directing groups present on the ring, allowing various modes of deprotonation to be employed by appropriate selection of base.[2] Because of these issues, metal–halogen exchange is most commonly employed to generate metalated pyridines[5,16,99] unless an ortho-directing group is present.[113,114] Deprotonation is also the method of choice if the metalation must be accomplished in the presence of an exchangeable halide.[115] Deprotonation can be achieved at C-2 using combinations of alkyl lithium reagents and an aminoalkoxy lithium reagent.[116]

DG=directing group

Selective metal–halogen exchange reactions have been explored on polyhalogenated pyridines, and in some cases very good selectivity for one position over the other can be achieved in the exchange process. In the case of magnesium–halogen exchange, selectivity for the C-3 bromide and C-5 bromide can be achieved over the C-2 bromide, and the C-3 position can be exchanged in the presence of the C-4 bromide.[6,114] For example, the 2-lithio or 5-lithio species from 2,5-dibromopyridine can be generated and reacted selectively depending on solvent selection and temperature, essentially determining whether the kinetic or thermodynamic anion will react.[117] Another way to ensure selective reaction is to use a pyridine substituted with different halogens, such as 2-iodo-5-bromopyridine.[118]

[113]Mongin, F.; Queguiner, G. *Tetrahedron* **2001**, *57*, 4059–4090.

[114]Mongin, F.; Trecourt, F.; Queguiner, G. *Tetrahedron Lett.* **1999**, *40*, 5483–5486.

[115]Fort, Y.; Gros, P.; Rodriguez, A. L. *Tetrahedron: Asymmetry* **2001**, *12*, 2631–2635.

[116]Gros, P.; Ben Younes-Millot, C.; Fort, Y. *Tetrahedron Lett.* **2000**, *41*, 303–306.

[117]Lee, J. C.; Lee, K.; Cha, J. K. *J. Org. Chem.* **2000**, *65*, 4773–4775.

[118]Mickel, S. J.; Sedelmeier, G. H.; Niederer, D.; Schuerch, F.; Seger, M.; Schreiner, K.; Daeffler, R.; Osmani, A.; Bixel, D.; Loiseleur, O.; Cercus, J.; Stettler, H.; Schaer, K.; Gamboni, R.; Bach, A.; Chen, G.-P.; Chen, W.; Geng, P.; Lee, G. T.; Loeser, E.; McKenna, J.; Kinder, F. R., Jr.; Konigsberger, K.; Prasad, K.; Ramsey, T. M.; Reel, N.; Repic, O.; Rogers, L.; Shieh, W.-C.; Wang, R.-M.; Waykole, L.; Xue, S.; Florence, G.; Paterson, I.; *Org. Process Res. Dev.* **2004**, *8*, 113–121.

Diazenes and triazenes can be deprotonated adjacent to nitrogen using amide bases; other positions can be metalated using directed metalation[119] or metal–halogen exchange. The anions become less stable as electrophilicity increases along with the number of nitrogens in the ring. Triazene anions are quite unstable and difficult to utilize.[118]

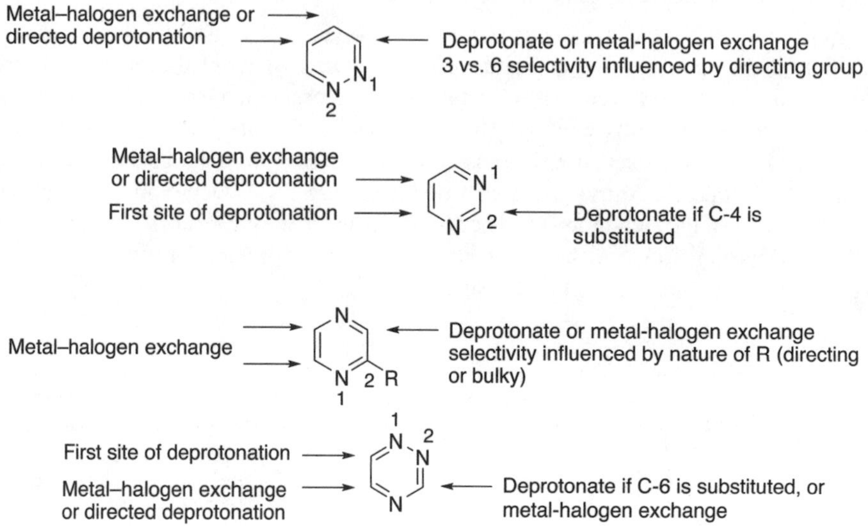

## 12.4   REACTIONS OF "NUCLEOPHILIC" ORGANOMETALLIC REAGENTS

In this section, the introduction of a heteroatom at a metalated position is summarized. An excellent overview of this topic has been published.[120]

### 12.4.1   Uncatalyzed Conversion of C–M to C–(X) Bond

Converting a carbon–metal bond to a carbon–oxygen bond is a difficult transformation. Some special cases are discussed elsewhere in this book: the oxidation of a carbon–boron bond resulting from hydroboration in Section 3.3.1.1.1, the Tamao oxidation (C-Si to C-O) in Section 10.9.2, the Sommelet reaction (benzylic halide to aldehyde) in Section 10.9.1, and the introduction of an oxygen $\alpha$- to a carbonyl via the enolate in Section 10.3.4. The oxidation of a carbon–metal bond to a carbon–oxygen bond is frequently employed in the process of converting an aryl halide to a phenol. One method is to proceed via the intermediacy of a borate, as shown below.[121]

[119]Turck, A.; Ple, N.; Mongin, F.; Queguiner, G. *Tetrahedron* **2001**, *57*, 4489–4505.
[120]Atkins, R. J.; Banks, A.; Bellingham, R. K.; Breen, G. F.; Carey, J. S.; Etridge, S. K.; Hayes, J. F.; Hussain, N.; Morgan, D. O.; Oxley, P.; Passey, S. C.; Walsgrove, T. C.; Wells, A. S. *Org. Process Res. Dev.* **2003**, *7*, 663–675.
[121]Kidwell, R. L.; Murphy, M.; Darling, S. D. *Org. Synth.* **1969**, *49*, 90–93.

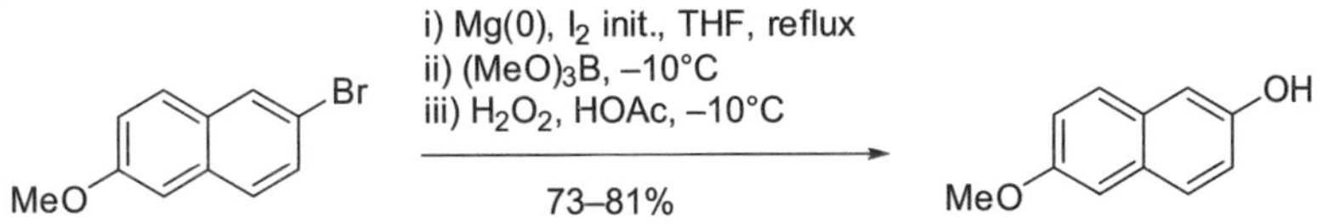
i) Mg(0), $I_2$ init., THF, reflux
ii) $(MeO)_3B$, –10°C
iii) $H_2O_2$, HOAc, –10°C
73–81%
Br
MeO
OH
MeO

Alternatively, a more reactive organometallic species such as a Grignard reagent can be reacted directly with a peroxide as shown below.[84]

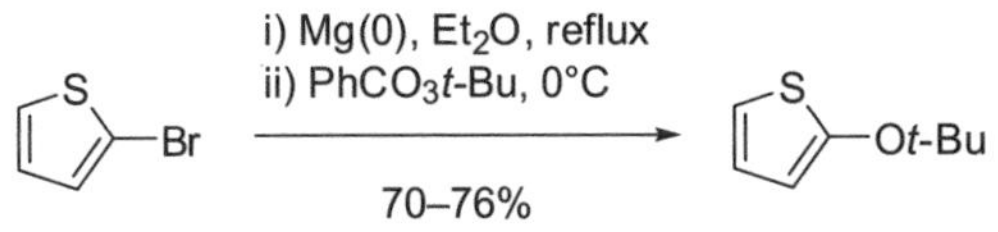
i) Mg(0), $Et_2O$, reflux
ii) $PhCO_3t$-Bu, 0°C
70–76%
S
Br
S
Ot-Bu

### 12.4.2 Uncatalyzed Conversion of C–M to C–(X) Bond

The conversion of a carbon–metal bond to a carbon–selenium bond or a carbon–sulfur bond can be achieved using a number of reagents. Selenium(0) or sulfur(0) can be used as the trapping agent, in combination with an organolithium or Grignard reagent, as shown in the two *Organic Syntheses* preparations below.[122,123]

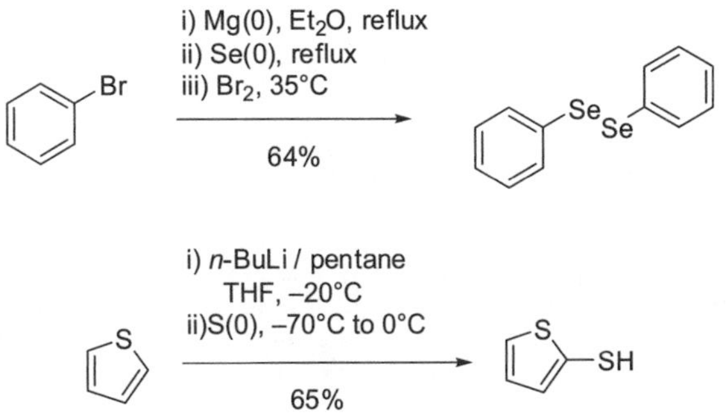
i) Mg(0), $Et_2O$, reflux
ii) Se(0), reflux
iii) $Br_2$, 35°C
64%
Br
Se
Se
i) n-BuLi / pentane
THF, –20°C
ii)S(0), –70°C to 0°C
65%
S
S
SH

Alternatively, disulfides, diselenides, selenylchlorides, sulfur chlorides, and reagents of the structure RS-$SO_2$R' can be used as trapping agents as shown in the 3 kg preparation below.[124]

[122]Reich, H. J.; Cohen, M. L.; Clark, P. S. *Org. Synth.* **1980**, *59*, 141–147.

[123]Jones, E.; Moodie, I. M. *Org. Synth.* **1970**, *50*, 979–980.

[124]Alcaraz, M.-L.; Atkinson, S.; Cornwall, P.; Foster, A. C.; Gill, D. M.; Humphries, L. A.; Keegan, P. S.; Kemp, R.; Merifield, E.; Nixon, R. A.; Noble, A. J.; O'Beirne, D.; Patel, Z. M.; Perkins, J.; Rowan, P.; Sadler, P.; Singleton, J. T.; Tornos, J.; Watts, A. J.; Woodland, I. A. *Org. Process Res. Dev.* **2005**, *9*, 555–569.

Organozirconium reagents can be reacted with diselenides, selenyl chlorides, or selenyl phthalimides to produce the selenated product.[125]

### 12.4.3   Uncatalyzed C−M to C−X

Most organometallic species can be converted to the corresponding halide after treatment with the elemental halide, or with reagents such as NCS, NBS, and NIS, as exemplified by the reaction of the Grignard reagent with iodine below.[28]

Carboalumination or hydroalumination of acetylenes can be followed by treatment with a halide such as iodine to make the vinyl halide product.[33]

Hydrozirconation products can be directly converted to the halide on treatment with $I_2$, $Br_2$, NBS, $PhICl_2$, or NCS. The stereochemistry of the C−Zr bond is retained in these processes. This is one of the most utilized methods for the conversion of acetylenes into vinyl halides.[78,126]

[125]Fryzuk, M. D.; Bates, G. S.; Stone, C. *J. Org. Chem.* **1991**, *56*, 7201–7211.
[126]Treilhou, M.; Fauve, A.; Pougny, J. R.; Prome, J. C.; Veschambre, H. *J. Org. Chem.* **1992**, *57*, 3203–3208.

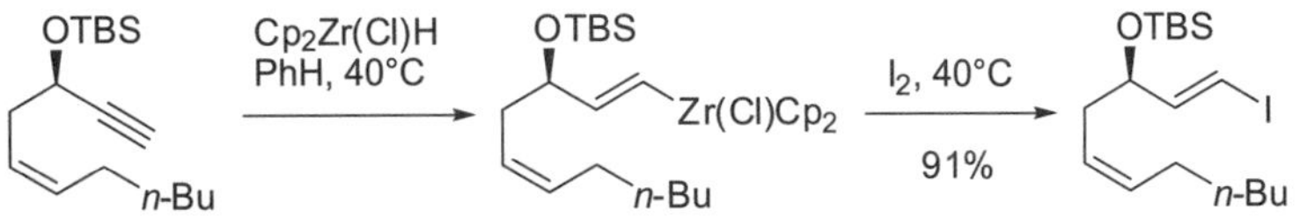

## 12.4.4    Uncatalyzed Conversion of C−M to C−(X) Bond

The conversion of a C−M bond to a C−N bond is a much less common process than the conversion of a C−X bond to a C−N bond as in S$_N$Ar reactions and transition-metal catalyzed or uncatalyzed aryl aminations. Some examples do exist and are synthetically useful, such as the case below, where an organozirconium reagent was reacted with an *O*-sulfonylated amine to produce an amine product, as shown below.[127]

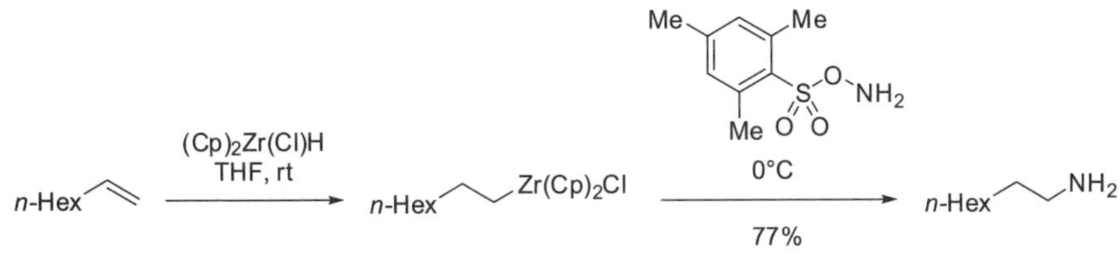

[127]Zheng, B.; Srebnik, M. *J. Org. Chem.* **1995**, *60*, 1912–1913.

# 13

# SYNTHESIS OF COMMON AROMATIC HETEROCYCLES

STÉPHANE CARON

## 13.1  INTRODUCTION

While heterocyclic chemistry was a significant contributor to organic chemistry in the early part of the twentieth century, the emphasis in current organic chemistry research is on catalytic stereocontrolled synthesis of acyclic compounds. However, much of the chemical industry continues to rely on the efficient preparation of heterocycles. This chapter will focus on the preferred methods for the preparation of several common aromatic heterocyclic ring systems. Specifically, only the heterocyclic ring formation reactions will be discussed and not the synthesis of complex molecules containing one or many heterocycles. In this context, the chapters on metalation of heterocycles (see Section 12.3) and metal-mediated cross couplings (Chapter 6) might provide additional useful information. Methods relying on either the oxidation or reduction of a heterocycle in the final steps are generally not covered. This chapter is organized first by the atom included in the heterocyclic (N, O, S, mixed), ring size, number of heteroatoms, and position of the heteroatom.

## 13.2  PYRROLES

The pyrrole ring system has been widely studied since it is a common fragment in several naturally occurring molecules and constitutes the backbone of porphyrins. This heterocycle is a core found in several active pharmaceutical ingredients, including atorvastatin. While several methods exist for the preparation of pyrroles, the reaction of a 1,4-dicarbonyl with an amine has been the most widely utilized.

---

*Practical Synthetic Organic Chemistry: Reactions, Principles, and Techniques*, First Edition.
Edited by Stéphane Caron.
© 2011 John Wiley & Sons, Inc. Published 2011 by John Wiley & Sons, Inc.

### 13.2.1 Condensation of 1,4-Dicarbonyls with a Primary Amine

The preferred method for the preparation of a pyrrole ring is the Paal–Knorr reaction. In this synthetic method, a 1,4-dicarbonyl starting material is treated with a primary amine. The reaction is generally high yielding[1] and is an effective way to protect anilines as the corresponding dimethyl pyrroles.[2]

### 13.2.2 Condensation of 1,3-Dicarbonyls with an α-Aminocarbonyl Compound

One of the best known procedures for the preparation of highly substituted pyrroles is the Knorr reaction, where an α-aminocarbonyl compound is condensed with a 1,3-dicarbonyl to provide the pyrrole, often in a modest yield.[3]

### 13.2.3 Dipolar Cycloaddition

Another popular method for the preparation of pyrroles is the 1,3-dipolar cycloaddition of an azomethine ylide with an alkyne. The dipole can be generated using a number of different methods such as N-alkylation of an azalactone[4] or through activation of an acylated amino

[1]Singer, R. A.; Caron, S.; McDermott, R. E.; Arpin, P.; Do, N. M. *Synthesis* **2003**, 1727–1731.
[2]Ragan, J. A.; Jones, B. P.; Castaldi, M. J.; Hill, P. D.; Makowski, T. W. *Org. Synth.* **2002**, *78*, 63–72.
[3]Lancaster, R. E., Jr.; VanderWerf, C. A. *J. Org. Chem.* **1958**, *23*, 1208–1209.
[4]Hershenson, F. M.; Pavia, M. R. *Synthesis* **1988**, 999–1001.

acid in the presence of an alkyne.[5] One advantage of this method is the high level of complexity that can be obtained in a single operation.

80%

Ac₂O, 90°C

43%

## 13.3 INDOLES

The indole nucleus has been one of the most studied heterocycles, and several efficient methods have been developed for its preparation. The desired substitution pattern will often dictate the choice of synthetic route. The indole nucleus forms the core of several active pharmaceutical ingredients, most notably in the "triptan" family such as eletriptan, sumatriptan, rizatriptan, and zolmitriptan, which are 5-hydroxytriptamine₁ receptor agonists for the treatment of migraines.

### 13.3.1 Fischer Indole Synthesis

The Fischer indole synthesis is undeniably the most widely used method for the preparation of indoles, especially in light of the fact that the reaction requires an arylhydrazine and a ketone or aldehyde, two very simple starting materials (see Section 7.3.2.5). The only drawback of the reaction is that a mixture of regioisomers can be obtained with *meta*-substituted hydrazines. In the first example below, the ammonia generated during the indole synthesis displaces a primary chloride in 88% overall yield.[6] A ketohydrazone can be also used for the formation of indoles.[7]

[5]Roth, B. D.; Blankley, C. J.; Chucholowski, A. W.; Ferguson, E.; Hoefle, M. L.; Ortwine, D. F.; Newton, R. S.; Sekerke, C. S.; Sliskovic, D. R.;et al. *J. Med. Chem.* **1991**, *34*, 357–366.
[6]Fleck, T. J.; Chen, J. J.; Lu, C. V.; Hanson, K. J. *Org. Process Res. Dev.* **2006**, *10*, 334–338.
[7]Hillier, M. C.; Marcoux, J.-F.; Zhao, D.; Grabowski, E. J. J.; McKeown, A. E.; Tillyer, R. D. *J. Org. Chem.* **2005**, *70*, 8385–8394.

### 13.3.2 Intramolecular Condensation of Anilines with Phenacyl Derivatives

Indoles can also be synthesized by the cyclodehydration of phenacyl derivatives with an *ortho*-substituted aniline. This method is especially attractive for the preparation of 2-substituted indoles.[8]

### 13.3.3 Cycloelimination of Enamines

An excellent method for the preparation of indoles unsubstituted at the 2- and 3-position is the Batcho–Leimgruber indole synthesis, where an aniline cyclizes onto an enamine under acidic conditions. The substrate for this reaction is easily obtained in a single step from the corresponding nitrotoluene and DMF•DMA.[9] Very often, the indole formation will proceed during the reduction of the nitro group to the aniline.[10]

[8]Peters, R.; Waldmeier, P.; Joncour, A. *Org. Process Res. Dev.* **2005**, *9*, 508–512.
[9]Vetelino, M. G.; Coe, J. W. *Tetrahedron Lett.* **1994**, *35*, 219–222.
[10]Ponticello, G. S.; Baldwin, J. J. *J. Org. Chem.* **1979**, *44*, 4003–4005.

### 13.3.4  Aldol and Michael Additions

Indoles acylated at the 2-position may be easily accessed via an intramolecular 1,2- or 1,4- nucleophilic addition reaction. In general, a mild base is sufficient to induce the cyclization. In the first example below, a second step was needed for the dehydration of the intermediate aldol adduct to generate the indole.[11] In the second case, the nitrogen-protecting group was eliminated by treatment with NaOH to afford the indole nucleus.[12]

### 13.3.5  Addition of Vinyl Grignard Reagents to Nitrobenzene Derivatives

The addition of a vinyl Grignard reagent to a nitrobenzene to generate an indole is known as the Bartoli indole synthesis. While the reaction proceeds in a single step from a nitroarene and is tolerant of some functionality that cannot be present with other methods, the necessity of low temperature renders this reaction less practical. The Bartoli indole synthesis is not the preferred method for preparation of indoles on large scale but can be considered a good alternative for laboratory scale synthesis.[13]

### 13.4  OXINDOLES (2-INDOLINONES)

Oxindoles are fairly straightforward to prepare and have been widely employed as pharmaceutical agents. For instance, ropinirole is a simple oxindole used for the treatment

[11]Jones, C. D. *J. Org. Chem.* **1972**, *37*, 3624–3625.

[12]Caron, S.; Vazquez, E.; Stevens, R. W.; Nakao, K.; Koike, H.; Murata, Y. *J. Org. Chem.* **2003**, *68*, 4104–4107.

[13]Faul, M. M.; Engler, T. A.; Sullivan, K. A.; Grutsch, J. L.; Clayton, M. T.; Martinelli, M. J.; Pawlak, J. M.; LeTourneau, M.; Coffey, D. S.; Pedersen, S. W.; Kolis, S. P.; Furness, K.; Malhotra, S.; Al-Awar, R. S.; Ray, J. E. *J. Org. Chem.* **2004**, *69*, 2967–2975.

of Parkinson's disease and restless leg syndrome, sunitinib is a multi-kinase inhibitor in oncology and ziprazidone is an atypical antipsychotic. Two main methods are utilized for their preparation.

### 13.4.1 Friedel–Crafts Alkylation

The Friedel–Crafts alkylation of α-chloroacyl anilines is a reliable method for the preparation of oxindoles. The advantage of this method is that anilines and α-chloroacetyl chloride derivatives are simple starting materials. Unfortunately, high temperatures are usually required for the cyclization to proceed.[14]

### 13.4.2 Lactamization

The other common method for the preparation of oxindoles is the lactamization of anilines onto a phenylacetic acid derivative.[15] The reaction often proceeds directly upon reduction of the corresponding nitroarene.[16]

### 13.5 ISATINS (2,3-INDOLINDIONES)

Isatins are very useful synthetic intermediates due to their high reactivity towards nucleophiles. There are two efficient methods for the preparation of these compounds.

[14]Acemoglu, M.; Allmendinger, T.; Calienni, J.; Cercus, J.; Loiseleur, O.; Sedelmeier, G. H.; Xu, D. *Tetrahedron* **2004**, *60*, 11571–11586.

[15]Moser, P.; Sallmann, A.; Wiesenberg, I. *J. Med. Chem.* **1990**, *33*, 2358–2368.

[16]Quallich, G. J.; Morrissey, P. M. *Synthesis* **1993**, 51–53.

### 13.5.1 Cyclization of Isonitrosoacetanilides

The most common method for the preparation of isatins is the reaction of an aniline with trichloroacetaldehyde (chloral) in the presence of hydroxylamine.[17] The resulting isonitroso intermediate undergoes cyclization under acidic conditions.[18]

### 13.5.2 Friedel–Crafts Acylation

The other method for the preparation of isatins is the Friedel–Crafts cyclization of amides of oxalic acid. This method is often higher yielding than the chloral method.[19]

## 13.6 PYRAZOLES

This heterocycle forms the core of several important pharmaceutical drugs, notably celecoxib, a 1,3,5-trisubstituted pyrazole which is a COX-2 inhibitor and rimonabant, a 1,3,4,5-tetrasubstituted pyrazole acting as a CB-1 receptor antagonist. Most syntheses of pyrazoles utilize the addition of a hydrazine to a 1,3-dicarbonyl compound or a Michael acceptor.

### 13.6.1 Condensation of a Hydrazine with a 1,3-Dicarbonyl Derivative

The most common method for preparation of a pyrazole is condensation of a hydrazine with a 1,3-dicarbonyl reagent. In the case of a symmetrical diketone[20] or when using hydrazine[20] the reaction is usually high yielding with no regiochemical issues. In cases where an unsymmetrical diketone or ketoester is treated with a substituted hydrazine, regioisomers of the pyrazole can be obtained.[21] It is noteworthy that appropriate safety precautions should be taken when working with hydrazines.

[17]Marvel, C. S.; Hiers, G. S. *Org. Synth.* **1925**, *V*, 71–74.

[18]Wakelin, L. P. G.; Bu, X.; Eleftheriou, A.; Parmar, A.; Hayek, C.; Stewart, B. W. *J. Med. Chem.* **2003**, *46*, 5790–5802.

[19]Soll, R. M.; Guinosso, C.; Asselin, A. *J. Org. Chem.* **1988**, *53*, 2844–2847.

[20a]Singer, R. A.; Caron, S.; McDermott, R. E.; Arpin, P.; Do, N. M. Synthesis, **2003**, *11*, 1727–1731.

[b]Patel, M. V.; Bell, R.; Majest, S.; Henry, R.; Kolasa, T. *J. Org. Chem.* **2004**, *69*, 7058–7065.

[21]Hanefeld, U.; Rees, C. W.; White, A. J. P.; Williams, D. J. *J. Chem. Soc., Perkin Trans. 1* **1996**, 1545–1552.

### 13.6.2 Condensation of a Hydrazine with a Michael Acceptor

The second method for construction of a pyrazole is the condensation of a hydrazine with a Michael acceptor. The product of the reaction depends on the initial regioselectivity of hydrazine addition (1,2 or 1,4). Michael acceptors[22] bearing a leaving group such as an amine[23] or an ether[24] at the $\beta$-position have been used for this transformation.

[22]Hamper, B. C.; Kurtzweil, M. L.; Beck, J. P. *J. Org. Chem.* **1992**, *57*, 5680–5686.

[23]Singh, R. K.; Sinha, N.; Jain, S.; Salman, M.; Naqvi, F.; Anand, N. *Tetrahedron* **2005**, *61*, 8868–8874.

[24]Flores, A. F. C.; Brondani, S.; Pizzuti, L.; Martins, M. A. P.; Zanatta, N.; Bonacorso, H. G.; Flores, D. C. *Synthesis* **2005**, 2744–2750.

## 13.7  INDAZOLES

Three methods are commonly utilized for the preparation of indazoles. The diazotization of *ortho*-toluidines is straightforward on small scale but can present safety challenges on scale-up. The nucleophilic aromatic substitution of hydrazone is an efficient method, usually starting from easily obtained reagents. Finally, catalytic metal-mediated cyclization of hydrazones has been demonstrated as a mild and efficient way to access this ring system.

### 13.7.1  Nucleophilic Aromatic Substitution of Arylhydrazones

An efficient method for construction of an indazole is the intramolecular $S_NAr$ reaction of an *ortho*-activated hydrazone. The hydrazone can be generated *in situ* from the corresponding ketone and hydrazine, followed by intramolecular cyclization.[25,26] Both *cis* and *trans* hydrazones convert to the desired product.

### 13.7.2  Diazotization of a Toluidine

Another popular method for the preparation of indazoles is the diazotization of *ortho*-toluidines. Upon generation of the diazonium species, the benzylic proton at the *ortho*-position is sufficiently acidic to be deprotonated by a mild base. Addition of the resulting anion to the diazonium group leads to the desired indazole.[27] Special precautions should be taken when working with diazonium compounds.

[25]Caron, S.; Vazquez, E. *Synthesis* **1999**, 588–592.
[26]Caron, S.; Vazquez, E. *Org. Process Res. Dev.* **2001**, *5*, 587–592.
[27]Sun, J.-H.; Teleha, C. A.; Yan, J.-S.; Rodgers, J. D.; Nugiel, D. A. *J. Org. Chem.* **1997**, *62*, 5627–5629.

### 13.7.3    Metal-Mediated Cyclization

A more recent method for the generation of indazole is the metal-mediated cyclization of hydrazones onto aryl halides. This transformation has been accomplished using either copper[28] or palladium[29] catalysts.

DPEphos = bis[2-(phenylphosphino)phenyl] ether

## 13.8    IMIDAZOLES AND BENZIMIDAZOLES

Imidazoles and benzimidazoles have been one of the most utilized classes of compounds in the pharmaceutical industry. This pharmacophore has led to several gastric-acid pump inhibitors, namely omeprazole and its enantiomer esomeprazole. Several excellent synthetic methods exist for the preparation of these heterocycles.

### 13.8.1    Condensation of a 1,2-Diamine with a Carboxylic Acid

Condensation of an *ortho*-bisaniline with a carboxylic acid derivative is the best method for preparation of benzimidazoles. Carboxylic acids,[30] acid chlorides,[31] and anhydrides[32] have been utilized successfully for this transformation.

[28]Watson, T. J.; Ayers, T. A.; Shah, N.; Wenstrup, D.; Webster, M.; Freund, D.; Horgan, S.; Carey, J. P. *Org. Process Res. Dev.* **2003**, *7*, 521–532.

[29]Lebedev, A. Y.; Khartulyari, A. S.; Voskoboynikov, A. Z. *J. Org. Chem.* **2005**, *70*, 596–602.

[30]Huff, J. R.; King, S. W.; Saari, W. S. *J. Org. Chem.* **1982**, *47*, 582–585.

[31]Mertens, A.; Mueller-Beckmann, B.; Kampe, W.; Hoelck, J. P.; Von der Saal, W. *J. Med. Chem.* **1987**, *30*, 1279–1287.

[32]Caron, S.; Do, N. M.; McDermott, R. E.; Bahmanyar, S. *Org. Process Res. Dev.* **2006**, *10*, 257–261.

### 13.8.2    Condensation of an Amidine with a Halocarbonyl Derivative

A good method for preparation of imidazoles is the reaction of an amidine with a carbonyl compound containing a leaving group at the α-position.[33] In the second example below, a bromoacetaldehyde equivalent is introduced in the form of the enol ether[34] while the third example shows the preparation of an N-substituted imidazole.[35]

[33]Tsunoda, T.; Tanaka, A.; Mase, T.; Sakamoto, S. *Heterocycles* **2004**, *63*, 1113–1122.

[34]Lipinski, C. A.; Blizniak, T. E.; Craig, R. H. *J. Org. Chem.* **1984**, *49*, 566–570.

[35]Shilcrat, S. C.; Mokhallalati, M. K.; Fortunak, J. M. D.; Pridgen, L. N. *J. Org. Chem.* **1997**, *62*, 8449–8454.

### 13.8.3   Condensation of 1,4-Dicarbonyls with an Amine

A 1,4-dicarbonyl compound will react with an amine to first generate an enamine that undergoes cyclodehydration. The acyclated α-aminoketone required for this transformation can easily be accessed using organocatalysis.[36]

### 13.8.4   Condensation of 1,2-Dicarbonyls with an Aldehyde in the Presence of Ammonia

Another method for the preparation of the imidazole ring is the condensation of a 1,2-dicarbonyl with an aldehyde in the presence of an ammonia source.[37]

## 13.9   1,2,3-TRIAZOLES

The 1,2,3-triazole ring system is a well-studied heterocycle. The most common method for its preparation is a dipolar cycloaddition.

### 13.9.1   Dipolar Cycloaddition of Azides with an Alkynes

There is only one reliable method for the preparation of 1,2,3-triazoles, namely the cycloaddition of an azide with an alkyne. The two most common reagents to obtain the unsubstituted 1,2,3-triazole are sodium azide[38] or trimethylsilyl azide.[39] Alkyl azides can also be employed, but the substrates must be selected carefully to avoid mixtures of regioisomers.[40] Special precautions should be taken when working with azides as they are highly energetic and often shock sensitive compounds.

[36]Frantz, D. E.; Morency, L.; Soheili, A.; Murry, J. A.; Grabowski, E. J. J.; Tillyer, R. D. *Org. Lett.* **2004**, *6*, 843–846.
[37]Krebs, F. C.; Jorgensen, M. *J. Org. Chem.* **2001**, *66*, 6169–6173.
[38]Trybulski, E. J.; Benjamin, L.; Vitone, S.; Walser, A.; Fryer, R. I. *J. Med. Chem.* **1983**, *26*, 367–372.
[39]Blass, B. E.; Coburn, K.; Lee, W.; Fairweather, N.; Fluxe, A.; Wu, S.; Janusz, J. M.; Murawsky, M.; Fadayel, G. M.; Fang, B.; Hare, M.; Ridgeway, J.; White, R.; Jackson, C.; Djandjighian, L.; Hedges, R.; Wireko, F. C.; Ritter, A. L. *Bioorg. Med. Chem. Lett.* **2006**, *16*, 4629–4632.
[40]Coats, S. J.; Link, J. S.; Gauthier, D.; Hlasta, D. J. *Org. Lett.* **2005**, *7*, 1469–1472.

### 13.9.2    Dipolar Cycloaddition of Azides with Enolates

An excellent procedure for the preparation of 4,5-disubstituted-1,2,3-triazoles is the addition of an enolates to an alkylazide, also known as the Dimroth reaction.[41] This procedure provides high levels of regiocontrol, which is not always the case for the dipolar cycloaddition on disubstituted alkynes as shown in the example below.

## 13.10    1,2,4-TRIAZOLES

The 1,2,4-triazole unit is far more commonly found in pharmaceutical agents then the 1,2,3-triazole moiety. The ring system is usually obtained through a cyclodehydration approach.

### 13.10.1    Cyclodehydration

Most 1,2,4-triazoles and derivatives thereof are prepared by intramolecular cyclodehydration of an amide or acylhydrazine. The desired substrates can be accessed in a number of

[41]Cottrell, I. F.; Hands, D.; Houghton, P. G.; Humphrey, G. R.; Wright, S. H. B. *J. Heterocycl. Chem.* **1991**, *28*, 301–304.

ways, either through addition of an acyl hydrazine to an imidate or a nitrile,[42] addition of hydrazine or an alkylhydrazine to an imine,[43] or addition to a diazo compound.[44] The cyclization itself proceeds under either acidic or basic conditions.

## 13.11 TETRAZOLES

Tetrazoles represent a carboxylic acid isostere and are by far the most common heterocycle introduced on the side chain of β-lactam antibiotics. The synthetic strategy for their preparation usually involves the cycloaddition of an azide onto a nitrile or activated amide.

### 13.11.1 Cycloaddition of an Azide on a Nitrile

The most common method to prepare a tetrazole is by cycloaddition of an azide with a nitrile.[45] While several of the literature procedures utilize sodium azide in the presence of a proton source at elevated temperature, this method generates hydrazoic acid, which is hazardous under the reaction conditions.[46] Me$_3$SnCl has been used instead of a proton source, however the reaction requires a full equivalent of the tin reagent, which is highly undesirable.[47] A method utilizing catalytic quantities of dibutyltin oxide in conjunction with trimethysilyl azide has been developed and is considered less hazardous than the sodium azide method, mainly because of the lower risk in handling TMSN$_3$.[48]

[42]Omodei-Sale, A.; Consonni, P.; Galliani, G. *J. Med. Chem.* **1983**, *26*, 1187–1192.

[43]Lange, J. H. M.; van Stuivenberg, H. H.; Coolen, H. K. A. C.; Adolfs, T. J. P.; McCreary, A. C.; Keizer, H. G.; Wals, H. C.; Veerman, W.; Borst, A. J. M.; de Looff, W.; Verveer, P. C.; Kruse, C. G. *J. Med. Chem.* **2005**, *48*, 1823–1838.

[44]Lin, Y.-I.; Lang, S. A., Jr.; Lovell, M. F.; Perkinson, N. A. *J. Org. Chem.* **1979**, *44*, 4160–4164.

[45]Butler, R. N. In *Comprehensive Heterocyclic Chemistry*; Katritzky, A. R., Rees, C. W., Scriven, E. F. V., Eds.; Pergamon: Oxford, U.K., **1996**; Vol. 4.

[46]Nakamura, T.; Sato, M.; Kakinuma, H.; Miyata, N.; Taniguchi, K.; Bando, K.; Koda, A.; Kameo, K. *J. Med. Chem.* **2003**, *46*, 5416–5427.

[47]Kerdesky, F. A. J.; Haight, A.; Narayanan, B. A.; Nordeen, C. W.; Scarpetti, D.; Seif, L. S.; Wittenberger, S.; Morton, H. E. *Synth. Commun.* **1993**, *23*, 2027–2039.

[48]Wittenberger, S. J.; Donner, B. G. *J. Org. Chem.* **1993**, *58*, 4139–4141.

## 13.11.2    Activation of an Amide and Addition of an Azide

Another less common method for the preparation of a tetrazole is the activation of an amide, mainly to an iminoyl chloride,[49] followed by addition of trimethylsilyl azide. This method has also been utilized directly on an amide under Mitsunobu-type conditions.[50]

[49]Meanwell, N. A.; Hewawasam, P.; Thomas, J. A.; Wright, J. J. K.; Russell, J. W.; Gamberdella, M.; Goldenberg, H. J.; Seiler, S. M.; Zavoico, G. B. *J. Med. Chem.* **1993**, *36*, 3251–3264.

[50]Nelson, D. W.; Gregg, R. J.; Kort, M. E.; Perez-Medrano, A.; Voight, E. A.; Wang, Y.; Grayson, G.; Namovic, M. T.; Donnelly-Roberts, D. L.; Niforatos, W.; Honore, P.; Jarvis, M. F.; Faltynek, C. R.; Carroll, W. A. *J. Med. Chem.* **2006**, *49*, 3659–3666.

## 13.12   DIHYDROPYRIDINES

The 1,4-dihydropyridine nucleus has been extensively studied because of its pharmacological activity as a calcium channel blocker. For instance, amlodipine has been a multi-billion dollar product for the treatment of hypertension and angina. Two major synthetic methods have provided access to the basic framework in a single step. It is noteworthy that dihydropyridines can easily be oxidized to the corresponding pyridines and can be used as intermediates in their synthesis.

### 13.12.1   Reaction of Ketoesters and Aldehydes in the Presence of Ammonia

The Hantzsch dihydropyridine synthesis is by far the most well known method for the preparation of this class of compounds. The reaction can be low yielding, but the simplicity of synthesis and low cost of the required starting materials counterbalance this issue. Two equivalents of a β-ketoester react with an aldehyde in the presence of ammonia. It is generally believed that ammonia reacts with one equivalent of the ketoester to generate an aminocrotonate. The second equivalent of the ketoester undergoes Knoevenagel condensation with the aldehyde to provide an enone. The two components undergo condensation followed by ring closure to the dihydropyridine. Using the classical Hantzch dihydropyridine synthesis, only symmetrical products are formed.[51]

### 13.12.2   Reaction of Aminocrotonates with Aldehydes and β-Ketoesters

The preferred method for the preparation of unsymmetrical dihydropyridines is a stepwise modification of the Hantzsch synthesis. Rather than utilizing two equivalents of the ketoester, only one equivalent is used in the presence of equimolar amounts of a previously prepared aminocrotonate to afford the desired dihydropyridine in high yields.[52]

[51]Boecker, R. H.; Guengerich, F. P. *J. Med. Chem.* **1986**, *29*, 1596–1603.
[52]Alker, D.; Denton, S. M. *Tetrahedron* **1990**, *46*, 3693–3702.

## 13.13  PYRIDINES

Pyridines are among the most common heterocycles and numerous methods for their preparation are available and have been reviewed extensively.[53] The three most common methods are presented below.

### 13.13.1  Condensation of a 1,3-Dicarbonyl Derivative in the Presence of a Cyanoacetamide

Reaction of a β-ketoester with a reagent such as cyanoacetamide, known as the Guareschi–Thorpe pyridine synthesis, is a proven method for the preparation of nicotinamide derivatives. The reaction is usually carried out in an alcoholic solvent in the presence of a mild base such as piperidine. This method leads to a 2,6-dihydroxypyridine (or 6-hydroxypyridone) with substitution possible at the 2-, 3-, 4- and 6-position.[54,55]

### 13.13.2  Condensation of Enolates with an Enaminoesters

One of the most common and reliable methods for the preparation of 2-hydroxypyridine is the Friedländer condensation, which involves the reaction of an enaminoester with an enolate. While the reaction works well with simple enolates, it is usually more efficient to start with a 1,3-dicarbonyl compound. When an unsubstituted malonate is used, an ester group remains at the 3-position of the pyridine. However, if a substituted ketoester is used, decarboxylation occurs either during the pyridine synthesis or as a subsequent step leading to a 2,3,4-trisubstituted product.[56]

when R = Me, R' = Me
when R = OEt, R' = CO$_2$Et

[53]Henry, G. D. *Tetrahedron* **2004**, *60*, 6043–6061.
[54]Holland, G. F.; Pereira, J. N. *J. Med. Chem.* **1967**, *10*, 149–154.
[55]Kutney, J. P.; Selby, R. C. *J. Org. Chem.* **1961**, *26*, 2733–2737.
[56]McElroy, W. T.; DeShong, P. *Org. Lett.* **2003**, *5*, 4779–4782.

### 13.13.3   Condensation of a 1,5-Dicarbonyl Compound with Ammonia

Condensation of a 1,5-dicarbonyl compound with ammonia is a less common method to prepare pyridines, mainly because of the complexity or lack of availability of the starting material. In order to obtain the pyridine, the starting material must contain a functional group poised for elimination, or hydroxylamine can be used in place of ammonia. The reaction has been demonstrated with a hydroxydialdehyde and ammonia.[57] It is worthy to note that a 1,5-dicarbonyl is usually the final intermediate prior to condensation with ammonium acetate in the Kröhnke pyridine synthesis that starts from an acylpyridinium salt.[58]

Another efficient method for the preparation of pyridines is the reaction of a ketone with an enone and a source of ammonia. Ketone enolates can be converted to bis(methylthio)-enones that undergo pyridine formation upon treatment with ammonium hydroxide.[59] An extension of this methodology was discovered at Merck through the use of vinamidinium salts. This method is especially useful for the formation of 2,3-disubstituted pyrdines.[60,61]

[57]Jiang, B.; Xiong, W.; Zhang, X.; Zhang, F. *Org. Process Res. Dev.* **2001**, *5*, 531–534.

[58]Kroehnke, F. *Synthesis* **1976**, 1–24.

[59]Potts, K. T.; Ralli, P.; Theodoridis, G.; Winslow, P. *Org. Synth.* **1986**, *64*, 189–195.

[60]Davies, I. W.; Marcoux, J.-F.; Corley, E. G.; Journet, M.; Cai, D.-W.; Palucki, M.; Wu, J.; Larsen, R. D.; Rossen, K.; Pye, P. J.; DiMichele, L.; Dormer, P.; Reider, P. J. *J. Org. Chem.* **2000**, *65*, 8415–8420.

[61]Marcoux, J.-F.; Marcotte, F.-A.; Wu, J.; Dormer, P. G.; Davies, I. W.; Hughes, D.; Reider, P. J. *J. Org. Chem.* **2001**, *66*, 4194–4199.

## 13.14   QUINOLINES

Quinolines are present in a wide variety of naturally occurring alkaloids as well as in many pharmaceuticals. Several agents of this class have been utilized as protozoacids for the treatment of malaria. The most common method for the preparation of quinolines is the Friedländer quinoline synthesis. Methods discussed for the preparation of structurally related quinolones and quinolinones are described in section 13.17 and 13.16.

### 13.14.1   Friedländer Quinoline Synthesis

The Friedländer quinoline synthesis involves the reaction of an *ortho*-aminoacetophenone with an enolizable aldehyde or ketone. One of the major advantages of this method is that highly functionalized quinolines can be readily obtained.[62]

### 13.14.2   Addition to Isatins

The best method to prepare a quinoline containing a carboxylic acid at the 4-position is the Pfitzinger reaction. This synthesis begins with an isatin that is opened to the corresponding isatoic acid. Further reaction with a carbonyl compound generates an imine that undergoes intramolecular condensation to provide the desired quinoline.[63]

### 13.14.3   Electrophilic Aromatic Substitution

A less popular method for the preparation of quinolines is the electrophilic cyclization of a ketone, known as the Combes reaction. The reaction proceeds under a variety of acidic or dehydrating conditions and is a good route to access polyaromatic systems.[64]

[62]Mizuno, M.; Inagaki, A.; Yamashita, M.; Soma, N.; Maeda, Y.; Nakatani, H. *Tetrahedron* **2006**, *62*, 4065–4070.
[63]Atwell, G. J.; Baguley, B. C.; Denny, W. A. *J. Med. Chem.* **1989**, *32*, 396–401.
[64]Atkins, R. J.; Breen, G. F.; Crawford, L. P.; Grinter, T. J.; Harris, M. A.; Hayes, J. F.; Moores, C. J.; Saunders, R. N.; Share, A. C.; Walsgrove, T. C.; Wicks, C. *Org. Process Res. Dev.* **1997**, *1*, 185–197.

### 13.14.4   Intramolecular Cyclization of an Iminium Ion

The Meth–Cohn quinoline synthesis is a very efficient method for the preparation of a 2-chloroquinoline. Reaction of an amide with $POCl_3$ leads to the imidoyl chloride that further reacts with the Vilsmeier reagent generated *in situ* from DMF. The iminium generated undergoes an intramolecular electrophilic aromatic substitution to furnish the desired quinoline.[65]

### 13.15   ISOQUINOLINES

Several methods are available for the preparation of isoquinolines. Electrophilic aromatic cyclization of an amide is the most common method to access this heterocycle. If the desired isoquinoline is a synthetic intermediate that will be further elaborated at the 1-position, chlorination of the corresponding quinolinone is often the best choice.[66]

### 13.15.1   Intramolecular Cyclization of Imidoyl Chlorides

The method of choice to prepare dihydroisoquinolines is the Bischler–Napieralski reaction, where an amide is treated with a dehydrating agent such as $POCl_3$ to induce cyclization. Electron-rich arenes cyclize readily.[67] This ring system can also be obtained through elimination of an ether, which is sometimes referred to as the Pictet–Gams reaction.[68]

[65]Mabire, D.; Coupa, S.; Adelinet, C.; Poncelet, A.; Simonnet, Y.; Venet, M.; Wouters, R.; Lesage, A. S. J.; Van Beijsterveldt, L.; Bischoff, F. *J. Med. Chem.* **2005**, *48*, 2134–2153.

[66]Tucker, S. C.; Brown, J. M.; Oakes, J.; Thornthwaite, D. *Tetrahedron* **2001**, *57*, 2545–2554.

[67]Geen, G. R.; Mann, I. S.; Mullane, M. V.; McKillop, A. *Tetrahedron* **1998**, *54*, 9875–9894.

[68]Poszavacz, L.; Simig, G. *Tetrahedron* **2001**, *57*, 8573–8580.

## 13.15.2  Intramolecular Cyclization of an Oxonium Ion

Another electrophilic cyclization that leads to an isoquinoline is the formation of a benzylic imine containing an oxonium ion precursor at the β-position. While this sequence is attractive due to the availability of the starting materials, it is often low yielding.[69]

## 13.15.3  Condensation of Phenacyl Derivatives with Ammonia

Condensation of ammonia with a putative 1,5-dicarbonyl compound will generate an imine that tautomerizes to an enamine and undergoes intramolecular cyclization to the isoquinoline.[70] The disadvantage of this method is that the requisite starting material is not always easily accessible.

## 13.16  QUINOLINONES AND 2-HYDROXYQUINOLINES

Quinolinones are very useful synthetic intermediates, especially for further functionalization at the 2-position. There are three main methods for the preparation of quinolinones and their tautomeric 2-hydroxyquinolines.

[69]Briet, N.; Brookes, M. H.; Davenport, R. J.; Galvin, F. C. A.; Gilbert, P. J.; Mack, S. R.; Sabin, V. *Tetrahedron* **2002**, *58*, 5761–5766.
[70]Flippin, L. A.; Muchowski, J. M. *J. Org. Chem.* **1993**, *58*, 2631–2632.

### 13.16.1 Electrophilic Cyclization

It is possible to generate the pyridine ring of a quiloninone through electrophilic cyclization although it is not the most general method. The advantage of this method is that simple anilines are used as the starting material. Cyclization of ketoamides, which are easily obtained by condensation of an aniline with diketene,[71] and enolethers[72] has been demonstrated.

### 13.16.2 Intramolecular Aldol Ring Closure

Another method to generate quinolinones is from a dicarbonyl intermediate under basic conditions. A malonamide can be utilized for condensation onto a ketone to provide the desired ring system after dehydration.[73]

An extension of this procedure is a modified Pfitzinger quinoline synthesis that utilizes the commonly accessible isatin as the starting material.[74]

[71]Davis, S. E.; Rauckman, B. S.; Chan, J. H.; Roth, B. *J. Med. Chem.* **1989**, *32*, 1936–1942.
[72]Janiak, C.; Deblon, S.; Uehlin, L. *Synthesis* **1999**, 959–964.
[73]Robl, J. A. *Synthesis* **1991**, 56–58.
[74]Cappelli, A.; Gallelli, A.; Manini, M.; Anzini, M.; Mennuni, L.; Makovec, F.; Menziani, M. C.; Alcaro, S.; Ortuso, F.; Vomero, S. *J. Med. Chem.* **2005**, *48*, 3564–3575.

### 13.16.3  Oxidation of a Quinoline

Oxidation of a quinoline is a very good method for the preparation of quinolinones, especially when the required quinoline is readily available. A quinoline can be oxidized to its N-oxide and rearranged to the quinolinone upon treatment with tosyl chloride.[75] Alternatively, a quinoline can be quaternized and oxidized with an oxidant such as $KMnO_4$.[76]

## 13.17  QUINOLONES (4-HYDROXYQUINOLINES)

Quinolones are an important class of compounds because of their antimicrobial activity. Fluoroquinolones have especially been targeted as antiinfectives.[77] There are three major reliable methods for the preparation of the 4-quinolone nucleus.

### 13.17.1  Electrophilic Cyclization

One of the oldest methods for the preparation of quinolones is the Gould–Jacobs reaction. In general, an aniline is condensed with diethyl(ethoxymethylene)malonate and cyclized under Friedel–Crafts conditions or elevated temperatures. When the aniline is *meta*-substituted, regiomeric mixtures may be obtained. The one drawback of this method is that the cyclization usually requires forcing conditions.[63] The vinylogous amide intermediate can be obtained by condensation of an aniline with a β-ketoester and cyclized under thermal conditions.[78] This reaction is also known as the Conrad–Limpach reaction.

[75]Payack, J. F.; Vazquez, E.; Matty, L.; Kress, M. H.; McNamara, J. *J. Org. Chem.* **2005**, *70*, 175–178.
[76]Venkov, A. P.; Statkova-Abeghe, S. M. *Tetrahedron* **1996**, *52*, 1451–1460.
[77]Radl, S.; Bouzard, D. *Heterocycles* **1992**, *34*, 2143–2177.
[78]Moyer, M. P.; Weber, F. H.; Gross, J. L. *J. Med. Chem.* **1992**, *35*, 4595–4601.

### 13.17.2    Nucleophilic Aromatic Substitution

A second and very popular method for preparation of quinolones is through a nucleophilic aromatic substitution. Because a leaving group is activated by the presence of the ketone at the *ortho*-position, a facile $S_NAr$ usually occurs. This has been the method of choice for the production of fluoroquinolones, since the additional fluorine atoms further activate the system toward cyclization.[79]

### 13.17.3    Intramolecular Aldol Ring Closure

The third method for the preparation of 4-quinolones is through a condensation of a ketone enolate with an amide followed by dehydration.[80]

## 13.18    PYRIMIDINES AND PYRIMIDONES

The pyrimidine ring system is a very prevalent heterocycle. It has been utilized in pharmaceutical agents as antifungals (voriconazole), in neuroscience (buspirone), and in oncology (imatinib). This heterocycle is usually obtained by condensation of an amidine with either a 1,3-dicarbonyl compound or a Michael acceptor.

### 13.18.1    Condensation of Amidines with 1,3-Dicarbonyl Derivatives

The most common method for preparation of pyrimidines is the reaction of an amidine with a 1,3-dicarbonyl derivative, also known as the Pinner pyrimidine synthesis. The

[79]Barnes, D. M.; Christesen, A. C.; Engstrom, K. M.; Haight, A. R.; Hsu, M. C.; Lee, E. C.; Peterson, M. J.; Plata, D. J.; Raje, P. S.; Stoner, E. J.; Tedrow, J. S.; Wagaw, S. *Org. Process Res. Dev.* **2006**, *10*, 803–807.
[80]Willemsens, B.; Vervest, I.; Ormerod, D.; Aelterman, W.; Fannes, C.; Mertens, N.; Marko, I. E.; Lemaire, S. *Org. Process Res. Dev.* **2006**, *10*, 1275–1281.

reaction is very general. When a β-ketoester is used, the 2-hydroxypyrimidine (pyrimidone) is obtained[81] while the 2-amino derivative is the product when starting from the β-cyanoester.[82] The third example below utilizes a masked dialdehyde.[83]

### 13.18.2 Condensation of Amidines with Michael Acceptors

A second popular method for the preparation of pyrimidines is the reaction of an amidine with a Michael acceptor. Vinylogous amides can be used as the electrophile for this transformation.[20,84]

[81]Tice, C. M.; Bryman, L. M. *Tetrahedron* **2001**, *57*, 2689–2700.

[82]Taylor, E. C.; Gillespie, P. *J. Org. Chem.* **1992**, *57*, 5757–5761.

[83]Zhichkin, P.; Fairfax, D. J.; Eisenbeis, S. A. *Synthesis* **2002**, 720–722.

[84]Reiter, L. A. *J. Org. Chem.* **1984**, *49*, 3494–3498.

## 13.19  QUINAZOLINES AND QUINAZOLINONES

Quinazolines have been prepared in a number of different ways and remain a highly studied heterocycle, especially since they often mimic purines as a pharmacophore. They comprise the main structural component of the "zosin" class of pharmaceutical agents for the treatment of the signs and symptoms of benign prostatic hyperplasia (BPH). 4-Aminoquinazoline derivatives such as erlotinib and gefitinib have also been used extensively as EGFR tyrosine kinase inhibitors in the treatment of cancer. Most of the methods for the preparation of this ring system begin from an anthranilic acid.

### 13.19.1  Derivatization of Anthranilic Acids

Several reagents are used for the direct conversion of an anthranilic acid to a quinazolinedione. The use of sodium or potassium cyanate under pH control is an excellent way to obtain the desired heterocycle.[85] Another commonly employed reagent is urea, although its use usually requires higher reaction temperatures.[86]

### 13.19.2  Reaction of Benzonitriles

One of the few methods to access a 2,4-disubstituted quinazoline in a single step is the addition of a Grignard reagent to a 2-aminobenzonitrile. This provides an imine intermediate that can react with an acid chloride to afford the desired quinazoline upon condensation.[87] The most common method for the direct preparation of a substituted 4-aminoquinazoline is the addition of an aniline to a benzonitrile.[88]

[85]Goto, S.; Tsuboi, H.; Kanoda, M.; Mukai, K.; Kagara, K. *Org. Process Res. Dev.* **2003**, *7*, 700–706.
[86]Lee, A. H. F.; Kool, E. T. *J. Org. Chem.* **2005**, *70*, 132–140.
[87]Bergman, J.; Brynolf, A.; Elman, B.; Vuorinen, E. *Tetrahedron* **1986**, *42*, 3697–3706.
[88]Wissner, A.; Floyd, M. B.; Johnson, B. D.; Fraser, H.; Ingalls, C.; Nittoli, T.; Dushin, R. G.; Discafani, C.; Nilakantan, R.; Marini, J.; Ravi, M.; Cheung, K.; Tan, X.; Musto, S.; Annable, T.; Siegel, M. M.; Loganzo, F. *J. Med. Chem.* **2005**, *48*, 7560–7581.

### 13.19.3 Intramolecular Condensations

A quinazoline can be accessed from a benzamide by acylation of an amine substituted at the *ortho*-position followed by cyclization under basic conditions. The condensation can proceed without loss of stereochemical integrity on the acylated sidechain.[89]

### 13.19.4 Electrophilic Aromatic Substitution

A method far less utilized for the preparation of a quinazoline is via electrophilic aromatic substitution. However, this method can be advantageous for electron-rich arene substrates, especially in light of the fact that the starting N-acylurea may be obtained by reaction of an isocyanate with an amide.[90]

### 13.20 PYRAZINES AND QUINOXALINES

Pyrazines and quinoxalines have been used extensively in the pharmaceutical industry. Two important examples of this class of compounds are eszopiclone for the treatment of insomnia and varenicline as an aid for smoking cessation.

[89]Bergman, J.; Brynolf, A. *Tetrahedron* **1990**, *46*, 1295–1310.
[90]Bandurco, V. T.; Schwender, C. F.; Bell, S. C.; Combs, D. W.; Kanojia, R. M.; Levine, S. D.; Mulvey, D. M.; Appollina, M. A.; Reed, M. S.;et al. *J. Med. Chem.* **1987**, *30*, 1421–1426.

### 13.20.1    Condensation of Dianilines

There is only one reliable method for the preparation of quinoxaline: the condensation of a dianiline with a 1,2-dicarbonyl compound. When one of the carbonyls is at a higher oxidation state, a quinoxalinone is obtained.[91] For the unsubstituted quinoxalines, aqueous glyoxal is generally utilized[92] but other glyoxal equivalents have also been employed.[93] For the preparation of pyrazines using 1,2-diaminoethane, an additional oxidation step is necessary after the ring-forming reaction.[94]

### 13.20.2    Condensation of Dicarbonyl Derivatives with Ammonia

An excellent method for the preparation of 2-hydroxypyrazine is the condensation of a 1,5-dicarbonyl compound with a source of ammonia. The substrate is easily obtained by an amide coupling of an α-ketoacid with an α-aminoketone.[95]

[91]Willardsen, J. A.; Dudley, D. A.; Cody, W. L.; Chi, L.; McClanahan, T. B.; Mertz, T. E.; Potoczak, R. E.; Narasimhan, L. S.; Holland, D. R.; Rapundalo, S. T.; Edmunds, J. J. *J. Med. Chem.* **2004**, *47*, 4089–4099.
[92]Marterer, W.; Prikoszovich, W.; Wiss, J.; Prashad, M. *Org. Process Res. Dev.* **2003**, *7*, 318–323.
[93]Venuti, M. C. *Synthesis* **1982**, 61–63.
[94]Heirtzler, F. R. *Synlett* **1999**, 1203–1206.
[95]Roberts, D. A.; Bradbury, R. H.; Brown, D.; Faull, A.; Griffiths, D.; Major, J. S.; Oldham, A. A.; Pearce, R. J.; Ratcliffe, A. H.;et al. *J. Med. Chem.* **1990**, *33*, 2326–2334.

## 13.21  FURANS AND BENZOFURANS

Furans and benzofurans have been used extensively as synthetic intermediates and also as pharmaceutical targets in antibiotics such as ceftiofur. Most of the syntheses of furans start from either a 1,3- or 1,4-dicarbonyl compound. In the case of benzofurans, metal-mediated approaches have gained widespread use in recent years.

### 13.21.1  Condensation of 1,3-Dicarbonyls with an α-Halocarbonyl

A well-precedented method for the synthesis of furans is the condensation of a 1,3-dicarbonyl compound with a α-halocarbonyl compound, known as the Feist–Bénary reaction. The resulting 1,3-dicarbonyl intermediate possesses a leaving group that undergoes cyclization to provide the furan. The advantage of this method is the wide availability of the two key starting materials. This method always affords a 3-acyl furan analog.[96] A variation of this method uses an α-hydroxyaldehyde, such as glyceraldehyde, in place of the haloaldehyde.[97]

### 13.21.2  Cyclodehydration of 1,4-Dicarbonyls

Another well-precedented method is the Paal–Knorr furan synthesis that is analogous to the better-known Paal–Knorr pyrrole synthesis. In this case, a 1,4-dicarbonyl is treated under acidic conditions to induce cyclodehydration. One of the disadvantages of this method is that the starting material is not as easily accessible as in the previous method.[98]

[96]Zambias, R. A.; Caldwell, C. G.; Kopka, I. E.; Hammond, M. L. *J. Org. Chem.* **1988**, *53*, 4135–4137.
[97]Toro, A.; Deslongchamps, P. *Synth. Commun.* **1999**, *29*, 2317–2321.
[98]Effland, R. C. *J. Med. Chem.* **1977**, *20*, 1703–1705.

### 13.21.3 Dehydration

A hydroxydihydrofuran can easily be dehydrated to the furan. This has been accomplished from an aldol adduct with sulfuric acid[99] or from the product of an organocerium addition to a ketone with $p$-TsOH.[100]

### 13.21.4 Metal-Mediated Cyclization

A more recent method for the preparation of benzofurans is the cyclization of a phenol onto an alkyne. One attractive feature of this process is that a readily available *ortho*-halophenol can be utilized in a tandem Sonogashira coupling and furan formation under the same reaction conditions.[101]

## 13.22 BENZOPYRAN-4-ONE (CHROMEN-4-ONE, FLAVONE) AND XANTHONE

The benzopyran-4-one heterocycle is abundant in natural products and has been studied for medicinal properties. It is usually prepared by intramolecular cyclization.

[99]Ragan, J. A.; Murry, J. A.; Castaldi, M. J.; Conrad, A. K.; Jones, B. P.; Li, B.; Makowski, T. W.; McDermott, R.; Sitter, B. J.; White, T. D.; Young, G. R. *Org. Process Res. Dev.* **2001**, *5*, 498–507.
[100]Ohno, M.; Miyamoto, M.; Hoshi, K.; Takeda, T.; Yamada, N.; Ohtake, A. *J. Med. Chem.* **2005**, *48*, 5279–5294.
[101]Pu, Y.-M.; Grieme, T.; Gupta, A.; Plata, D.; Bhatia, A. V.; Cowart, M.; Ku, Y.-Y. *Org. Process Res. Dev.* **2005**, *9*, 45–50.

### 13.22.1  Condensation of *ortho*-Phenoxy-1,3-dicarbonyl Derivatives

The most reliable method for the synthesis of a benzopyran-4-one is the cyclization of a phenol onto a 1,3-dicarbonyl substituent at the *ortho*-position, usually under acidic conditions.[102] An extension of this methodology is the cyclization onto a vinylogous amide under acidic conditions.[103]

### 13.22.2  Condensation of *ortho*-Acylcarbonyl Derivatives

A common method for the preparation of chromen-4-ones is the Kostanecki–Robinson reaction.[104] Depending on the substrate, this reaction will sometimes provide mixtures of the chromen-4-one and the coumarin.

### 13.22.3  Electrophilic Cyclization

The preferred method for the synthesis of xanthones is the electrophilic cyclization of an *ortho*-aryloxy carboxylic acid in the presence of Eaton's reagent (phosphorous pentoxide in MsOH). The reaction usually proceeds at room temperature but requires a careful workup to quench the residual $P_2O_5$.[105]

[102]Geen, G. R.; Giles, R. G.; Grinter, T. J.; Hayler, J. D.; Howie, S. L. B.; Johnson, G.; Mann, I. S.; Novack, V. J.; Oxley, P. W.; Quick, J. K.; Smith, N. *Synth. Commun.* **1997**, *27*, 1065–1073.

[103]Yoshimura, H.; Nagai, M.; Hibi, S.; Kikuchi, K.; Abe, S.; Hida, T.; Higashi, S.; Hishinuma, I.; Yamanaka, T. *J. Med. Chem.* **1995**, *38*, 3163–3173.

[104]Adam, W.; Rao, P. B.; Degen, H.–G.; Levai, A.; Patonay, T.; Saha-Moeller, C. R. *J. Org. Chem.* **2002**, *67*, 259–264.

[105]Sawyer, J. S.; Schmittling, E. A.; Palkowitz, J. A.; Smith, W. J., III *J. Org. Chem.* **1998**, *63*, 6338–6343.

### 13.22.4   Nucleophilic Aromatic Substitution

If one of the aromatic rings in the xanthone contains an appropriately substituted electron-withdrawing group, a nucleophilic aromatic substitution is an appropriate synthetic strategy for the preparation of these compounds from a diarylketone. The example below is operationally simple and was conducted on 140 g.[106]

### 13.23   COUMARINS

Coumarins are very well-studied natural products and pharmaceutical agents. Warfarin is probably the most recognizable member of the class of compounds, which remain as very commonly used anticoagulants despite having been on the market in the United States since 1954. The best method for their preparation is an intramolecular condensation.

### 13.23.1   Condensation of *ortho*-Acylcarbonyl Derivatives

The best method for preparation of coumarins is the Perkin reaction or its modifications. The reaction proceeds by acylation of a phenol to provide an enolizable ester, which condenses with an *ortho*-carbonyl substituent. The acylation can be accomplished using standard conditions with an acid chloride[107] or an anhydride.[108] This reaction can also lead to the chromen-4-one, especially for unactivated esters with a higher $pK_a$.

[106]Greco, M. N.; Rasmussen, C. R. *J. Org. Chem.* **1992**, *57*, 5532–5535.
[107]Rao, P. P.; Srimannarayana, G. *Synthesis* **1981**, 887–888.
[108]Li, X.; Jain, N.; Russell, R. K.; Ma, R.; Branum, S.; Xu, J.; Sui, Z. *Org. Process Res. Dev.* **2006**, *10*, 354–360.

## 13.24    THIOPHENES AND BENZOTHIOPHENES

Thiophenes and benzothiophenes have been studied extensively as pharmaceutical agents and have proven to be an important pharmacophore in several therapeutic areas. Examples of this class of compounds are clopidogrel (anticoagulant), duloxetine (antidepressive), olanzapine (antipsychotic), raloxifene (treatment and prevention of osteoporosis), and tiotropium (management of chronic obstructive pulmonary disease [COPD]). All reliable methods usually involve a cyclodehydration.

### 13.24.1    Cyclodehydration of 1,4-Dicarbonyl Derivatives

Thiophenes and benzothiophenes can be obtained by cyclodehydration of 1,4-dicarbonyl compounds in the presence of a sulfur source.[109]

### 13.24.2    Knoevenagel Condensation

Thiophenes and benzothiophenes can be obtained by Knoevenagel condensation of an enolate with a carbonyl compound under acidic[110] or basic[111] conditions. The starting material can be easily obtained and the reaction usually proceeds in high yield. The drawback of this approach resides in its specificity to the formation of 2-acylthiophenes.

### 13.24.3    Nucleophilic Addition to Sulfur Followed by Cyclocondensation

A very efficient method for the preparation of 2-aminothiophene is the Gewald reaction. An anion is generated and trapped with elemental sulfur. The resulting thiolate provides the desired aminothiophene upon condensation on the nitrile.[112]

[109]Ibrahim, Y. A.; Al-Saleh, B.; Mahmoud, A. A. A. *Tetrahedron* **2003**, *59*, 8489–8498.

[110]LaLonde, R. T.; Florence, R. A.; Horenstein, B. A.; Fritz, R. C.; Silveira, L.; Clardy, J.; Krishnan, B. S. *J. Org. Chem.* **1985**. *50*, 85–91.

[111]Hsiao, C. N.; Bhagavatula, L.; Pariza, R. J. *Synth. Commun.* **1990**, *20*, 1687–1695.

[112]Barnes, D. M.; Haight, A. R.; Hameury, T.; McLaughlin, M. A.; Mei, J.; Tedrow, J. S.; Dalla Riva Toma, J. *Tetrahedron* **2006**, *62*, 11311–11319.

## 13.25   ISOXAZOLES AND BENZISOXAZOLES

Isoxazoles and benzisoxazoles are a common class of compounds in the pharmaceutical industry. They have been used extensively as side chains for a number of β-lactams (oxacillin), for the treatment of schizophrenia (risperidone), and as an anti-inflamatory agent (leflunomide). Several methods exist for their preparation, but the most common is the condensation of hydroxylamine with a 1,3-dicarbonyl or Michael acceptor.

### 13.25.1   Addition of Hydroxylamine to 1,3-Dicarbonyl Derivatives or Enaminoketones

Condensation of hydroxylamine to a 1,3-dicarbonyl compound or a Michael acceptor containing a β-leaving group is a common way to prepare isoxazoles. This method is especially useful for symmetrical 1,3-diketones[113] substrates with a clear regiochemical preference for the amine addition.[114] Unsymmetrical 1,3-diketones typically yield a mixture of regioisomers unless one of the carbonyl is converted to an enaminoketone to induce regioselectivity.[115]

[113]Mashraqui, S. H.; Keehn, P. M. *J. Org. Chem.* **1983**, *48*, 1341–1344.

[114]Wiles, C.; Watts, P.; Haswell, S. J.; Pombo-Villar, E. *Org. Process Res. Dev.* **2004**, *8*, 28–32.

[115]Ohigashi, A.; Kanda, A.; Tsuboi, H.; Hashimoto, N. *Org. Process Res. Dev.* **2005**, *9*, 179–184.

### 13.25.2   Dipolar Cycloaddition

The reaction of a nitrone[116] or hydroximoyl halide[117] with an acetylene is a method for the rapid and highly convergent preparation of highly substituted isoxazoles. However, the yield and regioselectivity can vary greatly depending on the substrate.

### 13.25.3   Nucleophilic Aromatic Substitution

Benzisoxazoles have been prepared by intramolecular nucleophilic aromatic substitution. A carbonyl compound is treated with hydroxylamine to provide an oxime that undergoes cyclization.[118] For this reaction to be productive, the Z-isomer of the oxime must be able to isomerize to the E-isomer that undergoes cyclization.

### 13.26   OXAZOLES AND BENZOXAZOLES

Oxazoles and benzoxazoles are a common class of compounds in the pharmaceutical industry. Oxaprozin is a nonsteroidal anti-inflammatory from this structural class. Two excellent methods are utilized for their syntheses depending on the desired substitution pattern in the product.

[116]Lee, C. K. Y.; Herlt, A. J.; Simpson, G. W.; Willis, A. C.; Easton, C. J. *J. Org. Chem.* **2006**, *71*, 3221–3231.

[117]Yao, C.-F.; Kao, K.-H.; Liu, J.-T.; Chu, C.-M.; Wang, Y.; Chen, W.-C.; Lin, Y.-M.; Lin, W.-W.; Yan, M.-C.; Liu, J.-Y.; Chuang, M.-C.; Shiue, J.-L. *Tetrahedron* **1998**, *54*, 791–822.

[118]Fink, D. M.; Kurys, B. E. *Tetrahedron Lett.* **1996**, *37*, 995–998.

### 13.26.1   Cyclodehydration

The cyclodehydration of an acylated $\alpha$-aminoketone is the most common method for the preparation of oxazoles. The reagent of choice for this preparation is $POCl_3$, although many other dehydrating agents have been utilized. One of the biggest advantages of the method is the ease of preparation of the required starting material.[119]

### 13.26.2   Dipolar Cycloaddition

The dipolar cycloaddition of an isonitrile, most commonly tosylmethyl isocyanide (TosMIC) with an aldehyde is a popular method for the preparation of 5-substituted oxazoles. The reaction proceeds in high yield on simple substrates[120] as well as on materials that might be sensitive to dehydrating conditions.[121]

## 13.27   ISOTHIAZOLES AND BENZISOTHIAZOLES

Isothiazoles and benzisothiazoles are usually prepared through an intramolecular cyclization using a number of different methods. The choice of method usually depends on the ease of preparation of the requisite starting material. An example of this class of compound is zaprasidone, an atypical antipsychotic.

[119]Godfrey, A. G.; Brooks, D. A.; Hay, L. A.; Peters, M.; McCarthy, J. R.; Mitchell, D. *J. Org. Chem.* **2003**, *68*, 2623–2632.
[120]Herr, R. J.; Fairfax, D. J.; Meckler, H.; Wilson, J. D. *Org. Process Res. Dev.* **2002**, *6*, 677–681.
[121]Anderson, B. A.; Becke, L. M.; Booher, R. N.; Flaugh, M. E.; Harn, N. K.; Kress, T. J.; Varie, D. L.; Wepsiec, J. P. *J. Org. Chem.* **1997**, *62*, 8634–8639.

### 13.27.1 Intramolecular Cyclization

An efficient method to generate the isothiazole ring is through an intramolecular cyclization to generate the N–S bond. This method has proven successful under oxidative conditions starting from thioamides in the preparation of 5-aminoisothiazoles.[122] This strategy also works in the preparation of benzothiazoles from benzophenone derivatives,[123] and iso-thiazoles from oximes.[124]

### 13.27.2 Addition to *ortho*-Thiobenzonitriles

One of the best synthetic methods for the preparation of a 3-aminobenzisothiazole is by addition of an amine to an *ortho*-thiobenzonitrile. The disulfide is generally used as the thiophenol derivative since it is easily prepared and has a good leaving group embedded in it.[125] This strategy is not as efficient for the preparation of carbon derivatives at the 3-position.[126]

[122]Etzbach, K. H.; Eilingsfeld, H. *Synthesis* **1988**, 449–452.

[123]Dehmlow, H.; Aebi, J. D.; Jolidon, S.; Ji, Y.-H.; Von Mark, E. M.; Himber, J.; Morand, O. H. *J. Med. Chem.* **2003**, *46*, 3354–3370.

[124]Dieter, R. K.; Chang, H. J. *J. Org. Chem.* **1989**, *54*, 1088–1092.

[125]Walinsky, S. W.; Fox, D. E.; Lambert, J. F.; Sinay, T. G. *Org. Process Res. Dev.* **1999**, *3*, 126–130.

[126]Chimichi, S.; Giomi, D.; Tedeschi, P. *Synth. Commun.* **1993**, *23*, 73–78.

## 13.28   THIAZOLES AND BENZOTHIAZOLES

Thiazoles and benzothiazoles have been studied extensively as pharmaceutical and agro-chemical agents. A few methods have proven to be reliable for their preparation, most notably alkylation of thioamide derivatives followed by cyclodehydration.

### 13.28.1   Condensation of Thioamides with Haloketones

A practical method for the preparation of a thiazole is the condensation of a thioacetamide with a haloketone, which generally proceeds in high yield.[127] This reaction also works on more complicated substrates such as thioureas to generate 2,4-diaminothiazoles.[128]

[127]Rooney, C. S.; Cochran, D. W.; Ziegler, C.; Cragoe, E. J., Jr.; Williams, H. W. R. *J. Org. Chem.* **1984**, *49*, 2212–2217.
[128]Masquelin, T.; Obrecht, D. *Tetrahedron* **2001**, *57*, 153–156.

### 13.28.2    Condensation of Carboxylic Acid Derivatives with 2-Aminothiols

The most common method for preparation of benzothiazoles is the reaction of 2-aminothiols with carboxylic acid derivatives. Reaction with an acid chloride is usually straightforward, and toluene has been demonstrated to be the preferred solvent for the transformation.[129]

### 13.28.3    Nucleophilic Aromatic Substitution

A less commonly used method for preparation of benzothiazoles is an intramolecular nucleophilic aromatic substitution. This method is especially attractive for electron-deficient substrates.[130]

[129]Rudrawar, S.; Kondaskar, A.; Chakraborti, A. K. *Synthesis* **2005**, 2521–2526.
[130]Zhu, L.; Zhang, M. *J. Org. Chem.* **2004**, *69*, 7371–7374.

# 14

## ACCESS TO CHIRALITY

Robert W. Dugger

## 14.1 INTRODUCTION

Synthesis of optically active compounds, whether natural products or synthetic molecules, is a significant challenge to the synthetic organic chemist.[1] Throughout this book are many examples of using chiral reagents, catalysts and auxiliaries to influence the creation of new chiral centers (hydroboration, hydrogenation, aldol reactions, etc.). In this chapter we will focus on other methods of obtaining molecules in enantiomerically pure form.

## 14.2 USING THE CHIRAL POOL

An obvious starting point for the synthesis of chiral molecules is to use a natural product that is readily available as a single enantiomer. Many sugars, terpenes, amino acids, and so on, have been used as the starting materials for the synthesis of a wide variety of chiral molecules. Obviously this method works best when there is a large degree of structural similarity between the chiral pool starting material and the final product. The greater the structural differences, the more complex the synthesis will become. Additionally, in most instances nature has not always seen fit to provide us with both enantiomers of a chiral molecule, thereby further limiting the availability of starting materials.

## 14.3 CLASSICAL RESOLUTIONS

Despite advances in enantioselective synthesis, classical resolution is still one of the most widely utilized methods for the production of enantiopure substances. The most common

---

[1]Corey, E. J.; Kurti, L. *Enantioselective Chemical Synthesis*; Direct Book Publishing: Dallas, TX, **2010**.

*Practical Synthetic Organic Chemistry: Reactions, Principles, and Techniques*, First Edition.
Edited by Stéphane Caron.
© 2011 John Wiley & Sons, Inc. Published 2011 by John Wiley & Sons, Inc.

variation is the formation of diastereomeric salts via combination of a racemate with a single enantiomer, producing a 1 : 1 mixture of diastereomeric salts. These salts can then be separated by physical means, usually by crystallization from an appropriate solvent or combination of solvents. Since the goal is to try to selectively crystallize one diastereomeric salt while leaving the other in solution, solvents of intermediate polarity, like the lower alcohols, often work the best. Since organic salts are highly soluble in polar organic solvents such as DMF, such solvents are rarely used since both salts will likely dissolve. Likewise very nonpolar solvents (heptane, toluene, etc.) will not dissolve most salts and thereby will not be useful for selective crystallization of one salt. A survey of diastereomeric salt resolutions[2] shows that over 50% of published resolutions are carried out in the lower alcohols (MeOH, EtOH and i-PrOH), sometimes with a small amount of water added. Other common solvents are EtOAc and acetone.

Fortunately, a variety of chiral acids and bases are available from the chiral pool or are inexpensive commodity chemicals. Acids such as tartaric, dibenzoyltartaric, ditoluoyltartaric, mandelic, and camphorsulfonic are often used to resolve basic molecules (75% of the cases in reference 1) and bases such as sec-phenethylamine, brucine, ephedrine, quinine, cinchonine, and cinchonidine are typically used to resolve acids (70% of the cases in reference 1).

For example, treatment of racemic phenethylamine with (+)-tartaric acid in methanol results in the precipitation of the (–)-phenethylamine-(+)-tartaric acid salt.[3] Conversion of the salt to the free amine using NaOH followed by distillation produces (S)-phenethylamine in 29% yield (58% of theory) and 98% ee.

In cases where the racemate is not acidic or basic, it may be possible to temporarily derivatize the molecule, to make it acidic or basic, conduct the resolution, and then remove the derivatization. For example, 1-phenyl-1-propanol was treated with maleic anhydride producing the monoester, which was efficiently resolved with cinchonidine (41% yield, >99% ee).[4]

[2]Kozma, D. *CRC Handbook of Optical Resolutions via Diastereomeric Salt Formation*; CRC Press: Boca Raton, FL, **2001**.

[3]Ault, A. *Org. Syn.* **1969**, *49*, 93–98.

[4]Kiss, V.; Egri, G.; Bálint, J.; Fogassy, E. *Chirality* **2006**, *18*, 116–120.

An empirical screening method that can be used to select a resolving agent and solvent to first attempt to scale has been published. It relies on slurrying the racemic compound with the resolving agent in various solvents and assessing the effectiveness of the resolution by HPLC analysis of the mother liquor.[5] Another method using differential scanning calorimetry (DSC) to determine whether or not resolving agents will form diastereomeric salts has also been published.[6]

### 14.3.1 The Family Approach (the Dutch Resolution)

In an attempt to accelerate the process of finding the best resolving agent, researchers at Syncom treated a racemic compound with a mixture of resolving agents in the hopes that the least soluble diastereomeric salt would precipitate, thereby self-selecting the best resolving agent from the mixture.[7] To their surprise, the salts that crystallized contained mixtures of the resolving agents and the ratio did not change appreciably upon recrystallization of the salt. Additionally, the optical purity of the resolved material was often higher than that obtained by forming salts with the individual resolving agents.

For example, treatment of $p$-methylphenethylamine with a 1 : 1 mixture of ($S$)-mandelic acid and ($S$)-$p$-methylmandelic acid produced crystals that contained amine of 87% ee and contained a 1 : 4 ratio of the two acids. Treating $p$-methylphenethylamine with only ($S$)-$p$-methylmandelic acid produced crystals in which the amine was only 57% ee.

### 14.3.2 Separation of Covalent Diastereomers

Another resolution strategy is to covalently derivatize a racemate with a single enantiomer of a chiral molecule, producing a mixture of diastereomers that can be separated by crystallization or chromatography. A particularly successful example of this process is the reaction of racemic timolol with dibenzoyltartaric acid anhydride in acetone.[8] A single diastereomer precipitates from the reaction mixture in 42% yield (50% maximum) and in high optical purity. Hydrolysis to the optically pure timolol is straightforward.

### 14.3.3 Kinetic Resolutions

A kinetic resolution occurs when a racemate undergoes reaction with a chiral agent (reagent, catalyst, etc.) and the enantiomers react at different rates. If the difference in reaction rates is high enough, one enantiomer of product and/or starting material can often be isolated in high

[5]Borghese, A.; Libert, V.; Zhang, T.; Alt, C. A. *Org. Process Res. Dev.* **2004**, *8*, 532–534.

[6]Dyer, U. C.; Henderson, D. A.; Mitchell, M. B. *Org. Process Res. Dev.* **1999**, *3*, 161–165.

[7]Vries, T.; Wynberg, H.; Van Echten, E.; Koek, J.; Ten Hoeve, W.; Kellogg, R. M.; Broxterman, Q. B.; Minnaard, A.; Kaptein, B.; Van der Sluis, S.; Hulshof, L.; Kooistra, J. *Angew. Chem., Int. Ed.* **1998**, *37*, 2349–2354.

[8]Varkonyi-Schlovicsko, E.; Takacs, K.; Hermecz, I. *J. Heterocycl. Chem.* **1997**, *34*, 1065–1066.

yield. As with classical salt-forming resolutions, the maximum yield is 50%. If the reaction follows first order kinetics, equations can be derived to calculate the enantioselectivity (s) of a reaction based on the percent conversion (C) and ee of the product (eq. 1). Likewise, the enantioselectivity can be calculated from the ee of the unreacted starting material (ee', eq. 2).

$$s = \frac{\ln[(1 - C)(1 - ee)]}{\ln[(1 - C)(1 + ee)]} \tag{1}$$

$$s = \frac{\ln[1 - C(1 + ee')]}{\ln[1 - C(1 - ee')]} \tag{2}$$

In order to obtain a high yield and high optical purity, the enantioselectivity of the reaction needs to be greater than 100. If the enantioselectivity is lower, it is still possible to obtain high ee material if one is willing to sacrifice yield. For example, if the unreacted starting material is the desired product and enantioselectivity is 10, running the reaction to 70% conversion should produce unreacted starting material in >99% ee. Although there are many examples of non-enzymatic kinetic resolution processes, this is an area in which enzymatic reactions are used extensively.

*14.3.3.1  Resolution of Alcohols*    A classic example of kinetic resolution can be found in the Sharpless epoxidation of chiral, racemic allylic alcohols. For example, epoxidation of 1-nonen-3-ol using ( + )-diisopropyl tartrate, Ti(O-*i*-Pr)$_4$ and *t*-butylhydroperoxide (TBHP) to 52% conversion produced the epoxide in 49% yield and >96% ee (erythro/threo ratio was 99 : 1).[9] The unreacted alcohol was the (R) enantiomer.

Enzymatically, alcohols can be resolved either by acylation of the alcohol or by hydrolysis of an ester of the alcohol. It is important to note that these are complementary processes.

[9]Martin, V. S.; Woodard, S. S.; Katsuki, T.; Yamada, Y.; Ikeda, M.; Sharpless, K. B. *J. Am. Chem. Soc.* **1981**, *103*, 6237–6240.

If an enzyme has a preference for reacting with the (S)-enantiomer then an acylation process will produce the ester of the (S)-enantiomer and the unreacted alcohol will be the (R)-enantiomer. If the same enzyme is used in a hydrolytic process the enzymatic reaction will produce the (S)-alcohol and the remaining ester will be the (R)-enantiomer. Typically, vinyl esters, isopropenyl esters or acid anhydrides are used as the acylating agent because they react irreversibly. Although used extensively to prepare small quantities of optically pure alcohols, the major problem with this approach is the separation of the alcohol and ester. Often the physical properties are similar and the only method for separation is chromatography. One approach to solving this problem is the use of cyclic anhydrides as the acylating agent. The product and unreacted starting material can then easily be separated by extraction. In the example below, the racemic hydroxynitrile was acylated with succinic anhydride catalyzed by the lipase CALB in an immobilized form (Novozym 435).[10] The hemisuccinate was extracted into a mildly basic aqueous phase and the unreacted hydroxynitrile remained in the MTBE layer. Addition of NaOH to the aqueous layer hydrolyzed the hemisuccinate and the desired product was extracted into MTBE.

Another interesting solution for the separation issue is to make the acylating agent highly lipophilic so that the acylated product is very soluble in nonpolar organic solvents such as heptane.[11] The unreacted alcohol can be washed out of the heptane layer using water or methanol.

[10]Vaidyanathan, R.; Hesmondhalgh, L.; Hu, S. *Org. Process Res. Dev.* **2007**, *11*, 903–906.
[11]ter Halle, R.; Bernet, Y.; Billard, S.; Bufferne, C.; Carlier, P.; Delaitre, C.; Flouzat, C.; Humblot, G.; Laigle, J. C.; Lombard, F.; Wilmouth, S. *Org. Process Res. Dev.* **2004**, *8*, 283–286.

*14.3.3.2  Resolution of Amines*   Like alcohols, some amines can be resolved by enzymatic acylation. A number of lipases are known to catalyze the acylation of amines by esters. In this case, the products are usually easy to separate by extraction since the product is neutral and the remaining unreacted amine is basic. One example is the acylation of *sec*-phenethylamine with ethyl methoxyacetate, catalyzed by a lipase from *B. plantarii*.[12] Running the reaction to 52% conversion affords the (*S*)-enantiomer of phenethylamine in 46% yield and the methoxyamide in 48% yield and 93% ee.

*14.3.3.3  Resolution of Epoxides*   Epoxides are a very useful functional group due to the ease with which they can be transformed into other functionality. Numerous strategies have been employed to produce optically enriched epoxides (Section 3.10.4). An alternative strategy would be a kinetic resolution approach. Jacobsen has discovered that (salen) Co complexes can effectively hydrolyze many terminal epoxides with very high enantioselectivity.[13]

## 14.4   DYNAMIC KINETIC RESOLUTIONS

Unlike classical resolutions that can produce a maximum 50% yield of the desired enantiomer, dynamic resolutions allow for the production of a 100% yield of a single enantiomer from a racemate. This can be accomplished by incorporating a racemization reaction into the resolution process.

[12]Balkenhohl, F.; Ditrich, K.; Hauer, B.; Ladner, W. *J. Prakt. Chem./Chem-Ztg* **1997**, *339*, 381–384.
[13]Schaus, S. E.; Brandes, B. D.; Larrow, J. F.; Tokunaga, M.; Hansen, K. B.; Gould, A. E.; Furrow, M. E.; Jacobsen, E. N. *J. Am. Chem. Soc.* **2002**, *124*, 1307–1315.

### 14.4.1  Dynamic Kinetic Resolutions via Chemical Reactions

One of the classic examples of dynamic kinetic resolution was published by Noyori in 1989.[14] He demonstrated that it was possible to reduce $\alpha$-substituted $\beta$-keto esters with a chiral catalyst to produce a single diastereomer in high ee. The $\beta$-ketoesters rapidly racemize via tautomerization to the enol form and one enantiomer of the keto form is selectively reduced by the chiral catalyst. One key example is the reduction of an $\alpha$-amidomethyl substrate to the corresponding $\beta$-hydroxyester in 98% ee. The product can be converted into the acetoxy azetidinone, a key starting point for many $\beta$-lactam antibiotics.

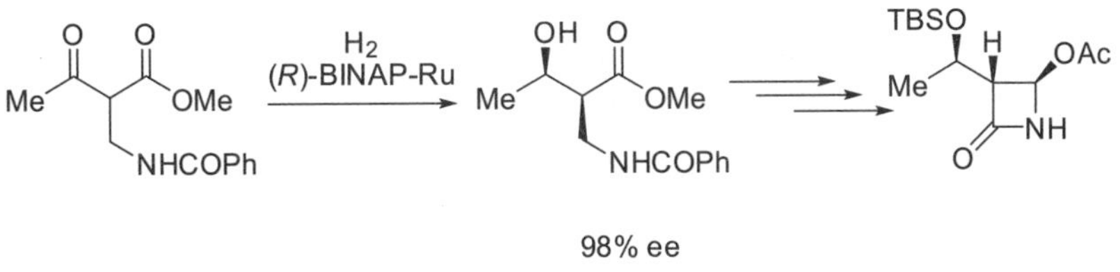

98% ee

A number of optically enriched amino acids can be produced from their corresponding racemates via an enzymatically catalyzed hydrolysis of the readily enolized azalactone derivatives.[15] This is a particularly useful way to make non-natural amino acids.

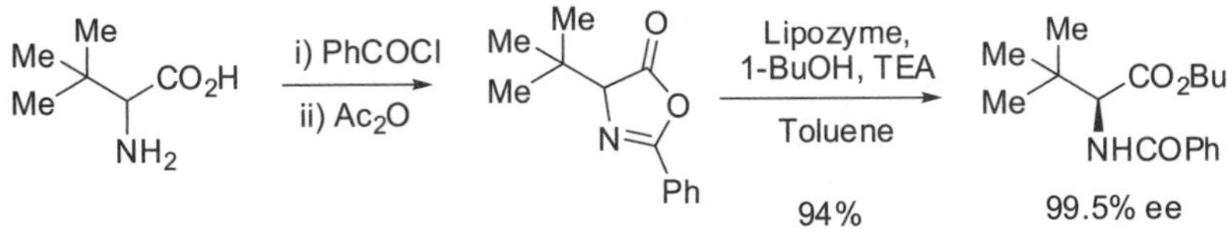

An interesting method for performing a dynamic kinetic resolution of secondary alcohols has evolved over the past decade. Several metals are known to racemize secondary alcohols via a reversible oxidation/reduction process. With readily oxidized alcohols, the redox process is rapid enough that it can be coupled with an enzymatic acylation to produce high yields of the resolved alcohol ester. For example, Bäckvall has published a procedure for resolving phenethyl alcohol on mole scale using only 0.05 mol% of the ruthenium catalyst shown below and the enzyme *Candida antarctica* lipase B (CALB).[16]

[14]Noyori, R.; Ikeda, T.; Ohkuma, T.; Widhalm, M.; Kitamura, M.; Takaya, H.; Akutagawa, S.; Sayo, N.; Saito, T.; et, a. *J. Am. Chem. Soc.* **1989**, *111*, 9134–9135.

[15]Turner, N. J.; Winterman, J. R.; McCague, R.; Parratt, J. S.; Taylor, S. J. C. *Tetrahedron Lett.* **1995**, *36*, 1113–1116.

[16]Bogár, K.; Martín-Matute, B.; Bäckvall, J.-E. *Beilstein J. Org. Chem.* **2007**, *3*, 50.

A similar process has been demonstrated with amines using a catalyst that induces the epimerization at lower temperatures.[17]

### 14.4.2   Dynamic Kinetic Resolutions via Crystallization

The dynamic kinetic resolution can also be driven by a physical process, such as crystallization, rather than by a chemical transformation. For example, a group at Glaxo intensively studied the resolution of phenylglycinate esters with L-tartaric acid.[18] A key finding of their work, which has been adopted by many others, was the observation that addition of a variety of carbonyl compounds accelerated the rate of the racemization process.

Presumably, the carbonyl additive reacts with the amine to form an imine that enhances the acidity of the α-proton, thereby increasing the rate of enolization and epimerization. Another interesting example was reported by a Hoechst group.[19] D-p-hydroxyphenylglycine is used in the side chain of a number of semisynthetic β-lactams. They found that heating racemic p-hydroxyphenylglycine with (+)-3-bromocamphor-8-sulfonic acid (BCSA) to 70°C in acetic acid with salicylaldehyde as the epimerization catalyst produced a 99% yield of the salt of the D isomer in 98–99% de.

[17]Blacker, A. J.; Stirling, M. J.; Page, M. I. *Org. Process Res. Dev.* **2007**.
[18]Clark, J. C.; Phillipps, G. H.; Steer, M. R. *J. Chem. Soc., Perkin Trans. 1* **1976**, 475–481.
[19]Bhattacharya, A.; Araullo-Mcadams, C.; Meier, M. B. *Synth. Commun.* **1994**, *24*, 2449–2459.

99%

98% de

This process can also be applied to other amino acids with less acidic α-protons. For example, (R)-proline can be obtained from the natural occurring (S)-proline or racemic proline by treatment with D-tartaric acid in butanoic acid at 80°C with 10 mol% butanal.[20] The (R)-proline • D-tartaric acid salt is obtained in 97% yield and 93–95% de.

97%

93–95% de

More complex molecules are also readily amenable to dynamic kinetic resolution. For example, the fluorophenyloxazine precursor to the antiemetic drug aprepitant undergoes a dynamic kinetic resolution with (–)-3-bromocamphor-8-sulfonic acid (BCSA) in 90% yield and 99% de.[21]

90%

99% de

In all of the cases above, the racemization occurs under acidic conditions. Merck chemists have reported a case where the epimerization occurs faster under basic conditions. Treatment of the 3-aminobenzodiazapinone with a full equivalent of (+)-10-camphorsulfonic acid (CSA) led to very slow racemization. If slightly less than one equivalent was used, racemization was substantially faster, providing a 91% yield of the salt (99% based on CSA) after 12 hours at 25°C.[22]

91%

>99.5% de

[20]Shiraiwa, T.; Shinjo, K.; Kurokawa, H. *Bull. Chem. Soc. Jpn.* **1991**, *64*, 3251–3255.
[21]Kolla, N.; Elati, C. R.; Arunagiri, M.; Gangula, S.; Vankawala, P. J.; Anjaneyulu, Y.; Bhattacharya, A.; Venkatraman, S.; Mathad, V. T. *Org. Process Res. Dev.* **2007**, *11*, 455–457.
[22]Reider, P. J.; Davis, P.; Hughes, D. L.; Grabowski, E. J. J. *J. Org. Chem.* **1987**, *52*, 955–957.

The vast majority of examples in the literature involve enolization at a chiral center α to a carbonyl, but there are also examples of atropisomer interconversion, Michael/retro-Michael reactions, and reversible additions to a carbonyl.[23,24]

## 14.5 DESYMMETRIZATION OF *MESO* COMPOUNDS

One intriguing method for producing a chiral compound involves the desymmetrization of a *meso* compound. A particularly useful aspect of this transformation is the potential to convert all of the *meso* compounds into a single enantiomer product. It is also possible to produce several chiral centers at once using this method. Typically the desymmetrization is accomplished by selectively derivatizing one of two pro-chiral functional groups present in the molecule. Diols, diesters, and cyclic anhydrides are the typical substrates. For example, 3-methyl glutaric anhydride was reacted with (S)-1-(1'-naphthyl)ethanol, followed by esterification with diazomethane to give a high yield of the diester.[25] Subsequent conversion to 3-methylvalerolactone and comparison to literature data showed that the (R)-enantiomer had been produced in approximately 86% ee.

Numerous other examples with various nucleophiles exist.[26] The major drawback of such an approach is the use of a stoichiometric amount of a chiral agent that will usually have to be removed later. Catalytic methods are preferable and enzymatic methods have been particularly valuable.[27] For example, the monoacid below can be obtained by resolution of the racemate with cinchonidine, but of course the maximum yield is 50%. Chemists at Bristol-Myers Squibb developed an enzymatic hydrolysis of the *meso* diester that provided the desired mono acid in >99% ee and 98% yield.[28]

[23]Pellissier, H. *Tetrahedron* **2003**, *59*, 8291–8327.
[24]Pellissier, H. *Tetrahedron* **2008**, *64*, 1563–1601.
[25]Theisen, P. D.; Heathcock, C. H. *J. Org. Chem.* **1993**, *58*, 142–146.
[26]Atodiresei, I.; Schiffers, I.; Bolm, C. *Chem. Rev.* **2007**, *107*, 5683–5712.
[27]Garcia-Urdiales, E.; Alfonso, I.; Gotor, V. *Chem. Rev.* **2005**, *105*, 313–354.
[28]Goswami, A.; Kissick, T. P. *Org. Process Res. Dev.* **2009**, *13*, 483–488.

In another example, a key building block for the synthesis of vitamin D analogs can be obtained in excellent ee by acylation of the *meso* diol.[29]

## 14.6   CHIRAL CHROMATOGRAPHY

Preparative chiral column chromatography is a valuable method for the resolution of enantiomers. For small amounts of material, it is often easier and faster to use a racemic synthesis followed by chiral column chromatography than to develop an asymmetric synthesis. This is particularly true in medicinal chemistry, where the individual enantiomers need to be assayed but the asymmetric synthesis of numerous analogs would be prohibitively time consuming.[30] Even on multi-kilogram scale, chiral column chromatography can be practical if the molecule to be resolved is relatively inexpensive and the chromatography is efficient in terms of having a good resolution ($\alpha$ value) and a high loading. Because of the potential high cost of the chiral stationary phase for large scale applications, it is often necessary to reuse the chiral stationary phase to lower the economic impact. One interesting example of the large scale application of chiral chromatography is in the resolution of the tetralone used in the manufacture of sertraline.[31]

The desired 4-(S)-tetralone is obtained in >949% yield and 99.7% ee. In this particular case the undesired enantiomer can be racemized by treatment with base, providing more racemic feed.

Another interesting example is the chromatographic resolution of cetirizine.[32] In this case, derivatives of the desired acid were found to be much more soluble in the mobile phase

[29]Hilpert, H.; Wirz, B. *Tetrahedron* **2001**, *57*, 681–694.

[30]Leonard, W. R.; Henderson, D. W.; Miller, R. A.; Spencer, G. A.; Sudah, O. S.; Biba, M.; Welch, C. J. *Chirality* **2007**, *19*, 693–700.

[31]Quallich, G. J. *Chirality* **2005**, *17*, S120–S126.

[32]Pflum, D. A.; Wilkinson, H. S.; Tanoury, G. J.; Kessler, D. W.; Kraus, H. B.; Senanayake, C. H.; Wald, S. A. *Org. Process Res. Dev.* **2001**, *5*, 110–115.

thereby increasing the potential loading and efficiency relative to chromatography of the acid. The efficiency was further increased by screening different derivatives to find the one with the highest resolution ($\alpha$ value), which led to the selection of the primary amide. Chromatography produced the desired enantiomer in 99.8% ee and 98% recovery.

Cetirizine

# 15

# SYNTHETIC ROUTE DEVELOPMENT OF SELECTED CONTEMPORARY PHARMACEUTICAL DRUGS

Stéphane Caron

## 15.1 INTRODUCTION

The identification of a synthetic route to a molecule can be influenced by many factors such as the availability of a starting material, the ability to derivatize a key intermediate to a larger number of synthetic targets, the realistic amount of time allocated for its demonstration or simply the chemists preference for a given synthetic sequence or specific reaction. This is true in both academia and industry. Chemists in academia developing a new synthetic methodology have often proceeded with a total synthesis of a natural product to showcase the power of the methodology even if the synthesis itself was not necessarily the most efficient way to construct the target. Medicinal chemists think about divergent synthesis that can provide several potential targets as opposed to the best synthesis of individual molecules. The process chemist, however, is given the task to ultimately deliver the best possible synthesis of the active pharmaceutical ingredient or drug substance. The development of a process to make a new drug is often evolutionary since the synthetic technology must be adequate to support a drug candidate to the next milestone as rapidly as possible. Often, the medicinal chemistry route will be enabled to allow the preparation of enough material to support pre-clinical safety evaluation and introduction of the candidate to the clinic. As demand for "bulk" increases, the route needs to improve to ultimately meet the objectives of the manufacturing division with regards to efficiency, robustness, environmental impact, cost, and other considerations.

*Practical Synthetic Organic Chemistry: Reactions, Principles, and Techniques*, First Edition.
Edited by Stéphane Caron.
© 2011 John Wiley & Sons, Inc. Published 2011 by John Wiley & Sons, Inc.

In this chapter, the process history of a selection of new drugs approved in the United States by the Food and Drug Administration (FDA) in recent years will be presented. This chapter should provide an idea of the concerns addressed by the process chemists to ensure that a chemical process is appropriate to meet the needs of a project at various stages of development. Told bold numbers between parentheses indicates compounds shown in the schemes at the end of each sections.

## 15.2  VAPRISOL® (CONIVAPTAN HYDROCHLORIDE)

### 15.2.1  Background

Conivaptan is a nonpeptide, dual antagonist of arginine vasopressin (AVP) $V_{1A}$ and $V_2$ receptors. The drug is indicated for the treatment of euvolemic hyponatremia, the syndrome of inappropriate secretion of antidiuretic hormone, or in the setting of hypothyroidism, adrenal insufficiency, pulmonary disorder, and so on. Conivaptan works by raising blood sodium levels in a patient who has too little sodium in the bloodstream but a normal amount of fluid in the body.

### 15.2.2  Medicinal Chemistry Synthesis

The medicinal chemistry route to conivaptan builds the target molecule through multiple amide bond formations and other highly precedented chemical methods. Starting from the commercially available 1,2,3,4-tetrahydro-benzo[*b*]azepin-5-one (**1**), acylation is conducted with *p*-nitrobenzoyl chloride (**2**) in dichloromethane in the presence of triethylamine. The nitro group is reduced to the aniline (**4**) by catalytic hydrogenation using either Pd/C or Raney Nickel in methanol.[1] In the third step of the synthesis, the acid chloride (**5**) obtained from the reaction of biphenyl-2-carboxylic acid (**6**) and oxalyl chloride in dichloromethane in the presence of catalytic DMF is reacted with the aniline using triethylamine as a base to provide amide **7**. The end game of the synthesis focuses on the installation of the imidazole heterocycle. α-Bromination using either bromine or $CuBr_2$ in chloroform provides the bromoketone (**8**), which is condensed with acetamidine hydrochloride (**9**) to provide the crude conivaptan free base (**10**). After chromatography, the free base is converted to the salt using HCl in ethyl acetate leading to conivaptan hydrochloride **11**.[2] The medicinal chemistry route suffered from a low overall yield, particularly due to the inefficient conversion of ketone **7** to the API proceeding in only 38% yield. However, the formation of the two amide bonds is straightforward as one would expect and probably allowed for a study of the SAR.

### 15.2.3  Second Route to Conivaptan

In the second synthesis reported, the steps are essentially the same as the discovery route but have been reordered. The imidazole portion of the molecule is introduced earlier in the synthesis and the most costly starting material, the carboxylic acid precursor to acid chloride

---

[1]Ogawa, H.; Yamashita, H.; Kondo, K.; Yamamura, Y.; Miyamoto, H.; Kan, K.; Kitano, K.; Tanaka, M.; Nakaya, K. et al. *J. Med. Chem.* **1996**, *39*, 3547–3555.
[2]Matsuhisa, A.; Taniguchi, N.; Koshio, H.; Yatsu, T.; Tanaka, A. *Chem. Pharm. Bull. (Tokyo).* **2000**, *48*, 21–31.

**5**, is introduced late in the synthesis. Bromination of ketone **3** is accomplished with bromine and the resulting bromoketone is reacted with acetamidine **9**. The condensation results in a low yield as a mixture of the desired imidazole (**13**) and undesired oxazole (**14**) is obtained. This reaction is highly sensitive to the water content and higher levels resulted in a 10:1 ratio favoring the desired imidazole.[3] Reduction of the nitroarene and amide formation proceeded uneventfully. Overall, this synthesis is superior since it is one less step and the chromatography of the penultimate intermediate was eliminated although the synthesis is still linear. The problem in the imidazole formation was addressed earlier in the synthesis and minimized the amount of oxazole generated.

### 15.2.4   Third Route to Conivaptan

The third route reported attempts to bring the remaining problematic step, the formation of the imidazole, early in the synthesis. It starts from tosylamide **16** that is easily brominated with pyridinum hydrobromide perbromide, a much more practical brominating agent than CuBr$_2$. Bromide **17** is alkylated with imidate **9** and after cyclodehydration, a 2:1 ratio of the imidazole **18** to the undesired oxazole is obtained. The protecting group is cleaved using H$_2$SO$_4$ in AcOH and the resulting aniline (**19**) is acylated with acid chloride **2**. From this point forward, a sequence similar to the medicinal chemistry route afforded the desired API in 13% overall yield.[3] This route suffered from its linearity since many steps were required for the installation of the bottom side chain.

### 15.2.5   Fourth Route to Conivaptan

The fourth route for the preparation of conivaptan hydrochloride was reported in 2005.[4] This final route presents the advantage of having a very high level of convergency, a more environmentally friendly selection of solvents and better process efficiency. Further evaluation of the imidazole formation from the previous route provided conditions leading to a 10:1 ratio in 69% yield. More importantly, the chlorinated solvent was eliminated and the rate of the reaction was greatly enhanced. The first two steps of the synthesis were telescoped by using toluene as the extraction solvent for the bromination and utilizing the toluene solution directly in the imidazole formation to provide the desired benzazepine. The entire side chain was efficiently prepared by utilization of 4-aminobenzoic acid (**20**), the previous reduction of the nitroarene was no longer necessary. Finally, the cyclic amine of **23** was acylated with the acid chloride in acetonitrile, which allows for direct formation of the desired HCl salt and isolation in ethanol. Overall, the final developed process has eliminated undesirable solvents, especially chloroform and pyridine, and utilizes toluene for several operations including multiple steps without isolation of the intermediates. The synthesis is highly convergent since the bottom side chain is prepared and coupled to the cyclic amine. Furthermore, the final step was cleverly designed to directly provide the desired final form.

[3]Tsunoda, T.; Tanaka, A.; Mase, T.; Sakamoto, S. *Heterocycles* **2004**, *63*, 1113–1122.
[4]Tsunoda, T.; Yamazaki, A.; Mase, T.; Sakamoto, S. *Org. Process Res. Dev.* **2005**, *9*, 593–598.

## SCHEME 1    MEDICINAL CHEMISTRY SYNTHESIS TO CONIVAPTAN

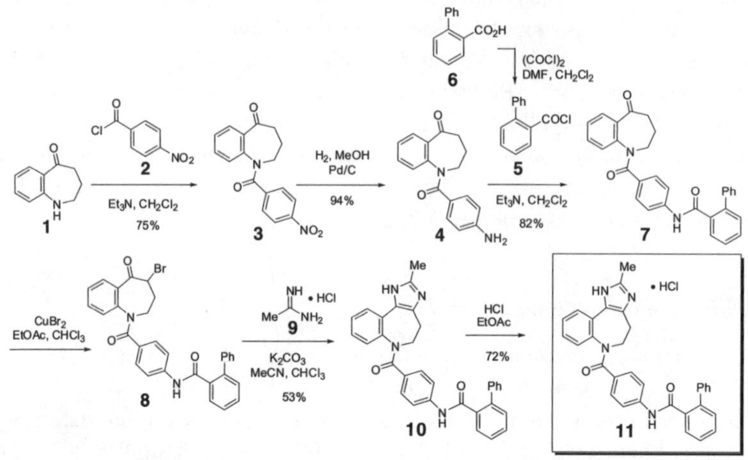

- Low overall yield, mainly due to linearity and poor imidazole formation.
- Required chromatography.
- Reliable methods used in the synthesis.
- Amenable to SAR.

## SCHEME 2    FIRST DEVELOPMENT ROUTE TO CONIVAPTAN

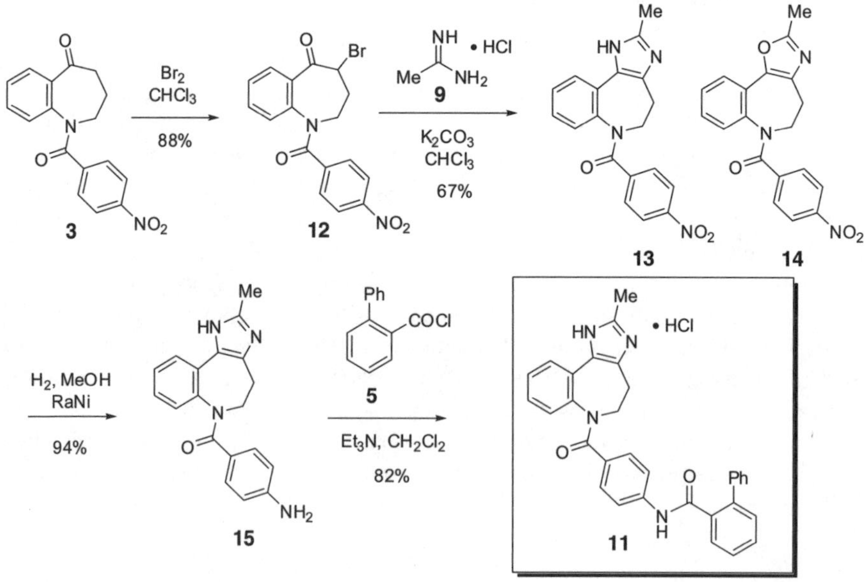

- Fairly linear synthesis.
- Construction of the imidazole on an advanced intermediate.
- Improved selectivity in imidazole formation.

## SCHEME 3   SECOND DEVELOPMENT ROUTE TO CONIVAPTAN

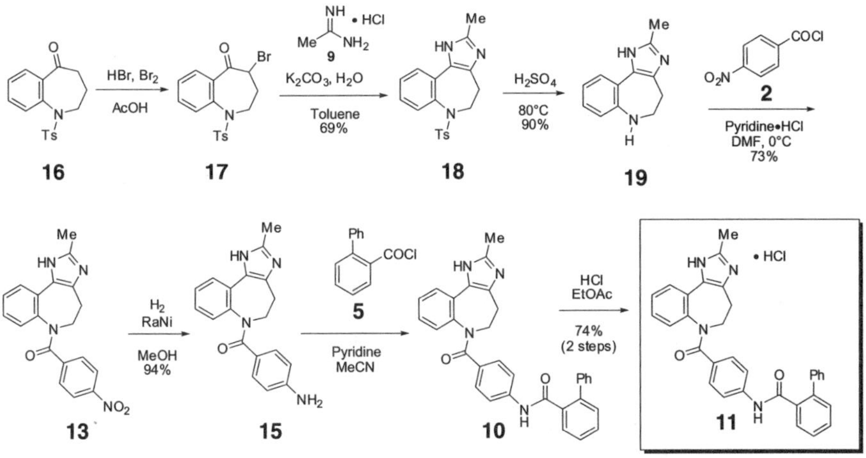

- Lengthy sequence to install bottom side chain.
- Imidazole formation completed on a very early intermediate.

## SCHEME 4   THIRD DEVELOPMENT ROUTE TO CONIVAPTAN

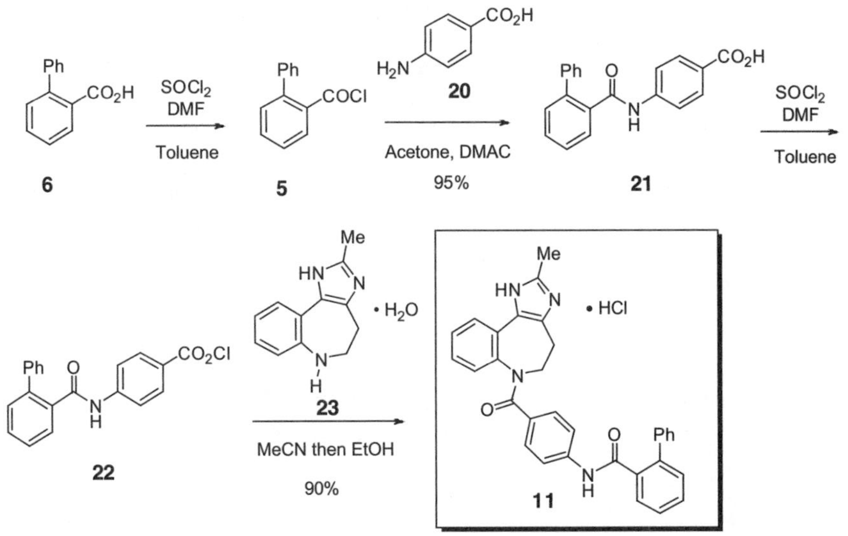

- Very convergent synthesis, poised for purchase of two key intermediates.
- Efficient solvent utilization and minimal process operations.
- High overall yield.

## 15.3   APTIVUS® (TIPRANAVIR)

### 15.3.1   Background

Tipranavir is a non-peptidic HIV-1 protease inhibitor for the treatment of AIDS.[5] It inhibits the virus-specific processing of the viral Gag and Gag-Pol polyproteins in HIV-1 infected cells, thus preventing formation of mature virions. The compound is a 4-hydroxy-5,6-dihydro-2-pyrone sulfonamide that presents a significant synthetic challenge, mainly due to the quaternary stereocenter and the remote chiral ethyl side chain.

### 15.3.2   Medicinal Chemistry Synthesis

The medicinal chemistry route to tipranavir (PNU-140690) was first reported in 1998.[6] The dianion of methyl acetoacetate (**2**) is added to ketone **1** to provide the tertiary alcohol that is deprotonated and cyclized to dihydropyrone **3**. Condensation with benzaldehyde **4** is catalyzed by aluminum trichloride at low temperature to afford **5**, which undergoes 1,4-addition with triethyl aluminum yielding **6**. Reduction of the nitroarene is followed by sulfonylation of the resulting aniline (**8**) providing racemic product **10**. The choice of the base in this transformation was important for chemoselective reaction at nitrogen. In order to access the chiral product, aniline **8** is converted to its Cbz derivative that can be separated to its four diastereomers by chiral chromatography and then hydrogenolyzed. While the medicinal chemistry route allowed for preparation of small quantities of material, it is not suitable for scale-up because of reagents such as NaH, triethyl aluminum, and several cryogenic reactions as well as multiple chromatographic separations.

### 15.3.3   Second Route to Tipranavir

The second route published on tripranavir firmly established the stereochemistry of the molecule.[7] Acylated oxazolidinone **13** undergoes an enantioselective copper-mediated 1,4-addition leading to **15** after dibenzylation of the aniline. Condensation of the titanium enolate with methoxydioxolane **16** provides the desired ketone after deprotection. It was hoped that the quaternary stereocenter would be established using Evans' keto-imide chemistry.[8] Unfortunately, a disappointing 3:2 ratio of **19** was obtained. The synthesis is completed through formation of dihydropyrone **20** under basic conditions and intercepts the intermediate of the previous route. While this route provides access to chirality through the use of an auxiliary, it remains lengthy, provides poor diastereoselectivity in the aldol step, and requires multiple protecting group interconversions.

### 15.3.4   Modification of the Second Route to Tipranavir

One important modification of the previous route allowed for excellent diastereoselectivity in the key titanium–aldol reaction. By switching from an *n*-propyl substituent to an

[5]Aristoff, P. A. *Drugs of the Future* **1998**, *23*, 995–999.
[6]Turner, S. R.; Strohbach, J. W.; Tommasi, R. A.; Aristoff, P. A.; Johnson, P. D.; Skulnick, H. I.; Dolak, L. A.; Seest, E. P.; Tomich, P. K.; Bohanon, M. J.; Horng, M.-M.; Lynn, J. C.; Chong, K.-T.; Hinshaw, R. R.; Watenpaugh, K. D.; Janakiraman, M. N.; Thaisrivongs, S. *J. Med. Chem.* **1998**, *41*, 3467–3476.
[7]Judge, T. M.; Phillips, G.; Morris, J. K.; Lovasz, K. D.; Romines, K. R.; Luke, G. P.; Tulinsky, J.; Tustin, J. M.; Chrusciel, R. A.; Dolak, L. A.; Mizsak, S. A.; Watt, W.; Morris, J.; Vander Velde, S. L.; Strohbach, J. W.; Gammill, R. B. *J. Am. Chem. Soc.* **1997**, *119*, 3627–3628.
[8]Evans, D. A.; Clark, J. S.; Metternich, R.; Novack, V. J.; Sheppard, G. S. *J. Am. Chem. Soc.* **1990**, *112*, 866–868.

acetylenic ketone (**21**), a 25:1 ratio was observed. Completion of the synthesis proceeded in the same fashion and the alkyne was reduced to the saturated chain during the removal of the two benzyl groups of the aniline. While the end-game of this synthesis is very appealing, preparation of the key keto-imide **17** was rather lengthy and required a chiral auxiliary.

### 15.3.5    The Third Route to Tipranavir

A third route to tipranavir was published in 1998 and incorporated many attractive elements.[9] Alcohol **24** is resolved by enzymatic resolution using isoprenyl acetate and an Amano lipase. Homologation of benzylic alcohol **25** is achieved by displacement of the mesylate with a malonate to afford **26**, which is decarboxylated and esterified to provide **27** as a key intermediate. For the other key piece, ketone **1** is added to the anion of ethyl acetate and the ester hydrolyzed to provide racemic hydroxy acid **28** in high yield. Classical resolution using norephedrine gives the chiral alcohol. After salt break, the alcohol is protected as a POM ether with concomitant ester formation. DIBAL-H reduction of **30** and TEMPO oxidation of the resulting primary alcohol provides aldehyde **31**, which undergoes a smooth aldol addition to alcohol **32**. Oxidation with PCC followed by deprotection of the POM ether under acidic conditions affords keto-ester **33**, which cyclizes to dihydropyrone **34** under basic conditions. While this synthesis allows for straightforward access to the two chiral fragments, it suffers from the need for frequent modification of the oxidation state of a single carbon, the carboxylic acid of **29**, which undergoes one un-needed protection, one reduction, and two oxidations. These issues were addressed in the fourth and final route.

### 15.3.6    The Fourth Route to Tipranavir

The last route to tipranavir can be found in the patent literature.[10,11] This synthesis builds on the strength of the previous route, especially in accessing the quaternary stereocenter through a classical resolution to **29**. The salt is broken with HCl and the carboxylic acid activated with CDI. Upon addition of the magnesium salt of monoethyl malonate, a smooth condensation occurs followed by decarboxylation to produce ketoester **35**. It was discovered that the cyclization to the dihydroparanone **36** could be achieved with NaOH followed by acidification. The key strategic change in this synthesis is to utilize the dihydropyranone as the nucleophile in the form of the titanium enolate and react it with ketone **37** in an aldol reaction that provides the alkene upon dehydration. The key step of this synthesis is control of the stereochemistry of the ethyl subsituent through a chiral hydrogenation of **38**. This was achieved using a ruthenium catalyst in the presence of DuPhos. Upon hydrogenation of the nitroarene, the same key intermediate (**39**) was obtained. This synthesis is remarkable given the complexity of the molecule. Very simple reagents and solvents are used and cryogenic reactions have been eliminated. Furthermore, crystalline intermediates are obtained allowing for facile purification. One stereocenter is obtained through a classical resolution while the second generated in an asymmetric hydrogenation.

[9]Fors, K. S.; Gage, J. R.; Heier, R. F.; Kelly, R. C.; Perrault, W. R.; Wicnienski, N. *J. Org. Chem.* **1998**, *63*, 7348–7356.

[10]Gage, J. R.; Kelly, R. C.; Hewitt, B. D. *Process to produce 4-hydroxy-2-oxo-pyran derivatives useful as protease inhibitors* WO9912919 A1 1999, 45 pp

[11]Hewitt, B. D.; Burk, M. J.; Johnson, N. B. *Asymmetric hydrogenation of arylpropenylpyrones using rhodium phosphine catalysts* WO2000055150 A1 2000, 35 pp

**SCHEME 5    MEDICINAL CHEMISTRY SYNTHESIS OF TIPRANAVIR**

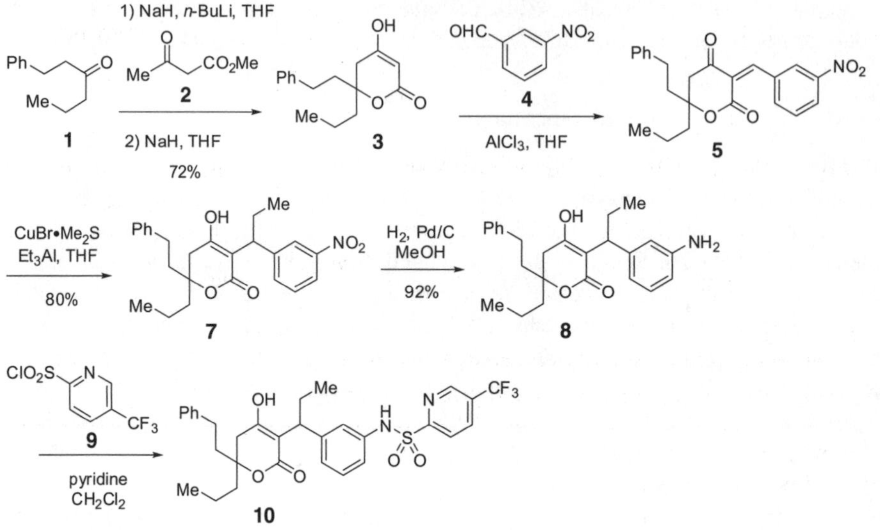

- Several undesirable reagents (NaH, Et$_3$Al) and multiple chromatographic separations.
- No control of the stereochemistry.

**SCHEME 6    FIRST ENANTIOSELECTIVE ROUTE TO TIPRANAVIR**

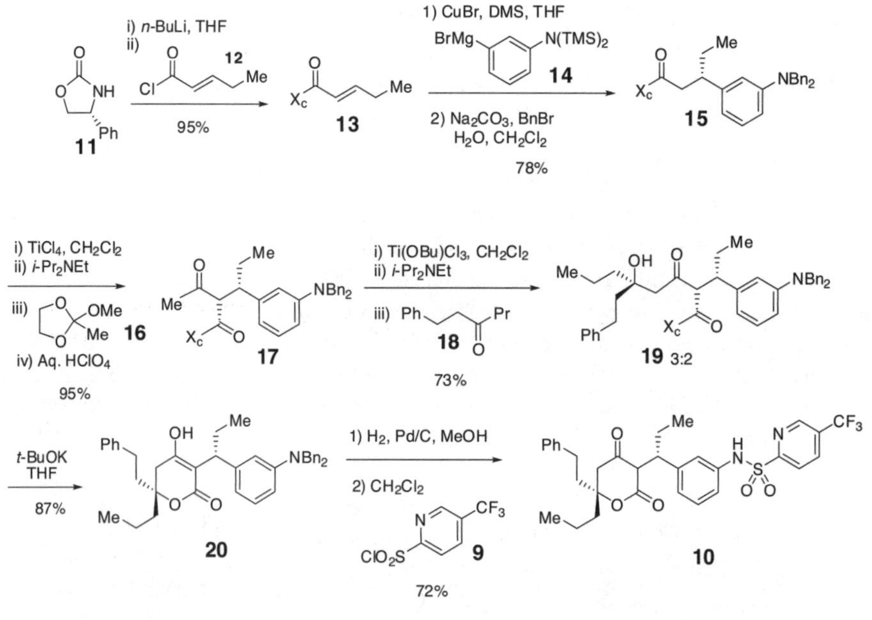

- Poor diastereoselectivity in the aldol condensation.
- Several cryogenic reactions and protecting groups.

## SCHEME 7    SECOND ENANTIOSELECTIVE ROUTE TO TIPRANAVIR

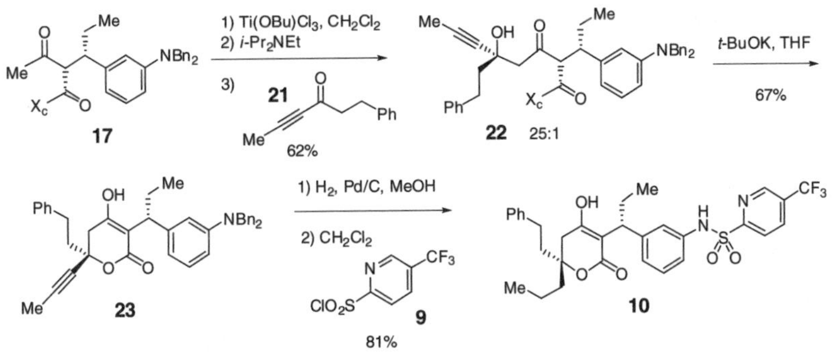

- High diastereoselectivity and enantioselectivity.
- Several cryogenic reactions and protecting groups.

## SCHEME 8    THIRD DEVELOPMENT ROUTE TO TIPRANAVIR

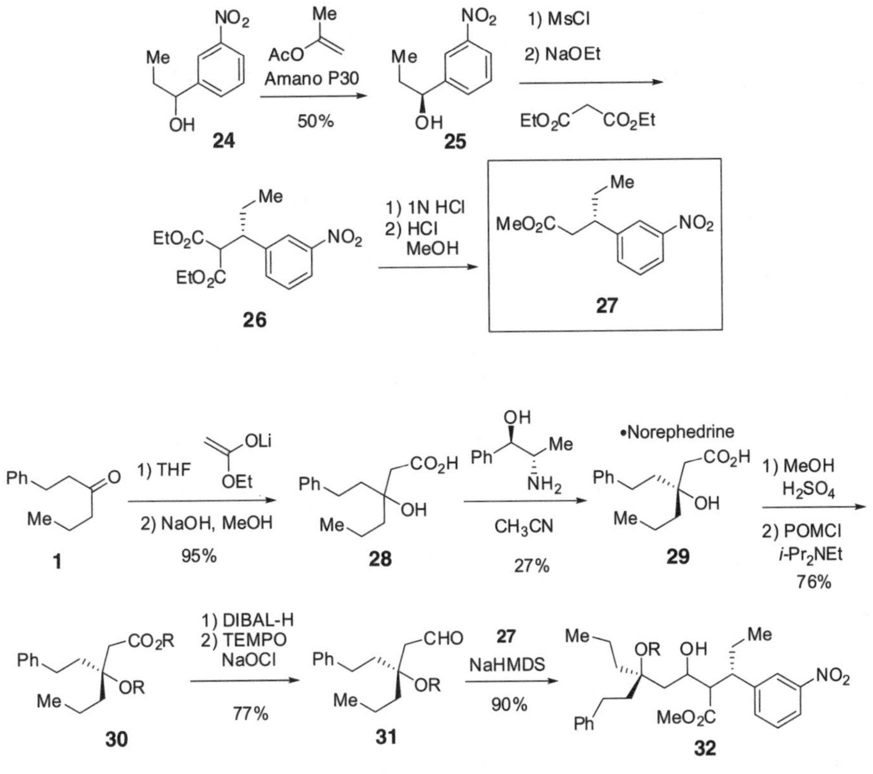

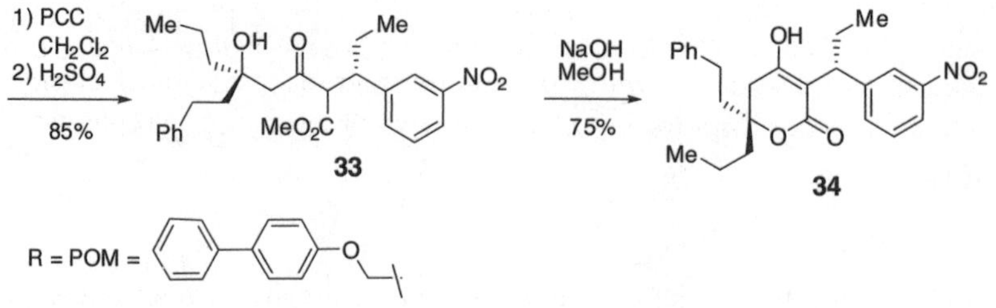

- Hydroxyacid **29** prepared efficiently.
- Multiple oxidation/reduction.
- Synthesis required inefficient use of a large protecting group.

**SCHEME 9   FINAL ROUTE TO TIPRANAVIR**

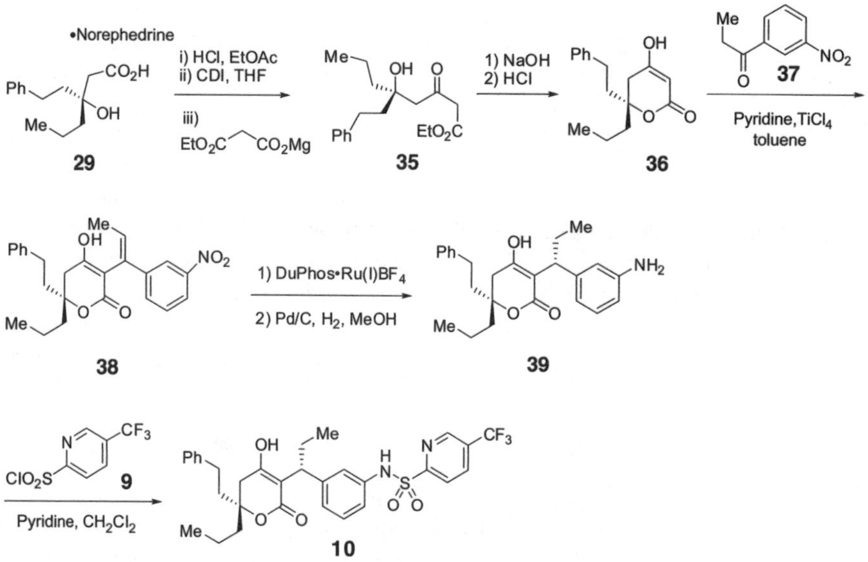

- High level of stereocontrol through a chiral hydrogenation
- Elimination of protecting groups and reduction/oxidation sequence.
- Elimination of cryogenic reactions.

## 15.4   EMEND® (APREPITANT)

### 15.4.1   Background

Aprepitant is a selective high-affinity antagonist of human substance P/neurokinin-1 (NK1) receptors. It is part of a new class of antiemetics for control of chemotherapy-induced nausea and vomiting.[12] The molecule is a chiral 2,3-substituted morpholine with an additional stereocenter on the ether side chain at the 2-position. The clear synthetic challenge for this pharmaceutical target is the control of the absolute and relative stereochemistry on the morpholine ring, especially at the glycosidic-like stereocenter of the morpholine.

### 15.4.2  Medicinal Chemistry Synthesis

The details around the first medicinal chemistry route to aprepitant, which is also known as L-754,939 or MK-869, were reported in the patent literature. A key intermediate in the synthesis was morpholinone **7**.[13] The synthesis starts from a phenylacetic acid derivative (**1**) that is activated and converted to an Evans' oxazolidinone (**3**).[14] Upon deprotonation with KHMDS and treatment with an arylsulfonyl azide, intermediate **4** is obtained enantioselectively. The chiral auxiliary is cleaved under standard conditions and the azide hydrogenated providing the 4-fluorophenyl glycine **5**. The primary amine is protected through reductive amination and the morpholinone generated through a bis-*N,O*-alkylation to afford **7**. While this synthesis allows for the preparation of the desired intermediate as a single enantiomer, the synthesis is lengthy and requires preparation of an azide that is undesirable in the long term.

The medicinal chemistry synthesis of aprepitant[15] was completed by reduction of lactone **7** to the hemiacetal followed by acylation with acid chloride **8** to afford ester **9**. This ester was converted to enol ether **10** using dimethyl titanocene. This reaction proved to be low yielding with the more commonly used Tebbe's reagent. A more detailed description around the issues with scaling up this olefination and how it could be accomplished on a kilomole scale has been published.[16] The enol ether (**10**) was diastereoselectively hydrogenated (8:1 selectivity) to afford the final stereocenter. The triazolone was incorporated in a two-step sequence, first by alkylation of morpholine **11** with **12** followed by cyclodehydration to provide the final product. While the medicinal chemistry route to aprepitant provides good level of stereocontrol for the three chiral centers, it is lengthy and requires several chromatographic purifications, which is common for a first route.

### 15.4.3  Second Generation Route to the Morpholine Core

One of the key issues to address for a more practical synthesis of aprepitant was the preparation of the morpholine nucleus of the molecule. A very attractive approach utilizing a dynamic resolution was published in 1997.[17] The synthesis initiated with

[12]Sorbera, L. A.; Castaner, J.; Bayes, M.; Silvestre, J. *Drugs of the Future* **2002**, *27*, 211–222.

[13]Dorn, C. P.; Hale, J. J.; Maccoss, M.; Mills, S. G.; Ladduwahetty, T.; Shah, S. K. *Morpholine and thiomorpholine tachykinin receptor antagonists* EP577394 A1 1994, 96 pp

[14]Evans, D. A.; Britton, T. C.; Ellman, J. A.; Dorow, R. L. *J. Am. Chem. Soc.* **1990**, *112*, 4011–4030.

[15]Hale, J. J.; Mills, S. G.; MacCoss, M.; Finke, P. E.; Cascieri, M. A.; Sadowski, S.; Ber, E.; Chicchi, G. G.; Kurtz, M.; Metzger, J.; Eiermann, G.; Tsou, N. N.; Tattersall, F. D.; Rupniak, N. M. J.; Williams, A. R.; Rycroft, W.; Hargreaves, R.; MacIntyre, D. E. *J. Med. Chem.* **1998**, *41*, 4607–4614.

[16]Payack, J. F.; Huffman, M. A.; Cai, D.; Hughes, D. L.; Collins, P. C.; Johnson, B. K.; Cottrell, I. F.; Tuma, L. D. *Org. Process Res. Dev.* **2004**, *8*, 256–259.

[17]Alabaster, R. J.; Gibson, A. W.; Johnson, S. A.; Edwards, J. S.; Cottrell, I. F. *Tetrahedron: Asymmetry* **1997**, *8*, 447–450.

commercially available benzaldehyde **14** that is converted to aminonitrile **15** by a Strecker reaction. The nitrile is successively hydrolysed to amide **16** and acid **17**, which undergoes a bis-alkylaton to racemic oxazinone **7** in acceptable overall yield. It was discovered that treatment of the racemate, in *i*-PrOAc at reflux with a seed of the resolved salt **18** followed by slow addition of additional chiral 3-bromocamphor-8-sulfonic acid, led to crystallization of the desired resolved salt with concomitant epimerization of the wrong enantiomer. This elegant dynamic resolution provides 90% yield from the racemate in >99% ee. The only drawback of this procedure is the high cost and limited availability of resolving agent.

### 15.4.4   Third Generation Route to the Morpholine Core

An alternative approach to the oxazinone addressed the previous concerns.[18] The synthesis begins with a inexpensive and readily available starting material, α-methyl benzylamine **19**. Reaction with ethyl oxalyl chloride under Schotten–Baumann conditions followed by hydrolysis provides amide **20**. The reduction of **20** to aminoalcohol **21** proved to be the most problematic step and was accomplished with borane generated *in situ*. A small amount of unreduced amide remained and was resistant to further reduction but could be taken crude in the next step. Compound **21** was condensed with glyoxal derivative **23** that was easily prepared from acetophenone **22**. This afforded a 2:1 mixture of diastereomers favoring the undesired isomer. Since it was known from the previous route that the stereocenter was easily racemized under acidic conditions, several salts were screened. The HCl salt (**25**) was found to crystallize preferentially, once again out of *i*-PrOAc, in high yield and diastereoselection. In this case, the stereocenter on the benzylamine dictates the diastereoselectivity rather than the nature of the counterion. This synthesis provided excellent level of stereocontrol through the use of a chiral auxiliary.

### 15.4.5   Preparation of the Chiral Benzylic Alcohol

Rather than relying on a straightforward ester formation at C-2 followed by a difficult methylenation/reduction sequence to obtain the methyl group, a more challenging diastereoselective glycosidation approach was selected to bring more convergency to the route. This strategy also provided a convenient starting material in (*R*)-3,5-bis(trifluoromethyl) phenyl] ethanol (**28**). Two attractive approaches for the preparation of **28** have been demonstrated. The first process relies on an asymmetric transfer-hydrogenation using (1*S*,2*R*)-*cis*-amino-2-indanol (**27**) as the chiral ligand.[19] The hydrogenation provides **28** in high yields and good enantioselectivity (91% ee). Moreover, it was found that the DABCO 2:1 inclusion complex could be crystallized and enrich the desired enantiomer to levels >99% ee.

[18]Zhao, M. M.; McNamara, J. M.; Ho, G.-J.; Emerson, K. M.; Song, Z. J.; Tschaen, D. M.; Brands, K. M. J.; Dolling, U.-H.; Grabowski, E. J. J.; Reider, P. J.; Cottrell, I. F.; Ashwood, M. S.; Bishop, B. C. *J. Org. Chem.* **2002**, *67*, 6743–6747.

[19]Hansen, K. B.; Chilenski, J. R.; Desmond, R.; Devine, P. N.; Grabowski, E. J. J.; Heid, R.; Kubryk, M.; Mathre, D. J.; Varsolona, R. *Tetrahedron: Asymmetry* **2003**, *14*, 3581–3587.

The second approach relies on an asymmetric hydride reduction, namely an oxazaborolidine-catalyzed borane reduction. Addition of acetophenone (**26**) to 2 mol % of (*S*)-catalyst **30** and the BH$_3$•PhNEt$_2$ complex in MTBE affords the desired alcohol in 97% yield and 93% ee.[20]

### 15.4.6    Preparation of the 1,2,4-Triazolin-5-one

Incorporation of the lower side chain of aprepitant is accomplished at the very end in the original route. This certainly made a lot of sense from a medicinal chemistry standpoint since it allows for a divergence point for the structure activity relationship evaluation. To bring in additional synthetic efficiency, it would be preferable to append the whole side chain in a single operation using an alkylation.[21] Starting from semicarbazide hydrochloride **31**, condensation with benzyloxyacetyl chloride **32** affords **33** that cyclizes directly under basic conditions. Deprotection of the benzyl ether via catalytic hydrogenation yields the alcohol that is converted to alkyl chloride **36** under standard conditions. Despite the fact that this sequence utilizes very inexpensive starting materials, it is rather lengthy. However, it was found that the desired triazolinone could be obtained from **31** in a single step by treatment with an orthoester in near quantitative yield. This became the method of choice for the preparation of this key intermediate.

### 15.4.7    Towards a Commercial Process

With all the key pieces in place, assembly of aprepitant could be completed.[18] Oxazinone **25** is reduced diastereoselectively at −20°C and the crude lactol converted to unstable trichloroacetamidate **37**. This intermediate undergoes a smooth glycosidation reaction with alcohol **28** providing a 90% yield of a 96:4 mixture of diastereomers that can be recrystallized to high purity and only 1.5–2.5% *cis*-isomer. Hydrogenolysis of the chiral auxiliary is accomplished under standard conditions. It is noteworthy that unlike the previous synthesis, the undesired *trans*-isomer is obtained. The stereochemistry is corrected by chlorination of the nitrogen with NCS and elimination to imine **40**. Hydrogenation of the double bond proceeds with greater than 99% selectivity favoring the desired *cis*-isomer. Morpholine **11** is isolated as the HCl salt. This synthesis proved the viability of the glycosidation strategy. Unfortunately, adjustment in the chirality of the fluorophenyl group required additional synthetic steps.

Final alkylation of the triazolinoyl chloride **36** was demonstrated on the salt of morpholine **11**. This alkylation is rapid and tolerant of either inorganic or organic bases to provide aprepitant in 99% yield.[21]

[20]Mathre, D. J.; Thompson, A. S.; Douglas, A. W.; Hoogsteen, K.; Carroll, J. D.; Corley, E. G.; Grabowski, E. J. J. *J. Org. Chem.* **1993**, *58*, 2880–2888.

[21]Cowden, C. J.; Wilson, R. D.; Bishop, B. C.; Cottrell, I. F.; Davies, A. J.; Dolling, U.-H. *Tetrahedron Lett.* **2000**, *41*, 8661–8664.

### 15.4.8 The Final Route

The final route to aprepitant is extremely elegant and built on the knowledge acquired throughout its development.[22] Starting from aminoalcohol **41**, which is easily obtained by reductive amination of aminoethanol, condensation with glyoxilic acid affords lactol **42** directly.[23] It was determined that for the glycosylation reaction, generation of the trifluoroacetate was sufficient to provide a more stable intermediate and higher yields. The fact that the newly generated stereocenter of **43** is not controlled is of little consequence since it can be epimerized readily using a hindered based to minimize the 1,2-Wittig rearrangement. The potassium salt of 3,7-dimethyl-3-octanol was found to be optimal for this transformation. Using this protocol, lactam **43** of greater than 99% ee is obtained. Introduction of the 4-fluorophenyl substituent is achieved through the use of the Grignard reagent providing the aminal after a quench with MeOH and TsOH. The aminal and the benzyl group are hydrogenated under conditions that provide greater than 300:1 selectivity at the newly created stereocenter. Finally, previously reported intermediate **11** is crystallized as the HCl salt in 91% overall yield. Details around the kinetics of the hydrogenation have been reported. It appears that debenzylation proceeds first followed by generation of the imine that is then reduced.[24] In this route, the chiral alcohol provides control for the desired stereochemistry of the acetal. This stereocenter now controls the regiochemical outcome of the hydrogenation of the imine therefore eliminating the need for a chiral auxiliary.

**SCHEME 10   MEDICINAL CHEMISTRY ROUTE TO APREPITANT**

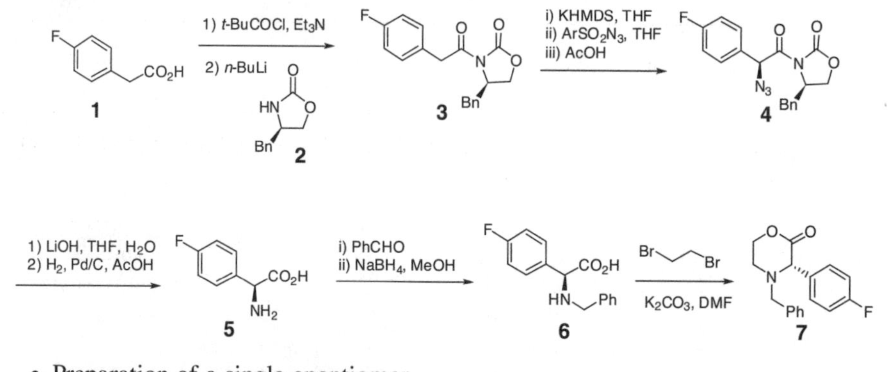

- Preparation of a single enantiomer.
- Amino acid prepared utilizing an azide.

[22]Brands, K. M. J.; Payack, J. F.; Rosen, J. D.; Nelson, T. D.; Candelario, A.; Huffman, M. A.; Zhao, M. M.; Li, J.; Craig, B.; Song, Z. J.; Tschaen, D. M.; Hansen, K.; Devine, P. N.; Pye, P. J.; Rossen, K.; Dormer, P. G.; Reamer, R. A.; Welch, C. J.; Mathre, D. J.; Tsou, N. N.; McNamara, J. M.; Reider, P. J. *J. Am. Chem. Soc.* **2003**, *125*, 2129–2135.
[23]Nelson, T. D.; Rosen, J. D.; Brands, K. M. J.; Craig, B.; Huffman, M. A.; McNamara, J. M. *Tetrahedron Lett.* **2004**, *45*, 8917–8920.
[24]Brands, K. M. J.; Krska, S. W.; Rosner, T.; Conrad, K. M.; Corley, E. G.; Kaba, M.; Larsen, R. D.; Reamer, R. A.; Sun, Y.; Tsay, F.-R. *Org. Process Res. Dev.* **2006**, *10*, 109–117.

## SCHEME 11 COMPLETION OF THE MEDICINAL CHEMISTRY ROUTE

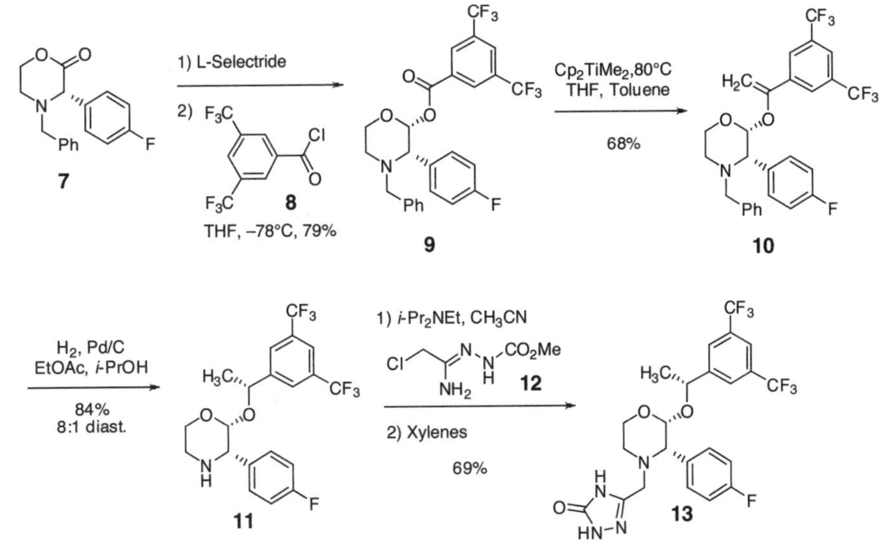

- Acceptable stereocontrol in hydrogenation.
- Linear synthesis
- Requires cryogenic reactions and several chromatographic purifications.

## SCHEME 12 SECOND GENERATION ROUTE TO MORPHOLINE CORE

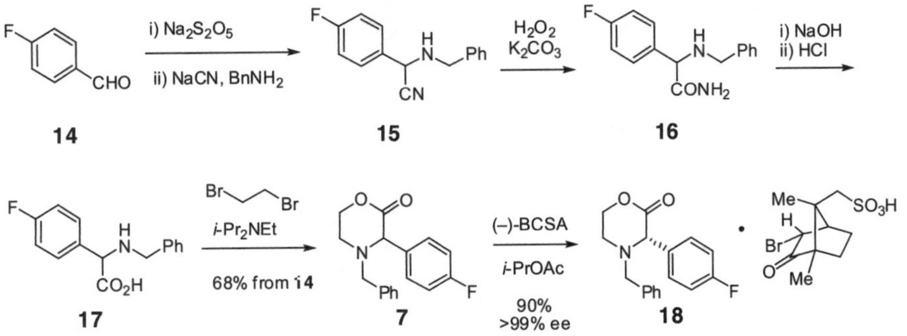

- Outstanding dynamic resolution.
- Limited commercial availability of the resolving agent.

**SCHEME 13 THIRD GENERATION ROUTE TO THE MORPHOLINE CORE**

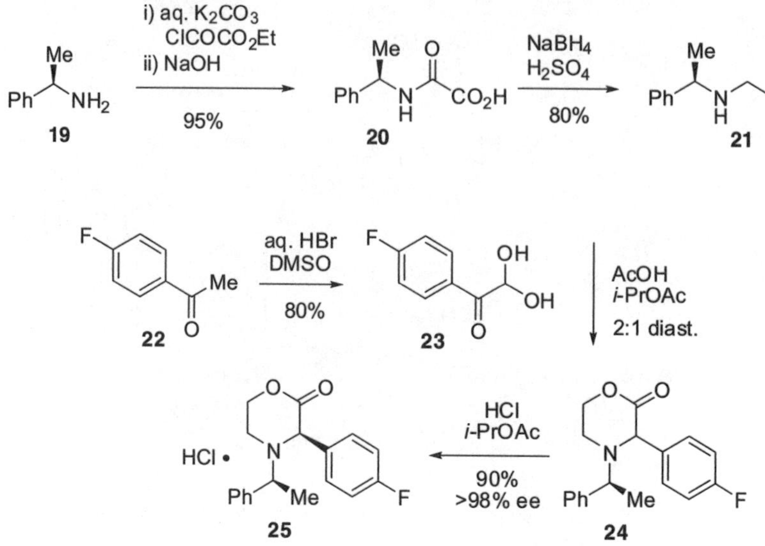

- Chirality obtained from the starting material as a chiral auxiliary.
- Epimerization provides desired stereochemistry on the ring.

**SCHEME 14 PREPARATION OF CHIRAL BENZYLIC ALCOHOL 28**

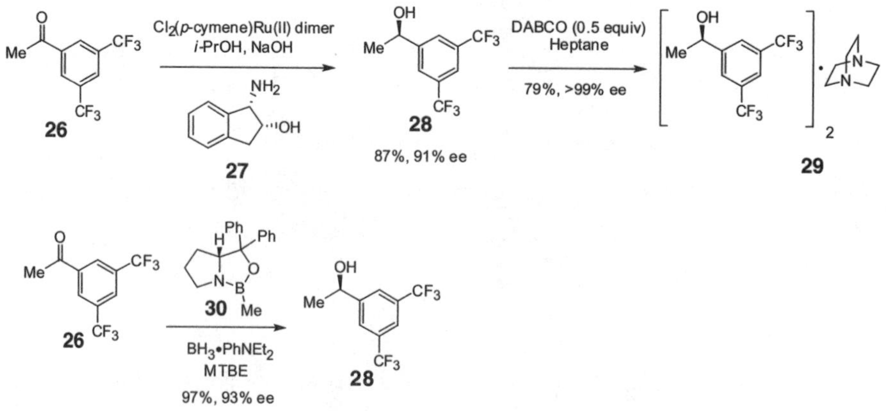

- Chiral hydrogenation is efficient and ee can be easily upgraded with DABCO.
- Oxazoborolidine reduction provides higher yield but is not as simple operationally.

## SCHEME 15   PREPARATION OF THE 1,2,4-TRIAZOLIN-5-ONE

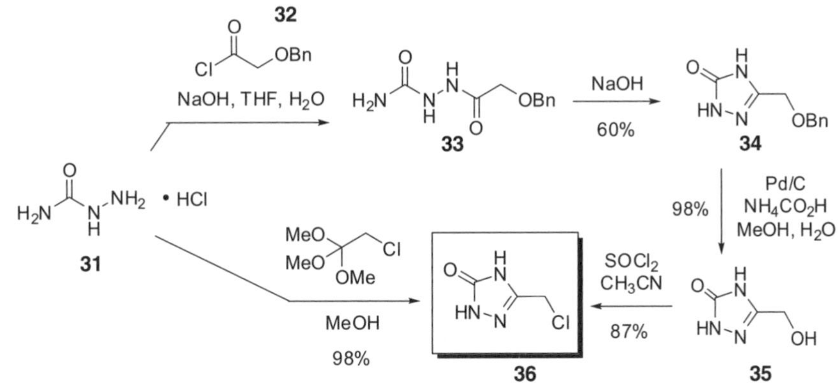

- One-step synthesis identified for preparation of the target.
- Allowed for a more convergent synthesis.

## SCHEME 16   ALCOHOL INTRODUCTION THROUGH GLYCOSYLATION

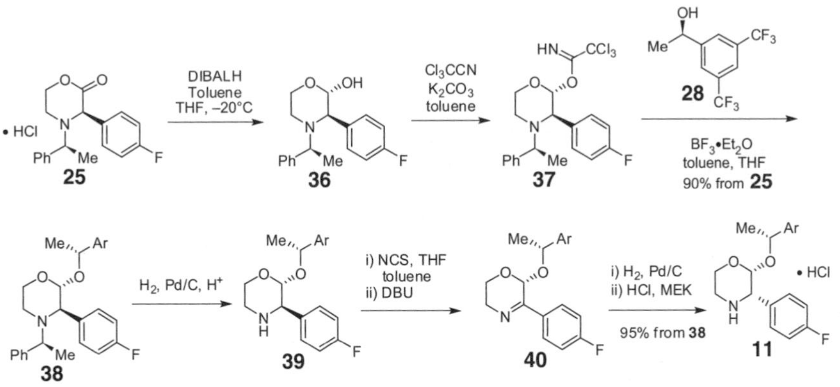

- Glycosylation provides more convergency than methylenation.
- Chiral auxiliary provides wrong stereochemistry, which is fixed through imine formation.
- Imine hydrogenation is stereoselective.

## SCHEME 17    COMPLETION OF THE SYNTHESIS BY ALKYLATION

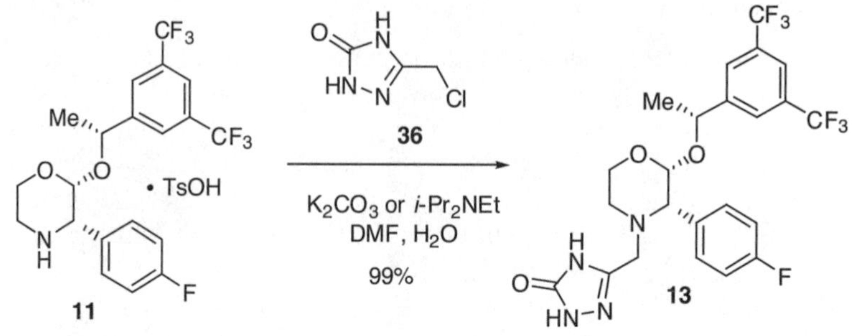

- High yielding and operationally simple last step.
- Provides convergency to the route.

## SCHEME 18    COMMERCIAL ROUTE TO APREPITANT

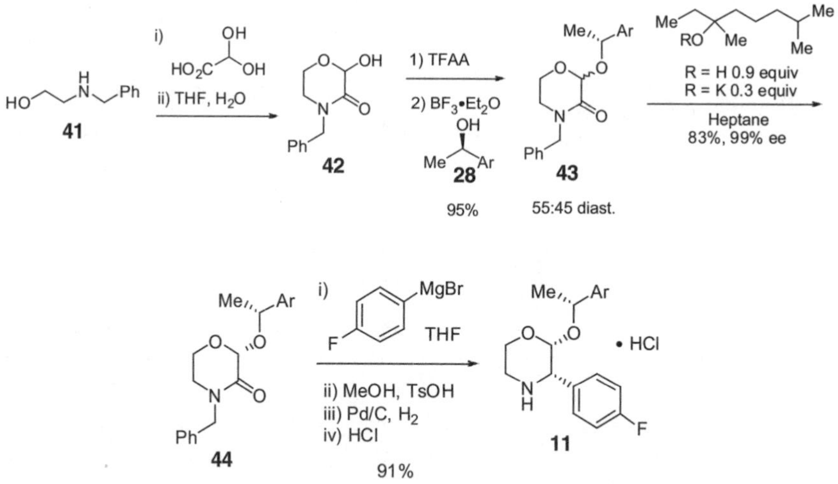

- Highly convergent synthesis.
- Chiral alcohol controls stereochemistry of the acetal epimerization and facial selectivity in the imine hydrogenation.
- No chiral auxiliary needed.

## 15.5  SUTENT® (SUNITINIB MALATE)

### 15.5.1  Background

Sunitinib malate is a multi-kinase inhibitor targeting several receptor tyrosine kinases (RTKs) that control signal transduction in diverse cellular processes. These kinases are transmembrane proteins containing extracellular ligand-binding domains and intracellular catalytic domains. RTKs have been shown to play a crucial role in tumor growth, progression, and metastasis. Sunitinib malate inhibits the vascular endothelial growth factor (VEGF) and the platelet-derived growth factor (PDGF) as well as Flt3 and Kit tyrosine kinases with potency in the nanomolar range.[25]

### 15.5.2  Medicinal Chemistry Synthesis

The details around the medicinal chemistry synthesis of sunitinib (SU011248) have been described in the patent literature[26] and the synthesis of a close analogue has been disclosed.[27] The synthesis is initiated by nitrosation of *t*-butyl acetoacetate (**1**) and the resulting oxime (**2**) undergoes a reductive cyclization with ethyl acetoacetate (**3**) to generate desired pyrrole **4** in good overall yield. The *t*-butyl ester is hydrolyzed selectively under acidic conditions followed by decarboxylation to afford **5**, which undergoes a formylation. After hydrolysis of the ethyl ester and acidification, carboxylic acid **6** is isolated in 93% yield. Activation of the acid with EDC in the presence of HOBt provides amide **7**, which is coupled with the oxindole **8** under mild basic conditions to provide sunitinib (**9**) in 58% yield. One interesting note on the amide formation is the fact that when the coupling reaction is performed with CDI, additional $CO_2$ leads to a substantial rate enhancement.[28]

Overall, this is a very good synthesis. From a medicinal chemistry standpoint, the two SAR points, formation of the amide and condensation with the oxindole, are late in the synthesis.

### 15.5.3  Process Chemistry Synthesis

While the discovery route is very good, it is clear that it might be more practical to bring in the appropriate side chain at the C-3 position of the pyrrole at the onset of its formation.[29,30] The prerequisite ketoamide **12** is prepared by condensation of diketene **10** with amine **11**. It was found that catalytic hydrogenation conditions are acceptable for the reductive cyclization to pyrrole **13** that is decarboxylated in quantitative yields. The Vilsmeier–Haack reaction affords iminium **16**, which is treated directly with the oxindole **8** in the presence of KOH to provide the desired drug substance (**9**).

[25]McIntyre, J. A.; Castaner, J. *Drugs of the Future* **2005**, *30*, 785–792.

[26]Tang, P. C.; Miller, T.; Li, X.; Sun, L.; Wei, C. C.; Shirazian, S.; Liang, C.; Vojkovsky, T.; Nematalla, A. S. *Preparation of pyrrole substituted 2-indolinone protein kinase inhibitors for treatment of cancer* WO2001060814 A2 2001, 225 pp

[27]Sun, L.; Liang, C.; Shirazian, S.; Zhou, Y.; Miller, T.; Cui, J.; Fukuda, J. Y.; Chu, J.-Y.; Nematalla, A.; Wang, X.; Chen, H.; Sistla, A.; Luu, T. C.; Tang, F.; Wei, J.; Tang, C. *J. Med. Chem.* **2003**, *46*, 1116–1119.

[28]Vaidyanathan, R.; Kalthod, V. G.; Ngo, D. P.; Manley, J. M.; Lapekas, S. P. *J. Org. Chem.* **2004**, *69*, 2565–2568.

[29]Manley, J. M.; Kalman, M. J.; Conway, B. G.; Ball, C. C.; Havens, J. L.; Vaidyanathan, R. *J. Org. Chem.* **2003**, *68*, 6447–6450.

[30]Havens, J. L.; Vaidyanathan, R. *Method of synthesizing indolinone compounds* US2006009510 A1 2006, 28 pp

This synthesis of sunitinib is highly efficient and convergent, highlighting the fact that only a minor change in the synthetic strategy can have a large impact on route efficiency.

---

**SCHEME 19   MEDICINAL CHEMISTRY ROUTE TO SUNITINIB**

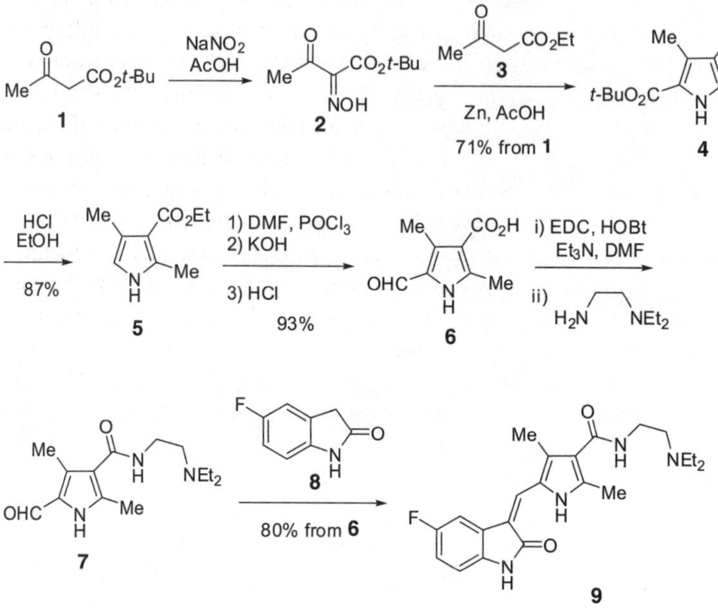

- Highly convergent synthesis.
- Poised for rapid SAR.

---

**SCHEME 20   DEVELOPED SYNTHESIS OF SUNITINIB**

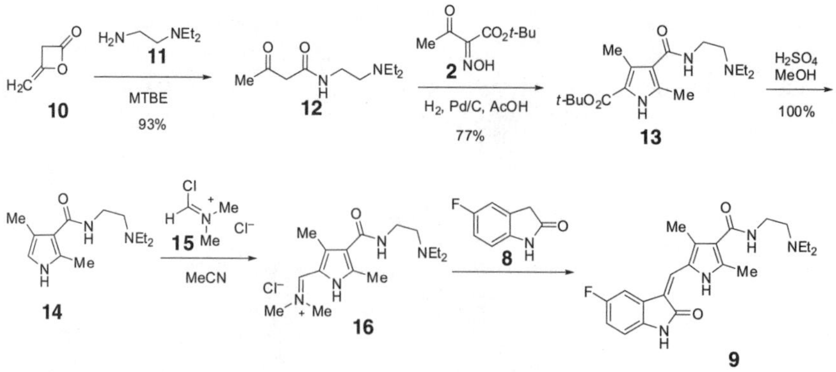

- Right side chain introduced as part of the starting material.
- Product of the Vilsmeier–Haack used directly in the coupling with oxindole.

## 15.6    CONCLUSION

The evolution of a chemical process varies from one drug candidate to another. Often, the medicinal chemist has no clear definition of what the ultimate structure of the final target should be but rather focuses on which class of compounds should be prepared. A key intermediate might be the most important target if it allows for the opportunity to prepare multiple analogues late in the synthesis and expand the SAR evaluation. The medicinal chemist usually does not have the opportunity to spend time on a speculative synthetic approach and will more often rely on proven chemistry, even if it means a few more synthetic steps or a costly reagent. Once a compound enters drug development, the synthetic target is now well defined. The early development chemist makes the assessment if the medicinal chemistry route can be enabled to a scaleable and safe process that could provide sufficient quantities of the drug substance to fuel the pre-clinical evaluation and entry into the clinic. As a drug candidate progresses through development and the demand for the API increases, a more efficient chemical process is necessary and often leads to a totally new synthetic route. Finally, once a drug candidate shows signs of clinical safety and efficacy, the commercial process must be identified. The choice of starting materials, reagents, solvent, and so on is important in addition to the development of a process incorporating practicality and robustness such that it performs reliably on large scale. Additionally, the commercial route must meet commercial objectives from an economical, environmental, and regulatory perspective. Even after registration of a new drug, continual improvement, either in terms of process optimization or development of a new route, never cease.

# 16

# GREEN CHEMISTRY

Juan C. Colberg

## 16.1 INTRODUCTION

The green chemistry movement gained popularity in the early 1990s. Since then, there have been major contributions from all around the globe, with more than 5,000 publications in this area. The evaluation of the environmental impact of a pharmaceutical process requires the consideration of many factors. Processing efficiency, safety, and cost effectiveness have historically been the primary consideration of development chemists as a new drug candidate progresses toward commercialization. The green chemistry principles also demand an in depth analysis of health and environmental impact. Each of these components must be evaluated across an average of 10 synthetic steps, across a wide range of functional group transformations and in light of hundreds of options for reagents and solvents.

In order to achieve an efficient, environmentally benign synthesis, a chemist needs to have the tools to guide the selection of reagents and metrics to assess the changes being made. Many tools have been developed during recent years for measuring how green a process is, from the early work done by Trost introducing the concept of atom economy,[1] which measures the efficiency of raw material use; to Sheldon, who introduced the E-Factor analysis[2] that quantitates waste generation. Following the principle that prevention is better than cure, Anastas and Warner formulated the widely spread 12 principles of green chemistry in 1998 (Table 16.1).[3] Following the publication of these principles, more specific guides were published for process chemists and engineers (Table 16.2).[4] Efforts have continued in

---

[1]Trost, B. M. *Science* **1991**, *254*, 1471–1477.
[2]Sheldon, R. A. *Chem. Ind.* **1998**, *75*, 273–288.
[3]Anastas, P.; Warner, J. *Green Chemistry: Theory and Practice*; Oxford Univ Press: Oxford, UK, **1998**.
[4]Anastas, P. T.; Zimmerman Julie, B. *Environ. Sci. Technol.* **2003**, *37*, 94A–101A.

---

*Practical Synthetic Organic Chemistry: Reactions, Principles, and Techniques*, First Edition.
Edited by Stéphane Caron.
© 2011 John Wiley & Sons, Inc. Published 2011 by John Wiley & Sons, Inc.

**TABLE 16.1   The 12 Green Chemistry Principles**

| | |
|---|---|
| 1. Prevent waste rather than treat. | 7. Use renewable raw materials. |
| 2. Maximize incorporation of all materials (atom economy). | 8. Minimize unnecessary derivatization. |
| 3. Design synthesis to use or generate least hazardous chemical substances. | 9. Use catalytic versus stoichiometric reagents. |
| 4. Design safer chemicals to do the desired function. | 10. Process-related products should be designed to be biodegradable. |
| 5. Minimize or use innocuous auxiliary agents (solvents, etc). | 11. Use online analytical process monitoring to minimize formation of hazardous byproducts. |
| 6. Minimize energy requirements. | 12. Choose safer reagents that minimize the potential for accidents. |

academia and industry to further develop metrics and educational tools that help chemists develop greener processes.[4–26] Some of the recently published tools include the development of a series of solvent and reaction selection guides to help chemists during route development, even as early as the first medicinal chemistry route.[5]

## 16.2   GREEN CHEMISTRY METRICS

In addition to the green chemistry and engineering principles that allow chemists to perform a qualitative assessment, quantitative metrics are needed for process comparison on both laboratory and larger scale. In general, this type of metric should provide a simple,

[5]Alfonsi, K.; Colberg, J.; Dunn, P. J.; Fevig, T.; Jennings, S.; Johnson, T. A.; Kleine, H. P.; Knight, C.; Nagy, M. A.; Perry, D. A.; Stefaniak, M. *Green Chem.* **2008**, *10*, 31–36.

[6]Auge, J. *Green Chem.* **2008**, *10*, 225–231.

[7]Trost, B. M. *Acc. Chem. Res.* **2002**, *35*, 695–705.

[8]Andraos, J. *Org. Process Res. Dev.* **2005**, *9*, 149–163.

[9]Andraos, J. *Org. Process Res. Dev.* **2005**, *9*, 404–431.

[10]Andraos, J. *Org. Process Res. Dev.* **2006**, *10*, 212–240.

[11]Andraos, J.; Sayed, M. *J. Chem. Educ.* **2007**, *84*, 1004–1010.

[12]Eissen, M.; Metzger, J. O. *Chem.-Eur. J.* **2002**, *8*, 3580–3585.

[13]Constable, D. J. C.; Curzons, A. D.; Cunningham, V. L. *Green Chem.* **2002**, *4*, 521–527.

[14]Pirrung, M. C. *Chem.-Eur. J.* **2006**, *12*, 1312–1317.

[15]Brummond, K. M.; Wach, C. K. *Mini-Rev. Org. Chem.* **2007**, *4*, 89–103.

[16]Dallinger, D.; Kappe, C. O. *Chem. Rev.* **2007**, *107*, 2563–2591.

[17]Guillena, G.; Hita, M. d. C.; Najera, C.; Viozquez, S. F. *Tetrahedron: Asymmetry* **2007**, *18*, 2300–2304.

[18]Hossain, K. A.; Khan, F. I.; Hawboldt, K. *Ind. Eng. Chem. Res.* **2007**, *46*, 8787–8795.

[19]Hayashi, Y.; Urushima, T.; Aratake, S.; Okano, T.; Obi, K. *Org. Lett.* **2008**, *10*, 21–24.

[20]Gani, R.; Jimenez-Gonzalez, C.; Constable, D. J. C. *Comput. Chem. Eng.* **2005**, *29*, 1661–1676.

[21]Jimenez-Gonzalez, C.; Curzons, A. D.; Constable, D. J. C.; Cunningham, V. L. *Clean Technol. Environ. Policy* **2005**, *7*, 42–50.

[22]Slater, C. S.; Savelski, M. *J. Environ. Sci. Health, Part A* **2007**, *42*, 1595–1605.

[23]Gani, R.; Jimenez-Gonzalez, C.; ten Kate, A.; Crafts, P. A.; Powell, M. J.; Powell, L.; Atherton, J. H.; Cordiner, J. L. *Chem. Eng.* **2006**, *113*, 30–43.

[24]Mikami, K.; Kotera, O.; Motoyama, Y.; Sakaguchi, H. *Synlett* **1995**, 975–977.

[25]Jimenez-Gonzalez, C.; Constable, D. J. C.; Curzons, A. D.; Cunningham, V. L. *Clean Technol. Environ. Policy* **2002**, *4*, 44–53.

[26]Hamada, T.; Manabe, K.; Kobayashi, S. *J. Am. Chem. Soc.* **2004**, *126*, 7768–7769.

**TABLE 16.2    The 12 Green Principles of Process Engineering**

1. Designers need to strive to ensure that all material and energy inputs and outputs are as inherently non-hazardous as possible.
2. It is better to prevent waste than to treat or clean up waste after it is formed.
3. Separation and purification operations should be designed to minimize energy consumption and materials use.
4. Products, processes, and systems should be designed to maximize mass, energy, space, and time efficiency.
5. Products, processes, and systems should be "output pulled" rather than "input pushed" through the use of energy and materials.
6. Embedded entropy and complexity must be viewed as an investment when making design choices on recycle, reuse, or beneficial disposition.
7. Targeted durability, not immortality, should be a design goal.
8. Design for unnecessary capacity or capability (e.g., "one size fits all") solutions should be considered a design flaw.
9. Material diversity in multi-component products should be minimized to promote disassembly and value retention.
10. Design of products, processes, and systems must include integration and interconnectivity with available energy and materials flows.
11. Products, processes, and systems should be designed for performance in a commercial "afterlife."
12. Material and energy inputs should be renewable rather than depleting.

measurable, and objective guidance that ultimately enables the development of the most efficient, environmentally benign process. Perhaps the simplest of all process metrics is overall yield. Yield is one measure of a step's or a process' efficiency but does not take into account byproduct formation or waste generation, which can carry significant cost and environmental impact. As a result, a number of other metrics have been proposed to more fully gauge the overall greenness of a process, the most common of which will be presented in the following sections.

### 16.2.1    Atom Economy (AE)

Atom economy (AE), first introduced by Barry Trost,[1] is a metric designed to encourage greater efficiency and lessened environmental impact when planning the synthesis of complex organic molecules. The classic definition of atom economy is the percentage of the molecular weight (MW) of the starting materials present in the molecular weight of the desired product after multiple chemical transformations.[1] AE can also be thought of as a comparison of the molecular weight of the product to the sum of the molecular weights of all the starting materials (salts, activating groups, removed protecting groups, etc). Catalysts, solvents, and stoichiometric excesses are typically not included in the analysis. The concept can also be used to calculate the efficiency of a single reaction. AE is expressed as follows for a single step:

$$A + B \rightarrow P$$

$$AE = \left[ \frac{MW \text{ of } P}{MW \text{ of } A + MW \text{ of } B} \right] \times 100$$

For a multi-step process, a more complex calculation is needed, including all new reagents (A, B, D, and F), but not the intermediate products (C and E), to avoid double counting weights.

$$A + B \rightarrow C$$

$$C + D \rightarrow E$$

$$E + F \rightarrow P$$

$$AE = \left[ \frac{MW \text{ of } P}{MW \text{ of } A + MW \text{ of } B + MW \text{ of } D + MW \text{ of } F} \right] \times 100$$

For convergent syntheses, the calculations are similar, and examples are available in the literature.[6] Trost has also published AE analyses for a variety of reactions.[7] The AE metric, which is solely based on the molecular weights of the reaction components, is limited by not taking into account yield, stoichiometric excess, catalyst or solvent usage, or workup materials.

### 16.2.2   Reaction Mass Efficiency (RME)

Reaction mass efficiency (RME)[8–11] is the percentage of the reactant mass going into a reaction that remains in the final product. This metric takes AE into account but also the stoichiometry of reagents and yield contributions from reaction steps. For the reaction:

$$A + B \rightarrow P$$

$$RME = \left[ \frac{\text{total mass of } P}{\text{total mass of } A + \text{total mass of } B} \right] \times 100$$

This calculation can be used to quantitate process improvements over time as conditions change. An increasing RME will result from increasing yield or smaller stoichiometric excess of reagents, as well as improvements in the AE associated with the reagent choice.

### 16.2.3   E-Factor

As neither AE nor RME account for chemical usage beyond reagents, Roger Sheldon developed the E-Factor metric.[2,6,8–12] The E-Factor is calculated as the ratio of total weight of waste generated to the total weight of product isolated.

$$\text{E-Factor} = \frac{\text{total waste (kg)}}{\text{kg of product}}$$

This metric allows for rapid comparison of many different routes to the same product or across multiple products. It can also serve as a metric across different organizations in a similar field. Because the E-Factor looks across the whole multi-step process, RME or AE may be more useful for process development on a single step.

### 16.2.4  Process Mass Intensity (PMI)

Process mass intensity (PMI)[6,8,9,11] is a variation of the E-Factor that calculates the ratio of the total mass used in a process relative to the mass of the product.

$$PMI = \frac{\text{total mass used in a process or process steps}}{\text{mass of final product}}$$

For this metric, "total mass used" includes: reactants, reagents, solvents, catalysts, acids, bases, salts, as well as workup solvents (for washes, extractions, crystallizations, solvent displacements, etc.). To use PMI to compare a set of processes,[13] consistency is needed in terms of what makes up the total mass used in the process. This makes it very difficult to compare PMI values from different laboratories or companies.

### 16.2.5  Additional Metrics

A number of other metrics, including carbon efficiency and mass productivity, have been developed to compare the efficiency and sustainability of processes. These metrics are less frequently used due to the complexity of calculation relative to those described above.[6]

## 16.3  SOLVENT AND REAGENT SELECTION

### 16.3.1  Organic Solvent Selection

Solvent selection is crucial to the outcome of many chemical reactions, and impacts the overall efficiency and sustainability of a process. Solvents are usually the biggest contributors to process waste, typically representing 85% of the total weight. Solvents are generally required during a reaction to enable heat and mass transfer, and during post-reaction processing and isolation of products.

The use of a number of common solvents has been questioned in recent years as their hazardous characteristics have come to light. As environmental, safety, and health issues from common solvents like dichloromethane, toluene, and dimethylsulfoxide have come to light, it has become critical that pharmaceutical and fine chemical businesses find suitable alternatives. This need was further increased by regulation of "undesired" solvents by regulatory agencies.

This raised the challenge of identifying substitutes with the appropriate physical characteristics (boiling point, vapor pressure, etc.) and molecular properties (dipolar moment, polarizability, etc.) without negatively impacting reactivity and ease of processing. As a result, several companies have developed solvent selection guides that suggest potential alternatives for commonly used non-green solvents.[5,14–23] One example is the guide developed by Pfizer[5] that presents a simple solvent selection tool that evaluates each solvent on three simple areas:

1. Worker safety, including carcinogenicity, mutagenicity, reproductive toxicity, skin absorption/sensitization, and toxicity.
2. Process safety, including flammability, potential for high emissions through high vapor pressure, static charge, and potential for peroxide formation and odor issues.

3. Environmental and regulatory considerations, including ecotoxicity and ground water contamination, potential EHS regulatory restrictions, ozone depletion potential, photoreactive potential.

A part of this tool is presented in Chapter 19. Most of the solvents in the "undesired" category have good options for replacements with the exception of polar aprotic solvents such as DMF, DMAc, and NMP.

Ideally, running a reaction in the absence of solvent presents the most environmentally friendly scenario, but as most reactions require some solvent, water or an aqueous mixture is the second best alternative.

### 16.3.2   Aqueous Systems

There has been a growing interest in using water as a solvent for organic reactions.[14,16,19,24–48] This has driven both by the increase in sustainability and greenness and by advantages in reactivity/selectivity for some classic reactions.[49–51] In some cases, rates and selectivity are increased due to hydrophobic effects when reactions are conducted in water.[19,26,28,33,38,40,41] The low solubility of oxygen in water can, in some cases, facilitate air-sensitive catalytic organic reactions to occur without extensive degassing procedures. In addition, using enzymatic reactions as the model, biotechnology groups have been investigating organic reactions in aqueous environments.[52,53]

*16.3.2.1   Example of Classic Reactions Run in Water*   MacMillan and coworkers[54] reported on organocatalyzed Diels–Alder reaction between unsaturated aldehydes and

[27]Anderson, K. W.; Buchwald, S. L. *Angew. Chem., Int. Ed. Engl.* **2005**, *44*, 6173–6177.

[28]Bhattacharya, S.; Srivastava, A.; Sengupta, S. *Tetrahedron Lett.* **2005**, *46*, 3557–3560.

[29]Shaughnessy, K. H.; DeVasher, R. B. *Curr. Org. Chem.* **2005**, *9*, 585–604.

[30]Azizi, N.; Torkiyan, L.; Saidi, M. R. *Org. Lett.* **2006**, *8*, 2079–2082.

[31]Hamada, T.; Manabe, K.; Kobayashi, S. *Chem.-Eur. J.* **2006**, *12*, 1205–1215.

[32]Itoh, J.; Fuchibe, K.; Akiyama, T. *Synthesis* **2006**, 4075–4080.

[33]Jiang, Z.; Liang, Z.; Wu, X.; Lu, Y. *Chem. Commun.* **2006**, 2801–2803.

[34]Lysen, M.; Koehler, K. *Synthesis* **2006**, 692–698.

[35]Ollevier, T.; Nadeau, E.; Guay-Begin, A.-A. *Tetrahedron Lett.* **2006**, *47*, 8351–8354.

[36]Raju, S.; Kumar, P. R.; Mukkanti, K.; Annamalai, P.; Pal, M. *Bioorg. Med. Chem. Lett.* **2006**, *16*, 6185–6189.

[37]Shimizu, S.; Shimada, N.; Sasaki, Y. *Green Chem.* **2006**, *8*, 608–614.

[38]Blackmond, D. G.; Armstrong, A.; Coombe, V.; Wells, A. *Angew. Chem., Int. Ed. Engl.* **2007**, *46*, 3798–3800.

[39]Fleckenstein, C.; Roy, S.; Leuthaeusser, S.; Plenio, H. *Chem. Commun.* **2007**, 2870–2872.

[40]Fringuelli, F.; Piermatti, O.; Pizzo, F.; Vaccaro, L. *Org. React. Water* **2007**, 146–184.

[41]Kobayashi, S. *Pure Appl. Chem.* **2007**, *79*, 235–245.

[42]Li, H.; Chai, W.; Zhang, F.; Chen, J. *Green Chem.* **2007**, *9*, 1223–1228.

[43]Narayan, S.; Fokin, V. V.; Sharpless, K. B. *Org. React. Water* **2007**, 350–365.

[44]Xiao, J.; Loh, T.-P. *Synlett* **2007**, 815–817.

[45]Xin, B.; Zhang, Y.; Cheng, K. *Synthesis* **2007**, 1970–1978.

[46]Ouach, A.; Gmouh, S.; Pucheault, M.; Vaultier, M. *Tetrahedron* **2008**, *64*, 1962–1970.

[47]Pan, C.; Wang, Z. *Coord. Chem. Rev.* **2008**, *252*, 736–750.

[48]Hailes, H. C. *Org. Process Res. Dev.* **2007**, *11*, 114–120.

[49]Lindstroem, U. M. *Chem. Rev.* **2002**, *102*, 2751–2771.

[50]Bonollo, S.; Fringuelli, F.; Pizzo, F.; Vaccaro, L. *Synlett* **2007**, 2683–2686.

[51]Bonollo, S.; Fringuelli, F.; Pizzo, F.; Vaccaro, L. *Green Chem.* **2006**, *8*, 960–964.

[52]Rong, L.; Wang, H.; Shi, J.; Yang, F.; Yao, H.; Tu, S.; Shi, D. *J. Heterocycl. Chem.* **2007**, *44*, 1505–1508.

[53]Fringuelli, F.; Piermatti, O.; Pizzo, F.; Vaccaro, L. *Eur. J. Org. Chem.* **2001**, 439–455.

[54]Ahrendt, K. A.; Borths, C. J.; MacMillan, D. W. C. *J. Am. Chem. Soc.* **2000**, *122*, 4243–4244.

conjugate dienes (cyclic or acyclic) in MeOH/water. In this case, the addition of water improved selectivity.

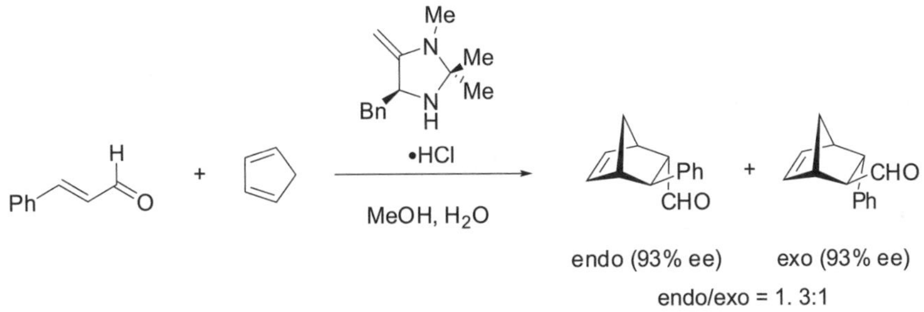

endo (93% ee)     exo (93% ee)

endo/exo = 1. 3:1

Mannich-type reactions have also been run successfully in water. Lu and coworkers[55] reported aqueous conditions for three-component Mannich reactions with good to excellent yields. The combination of the benzyl protecting group, the utilization of a threonine-derived organocatalyst, and an aqueous environment provided the best diastereoselectivity and enantioselectivity for this class of substrates.

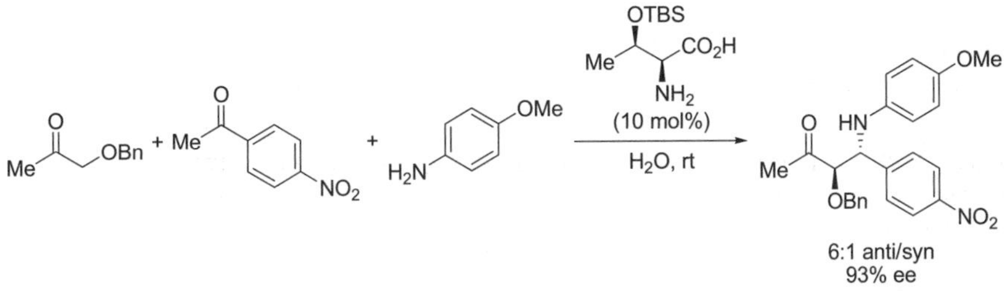

6:1 anti/syn
93% ee

Hayashi and coworkers[19] also reported a proline-derived catalyst for the asymmetric three-component Mannich reaction providing excellent yields and enantioselectivity.

4.6:1 syn/anti
93% ee

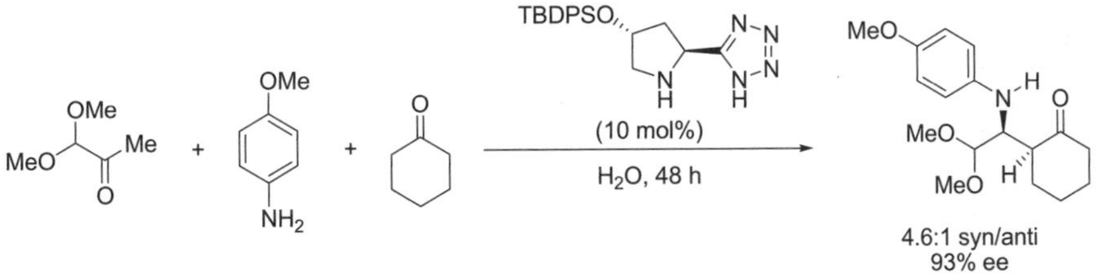

[55]Cheng, L.; Wu, X.; Lu, Y. *Org. Biomol. Chem.* **2007**, *5*, 1018–1020.

Several other successful Mannich and Mannich-type reactions in water have been reported.[26,30–32,35,37,41,44,47,49,55,56] Palladium coupling reactions, such as the Suzuki–Miyaura variant, have also been extensively explored with water as the solvent.[27,28,34,39,57–63] Lysén and Köhlera[34] reported successful Suzuki couplings in water with palladium on carbon as the catalyst. In this case, the catalyst can be recovered via filtration and reused for four more cycles. This method was applied to the reaction of arylboronic esters and potassium trifluoroborates with aryl halides.

In cases where homogeneous catalysts are needed, the typical approach involves modification of the ligands with groups that result in sufficient aqueous solubility. Plenio and coworkers[39] reported the synthesis of disulfonated N-heterocyclic imidazolium and imidazolinium salt ligands and their use in aqueous Suzuki–Miyaura coupling reactions. These conditions provided high yields when tested with aryl halides and aryl boronic acids, even at 0.1% catalyst loading.

[56]Li, P.; Wang, L. *Tetrahedron* **2007**, *63*, 5455–5459.

[57]Yin, L.; Liebscher, J. *Chem. Rev.* **2007**, *107*, 133–173.

[58]Mason, B. P.; Price, K. E.; Steinbacher, J. L.; Bogdan, A. R.; McQuade, D. T. *Chem. Rev.* **2007**, *107*, 2300–2318.

[59]Phan, N. T. S.; Van Der Sluys, M.; Jones, C. W. *Adv. Synth. Cat.* **2006**, *348*, 609–679.

[60]Marion, N.; Navarro, O.; Mei, J.; Stevens, E. D.; Scott, N. M.; Nolan, S. P. *J. Am. Chem. Soc.* **2006**, *128*, 4101–4111.

[61]Demchuk, O. M.; Yoruk, B.; Blackburn, T.; Snieckus, V. *Synlett* **2006**, 2908–2913.

[62]Nicolaou, K. C.; Bulger, P. G.; Sarlah, D. *Angew. Chem., Int. Ed. Engl.* **2005**, *44*, 4442–4489.

[63]Li, C.-J. *Chem. Rev.* **2005**, *105*, 3095–3165.

Other types of metal-catalyzed reactions that have been explored in water include: Sonogashira,[27,36,57,62,63] Heck,[28,57,59,61–64] carbonylation,[57] Fukuyama,[65] Ullmann,[42,57,63] Claisen rearrangements,[14,49,63,66,67] asymmetric aldol,[14,17,33,47,49,68] olefin oxidation and catalytic asymmetric hydrogenation.[39,49]

### 16.3.3 Green Reactions/Reagents

Once a synthetic strategy has been chosen, selection of reaction conditions and reagents becomes critical for the economical, environmental, and safety impact of the process. In 2005, Dugger, Ragan, and Ripin[69] published a review of reactions scaled in the GMP facilities at Pfizer's research facilities. They found that carbon–carbon and carbon–nitrogen bond formation were two of the most common reactions in API synthesis. The methodology used for carbon–carbon bond formation varied from the use of organolithium reagents to metal-catalyzed coupling reactions. For carbon–nitrogen bond formation, reductive amination and metal-catalyzed coupling were the most common. A cross-site analysis among Pfizer, GlaxoSmithKline, and AstraZeneca[70] also found that carbon–carbon and carbon–heteroatom bond formation were the most common. With the variety of reagents/conditions available to perform these transformations, it can be challenging for a chemist to decide which one to use. Metrics, such as those introduced in Section 16.2, can assist in comparisons between methods, but do not provide guidance for reagent selection.

Many companies have developed tools, similar to the solvent selection tool, to help their scientists choose green reagents during their route selection process. As an example, Pfizer published a reagent guide as part of an extensive publication to target both greener solvent and reagent selections[5] by providing alternative chemical methods for several reaction types. These methods were chosen because of established performance, scalability, and greenness. The guide is intended to help introduce green reagents earlier in the discovery/development process.

Although green alternatives have been identified for several transformations, gaps exist where there is no useful green substitute. The ACS Green Chemistry Institute Pharmaceutical Roundtable in 2007[71] identified the following gaps: reduction of amides, bromination, sulfonation, atom economical amide formations, nitrations, demethylations, Friedel–Crafts reactions, ester hydrolysis, alcohol substitution for displacement, epoxidation, and carbonyl olefinations.

[64]Herrmann, W. A.; Reisinger, C.-P.; Haerter, P. *Multiphase Homogeneous Catal.* **2005**, *1*, 230–238.

[65]Seki, M.; Hatsuda, M.; Mori, Y.; Yoshida, S.-i.; Yamada, S.-i.; Shimizu, T. *Chem.-Eur. J.* **2004**, *10*, 6102–6110.

[66]Kuzemko, M. A.; Van Arnum, S. D.; Niemczyk, H. J. *Org. Process Res. Dev.* **2007**, *11*, 470–476.

[67]Rossi, F.; Corcella, F.; Caldarelli, F. S.; Heidempergher, F.; Marchionni, C.; Auguadro, M.; Cattaneo, M.; Ceriani, L.; Visentin, G.; Ventrella, G.; Pinciroli, V.; Ramella, G.; Candiani, I.; Bedeschi, A.; Tomasi, A.; Kline, B. J.; Martinez, C. A.; Yazbeck, D.; Kucera, D. J. *Org. Process Res. Dev.* **2008**, *12*, 322–338.

[68]Tanaka, K.; Toda, F. *Chem. Rev.* **2000**, *100*, 1025–1074.

[69]Dugger, R. W.; Ragan, J. A.; Brown Ripin, D. H. *Org. Process Res. Dev.* **2005**, *9*, 253–258.

[70]Carey, J. S.; Laffan, D.; Thomson, C.; Williams, M. T. *Org. Biomol. Chem.* **2006**, *4*, 2337–2347.

[71]Constable, D. J. C.; Dunn, P. J.; Hayler, J. D.; Humphrey, G. R.; Leazer, J. L., Jr.; Linderman, R. J.; Lorenz, K.; Manley, J.; Pearlman, B. A.; Wells, A.; Zaks, A.; Zhang, T. Y. *Green Chem.* **2007**, *9*, 411–420.

## 16.4   EXAMPLES OF GREEN METHODS AND REAGENTS FOR COMMON REACTION TYPES

### 16.4.1   Formation of Aryl Amines and Aryl Amides

Aryl amines and aryl amides can be formed from a coupling of an amine or amide with an aryl halide, with or without metal catalysis (see Chapter 6).[72,73] Direct displacement of an aryl halide with an amine is very atom efficient and leads to virtually no side products of concern but typically requires harsh conditions. As a result, the substitution is more commonly conducted in the presence of small amounts of a heavy metal catalyst under milder conditions. This method is effective for coupling to aryl bromides, iodides, and triflates. Several reviews have been published in the aryl amination area, including some that address large scale reactions for both ligand preparation and their use in multi-kilo scale aryl aminations.[72]

Aryl chlorides are typically less expensive relative to the other aryl halides but are generally less reactive. A number of catalyst–ligand combinations have been shown to facilitate this coupling reaction. Although many have been demonstrated on small scale, there are limited examples of these couplings on larger scale, most likely due to the difficulty associated with preparing large quantities of the ligands and catalysts.

The Buchwald and Hartwig groups have pioneered the use of metal–phosphine ligand complexes to catalyze the amination of aryl chlorides.[38,55,70–75] In 2007, Buchwald[74] reported the use of a hybrid ligand (Buchwald–Hartwig type) that promoted with the coupling reaction between aryl chlorides and amides in excellent yields.

Ligand =

Singer and coworkers[75] reported the palladium-catalyzed aryl amination reaction using pyrazole ligands. These ligands were successfully prepared at large scale and were used to couple amines and aryl chlorides in high yields under mild conditions in benign solvents.

[72]Buchwald, S. L.; Mauger, C.; Mignani, G.; Scholz, U. *Adv. Synth. Cat.* **2006**, *348*, 23–39.

[73]Hartwig, J. F. *Synlett* **2006**, 1283–1294.

[74]Altman, R. A.; Fors, B. P.; Buchwald, S. L. *Nat. Protoc.* **2007**, *2*, 2881–2887.

[75]Singer, R. A.; Dore, M.; Sieser, J. E.; Berliner, M. A. *Tetrahedron Lett.* **2006**, *47*, 3727–3731.

The successful use of copper as metal has also been reported for several types of coupling reactions with aryl chlorides.[60,76–80]

### 16.4.2   Carbon–Carbon Bond Formation

When forming carbon–carbon bonds, palladium-catalyzed coupling reactions (Heck, Suzuki, Sonogashira, etc.) can serve as a green alternative to more traditional methods. For instance, the Heck reaction can be used to produce Wittig-type products without the phosphorus oxide waste. For the Heck, as with other palladium-catalyzed coupling reactions, bulky phosphorus ligands tend to work the best.[72,74,80–82] Ligands such as the type developed by Buchwald and Hartwig have shown remarkable results for this type of reaction under mild conditions.[80–82] Several other ligands, also used for aryl amination reactions, have been found to be efficient for a variety of Heck reactions[28,57,59,61–64] including the Nolan imidazolinium type[60,83,84] and modified ureas.[85] As the availability of boronic acid and ester derivatives increases at commercial scale, the Suzuki reaction has become a feasible alternative to organometallic chemistry such as organomagnesium or organolithium reactions. These developments make the use of economical substrates such as aryl chlorides become more practical.[27,28,34,39,57–63] Reaction methodology to couple hindered substrates

[76]Altman, R. A.; Shafir, A.; Choi, A.; Lichtor, P. A.; Buchwald, S. L. *J. Org. Chem.* **2008**, *73*, 284–286.

[77]Verma, A. K.; Singh, J.; Sankar, V. K.; Chaudhary, R.; Chandra, R. *Tetrahedron Lett.* **2007**, *48*, 4207–4210.

[78]Altman, R. A.; Buchwald, S. L. *Org. Lett.* **2007**, *9*, 643–646.

[79]Ghosh, A.; Sieser, J. E.; Caron, S.; Couturier, M.; Dupont-Gaudet, K.; Girardin, M. *J. Org. Chem.* **2006**, *71*, 1258–1261.

[80]Ikawa, T.; Barder, T. E.; Biscoe, M. R.; Buchwald, S. L. *J. Am. Chem. Soc.* **2007**, *129*, 13001–13007.

[81]Beccalli, E. M.; Broggini, G.; Martinelli, M.; Sottocornola, S. *Chem. Rev.* **2007**, *107*, 5318–5365.

[82]Shekhar, S.; Ryberg, P.; Hartwig, J. F.; Mathew, J. S.; Blackmond, D. G.; Strieter, E. R.; Buchwald, S. L. *J. Am. Chem. Soc.* **2006**, *128*, 3584–3591.

[83]Marion, N.; de Fremont, P.; Puijk, I. M.; Ecarnot, E. C.; Amoroso, D.; Bell, A.; Nolan, S. P. *Adv. Synth. Cat.* **2007**, *349*, 2380–2384.

[84]Clavier, H.; Nolan, S. P. *Annu. Rep. Prog. Chem., Sect. B: Org. Chem.* **2007**, *103*, 193–222.

[85]Sergeev, A. G.; Artamkina, G. A.; Beletskaya, I. P. *Russ. J. Org. Chem.* **2003**, *39*, 1741–1752.

or those with asymmetric centers[86,87] with retention of configuration have also been developed.

### 16.4.3 Oxidation

The oxidation of alcohols is a common transformation for both industrial and laboratory scale synthesis,[88] but traditional methods present efficiency and safety limitations (see Section 10.4.1). Alcohol oxidation has frequently been performed with stoichiometric inorganic oxidants, such as chromium(VI) reagents,[89,90] which have a number of safety concerns, or by Swern-type oxidations. The Swern oxidation [91,92] is challenging on large scale due to, in most cases, the need for cryogenic temperatures and the extremely reactive reagent oxalyl chloride, which can present handling and safety issues. In addition, large amounts of gas byproducts ($CO_2$ and CO) are produced, presenting a potential safety and environmental issue. Recently, the use of 1-oxy-2,2,6,6-tetramethylpiperidinyl (TEMPO) for oxidation has been explored by various groups as a green alternative.[67,93–95] Other greener oxidation alternatives have been explored and reported in the literature.[65,90,94–98]

### 16.5   GREEN CHEMISTRY IMPROVEMENTS IN PROCESS DEVELOPMENT

### 16.5.1   Introduction

Throughout the development of a commercial route for an active pharmaceutical ingredient (API), process chemists face numerous challenges, including the need to select the best route, starting materials, synthetic methods, and technologies that provide the desired target under a timeline dictated by project needs and loss of exclusivity due to patent lifetimes.

The overall goal of a process chemist is to develop the most cost-effective process for large scale synthesis of an API. A cost-effective process typically has high overall yield, little waste (cost for disposal and cost of material use), and good reagent/reactant utilization factors, among others. As development progresses and these factors improve, the process becomes greener. In terms of metrics, the RME increases as the yield increases and the

[86]Bringmann, G.; Ruedenauer, S.; Bruhn, T.; Benson, L.; Brun, R. *Tetrahedron* **2008**, *64*, 5563–5568.

[87]Matos, K.; Soderquist, J. A. *J. Org. Chem.* **1998**, *63*, 461–470.

[88]Caron, S.; Dugger, R. W.; Ruggeri, S. G.; Ragan, J. A.; Ripin, D. H. B. *Chem. Rev.* **2006**, *106*, 2943–2989.

[89]Luzzio, F. A.; Guziec, F. S., Jr. *Org. Prep. Proced. Int.* **1988**, *20*, 533–584.

[90]Hayashi, M.; Kawabata, H. *Adv. Chem. Res.* **2006**, *1*, 45–62.

[91]Tidwell, T. T. *Organic Reactions (New York)* **1990**, *39*, 297–572.

[92]Ahmad, N. M.In *Name Reactions for Functional Group Transformations*; Wiley: Hoboken, N. J., **2007**, pp 291–308.

[93]Breton, T.; Bashiardes, G.; Leger, J.-M.; Kokoh, K. B. *Eur. J. Org. Chem.* **2007**, 1567–1570.

[94]Holczknecht, O.; Cavazzini, M.; Quici, S.; Shepperson, I.; Pozzi, G. *Adv. Synth. Cat.* **2005**, *347*, 677–688.

[95]Sheldon, R. A.; Arends, I. W. C. E.; ten Brink, G.-J.; Dijksman, A. *Acc. Chem. Res.* **2002**, *35*, 774–781.

[96]Ruddy, D. A.; Tilley, T. D. *Chem. Commun.* **2007**, 3350–3352.

[97]Guo, M.-L.; Li, H.-Z. *Green Chem.* **2007**, *9*, 421–423.

[98]Merbouh, N.; Bobbitt, J. M.; Brueckner, C. *Org. Prep. Proced. Int.* **2004**, *36*, 3–31.

reagents are used more sparingly. The E-Factor will also decrease over time as the waste is decreased with more efficient workups, telescoping, and reduced use of solvent. The process can then be further improved when "greener" methods/reagents are introduced. In the following sections, a number of examples will be presented that combine both green reagent/method selections with good process chemistry to create green processes.

### 16.5.2 Sildenafil Citrate

Sildenafil citrate[99,100] is a selective inhibitor of phosphodiesterase 5 (PDE5) developed by Pfizer for the treatment of male erectile dysfunction. As this drug rapidly reached more than $1 billion in annual sales a highly efficient process was required to provide several metric tons of the active pharmaceutical ingredient.

The medicinal chemistry route for sildenafil presented several issues for commercialization, including the use of stannous chloride in the reduction of a nitro group and the use of thionyl chloride as a reaction solvent. In addition, the linear approach reduced overall throughput and increased the impact of both stannous chloride and thionyl chloride per kilogram of API prepared.

DISCOVERY ROUTE TO SILDENAFIL

[99]Dunn, P. J. *Process Chem. Pharm. Ind.* **2008**, 267–277.
[100]Dunn, P. J.; Galvin, S.; Hettenbach, K. *Green Chem.* **2004**, *6*, 43–48.

In their first route, the process group replaced the tin reduction with a catalytic hydrogenation. This eliminated the tin from the synthesis and significantly reduced the amount of metal needed. To limit the use of thionyl chloride, toluene was selected as the reaction solvent. This allowed reduction of thionyl chloride from 13 to 1.6 equivalents. By repositioning selected bond connections, the synthesis was also made more convergent by a different connection strategy. An analysis of this work detailing how the modifications improved the process and limited solvent use was published by Dunn et al.[100]

**COMMERCIAL ROUTE TO SILDENAFIL**

### 16.5.3   Sertraline Hydrochloride

Sertraline hydrochloride is the active ingredient of Zoloft, a selective serotonin reuptake inhibitor that is used to treat depression as well as dependency and other anxiety-related disorders. The commercial process for the synthesis of sertaline hydrochloride has gone through multiple iterations since the original discovery synthesis.[101–103] In each approach,

[101] Spavins, J. C. *Process for preparing a ketimine, N-[4-(3,4-dichlorophenyl)-3,4-dihydro-1(2H)-naphthalenylidene]methanamine.* US4855500A, **1989**, 4 pp.
[102] Welch, W. M.; Kraska, A. R.; Sarges, R.; Koe, B. K. *J. Med. Chem.* **1984**, *27*, 1508–1515.
[103] Taber, G. P.; Pfisterer, D. M.; Colberg, J. C. *Org. Process Res. Dev.* **2004**, *8*, 385–388.

a classical resolution with *D*-mandelic acid has been used to separate the desired enantiomer.[101,102,104] The mandelate salt can then be converted to the hydrochloride salt for final crystallization.

**DISCOVERY ROUTE TO SERTRALINE HYDROCHLORIDE**

In the original discovery route, the sertraline ketone was reacted with excess methylamine and titanium tetrachloride to produce the corresponding sertraline imine, titanium oxide, and monomethylamine hydrochloride as byproducts. The reduction of the imine with sodium borohydride resulted in a 1:1 mixture of the *cis* and *trans* diastereoisomers. Later, this process was improved by using a hydrogenation in the presence of Pd/C instead of the borohydride.[101,105] Although this increased the yield of the desired *cis* isomer, it also resulted in the formation of undesired byproducts such as mono- and dideschloro racemic sertraline. Lengthy and yield-consuming crystallizations using large volumes of solvent were needed to effectively remove these impurities.

[104]Vukics, K.; Fodor, T.; Fischer, J.; Fellegvari, I.; Levai, S. *Org. Process Res. Dev.* **2002**, *6*, 82–85.
[105]Williams, M.; Quallich, G. *Chem. Ind.* **1990**, 315–319.

## FIRST COMMERCIAL ROUTE TO SERTRALINE HYDROCHLORIDE

**Both reagent & solvent**

**Tetralone**

70%

1) H₂,Pd/C THF, rt
2) HCl

+ TiO₂
+ MeNH₄Cl

TiCl₄
MeNH₂

Toluene
hexanes
22–25°C

AlCl₃ (2 eq.)
0–5°C

cis/trans = 6:1

1) NaOH
2) (D)-mandelic acid (1.1 eq.) EtOH, rt
75%

**Sertraline mandelate**

1) NaOH, 65°C H₂O, EtOAc
2) EtOAc/HCl
96%

**Sertraline**

Since an efficient large scale synthesis of the racemic tetralone was available,[106] it was used as an early intermediate in later synthetic routes. In 1997 it was reported that the use of simple solubility differences between the tetralone and corresponding imine in alcoholic solvent could be used to drive the equilibrium to the imine to 95% conversion.[103] This approach eliminated the need for a Lewis acid such as titanium tetrachloride and provided the opportunity to telescope this step into the reduction step, as it was performed in the same type of solvent. By changing the type of catalyst support from carbon to calcium carbonate, the reduction selectivity increased from 6:1 to >20:1 *cis/trans*. With over 200 tons of sertraline hydrochloride produced in 2002, the improvements resulted in the elimination of 140 metric tons of $TiCl_4$ and 440 tons of $TiO_2MeNH_2 \cdot HCl$ wet cake waste per year. By increasing the diastereoselectivity of the step 2 reduction, over 40 tons of *trans*-isomer waste was also eliminated.

This process did not eliminate the late stage resolution using mandelic acid nor the resulting loss of material. To eliminate this problem, a process for the synthesis of the optically active sertraline ketone was reported.[107] This was accomplished by using a

[106]Adrian, G. *Preparation of 4-(disubstituted aryl)-1-tetralones as intermediates for serotonin antagonists.* EP346226 A1, **1989**, 6 pp.

[107]Dapremont, O.; Geiser, F.; Zhang, T.; Guhan, S. S.; Guinn, R. M.; Quallich, G. J. *Process for the production of enantiomerically pure or optically enriched sertraline-tetralone using continuous chromatography.* WO9957089 A1, **1999**, 16 pp.

novel multi-column chromatography (MCC) system to separate the racemic ketone. In this case, the unwanted enantiomer could be epimerized with base to generate more racemate and be recycled to maximize material use and reduce waste. By eliminating the classical resolution, over 92 kg of waste for every kg of API was eliminated. The resulting process has an E-Factor of 8 kg waste/kg of product, a level rarely achieved for pharmaceutical processes.

**COMMERCIAL ROUTE TO SERTRALINE HYDROCHLORIDE**

### 16.5.4 Torcetrapib

Torcetrapib[108,109] is a potent inhibitor of cholesteryl ester transfer protein (CETP) developed by Pfizer for the treatment and prevention of dyslipidemia and atherosclerosis. The compound development was halted by the company during phase III studies. Due to the large amount of API needed to supply clinical trials, over five metric tons were produced throughout process development. As the need for this compound was expected to be over 200 metric tons per year, a careful green chemistry analysis was performed seeking the most environmentally friendly route.

The optimized discovery route (named the "Resolution Route") was used for the synthesis of the API supporting early drug safety studies and initial clinical evaluation, but was not able to efficiently provide the necessary quantities for the larger volume batches.

[108]Damon, D. B.; Dugger, R. W.; Hubbs, S. E.; Scott, J. M.; Scott, R. W. *Org. Process Res. Dev.* **2006**, *10*, 472–480.
[109]Damon, D. B.; Dugger, R. W.; Magnus-Aryitey, G.; Ruggeri, R. B.; Wester, R. T.; Tu, M.; Abramov, Y. *Org. Process Res. Dev.* **2006**, *10*, 464–471.

## OPTIMIZED DISCOVERY ROUTE (RESOLUTION ROUTE) TO TORCETRAPIB

This sequence raised two specific concerns associated with this chemistry included the use of 4-trifluoromethylaniline, which can undergo exothermic degradation releasing HF gas,[110] and the use of a low molecular weight azide generated *in situ*, as it was found to be thermally unstable,[111] with the potential to violently decompose at low temperatures. The special equipment needed to safely contain these reactions could have limited the number of long-term external suppliers as the bulk volume increased.

As this route was developed, it also became apparent that if the N-vinyl-O-benzyl urethane did not react quickly, a new impurity was formed that was very difficult to purge downstream.[109] Finally, the late stage resolution also offered a challenge since it would also generate hundreds of tons of the undesired enantiomer salt as waste as the scale increased.

---

[110]Tickner, D.; Kasthurikrishnan, N. *Org. Process Res. Dev.* **2001**, *5*, 270–271.

[111]am Ende, D. J.; DeVries, K. M.; Clifford, P. J.; Brenek, S. J. *Org. Process Res. Dev.* **1998**, *2*, 382–392.

## CHIRAL POOL ROUTE TO TORCETRAPIB

In face of the challenges described above, a new route based on chiral pool (see Chapter 14) was developed to increase the process efficiency.[108] In this new approach, (R)-3-aminopentanenitrile was reacted with p-chloro- or p-bromotrifluoromethylbenzene under palladium-catalyzed conditions. The resulting product was not isolated, but was directly moved into the second step as a toluene solution. After hydrolysis of the nitrile functionality, the product was reacted with lithium t-butoxide and methylchloroformate in THF to form the corresponding imide.

The key step in the synthesis was the reduction of the imide with sodium borohydride and subsequent diastereoselective cyclization to form the tetrahydroquinoline ring. In the final intermediate step, ethylchloroformate was coupled in dichloromethane, using pyridine as the base. The product was then reacted with the benzylbromide in the presence of potassium t-butoxide in dichloromethane to produce the corresponding API.

Further development was done to optimize the last two steps of the synthesis. Improvements focused on eliminating the use of chlorinated solvents, the use of pyridine as the base for the reaction with ethylchloroformate, and the use of potassium t-butoxide. After screening other bases and solvents, it was found that tribasic sodium phosphate or sodium carbonate in THF at reflux drove the reaction with ethylchloroformate to completion. This replaced pyridine and eliminated the need for dichloromethane. In addition, ethanol/water was introduced as the isolation solvent matrix, removing isopropyl ether and related peroxide hazards.

The potassium *t*-butoxide was eliminated from the alkylation in the last step of the synthesis by using biphasic conditions with 50% caustic and a phase-transfer catalyst, tetrabutylammonium bromide.

For the projected volume of 200 metric tons of API at peak sales, these improvements would have accomplished the following: elimination of isopropyl ether (2400 metric tons/year); reduction of dichloromethane (5000 metric tons/year); elimination of pyridine (160 metric tons/year); elimination of DABCO, originally used as a scavenging agent for the benzyl bromide, (50 metric tons/year); and elimination of potassium *t*-butoxide (50 metric tons). In addition, the chiral pool route reduced the process volume by 40% relative to the original process, as well as the number of solvents needed. Significant reductions were also obtained for chlorinated solvents (about 50%) and other solvents. Both RME and yield data showed significant improvement and the overall yield doubled as the route changed from the resolution to the chiral pool route.

# 17

# NAMING CARBOCYCLES AND HETEROCYCLES

Heather N. Frost and David H. Brown Ripin

## 17.1 INTRODUCTION

The naming of heterocycles and the determination of a ring system's structure based on its name are critical to the synthetic organic chemist. Ring systems are sometimes called by their common names, and sometimes by their more complex IUPAC names. In this chapter, the IUPAC naming conventions for heterocyclic and carbocyclic ring systems are summarized, and tables are provided listing common ring systems and their common and IUPAC names.[1] Finally, the names are indexed alphabetically at the end of the chapter to simplify the conversion of common names into structure. This chapter is a summary of the methods used to name most ring systems, but for complex systems the IUPAC guidance should be consulted. The flow chart shown below can serve as a decision tree in the naming of a specific compound. A compound may have multiple components that must be named using different strategies. To further complicate matters, some polycyclic systems are named based on the common name of the all-carbon ring analogies.

---

[1] http://www.chem.qmul.ac.uk/iupac/ This HTML reproduction is as close as possible to the published version (see **IUPAC, Commission on Nomenclature of Organic Chemistry.** *A Guide to IUPAC Nomenclature of Organic Compounds [Recommendations 1993]*, 1993, Blackwell Scientific publications, Copyright 1993 IUPAC]). If you need to cite these rules please quote this reference as their source.This HTML reproduction of Sections A, B, and C of IUPAC "Blue Book" is as close as possible to the published version (see *Nomenclature of Organic Chemistry, Sections A, B, C, D, E, F, and H*, Pergamon Press, Oxford, 1979. Copyright 1979 IUPAC). If you need to cite these rules please quote this reference as their source.The HTML reproduction of the IUPAC Nomenclature of Organic Chemistry is published by Advanced Chemistry Development, Inc. with permission of the IUPAC. The following IUPAC publications were taken as a source and should be used as primary references: *Nomenclature of Organic Chemistry, Sections A, B, C, D, E, F, and H;* Pergamon Press: Oxford, 1979. Copyright 1979 IUPAC. *A Guide to IUPAC Nomenclature of Organic Compounds (Recommendations 1993)*, 1993, Blackwell Scientific publications, Copyright 1993 IUPAC.

---

*Practical Synthetic Organic Chemistry: Reactions, Principles, and Techniques*, First Edition.
Edited by Stéphane Caron.
© 2011 John Wiley & Sons, Inc. Published 2011 by John Wiley & Sons, Inc.

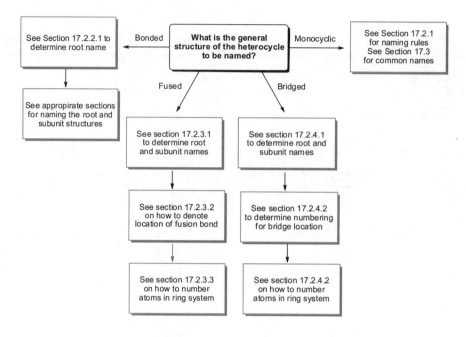

## 17.2   IUPAC RULES FOR NAMING HETEROCYCLES

### 17.2.1   Monocyclic Heterocycles

Heteroatoms in a ring are prioritized in the order as listed in Table 17.1.

**TABLE 17.1   Heteroatom Numbering Prioritization**

| Element | Prefix |
| --- | --- |
| O | oxa |
| S | thia |
| Se | selena |
| Te | tellura |
| N | aza |
| P | phospha |
|  | phosphor when followed by in or ine |
| As | arsa |
|  | arsen when followed by in or ine |
| Sb | stiba |
|  | antimon when followed by in or ine |
| Bi | bisma |
| Si | sila |
| Ge | germa |
| Sn | stanna |
| Pb | plumba |
| B | bora |

The suffix for the name is determined by the size of the ring. Compounds with an asterisk (*) note exceptions to the convention. The last letter of the prefix is dropped prior to the suffix being added, for example, an unsaturated 3-membered nitrogen containing ring is "az" + "irine," not "aza" + "irine."

**TABLE 17.2   Naming Conventions–Suffixes for Nitrogen-Containing Heterocycles**

| Ring Size | Unsaturated Ring Containing Nitrogen | Examples | Saturated Ring Containing Nitrogen | Examples |
|---|---|---|---|---|
| 3 | -irine | 2-azirine | -iridine | aziridine |
| 4 | -ete | azete        2-azetine* | -etidine | azetidine |
| 5 | -ole | Azole (1H-Pyrrole) | -olidine | Azolidine (Pyrrolidine) |
| 6 | -ine | Azine (Pyridine) | hexahydro-__-ine | hexahydroazine (piperidine) |
| 7 | -epine | azepine | hexahydro-__-epine | hexahydroazepine |
| 8 | -ocene | azocene | octahydro-__-ocene | octahydroazocene |
| 9 | -onine | azonine | octahydro-__-onine | octahydroazonine |
| 10 | -e'cine | azecine | decahydro-__-ecine | decahydroazecine |

Different suffixes are used for rings that do not contain nitrogen. The same rules apply for a non-nitrogen containing ring with regard to joining the prefix and suffix, that is, an unsaturated 3-membered non-nitrogen containing ring is "ox" + "irene," not "oxa" + "irene."

**TABLE 17.3  Naming Conventions–Suffixes for Oxygen-Containing Heterocycles**

| Ring Size | Unsaturated Ring Containing no Nitrogen | Examples | Saturated Ring Containing no Nitrogen | Examples |
|-----------|------------------------------------------|----------|----------------------------------------|----------|
| 3 | -irene | oxirene | -irane | oxirane |
| 4 | -ete | oxete | -etane |  oxetane |
| 5 | -ole |  oxole (furan) | -olane |  oxolane (tetrahydrofuran) |
| 6 | -in | 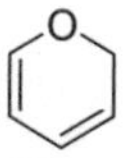 2*H*-oxin 2*H*-pyran | -ane | 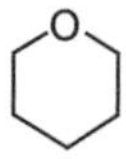 oxane |
| 7 | -epin |  oxepin | -epane | 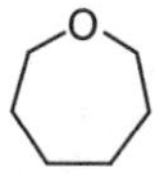 oxepane |
| 8 | -ocin | 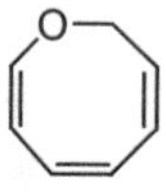 2*H*-oxocin | -ocane | 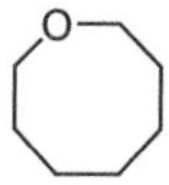 oxocane |
| 9 | -onin | 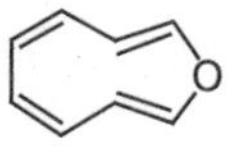 oxonin | -onane | 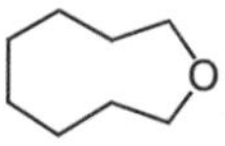 oxonane |
| 10 | -ecin | 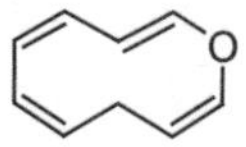 4*H*-oxecin | -ecane | 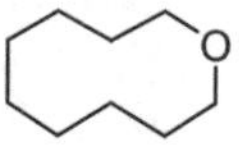 oxecane |

Heterocycles containing more than one heteroatom are named with the highest priority heteroatom listed first.

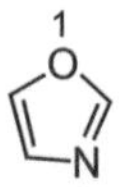

3-oxazole
not
3-azoxole

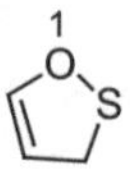

3*H*-2-oxathiole
not
5*H*-2-thiooxole

When a heterocycle is a substituent on a molecule, the heterocycle is named as a radical. If the heterocycle is bonded to the molecule with a single bond, the name is modified by dropping the letter "e" from the end of the name and replacing it with "yl." The connection point on the substituent heterocycle is denoted prior to the "yl" ending as shown in the following example.

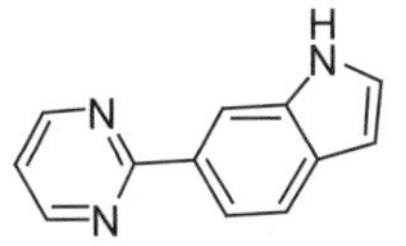

6-(pyrimidin-2-yl)indole

A few common exceptions, where a different modification is made to the root name, are furyl, pyridyl, piperidyl, quinolyl, isoquinolyl, and thienyl.

*17.2.1.1  Atom Numbering*    Heterocycles are numbered according to the following rules, applied in the order listed.

1.  Numbering is started with the highest priority heteroatom.

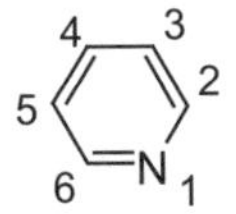

2.  For heterocycles with only one heteroatom, the numbering goes around the ring in the direction that results in the lowest locants for the substituents on the ring.

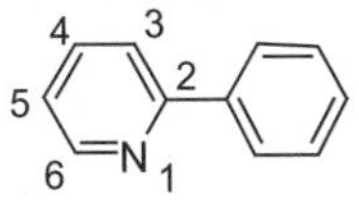

2-phenylpyridine

3.  If the heterocycle contains more than one heteroatom, the ring is numbered in the direction that results in the lowest locants for the heteroatoms.

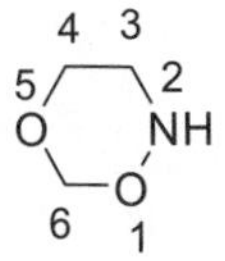

2-aza-1,5-dioxane
not
6-aza-1,3-dioxane
or
4-aza-1,3-dioxane

### 17.2.2  Bonded Heterocycles

*17.2.2.1  Root Name*    In the case of attached heterocycles, the base name used is the ring system with the highest priority. Determining highest priority is complex, but some more commonly applicable rules are discussed here.

1. Any heterocycle has a higher priority than an all-carbon ring system.

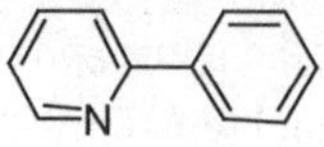

2-phenylpyridine
not
pyrid-2-ylbenzene

2. Heterocycles are ranked according to the priority of the heteroatoms in the ring (e.g., oxepane is higher priority than thiopane is higher than hexahydroazepine).

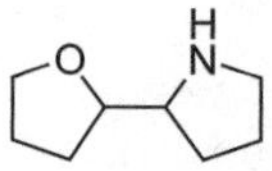

2-(pyrrolidin-2-yl)tetrahydrofuran
not
2-(tetrahydrofuran-2-yl)pyrrolidine

3. The ring system with the highest number of rings has highest priority.

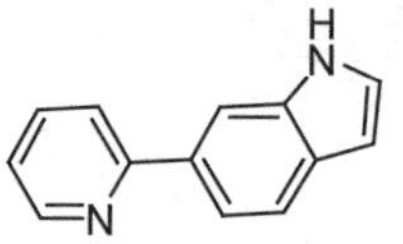

6-(pyrid-2-yl)indole
not
2-(indol-6-yl)pyridine

4. Larger rings are prioritized higher than smaller rings.

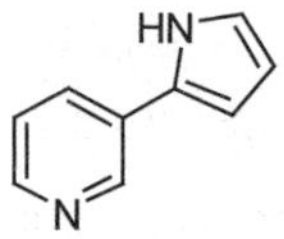

3-(pyrrol-2-yl)pyridine
not
2-(pyrid-3-yl)pyrrole

5. The ring with the highest degree of unsaturation gets priority.

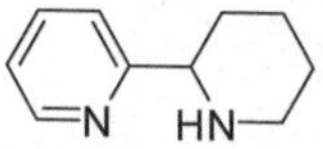

2-(piperid-2-yl)pyridine
not
2-(pyridin-2-yl)piperidine

### 17.2.3  Fused Heterocycles

*17.2.3.1  Root Name*   In the case of fused heterocycles, the root name in the ring system is determined by the following rules. In the case of multiple options, proceed to the next rule.

(See Section 17.2.3.3 for the appropriate order for numbering rings and Section 17.2.3.2 for the rules for lettering bonds. Note that prioritization of polycycles does not necessarily follow the prioritization of monocycles.) The name of the appended ring is modified by dropping the final "e" from the name and replacing it with "o." Examples are shown above and below. Some common exceptions to this rule are benzo, furo, imidazo, isoquino, pyrido, pyrimido, quino, and thieno.

1. A nitrogen-containing component takes precedence over other ring systems.

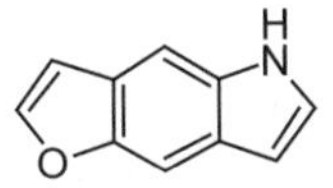

furo[2,3-*f*]indole
not
pyrrolo[2,3-*f*]benzofuran

2. A component containing a heteroatom (other than nitrogen) follows the prioritization rules for monocycles.

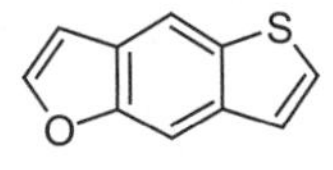

thiazolo[2,3-*f*]benzofuran
not
furo[2,3-*f*]benzthiazole

3. A component containing the greatest number of heteroatoms, regardless of ring size, has priority.

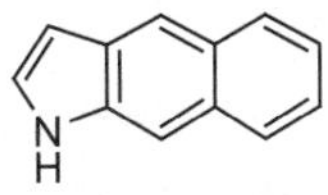

benzo[*f*]indole
not
naphtho[*b*]pyrrole

4. A component containing the largest possible heterocyclic ring has priority.

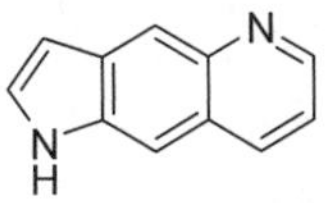

pyrrolo[2,3-*g*]quinoline
not
pyrido[2,3-*f*]indole

5. A component containing the greatest number of hetero atoms of any kind has priority.

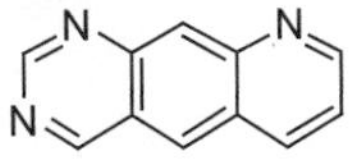

pyrido[3,2-*g*]quinazoline
not
pyrimidino[5,6-*g*]quinoline

6. A component containing the greatest variety of hetero atoms has priority.

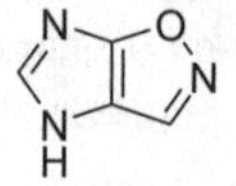

imidazolo[5,4-*d*]isoxazole
not
isoxazolo[5,4-*d*]imidazole

7. The component with the lower numbers for the hetero atoms before fusion has priority.

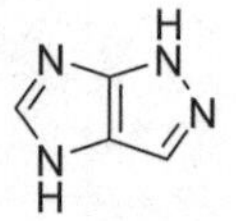

imidazolo[5,4-*d*]pyrazole
not
pyrazolo[5,4-*d*]imidazole

8. For ring systems where a heteroatom is shared between two rings, both rings are named to include the heteroatom.

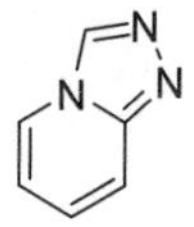

pyrido[2,1-*c*]-1,2,4-triazole

***17.2.3.2   Fusion Bond Numbering System***   In the case of fused heterocycles, the root name is the ring system with the highest number of rings. On the root heterocycle, the bonds are lettered "a" for the 1–2 bond, "b" for the 2–3 or 2–2a bond, and so on around the exterior of the ring system. The name is denoted "fused ring[x,y-*n*]parent ring" where x is the substitution on the fused ring attached to the lowest number on the parent ring, y is the substitution on the fused ring attached to the higher number on the parent ring, and *n* is the letter of the fusion bond. In the case of rings fused at 3 atoms, the denotation is "fused ring[x,y,z-*nm*]parent ring."

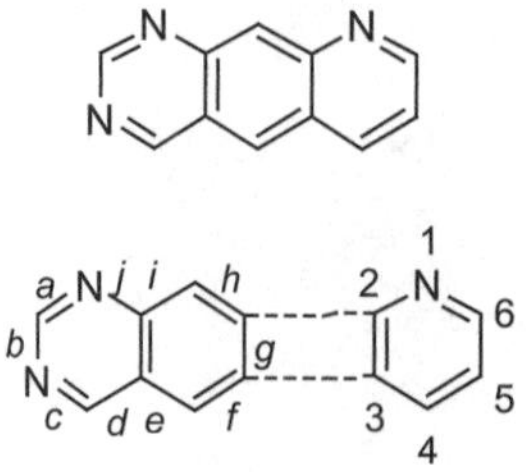

pyrido[3,2-*g*]quinazoline

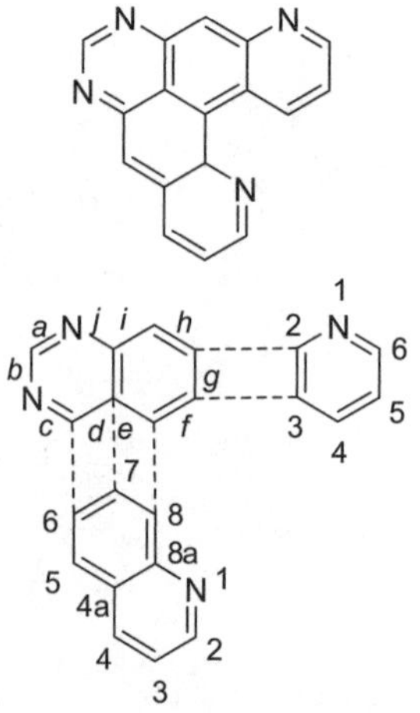

quinolino[6,7,8-*de*]pyrido[3,2-*g*]quinazoline

*17.2.3.3  Peripheral Atom Numbering*    Once the ring system is named, the atoms need to be numbered to denote the location of substituents. Numbering fused heterocycles is a complex operation, and the atom numbers do not have to match the numbers used when denoting the name of the ring system. The first step is to lay the heterocycle out in the appropriate orientation, which is:

1. The greatest number of rings is in a horizontal row (this does not need to be the ring from which the base name is derived).
2. The maximum number of rings is above and to the right of the horizontal row.
3. For heterocycles, give low numbers to heteroatoms. Where there are multiple heteroatoms, the lowest number goes to the highest priority heteroatom.
4. Where there is a choice, carbon atoms common to two or more rings follow the lowest possible numbers.
5. Where there is a choice, atoms that carry an indicated hydrogen atom ("*H*") are numbered as low as possible.
6. For heterocycles, carbon atoms common to two or more rings are given the lowest possible number.

Next, the atoms are numbered in a clockwise direction, starting with the first atom not at a fusion position of the ring at the top right-most position in the molecule. Carbons along the exterior, common to two or more rings are given the number of the preceding numbered atom followed by "a," "b," "c." Heteroatoms common to two or more rings are given a distinct number. Interior atoms are numbered following the highest adjacent atom followed by "a," "b," "c," in a clockwise direction. Some examples are shown in the box below:

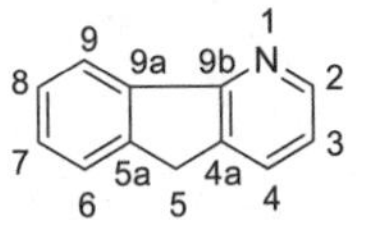

5*H*-indeno[1,2-*b*]pyridine
4-azafluorene

NOT:

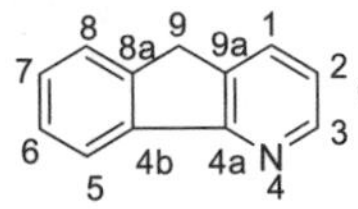

OR

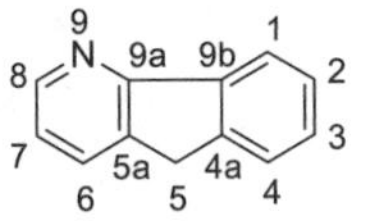

Rule 3—lowest possible locant for heteroatom

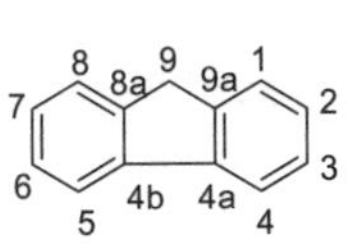

9*H*-fluorene

NOT:

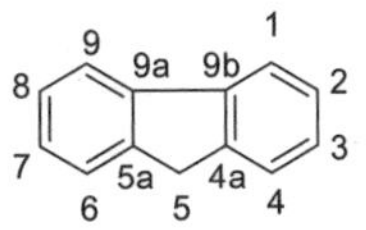

Rule 4—carbon atoms common to two or more rings follow the lowest possible numbers *predominates over* Rule 5—atoms which carry an indicated hydrogen atom ("H") are numbered as low as possible

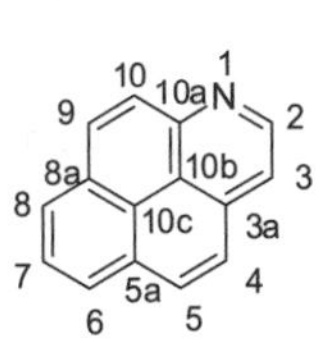

naphtho[2,1,8-*def*]quinoline

NOT:

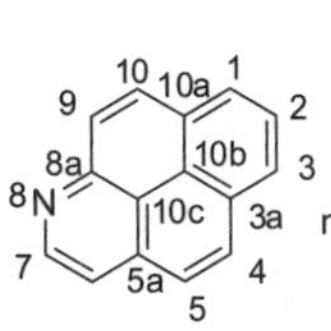

Locant for nitrogen not minimized

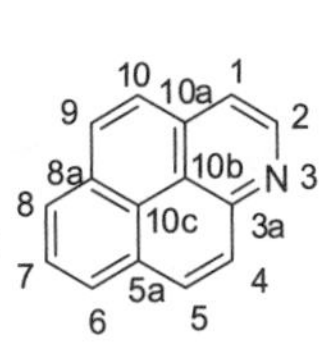

cyclopenteno[*d,e*]isoquinoline

NOT:

Rule 4—carbonatoms common to two or more rings follow the lowest possible numbers

indeno[3,3a,4,5-*bcd*]indole    6-bromo-thieno[3,2-*c*]-*s*-triazolo[4,3-*a*]pyridine

**Practice Examples**

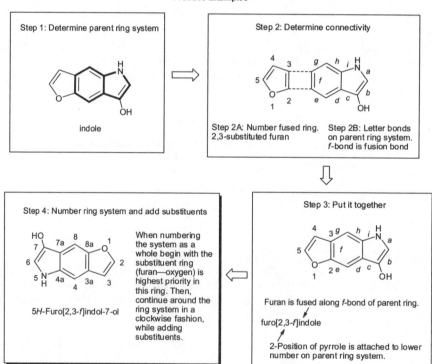

Step 1: Determine parent ring system

indole

Step 2: Determine connectivity

Step 2A: Number fused ring. 2,3-substituted furan

Step 2B: Letter bonds on parent ring system. *f*-bond is fusion bond

Step 3: Put it together

Furan is fused along *f*-bond of parent ring.

furo[2,3-*f*]indole

2-Position of pyrrole is attached to lower number on parent ring system.

Step 4: Number ring system and add substituents

When numbering the system as a whole begin with the substituent ring (furan—oxygen) is highest priority in this ring. Then, continue around the ring system in a clockwise fashion, while adding substituents.

5*H*-Furo[2,3-*f*]indol-7-ol

For the isomer:

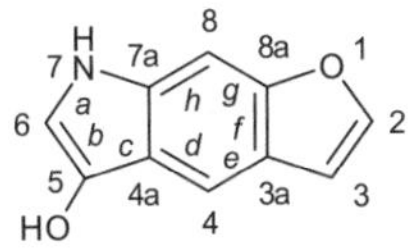

Begin with furan ringand number as you would in step 4 above, noting position of substituents.

7*H*-Furo[3,2-*f*]indol-5-ol

↑

3-Position of pyrrole(from step2A above) is attached to lower number on parent ring system.

Step 1: Determine root name

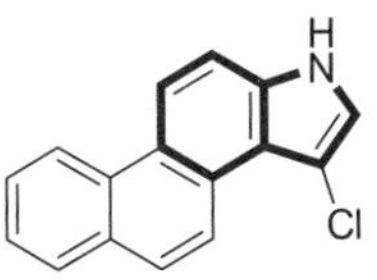

Step 2: Determine connectivity

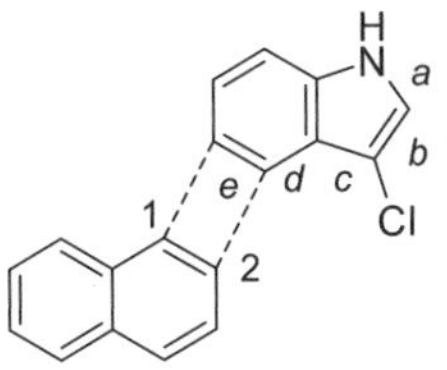

naphth[2,1-e]indole

step 3: Number substituents

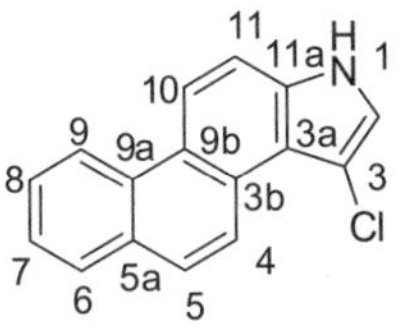

3-chloro-1*H*-naphth[2,1-e]indole

↑

2,1 is naphthalene numbering not compound numbering; nitrogen position is indicated by fusion bond

For the isomer:

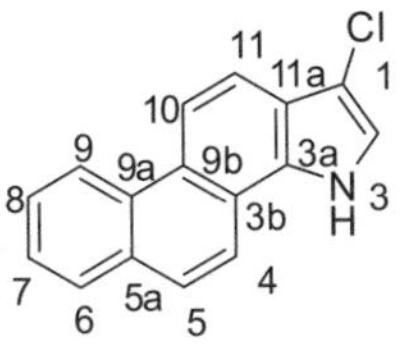

1-chloro-3*H*-naphth[1,2-g]indole

Numbering begins at the same position

## 17.2.4  Bridged Heterocycles

Bridged heterocycles are named in accordance with the rules for naming bridged all-carbon ring systems, with the heterocycle position denoted as discussed below. The rules for naming bridged all-carbon ring systems are also summarized below.

*17.2.4.1  Root Name: Bicyclic Compounds*  For bridged bicyclic compounds, the root name is based on the hydrocarbon name representing the total number of atoms in the bicyclic ring system, prefixed by "bicyclo." Between the bicyclo prefix and the hydrocarbon name, the number of atoms in each bridge, from greatest to least, are designated in brackets and separated by periods. Thus, the name will have the form bicyclo [x.y.z]hydrocarbon name where x, y, and z are the number of atoms in the bridges. An example is below.

bicyclo[2.2.1]heptane

***17.2.4.2  Bridged Ring System Numbering***    The ring system is numbered starting at a bridgehead, and going around the longest bridge first, then the second longest, then the shortest. In the event that there are bridges of equal length, atoms are numbered such that substituents have the lowest possible locant.

2-azabicyclo[3.2.1]octane            2-azabicyclo[2.2.1]heptane

***17.2.4.3  Root Name: Polycyclic Compounds***    The naming of a polycyclic-bridged ring system uses the same system as for a bicyclic ring system. For example:

tricyclo[4.2.2.2]dodecane            tricyclo[2.2.1.0]heptane

***17.2.4.4  Atom Numbering***    The polycyclic ring system is numbered based on the above rules for numbering a bicyclic ring system. The largest ring and the largest bridge make up the bicyclic ring system to be numbered. The atom connectivity of the minor bridges is denoted as superscripts after the corresponding number in the brackets with the best depicted in the examples below:

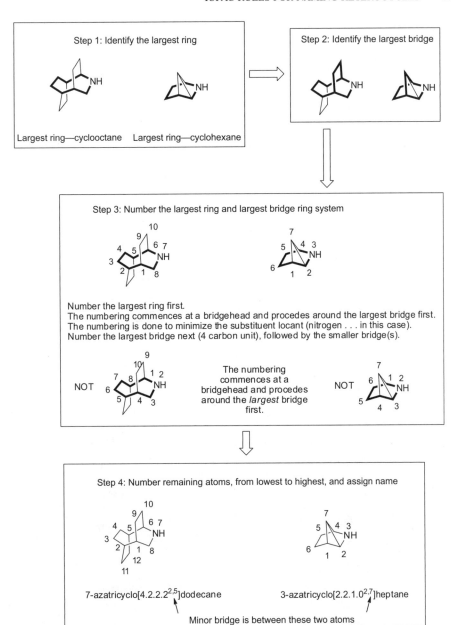

## 17.2.5   Fused and Bridged Ring Nomenclature

If a fused heterocyclic ring system is also bridged, the naming is based on the fused ring system name. The name will take the form x,y-bridge name heterocycle name, where x and y are the fusion points for the bridge on the fused ring system. The bridge name is indicated by taking the name of the corresponding hydrocarbon, and changing the final "e" into an "o." For example:

4,7-dihydro-4,7-ethanoindole     5,8-dihydro-5,8-[2,3-furano]quinoxaline

For bridges that do not contain carbon and a few hydrocarbon examples, names are listed below.

**TABLE 17.4 Naming Conventions–Prefixes for Bridged Heterocycles**

| Bridge | Prefix | Bridge | Prefix | Bridge | Prefix |
|---|---|---|---|---|---|
| HN- | imino | -N=N- | diazeno | -NH-NH- | diazano |
| | epimino | | azo | | biimino |
| -O-O- | epidioxy | -S-S- | epidithio | -S- | epithio |
| -S-O-NH- | epithoximino | -O- | epoxy | -O-NH- | epoxyimino |
| -O=N- | epoxynitilo | -O-S- | epoxythio | -O-S-O- | epoxythioxy |
| -C$_4$H$_2$O | furano | -N=C- | nitrilo | -CH$_2$- | methano |
| -C$_6$H$_4$- | [1,2]benzeno | -CH=CH-CH$_2$- | [1]propeno | | |

**TABLE 17.5 Monocyclic Heterocycles Containing One Nitrogen**

| # | Ring System | IUPAC Name (Common Name) |
|---|---|---|
| 1 | | aziridine |
| 2 | | 2-azirine |
| 3 | | azetidine |
| 4 | | 2-azetine |
| 5 | | azete |
| 6 | | azolidine (pyrrolidine) |
| 7 | | 2-azoline (2-pyrroline) |
| 8 | | azole (pyrrole) |
| 9 | | 2*H*-azole (2*H*-pyrrole) |
| 10 | | pyrrolidine-2,5-dione (succinimide) |

**TABLE 17.5**   (*Continued*)

| # | Ring System | IUPAC Name (Common Name) |
|---|---|---|
| 11 | | 1*H*-pyrrole-2,5-dione (maleimide) |
| 12 | | hexahydroazine (piperidine) |
| 13 | | 1,4*H*-azine (1,4*H*-pyridine) |
| 14 | | azine (pyridine) |
| 15 | | 3-pyridinecarboxylic acid (nicotinic acid) |
| 16 | | 4-pyridinecarboxylic acid (isonicotinic acid) |
| 17 | | hexahydroazepine |
| 18 | | azepine |
| 19 | | octahydroazocene |
| 20 | | azocene |
| 21 | | octahydroazonine |
| 22 | | azonine |
| 23 | | decahydroazecine |
| 24 | | azecine |

**TABLE 17.6   Monocyclic Heterocycles Containing More Than One Nitrogen**

| # | Ring System | IUPAC Name (Common Name) |
|---|---|---|
| 1 | | diaziridine |
| 2 | | 1-diazirine |
| 3 | | 2-diazetidine |
| 4 | | 2-diazete |
| 5 | | (imidazolidine) |
| 6 | | (pyrazolidine) |
| 7 | | 4,5-dihydro-1*H*-imidazole (2-imidazoline) |
| 8 | | 3,4*H*-2-pyrazol-2-ine (2-pyrazoline) |
| 9 | | 2,3-dihydro-1*H*-pyrazole (3-pyrazoline) |
| 10 | | 3-diazole (imidazole) |
| 11 | | 2*H*-1,3-diazole (2*H*-imidazole) |
| 12 | | 2-diazole (pyrazole) |
| 13 | | imidazolidine-2,4-dione (hydantoin) (glycolylurea) |
| 14 | | 1,2,3-triazole (2,5-diazapyrrole) (osotriazole) |

**TABLE 17.6**   (*Continued*)

| # | Ring System | IUPAC Name (Common Name) |
|---|---|---|
| 15 | | 1,2,4-triazole<br>(*s*-triazole; 3,4-diazapyrrole) |
| 16 | | 1*H*-tetrazole |
| 17 | | tetrahydro-1,4-diazine<br>(piperazine) |
| 18 | | pyridazine<br>1,2-diazine, *o*-diazine |
| 19 | | pyrimidine<br>1,3-diazine, *m*-diazine |
| 20 | | pyrazine<br>(1,4-diazine, *p*-diazine) |
| 21 | | 2,4-dihydroxypyrimidine<br>2,4-pyrimidinedione<br>(uracil) |
| 22 | | 1-β-D-ribosyluracil<br>(uridine) |
| 23 | | 4-amino-2(1*H*)-pyrimidinone<br>4-amino-2-hydroxypyrimidine<br>4-aminouracil<br>(cytosine) |

(*continued*)

**TABLE 17.6** (*Continued*)

| # | Ring System | IUPAC Name (Common Name) |
|---|---|---|
| 24 | | 1-β-D-ribosylcytosine (cytidine) |
| 25 | | 1-(2-deoxy-β-D-ribofuranosyl)cytosine (2′-deoxycytidine) |
| 26 | | 5-methyl-2,4(1*H*,3*H*)-pyrimidinedione 2,4-dihydroxy-5-methylpyrimidine (thymine) |
| 27 | | 1-(2-deoxy-β-D-ribofuranosyl)thymine (thymidine) |
| 28 | | 2,4,6(1*H*,3*H*,5*H*)-pyrimidinetrione (barbituric acid) |
| 29 | | 1,3,5-hexahydrotriazine (hexahydro-*s*-triazine) |
| 30 | | 1,2,3-triazine (*v*-triazine) |
| 31 | | 1,2,4-triazine (*as*-triazine) |

**TABLE 17.6** (*Continued*)

| # | Ring System | IUPAC Name (Common Name) |
|---|---|---|
| 32 | 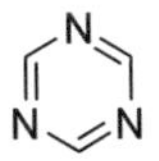 | 1,3,5-triazine<br>(*s*-triazine) |
| 33 | 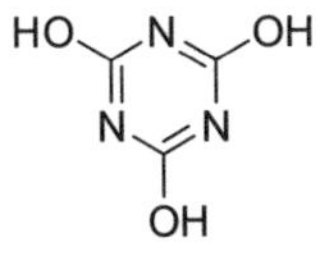 | 1,3,5-triazine-2,4,6(1*H*,3*H*,5*H*)-trione<br>(isocyanuric acid)<br>2,4,6-trihydroxy-*s*-triazine<br>(cyanuric acid) |
| 34 | 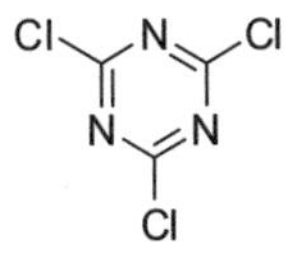 | 2,4,6-trichloro-1,3,5-triazine<br>(cyanuric chloride) |
| 35 | 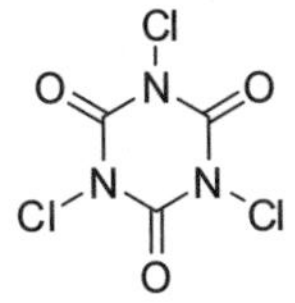 | (1*H*,3*H*,5*H*)-1,3,5-trichloro-1,3,5-triazine-2,4,6-trione,<br>(isocyanuric chloride) |
| 36 | 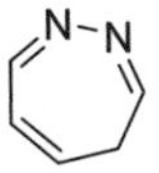 | 4*H*-1,2-diazepine |
| 37 | 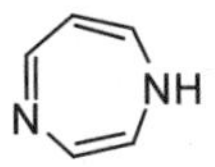 | 4-diazepine<br>1*H*-4-diazepine |

**TABLE 17.7  Polycyclic Heterocycles Containing Nitrogen**

| # | Ring System | IUPAC Name (Common Name) |
|---|---|---|
| 1 | 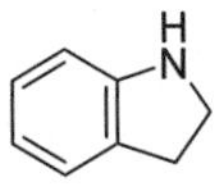 | indoline |
| 2 | 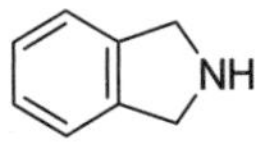 | isoindoline |
| 3 | 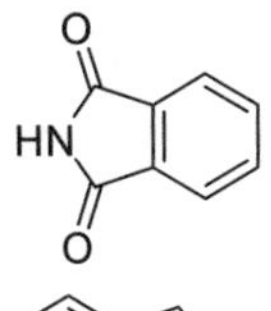 | 1*H*-isoindole-1,3(2*H*)-dione<br>(phthalimide) |
| 4 | 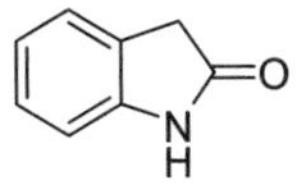 | 1,3-dihydro-2*H*-indol-2-one<br>(indolin-2-one)<br>(oxindole) |

(*continued*)

**TABLE 17.7** (*Continued*)

| # | Ring System | IUPAC Name (Common Name) |
|---|---|---|
| 5 | | 1*H*-indole-2,3-dione (isatin) |
| 6 | | indolizine |
| 7 | | isoindole |
| 8 | | 3*H*-indole |
| 9 | | indole |
| 10 | | 5H-cyclopenta[*b*]pyridine (7H-1-pyrindine) |
| 11 | | 1*H*-indazole |
| 12 | | benzimidazole |
| 13 | | 7H-pyrrolo[2,3-*d*]pyrimidine (1,5,7-triazaindene) |
| 14 | | 5*H*-pyrrolo[3,2-*d*]pyrimidine (1,4,6-triazaindene) |
| 15 | | 1*H*-pyrrolo[3,2-*b*]pyridine (4-azaindole) |
| 16 | | 7*H*-purine |

**TABLE 17.7**    (*Continued*)

| # | Ring System | IUPAC Name (Common Name) |
|---|---|---|
| 17 | | 6-amino-1*H*-purin (adenine) (vitamin B4) |
| 18 | | 9-β-D-ribofuranosyladenine (adenosine) (β-adenosine) |
| 19 | | 9-(2-deoxy-β-D-ribofuranosyl)adenosine (2′-deoxyadenosine) |
| 20 | | 2-amino-6-hydroxy-1*H*-purine (guanine) |
| 21 | | 9-β-D-ribofuranosylguanine (guanosine) |
| 22 | | 9-(2-deoxy-β-D-ribofuranosyl)guanine (2′-deoxyguanosine) |
| 23 | | 2-amino-4-hydroxy-7*H*-pyrrolo[2,3-*d*]pyrimidine (7-deazaguanine) |

(*continued*)

**TABLE 17.7** (*Continued*)

| # | Ring System | IUPAC Name (Common Name) |
|---|---|---|
| 24 | | 1,2,4-triazolo[4,3-*a*]pyridine |
| 25 | | pyrazolo[1,5-*a*]pyrimidine (1,4,7a-triazaindene) |
| 26 | | 5,6,7,8-tetrahydro-1,2,4-triazolo[4,3-*a*]pyrazine |
| 27 | | tetrahydroquinoline |
| 28 | | tetrahydroisoquinoline |
| 29 | | 4*H*-quinolizine |
| 30 | | isoquinoline |
| 31 | | quinoline |
| 32 | | phthalazine |
| 33 | | 1,8-naphthyridine |
| 34 | | quinoxaline |
| 35 | | quinazoline |
| 36 | | cinnoline |
| 37 | | pteridine |

**TABLE 17.7**    (*Continued*)

| # | Ring System | IUPAC Name (Common Name) |
|---|---|---|
| 38 | | 1,4-dihydro-4-oxoquinoline<br>(quinolone) |
| 39 | | 1,3,4,5-tetrahydrobenzo[*b*]azepin-2-one |
| 40 | | 1,8-diaza-7-bicyclo[5.4.0]undecene<br>(DBU) |
| 41 | | 3*H*-1,4-benzo[*f*]diazepine<br>(benzodiazepine) |
| 42 | | 1-azabicyclo[2.2.2]octane<br>(quinuclidine) |
| 43 | | 1,4-diazabicyclo[2.2.2]octane<br>(DABCO) |
| 44 | | 2-azabicyclo[2.2.2]octane<br>(isoquinuclidine) |
| 45 | | 8-methyl-8-azabicyclo[3.2.1]octane<br>(tropane) |
| 46 | | carbazole |
| 47 | | 4a*H*-carbazole |
| 48 | | 9*H*-pyrido[3,4-*b*]indole<br>(β-carboline)<br>(norharman) |

(continued)

**TABLE 17.7**    (*Continued*)

| # | Ring System | IUPAC Name (Common Name) |
|---|---|---|
| 49 | | phenanthridine |
| 50 | | acridine |
| 51 | | perimidine |
| 52 | | 1,10-phenanthroline |
| 53 | | phenazine |
| 54 | | 2,3,6,7-tetrahydro-1*H*,5*H*-benzo[*i,j*]quinolizine (julolidine) |
| 55 | | 10,11-dihydro-5*H*-dibenzo[*b,f*]azepine iminodibenzyl |
| 56 | | 5*H*-dibenzo[*b,f*]azepine |
| 57 | | (5*S*,11*S*)- 2,8-dimethyl-6*H*,12*H*-5, 11-methanodibenzo[*b,f*][1,5]diazocine ((+)-Troger's base) |
| 58 | | 1,3,5,7-tetraazatricyclo[3.3.1.1³,⁷]decane (hexamethylenetetramine) (HMTA) |

**TABLE 17.8   Monocyclic Heterocycles Containing One Oxygen**

| # | Ring System | IUPAC Name (Common Name) |
|---|---|---|
| 1 | | oxirane<br>(ethylene oxide) |
| 2 | | 2,3-epoxy-1-propanol<br>(glycidol) |
| 3 | | oxirene |
| 4 | | oxetane |
| 5 | | oxete |
| 6 | | tetrahydrofuran |
| 7 | | 2,3-dihydrofuran |
| 8 | | furan |
| 9 | | 2*H*-tetrahydropyran<br>oxane |
| 10 | | 3,4-2*H*-dihydropyran |
| 11 | | 2*H*-pyran |
| 12 | | 4*H*-pyran |
| 13 | | 2*H*-pyran-2-one<br>(2-pyrone) |
| 14 | | 4*H*-pyran-2-one<br>(4-pyrone) |
| 15 | | oxepane |
| 16 | | oxepin |

(*continued*)

**TABLE 17.8** (*Continued*)

| # | Ring System | IUPAC Name (Common Name) |
|---|---|---|
| 17 | | oxocane |
| 18 | | 2*H*-oxocin |
| 19 | | oxonane |
| 20 | | oxonin |
| 21 | | oxecane |
| 22 | | 4*H*-oxecin |

**TABLE 17.9** **Monocyclic Heterocycles Containing More Than One Oxygen**

| # | Ring System | IUPAC Name (Common Name) |
|---|---|---|
| 1 | | 1,3-dioxolane |
| 2 | | 1,2-dioxin |
| 3 | | 4*H*-1,3-dioxin |
| 4 | | 1,4-dioxane |
| 5 | | 2,2-dimethyl-1,3-dioxane-4,6-dione<br>Meldrum's acid |
| 6 | | 1,4,7,10,13,16-hexaoxacyclooctadecane<br>18-crown-6 |

**TABLE 17.10   Polycyclic Heterocycles Containing Oxygen**

| # | Ring System | IUPAC Name (Common Name) |
|---|---|---|
| 1 | | hexahydro-2H-cyclopenta[b]furan-2-one |
| 2 | | phthalan |
| 3 | | benzofuran |
| 4 | | isobenzofuran |
| 5 | | 1,3-benzodioxole |
| 6 | | 5-hydroxy-1,3-benzodioxole (sesamol) |
| 7 | | 1,3-benzodioxole-5-methanol (piperonyl alcohol) |
| 8 | | 5-(2-propenyl)-1,3-benzodioxole (safrole) |
| 9 | | 5-(1-propenyl)-1,3-benzodioxole (isosafrole) |
| 10 | | chroman |
| 11 | | isochroman |
| 12 | | 2H-1-benzopyran (2H-chromene) |

(*continued*)

**TABLE 17.10** (*Continued*)

| # | Ring System | IUPAC Name (Common Name) |
|---|---|---|
| 13 | | 2,2-dimethyl-7-methoxy-2*H*-1-benzopyran (precocene I) |
| 14 | | 6,7-dimethoxy-2,2-dimethyl-2*H*-1-benzopyran (precocene II) |
| 15 | | 1*H*-2-benzopyran-1-one (isocoumarin) |
| 16 | | 2*H*-benzopyran-2-one (coumarin) |
| 17 | | 3*H*-2-benzopyran-1-one (isochromen-3-one) |
| 18 | | 4*H*-1-benzopyran-4-one (chromene-4-one) |
| 19 | | flavone |
| 20 | | flavanone |
| 21 | | xanthene |

**TABLE 17.11    Pentose Sugars**

| # | Straight-chain Projection | α/β | Furanose | Pyranose |
|---|---|---|---|---|
| 1 | CHO / H—OH / H—OH / H—OH / CH$_2$OH <br><br> D-ribose | α <br><br> α | <br> α-D-ribofuranose <br> α-D-ribose | <br> α-D-ribopyranose |
| 2 | <br> D-ribose | β <br><br> β | <br> β-D-ribofuranose <br> β-D-ribose | <br> β-D-ribopyranose |
| 3 | CH$_2$OH / =O / H—OH / H—OH / CH$_2$OH <br><br> D-erythro-pentulose <br><br> D(−)-ribulose <br> D-adonose <br> D-arabinulose <br> D-araboketose <br> D-erythropentulose | α <br><br> α | <br> α-D-erythro-2-pentulofuranose <br> α-D-ribulose | |

(*continued*)

**TABLE 17.11** *(Continued)*

| # | Straight-chain Projection | α/β | Furanose | Pyranose |
|---|---|---|---|---|

| 4 | | β | | |

| | D-erythro-pentulose | β | β-D-erythro-2-pentulofuranose | |
| | | | β-D-ribulose | |
| | D(−)-ribulose | | | |
| | D-adonose | | | |
| | D-arabinulose | | | |
| | D-araboketose | | | |
| | D-erythropentulose | | | |

| 5 | | α | | |

| | 2-deoxy-D-ribose | α | 2-deoxy-α-D-erythro-pentofuranose | 2-deoxy-α-D-erythro-pentopyranose |
| | | | 2-deoxy-α-D-ribose | |

| 6 | | β | | |

| | 2-deoxy-D-ribose | β | 2-deoxy-β-D-erythro-pentofuranose | 2-deoxy-β-D-erythro-pentopyranose |
| | | | 2-deoxy-β-D-ribose | |

**TABLE 17.11**    (*Continued*)

| # | Straight-chain Projection | α/β | Furanose | Pyranose |
|---|---|---|---|---|
| 7 | D-arabanose | α<br><br>α | α-D-arabanofuranose | α-D-arabanopyranose<br>α-D-arabanose |
| 8 | D-arabanose | β<br><br>β | β-D-arabanofuranose | β-D-arabanopyranose<br>β-D-arabanose |
| 9 | D-xylose | α<br><br>α | α-D-xylofuranose | α-D-xylopyranose<br>D-xylose |
| 10 | D-xylose | β<br><br>β | β-D-xylofuranose | β-D-xylopyranose<br>D-xylose |

(*continued*)

**TABLE 17.11** (*Continued*)

| # | Straight-chain Projection | α/β | Furanose | Pyranose |
|---|---|---|---|---|
| 11 | | α | | |
| | D-threo-2-pentulose | α | α-D-threo-2-pentulofuranose | |
| | D-xylulose | | α-D-xylulose | |
| 12 | | β | | |
| | D-threo-2-pentulose | β | β-D-threo-2-pentulofuranose | |
| | D-xylulose | | β-D-xylulose | |
| 13 | | α | | |
| | D-lyxose | α | α-D-lyxofuranose | α-D-lyxopyranose D-lyxose |
| 14 | | β | | |
| | D-lyxose | β | β-D-lyxofuranose | β-D-lyxopyranose D-lyxose |

**TABLE 17.12  Hexose Sugars**

| # | Straight-chain Projection | α/β | Furanose | Pyranose | Heptanose |
|---|---|---|---|---|---|
| 1 | D-allose | α | α-D-allofuranose | α-D-allopyranose α-D-allose | α-D-alloseptanose |
| 2 | D-allose | β | β-D-allofuranose | β-D-allopyranose β-D-allose | β-D-alloseptanose |

*(continued)*

**TABLE 17.12** (*Continued*)

| # | Straight-chain Projection | α/β | Furanose | Pyranose | Heptanose |
|---|---|---|---|---|---|
| 3 | D-altrose | α | α-D-altrofuranose | α-D-altropyranose<br>α-D-altrose | α-D-altroseptanose |
| 4 | D-altrose | α<br>β | β-D-altrofuranose | β-D-altropyranose<br>β-D-altrose | β-D-altroseptanose |

CH₂OH

α-D-fructopyranose
α-D-fructose

β-D-fructopyranose
β-D-fructose

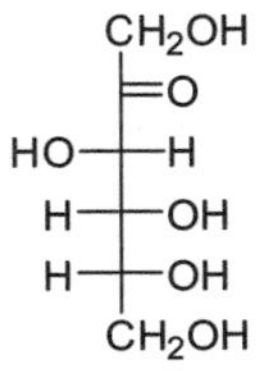

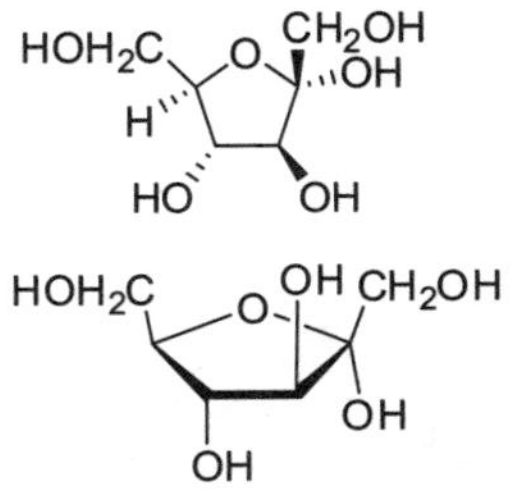

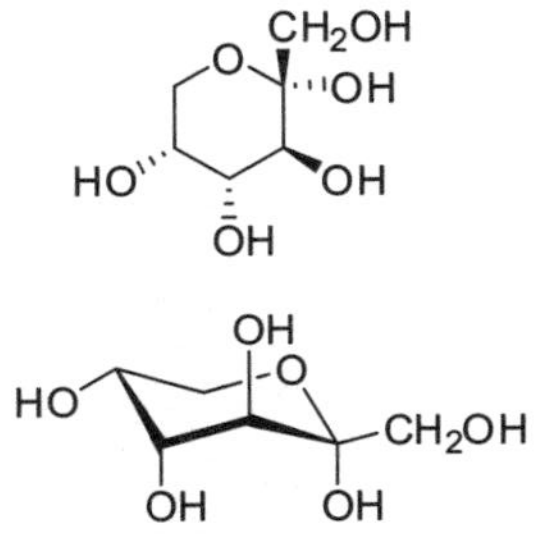

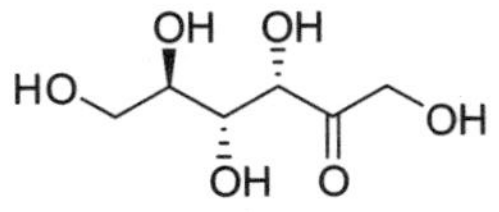

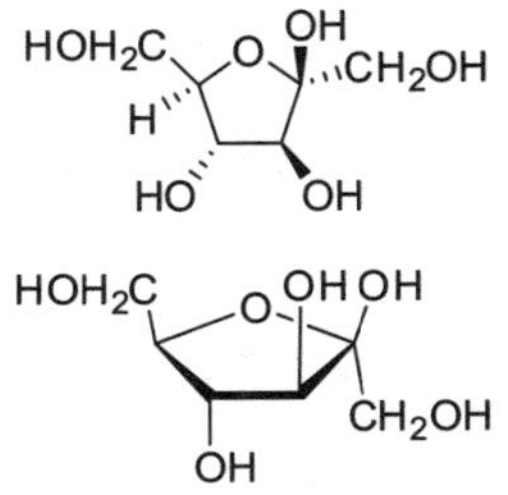

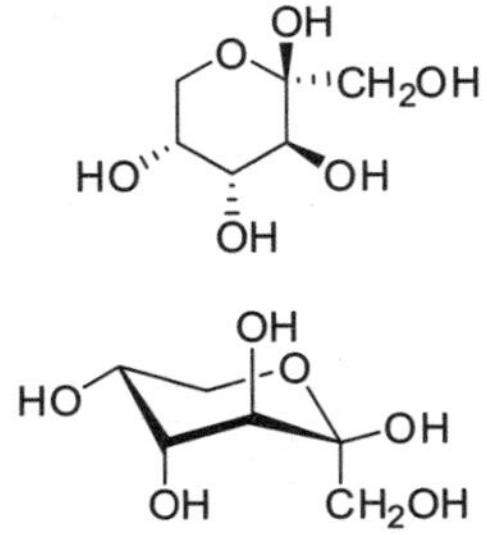

α-D-fructofuranose

β-D-fructofuranose

α

α

β

β

5

CH₂OH

D-fructose

6

D-fructose

**TABLE 17.12** (*Continued*)

| # | Straight-chain Projection | α/β | Furanose | Pyranose | Heptanose |
|---|---|---|---|---|---|
| 7 | D-galactose | α | α-D-galactofuranose | α-D-galactopyranose<br>α-D-galactose | α-D-galactoseptanose |
|  |  | α |  |  |  |
|  |  | β |  | β-D-galactopyranose<br>β-D-galactose | β-D-galactoseptanose |
| 8 | D-galactose | β | β-D-galactofuranose |  |  |

9

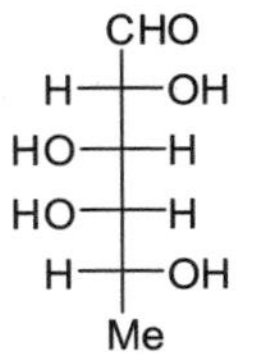

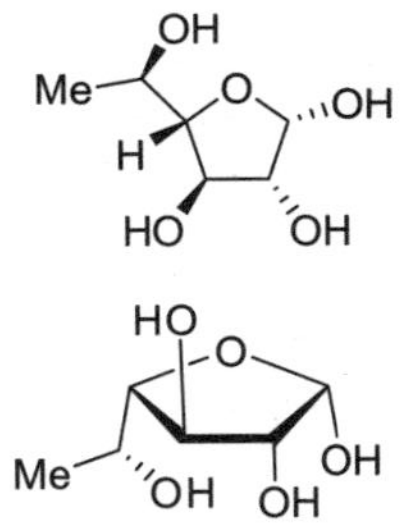

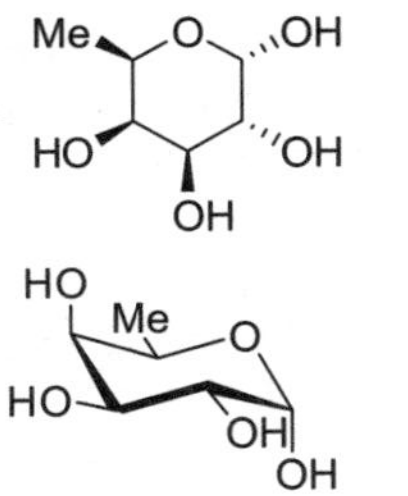

D-fucose
6-deoxy-D-galactose

α

α-D-fucofuranose
α-6-deoxy-
D-galactofuranose

α-D-fucopyranose
α-6-deoxy-
D-galactopyranose

β

β-D-fucofuranose
β-6-deoxy-D-galactofuranose

β-D-fucopyranose
β-6-deoxy-D-galactopyranose

10

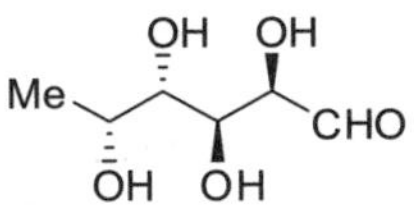

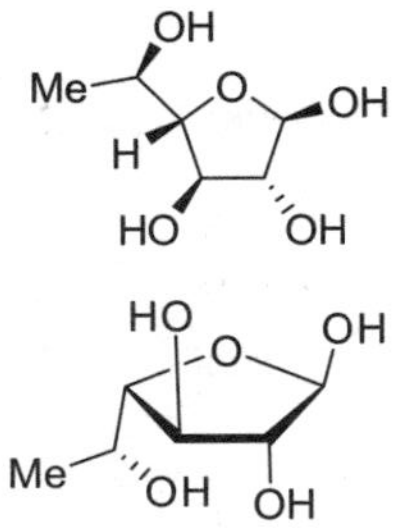

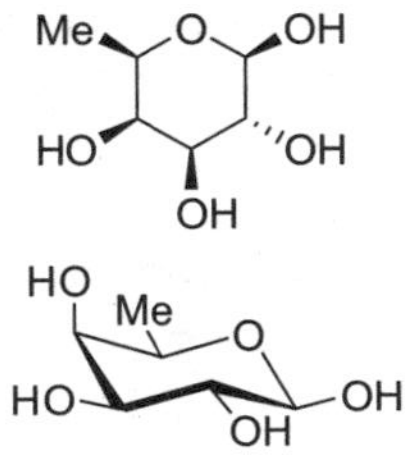

D-fucose
6-deoxy-D-galactose

*(continued)*

**TABLE 17.12** (*Continued*)

| # | Straight-chain Projection | α/β | Furanose | Pyranose | Heptanose |
|---|---|---|---|---|---|
| 11 | D-glucose | α | α-D-glucofuranose | α-D-glucopyranose<br>α-D-glucose | α-D-glucoseptanose |
| 12 | D-glucose | β | β-D-glucofuranose | β-D-glucopyranose<br>β-D-glucose | β-D-glucoseptanose |

α-D-guloseptanose

β-D-guloseptanose

α-D-gulopyranose
α-D-gulose

β-D-gulopyranose
β-D-gulose

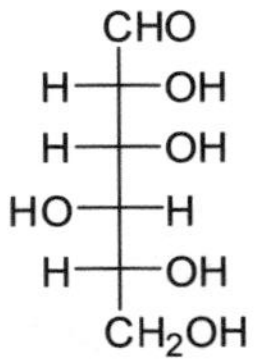

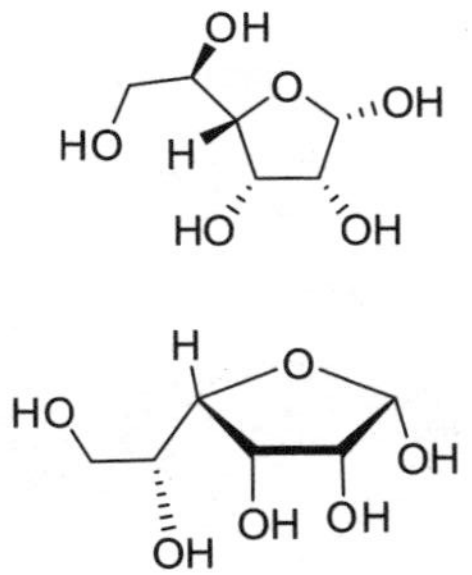

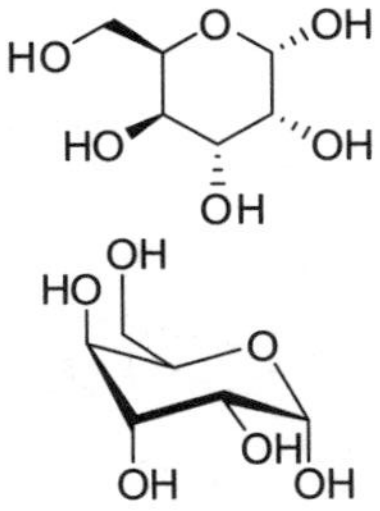

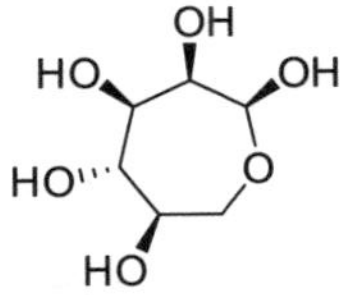

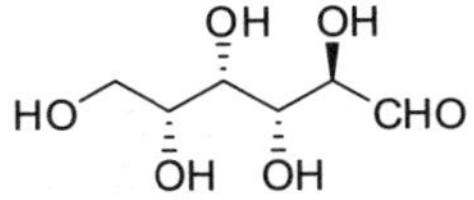

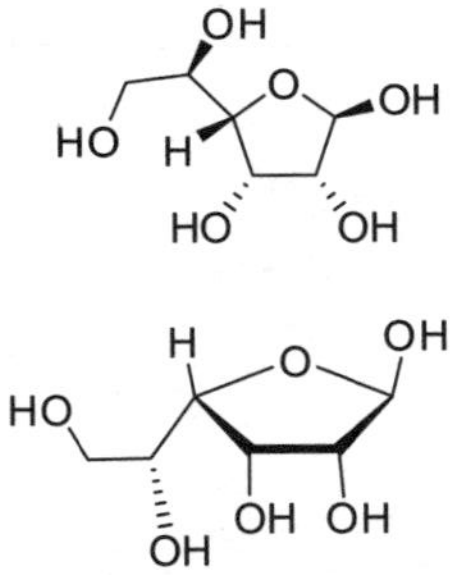

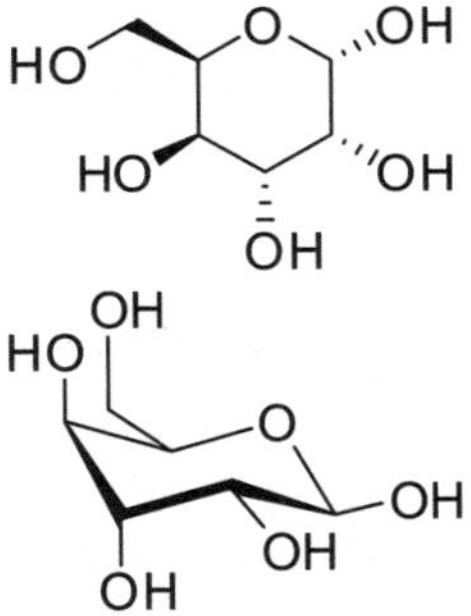

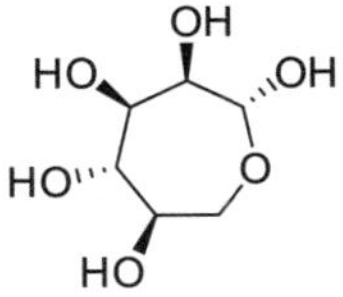

α

α-D-gulofuranose

β

β-D-gulofuranose

α

α

β

β

D-gulose

D-gulose

13

14

*(continued)*

**TABLE 17.12** (*Continued*)

| # | Straight-chain Projection | α/β | Furanose | Pyranose | Heptanose |
|---|---|---|---|---|---|
| 15 | D-idose | α | α-D-idofuranose | α-D-idopyranose<br>α-D-idose | α-D-idoseptanose |
| 16 | D-idose | β | β-D-idofuranose | β-D-idopyranose<br>β-D-idose | β-D-idoseptanose |

α-D-sorbopyranose
α-D-sorbose

β-D-sorbopyranose
β-D-sorbose

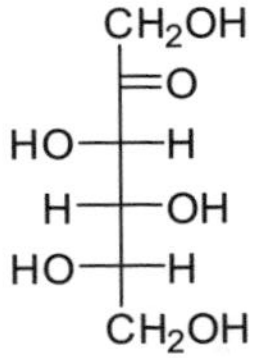

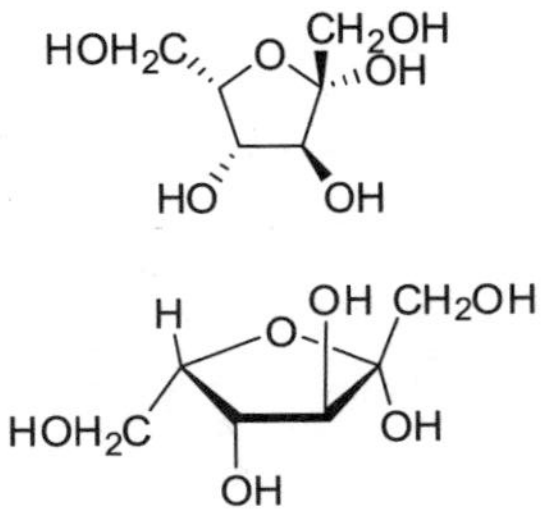

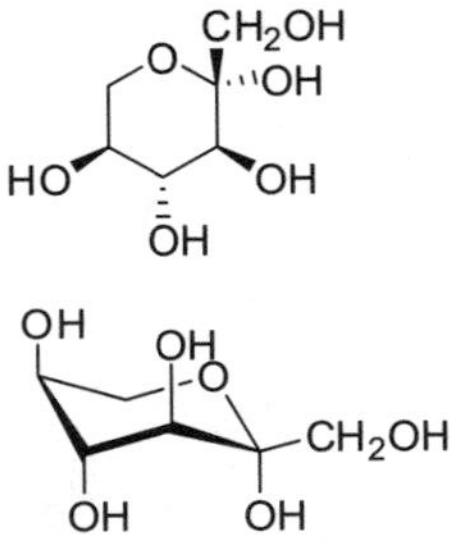

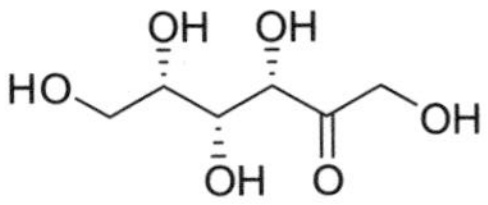

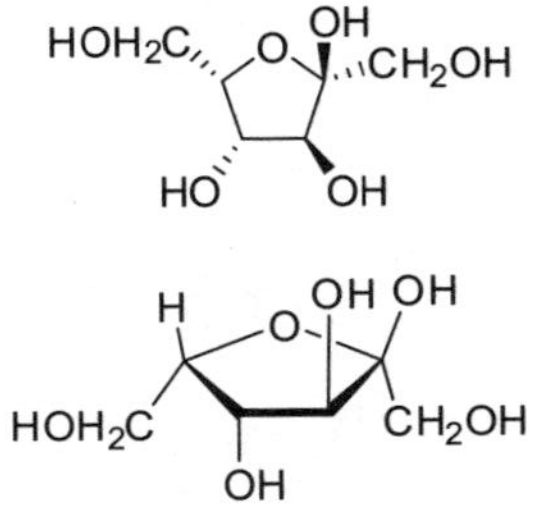

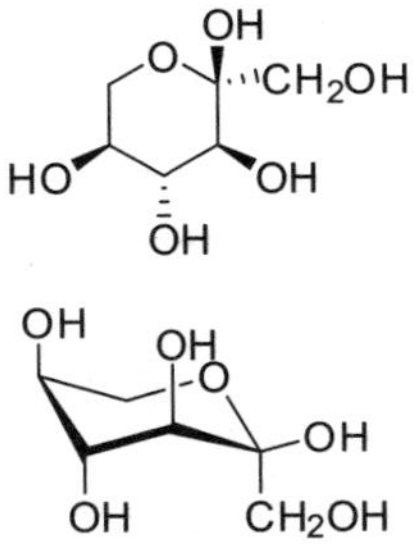

α-D-sorbofuranose

β-D-sorbofuranose

α

α

β

β

D-sorbose

D-sorbose

17

18

(*continued*)

**TABLE 17.12** (*Continued*)

| # | Straight-chain Projection | α/β | Furanose | Pyranose | Heptanose |
|---|---|---|---|---|---|
| 19 | D-mannose | α | α-D-mannofuranose | α-D-mannopyranose α-D-mannose | α-D-mannoseptanose |
| 20 | D-mannose | β | β-D-mannofuranose | β-D-mannopyranose β-D-mannose | β-D-mannoseptanose |

α-D-tagatopyranose
α-D-tagatose

β-D-tagatopyranose
β-D-tagatose

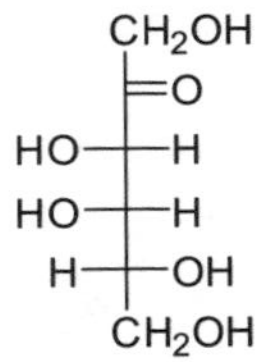

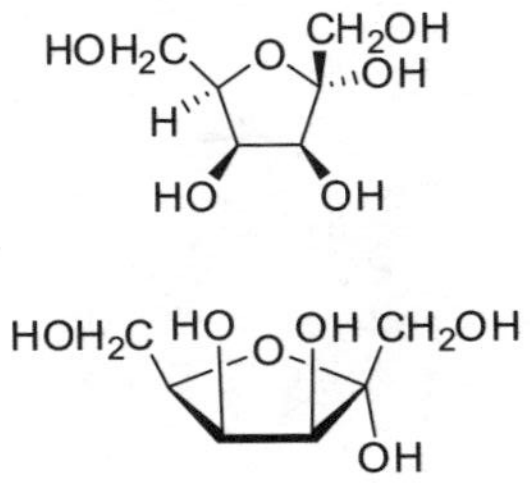

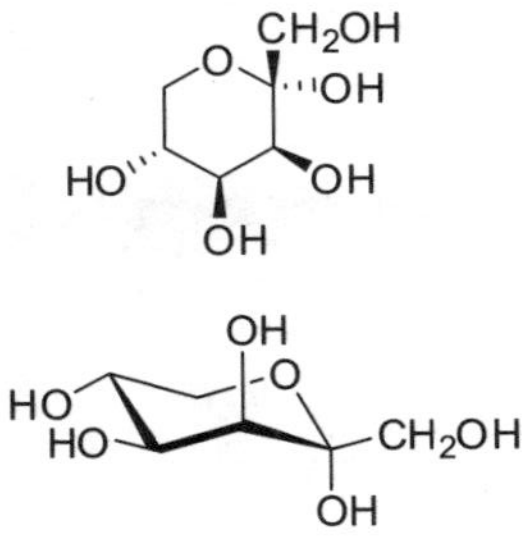

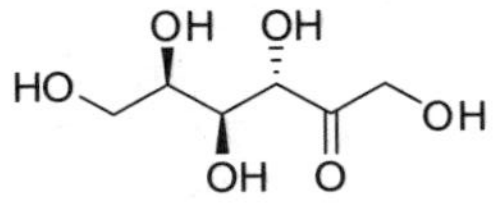

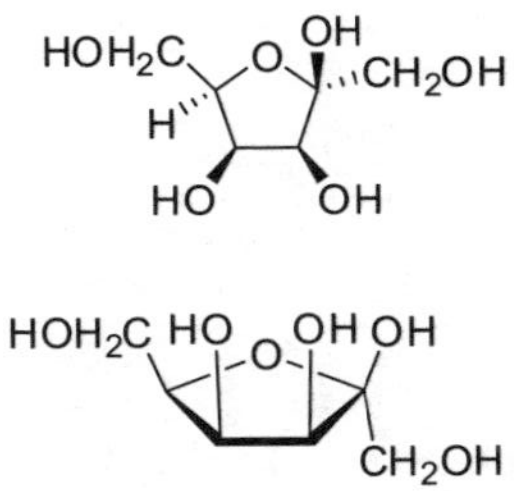

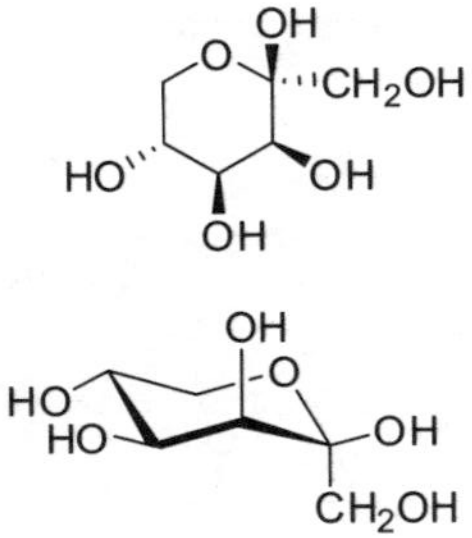

α-D-tagatofuranose

β-D-tagatofuranose

α

α

β

β

D-tagatose

D-tagatose

21

22

(*continued*)

**TABLE 17.12** (*Continued*)

| # | Straight-chain Projection | α/β | Furanose | Pyranose | Heptanose |
|---|---|---|---|---|---|
| 23 | D-rhamnose | α | α-D-rhamnofuranose | α-D-rhamnopyranose<br>α-D-rhamnose | |
| | | α | | | |
| | | β | β-D-rhamnofuranose | β-D-rhamnopyranose<br>β-D-rhamnose | |
| 24 | D-rhamnose | β | | | |

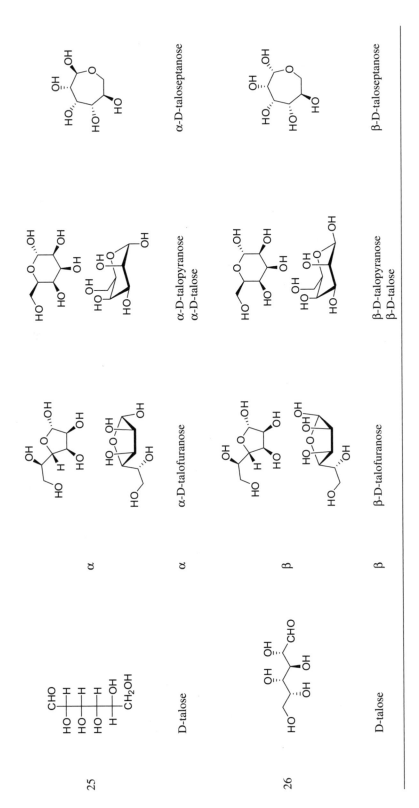

α-D-taloseptanose

β-D-taloseptanose

α-D-talopyranose
α-D-talose

β-D-talopyranose
β-D-talose

α-D-talofuranose

β-D-talofuranose

α

α

β

β

D-talose

D-talose

25

26

**TABLE 17.13   Monocyclic Heterocycles Containing One Sulfur**

| # | Ring System | IUPAC Name (Common Name) |
|---|---|---|
| 1 | | thiirane (ethylene sulfide) (ethylene episulfide) |
| 2 | | thiirene |
| 3 | | thietane (trimethylene sulfide) (1,3-epithio-propane) (thiacyclobutane) |
| 4 | | thiete 2*H*-thiete |
| 5 | | tetrahydrothiophene |
| 6 | | 2,3-dihydrothiophene |
| 7 | | thiophene |
| 8 | | 2*H*-tetrahydropyran oxane |
| 9 | | 3,4-2*H*-dihydrothiopyran |
| 10 | | 2*H*-thiopyran |
| 11 | | 4*H*-thiopyran |
| 12 | | 2*H*-thiopyran-2-one |
| 13 | | thiepin |

**TABLE 17.14   Monocyclic Heterocycles Containing More Than One Sulfur**

| # | Ring System | IUPAC Name (Common Name) |
|---|---|---|
| 1 | | 1,2-dithiolane |
| 2 | | 1,3-dithiolane |
| 3 | | 1,2-dithiole |
| 4 | | 1,3-dithiole |
| 5 | | 1,4-dithiane |
| 6 | | 1,3,5-trithiane |

**TABLE 17.15   Polycyclic Heterocycles Containing Sulfur**

| # | Ring System | IUPAC Name (Common Name) |
|---|---|---|
| 1 | | thioindene |
| 2 | | benzo(*b*)thiophene |
| 3 | | benzo(*c*)thiophene |
| 4 | | 2*H*-thiochromene |
| 5 | | thianthrene |
| 6 | | dibenzothiophene |

**TABLE 17.16  Monocyclic Heterocycles Containing Multiple Heteroatoms**

| # | Ring System | IUPAC Name (Common Name) |
|---|---|---|
| 1 | | oxaziridine |
| 2 | | oxazirine |
| 3 | | 2-thiazoline |
| 4 | | thiazole |
| 5 | | isothiazole |
| 6 | | oxazolidine |
| 7 | | oxazoline |
| 8 | | oxazole |
| 9 | | 1,2-isoxazole |
| 10 | | 3*H*-1,2-oxathiole |
| 11 | | 2*H*-1,3-oxathiole |
| 12 | | 1,3,4-thiadiazole |
| 13 | | 1,2,3-oxadiazole |
| 14 | | 1,2,4-oxadiazole |
| 15 | | 1,2,5-oxadiazole (furazan) |

**TABLE 17.16**    (*Continued*)

| # | Ring System | IUPAC Name (Common Name) |
|---|---|---|
| 16 | | 1,3,4-oxadiazole |
| 17 | | 1,2,3,4-oxatriazole |
| 18 | | 1,2,3,5-oxatriazole |
| 19 | | 3*H*-1,2,3-dioxazole |
| 20 | | 1,2,4-dioxazole |
| 21 | | 1,3,2-dioxazole |
| 22 | | 1,3,4-dioxazole |
| 23 | | 5*H*-1,2,5-oxathiazole |
| 24 | | 5-oxazolone |
| 25 | | oxazolidinone |
| 26 | | imidazoline-2-thione |
| 27 | | 2-methyl-1,3,2-dioxaborolane |
| 28 | | 4*H*-1,2-oxazine |
| 29 | | 2*H*-1,3-oxazine |
| 30 | | 6*H*-1,3-oxazine |

(*continued*)

**TABLE 17.16**   (*Continued*)

| # | Ring System | IUPAC Name (Common Name) |
|---|---|---|
| 31 | | 6*H*-1,2-oxazine |
| 32 | | 1,4-oxazine |
| 33 | | 2*H*-1,2-oxazine |
| 34 | | 4*H*-1,4-oxazine |
| 35 | | 1,2,5-oxathiazine |
| 36 | | 2*H*-1,2,4-oxadiazine |
| 37 | | 1,3,5-oxadiazine |
| 38 | | morpholine |
| 39 | | thiomorpholine |
| 40 | | 2-aza-1,5-dioxane |
| 41 | | 1*H*-pyridin-2-one |
| 42 | | 4-oxazepine |

**TABLE 17.17   Polycyclic Heterocycles Containing Multiple Heteroatoms**

| # | Ring System | IUPAC Name (Common Name) |
|---|---|---|
| 1 | | (penicillin nucleus) |
| 2 | | (cephalosporin nucleus) |
| 3 | | (penem nucleus) |
| 4 | | (carbapenem nucleus) |
| 5 | | indolo[5,4-*d*]isoxazole |
| 6 | | (tetramisole) |
| 7 | | (biotin) |
| 8 | | pyrano[3,4-*b*]-pyrrole |
| 9 | | benzisoxazole (indoxazine) |
| 10 | | benzoxazole |

*(continued)*

**TABLE 17.17**  (*Continued*)

| # | Ring System | IUPAC Name (Common Name) |
|---|---|---|
| 11 | | 2,1-benzisoxazole |
| 12 | | benzofurazan |
| 13 | | furo[3,2-*b*]pyridine |
| 14 | | oxazolo[4,5-*b*]pyridine |
| 15 | | 2*H*-1,3-benzoxazine |
| 16 | | 2*H*-1,4-benzoxazine |
| 17 | | 1*H*-2,3-benzoxazine |
| 18 | | 2*H*-benzo[*e*][1,2,4]oxadiazine |
| 19 | | 4*H*-3,1-benzoxazine |
| 20 | | 4*H*-1,4-benzoxazine |
| 21 | | 1*H*-benzo[*d*][1,3]oxazine-2,4-dione (isatoic anhydride) |
| 22 | | 2,3-dioxo-1,4-benzoxazine |
| 23 | | 1,4-dihydro-2*H*-3,1-benzoxazin-2-one |

**TABLE 17.17**    (*Continued*)

| # | Ring System | IUPAC Name (Common Name) |
|---|---|---|
| 24 | | 1,3-(4*H*)-benzothiazine |
| 25 | | 4-oxo-2,3,4,5-tetrahydrobenzo [*b*][1,4]thiazepine |
| 26 | | furo[2,3-*f*]indole |
| 27 | | thiazolo[2,3-*f*]benzofuran |
| 28 | | phenoxathiine |
| 29 | | thianthrene |
| 30 | | phenarsazine |
| 31 | | phenoxazine |
| 32 | | phenothiazine |
| 33 | | 2,6-bis(oxazolin-2-yl)pyridine (pyridine bisoxazoline) (pybox) |

**TABLE 17.18   Monocycles**

| # | Ring System | Common Name |
|---|---|---|
| 1 | | *o*-xylene |
| 2 | | mesitylene |
| 3 | | cumene |
| 4 | | styrene |
| 5 | | cinnamaldehyde |
| 6 | | stilbene |
| 7 | | chalcone |
| 8 | | benzoin |
| 9 | | benzil |
| 10 | | *o*-cymene |

**TABLE 17.18**    (*Continued*)

| # | Ring System | Common Name |
|---|---|---|
| 11 | OMe | anisole |
| 12 | OEt | phenetole |
| 13 | OMe, OMe | veratrole |
| 14 | OH, OMe | guaiacol |
| 15 | OH, OH | catechol (1,2-dihydroxybenzene) |
| 16 | OH, OH | resorcinol (1,3-dihydroxybenzene) |
| 17 | OH, OH | hydroquinone (1,4-dihydroxybenzene) |
| 18 | O, O | quinone |
| 19 | OH, CH₃ | 1,4-cresol |

(*continued*)

**TABLE 17.18**    *(Continued)*

| # | Ring System | Common Name |
|---|---|---|
| 20 | | *p*-anisaldehyde |
| 21 | | salicylic acid |
| 22 | | anthranilic acid |
| 23 | | phthalic acid |
| 24 | | isophthalic acid |
| 25 | | terephthalic acid |
| 26 | | 2,3-xylenol |
| 27 | | carvacrol |
| 28 | | thymol |

**TABLE 17.18**    (*Continued*)

| # | Ring System | Common Name |
|---|---|---|
| 29 | | eugenol |
| 30 | | isoeugenol |
| 31 | | coniferyl alcohol |
| 32 | | menthol |
| 33 | | pyrogallol (1,2,3-trihydroxybenzene) |
| 34 | | phloroglucinol (1,3,5-trihydroxybenzene) |
| 35 | | orcinol |

(*continued*)

**TABLE 17.18**    (*Continued*)

| # | Ring System | Common Name |
|---|---|---|
| 36 | | olivetol |
| 37 | | vanillin |
| 38 | | aniline |
| 39 | | 1,2-phenylene diamine |
| 40 | | *o*-toluidine |
| 41 | | 2,3-xylidine |
| 42 | | *o*-anisidine |
| 43 | | *o*-phenetidine |
| 44 | | picric acid (2,4,6-trinitrophenol) |

**TABLE 17.18** (*Continued*)

| # | Ring System | Common Name |
|---|---|---|
| 45 | 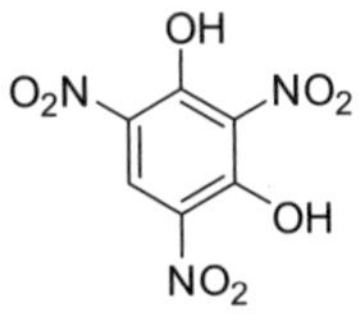 | styphnic acid |
| 46 | 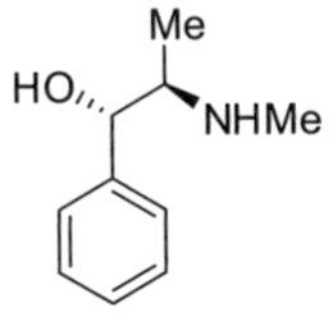 | ephedrine |
| 47 | 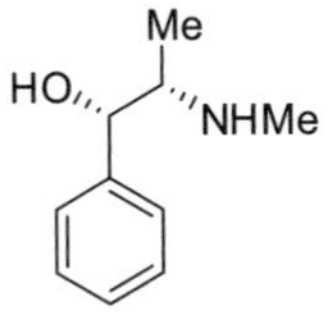 | pseudoephedrine |
| 48 | 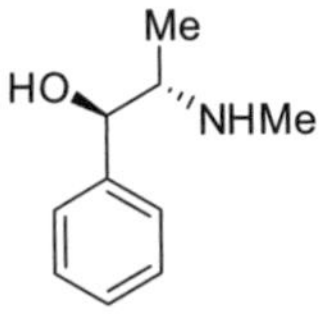 | norpseudoephedrine |
| 49 | 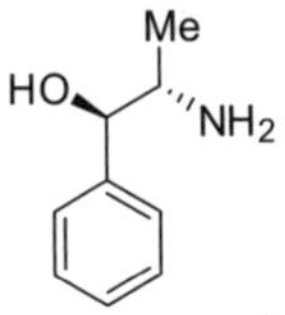 | norephedrine |
| 50 | 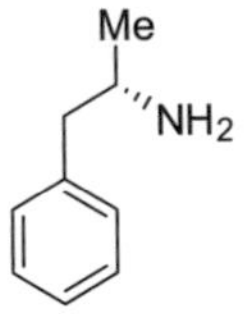 | amphetamine |
| 51 | 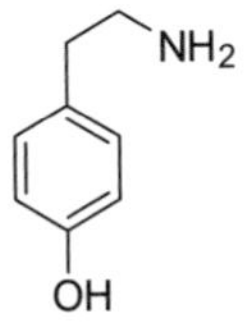 | tyramine |
| 52 | 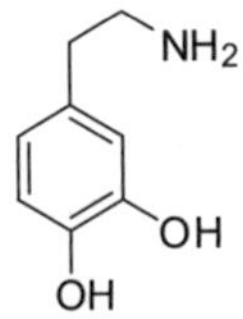 | dopamine |

(*continued*)

**TABLE 17.18**   (*Continued*)

| # | Ring System | Common Name |
|---|---|---|
| 53 | | octopamine |
| 54 | | synephrine |
| 55 | | epinephrine |
| 56 | | norepinephrine |
| 57 | | metanephrine |
| 58 | | isoproterenol |
| 59 | | inositol |

**TABLE 17.18**    (*Continued*)

| # | Ring System | Common Name |
|---|---|---|
| 60 | 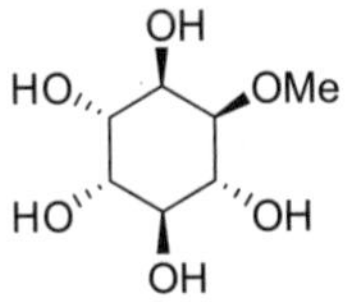 | (-)-quebrachitol<br>L-2-O-methyl-*chiro*-inositol |
| 61 | 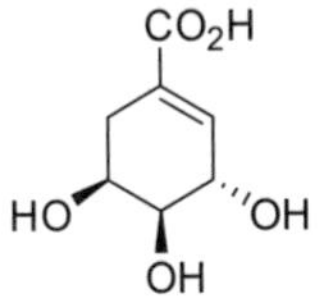 | L-(-)-shikimic acid |
| 62 | 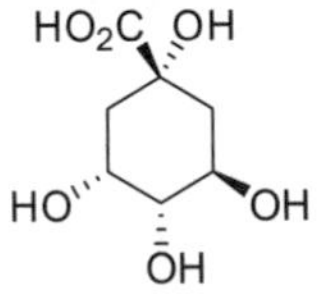 | (1α,3R,4α,5R)-1,3,4,5-tetrahydroxy-<br>D-(-)-quinic acid |
| 63 | 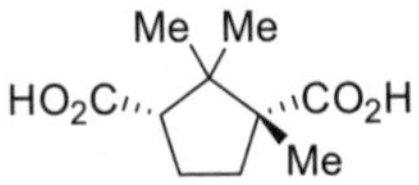 | (R)-( + )-camphoric acid<br>(1R,3S)-1,2,2-trimethyl-1,3- |
| 64 | 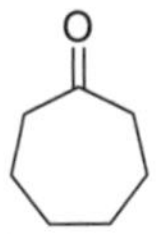 | subarone<br>cycloheptanone |

**TABLE 17.19    Polycycles**

| # | Ring System | Common Name |
|---|---|---|
| 1 |  | spiropentane |
| 2 | 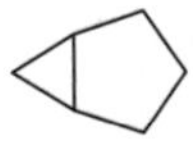 | bicyclo[3.2.1]hexane |
| 3 |  | norbornane |

(*continued*)

**TABLE 17.19** (*Continued*)

| # | Ring System | Common Name |
|---|---|---|
| 4 | | norpinane |
| 5 | | (*R*)-(+)-camphor |
| 6 | | L-(-)-fenchone |
| 7 | | (-)-α-thujone |
| 8 | | (-)-pinene |
| 9 | | (-)-isopinocampheol |
| 10 | | S-(-)-*cis*-verbenol |
| 11 | | 6,6-dimethyl-2-norpinene-2-ethanol nopol |

**TABLE 17.19** (*Continued*)

| # | Ring System | Common Name |
|---|---|---|
| 12 | 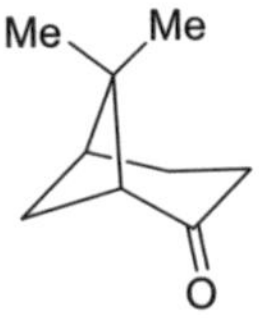 | (+)-nopinone |
| 13 | 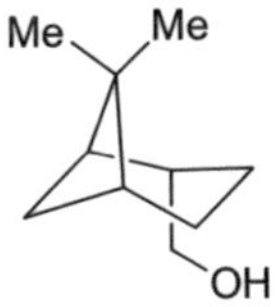 | myrtanol |
| 14 | 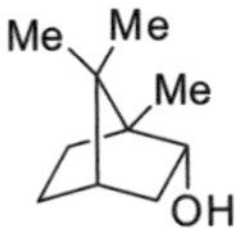 | L-(-)-borneol |
| 15 | 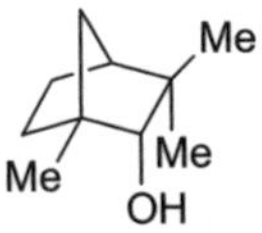 | *endo*-fenchyl alcohol<br>(+)-α-fenchyl alcohol |
| 16 | 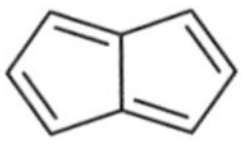 | pentalene |
| 17 | 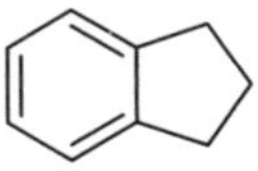 | indane |
| 18 | 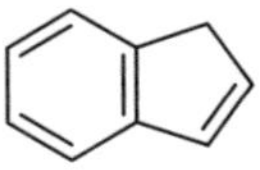 | indene |
| 19 | 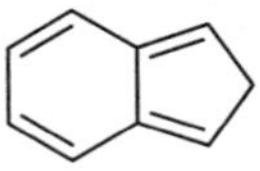 | 2*H*-indene<br>isoindene |
| 20 | 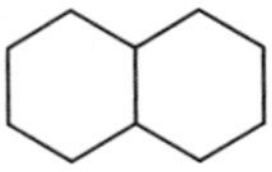 | decalin<br>octahydronapthalene |
| 21 | 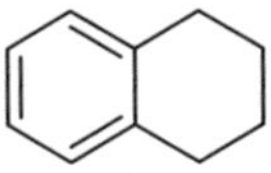 | tetralin<br>1,2,3,4-tetrahydronapthalene |

(*continued*)

**TABLE 17.19**   (*Continued*)

| # | Ring System | Common Name |
|---|---|---|
| 22 | | 2-tetralone |
| 23 | | napthalene |
| 24 | | 2-napthol<br>2-hydroxynaphthalene |
| 25 | | azulene |
| 26 | | heptalene |
| 27 | | biphenyl |
| 28 | | *trans*-stilbene |
| 29 | | *o*-terphenyl |
| 30 | | 1,1′-binaphthylene |
| 31 | | biphenylene |

**TABLE 17.19**    (*Continued*)

| # | Ring System | Common Name |
|---|---|---|
| 32 | | *asym*-indacene |
| 33 | | *sym*-indacene |
| 34 | | acenapthene |
| 35 | | acenapthylene |
| 36 | | fluorene |
| 37 | | phenalene |
| 38 | | phenanthrene |
| 39 | | anthracene |
| 40 | | dibenzosubarane |
| 41 | | dibenzosubarene |

(*continued*)

**TABLE 17.19**    (*Continued*)

| # | Ring System | Common Name |
|---|---|---|
| 42 | | fluoranthene |
| 43 | | acephenanthrylene |
| 44 | | aceanthrylene |
| 45 | | triphenylene |
| 46 | | pyrene |
| 47 | | chrysene |
| 48 | | naphthacene |
| 49 | | [2.2]paracyclophane |

**TABLE 17.19**    (*Continued*)

| # | Ring System | Common Name |
|---|---|---|
| 50 | | triptycene |
| 51 | | adamantane |
| 52 | | Steroid ring system:<br>R = nearly always methyl<br>R′ = usually methyl<br>R″ = various groups |

**CHAPTER 17    Table Index**

# 18

## pKa

DAVID H. BROWN RIPIN

## 18.1 INTRODUCTION

### 18.1.1 Deprotonation and Protonation of C−H

Deprotonation and protonation of organic compounds is a large and multidimensional topic. Evaluating the relative pKa of the base and acid to be reacted provides a thermodynamic perspective on where the equilibrium between the two and their conjugate acid and base will lie, but frequently this is only part of the story. In the case of metalated bases, varying the metal counterion frequently alters the kinetic basicity of the compound. A good example of this is the relative stability and functional group compatibility of an organozinc reagent to that of the corresponding organolithium. In cases where regioselectivity is an issue, the sterics or coordinating ability of the base can alter the outcome of the reaction. These two effects are a result of the kinetic acidity of a given proton when reacting with a specific base. Additionally, while some reactions require complete deprotonation of a substrate in order to succeed, others require access to low levels of the anion in an equilibrium process, directing the choice of base.

*Practical Synthetic Organic Chemistry: Reactions, Principles, and Techniques*, First Edition.
Edited by Stéphane Caron.
© 2011 John Wiley & Sons, Inc. Published 2011 by John Wiley & Sons, Inc.

## 18.1.2   pKa

The pKa$^{1-18}$ of a compound is a measure of the ability of a compound to lose a proton to the solvent under equilibrium. The pKa of organic compounds are typically measured in water for relatively acidic compounds, and in DMSO for the strongly basic compounds. The solvent can have a dramatic affect on the acidity of a compound, and therefore the pKa of a compound in water cannot be compared to the pKa of another compound in DMSO to estimate where the equilibrium between the two may lie. A table of pKa values is useful to compare the relative acidity of compounds in the same solvent, and from which to approximate where an equilibrium may lie in order to select a base for a deprotonation.

Since the pKa of the majority of organic compounds has not been measured, values can be estimated by using trends relating to electronegativity of a substituent, conjugative stabilization of an anion, aromaticity or antiaromaticity, sterics, and ring constraints. The tables below evaluate the effect of various constraints of related compounds to serve this purpose. Unless otherwise noted, the values given are measured or estimated in the solvent DMSO.

Tables 18.1 and 18.2 show the effect of various substituents on the pKa of a hydrogen on a carbon to which they are appended. These averages are based on data available; some are averages of 10 or more pKa values, while some are from a single data point. *Use care when applying these numbers to a pKa estimate you are preparing.* Some functional groups are more affected by substituents in general than are others; examples are highlighted in some tables. Additionally, the effects of solvation, conformation, and sterics should not be overlooked; for example, a phenyl group will have a radically different effect on the pKa of a proton depending on whether or not the geometry of the molecule allows for the phenyl group to adopt a conformation in which the anion is stabilized by resonance. This can be seen in the Tables 18.1 and 18.2, which shows that the effect of more than one of the same substituent is not additive (2 methyl groups raise pKa an average of 0.8 units vs. 0.2 units for one methyl group and 2 fluorines lower pKa an average of 4.5 units vs. 1.8 for one fluorine).

[1]Alnajjar, M. S.; Zhang, X.-M.; Franz, J. A.; Bordwell, F. G. *J. Org. Chem.* **1995**, *60*, 4976–4977.

[2]Bordwell, F. G. *Acc. Chem. Res.* **1988**, *21*, 456–463.

[3]Bordwell, F. G.; Algrim, D. *J. Org. Chem.* **1976**, *41*, 2507–2508.

[4]Bordwell, F. G.; Algrim, D.; Vanier, N. R. *J. Org. Chem.* **1977**, *42*, 1817–1819.

[5]Bordwell, F. G.; Bartmess, J. E.; Hautala, J. A. *J. Org. Chem.* **1978**, *43*, 3095–3101.

[6]Bordwell, F. G.; Bausch, M. J. *J. Am. Chem. Soc.* **1983**, *105*, 6188–6189.

[7]Bordwell, F. G.; Branca, J. C.; Hughes, D. L.; Olmstead, W. N. *J. Org. Chem.* **1980**, *45*, 3305–3313.

[8]Bordwell, F. G.; Drucker, G. E.; Fried, H. E. *J. Org. Chem.* **1981**, *46*, 632–635.

[9]Bordwell, F. G.; Drucker, G. E.; McCollum, G. J. *J. Org. Chem.* **1982**, *47*, 2504–2510.

[10]Bordwell, F. G.; Fried, H. E. *J. Org. Chem.* **1991**, *56*, 4218–4223.

[11]Bordwell, F. G.; Fried, H. E.; Hughes, D. L.; Lynch, T. Y.; Satish, A. V.; Whang, Y. E. *J. Org. Chem.* **1990**, *55*, 3330–3336.

[12]Bordwell, F. G.; Liu, W.-Z. *J. Phys. Org. Chem.* **1998**, *11*, 397–406.

[13]Bordwell, F. G.; Matthews, W. S.; Vanier, N. R. *J. Am. Chem. Soc.* **1975**, *97*, 442–443.

[14]Bordwell, F. G.; Singer, D. L.; Satish, A. V. *J. Am. Chem. Soc.* **1993**, *115*, 3543–3547.

[15]Bordwell, F. G.; Zhang, X. M.; Cheng, J. P. *J. Org. Chem.* **1993**, *58*, 6410–6416.

[16]Jencks, W. P., Unpublished results.

[17]Perrin, D. D. *Aust. J. Chem.* **1964**, *17*, 484–488.

[18]Smith, M. B.; March, J. *March's Advanced Organic Chemistry: Reactions, Mechanisms, and Structure*, 5th ed; Wiley: Chichester, UK, 2000.

**TABLE 18.1   Average Change in pKa of a Proton Attached to an Alkyl
Carbon Bearing the Following Substituents**

| Substituent | Δ pKa | Substituent | Δ pKa |
|---|---|---|---|
| 2 X Me | 0.8 | SPh | −9.1 |
| Me | 0.2 | CH=CH$_2$ | −9.3 |
| **H** | 0.0 | CONMe$_2$ | −10.0 |
| OMe | −0.1 | dithiane | −10.3 |
| Bn | −1.0 | Me$_3$N$^+$ | −11.1 |
| F | −1.8 | C≡CPh | −11.2 |
| TMS | −2.9 | 2 X Ph | −11.8 |
| OPh | −3.7 | PO(OEt)$_2$ | −13.3 |
| NPh$_2$ | −4.4 | CONMe$_2$ | −14.7 |
| 2 X F | −4.5 | C$_6$F$_5$ | −15.5 |
| 3-thiophene | −4.8 | POPh$_2$ | −15.8 |
| CF$_3$ | −4.9 | SOPh | −16.2 |
| Cl | −5.2 | PyrN$^+$ | −16.8 |
| SePh | −6.1 | SO$_2$Ph | −17.4 |
| 2-thiophene | −6.6 | CO$_2$R | −17.5 |
| 2-furanyl | −6.7 | CN | −17.6 |
| C≡CH | −6.9 | COMe | −18.2 |
| Ph | −7.6 | COPh | −18.3 |
| TPS (triphenylsilyl) | −7.7 | SO$_2$CF$_3$ | −22.4 |
| 2 X TMS | −8.6 | Ph$_3$P$^+$ | −23.2 |
| PPh$_2$ | −8.8 | NO$_2$ | −27.0 |
| CH=CHPh | −8.8 | 2 X C$_6$F$_5$ | −30.3 |

## 18.2   pKa DATA

### 18.2.1   Effect of Substituents on pKa of C–H

Tables 18.1 and 18.2 below can be used as exemplified. To estimate the pKa of the molecule
below, the relative effect of the various substituents could be considered as follows:

In DMSO:

| 26.5 (pKa of acetone) | 30.3 (pKa of ethyl acetate) | 14.2 (pKa of ethyl acetoacetate) |
|---|---|---|
| −1.8 (fluorine substituent) | −1.8 (fluorine substituent) | −1.8 (fluorine substituent) |
| −17.5 (ethyl ester substituent) | −18.2 (methyl ketone substituent) | |
| **7.2 (estimate)** | **10.3 (estimate)** | **12.4 (estimate)** |

As can be discerned from the example above, the estimates are very rough and should only
be used to determine the order of magnitude of a pKa.

**TABLE 18.2  Average Change in pKa of a Proton Attached to an Alkyl Carbon Bearing the Following Categories of Substituent—A Different Organization to Table 18.1**

| Alkyl and Aryl | Δ pKa | Halide | Δ pKa | O, N, P | Δ pKa | S, Se | Δ pKa | Carbonyl and Nitrile | Δ pKa |
|---|---|---|---|---|---|---|---|---|---|
| 2 X Me | 0.8 | F | −1.8 | OMe | −0.1 | SePh | −6.1 | CONMe$_2$ | −10.0 |
| Me | 0.2 | 2 X F | −4.5 | OPh | −3.7 | SPh | −9.1 | CO$_2$R | −17.5 |
| H | 0.0 | CF$_3$ | −4.9 | NPh$_2$ | −4.4 | dithiane | −10.3 | CN | −17.6 |
| Bn | −1.0 | Cl | −5.2 | PPh$_2$ | −8.8 | SOPh | −16.2 | COMe | −18.2 |
| 3-thiophene | −4.8 | | | Me$_3$N$^+$ | −11.1 | SO$_2$Ph | −17.4 | COPh | −18.3 |
| 2-thiophene | −6.6 | | | PO(OEt)$_2$ | −13.3 | SO$_2$CF$_3$ | −22.4 | | |
| 2-furanyl | −6.7 | | | POPh$_2$ | −15.8 | | | | |
| C≡CH | −6.9 | | | PyrN$^+$ | 16.8 | | | | |
| Ph | −7.6 | | | Ph$_3$P$^+$ | 23.2 | | | | |
| CH=CHPh | −8.8 | | | NO$_2$ | 27.0 | | | | |
| CH=CH$_2$ | −9.3 | | | | | | | | |
| C≡CPh | −11.2 | | | | | | | | |
| 2 X Ph | −11.8 | | | | | | | | |

## 18.2.2   pKa of Hydrocarbons and Heterocycles

The pKa values of a number of hydrocarbons are presented in Tables 18.3 and 18.4. Note that pKa values of >35 in DMSO and >14 in water are estimated. Noteworthy is the dramatic increase in the acidity of an acetylenic proton when compared to a vinylic proton. Also interesting is the increase in pKa from cyclopentadiene to indene demonstrating a decrease in the benefit of aromatization in the presence of the conjugated phenyl ring. This is further evident on examination of Table 18.5.

In Table 18.4, some trends are highlighted from Table 18.3 below. The substitution of methyl groups on methane has an almost linear effect in decreasing the acidity of the C−H. Changing from an $sp^3$-bonded hydrogen (ethane) to an $sp^2$-bonded hydrogen (ethylene) results in no pKa change, whereas the sp-bonded hydrogen of acetylene is significantly more

**TABLE 18.3   pKa of Hydrocarbons and Hydrogen**

| Hydrocarbon | pKa in DMSO | pKa in Water | Hydrocarbon | pKa in DMSO | pKa in Water |
|---|---|---|---|---|---|
| Me–C(Me)(Me)H | | 53 | H–H | | 35 |
| Me–CH(Me) | | 51 | Ph–CH(Ph)H | 32.2 | 33.5 |
| CH2=CH–H | | 50 | Ph–C(Ph)(H)Ph | 30.6 | 31.5 |
| Me–H | 56 | 48 | H–≡–H | | 24 |
| (cyclopropyl)–H | | 46 | Ph–≡–H | 28.8 | 23 |
| CH2=CH–CH2–H | 44 | 43 | (tetramethylcyclopentadiene Me,Me,Me,Me,H) | 26.1 | |
| (phenyl)–H | | 43 | (indene, H) | 20.1 | 20 |
| Ph–CH2–H | 43 | 41 | (cyclopentadiene, H) | 18.0 | 15 |

**TABLE 18.4  Interesting Trends in Hydrocarbon pKa in DMSO (pKa in water)**

| | | | |
|---|---|---|---|
| Me–H | Me $\diagup$ H | Me $\diagup$ Me H | Me $\diagdown$ Me $\diagup$ Me H |
| 56 (48) | (49.5[a]) | (51) | (53) |
| Ph $\diagup$ H | $\diagup\!\!\!\diagup$ H | | |
| 43 (41) | (50) | | |
| Ph $\diagup$ Ph $\diagup$ H | H——H | | |
| 32.2 (33.5) | (24) | | |
| Ph $\diagup$ Ph $\diagdown$ Ph H | | | |
| 30.6 (31.5) | | | |

[a]extrapolated.

**TABLE 18.5  pKa Values of Carbocycles Resulting in Aromatic Anions**

| | pKa in DMSO (water) | | pKa in DMSO (water) |
|---|---|---|---|
| (cyclopentadiene) | 18.0 (15) | (pentamethylcyclopentadiene) | 26.1 |
| Ph–(cyclopentadiene)–Ph | 14.3 | (pentaphenylcyclopentadiene) | 12.5 |
| (indene) | 20.1 (20) | (indene)–Ph | 19.4 |
| (indene)–Me | 21.8 | (triphenylindene) Ph, Ph, Ph | 15.2 |
| (indene) Me | 22.5 | | |
| (fluorene) | 22.6 | (9-TMS-fluorene) TMS | 21.7 |
| (octafluorofluorene) F F F F F F F F | 10.8 | (9-Me-fluorene) Me | 22.3 |

**TABLE 18.6    pKa of Benzylic Protons in Aromatic and Heteroaromatic Systems**

| pKa in DMSO (water) | pKa in DMSO (water) |
|---|---|
| Me (structure) — 43 | Me (naphthalene structure) — 42 |
| Me (pyridine structure) — 35 | (pyridine-CH₂-Ph structure) — 30.2 |
| Ph (pyridine structure) — 26.7 | (pyridine-CH₂-Ph structure) — 28.2 |
| Ph (pyridine N-oxide structure) — 25.2 | SO₂Ph (pyridine structure) — 16.7 |
| O—Me (furan structure) — 43 | |
| S—Me (thiophene structure) — 43 | (thiophene-CH₂-Ph structure) — 29.9 |
| Me (benzothiazole structure) — 27.6 | Ph (benzothiazole structure) — 20.8 |

acidic. Substituting methane with phenyl groups results in a nonlinear acidification of the C–H due to steric constraints affecting the ability of the phenyl groups to conjugate to the resulting anion.

The pKa values of protons attached to a heterocycle are presented in Table 18.7. The use of selective deprotonations of heterocycles is a key strategy in the synthesis of substituted systems. See Section 12.3. on strategies for metalating heterocycles for a more empirical treatment of this data.

The effect on pKa of benzylic protons by substituents attached to the aromatic ring is presented in Table 18.8. Aryl substituent effects on other systems can be found in Tables 18.10, 18.27, 18.34, 18.36, 18.40, 18.41 and 18.44.

The effect on pKa of benzylic protons by substituents attached to the aromatic ring is presented in Table 18.6 and Table 18.8.

**TABLE 18.7   pKa Values of Heterocyclic C−H**

| | pKa in DMSO (water) | | pKa in DMSO (water) |
|---|---|---|---|
| | 28 | | 24.4 |
| | 30.3 | | 29.4 |
| | 18.2 | | 18.6 |
| | 18.5 | | 13.8 |

**TABLE 18.8   pKa Values of Toluenes**

| X | pKa in DMSO (water) | X | pKa in DMSO (water) |
|---|---|---|---|
| H | 43 | H | 23.4 |
| NO$_2$ | 20.4 | Me | 25.0 |
| CN | 30.8 | Br | 22.3 |
| SO$_2$Ph | 29.8 | | |
| SO$_2$CF$_3$ | 24.1 | | |

## 18.2.3   pKa Values of Carbonyl Compounds

The pKa values of substituted esters, ketones, aldehydes, and amides are presented in Tables 18.9 to 18.15. The substituents are listed in the same order as in Table 18.1. Compare the DMSO pKa of $t$-BuOH (29.4), (TMS)$_2$NH (30), and $i$-Pr$_2$NH (36), the conjugate bases of which are frequently utilized for enolization. Ester values are for the ethyl or methyl ester.

Conformational changes resulting from loss of a proton, and subsequent addition or relief of strain in the molecule, can affect pKa. The effect of a ring on the pKa of a series of cyclic ketones (in which endocyclic stabilization will occur) and phenyl ketones (in which exocyclic stabilization will occur) are presented in Tables 18.11 and 18.12. Note that the endocyclic stablilization in the 4-membered ring and 6-membered ring is larger than in the case of exocyclic stabilization. Table 18.13, listing the pKa values of some bicyclic ketones, more dramatically demonstrates the effect of conformation on pKa. The effects of rings on some other systems are shown in Table 18.21.

**TABLE 18.9  pKa Values of Aldehydes, Ketones, and Esters in DMSO (pKa in Water)**

|  | $\underset{H}{\overset{O}{\parallel}}\!\!-\!X$ | $\underset{Me}{\overset{O}{\parallel}}\!\!-\!X$ | $\underset{Ph}{\overset{O}{\parallel}}\!\!-\!X$ | $\underset{RO}{\overset{O}{\parallel}}\!\!-\!X$ | $\underset{Me_2N}{\overset{O}{\parallel}}\!\!-\!X$ |
|---|---|---|---|---|---|
| 2 X Me |  |  | 26.3 |  |  |
| Me |  | 24.4 |  |  |  |
| H | 24.5 | 26.5 | 24.7 | 30.3 | 35 |
| OMe |  |  | 22.8 |  |  |
| F |  |  | 21.6 |  |  |
| OPh |  |  | 21.1 |  |  |
| NPh$_2$ |  |  | 20.3 |  |  |
| 2 X F |  |  | 20.2 |  |  |
| SePh |  |  | 18.6 |  |  |
| Ph |  | 19.8 | 17.7 | 22.6 | 26.6 |
| SPh |  | 18.7 | 16.9 | 21.2 | 25.9 |
| dithiane |  | 18.4 | 17.7 | 20.9 | 27 |
| Me$_3$N$^+$ |  | 16.3 | 14.6 | 20.6 | 24.9 |
| 2 X Ph |  |  | 18.8 | 21.9 |  |
| PO(OEt)$_2$ |  |  |  | 18.6 |  |
| SOPh |  |  | 14 |  |  |
| PyrN$^+$ |  | 11.8 | 10.7 | 14.1 |  |
| SO$_2$Ph |  | 12.5 | 11.4 |  |  |
| CO$_2$R |  | 14.2 |  | 15.7 |  |
| COMe |  | 13.3 (8.9) | 12.7 | 14.2 | 18.2 |
| COPh |  | 12.7 | 13.3 |  |  |
| SO$_2$CF$_3$ |  |  | 5.1 |  |  |
| Ph$_3$P$^+$ | 6.1 | 7.1 | 6.2 | 8.5 | 12.6 |
| NO$_2$ |  |  | 7.7 |  |  |

**TABLE 18.10  pKa Values of Acetophenones in DMSO**

|  | pKa |
|---|---|
| H | 24.7 |
| OMe | 25.7 |
| CN | 22.0 |
| Br | 23.8 |

|  | pKa |
|---|---|
|  | 21.8 |

**TABLE 18.11   Effect of Ring Size on pKa Values of Ketones and Esters in Aliphatic Systems**

| Ring size | | pKa | | pKa | | pKa |
|---|---|---|---|---|---|---|
| No ring | Me–C(=O)–Me | 26.5 | Me–CH(Me)–C(=O)–Ph | 26.3 | MeO–C(=O)–CH$_2$–Me | 30.0 |
| 3 | | | cyclopropyl–C(=O)–Ph | 28.2 | | |
| 4 | cyclobutanone | 25.0 | cyclobutyl–C(=O)–Ph | 26.2 | | |
| 5 | cyclopentanone | 25.8 | cyclopentyl–C(=O)–Ph | 25.8 | | |
| 6 | cyclohexanone | 26.4 | cyclohexyl–C(=O)–Ph | 26.7 | δ-valerolactone | 25.2 |
| 7 | cycloheptanone | 27.8 | | | | |
| 8 | cyclooctanone | 27.4 | | | | |
| 9 | cyclononanone | N/A | | | | |
| 10 | cyclodecanone | 26.8 | | | | |

**TABLE 18.12   Effect of Ring Size on pKa Values of Ketones and Esters in Aromatic Systems**

| No Ring | pKa | 5-Membered Ring | pKa | 6-Membered Ring | pKa |
|---|---|---|---|---|---|
| | $25.1^a$ | | 23.0 | | |
| | 18.7 | | 16.9 | | |
| | 11.4 | | 10.1 | | |
| | 22.6 | | 13.5 | | 18.8 |

$^a$extrapolated.

**TABLE 18.13   pKa Values of Bicyclic Ketones**

| | | | |
|---|---|---|---|
| 28.1 | 29.0 | 25.5 | 32.4 |

**TABLE 18.14   pKa Values of Cyclic 1,3-Diketones and 1,3-Diesters**

| | | | |
|---|---|---|---|
| 10.3 | 11.2 | 7.3 (4.8) | 7.4 |

### TABLE 18.15   pKa Value of 2-Substituted Malonate Esters

EtO$_2$C  CO$_2$Et
     |
     X

| 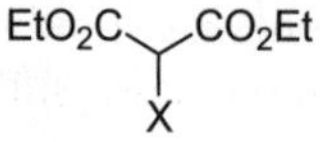  | pKa |
|---|---|
| H | 15.7 |
| *i*-Pr | 20.5 |
| *t*-Bu | 24.7 |
| TMS | 19.0 |
| CF$_3$ | 10.8 |

### 18.2.4   pKa of Nitriles and Nitroalkyls

The pKa values of various substituted acetonitriles and nitromethanes are listed in Table 18.16, along with the average effect of the substituent on the pKa. The pKa of an imine is listed in Table 18.17.

### TABLE 18.16   pKa Values of Nitriles and Nitroalkyls

|  | 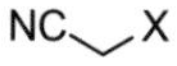 NC$\diagdown$X | 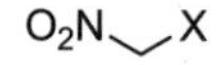 O$_2$N$\diagdown$X | Average ΔpKa |
|---|---|---|---|
| 2 X Me |  | 16.9 | −0.8 |
| Me | 32.5 | 16.7 | −0.2 |
| H | 31.3 | 17.2 (10.0) | 0.0 |
| Bn |  | 16.2 | 1.0 |
| OPh | 28.1 |  | 3.7 |
| Ph | 21.9 | 12.2 | 7.6 |
| CH=CH$_2$ |  | 7.7 | 9.3 |
| SPh | 20.8 | 11.8 | 9.1 |
| dithiane | 19.1 |  | 10.3 |
| Me$_3$N$^+$ | 20.6 |  | 11.1 |
| 2 X Ph | 17.5 |  | 11.8 |
| PO(OEt)$_2$ | 16.4 |  | 13.3 |
| C$_6$F$_5$ | 15.8 |  | 15.5 |
| POPh$_2$ | 16.9 |  | 15.8 |
| PyrN$^+$ | 16.5 |  | 16.8 |
| SO$_2$Ph | 12 | 7.1 | 17.4 |
| CO$_2$R | 13.1 |  | 17.5 |
| CN | 11.1 (11.0) |  | 17.6 |
| COPh | 10.2 | 7.7 | 18.3 |
| Ph$_3$P$^+$ | 7 |  | 23.2 |
| 2 X C$_6$F$_5$ | 8 |  | 30.3 |

### TABLE 18.17   pKa Value of Imines

Ph
  \
Ph  N  Ph

| 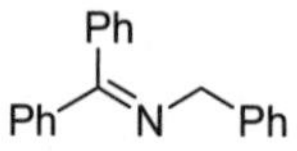 24.3 |
|---|

### 18.2.5 pKa of C−H with a Sulfur Substituent

The pKa of various substituted sulfides, sulfoxides, and sulfones are listed in Table 18.18, along with the average effect of the substituent on the pKa. Sulfonimides, sulfoximides, and additional sulfones are examined in Table 18.19. The structure of the sulfide substituent and the constraint of the sufide in a ring and its effect on pKa is evident from the data in Tables 18.20 and 18.21. It is apparent that an arylsulfide substituent is substantially more acidifying than an alkyl sulfide. Some data on cyclic nitroalkyls is also included in Table 18.21. The trends in this data do not correlate any trends discernable in Table 18.11 above (with the exception of the acidity lowering effect of the 3-membered ring), but with the exception of the 3-membered ring effect, the ring effect is minimal.

**TABLE 18.18  pKa Values of Sulfides, Sulfoxides, and Sulfones**

| X | PhS-CH2-X | PhS(O)-CH2-X | PhSO2-CH2-X | F3C-SO2-CH2-X | Average ΔpKa |
|---|---|---|---|---|---|
| Me | | | 31 | | −0.2 |
| **H** | **42** | **33** | **29** | **18.8** | **0.0** |
| OMe | | | 30.7 | | 0.1 |
| F | | | 28.5 | | 1.8 |
| TMS | | | 26.1 | | 2.9 |
| OPh | | | 27.9 | | 3.7 |
| 3-thiophenyl | | | 24.2 | | 4.8 |
| Cl | | | 23.8 | | 5.2 |
| 2-thiophenyl | | | 22.4 | | 6.6 |
| 2-furanyl | | | 22.3 | | 6.7 |
| C≡CH | | | 22.1 | | 6.9 |
| TPS | | | 21.3 | | 7.7 |
| Ph | 30.8 | 27.2 | 23.4 | 14.6 | 7.6 |
| TMS$_2$ | | | 20.4 | | 8.6 |
| PPh$_2$ | | | 20.2 | | 8.8 |
| CH=CHPh | | | 20.2 | | 8.8 |
| CH=CH$_2$ | | | 22.5 | | 9.3 |
| SPh | 30.8 | | 20.5 | 11 | 9.1 |
| Me$_3$N$^+$ | 28 | | 19.4 | | 11.1 |
| C≡CPh | | | 17.8 | | 11.2 |
| Ph$_2$ | 26.8 | 24.5 | 22.3 | | 11.8 |
| POPh$_2$ | 24.9 | | | | 15.8 |
| SOPh | | 18.2 | | | 16.2 |
| PyrN$^+$ | 17.7 | | | | 16.8 |
| SO$_2$Ph | 20.5 | | 12.2 | | 17.4 |
| COMe | 18.7 | | 12.5 | | 18.2 |
| COPh | 16.9 | 14 | 11.4 | | 18.3 |
| CN | 20.8 | | 12 | | 17.6 |
| Ph$_3$P$^+$ | 14.9 | | | | 23.2 |
| NO$_2$ | 11.8 | | 7.1 | | 27.0 |
| SO$_2$CF$_3$ | 11 | | | 2.1 | 22.4 |

**TABLE 18.19  pKa Values of Other Sulfones, Sulfoximines, Sulfoxides, and Sulfilimines**

| Sulfone | pKa | Sulfoximine | pKa | $X{-}SO_2{-}CH_2Ph$ (X) | pKa | Sulfimine/ Sulfoxide | pKa |
|---|---|---|---|---|---|---|---|
| $PhSO_2Me$ | 29.0 | Ph, $S(O)(=N^+Me_2)Me$ | 14.4 | Ph | 23.4 | Ts-tolyl sulfonimidoyl, $=N{-}Ph$, H | 27.6 |
| $Me{-}SO_2{-}Me$ | 31.1 | Ph, $S(O)(=NMe)Me$ | 33 | Bn | 23.9 | Ts-tolyl sulfonimidoyl, $=N{-}Ph$, $Me_2$/H | 30.7 |
| $Me{-}CH_2{-}SO_2{-}Me$ | 32.8 | Ph, $S(O)(=NTs)Me$ | 24.5 | $NMe_2$ | 25.2 | $Me{-}S(O){-}Me$ | 35 |
| $F_3C{-}SO_2{-}Me$ | 18.8 | Ph, $S(O)(=NTs)CH_2Cl$ | 20.7 | OPh | 19.9 | | |
| $F_3C{-}SO_2{-}$cyclopropyl | 32.8 | Ph, $S(O)(=NTs)CHCl_2$ | 17.7 | SBn | 19.1 | | |
| | | | | F | 16.9 | | |

**TABLE 18.20  Effect of Sulfide Structure on pKa**

| NC–SR | pKa | PhSO₂CH₂SR | pKa | RS–CH(Ph)–RS H | pKa | RS–C(SR) RS H | pKa |
|---|---|---|---|---|---|---|---|
| Me | 24.3 | Me | 23.4 | Ph | 23.0 | Ph | 22.8 |
| Et | 24.0 | Ph | 20.5 | dithiane | 30.7 | Pr | 31.3 |
| *i*-Pr | 23.6 | | | | | (bicyclic S) | 30.5 |
| *t*-Bu | 22.9 | | | | | | |

**TABLE 18.21  Effect of Ring Size on Sulfides and Nitroaryls**

| Ring size | | pKa | | pKa | | pKa |
|---|---|---|---|---|---|---|
| No ring | (MeS)₂CH–COPh | 17.8 | (PhS)₂CH–COPh | 12.0 | Me₂CH–NO₂ | 16.9 |
| 3 | | | | | ▷–NO₂ | 26.9 |
| 4 | | | | | ◇–NO₂ | 17.8 |
| 5 | (1,3-dithiolane)–COPh | 18.0 | (benzodithiole)–COPh | 15.2 | cyclopentyl–NO₂ | 16.0 |
| 6 | (1,3-dithiane)–COPh | 17.7 | | | cyclohexyl–NO₂ | 17.9 |
| 7 | (1,3-dithiepane)–COPh | 16.5 | | | cycloheptyl–NO₂ | 15.8 |

The pKa of various thiocarbonyl containing compounds are listed in Table 18.22, and cationic sulfur containing compounds in Table 18.23.

**TABLE 18.22  Effect of a Thiocarbonyl on pKa**

| Compound | X = O | X = S |
|---|---|---|
| Me–C(=X)–NMe₂ | 35.0 | 25.6 |
| Ph–CH₂–C(=X)–NMe₂ | 26.6 | 21.3 |
| Me–C(=X)–NH₂ | 25.5 (15.1) | 18.4 |

*(continued)*

**TABLE 18.22**   (*Continued*)

| Compound | X = O | X = S |
|---|---|---|
| Ph–C(=X)–NH$_2$ | 23.3 | 16.9 |
| Me–C(=X)–N(H)–Ph | 21.4 | 14.7 |
| Ph–N(H)–C(=X)–N(H)–Ph | 19.6 | 13.4 |
| (piperidin-2-one, =X) | 26.4 | 20.1 |
| H$_2$N–C(=X)–NH$_2$ | 26.9 | 21.2 |

**TABLE 18.23   pKa of Other Sulfur Containing Compounds**

| Compound | pKa |
|---|---|
| Ph–S$^+$(Me)–CH(H)–Ph | 16.3 |
| Me–S$^+$(Me)(Me)=O (with Me groups) | 18.2 |

## 18.2.6   pKa of C–H with Other Heteroatomic Substituents

Phosphonium salts, phosphines, phosphine oxides, and phosphonates significantly acidify the protons on the carbon to which they are affixed. Table 18.24 lists the pKa values of phosphorus-containing compounds.

**TABLE 18.24   pKa Values of Phosphines, Phosphine Oxides, and Phosphonates**

| Phosphonium salt | pKa | Phosphine/Phosphine oxide | pKa | Phosphonate | pKa |
|---|---|---|---|---|---|
| Ph$_3$P$^+$CH$_3$ | 22.4 | Ph–S(=O)$_2$–CH$_2$–PPh$_2$ | 20.2 | TMS–CH$_2$–P(=O)(OEt)$_2$ | 28.8 |
| Me–CH(Me)–P$^+$Ph$_3$ | 21.2 | Ph$_2$P–CH$_2$–PPh$_2$ | 29.9 | Cl–CH$_2$–P(=O)(OEt)$_2$ | 26.2 |

**TABLE 18.24** (*Continued*)

| Phosphonium salt | pKa | Phosphine/Phosphine oxide | pKa | Phosphonate | pKa |
|---|---|---|---|---|---|
| NC$\diagup$P$^+$Ph$_3$ | 7.0 | NC$\diagup$PPh$_2$ (O) | 16.9 | NC$\diagup$P(OEt)$_2$ (O) | 16.4 |
| PhS$\diagup$P$^+$Ph$_3$ | 14.9 | PhS$\diagup$PPh$_2$ (O) | 24.9 | Ph$\diagup$P(OEt)$_2$ (O) | 27.6 |
| RO$\diagup$C(O)$\diagup$P$^+$Ph$_3$ | 8.5 | | | RO$\diagup$C(O)$\diagup$P(OEt)$_2$ (O) | 18.6 |
| Me$_2$N$\diagup$C(O)$\diagup$P$^+$Ph$_3$ | 12.6 | | | | |
| Ph$\diagup$C(O)$\diagup$P$^+$Ph$_3$ | 6.2 | | | | |
| Me$\diagup$C(O)$\diagup$P$^+$Ph$_3$ | 7.1 | | | | |
| H$\diagup$C(O)$\diagup$P$^+$Ph$_3$ | 6.1 | | | | |

The pKa of various selenium-containing compounds is listed in Table 18.25

**TABLE 18.25  pKa Values of Selenides**

| Compound | pKa |
|---|---|
| PhSe$\diagup$SePh with CHPh, H | 16.2 |
| PhSe$\diagup$Ph | 31.0 |
| PhSe$\diagup$Ph with Ph | 27.5 |

## 18.2.7  pKa of O−H

The pKa values of some alcohols and phenols are listed in Table 18.26. The solvation of the resultant anion can explain the increase in pKa as the alcohol is more sterically hindered. This is dramatically evident in the difference in the pKa values of *t*-butanol and tricyclo-hexylmethanol measured in water. Also noteworthy is the significantly higher acidity of

**TABLE 18.26   pKa Values of O−H**

| Alcohol | pKa (water) | Phenol | pKa (water) | Hydroxypyridine | pKa (water) |
|---|---|---|---|---|---|
| $H_2O$ | 27.5 (15.7) | Phenol | 18.0 (10.0) | | 14.8 |
| MeOH | 27.9 (15.54) | $p$-$N^+Me_3$ | 16.8 | | 15.8 |
| $i$-PrOH | 29.3 (16.5) | $p$-$NO_2$ | 10.8 (7.14) | | 17.0 |
| $t$-BuOH | 29.4 (17) | $m$-$NO_2$ | (8.35) | | 20.7 |
| ($c$-hex)$_3$COH | (24) | $o,p$-$NO_2$ | 5.1 | | |
| $CF_3CH_2OH$ | 23.5 (12.5) | $o,o$-$NO_2$ | 4.9 | | |
| $(CF_3)_2CHOH$ | 17.9 | $p$-OMe | 19.1 (10.20) | | |
| $(CF_3)_3COH$ | 10.7 | | 17.1 | | |
| MeOOH | (11.5) | | 13.7 | | |
| HOOH | (11.6) | | | | |

methanol and water and their corresponding peroxides. In the case of hydroxypyridines, the pKa difference between 2- and 4-hydroxypyridine is noteworthy. The difference in acidity between 2-hydroxynaphthalene and 2,3-dihydroxynaphthalene can be explained by a hydrogen bonding stabilization by the adjacent alcohol.

The pKa values of carboxylic acids and benzoic acids are listed below. The majority of the data in this series is measured in water. The pKa of peracetic acid is significantly higher than that of acetic acid, opposite to the trend seen in the cases of alcohols and their corresponding peroxides. The effect of hydrogen bond stabilization acidifying a proton and charge–charge repulsion decreasing the acidity of a proton can be seen by the difference in the first and second pKa values of fumaric and maleic acid.

**TABLE 18.27   pKa Values of Carboxylic Acids**

| Carboxylic Acid | pKa (water) | Benzoic Acid | pKa (water) |
|---|---|---|---|
| $H_3CO_2H$ | 12.3 (4.76) | H (Benzoic acid) | 11.0 (4.2) |
| $HCO_2H$ | (3.77) | $p$-$NO_2$ | 9.0 (3.44) |
| $FH_2CCO_2H$ | (2.66) | $m$-$NO_2$ | (2.45) |
| $BrH_2CCO_2H$ | (2.86) | $o$-$NO_2$ | (2.17) |
| $IH_2CCO_2H$ | (3.12) | $p$-OMe | (4.47) |
| $ClH_2CCO_2H$ | (2.86) | $p$-Cl | (3.99) |
| $Cl_2HCCO_2H$ | 6.4 (1.29) | $m$-Cl | (3.83) |
| $Cl_3CCO_2H$ | (0.65) | $o$-Cl | (2.94) |
| $CF_3CO_2H$ | 3.4 (−0.25) | $p$-$N^+Me_3$ | (3.43) |
| $NO_2H_2CCO_2H$ | (1.68) | $o$-$N^+Me_3$ | (1.37) |
| $HOCO_2H$ | (3.6, 10.3) | | |
| $MeCO_3H$ | (8.2) | | |
| $\diagup\!\!\diagdown CO_2H$ | (4.25) | | |
| $HO_2C\diagdown\!\!\diagup CO_2H$ | (3.02, 4.38) | | |
| $\diagup\!\!\diagdown CO_2H$ / $CO_2H$ | (1.92, 6.23) | | |

The pKa of other O-H bonds is listed in Table 18.28, and protonated oxygen in Table 18.29.

**TABLE 18.28   pKa Values of other O−H**

| Compound | pKa | Compound | pKa |
|---|---|---|---|
| (cyclohexanone oxime) | 24.3 | (benzophenone oxime) | 20.1 (11.3) |
| $Ph$ C(=O) N(Me)OH | 18.5 | $Ph$ C(=O) NH OH | 13.7 (8.88) |
| $MeSO_3H$ | 1.6 (−2.6) | $CF_3SO_3H$ | 0.3 (−14) |
| HONO | 7.5 | $PhSO_2H$ | 7.1 (3.5) |

**TABLE 18.29   pKa Values of Protonated Oxygen**

| Compound | pKa | Compound | pKa | Compound | pKa |
|---|---|---|---|---|---|
| $Ph\!-\!O^+(H)\!-\!Me$ | (−6.5) | $Me\!-\!O^+(H)\!-\!H$ | (−2.2) | $Ph\!-\!N^+(=O)\!-\!OH$ | (−12.4) |
| $Me\!-\!O^+(H)\!-\!Me$ | (−3.8) | $Me_2S=O^+\!-\!H$ | 1.8 (−1.8) | $Ph\!-\!C(O^+H)\!-\!OH$ | (−7.8) |
| (tetrahydrofuran $O^+$·H) | (−2.0) | (pyridinium $N^+$-OH) | (0.79) | $Ph\!-\!C(O^+H)\!-\!Me$ | (−6.2) |

### 18.2.8 pKa of N−H

The pKas of a number of N-H protons is listed in Tables 18.30 to 18.42. Note the similar pKa values of TMS$_2$NH and *t*-butanol (30 vs. 29.4), the conjugate bases of which are frequently used in deprotonation reactions.

The pKa values of a number of anilines are listed in Table 18.31. The trend seen for 2-, 3-, and 4-hydroxypyridines (pKa decreases with distance from the pyridine nitrogen) is not replicated in the aniline series (See Table 18.26).

**TABLE 18.30  pKa Values of Amines**

| Compound | pKa | Compound | pKa |
|---|---|---|---|
| HN$_3$ | 7.9 (4.7) | H$_2$NCN | 16.9 |
| NH$_3$ | 41 (38) | TMS$_2$NH | 30 |
| *i*-Pr$_2$NH | 36 | | |
| | 44 | | 37 |

**TABLE 18.31  pKa Values of Anilines**

| Aniline | pKa |
|---|---|
| PhNH$_2$ | 30.6 |
| Ph$_2$NH | 25.0 |
| | 26.5 |
| | 28.5 |
| | 27.7 |
| | 23.3 |
| | 25.3 |

The pKa of various acylated and sulfonylated NH protons is listed in Tables 18.32 – 18.36.

**TABLE 18.32  pKa Values of Amide, Urea, and Carbamate N−H**

| R (RC(O)NH₂) | pKa | R (RC(O)NHR) | pKa |
|---|---|---|---|
| H | 23.4 | H | 23.4 |
| Me | 25.5 (15.1) | Me | 25.9 |
| t-Bu | 25.5 | t-Bu | 28.0 |
| CF₃ | 17.2 | | |
| NH₂ | 26.9 | | |
| OMe | 24.8 | | |

**TABLE 18.33  pKa values of Imides, Lactams, and Cyclic Carbamates**

| Compound | pKa | Compound | pKa |
|---|---|---|---|
| | 17.9 | | 24.1 |
| | 14.7 | | 26.4 |
| | 8.3 | | 20.5 |

**TABLE 18.34  pKa Values of Aryl Amide N−H**

| MeC(O)NH-Ar-X | pKa | X-Ar-C(O)NH₂ | pKa |
|---|---|---|---|
| H | 21.4 | H | 23.4 |
| p-OMe | 22.1 | p-OMe | 24.0 |
| p-CN | 18.6 | o-CF₃ | 21.8 |
| | | PhC(O)NHPh | 18.8 |

**TABLE 18.35   pKa Values of acyl Hydrazines, Sulfonyl Hydrazines, and Sulfonamides**

| R | pKa | | pKa | R | pKa |
|---|-----|---|-----|---|-----|
| Me | 21.8 | H | 17.1 | Me | 17.5 |
| Ph | 18.9 | Me | 15.8 | Ph | 16.1 |
| 3-pyr | 17.5 | | | $CF_3$ | 9.7 (6.3) |
| 4-pyr | 16.8 | | | | |
| $NHNH_2$ | 23.3 | | | | |

**TABLE 18.36   pKa Values of Aryl Sulfonamides**

| X | pKa | X | pKa | X | pKa |
|---|-----|---|-----|---|-----|
| H | 22.6 | H | 12 | H | 21.4 |
| OMe | 23.9 | OMe | 13 | OMe | 22.1 |
| CN | 18.5 | CN | 9.3 | CN | 18.6 |
| Br | 21.6 | Br | 11.3 | | |

The pKa values of a number of heterocycles are listed in Table 18.37. On the right-hand side, some pKa values of some comparable compounds are shown.

**TABLE 18.37   pKa Values of Heterocyclic N–H**

| Heterocycle | pKa | Comparable Heterocycle | pKa |
|-------------|-----|------------------------|-----|
| | 23.0 | | 21.0 |
| | 19.8 | | |
| | 18.6 | | 16.4 |
| | 14.8 | | |
| | 13.9 | | 11.9 |
| | | | |

**TABLE 18.37**    (*Continued*)

| Heterocycle | pKa | Comparable Heterocycle | pKa |
|---|---|---|---|
| | 15.0 | | 12.1 |
| | 20.5 | | 24.6 |
| | 14.1 | | 12.7 |
| | 13.0 | | 8.4 |
| | 14.8 | | 0.8 |
| | 4.0 | | |
| | 14.2 | | |
| | 19.9 | | 25.5 |
| | 17.7 | | |
| | 24.9 | | 26.1 |
| | 21.6 | | 22.7 |

**TABLE 18.38  pKa Values of Acyclic, Aliphatic, Protonated Nitrogen**

| Compound | pKa | Comparable Compound | pKa | Comparable Compound | pKa |
|---|---|---|---|---|---|
| $NH_4^+$ | 10.5 (9.24) | $H_2NNH_3^+$ | (8.12) | $HONH_3^+$ | (5.96) |
| $EtNH_3^+$ | (10.65) | $MeNH_3^+$ | (10.66) | | |
| $Et_2NH_2^+$ | (10.94) | $Me_2NH_2^+$ | (10.78) | $i\text{-}Pr_2NH_2^+$ | (11.05) |
| $Et_3NH^+$ | 9.0 (10.75) | $Et_3NH^+$ | (9.80) | | |
| HO⌢$N^+H_3$ | (9.50) | ephedrine: Me/Ph, HO⌢$N^+H_2Me$ (ephedrine) | (9.55) | | |
| HO⌢$N^+H_2$ (×2) | (8.88) | HO⌢Me⌢$H_3^+N$⌢OH | (8.80) | | |
| HO⌢$N^+H$ (×3) | (7.76) | | | | |
| $H_3^+N$⌢$N^+H_3$ | (6.90, 9.93) | cyclohexane-$N^+H_3$, $NH_2$ | (9.80) | cyclohexane-$N^+H_3$, $NH_2$ (trans) | (9.77) |
| $H_3^+N$(⌢)$_5$$NH_2$ | (10.93) | | | | |
| EtO₂C⌢$N^+H_3$ | 8.7 | $^-O_2C$⌢$N^+H_3$ | 7.5 | | |

**TABLE 18.39    pKa Values of Protonated Cyclic Aliphatic Nitrogen**

| Compound | pKa | Compound | pKa |
|---|---|---|---|
| ▷N$^+$H$_2$ | (8.04) | ⬠N$^+$H$_2$ | (11.30) |
| ◇N$^+$H$_2$ | (11.29) | ⬡N$^+$H$_2$ | (11.12) |
| (quinuclidine-like) N$^+$H | 9.8 (11.0) | H Me N$^+$ morpholine | (7.38) |
| (diazabicyclo) N$^+$·N$^+$H | (2.97, 8.82) | O morpholine N$^+$H$_2$ | (8.36) |
| DBU | (11–12)* | HN⌒N$^+$H$_2$ | (9.79) |

**TABLE 18.40    pKa Values of Protonated, Aryl Substituted Nitrogen**

| Compound | pKa | Compound | pKa |
|---|---|---|---|
| PhN$^+$H$_3$ | 3.6 (4.61) | naphthalene-N$^+$H$_3$ | (4.16) |
| Me—C$_6$H$_4$—N$^+$H$_3$ | (5.07) | H$_3^+$N  N$^+$H$_3$ (naphthalene) | "Proton sponge" 7.5 (−9.0, 12.0) |
| PhN$^+$HMe$_2$ | 2.5 (5.20) | | |
| Ph$_2$N$^+$H$_2$ | (0.78) | | |
| NO$_2$ (ortho) —N$^+$H$_3$ | (−0.26) | O$_2$N (meta) —N$^+$H$_3$ | (2.46) |
| O$_2$N—C$_6$H$_4$—N$^+$H$_3$ | (1.00) | | |
| F—C$_6$H$_4$—N$^+$H$_3$ | (4.65) | BzO—C$_6$H$_4$—N$^+$H$_3$ | (2.17) |
| Br (meta) —N$^+$H$_3$ | (3.53) | MeO (meta) —N$^+$H$_3$ | (4.23) |
| Cl (ortho) —N$^+$H$_3$ | (2.64) | OMe (ortho) —N$^+$H$_3$ | (4.52) |

**TABLE 18.41   pKa Values of Protonated Aromatic Nitrogens**

| Compound | pKa | Compound | pKa |
|---|---|---|---|
| | 3.4 (5.2) | | (4.33) |
| | (5.74) | | (5.81) |
| | 4.5 (6.75) | | 0.9 (4.95) |
| | (6.18) | | |
| | (0.72) | | (2.89) |
| | (9.2) | | (6.45) |
| | (7.37) | | (6.95) |
| | (7.29) | | (4.86) |
| | (5.41) | | (5.60) |

**TABLE 18.41**    (*Continued*)

| Compound | pKa | Compound | pKa |
|---|---|---|---|
| | (3.96) | | (3.48) |
| | (6.95) | | (4.15) |

**TABLE 18.42    pKa Values of other N—H**

| Compound | pKa | Compound | pKa |
|---|---|---|---|
| | 21.6 | $H_2NCN$ | 16.9 |
| | 17.3 | | 15.0 |
| | 13.7 (8.88) | $PhCN^+H$ | (−10) |

## 18.2.9    pKa Values of Amino Acids

The pKAs of common amino acids are listed in Table 18.43.

**TABLE 18.43    pKa Values of Amino Acids**

| Name | Side Chain | pKa ($CO_2H$) | pKa ($NH_3$) | pKa (side chain) | pI[a] |
|---|---|---|---|---|---|
| Glycine (Gly, G) | H | 2.4 | 9.8 | | 6.10 |
| Alanine (Ala, A) | Me | 2.4 | 9.9 | | 6.15 |
| Valine (Val, V) | *i*-Pr | 2.3 | 9.7 | | 6.00 |
| Leucine (Leu, L) | | 2.3 | 9.7 | | 6.00 |
| Isoleucine (Ile, I) | | 2.3 | 9.8 | | 6.05 |

(*continued*)

**TABLE 18.43** (*Continued*)

| Name | Side Chain | pKa (CO$_2$H) | pKa (NH$_3$) | pKa (side chain) | pI$^a$ |
|---|---|---|---|---|---|
| Proline (Pro, P) | | 2.0 | 10.6 | | 6.30 |
| Serine (Ser, S) | | 2.2 | 9.2 | | 5.70 |
| Threonine (Thr, T) | | 2.1 | 9.1 | | 5.60 |
| Cysteine (Cys, C) | | 1.9 | 10.7 | 8.4 | 5.15 |
| Methionine (Met, M) | | 2.1 | 9.3 | | 5.70 |
| Asparagine (Asn, N) | | 2.1 | 8.7 | | 5.40 |
| Glutamine (Gln, Q) | | 2.2 | 9.1 | | 5.65 |
| Phenylalanine (Phe, F) | Bn | 2.2 | 9.3 | | 5.75 |
| Tyrosine (Tyr, Y) | | 2.2 | 9.2 | 10.5 | 5.70 |
| Tryptophan (Trp, W) | | 2.5 | 9.4 | | 5.95 |
| Lysine (Lys, K) | | 2.2 | 9.1 | 10.5 | 9.80 |
| Arginine (Arg, R) | | 1.8 | 9.0 | 12.5 | 10.75 |
| Histidine (His, H) | | 1.8 | 9.3 | 6.0 | 7.65 |
| Aspartate (Asp, D) | | 2.0 | 9.9 | 3.9 | 2.95 |
| Glutamate (Glu, E) | | 2.1 | 9.5 | 4.1 | 3.10 |

$^a a$ = isoelectric point.

## 18.2.10    pKa Values of S–H, Se–H, and P–H

The pKa of various S-H, Se-H, and P-H bonds are listed in Tables 18.44–18.46, respectively.

**TABLE 18.44    pKa Values of S–H**

| X | pKa | RSH | pKa |
|---|---|---|---|
| H | 10.3 (6.5) | Bu | 17.0 |
| *p*-OMe | 11.2 | Bn | 15.4 |
| *p*-NH$_2$ | 12.5 | PhCO | 5.2 |
| *p*-Me | 10.8 | Ph | 10.3 |
| *m*-Me | 10.6 | | |
| *o*-Me | 10.7 | | |
| *p*-Br | 9.0 | | |
| *p*-Cl | 9.3 | | |
| *m*-Cl | 8.6 | | |
| *o*-Cl | 8.6 | | |
| *m*-CF$_3$ | 8.1 | | |
| *o*-CO$_2$Me | 7.8 | | |
| *o*-NO$_2$ | 5.5 | | |

**TABLE 18.45    pKa Values of Se–H**

| Compound | pKa |
|---|---|
| PhSeH | 7.1 |

**TABLE 18.46    pKa Values of P–H**

| Compound | pKa |
|---|---|
| P$^+$H$_4$ | (−14) |
| MeP$^+$H$_3$ | (2.7) |
| Et$_3$P$^+$H | (9.1) |

## 18.2.11    pKa of Inorganic Acids

**TABLE 18.47    pKa Values of Inorganic Acids**

| Compound | pKa | Compound | pKa | Compound | pKa |
|---|---|---|---|---|---|
| H$_2$O | 32 (15.75) | HBr | 0.9 (−9.00) | HCN | 12.9 (9.1) |
| H$_3$O$^+$ | (−1.7) | HCl | 1.8 (−8.0) | HN$_3$ | 7.9 (4.72) |
| H$_2$S | (7.00) | HF | 15 (3.17) | HSCN | (4.00) |
| HOCl | (7.5) | HNO$_3$ | (−1.3) | MeSO$_3$H | 1.6 (−2.6) |
| HClO$_4$ | (−10) | HONO | 7.5 (3.29) | CF$_3$SO$_3$H | 0.3 (−14) |
| H$_2$SO$_3$ | (1.9, 7.21) | H$_2$CrO$_4$ | (−0.98, 6.50) | PhSO$_2$H | 7.1 (3.5) |
| H$_2$SO$_4$ | (−3.0, 1.99) | B(OH)$_3$ | (9.23) | N$^+$H$_4$ | 10.5 (9.24) |
| H$_3$PO$_4$ | (2.12, 7.21, 12.32) | HOOH | (11.6) | | |

**TABLE 18.48  Commonly Used Bases**

| Base | Estimated pKa of Conjugate Acid | Some Useful Applications |
|---|---|---|
| NaHCO₃, KHCO₃, NaH₂PO₄, KH₂PO₄, KF, NaOAc, pyridine | 2–5 (water) | Scavenging acid generated in an acylation reaction<br>Deprotonation of acidic organics such as carboxylic acids |
| Na₂HPO₄, K₂HPO₄, N-methylmorpholine, imidazole morpholine, DABCO | 7–8 (water)<br>8–9 (water) | Scavenging acid generated in an alkylation reaction<br>Scavenging acid generated in an alkylation reaction |
| ethylenediamine, NH₃, DMAP | 9–10 (water) | Scavenging acid generated in an alkylation reaction |
| NaCN, KCN | 9–10 (water)<br>12 (DMSO) | Introduction of nitrile |
| Na₂CO₃, K₂CO₃, Cs₂CO₃, i-Pr₂NEt, Et₃N, i-Pr₂NH, quinuclidine | 10–11 (water)<br>9–10 (DMSO) | Scavenging acid generated in an alkylation reaction |
| DBU | 11–12 (DMSO) | Epimerization of α-chiral ketones |
| tetramethylguanidine (TMG) | 13.6 (DMSO) | Epimerization of α-chiral ketones |
| NaOH, KOH, LiOH | 15.7 (water)<br>27 (DMSO) | Base in a biphasic reaction<br>Hydrolysis reaction |
| KOt-Bu, NaOEt, NaOMe, SEH (sodium ethylhexanoate) | 15–17 (water)<br>27–29 (DMSO) | Reversible deprotonation of alcohols or ketones<br>Irreversible deprotonation of phenols, β-ketoesters |
| NaHMDS, LiHMDS, KHMDS | 30 (DMSO) | Enolate generation, reversible in some cases<br>Irreversible deprotonation of alcohols |
| NaH, KH | 35 (DMSO) | Irreversible deprotonation of alcohols<br>Irreversible enolate formation |
| LDA, LTMP | 36–37 (DMSO) | Irreversible alcohol deprotonation<br>Differing size of bases results in different selectivities |
| MeLi, MeMgX, EtMgX, n-BuLi, n-HexLi | 56 (DMSO) | Generation of LDA, LTMP<br>Irreversible deprotonation of amide<br>Deprotonation of substrates where the presence of an amine (from LDA) would be harmful<br>Deprotonation of aryl C–H<br>Metal halogen exchange |
| t-BuLi, s-BuLi | >60 (DMSO) | Metal halogen exchange<br>Deprotonation of protic species with a very high pKa<br>Deprotonation of aryl C–H |

## 18.3 KINETIC BACISITY

While the use of pKa differences allows the calculation of where the equilibrium between an acid and a base may lie, this measure may not be predictive of the outcome of a kinetically controlled deprotonation. The kinetic acidity of a compound is proportional to the activation energy required for a deprotonation to occur. Thus, in many cases a thermodynamically less acidic proton (as determined by pKa) can be selectively deprotonated in the presence of a thermodynamically more acidic proton, and a "stronger base" may be stable in the presence of a "weaker base." The effect of sterics on a deprotonation is a kinetic, and not thermodynamic phenomenon. Another example of kinetic acidity is the directed deprotonation, where coordination of the base to a functional group in the molecule results in the deprotonation of a specific proton, even if other protons of equal or greater acidity are present in the molecule. A good demonstration of kinetic acidity is the ability of PhMgBr (pKa~43) to add to an ester in the presence of an amide NH (pKa~25) without deprotonation.[19]

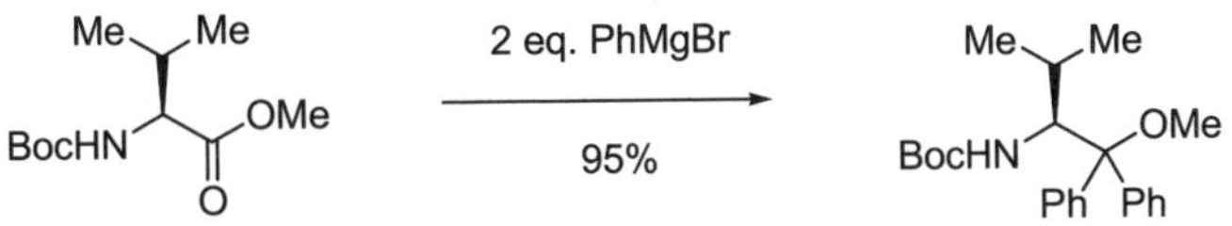

A list of synthetically useful bases is presented in Table 18.48, along with the types of substrates they are useful for deprotonating. Specific examples of bases useful for a given transformation can be found in the appropriate section of this book covering that transformation. This is in no way comprehensive, and does not take into consideration the use of Lewis acids to coordinate to an organic molecule and acidify specific protons, as happens in a number of enolization protocols.

The formation of kinetic versus thermodynamic enolates is a large topic and is covered in Section 1.2.4.1. Outside of a simple deprotonation, there are a few types of deprotonation that are noteworthy: directed deprotonation, enantioselective deprotonation reactions.

## 18.4 DIRECTED DEPROTONATIONS

Accomplishing a directed deprotonation requires a base that will coordinate with the substrate, and a directing group on the substrate to be deprotonated. Generally, a lithium base such as LDA, n-BuLi, s-BuLi, or t-BuLi is used, although other bases can be utilized. Directing groups include amines, amides, carbamates, heterocycles such as pyridine and oxazolines, enolates, and others. Some examples are shown below.[20–23]

[19]Brenner, M.; la Vecchia, L.; Leutert, T.; Seebach, D. *Org. Synth.* **2003**, *80*, 57–65.
[20]Hay, J. V.; Harris, T. M. *Org. Synth.* **1973**, *53*, 56–59.
[21]Keen, S. P.; Cowden, C. J.; Bishop, B. C.; Brands, K. M. J.; Davies, A. J.; Dolling, U. H.; Lieberman, D. R.; Stewart, G. W. *J. Org. Chem.* **2005**, *70*, 1771–1779.
[22]Iwao, M.; Kuraishi, T. *Org. Synth.* **1996**, *73*, 85–93.
[23]Borror, A. L.; Chinoporos, E.; Filosa, M. P.; Herchen, S. R.; Petersen, C. P.; Stern, C. A.; Onan, K. D. *J. Org. Chem.* **1988**, *53*, 2047–2052.

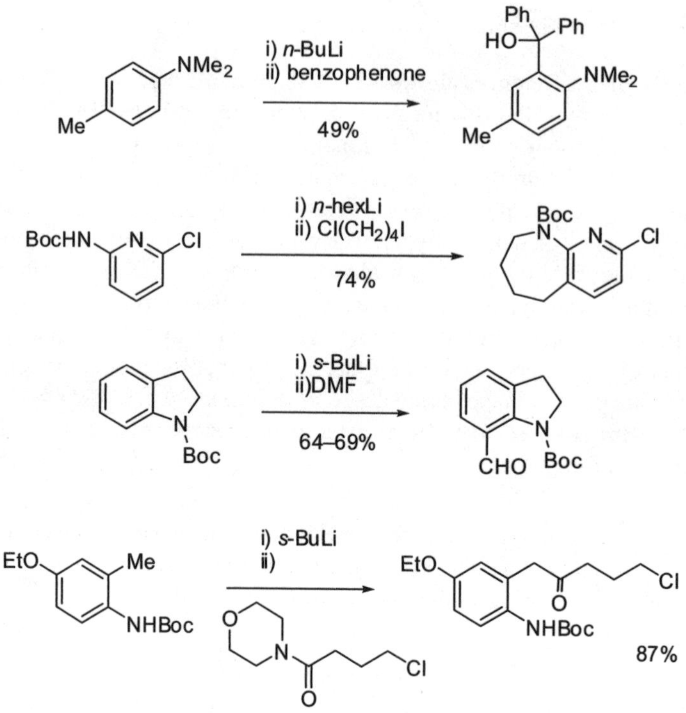

## 18.5   ENANTIOSELECTIVE DEPROTONATIONS

Chiral bases and bases complexed to chiral ligands can be used in enantioselective deprotonation reactions. This technology has been used to abstract pro-chiral protons or to abstract hydrogen at pro-chiral carbon centers in a desymmetrization process. A commonly used base is an organolithium reagent complexed with sparteine. Some examples of desymmetrization are shown below.[24] The third example demonstrates the power of using asymmetric amplification to improve a moderate ee to an excellent one.[25]

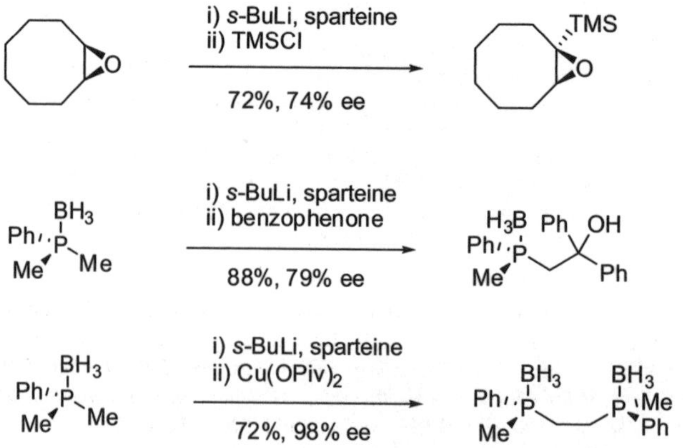

[24]Hodgson, D. M.; Gras, E. *Angew. Chem., Int. Ed. Engl.* **2002**, *41*, 2376–2378.
[25]Muci, A. R.; Campos, K. R.; Evans, D. A. *J. Am. Chem. Soc.* **1995**, *117*, 9075–9076.

Other bases commonly employed in asymmetric desymmetrizations are shown below with representative examples.[26–29]

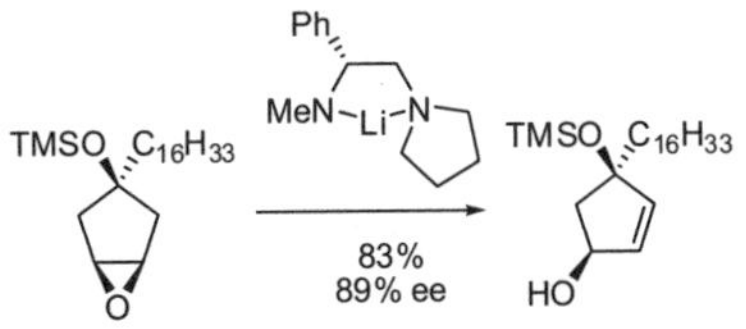
Ph
MeN
Li
N
TMSO
$C_{16}H_{33}$
O
83%
89% ee
TMSO
$C_{16}H_{33}$
HO

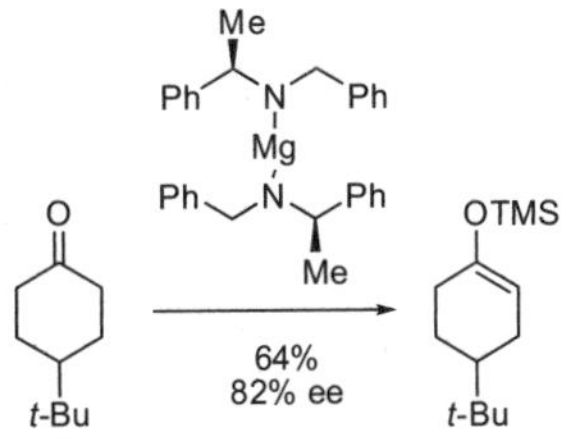
Me
Ph
N
Ph
Mg
Ph
N
Ph
Me
O
t-Bu
64%
82% ee
OTMS
t-Bu

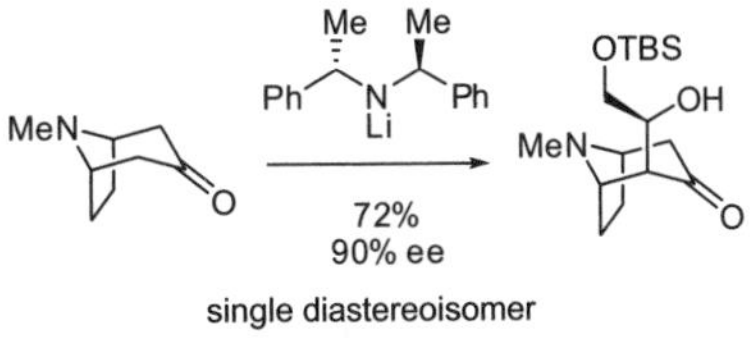
Me Me
Ph
N
Ph
Li
MeN
O
72%
90% ee
OTBS
OH
MeN
O
single diastereoisomer

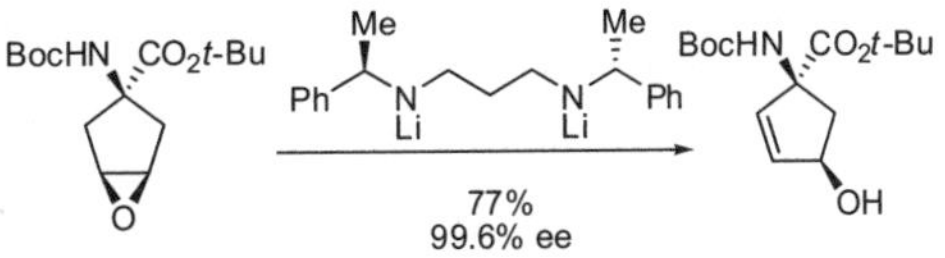
BocHN
$CO_2t$-Bu
O
Me
Me
Ph
N
N
Ph
Li
Li
77%
99.6% ee
BocHN
$CO_2t$-Bu
OH

Enantioselective abstraction of a pro-chiral proton can also be achieved in some cases, resulting in a chiral alkyl lithium intermediate. The proton α to the nitrogen of a carbamate is most commonly deprotonated in this type of reaction.[30–32]

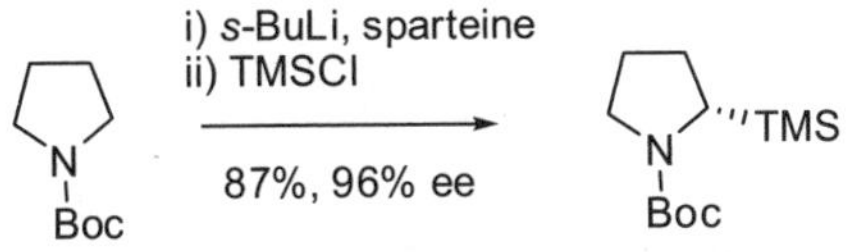
i) s-BuLi, sparteine
ii) TMSCl
87%, 96% ee
N
Boc
TMS
N
Boc

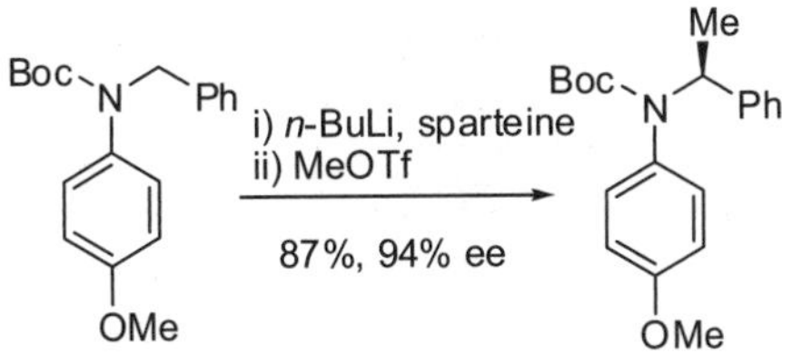
Boc
N
Ph
OMe
i) n-BuLi, sparteine
ii) MeOTf
87%, 94% ee
Me
Boc
N
Ph
OMe

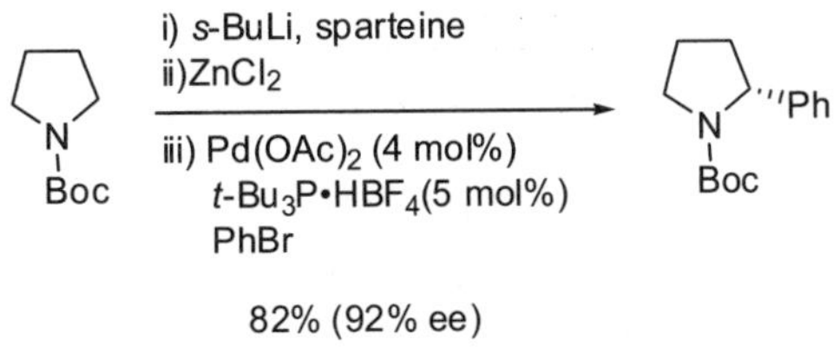
i) s-BuLi, sparteine
ii)$ZnCl_2$
iii) $Pd(OAc)_2$ (4 mol%)
t-$Bu_3P$•$HBF_4$(5 mol%)
PhBr
82% (92% ee)
N
Boc
Ph
N
Boc

[26]Varie, D. L.; Beck, C.; Borders, S. K.; Brady, M. D.; Cronin, J. S.; Ditsworth, T. K.; Hay, D. A.; Hoard, D. W.; Hoying, R. C.; Linder, R. J.; Miller, R. D.; Moher, E. D.; Remacle, J. R.; Rieck, J. A., III; Anderson, D. D.; Dodson, P. N.; Forst, M. B.; Pierson, D. A.; Turpin, J. A. *Org. Process Res. Dev.* **2007**, *11*, 546–559.

[27]Saravanan, P.; Bisai, A.; Baktharaman, S.; Chandrasekhar, M.; Singh, V. K. *Tetrahedron* **2002**, *58*, 4693–4706.

[28]Henderson, K. W.; Kerr, W. J.; Moir, J. H. *Tetrahedron* **2002**, *58*, 4573–4587.

[29]Lee, J. C.; Lee, K.; Cha, J. K. *J. Org. Chem.* **2000**, *65*, 4773–4775.

[30]Beak, P.; Kerrick, S. T.; Wu, S.; Chu, J. *J. Am. Chem. Soc.* **1994**, *116*, 3231–3239.

[31]Faibish, N. C.; Park, Y. S.; Lee, S.; Beak, P. *J. Am. Chem. Soc.* **1997**, *119*, 11561–11570.

[32]Campos, K. R.; Klapars, A.; Waldman, J. H.; Dormer, P. G.; Chen, C. *J. Am. Chem. Soc.* **2006**, *128*, 3538–3539.

# 19

# GENERAL SOLVENT PROPERTIES

Stéphane Caron

## 19.1  INTRODUCTION

Solvent selection is one of the most important aspects of a chemical reaction.[1] The solvent affects the solubility of starting materials, reagents and products. It will impact the isolation of the product either through precipitation/crystallization, extraction, or adsorption. More importantly, the choice of solvent will often impact how a reaction proceeds through stabilization of either the ground state or transition state of reagents and intermediates. Solvent properties can influence the rate of most reactions, especially non-unimolecular transformations, which are affected by the reaction concentration. Another important aspect of the solvent is that it provides a heat sink for exothermic reactions, which allows for additional process safety.[2]

Tables in this chapter provide useful information about common solvents. Tables 19.2 and 19.3, list solvents in alphabetical order along with a list of general properties. For a more comprehensive list of solvents and solvent properties, the book entitled *Organic Solvents; Physical Properties and Methods of Purification* is an excellent source of information.[3] Many azeotropes can be used to facilitate the displacement of a solvent from a mixture or to achieve drying by removal of water. For a more extensive list of azeotropes, the two volume set *Azeotropic Data* is highly recommended.[4]

[1]Lathbury, D. *Org. Process Res. Dev.* **2007**, *11*, 104.
[2]Laird, T. *Org. Process Res. Dev.* **2001**, *5*, 543.
[3]Riddick, J. A.; Bunger, W. B. *Techniques of Chemistry, Vol. 2: Organic Solvents: Physical Properties and Methods of Purification. 3rd ed*; Interscience: New York, N. Y., 1971.
[4]Gmehling, J. *Azeotropic Data*; VCH: Weinheim, Germany, 1994.

*Practical Synthetic Organic Chemistry: Reactions, Principles, and Techniques*, First Edition.
Edited by Stéphane Caron.
© 2011 John Wiley & Sons, Inc. Published 2011 by John Wiley & Sons, Inc.

**TABLE 19.1 Important Factors in Solvent Choice**

| Factor | Impact |
|---|---|
| Worker exposure | Known or potential toxicological impact. |
| Safety margin | Will the solvent serve as a heat sink or boiling point barrier? |
| Ease of handling | Does the solvent dissipate static charge effectively, form peroxides, have a high viscosity or low melting point, etc |
| Environmental impact | Will the solvent persist in the environment or have a negative effect in the atmosphere (i.e., ozone depletion), soil, or aquatic environment? |

Table 19.4 lists solvents in order of preference from a manufacturing point of view. Consideration of the choice of solvent for worker exposure, safety margins in reaction processing, ease of handling, and potential for environmental impact is of importance (see Table 19.1).[5] For example, *n*-hexane is very prone to generating a static charge upon stirring, to a much larger extent than *n*-heptane. Therefore, *n*-heptane is preferable from a process perspective.

The classification based on regulatory restrictions is provided in Table 19.4. When selecting a solvent for a given reaction, the solvent listed first in this table should be preferred based on safety and environmental impact.

## 19.2 DEFINITIONS AND ACRONYMS

**Azeotrope**     A liquid mixture of two or more components that retains the same composition in the vapor state as in the liquid state when distilled at a given pressure.

**Boiling point (bp)**     Temperature at which a liquid becomes a gas at standard atmospheric pressure.

**CMR**     Carcinogenic/Mutagenic/Reprotox hazard potential. CMR category 1 is considered a substance with sufficient evidence to establish causal relationship to be carcinogenic, mutagenic or reprotoxic hazard. Category 2 is considered a substance with sufficient evidence to establish a strong presumption of carcinogenic, mutagenic or reprotoxic properties. Category 3 is considered a substance that is under suspicion of having carcinogenic, mutagenic, or reprotoxic properties.

**Density (d)**     Concentration of a substance, measured by the mass per unit volume.

**Dielectric constant ($\varepsilon$)**     A measure of the effect of a medium on the potential energy of interaction between two charges. It is a measure of the relative effect a solvent has on the force with which two oppositely charged plates attract each other.

[5]Capello, C.; Fischer, U.; Hungerbuehler, K. *Green Chem.* **2007**, *9*, 927–934.

| | |
|---|---|
| **Flash point (fp)** | The lowest temperature at which a liquid or solid gives off enough vapor to form a flammable air–vapor mixture near its surface. |
| **ICH** | International Council on Harmonization is an alliance developed through collaboration between the U.S. Food and Drug Administration (FDA) and regulatory agencies in Japan and the European Union, to "harmonize" regulatory requirements to produce marketing applications acceptable in those jurisdictions. This alliance was formed to ensure that good-quality, safe, and effective medicines are developed and registered in the most efficient and cost-effective ways. These activities are pursued to prevent unnecessary duplication of clinical trials and to minimize the use of animal testing without compromising the regulatory obligations of safety and effectiveness. |
| **Heat Capacity ($C_p$)** | The specific heat capacity of a substance is the amount of heat required to raise the temperature of 1 gram of the substance 1°C at a constant pressure without change of phase. |
| **Heat of Vaporization ($\Delta H_v$)** | The amount of heat required to vaporize a defined quantity of a liquid. |
| **Melting point (mp)** | The temperature at which a solid becomes a liquid at standard atmospheric pressure. |
| **Molecular Weight (MW)** | The sum of the atomic weights of all the atoms in a molecule. |
| **Not. Est.** | Not established. |
| **TLV** | Threshold Limit Value is an exposure standard set by a committee of the American Conference of Governmental Industrial Hygienists (ACGIH). The value defines the level of exposure and is not intended as a line between safe and unsafe exposure. The objective is to minimize workers' exposure to hazardous concentrations as much as possible. The occupational exposure limit often considers the TLV for a time-weighed average (TWA) of 8 hours per day. |

## 19.3   SOLVENT PROPERTIES TABLE 19.2 PROVIDES IMPORTANT PROPERTIES OF COMMON SOLVENTS

**TABLE 19.2  Solvent Properties**

| Solvent | Acronym/ Shorthand | CAS # | Formula | MW | mp °C | bp °C | d g/ml | fp °C | ε | ΔH$_v$ kJ/mol | C$_P$ J/mol-K (25°C) | Solubility in H$_2$O (g/100 ml) | Azeotropes (at atm. Pressure) (Temp, % Solvent in 1$^{st}$ Column) |
|---|---|---|---|---|---|---|---|---|---|---|---|---|---|
| acetic acid | AcOH | 64-19-7 | C$_2$H$_4$O$_2$ | 60.05 | 17 | 118 | 1.049 | 39 | 6.15 | 23.36 | 122.3 | miscible | cyclohexane (79°C, 22%) toluene (105°C, 44%) heptane (92°C, 45%) |
| acetone | | 67-64-1 | C$_3$H$_6$O | 58.08 | −95 | 56 | 0.791 | −19 | 20.7 | 31.3 | 124.9 | miscible | MTBE (51°C, 44%) |
| acetonitrile | ACN MeCN | 75-05-8 | C$_2$H$_3$N | 41.05 | −45 | 82 | 0.786 | 5 | 36.6 | 32.94 | 91.46 | miscible | EtOH (73°C, 48%) MeOH (63°C, 19%) i-PrOH (75°C, 32%) THF (66°C, 8%) heptane (69°C, 62%) H$_2$O (76°C, 23%) |
| t-amyl alcohol [2-methyl-2-butanol] | | 75-85-4 | C$_5$H$_{12}$O | 88.15 | −11.9 | 102 | 0.805 | 3 | 5.82 | 50.2 | 244.3 | 12.36 at 25°C | cyclohexane (79°C, 11%) i-Pr$_2$O (89°C, 19%) H$_2$O (87°C, 38%) |
| benzene | PhH | 71-43-2 | C$_6$H$_6$ | 78.11 | 5.5 | 80 | 0.874 | −11 | 2.3 | 33.84 | 135.76 | 0.18 at 25°C | cyclohexane (78°C, 54%) H$_2$O (69°C, 70%) |
| 1-butanol | n-BuOH | 71-36-3 | C$_4$H$_{10}$O | 74.12 | −90 | 116 | 0.811 | 35 | 17.8 | 52.34 | 177.08 | 7.7 at 20°C | cyclohexane (79°C, 10%) toluene (106°C, 12%) heptane (94°C, 20%) H$_2$O (93°C, 24%) |
| t-butanol [2-methyl-2-propanol] | t-BuOH | 75-65-0 | C$_4$H$_{10}$O | 74.12 | 28 | 83 | 0.775 | 11 | 10.9 | 46.82 | 220.33 | miscible | cyclohexane (71°C, 27%) H$_2$O (80°C, 62%) |
| chloroform | | 67-66-3 | CHCl$_3$ | 119.38 | −63 | 61 | 1.492 | none | 4.81 | 33.35 | 116.90 | 0.8 at 20°C | MeOH (53°C, 65%) EtOH (53°C, 84%) H$_2$O (56°C, 85%) |
| cyclohexane | | 110-82-7 | C$_6$H$_{12}$ | 84.16 | 7 | 81 | 0.779 | −18 | 2.02 | 32.89 | 156.01 | insoluble | H$_2$O (69°C, 30%) EtOAc (70°C, 49%) i-PrOH (75°C, 49%) |
| 1,2-dichloroethane | DCE | 107-06-2 | C$_2$H$_4$Cl$_2$ | 98.96 | −35 | 83 | 1.256 | 15 | 16.7 | 35.15 | 128.99 | 0.87 | cyclohexane (74°C, 46%) H$_2$O (71°C, 68%) |

| | | | | | | | | | | | | | |
|---|---|---|---|---|---|---|---|---|---|---|---|---|---|
| dichloromethane | DCM | 75-09-2 | $CH_2Cl_2$ | 84.93 | −97 | 40 | 1.320 | none | 9.08 | 28.56 | 100.88 | 1.3 at 20°C | MeOH (38°C, 86%), EtOH (39°C, 98%), H$_2$O (38°C, 93%) |
| diethyl ether | Et$_2$O | 60-29-7 | $C_4H_{10}O$ | 74.12 | −116 | 35 | 0.706 | −40 | 4.34 | 27.2 | 172.5 | 6.9 at 20°C | H$_2$O (34°C, 95%) |
| diisopropyl ether | IPE, i-Pr$_2$O | 108-20-3 | $C_6H_{14}O$ | 102.17 | −87 | 68 | 0.725 | −12 | 3.88 | 32.0 | 216.1 | low | H$_2$O (63°C, 78%) |
| dimethoxyethane [ethylene elycol dimethyl ether] | DME | 110-71-4 | $C_4H_{10}O_2$ | 90.12 | −58 | 85 | 0.867 | 0 | 7.2 | 36.39 | 193.3 | miscible | H$_2$O (77°C, 90%) |
| dimethylacetamide | DMAC DMAc DMAA | 127-19-5 | $C_4H_9NO$ | 87.12 | −20 | 165 | 0.937 | 77.2 | 47.8 | 49.15 | 176 | miscible | |
| dimethylformamide | DMF | 68-12-2 | $C_3H_7NO$ | 73.09 | −61 | 152 | 0.944 | 58 | 38.3 | 47.51 | 148.36 | miscible | heptane (97°C, 8%) |
| dimethyl sulfoxide | DMSO | 67-68-5 | $C_2H_6OS$ | 78.13 | 6 | 189 | 1.101 | 95 | 47.2 | 52.88 | 153.18 | miscible | |
| 1,4-dioxane | | 123-91-1 | $C_4H_8O_2$ | 88.11 | 11.8 | 100 | 1.034 | 12 | 2.21 | 35.59 | 150.65 | miscible | heptane (92°C, 47%), EtOH (77°C, 28%), H$_2$O (88°C, 48%) |
| ethanol | EtOH | 64-17-5 | $C_2H_6O$ | 46.07 | −144 | 78 | 0.790 | 12 | 24.3 | 42.309 | 112.34 | miscible | THF (66°C, 9%), EtOAc (71°C, 46%), cyclohexane (65°C, 44%), H$_2$O (78°C, 90%) |
| ethyl acetate | EtOAc | 141-78-6 | $C_4H_8O_2$ | 88.11 | −83 | 77 | 0.902 | −4 | 6.02 | 35.62 | 169.0 | 9.0 | cyclohexane (72°C, 55%), hexane (66°C, 34%), H$_2$O (71°C, 30%) |
| ethylene glycol | | 104-21-1 | $C_2H_6O_2$ | 62.07 | −13 | 196 | 1.113 | 110 | 37.0 | 67.8 | 149.4 | miscible | toluene (110°C, 9%) |
| formic acid | | 64-18-6 | $CH_2O_2$ | 46.03 | 8.2 | 101 | 1.220 | 65 | 58 | 20.10 | 99.04 | miscible | 1,2-dichloroethane (77°C, 43%), heptane (78°C, 63%) |
| heptane | | 142-82-5 | $C_7H_{16}$ | 100.20 | −91 | 98 | 0.684 | −3 | 1.9 | 36.55 | 224.98 | insoluble | H$_2$O (79°C, 55%) |
| hexane | | 110-54-3 | $C_6H_{14}$ | 86.18 | −95 | 69 | 0.659 | −21 | 2.02 | 31.552 | 195.48 | insoluble | H$_2$O (62°C, 78%) |
| isoamyl alcohol [3-methyl-1-butanol] | | 123-51-3 | $C_5H_{12}O$ | 88.15 | −117 | 131 | 0.809 | 43 | 14.7 | 55.6 | 211.3 | 3 at 30°C | toluene (110°C, 12%), heptane (98°C, 8%), H$_2$O (95°C, 17%) |
| isopropyl acetate | i-PrOAc | 108-21-4 | $C_5H_{10}O_2$ | 102.13 | −73 | 89 | 0.874 | 2 | NA | 37.20 | 222.6 | 4.3 at 27°C | hexane (68°C, 8%), H$_2$O (76°C, 59%) |

(continued)

## TABLE 19.2 (Continued)

| Solvent | Acronym/ Shorthand | CAS # | Formula | MW | mp °C | bp °C | d g/ml | fp °C | ε | ΔH$_v$ kJ/mol | C$_P$ J/mol-K (25°C) | Solubility in H$_2$O (g/100 ml) | Azeotropes (at atm. Pressure) (Temp, % Solvent in 1st Column) |
|---|---|---|---|---|---|---|---|---|---|---|---|---|---|
| isopropyl alcohol [2-propanol] | i-PrOH, IPA, IPO, i-PrOH | 67-63-0 | C$_3$H$_8$O | 60.10 | −86 | 83 | 0.785 | 12 | 18.3 | 45.52 | 154.60 | miscible | EtOAc (75°C, 34%) cyclohexane (69°C, 41%) hexane (62°C, 28%) heptane (76°C, 60%) H$_2$O (80°C, 69%) |
| methanol | MeOH | 67-56-1 | CH$_4$O | 32.04 | −98 | 65 | 0.791 | 11 | 33 | 37.43 | 81.47 | miscible | acetone (56°C, 20%) THF (59°C, 51%) MTBE (51°C, 35%) cyclohexane (54°C, 60%) hexane (50°C, 49%) heptane (59°C, 73%) |
| methyl-t-butyl ether | MTBE | 1634-04-4 | C$_5$H$_{12}$O | 88.15 | −4 | 55.2 | 0.740 | −28 | | | | 4.2 at 20°C | H$_2$O (52°C, 83%) |
| methyl ethyl ketone [2-butanone] | MEK | 78-93-3 | C$_4$H$_8$O | 72.11 | −86 | 80 | 0.805 | −1 | 18.5 | 34.51 | 158.91 | 29 at 20°C | EtOAc (76°C, 17%) cyclohexane (71°C, 50%) hexane (64°C, 33%) H$_2$O (673°C, 66%) |
| 2-methyl tetrahydrofuran[6] | MeTHF | 96-47-9 | C$_5$H$_{10}$O | 86.13 | | 78 | 0.860 | −11 | 6.97 | 32.34 | | 16 at 25°C | H$_2$O (84°C, 76%) |
| N-methyl pyrrolidinone | NMP | 872-50-4 | C$_5$H$_9$NO | 99.13 | −24 | 202 | 1.027 | 86 | 32.2 | 52.80 | 166.1 | miscible | toluene (110°C, 22%) heptane (95°C, 30%) H$_2$O (94°C, 25%) |
| nitromethane | MeNO$_2$ | 75-52-5 | C$_3$NO$_2$ | 82.04 | −29 | 101 | 1.127 | 35 | 35.9 | 38.27 | 105.77 | 12 at 25°C | |
| pyridine | Py, pyr | 110-86-1 | C$_5$H$_5$N | 79.10 | −42 | 115 | 0.978 | 20 | 12.3 | 40.41 | 135.6 | miscible | |
| sulfolane | | 126-33-0 | C$_4$H$_8$O$_2$S | 120.17 | 28 | 285 | 1.261 | 165 | 43.3 | 62.8 | 180 | miscible | |
| tetrahydrofuran | THF | 109-99-9 | C$_4$H$_8$O | 72.11 | −108 | 64 | 0.889 | −17 | 7.52 | 32.0 | 123.9 | miscible | H$_2$O (63°C, 82%) |
| toluene | PhMe | 108-88-3 | C$_7$H$_8$ | 92.14 | −95 | 111 | 0.865 | 6 | 2.44 | 37.99 | 157.29 | insoluble | H$_2$O (85°C, 46%) |
| water | | 7732-18-5 | H$_2$O | 18.02 | 0 | 100 | 1.000 | none | 80 | 45.04 | 75.98 | miscible | |
| xylenes (60/14/9/17 mixture of m/p/o/Ethyl benzene) | | 1330-20-7 | C$_8$H$_{10}$ | 106.16 | | 137 | 0.860 | 29 | 2.37 | ~42.6 | ~183.4 | insoluble | H$_2$O (95°C, 80%) |

[6]Aycock, D. F. Org. Process Res. Dev. 2007, 11, 156–159.

## 19.4 MUTUAL SOLUBILITY OF WATER AND ORGANIC SOLVENTS TABLE 19.3 DESCRIBES THE SOLUBILITY PROPERTIES OF COMMONLY USED SOLVENTS WITH WATER

**TABLE 19.3  Mutual Solubility of Water and Organic Solvents**

| Solvent | CAS # | $H_2O$ Solubility %w/w | Solubility in $H_2O$ (g/100 ml) |
|---|---|---|---|
| acetic acid | 64-19-7 | miscible | miscible |
| acetone | 67-64-1 | miscible | miscible |
| acetonitrile | 75-05-8 | miscible | miscible |
| t-amyl alcohol [2-methyl-2-butanol] | 75-85-4 | 23.47% | 12.36 at 25°C |
| benzene | 71-43-2 | 0.06 | 0.18 at 25°C |
| 1-butanol | 71-36-3 | 20.5 | 7.7 at 20°C |
| t-butanol [2-methyl-2-propanol] | 75-65-0 | miscible | miscible |
| chloroform | 67-66-3 | 0.09 | 0.8 at 20°C |
| cyclohexane | 110-82-7 | 0.01 | insoluble |
| 1,2-dichloroethane | 107-06-2 | 0.81 | 0.87 |
| dichloromethane | 75-09-2 | 1.30 | 1.3 at 20°C |
| diethyl ether | 60-29-7 | 6.05 | 6.9 at 20°C |
| diisopropyl ether | 108-20-3 | 1.2 | low |
| dimethoxyethane [ethylene elycol dimethyl ether] | 110-71-4 | miscible | miscible |
| dimethylacetamide | 127-19-5 | miscible | miscible |
| dimethylformamide | 68-12-2 | miscible | miscible |
| dimethyl sulfoxide | 67-68-5 | miscible | miscible |
| 1,4-dioxane | 123-91-1 | miscible | miscible |
| ethanol | 64-17-5 | miscible | miscible |
| ethyl acetate | 141-78-6 | 8.08 | 9.0 |
| ethylene glycol | 104-21-1 | miscible | miscible |
| formic acid | 64-18-6 | miscible | miscible |
| heptane | 142-82-5 | <0.01 | insoluble |
| hexane | 110-54-3 | <0.01 | insoluble |
| isoamyl alcohol [3-methyl-1-butanol] | 123-51-3 | 2.67 | 3 at 30°C |
| isopropyl acetate | 108-21-4 | 2.9 | 4.3 at 27°C |
| isoropyl alcohol [2-propanol] | 67-63-0 | miscible | miscible |
| methanol | 67-56-1 | miscible | miscible |
| methyl-t-butyl ether | 1634-04-4 | 1.5 | 4.2 at 20°C |
| methyl ethyl ketone [2-butanone] | 78-93-3 | 10 | 29 at 20°C |
| 2-methyl tetrahydrofuran | 96-47-9 | 6.08 | 16 at 25°C |
| N-methyl pyrrolidinone | 872-50-4 | miscible | miscible |
| nitromethane | 75-52-5 | 2.09 | 12 at 25°C |
| pyridine | 110-86-1 | miscible | miscible |
| sulfolane | 126-33-0 | miscible | miscible |
| tetrahydrofuran | 109-99-9 | miscible | miscible |
| toluene | 108-88-3 | 0.03 | insoluble |
| xylenes | 1330-20-7 | 0.02 | 0.04 |

## 19.5  OTHER USEFUL INFORMATION ON SOLVENTS

Ethanol is available in several grades:[7]

[7]Budavari, S.; O'Neil, M.; Smith, A.; Heckelman, P.; Kinneary, J.;Editors *The Merck Index*, *12th ed*; Chapman & Hall: London, UK, 1996.

Absolute: 200 proof, >99.5% purity.

Alcohol: 190 proof, 95.0% containing ~5% water (which is the azeotrope).

For the following, a denaturant is added to a grade of ethanol. For example, 2B ethanol is 190 proof ethanol to which was added 0.5% of benzene by volume.

Denatured with 1–10% MeOH. Also known as industrial methylated spirits (IMS).

Denatured with 0.5% benzene in 95% EtOH is known as 2B ethanol.

Denatured with 5% MeOH in 95% EtOH is known as 3A ethanol.

Denatured with 0.5% pyridine in 95% EtOH is known as 6B ethanol.

Denatured with 5% benzene in 95% EtOH is known as 12A ethanol.

Denatured with 10% $Et_2O$ in 95% EtOH is known as 13A ethanol.

Denatured with 4% methyl isobutyl ketone + 1% kerosene in 95% EtOH is known as 19 ethanol.

Denatured with 5% $CHCl_3$ in 95% EtOH is known as 20 ethanol.

Denatured with 10% acetone in 95% EtOH is known as 23A ethanol.

Denatured with 10% benzene in 95% EtOH is known as 28 ethanol.

Denatured with 1% gasoline in 95% EtOH is known as 28A ethanol.

Denatured with 10% MeOH in 95% EtOH is known as 30 ethanol.

Denatured with 5% $Et_2O$ in 95% EtOH is known as 32 ethanol.

Denatured with 5% EtOAc in 95% EtOH is known as 35A ethanol.

Denatured with 20% n-BuOH in 95% EtOH is known as 44 ethanol.

Glyme is 1,2-dimethoxyethane (also known as monoglyme).

Diglyme is bis(2-methoxyethyl) ether.

Ligroin is a refined saturated hydrocarbon fraction usually composed of $C_7$–$C_{11}$ compounds with a boiling range of about 60–80°C.

Mineral spirits is also known as petroleum spirits. Like petroleum ether and ligroin, it is a hydrocarbon mixture obtained from petroleum refineries. This is a higher boiling fraction with a boiling range usually starting above 150°C.

Petroleum ether is a hydrocarbon mixture obtained from petroleum refineries. The mixture as a boiling range of ~30–60°C. It is also sometimes referred to as benzine.

Several solvents contain stabilizers for added safety. Most ethers contain 2,6-di-*tert*-butyl-4-methylphenol (BHT) as a radical scavenger to minimize the risk of generation of peroxides (THF usually contains 250 ppm). Dichloromethane is often stabilized with an olefin such as amylene (50–150 ppm). Chloroform is usually stabilized with ethanol (0.5–1%).

Xylenes is a very common industrial solvent. It is composed of 60% of the *meta* isomer, 14% of the *para* isomer, 9% of the *ortho* isomer, and 17% of ethylbenzene.

## 19.6   SOLVENT SAFETY

Table 19.4 lists solvents in order of preference for manufacture. It does not include solvents that are known ozone depleters (such as CFCs), solvents that are often reagents (such as acetic anhydride, orthoformates, etc.), or mineral acids.

**TABLE 19.4  Solvents in Order of Preference for Manufacture**

| Solvent | Formula | CMR Category | Skin Absorption (1) or Sensitizing Potential (2) | TLV ppm | Potential for Static Charge Buildup | Potential Reactivity | EU Classification | ICH Q3C Classification | [ ] Limit in API ppm |
|---|---|---|---|---|---|---|---|---|---|
| water | $H_2O$ | | | | | Active $H^+$ | | | |
| isoamyl alcohol | $C_5H_{12}O$ | | 1 | 100 | | Active $H^+$ | R10-20-37-66 | 3 | >5000 |
| 3-methyl-1-butanol | | | | | | | | | |
| t-butanol | $C_4H_{10}O$ | | | 100 | | Reacts with acids, active $H^+$ | R11-20 | | |
| 2-methyl-2-propanol | | | | | | | | | |
| isopropyl alcohol | $C_3H_8O$ | | 1 | 200 | | Active $H^+$ | R11-36-67 | 3 | >5000 |
| 2-propanol | | | | | | | | | |
| acetone | $C_3H_6O$ | | 1 | 500 | | Enolizable ketone | R11-36-66-67 | 3 | >5000 |
| 1-butanol | $C_4H_{10}O$ | | 1 | 20 | | Active $H^+$ | R10-22-37/38-41-67 | 3 | >5000 |
| ethanol | $C_2H_6O$ | | 1 | 1000 | | Active $H^+$ | R11 | 3 | >5000 |
| ethyl acetate | $C_4H_8O_2$ | | | 400 | | Enolizable ester | R11-36-66-67 | 3 | >5000 |
| heptane | $C_7H_{16}$ | | 1 | 400 | X | | R11-38-50/53-65-67 | 3 | >5000 |
| methanol | $CH_4O$ | | 1 | 200 | | Active $H^+$ | R11-23/24/25-39/23/24/25 | 2 | 3000 |
| methyl ethyl ketone | $C_4H_8O$ | | 1 | 200 | | Enolizale ketone | R11-36-66-67 | 3 | >5000 |
| 2-butanone | | | | | | | | | |
| N-methyl pyrrolidinone | $C_5H_9NO$ | | 1 | not est. | | Enolizable amide | R36/38 | 2 | 530 |
| 2-methyl tetrahydrofuran | $C_5H_{10}O$ | | 1 | | X | Can form peroxides | | | not est. |
| isopropyl acetate | $C_5H_{10}O_2$ | | 1 | 100 | | Enolizable ester | R11-36-66-67 | 3 | >5000 |
| tetrahydrofuran | $C_4H_8O$ | | 1 | 50 | X | Can form peroxides | R11-19-36/37 | 2 | 720 |
| acetic acid | $C_2H_4O_2$ | | 1 | 10 | | Active $H^+$ | R10-35  Harmful to aquatic organisms | 3 | >5000 |
| acetonitrile | $C_2H_3N$ | | 1 | 20 | | Enolizable nitrile | R11-20/21/22-36 | 2 | 410 |
| dimethyl sulfoxide | $C_2H_6OS$ | | 1 & 2 | not est. | | Enolizable sulfoxide | | 3 | >5000 |
| cyclohexane | $C_6H_{12}$ | | 1 | 100 | X | | R11-38-65-67-50/53 | 2 | 3880 |
| ethylene glycol | $C_2H_6O_2$ | | 1 & 2 | 100 | | Active $H^+$ | R22 | 2 | 620 |
| methyl-t-butyl ether | $C_4H_{12}O$ | | 1 | 50 | X | | R11-38  Persists in environment | 3 | not est. |

(continued)

**TABLE 19.4** (*Continued*)

| Solvent | Formula | CMR Category | Skin Absorption (1) or Sensitizing Potential (2) | TLV ppm | Potential for Static Charge Buildup | Potential Reactivity | EU Classification | ICH Q3C Classification | [ ] Limit in API ppm |
|---|---|---|---|---|---|---|---|---|---|
| toluene | $C_7H_8$ | | 1 | 50 | | Reactive arene | R11-38-48/20-63-65-67 | 2 | 890 |
| xylenes (mixt. of isomers) | $C_8H_{10}$ | | 1 & 2 | 100 | X | Reactive arene | R10-20/21-38 | 2 | 2170 |
| chloroform | $CHCl_3$ | 3 | 1 | 10 | | Reacts with base | R22-38-40-48/20/22 | 2 | 60 |
| 1,2-dichloroethane | $C_2H_4Cl_2$ | | 1 | 10 | | Alkylating agent | R45-11-22-36/37/38 | 1 | 5 |
| dichloromethane | $CH_2Cl_2$ | 3 | 1 | 50 | | Reacts with base, alkylating agent | R40<br>Potential ground water contaminant | 2 | 600 |
| diisopropyl ether | $C_6H_{14}O$ | | | 250 | X | Can form peroxides | R11-19-66-67 | | not est. |
| dimethoxyethane ethylene glycol dimethyl ether | $C_4H_{10}O_2$ | 2 | 1 | not est. | | Can form peroxides | R60-61-11-19-20 | 2 | 100 |
| dimethylacetamide | $C_4H_9NO$ | 2 | 1 | 10 | | Enolizable amide | R61-20/21 | 2 | 1090 |
| dimethylformamide | $C_3H_7NO$ | 2 | 1 | 10 | | Reacts with strong base | R61-20/21-36 | 2 | 380 |
| hexanes | $C_6H_{14}$ | 3 | 1 | 50 | X | | R11-38-48/20-62-65-67-51/53 | 2 | 290 |
| pyridine | $C_5H_5N$ | 3 | 2 | 5 | | Reacts with acids and electrophiles | R11-20/21/22 | 2 | 200 |
| sulfolane | $C_4H_8O_2S$ | | 1 | not est. | | Enolizable sulfone | R22-36-60 | 2 | 160 |
| benzene | $C_6H_6$ | 1 | 2 | 0.5 | | Reactive arene | R45-46-11-36/38-48/23/24/25-65 | 1 | 2 |
| 1,4-dioxane | $C_4H_8O_2$ | 1 | 1 & 2 | 20 | X | Can form peroxides | R11-19-36/37-40-66 | 2 | 380 |
| diethyl ether | $C_4H_{10}O$ | | 1 | 400 | X | Can form peroxides | R12-19-22-66-67 | 3 | |
| nitromethane | $CH_3NO_2$ | | 1 | 20 | | Shock sensitive. Reactive $h^+$ | R5-10-22 | 2 | |

## 19.7  RISK PHRASES USED IN THE COUNTRIES OF THE EUROPEAN UNION

**TABLE 19.5  Nature of Special Risks Attributed to Dangerous Substances and Preparations[8]**

| Category | Nature of Risk |
|---|---|
| R1 | Explosive when dry |
| R2 | Risk of explosion by shock, friction, fire, or other sources of ignition. |
| R3 | Extreme risk of explosion by shock, friction, fire, or other sources of ignition. |
| R4 | Forms very sensitive explosive metallic compounds. |
| R5 | Heating may cause an explosion. |
| R6 | Explosive with or without contact with air. |
| R7 | May cause fire. |
| R8 | Contact with combustible material may cause fire. |
| R9 | Explosive when mixed with combustible material. |
| R10 | Flammable. |
| R11 | Highly flammable. |
| R12 | Extremely flammable. |
| R14 | Reacts violently with water. |
| R15 | Contact with water liberates highly flammable gases. |
| R16 | Explosive when mixed with oxidizing substances. |
| R17 | Spontaneously flammable in air. |
| R18 | In use, may form flammable/explosive vapour–air mixture. |
| R19 | May form explosive peroxides. |
| R20 | Harmful by inhalation. |
| R21 | Harmful in contact with skin. |
| R22 | Harmful if swallowed. |
| R23 | Toxic by inhalation. |
| R24 | Toxic in contact with skin. |
| R25 | Toxic if swallowed. |
| R26 | Very toxic by inhalation. |
| R27 | Very toxic in contact with skin. |
| R28 | Very toxic if swallowed. |
| R29 | Contact with water liberates toxic gases. |
| R30 | Can become highly flammable in use. |
| R31 | Contact with acids liberates toxic gas. |
| R32 | Contact with acids liberates very toxic gas. |
| R33 | Danger of cumulative effects. |
| R34 | Causes burns |
| R35 | Causes severe burns. |
| R36 | Irritating to eyes. |
| R37 | Irritating to respiratory system. |
| R38 | Irritating to skin. |
| R39 | Danger of very serious irreversible effects. |
| R40 | Possible risks of irreversible effects. |
| R41 | Risk of serious damage to eyes. |

*(continued)*

[8]European Communities (EC) (1996). Council Directive 96/56/EC, 8th. Amendment of Directive 67/548/EEC. EU OJ L236, 18.09.1996.

**TABLE 19.5**   (*Continued*)

| Category | Nature of Risk |
| --- | --- |
| R42 | May cause sensitization by inhalation. |
| R43 | May cause sensitization by skin contact. |
| R44 | Risk of explosion if heated under confinement. |
| R45 | May cause cancer. |
| R46 | May cause heritable genetic damage. |
| R48 | Danger of serious damage to health by prolonged exposure. |
| R49 | May cause cancer by inhalation. |
| R50 | Very toxic to aquatic organisms. |
| R51 | Toxic to aquatic organisms. |
| R52 | Harmful to aquatic organisms. |
| R53 | May cause long-term adverse effects in the aquatic environment. |
| R54 | Toxic to flora. |
| R55 | Toxic to fauna. |
| R56 | Toxic to soil organisms. |
| R57 | Toxic to bees. |
| R58 | May cause long-term adverse effects in the environment. |
| R59 | Dangerous for the ozone layer. |
| R60 | May impair fertility. |
| R61 | May cause harm to the unborn child. |
| R62 | Possible risk of impaired fertility. |
| R63 | Possible risk of harm to the unborn child. |
| R64 | May cause harm to breastfed babies. |
| R14/15 | Reacts violently with water liberating highly flammable gases. |
| R15/29 | Contact with water liberates toxic, highly flammable gas. |
| R20/21 | Harmful by inhalation and in contact with skin. |
| R20/22 | Harmful by inhalation and if swallowed. |
| R20/21/22 | Harmful by inhalation, in contact with skin, and if swallowed. |
| R21/22 | Harmful in contact with skin and if swallowed. |
| R23/24 | Toxic by inhalation and in contact with skin. |
| R23/25 | Toxic by inhalation and if swallowed. |
| R23/24/25 | Toxic by inhalation, in contact with skin, and if swallowed. |
| R24/25 | Toxic in contact with skin and if swallowed. |
| R26/27: | Very toxic by inhalation and in contact with skin. |
| R26/28: | Very toxic by inhalation and if swallowed. |
| R26/27/28 | Very toxic by inhalation, in contact with skin, and if swallowed. |
| R27/28 | Very toxic in contact with skin and if swallowed. |
| R36/37 | Irritating to eyes and respiratory system. |
| R36/38 | Irritating to eyes and skin. |
| R36/37/38 | Irritating to eyes, respiratory system and skin. |
| R37/38 | Irritating to respiratory system and skin. |
| R39/23 | Toxic: danger of very serious irreversible effects through inhalation. |
| R39/24 | Toxic: danger of very serious irreversible effects in contact with skin. |
| R39/25 | Toxic: danger of very serious irreversible effects if swallowed. |
| R39/23/24 | Toxic: danger of very serious irreversible effects through inhalation and in contact with skin. |
| R39/23/25 | Toxic: danger of very serious irreversible effects through inhalation and if swallowed. |
| R39/24/25 | Toxic: danger of very serious irreversible effects in contact with skin and if swallowed. |

**TABLE 19.5**    (*Continued*)

| Category | Nature of Risk |
|---|---|
| R39/23/24/25 | Toxic: danger of very serious irreversible effects through inhalation, in contact with skin, and if swallowed. |
| R39/26 | Very toxic: danger of very serious irreversible effects through inhalation. |
| R39/27 | Very toxic: danger of very serious irreversible effects in contact with skin. |
| R39/28 | Very toxic: danger of very serious irreversible effects if swallowed. |
| R39/26/27 | Very toxic: danger of very serious irreversible effects through inhalation and in contact with skin. |
| R39/26/28 | Very toxic: danger of very serious irreversible effects through inhalation and if swallowed. |
| R39/27/28 | Very toxic: danger of very serious irreversible effects in contact with skin and if swallowed. |
| R39/26/27/28 | Very toxic: danger of very serious irreversible effects through inhalation, in contact with skin, and if swallowed. |
| R40/20 | Harmful: possible risk of irreversible effects through inhalation. |
| R40/21 | Harmful: possible risk of irreversible effects in contact with skin. |
| R40/22 | Harmful: possible risk of irreversible effects if swallowed. |
| R40/20/21 | Harmful: possible risk of irreversible effects through inhalation and in contact with skin. |
| R40/20/22 | Harmful: possible risk of irreversible effects through inhalation and if swallowed. |
| R40/21/22 | Harmful: possible risk of irreversible effects in contact with skin and if swallowed. |
| R40/20/21/22 | Harmful: possible risk of irreversible effects through inhalation, in contact with skin, and if swallowed. |
| R42/43 | May cause sensitization by inhalation and skin contact. |
| R48/20 | Harmful: danger of serious damage to health by prolonged exposure through inhalation. |
| R48/21 | Harmful: danger of serious damage to health by prolonged exposure in contact with skin. |
| R48/22 | Harmful: danger of serious damage to health by prolonged exposure if swallowed. |
| R48/20/21 | Harmful: danger of serious damage to health by prolonged exposure through inhalation and in contact with skin. |
| R48/20/22 | Harmful: danger of serious damage to health by prolonged exposure through inhalation and if swallowed. |
| R48/21/22 | Harmful: danger of serious damage to health by prolonged exposure in contact with skin and if swallowed. |
| R48/20/21/22 | Harmful: danger of serious damage to health by prolonged exposure in contact with skin and if swallowed. |
| R48/20/21/22 | Harmful: danger of serious damage to health by prolonged exposure through inhalation, in contact with skin, and if swallowed. |
| R48/23 | Toxic: danger of serious damage to health by prolonged exposure through inhalation. |
| R48/24 | Toxic: danger of serious damage to health by prolonged exposure in contact with skin. |
| R48/25 | Toxic: danger of serious damage to health by prolonged exposure if swallowed. |
| R48/23/24 | Toxic: danger of serious damage to health by prolonged exposure through inhalation and in contact with skin. |
| R48/23/25 | Toxic: danger of serious damage to health by prolonged exposure through inhalation and if swallowed. |

(*continued*)

**TABLE 19.5**   (*Continued*)

| Category | Nature of Risk |
| --- | --- |
| R48/24/25 | Toxic: danger of serious damage to health by prolonged exposure in contact with skin and if swallowed. |
| R48/23/24/25 | Toxic: danger of serious damage to health by prolonged exposure through inhalation, in contact with skin, and if swallowed. |
| R50/53 | Very toxic to aquatic organisms, may cause long-term adverse effects in the aquatic environment. |
| R51/53 | Toxic to aquatic organisms, may cause long-term adverse effects in the aquatic environment. |
| R52/53 | Harmful to aquatic organisms, may cause long-term adverse effects in the aquatic environment. |

# 20

# PRACTICAL CHEMISTRY CONCEPTS: TIPS FOR THE PRACTICING CHEMIST OR THINGS THEY DON'T TEACH YOU IN SCHOOL

SALLY GUT RUGGERI

## 20.1 INTRODUCTION

During the course of undergraduate study, chemists are taught the basics of carrying out organic reactions through lab courses and, sometimes, by carrying out undergraduate research. The fundamentals absorbed at this level generally include how to set up a simple reaction safely, how to work it up (usually by extraction) and how to isolate the product by crystallization, distillation, or chromatography. As a chemist progresses through graduate school, these techniques are refined and more technically challenging ones are introduced. Given the nature of academic research, most reactions are run on small to very small scale, and may require increasingly skilled techniques during the course of study. This approach is at the opposite end of the spectrum from what is considered practical in an industrial setting, especially in process groups. Here the intent is to design processes that are safe, robust, reproducible, and flexible enough to be carried out in any type of manufacturing equipment by operators not trained as chemists.[1] This chapter describes the collective learnings of the Pfizer process group on how to approach reaching this ideal, and common pitfalls or issues to be avoided.

---

[1] The term "bucket chemistry" has been used, often in a pejorative sense, to describe chemistry with these attributes; from a process chemist's perspective, there is no higher praise.

---

*Practical Synthetic Organic Chemistry: Reactions, Principles, and Techniques*, First Edition.
Edited by Stéphane Caron.
© 2011 John Wiley & Sons, Inc. Published 2011 by John Wiley & Sons, Inc.

## 20.2   REACTION EXECUTION

### 20.2.1   Heat Transfer

The success of a reaction that is influenced by temperature is dependent on the ability to control the temperature or heat flow. This is especially true for reactions where heat needs to be removed. The ability to control the temperature is relatively easy on the small scale normal to an academic setting, where ice baths are used for cooling and oil baths or heating mantles for heating. As a reaction is scaled up in glassware, control of the heat transfer becomes increasingly difficult. Assuming all spherical reactors, for every ten-fold increase in volume, the surface area/volume ratio is roughly halved.[2] This means that when scaling up a reaction 1000-fold (10 mL to 10 L), the surface area/volume ratio is approximately one tenth its original value, and equivalent means of temperature control will be much less efficient. Larger scale equipment is invariably jacketed, with internal or external circulating temperature control, which affords much more efficient heat transfer. The reactors are usually non-spherical as well, to increase the ability to transfer heat. Internal temperature probes should be used to monitor reaction temperature; external readings can often be $>10°C$ at variance with the actual temperature.

### 20.2.2   Heat Profiles

Monitoring the internal temperature of a reaction is the only way to ensure that the desired temperature control is being maintained (see above). The data generated can also be mined for valuable insights into the reaction mechanism, reaction completion, and safety margins. For example, a good understanding of temperature-induced impurity formation can be garnered by closely monitoring the internal reaction temperature. Polymorph control or particle size can sometimes be influenced by the precipitation temperature and the rate of cooling—parameters that are best measured by internal temperature probe. Important safety data can also be collected by monitoring for spikes in the heat profile.

### 20.2.3   Stirring

Mixing can have a large influence on reactions, especially heterogeneous ones. How the mixing is performed can also have a large impact on the reaction outcome. For example, the stirring achieved with a magnetic stir bar versus an overhead stirrer is not the same. In most cases, overhead stirring is more efficient, and is far more representative of a large scale reaction. However, stir bars can have a grinding effect on heterogeneous reactions, and an accelerating effect may be observed that is not reproducible on different scales. For example, when using an insoluble base such as cesium carbonate in THF, the rate may be accelerated with a magnetic stir bar because its action reduces the particle size of the solid, thereby increasing its surface area. This effect is not observed when using an overhead stirrer. A special case where different mixing methods can have striking effects occurs during hydrogenations: results can vary drastically depending on whether the reactor is shaken or stirred, and in extreme examples will give no product in one case and a good yield in the other. Again, the stirred reactor is far more predictive of large scale reactors.

---

[2] This is due to reaction mass (or volume) increasing as a cubic function $(x^3)$, whereas surface area increases as a square function $(x^2)$. For a 10-fold increase in volume, a 4.6-fold increase in surface area is observed for a spherical reactor $(4.6 = [(10)^{1/3}]^2)$.

### 20.2.4    Homogeneous versus Heterogeneous Reactions

Homogeneous reactions are generally less sensitive to scale than heterogeneous reactions, for obvious reasons: the rate of homogeneous reactions is controlled by diffusion, while heterogeneous reactions are dependent on mixing, surface area (for solids), absorption (for gases), and a host of other factors. Stir rate can have an enormous influence on heterogeneous reactions, and its effect should be studied to fully understand a reaction. The effective stir rate can also be affected by the size and shape of the reaction vessel. The particle size and shape of solids also has a large influence on the reaction rate. For slow reactions, the surface area of the solids should be maximized. A typical situation arises in the use of carbonate bases in organic solvents, where particle size reduction will greatly accelerate the reactions. In the lab this can be accomplished with a mortar and pestle; in an industrial setting, milling is a more standard technique.

### 20.2.5    Electrophilic versus Nucleophilic Reactions

As a very gross generalization, electrophilic reactions are often more reproducible, and will scale up better than nucleophilic reactions. In this context, an electrophilic reaction is defined as one in which the high-energy species is cationic, and in a nucleophilic reaction, it is anionic. One explanation that may account for this is a consideration of the decomposition pathways of the high-energy intermediate. As an example, consider the acylation of a phenyl ring. This can be accomplished in the electrophilic sense by a Friedel–Crafts acylation, and in the nucleophilic sense by a metalation/acylation. In the Friedel–Crafts acylation, the reagent being used to generate the acylium species is the high-energy intermediate, and is often used in excess to account for its decomposition. Note that in this case the "decomposition" pathway may be hydrolysis to the acid, and since this may have been the precursor to the acylium ion, it may not be lost from the productive pathway. Moreover, since the acylium ion is generated in the presence of the aromatic trap, there is little buildup during the course of the reaction. The substrate being acylated is by far the less reactive species of the two components, and is usually stable to decomposition pathways. That these reactions usually need to be carried out at elevated temperatures is a further indication of component stability.

In the corresponding nucleophilic reaction, the substrate being metalated is now high-energy, and often highly unstable. The reactions are usually carried out at low to cryogenic temperatures ($<-20°C$) to avoid decomposition of the metalated intermediate, which is usually fully generated prior to addition of the trap. Control of temperature on larger scale can be extremely challenging, making these types of reactions more difficult to reproduce. In addition, the decomposition pathway of the reaction almost never regenerates the starting material; in the case discussed, protonation is most likely, giving a product that cannot re-enter the reaction pathway.

### 20.2.6    Gas Generation

Reactions that generate gas have obvious safety implications, and care should be taken to ensure that they are properly vented and scrubbed. It is not good strategy to rely on needles to bleed pressure off a reaction. At a minimum, a gas inlet adapter that fits the largest neck on the flask should be used. It is less obvious that the method of removing the gas can influence the effectiveness of the reaction. In general, gas can be removed passively, under a positive pressure of nitrogen with a bubbler attached, or aggressively, with a strong nitrogen sweep. The former conditions are usually carried out on small-scale laboratory reactions, while larger reactions in kilo labs or pilot plants use the latter conditions. Although rare, this can make a difference, as in the example reported for the synthesis of an amide via the corresponding acyl imidazole.[3] It was found that efficient removal of the $CO_2$ being generated in the reactions substantially retarded the conversion to product. If the reaction was carried out retaining the evolved gas within the vessel, the reaction proceeded as expected. This result highlights the importance of testing different methods of inerting a reaction to see if they have an effect.

### 20.2.7 Execution of Energetic Reactions

Many reactions in organic chemistry are extremely energetic, usually evidenced by the generation of large exotherms and/or pressure development. The classic way that most chemists have been taught to deal with this situation is to cool the reaction down. This is one of the riskier ways to carry out such reactions, since it increases the potential to build up reactive intermediates that are the source of the high energy. Additionally, as noted above, the ability to control heat transfer decreases as the scale increases, leading to the situation where a reaction that seemed safe in a 10 mL flask can be disastrous at 1 L. The preferred way to carry out energetic reactions is to run them at the highest temperature possible, so that high-energy intermediates are consumed as soon as they are formed. In the reaction shown below, acryloyl azide reacts via a Curtius rearrangement and is trapped with benzyl alcohol to give the corresponding urethane.[4] Both the starting azide and intermediate isocyanate are thermally unstable and decompose at low onset temperatures. To address the safety issues, the reaction is carried out at a high enough temperature that the rearrangement and trap occur instantaneously as the azide is added to the rest of the reagents. There is no buildup of the isocyanate, and the generation of gas is controlled by the rate of addition.

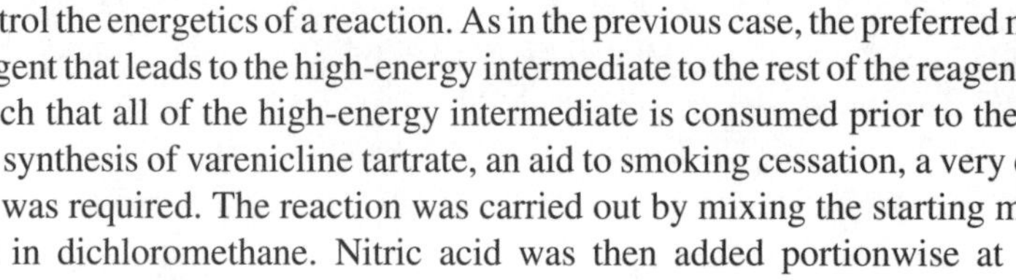

Even if heating is not possible, dose-control of one of the reactive components can still be used to control the energetics of a reaction. As in the previous case, the preferred method is to add the reagent that leads to the high-energy intermediate to the rest of the reagents in the pot at a rate such that all of the high-energy intermediate is consumed prior to the next dose. During the synthesis of varenicline tartrate, an aid to smoking cessation, a very exothermic dinitration was required. The reaction was carried out by mixing the starting material and triflic acid in dichloromethane. Nitric acid was then added portionwise at a rate that maintained the internal temperature slightly above ambient. Under these conditions, the nitronium ion being generated was consumed as it was formed.[5]

[3]Vaidyanathan, R.; Kalthod, V. G.; Ngo, D. P.; Manley, J. M.; Lapekas, S. P. *J. Org. Chem.* **2004**, *69*, 2565–2568.
[4]am Ende, D. J.; DeVries, K. M.; Clifford, P. J.; Brenek, S. J. *Org. Process Res. Dev.* **1998**, *2*, 382–392.
[5]Coe, J. W.; Watson, H. A., Jr.; Singer, R. A. *Varenicline: Discovery Synthesis and Process Chemistry Developments*; CRC Press: Boca Raton, Fl, **2008**.

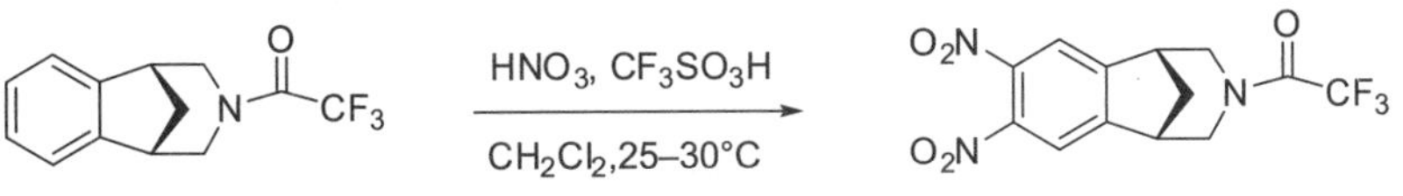

A completely different approach to running energetic reactions is based on a strategy of minimizing the quantity of the reaction at a given time, thus the total energy output possible, by running it under flow conditions. This approach is somewhat reminiscent of the earlier flash vacuum pyrolysis (FVP) method that was used to achieve ultra-high temperatures not otherwise attainable in normal glassware. Several reviews are available on the use of flow reactors,[6,7] and this technique has been used in the successful scale-up of many energetic[8] or otherwise problematic to scale reactions, such as the Newman–Kwart rearrangement shown below.[9]

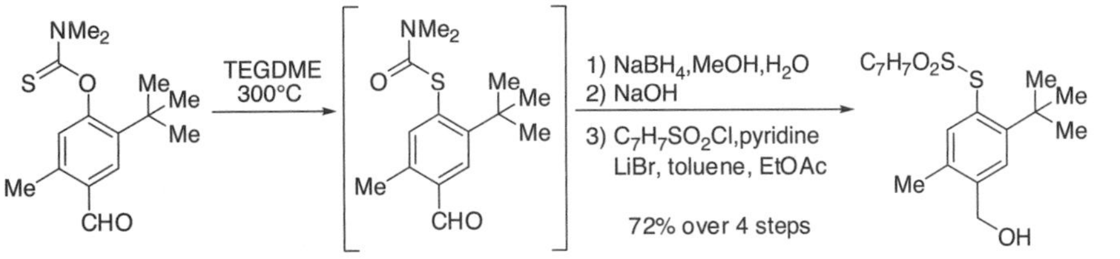

TEGDME = tetraethylene glycol dimethyl ether

## 20.3  SOLVENTS AND REAGENTS

### 20.3.1  Solvent Selection

Be creative and thoughtful when selecting an appropriate solvent for your reaction. Choice of the proper solvent can enhance desired reactivity, minimize byproduct formation, and simplify or even eliminate workups.[10] Following are some general guidelines and things to keep in mind:

1. Avoid solvents that react with your materials
   a. Ethyl acetate frequently reacts with amines to form the corresponding acyl amides; use of isopropyl acetate can reduce this side reaction. Ethyl acetate also decomposes very rapidly in the presence of hydroxide to acetic acid and ethanol. Therefore, it is a poor choice of organic solvent when an extraction is planned with sodium or potassium hydroxide. Aqueous solutions of carbonate bases can be used if a basic extraction is needed, with minimal decomposition of the ethyl acetate. It should be noted that $CO_2$ may be liberated in this approach, and proper venting should be ensured.

[6]Anderson, N. G. *Org. Process Res. Dev.* **2001**, *5*, 613–621.
[7]Pennemann, H.; Watts, P.; Haswell, S. J.; Hessel, V.; Loewe, H. *Org. Process Res. Dev.* **2004**, *8*, 422–439.
[8]Proctor, L. D.; Warr, A. J. *Org. Process Res. Dev.* **2002**, *6*, 884–892.
[9]Lin, S.; Moon, B.; Porter, K. T.; Rossman, C. A.; Zennie, T.; Wemple, J. *Org. Prep. Proced. Int.* **2000**, *32*, 547–555.
[10]Chen, C.-K.; Singh, A. K. *Org. Process Res. Dev.* **2001**, *5*, 508–513.

b. Ester exchange in alcohol solvents can be very rapid. For example, if a substrate containing a methyl ester is allowed to react in a solvent such as ethanol, a mixture of methyl and ethyl esters may be obtained. The exchange can occur under acidic, basic, or neutral conditions. From a strategic view, this may not matter to the overall sequence being pursued, but it can cause analytical issues. Whenever possible, the alcohol solvent should be chosen to match the alkyl group of the ester.

c. Dichloromethane will react with nucleophilic amines or alcohols to form chloromethyl amines and ethers or methylene-bridged dimers.[11] This can be a very competitive process, and it is generally prudent to avoid the combination. Dichloromethane should also be avoided when using azide ions, as it can form diazidomethane, an explosive compound.[12]

d. Ethereal solvents can form peroxides in oxidation reactions and should not be used.[13,14] Toluene, water, dichloromethane, and other related halogenated solvents are safer alternatives.

e. Ketone solvents can react with many different kinds of reagents, and should generally be avoided with amines or reagents like $POCl_3$.[15] Acetone can condense or polymerize in the presence of strong acids or bases.

f. Alcohol solvents have the potential to react with strong acids such as HCl or *p*-TsOH to form the corresponding chlorides or sulfonate esters. While these are not high-yielding reactions, it can become an issue if a nucleophile is present in the reaction. In the pharmaceutical industry, the presence of alkyl sulfonates or chlorides is a potential issue from a regulatory perspective because of their potential mutagenicity or carcinogenicity.

g. Cyclic ethers such as THF can undergo ring opening in the presence of strong protic or Lewis acids, and byproducts from the corresponding open-chain alcohols may be observed.

h. Polyether solvents such as glyme or diglyme can sometimes be used in place of crown ethers, and are usually much easier to handle.

2. Consider safety first. Some combinations of reagents and solvents are known to be unstable. For example, sodium hydride is known to decompose violently and auto-catalytically in the presence of solvents such as DMF, DMPU, or DMSO,[16] so avoid the combination. DMSO is also incompatible with a number of other reagents, such as perchloric acid and bromides.[17]

3. *t*-Amyl alcohol is a good choice for a high-polarity, water-immiscible solvent. It is also an excellent solvent to screen in palladium-catalyzed couplings.

4. If a reaction is exothermic, consider the boiling point of the solvent you select: A lower boiling point offers a potential barrier to a runaway reaction, or acts as a buffer to absorb the heat being produced.

---

[11]Mills, J. E.; Maryanoff, C. A.; McComsey, D. F.; Stanzione, R. C.; Scott, L. *J. Org. Chem.* **1987**, *52*, 1857–1859.
[12]Hassner, A.; Stern, M.; Gottlieb, H. E.; Frolow, F. *J. Org. Chem.* **1990**, *55*, 2304–2306.
[13]Williams, E. C. *Chem. Ind.* **1936**, 580–581.
[14]Aycock, D. F. *Org. Process Res. Dev.* **2007**, *11*, 156–159.
[15]Brenek, S. J.; am Ende, D. J.; Clifford, P. J. *Org. Process Res. Dev.* **2000**, *4*, 585–586.
[16]French, F. A. *Chem. Eng. News* **1966**, *44*, 48.
[17]Urben, P. G. *Chem. Health Saf.* **1994**, *1*, 30, 47.

5. If a solvent change is needed in the course of a reaction, either for isolation purposes or to telescope two reactions, consider the relative boiling points or existence of azeotropes of the solvents being used. When possible, a higher-boiling or favorable azeotrope solvent should be used to displace the first solvent.

6. Wet solvents can have much better solubilizing properties than dry solvents, so addition of a small amount of water can greatly increase the solubility of compounds. For example, wet 2-methyltetrahydrofuran (2-MeTHF) is much better at solubilizing organic substrates than when it is dry.

7. Some higher homologues of water-miscible solvents are immiscible, such as 1-butanol and 2-MeTHF. This may make them preferable for reactions in which aqueous extractions are planned. Even some normally water-miscible solvents can become immiscible depending on conditions such as pH, temperature, and the ionic strength of the medium. For example, THF is immiscible with water at high pH.

8. Some organic solvents are immiscible with other organics. For example, hexanes are immiscible with acetonitrile or DMAc, and can be used to extract non-polar impurities.

9. Use less solvent than you think you need; unless there is a reason for higher dilution, keep reactions concentrated. A rough rule of thumb is to try reactions first at 5–7 volumes (mL/g). In addition to being a more environmentally conscientious approach, your reaction may be accelerated as long as it is greater than first-order.

10. Try to avoid solvents that are chlorinated, toxic, or have very low flash points. Use heptane instead of hexanes. Older literature syntheses that use benzene usually work with toluene. If a chlorinated solvent is necessary, use dichloromethane or 1,2-dichloroethane instead of chloroform or carbon tetrachloride.

11. Try water as a reaction solvent; it often gives surprisingly good reactivity,[18] especially when the reaction mixture is not soluble in it.

12. If the solubility of the starting material is much higher than the product in a given solvent, the reaction can be designed such that the product precipitates in the course of the reaction therefore avoiding the need for a complicated workup and purification. This method is sometimes referred to as a direct drop process.

13. When transferring large amounts of solvent, a grounding strap or line should be used. It is very easy to generate "lightning" in a reaction vessel via static discharge, a situation obviously to be avoided when using flammable solvents.[19]

14. When using mixed solvents, choose those that can be recycled by distillation in a solvent recovery unit.

### 20.3.2    Removal of Water from Solvents or Reactions

Many types of reactions can be sensitive to the presence of water. During screening of reaction conditions, it is prudent to exclude water when its effect may be an issue. The standard method for removing water on an academic scale is to use solvent that has been distilled over a drying agent or passed through a solvent drying system. This process is

---

[18] Auge, J.; Lubin, N.; Lubineau, A. *Tetrahedron Lett.* **1994**, *35*, 7947–7948.
[19] Giles, M. R. *Org. Process Res. Dev.* **2003**, *7*, 1048–1050.

convenient on small scale, but is impractical, wasteful, and costly as the scale rises. The easiest and most economical method to achieve a dry solvent is to azeotropically remove the water. For a reaction mixture, this is achieved by addition of excess reaction solvent, followed by vacuum or atmospheric distillation to the desired volume. For lab use, fresh, anhydrous solvents poured from a bottle contain little or no more water than distilled solvents handled with syringe and needle. A practical way to dry a flask prior to use is to rinse once with anhydrous solvent. After screening has been completed, and a given reaction is being optimized, the assumed negative impact of water should be challenged. Many reactions, such as palladium-mediated couplings, will tolerate or even be enhanced by low levels of water. Experiments should be carried out during optimization to determine allowable and optimal ranges for water content.

To dry an organic solution after an aqueous workup, azeotropic removal of the water is again the preferred method. The use of drying agents, such as magnesium or sodium sulfate, is unwieldy and costly on large scale.

### 20.3.3   Solvent Contamination

Several commercially available solvents may be contaminated with compounds that can interfere or react competitively with desired transformations.

- DMF often contains some level of dimethylamine, or will decompose under reaction conditions to produce it. In nucleophilic substitution reactions, it can add competitively to give the dimethylamino product. DMAc or NMP are less prone to decomposition and are often comparable to DMF as a solvent.

- Solvents that are prone to air oxidation, such as ethers, are often stabilized with 2,6-di-*t*-butyl-4-methylphenol (BHT). If large volumes of solvents are used in a process and then removed by distillation, the residual BHT can sometimes be a significant contaminant in the isolated product.

- There are many grades of ethanol, most of which contain a co-solvent to decrease the potential for misuse. For example, 2B ethanol contains ~5% toluene and 3A ethanol contains ~5% methanol. During concentrations, these co-solvents can have a large effect on crystallizations or purifications, and should be taken into consideration.

- Chloroform is stabilized with either ethanol or amylenes. The amylenes rarely cause a compatibility issue in reactions and are easily removed, but the ethanol can cause problems.

- Alkane solvents are often contaminated with higher-boiling hydrocarbons that can become an issue during concentrations.

- Some solvents readily leach plasticizers or other organic-soluble components from plastic or rubber. For example, the phthalate esters in Tygon™ tubing readily dissolve in dichloromethane, and can easily contaminate reaction mixtures. Check the manufacturers' websites for solvent compatibility.

### 20.3.4   Reagent Selection and Compatibility

*20.3.4.1   Replacements for NaH*    As mentioned in the solvent selection section, NaH can react violently with some organic solvents. While this risk can be lessened by the proper choice of solvent, it is better to avoid its use when possible. If the product of the reaction is to be isolated by crystallization, adding the NaH as a mineral oil suspension might not cause problems in the isolation and is the safest handling alternative. Potassium *t*-butoxide in THF is a good combination to try as a replacement, and usually gives equivalent results. It can be conveniently purchased as an anhydrous solution. If sodium is desired as a counterion, sodium *t*-amylate is a good choice as a replacement. Phase-transfer conditions are often successful replacements as well.

### 20.3.4.2    *Use of Hydrides*    (See Section 20.4.1.3 for workups.)

Sodium borohydride can selectively reduce ketones and aldehydes in the presence of esters, but it will reduce esters at room temperature given sufficient time (and no competing functionality). Its reactivity in alcohol solvents follows the order MeOH > EtOH > 2-PrOH, however the rate of decomposition is also fastest in MeOH, so a larger excess may be required. As with other hydrides, it should not be used with polar, high-boiling solvents such as DMF.[20] Sodium triacetoxyborohydride (STAB) is also an excellent source of hydride for appropriate reactions, such as reductive aminations, and is easier to handle and quench.

LiAlH$_4$ (LAH) is more stable to air in THF-toluene mixtures than THF alone, and is sometimes more reactive toward organic substrates in them as well. A minimum of two moles of THF are required to solubilize the LAH, and a solution of this complex is commercially available. The reactivity can also be tuned by the addition of other ethers.

Borane is usually purchased as a complex with THF or SMe$_2$ to avoid having to crack the dimer. The THF complex is thermally unstable, especially as the concentration rises,[21] and is not recommended above 1 M for safety reasons, especially at large scale. The stability of the THF complex is claimed to be affected by the choice of stabilizer, with amines being preferred to NaBH$_4$.[22]

### 20.3.4.3    *Metal Catalysts*    Dry metal catalysts such as Pd/C are extremely hazardous, and can ignite in air and/or the presence of organic solvents. Water-wet catalysts should be used instead, and are typically available as 1:1 mixtures by weight with water to stabilize them. In most cases, the water-wet catalysts perform equally well as the dry catalysts, and do not spark when mixed with organic solvents. If the reaction does not tolerate water, a suspension of the catalyst can be dried azeotropically under nitrogen to minimize the fire hazard, prior to introduction of hydrogen. The reaction should be kept under an inert atmosphere until the catalyst has been removed by filtration, at which point it should be suspended in water for disposal. In addition to the safety issue, many substrates are efficiently oxidized in the presence of a palladium catalyst and oxygen, giving another reason for keeping the reaction inert.

Many different types of the same catalyst are available. The difference in the catalyst often stems from the source of the solid support. For example, the carbon in Pd/C can come from plant or animal sources, which can have a profound effect on reactivity. The catalysts can be acidic or basic enough to affect labile groups in the substrate. Different lots of the same catalyst can also vary in reactivity, and should be checked in a test reaction prior to large scale use.

---

[20]Liu, Y.; Schwartz, J. *J. Org. Chem.* **1993**, *58*, 5005–5007.

[21]Laird, T. *Org. Process Res. Dev.* **2003**, *7*, 1028.

[22]Vogt, P. F.; Am Ende, D. J. *Org. Process Res. Dev.* **2005**, *9*, 952–955.

The reactivity of homogeneous metal catalysts varies widely depending on the source and purity. The most reliable results will often be obtained by freshly preparing the catalyst, but this may not always be feasible for scale-up. If a commercial catalyst is used, it should always be use-tested prior to scaling up. It can be almost impossible to analytically detect small changes that result in gross effects on the reaction.

Raney nickel, sometimes referred to as sponge nickel, is an excellent reagent for reducing a variety of functional groups, but there are issues with its use, such as its pyrophoric nature. Nickel is a carcinogen, and most reagents based on it are assumed to be the same. As with many other catalysts, there are different types of Raney nickel that have varying reactivity. It is usually manufactured in the presence of hydroxide, and residual base may need to be washed out prior to use. Raney nickel loses activity with time, which usually necessitates the use of large excesses. Nickel boride[23] can be used as a replacement for Raney nickel for sufficiently reactive substrates.

*20.3.4.4  Generation of HCl*   For reactions requiring anhydrous HCl, gaseous acid is often used. This may not be easy or convenient if the stoichiometry of the acid needs to be carefully controlled. A more convenient and safer practical alternative is to generate the HCl *in situ* from an alcohol and acetyl chloride.[24]

*20.3.4.5  HF*   The hazards of HF (reactivity with bone, etching of glass) are well known, and its use should be avoided when possible. If HF must be used, reactions should be carried out in compatible plastics, such as polyethylene, or Hastelloy reactors. For additional worker safety, care should be taken to keep the pH basic after the reaction is complete. It should be realized that HF or its equivalent can be generated during reactions such as nucleophilic aromatic displacements of fluorides, and that these reactions can be as hazardous as using the reagent itself. When using HF, a solution of calcium gluconate should be available since this will readily deactivate HF by the formation of insoluble $CaF_2$. The antidote for skin contact is subcutaneous injection of the solution.

*20.3.4.6  Use of POCl₃*   Phosphorus oxychloride ($POCl_3$) is a reagent commonly used for dehydrations, chlorinations, and so on, and is often used in combination with DMF to generate the Vilsmeier reagent. Less well known is that this combination also generates dimethylcarbamoyl chloride (DMCC) as a byproduct.[25] DMCC is a known animal carcinogen. An assessment to a very low level is required if this combination is used to make material of pharmaceutical grade.

*20.3.4.7  KOH Contamination*   Solid KOH may be contaminated with nickel and other metals. This can be an issue if it is used as a base in palladium- or other metal-mediated reactions. Check the label for its presence.

*20.3.4.8  Stainless Steel Compatibility*   Stainless steel is not inert in the presence of halides. This can be a problem when drying compounds such as HCl salts in drying ovens, which are frequently constructed from stainless steel. The decomposition of the stainless steel can result in contamination of the solids being dried.

[23]Back, T. G.; Baron, D. L.; Yang, K. *J. Org. Chem.* **1993**, *58*, 2407–2413.
[24]Yadav, V. K.; Babu, K. G. *Eur. J. Org. Chem.* **2005**, 452–456.
[25]Levin, D. *Org. Process Res. Dev.* **1997**, *1*, 182.

## 20.4  ISOLATION

The workup and purification of a newly made compound can often be as challenging as the synthesis. This is especially true for large scale chemistry, where multiple extractions are costly and chromatographic purifications are discouraged or even prohibitive. The art of crystallization is the preferred method for isolation and purification of compounds that are solids. For compounds that are oils, formation of a salt to render them crystalline is the preferred option. If no salt can be formed, carrying forward the intermediate as a solution is the next best option, if the subsequent chemistry will tolerate the potential contaminants. Listed below is some general and specific guidance on workup and purification.

### 20.4.1  Reaction Workups

In the ideal process, the product precipitates directly from the reaction mixture as it is formed, and is isolated by simple filtration. This ideal is not often realized, however. Many reactions are subjected to aqueous workups to quench reactive intermediates and reagents or to wash byproducts away from the desired material. While these workups often cannot be avoided, they should be kept to the absolute minimum, that is, not done simply because they seem like a good idea, but because there is hard data that they are necessary. This is especially true if the product of the reaction is a solid, in which case precipitation of the solid by addition of an anti-solvent or displacement into a non-solubilizing solvent is preferred.

*20.4.1.1  Extractions*  As stated in the introduction, extractions should be minimized in any workup, and only carried out because there is data that suggests they are necessary. If extractions are required, the options for acid and base solutions should be considered. The most commonly used are probably aqueous HCl and NaOH. These are certainly sensible choices for cost and availability reasons, but may not be suitable for all reactions. If aqueous HCl is incompatible with a reaction stream, a solution of 10% citric acid is a very mild way of washing out basic impurities. It is also useful for extracting amines whose HCl salts are insoluble in water, and tend to precipitate. Carbonate or bicarbonate washes can be substituted for NaOH extractions, depending on the pH required to remove the impurities. These are especially useful when extracting from organic solvents that decompose in the presence of NaOH, such as EtOAc. A caution must be noted with their use, as they can generate $CO_2$ rapidly when quenching strongly acidic reaction mixtures.

*20.4.1.2  Emulsions*  If an emulsion forms during an aqueous workup, the simplest way to resolve it is to add more water and organic solvent. If this technique does not work, brine may be added to encourage separation of the layers. This approach works only with organic solvents that are less dense than water; with higher-density solvents, the addition of brine worsens the emulsion. If dilution is not desired or possible, the emulsion can be filtered through diatomaceous earth, which often resolves the layers. Heating the mixture can also break up an emulsion.

*20.4.1.3  Workup of Hydrides*  Hydride reagents based on aluminum, boron, or titanium can be challenging to work up. These metals tend to result in gels and slow-to-filter solids that can cause hopeless emulsions during extractive workups on large scale. Some strategies for removing these metals include:

- Non-extractive workups that rely on precipitation of salts. Control of the stoichiometry of water and/or hydroxide can result in a well-filtering solid. For aluminum reagents, this can be carried out by simple addition of the correct amount of water (4–5 molar equivalents per aluminum) to precipitate aluminum oxide. It is preferable to premix the water for quenching with a water-miscible solvent to more accurately control the rate of addition and dispersal into the reaction. If the solids are somewhat gelatinous, heating and cooling cycles in the organic reaction solvent can "dry" the solids.
- Alternatively, the Fieser workup[26] also usually gives filterable solids. A delayed exotherm is sometimes observed in these quenches, so care should be employed when running them for the first time.
- Another trick to avoid the gelatinous solids that can form is the addition of diatomaceous earth or another solid support to the reaction mixture. The solids adsorb onto the support without the usual blinding effect observed when the gelatinous solids are poured directly onto a filter aid. This works very well for quenches that generate titanium dioxide.
- If quenching with water is too exothermic, try 2-propanol instead. However, be watchful for a delayed exotherm. Other solvents that have been used include acetone or ethyl acetate, especially for DIBAL-H reactions.
- For extractive workups, try varying the pH of the aqueous phase. For some reactions, there is an optimal pH where the inorganic byproducts dissolve more readily, and emulsions can sometimes be avoided by identifying it. For example, when working up mixtures containing zinc salts, a gelatinous white precipitate forms around pH 9. This issue may be resolved by adding more base to a pH of 14
- Borane can be quenched with aqueous HCl, amines, or acetone. Methanol can also be used, and is useful when a distillation is planned, since the resulting trimethylborate ester has a very low boiling point (68–69°C).

***20.4.1.4  Workup of POCl₃*** While most chemists are aware that $POCl_3$ needs to be quenched cautiously, it is not as widely known that the reagent will react with some organic solvents in a potentially hazardous fashion. The greatest risk probably occurs with acetone, since it is commonly used as a rinse or wash solvent. Acetone and $POCl_3$ react at room temperature to produce heat and gas in a runaway reaction that can lead to a dangerous situation.[15] The $POCl_3$ should always be quenched after the desired reaction, and prior to mixing with other organic waste (this includes reagent that may have been distilled into rotavap traps). There is a long induction period when water is used as a quench, as they are not miscible; ammonium hydroxide is a better choice, since it reacts much more quickly.

***20.4.1.5  Reduction of Iodine*** Several inorganic salts are known to reduce excess iodine in a reaction, such as sodium bisulfite or sodium thiosulfate. However, the latter reagents may cause sulfur contamination of the product, which can lead to issues downstream.[27] Ascorbic acid can also be used as a reducing agent.

[26]Fieser and Fieser, *Reagents for Organic Synthesis* Wiley-Interscience, 1967 ,Vol. 1, p. 584.
[27]Xiang, Y.; Caron, P.-Y.; Lillie, B. M.; Vaidyanathan, R. *Org. Process Res. Dev.* **2008**, *12*, 116–119.

***20.4.1.6    Removal of Ph₃PO*** Triphenylphosphine oxide (TPPO) can be very difficult to purge from reaction mixtures by non-chromatographic methods. Occasionally, crystallization solvents can be identified that separate it from the product, but this is not a reliable approach. TPPO can be selectively extracted into MTBE if the product can be extracted into an aqueous phase, either through inherent solubility or by pH adjustment.

## 20.4.2    Purification

***20.4.2.1    Crystallization*** How many times have chemists been heard to curse because a compound starts to precipitate while they are preparing it for chromatographic purification? This is the best possible sign that the compound "wants" to crystallize, which is by far the preferred method for isolation and purification on a larger scale. Until the advent of flash chromatography, crystallization was the practice of choice for isolation. Since that time, the over-reliance on chromatographic purifications in the academic setting has made this somewhat of a lost art, and one that has to be relearned in industry. Chromatography is still the method of choice for small (<1 g) amounts of material or should be considered if a small pad filtration is very effective at removing polar impurities.

The simplest way to screen for crystallization solvents is to attempt to dissolve solids in a variety (typically 10–20) of different solvents; for pharmaceutical intermediates or candidates, Class 3 solvents (see Section 19.6) should be the first choice.

- Crystallizations should always be carried out with stirring, since it increases the chance of nucleation and ensures an even seed bed. This in turn helps reduce the likelihood of multi-modal particle size distribution. If a larger crystal is desired, such as for single crystal X-ray crystallography, it is better to try to enlarge the size of the crystal through processes such as Ostwald ripening[28] (i.e., repeated heating and cooling cycles).

- Crystallization attempts in which little dissolution appears to occur should be warmed to reflux. More solvent can be added if little or no sign of dissolution is observed. Those that become partially (see Section 20.4.2.2) or fully dissolved should be cooled; if the solid reprecipitates, it is a potential recrystallization solvent. Another method is to slowly evaporate a solution to see if solids appear. If the solids fail to redissolve when solvent is added, the system may be a candidate for the desired crystallization. The profile of the cooling cycle (i.e., rate of cooling, holding periods) can also have a profound effect on the characteristics of the crystals.

- The solvents that readily dissolve the substrate are not suitable unless an antisolvent can be identified. The two most common types of antisolvents are hydrocarbons or water. The latter is only practical for water-miscible solvents. The order of addition can have a large impact on whether the substrate precipitates as a solid or oil, as does the rate of addition.

- When attempting to crystallize a compound or salt, hygroscopicity (the tendency to pick up water) is often an issue, resulting in gumming of the solid. This may indicate

---

[28]Ratke, L. V.,,P. W. *Growth and Coarsening: Ostwald Ripening in Material Processing*; Springer: New York, 2002.

that a hydrate of the compound is a more stable species than the anhydrous form, and can be tested by adding small amounts (1–10%) of water to the solvent system to see if a stable hydrate is formed.

- It is preferable to identify a single solvent for crystallization. *n*-Butyl acetate may be a good alternative for ethyl acetate/hexanes. A single solvent gives greater control and facilitates a recycling of solvent.

***20.4.2.2   Reslurries or Repulping***   The classical method for purifying a crystalline solid is to identify the appropriate recrystallization solvent, that is, one in which it dissolves, usually with heating, then recrystallizes with cooling, simultaneously purging some or all of the impurities. Another method that can be equally effective is called reslurrying or repulping. In this case, the solid is suspended in a solvent that preferentially dissolves the impurities with minimal dissolution of the desired material. The method is based on the premise that impurities can be leached out of solids by taking advantage of a low equilibrium solubility of the substrate; given sufficient time, all of the solids will dissolve and recrystallize until the impurity is removed from the solid state. The reslurry can be carried out at room or elevated temperature. It is often easier to identify an appropriate reslurry solvent than a true recrystallization solvent.

***20.4.2.3   Formation of Salts or Adducts***   Salt formation is often a very effective method for isolation and purification of compounds, especially if they are not solids. The best way to attempt a salt formation is to mix a solution of the substrate with an equimolar amount of a solution of the acid/base and look for solid formation. Neither the substrate nor acid/base needs to be soluble for salt formation, but it is harder to detect if the mixture starts out as a slurry. For reactions that never become fully homogeneous, heating may be used to encourage dissolution and salt formation. For diacids or dibases, less than a full stoichiometric amount may be sufficient or optimal; in these cases, a range of stoichiometries should be investigated.

For formation of sodium salts in which anhydrous conditions need to be maintained, sodium 2-ethylhexanoate may be a good source of the counterion. The salt formation is driven by precipitation of the sodium salt of the substrate, since both the reagent and its counteracid are soluble in many organic solvents. Ethyl acetate is a particularly good solvent to try for this procedure.

When an amine counterion of an organic acid is desired, dicyclohexylamine is a good choice for inclusion in the screening, since it has a high tendency to form solid salts.

Aldehydes react with bisulfite to form adducts, which are often solids (see Section 2.3). This technique can be especially useful in the isolation and purification of aldehydes after an ozonolysis, since the bisulfite serves the double task of reducing ozonides and precipitating the product.[29] In some cases the adduct is not a solid, but it is usually water soluble, so purification of the substrate may still be effected by aqueous/organic extractions.

[29]Ragan, J. A.; am Ende, D. J.; Brenek, S. J.; Eisenbeis, S. A.; Singer, R. A.; Tickner, D. L.; Teixeira, J. J., Jr.; Vanderplas, B. C.; Weston, N. *Org. Process Res. Dev.* **2003**, *7*, 155–160.

*20.4.2.4  Purging Heavy Metals*    As the use of metals, particularly transition metals, in the use of organic reactions has risen, so have the approaches to purging residual metal from products.[30–33] Following are some of the most commonly used or effective ways of purging heavy metals, especially palladium:

- Activated carbon treatment is the most effective and inexpensive method for removing many metals, especially palladium, provided the substrate does not bind to it. In cases where the metal is tightly bound to the product, prolonged stirring at elevated temperatures may be required.
- Treatment with a small amount of silica gel can effectively remove many metals. This is most easily carried out by addition of one or two weight equivalents of silica, followed by stirring and filtration. Celite or other diatomaceous earth sources have also been used in an analogous fashion, but the colloidal metal in suspension tends to pass through during filtration.
- Treatment with amines such as triethylamine or 1,2-diaminopropane can reduce palladium levels and can be easier to separate from product than other methods.[34]
- For substrates with appropriate solubility, recrystallization from acetonitrile may purge palladium.

*20.4.2.5  Silica Gel Purification*    In some cases, there is no alternative to purification by a solid support, usually silica. The usual methods are well known, but in some situations are not fully needed. For removal of very polar impurities, a simple pad filtration may be sufficient. This can be easily carried out by adding one or two weight equivalents of silica to a reaction and filtering after stirring for an appropriate interval.

## 20.5  ANALYSIS

Several methods of analysis are available for tracking the course of a reaction and assessing its products. For following reaction completion, the most common are thin layer chromatography (TLC), high performance liquid chromatography (HPLC), and gas chromatography (GC); the latter two may also be coupled to a mass spectrometer. Others include infrared (IR), especially useful when used *in situ*, near IR, and Raman. For isolated material, organic chemists rely on these and NMR for confirmatory evidence of the product structure, along with elemental anaylsis, powder diffraction and single-crystal X-ray diffraction. Each of these methods has its advantages and disadvantages, and because none is ideal, orthogonal techniques should be used whenever evaluating a reaction or its product. This is especially important early in the development of the chemistry, to ensure that nothing is being missed.

[30]Pink, C. J.; Wong, H.-T.; Ferreira, F. C.; Livingston, A. G. *Org. Process Res. Dev.* **2008**, *12*, 589–595.

[31]Flahive, E. J.; Ewanicki, B. L.; Sach, N. W.; O'Neill-Slawecki, S. A.; Stankovic, N. S.; Yu, S.; Guinness, S. M.; Dunn, J. *Org. Process Res. Dev.* **2008**, *12*, 637–645.

[32]Koenigsberger, K.; Chen, G.-P.; Wu, R. R.; Girgis, M. J.; Prasad, K.; Repic, O.; Blacklock, T. J. *Org. Process Res. Dev.* **2003**, *7*, 733–742.

[33]Bullock, K. M.; Mitchell, M. B.; Toczko, J. F. *Org. Process Res. Dev.* **2008**, *12*, 896–899.

[34]Li, B.; Buzon, R. A.; Zhang, Z. *Org. Process Res. Dev.* **2007**, *11*, 951–955.

### 20.5.1  Thin Layer Chromatography

TLC using silica gel is a very old and well-studied method for following a reaction. UV-active compounds can be visualized with a UV lamp, usually available with at least two different wavelengths, and non-UV-active compounds can often be visualized with stains. It is optimal for compounds of medium polarity, since very polar compounds tend to stick to the baseline and very non-polar compounds often run with the solvent front. Following are a few tips or reminders:

- Addition of small amounts (1–5%) of $NH_4OH$ or TEA to the mobile phase can be helpful to resolve amines and moving them off the baseline.
- Addition of small amounts of HOAc is similarly useful for acids.
- For very polar compounds, up to 10% MeOH can be used in the mobile phase; higher levels result in dissolution of the silica stationary phase. If the compound is very insoluble in organic solvents, 10% MeOH in $CH_2Cl_2$ is a good choice for sample preparation. The MeOH should be evaporated prior to elution.
- For reactions run in DMF, DMSO, or other high-boiling solvents, evacuation on a high-vacuum pump for 10 minutes or so removes enough solvent to process the plate normally.
- Staining with iodine can be slow. Prepare a chamber of iodine-saturated silica by filling a TLC chamber with silica and add enough iodine to stain it dark brown. TLC plates dipped in this mixture stain very rapidly, especially those containing alkenes and aromatics, and amines give a negative stain. Iodine is also useful because it is reversible, and the same plate can be used in conjunction with other stains.
- The most general stains are p-anisaldehyde, phosphomolybdic acid, cerium sulfate, and potassium permanganate.
- Bromocresol Green is an effective stain for carboxylic acids.
- Ninhydrin is an excellent stain for amines, amino acids, sulfides, and phosphines.
- If your compounds do not separate or run well on silica plates, try alternative supports, such as alumina (acidic, basic, or neutral) or florisil.

### 20.5.2  High Performance Liquid Chromatography

HPLC is probably the most widely used method for reaction analysis in the pharmaceutical industry, and unlike TLC, can be coupled to a mass spectrometer. UV detection is the most common form of visualization, but other detector types may be substituted, such as refractive index, diode array, light scattering, or fluorescence. Most columns are reverse-phase relative to TLC systems, that is, more polar compounds elute first, however, normal phase columns also exist, particularly for chiral separations. The greatest danger when using HPLC, especially with a UV detector, is not accounting for differences in UV absorption among the components being analyzed. When used for quantitative analysis, it is vital to know the UV maximum of each component to pick an appropriate wavelength.

### 20.5.3  Gas Chromatography

GC coupled to a mass spectrometer is an excellent method for analyzing volatile compounds, especially those lacking a chromophore. The method relies on the degree of ionization to give information on relative percents of components, which is usually fairly accurate for organic reactions where the reactants and products are similar. Carboxylic acids can be problematic to analyze using GC, and other methods should be used to cross-check results. For some acids, formation of a silyl ester can resolve this issue.[35]

### 20.5.4   Nuclear Magnetic Resonance Spectroscopy

Most organic chemists are very familiar with the use of NMR. A relatively recent occurrence in the pharmaceutical industry is the use of $^1$H NMR to quantify the purity of compounds.[36,37] The method relies on addition of a weighed internal standard and proper adjustment of acquisition parameters, and can be used to detect impurities at or below 0.1%.[38] This can be a quick way to determine the potency of a substrate, especially when the suspected contaminants do not show a signal in the NMR, and is non-destructive of the sample.

[35]Haas, L. W.; DuBois, R. J. *Anal. Chem.* **1976**, *48*, 385–387.
[36]Wells, R. J.; Hook, J. M.; Al-Deen, T. S.; Hibbert, D. B. *J. Agric. Food Chem.* **2002**, *50*, 3366–3374.
[37]Malz, F.; Jancke, H. *J. Pharm. Biomed. Anal.* **2005**, *38*, 813–823.
[38]Soininen, P.; Haarala, J.; Vepsaelaeinen, J.; Niemitz, M.; Laatikainen, R. *Anal. Chim. Acta* **2005**, *542*, 178–185.

# FUNCTIONAL GROUP INTERCONVERSION INDEX

*Practical Synthetic Organic Chemistry: Reactions, Principles, and Techniques*, First Edition.
Edited by Stéphane Caron.
© 2011 John Wiley & Sons, Inc. Published 2011 by John Wiley & Sons, Inc.

# INDEX

It is also recommended to consult the Functional Group Interconversion Index for specific chemical transformations.

*Practical Synthetic Organic Chemistry: Reactions, Principles, and Techniques*, First Edition.
Edited by Stéphane Caron.
© 2011 John Wiley & Sons, Inc. Published 2011 by John Wiley & Sons, Inc.